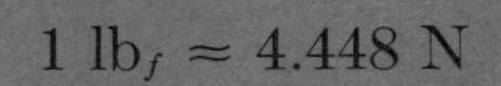
1 lb$_f$ ≈ 4.448 N

MW01004219

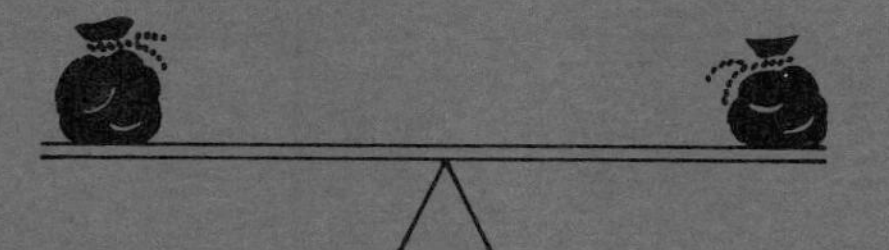

or

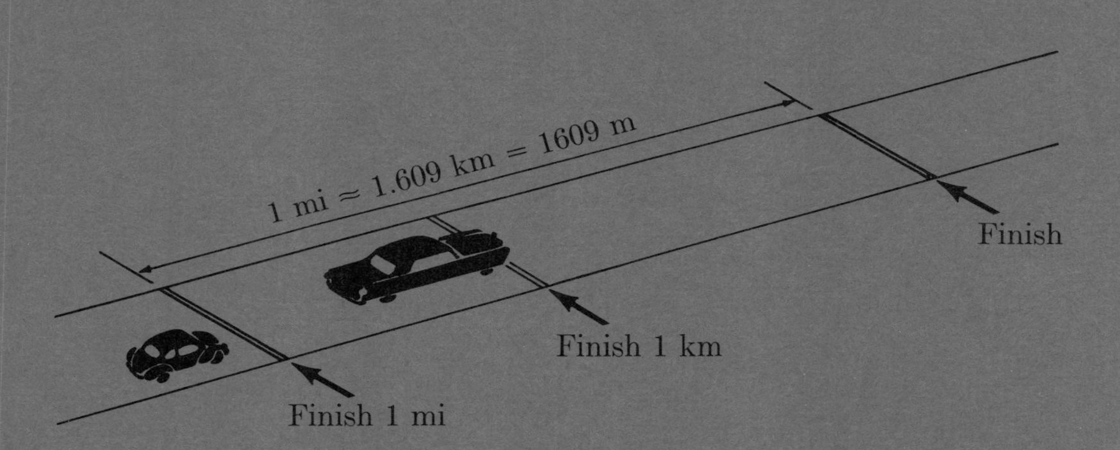
1 mi ≈ 1.609 km = 1609 m
Finish
Finish 1 km
Finish 1 mi

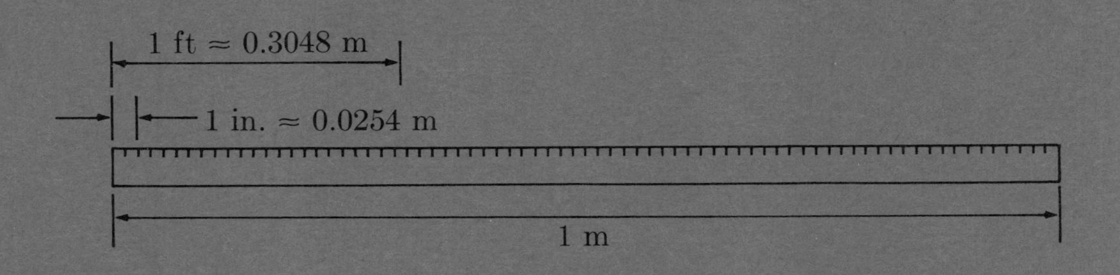
1 ft ≈ 0.3048 m
1 in. ≈ 0.0254 m
1 m

ELEMENTS OF
ENGINEERING MECHANICS

ELEMENTS OF ENGINEERING MECHANICS

PETER W. LIKINS
Mechanics and Structures Department
University of California, Los Angeles

McGraw-Hill Book Company
New York San Francisco St. Louis
Düsseldorf Johannesburg Kuala Lumpur
London Mexico Montreal
New Delhi Panama Rio de Janeiro
Singapore Sydney Toronto

This book was set in Monotype Modern 8A by The Maple Press Company. The editors were B. J. Clark and Marge Woodhurst, the designer was Janet Durey Bollow, and the production supervisor was Michael A. Ungersma. The drawings were done by Judith L. McCarty.

The printer and binder was Kingsport Press, Inc.

ELEMENTS OF ENGINEERING MECHANICS

Copyright © 1973 by McGraw-Hill, Inc. All rights reserved. No part of this publication may be reproduced, stored in a retrieval system, or transmitted, in any form or by any means, electronic, mechanical, photocopying, recording, or otherwise, without the prior written permission of the publisher.

Printed in the United States of America.

Library of Congress Cataloging in Publication Data

Likins, Peter W.
Elements of engineering mechanics.

1. Mechanics, Applied. I. Title.
TA350.L53 531 72-7423
ISBN 0-07-037852-5

1234567890 KPKP 79876543

To Teresa, Lora, Paul, Linda, Krista, and John
and
To Patricia
who makes it all possible

CONTENTS

CHAPTER FOUR PARTICLE DYNAMICS: FORMULATIONS

PART TWO MECHANICS OF RIGID BODIES

CHAPTER SEVEN STATICS: FORCE AND MOMENT EQUILIBRIUM

CHAPTER EIGHT RIGID-BODY DYNAMICS: FORMULATIONS

PREFACE

We are at the threshold of a period of radical change in undergraduate engineering instruction, and the pressures of educationally revolutionary forces will be felt most keenly in the tradition-bound provinces of mechanics. Although the teaching of the various academic subjects that comprise the engineering mechanics curriculum has changed significantly in the past 20 years, reflecting the national trend toward increasing the mathematical content of all courses in science and engineering, the practice of dividing mechanics into the traditional subdisciplines (statics, strength of materials, dynamics, and fluid mechanics) remains for the most part either unaltered or superficially modified by changes in nomenclature. While the depth of exploration and the level of presentation in mechanics courses have increased, in many modern engineering curricula the *number* of required courses in mechanics has been *reduced* in recent years, in order to make way for computer-science courses and other newborn subjects considered essential to the education of every young engineer. We have responded to the conflicting pressures to cover more material and to require fewer courses by simply accelerating the pace with which we cover the same old tracks. As a result we are conditioning our students to accept as normal a confused state of semiknowledge of the basic concepts upon which the practice of engineering is founded.

Introspection should provide sufficient stimulus for change, but if this fails we will soon be forced by external pressures to change our approach. The voices of other disciplines, clamoring for equal time, and the voices of the students themselves, demanding greater freedom of choice, will eventually be heard and heeded; fundamental changes in the teaching of engineering mechanics will be implemented. The queston is not "If," but "When" and "How."

We must respond to the pressures from our professorial colleagues to make our presentation of elementary mechanics more compact, and at the same time we must be responsive to student pressures to make our basic courses more stimulating intellectually. We must also heed our own internal demands for an honest treatment of our subject. These three constraints can be reconciled only by a very substantial restructuring of our curricula in engineering mechanics, and a deliberate sacrifice of previously sacrosanct subjects from our undergraduate courses, in a reasoned exchange of quantity for quality.

The traditional statics course is growing moribund and will soon be dead. The intellectual content of the old statics course will remain a necessary part of our educational program, but the traditional role of that course as the vehicle for the development of problem-solving skills must be distributed over other courses.

The most obvious response to the external pressure on mechanics faculty to dispense with the statics course is to combine it with the next course in the sequence, which is usually the old *strength of materials* course (often living under an alias nowadays).

Opportunity for more comprehensive reform is afforded by more complete integration of the subject matter of applied mechanics. In the various applications of Newton's laws, the most fundamental distinction to be made is in the range of *mathematical models* adopted for physical systems. In the purest sense, only two models are used: *particle models* and *continuum models*. The *rigid body* may be considered either as a highly constrained set of particles or as a rigid continuum. Practical considerations, however, justify treatment of the rigid body as a mathematical model distinct from the particle and the continuum, and further justify the separate treatment of fluid continua and deformable solid continua.

If mechanics must be subdivided into discrete segments for presentation in formal courses, it seems most logical to make the subdivisions on the basis of the mathematical model, treating first the mechanics (statics and dynamics) of particles, then the mechanics of rigid bodies, and then in turn the mechanics of deformable solids and the mechanics of fluids. This procedure entails an orderly progression from the simplest to the most difficult concepts, and yet offers a more unified treatment than is presently available. Such a *system-model sequence* for instruction in the elements of engineering mechanics is not only logically superior to the present pattern of instruction, but also easier and more enjoyable for both student and teacher.

In focusing first on the mechanics of a particle, the student is free to concentrate on the most fundamental implications of Newton's laws, including the concept of *static equilibrium* or rest in inertial space. With but one particle in his dynamical system, the student can more easily unravel the mathematical and kinematical mysteries that often becloud the subject of dynamics. By extending his scope to a system of particles, he can explore the advantages of most of the methods of statics and dynamics, and still remain within the familiar framework of particle mechanics.

The rigid body forms a natural link between the particle-system model and the continuum model since it can be treated as a special case of either, but it deserves attention in its own right as an idealization permitting particularly attractive formulation of equations and interpretation of results.

Extension to a nonrigid continuum is a very substantial step, involving many new concepts such as stress, strain, and compatibility requirements. Unlike the particle model, the continuum presents major obstacles to a formulation that permits both unlimited relative motions of constituent parts and arbitrary interaction forces; thus it is generally presented in one restricted form or another.

This book is designed to facilitate the transformation from the traditional pattern of instruction in elementary engineering mechanics to the system-model sequence in which the mechanics (statics and dynamics) of particles, rigid bodies, deformable solids, and fluids are considered in turn. This volume is intended to help accomplish the first step, being suitable for one or two courses on the mechanics of particles and rigid bodies. Particularly recommended for the second step are texts on deformable solid mechanics written for students who have not had a traditional statics course. Although few of these books are written specifically for the student with a course on the mechanics of particles and rigid bodies behind him, several of them serve such students very well, due to their integrated perspective. The system-model sequence of courses leaves the traditional course in fluid mechanics virtually intact, and presents no problems in text selection.

The detailed composition of this book is best evaluated by scanning its table of contents, and the peculiarities of its style and language can best be judged by the students who use it. The text is often deliberately redundant since I view multiple exposure from various perspectives to be necessary for effective learning at the basic level of this book. These redundancies also permit flexibility in the use of various portions of the book without covering all preceding material. I believe that it is more important in a basic mechanics course to learn certain thought processes than to learn any particular body of material, so I have tried to leave some latitude for the instructor in his selection of text coverage. Double asterisks are employed in the table of contents to identify the chapters, sections, or articles that might comprise my personal concept of a minimal course in the mechanics of particles and rigid bodies, single asterisks define the somewhat larger framework within which various instructors at U.C.L.A. teach this material.

Throughout this book there is a sharper demarcation than is customary between the business of formulating equations of motion and the task of solving them. This emphasis is dictated by the enormous influence of high-speed electronic computers on the modern practice of dynamics. Engineers today need advanced skills to formulate equations that only the digital computer can solve, and it is anachronistic to treat the derivation of equations of motion and their solution as one continuous operation. Thus Chap. 4 is devoted to methods of formulating equations of motion of a single particle, with considerable attention to the use of concepts of momentum, energy, etc., to facilitate their solution in special cases, while Chap. 5 is concerned solely with the mathematical solution of these equations. In an abbreviated treatment of the mechanics of particles and rigid bodies, Chap. 5 can be excluded entirely or used only for illustrative material selected by the instructor.

This book has been influenced much more by other people than by other books. The impact of Professor T. R. Kane of Stanford on my mode of thinking about kinematics and mechanics has been very substantial, and yet to some readers this book may appear to have little resemblance to those he has written. Professor R. E. Roberson of the University of California, San Diego, and Professor D. L. Mingori of the University of California, Los Angeles, have also contributed in significant ways to my approach to instruction in dynamics. The inspirational influence of Dean C. H. Norris of the University of Washington (formerly of M.I.T.) has been remote both in time and place, but nonetheless important to me. My secretary, Derfla Guthrie, has prepared the manuscript with the extraordinary cheerful competence that marks all of her activities, and I am equally appreciative of her fine work and her good nature. My greatest debt, however, is to my wife Patricia and our children (Teresa, Lora, Paul, Linda, Krista, and John). This book has been in gestation for more than 7 years, and its birth will be a time of celebration for the entire family, whose patience and forbearance have made it possible.

Peter W. Likins

INTRODUCTION

Mechanics is the study of the physical behavior of material bodies, with interest generally focused on the motions induced by forces. Because the quantities of interest in mechanics are amenable to primitive measurement, this was the first of the physical sciences to yield to mathematical formulation. Indeed, most of the mathematics used today in all branches of science and engineering developed originally under the stimulus of the growing demands of mechanics.

Modern science depends very fundamentally on the interaction of mathematical analysis and the observation of nature. In any rigorous scientific discipline, mathematics must play a crucial role, and this is particularly true in a science as mature and highly developed as mechanics. It should be recognized at the outset, however, that mathematics cannot be applied to nature directly, but only to a *model* of nature, suitably defined in mathematical terms.

It is the fundamental business of scientists and engineers to develop models that can be analyzed in order to obtain meaningful predictions of the behavior of real physical systems. The job of the scientist is to seek out the *truest* model, the model corresponding most closely to nature. The engineer's function, on the other hand, is to predict and ultimately to control the behavior of complex physical systems in order to accomplish a human objective. The engineer therefore adopts a mathematical model of nature that is no more complex than that required by the accuracy demanded of his predictions.

The earth, for example, may be idealized by the celestial mechanician as a particle (a point mass) for trajectory analysis of a space probe, but he may model the earth as a rigid oblate spheroid for orbital studies of earth satellites. The mechanician concerned with blast detection or earthquake analysis might model the same earth as a viscoelastic continuum, or perhaps as an elastic solid with a fluid core. The civil engineer building a dam of the earth's materials might require quite a different model of the substance of the earth, accommodating its inhomogeneity and representing its permeability to water flow as well as its complex structural properties. None of these models is *true* to nature, but each is *correct* for a given purpose. By working with the simplest mathematical model permitted by his problem, the engineer accomplishes analysis that might be impossible with a more precise model.

Thus this entire book is concerned with the mechanics of severely idealized material systems: particles and rigid bodies. The analyst must in each case determine the relevance to his problem of the results of analysis of these simple models.

Mastery of analytical methodology is of little value to an engineer who fails to exercise the judgment required for proper mathematical modeling of a physical system. The engineer must always ask not only "Is my analysis free of mathematical error?" but also "Am I analyzing an acceptable mathematical model of my physical system?"

Explorer I, the first artificial earth satellite successfully launched by the United States, provides an instructive and relevant case history. The satellite was a slender cylindrical body with four flexible wire antennae, as illustrated in Fig. I·1. The vehicle was designed to spin about the cylinder axis, with the intention that the spin would stabilize the attitude of the body; thus the spin axis was to remain in more or less the same orientation in space, even in the presence of small disturbances. As will be shown in §9·1·5, an elongated free rigid body is *stable* in rotation about its longitudinal axis. When such a body is disturbed slightly from its nominal rotational motion, the vehicle longitudinal axis executes a coning motion on a cone with a small apex angle (see Fig. I·1*b*), while the body continues to spin about this axis.

It is not surprising that in the press of a difficult schedule, amidst a multitude of uncertainties associated with this initial satellite launching, no one thought to ask, "Are we analyzing an acceptable mathematical model?" Not until the vehicle was in orbit was it apparent that the answer would have to be "No." The anticipated coning motion developed, but the apex angle grew rapidly with time, so that after a single orbit it was approximately 120°! (Compare Fig. I·1*c* and *b*.) Soon the satellite was hurtling through space tumbling end over end, and

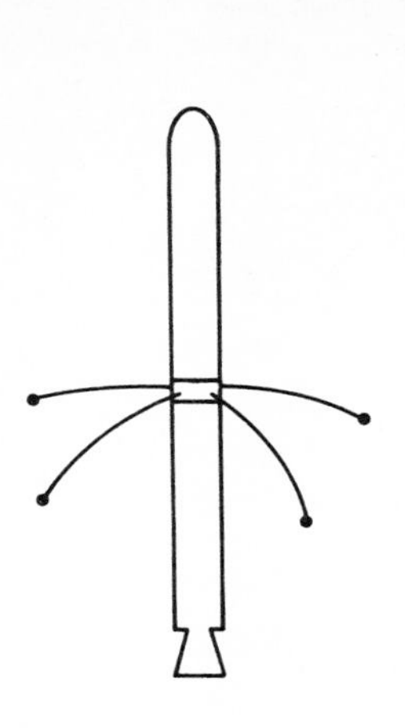

(*a*) *Explorer 1* satellite

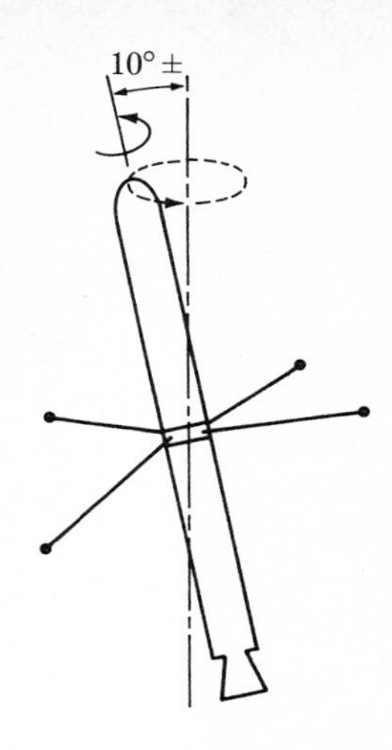

(*b*) Expected motion after one orbit

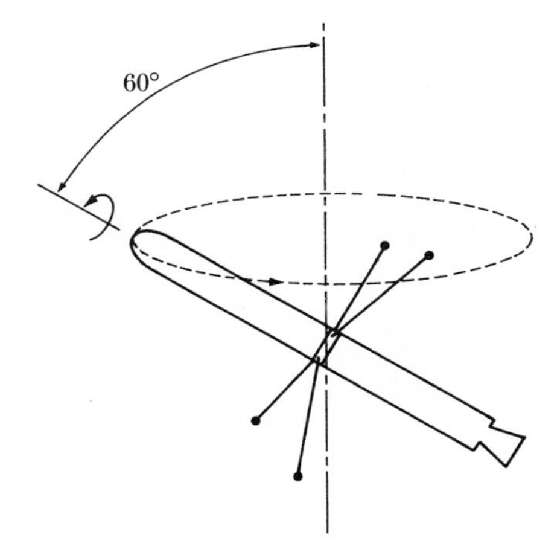

(*c*) Observed motion after one orbit

FIGURE 1·1

the resulting disturbance to the satellite radio signals made their interpretation difficult.

The postflight analysis of *Explorer I* revealed the explanation: The mathematical model of the vehicle was not good enough to reveal its fundamentally unstable pattern of behavior. Instability was evident analytically only after the analyst incorporated the flexibility and consequent energy dissipation of the lightweight wire antennae into the vehicle model.

Just as the engineer works with mathematical models of physical materials, deliberately choosing an imprecise model to facilitate analysis, so also he assumes the validity of physical laws that the modern student of physics knows to be inexact. Most potential engineers today learn in high school that the classical laws of physics attributed to Newton do not always conform in their predictions to experimental evidence; most students recognize that there are corrections to newtonian theory embodied in those branches of modern physics referred to as *relativistic mechanics* and *quantum mechanics*. They may not know, however, that in most modern enginerring applications the corrections of relativity theory to the predictions of newtonian mechanics can be ignored, and that in the study of macroscopic mechanical phenomena (visible to the naked eye) classical theory usually does not have to be modified to accommodate quantum mechanics either. The accuracy with which spacecraft motions can be measured under optimum conditions is just beginning to match the limits of newtonian mechanics, but in virtually every other macroscopic application the engineer may proceed safely within the classical framework. When such simplifications give acceptable predictions, they often *must* be elected, for without them most engineering problems would remain unsolved.

It is apparent, then, that engineering mechanics involves the study of *abstract idealizations* (such as particles, rigid bodies, and continua) under a set of *postulated laws*. The validity of the idealization and of the laws in application to nature is always an important consideration for the engineer, but the study of newtonian mechanics as a discipline can proceed without this encumbrance. This approach to mechanics makes it much easier to see the subject as a well-ordered whole, and this perspective contributes to confidence as well as competence in the application of the basic principles of mechanics to well-defined mathematical models. The engineering student who masters this formal discipline can then expect to spend the rest of his life developing his expertise in applying his knowledge to physical problems.

PART ONE
MECHANICS OF PARTICLES

CHAPTER ONE NEWTON'S LAWS

1·1 BACKGROUND

1·1·1 The origins of mechanics

Rarely is a fundamental "law" of science the fresh creation of any one man. Insight into the basic relationships that appear to govern the physical world almost always arises from the observations and speculations of many men. These ideas gradually become both more comprehensive in scope and more definitive in application, until finally a man of deep intellect carefully defines these relationships as "physical laws" and systematically develops their consequences.

The origins of the fundamental concepts of mechanics are historically obscure. The earliest attempts to explain natural phenomena were based on *animism*, the interpretation of the behavior of natural objects and events (rocks, rivers, thunderstorms, etc.) as conscious and living things. (If you've ever kicked a door or cursed a balky machine, you will understand the appeal of this primitive outlook.) Other explanations considered all physical events to be the conscious act of one or more deities.

We should understand that these attempts to explain nature differ in a very fundamental way from the efforts of the modern scientist. The ancients were preoccupied with the philosophical question "Why," whereas today we concentrate on the more pragmatic questions "What are the external manifestations of a physical event, and how can we predict or control it." Not until the time of Galileo (1564–1642) was this shift in basic outlook clearly evident, and it is for this reason that Galileo is so often called the founder of modern science.

It should not be imagined that all pregalilean mechanics was clouded by superstition and misconception. Nor was it limited to *qualitative* formulation of physical laws. Nearly 2000 yr before Galileo, Archimedes observed that a floating object displaces a volume of fluid weighing the same as the object. Aristotle (384–322 B.C.) wrote extensively about mechanics even earlier, proposing quantitative laws of motion that dominated thought on the subject of dynamics until the time of Galileo.

1·1·2 Aristotelian mechanics

Although Aristotle's laws of dynamics were fundamentally incorrect, they reveal in their reasonableness the subtle brilliance of the improvements wrought by Galileo and his successors. Aristotle believed that a moving agent or *force* must be applied in order to make an object deviate from its "natural motion," and that in most cases the natural motion is "rest." To Aristotle, forces could be applied only by contact, so the influence of gravity was explained as the natural tendency of heavy objects to fall and light objects to rise.[1] Thus in Aristotle's scheme the natural motion of a stone released in space is a falling motion, but the natural motion of the stone on a smooth table is a state of rest. A stone initially sliding across such a table was understood to come to rest because this was the "natural state" of the stone. In order for the stone to move from one point to another on the table, some force or moving agent must come into play. Aristotle argued that if force $\mathbf{F}$ will move body B a distance D in time T, then $\mathbf{F}$ will move half of B twice as far in the same time, or $\mathbf{F}$ will move half of B the same distance D in half the time. This comes very close to saying that force is proportional to mass times (average) velocity; but Aristotle realized that this proposition would conflict with his observation that a very small force won't move a large object at all, so he stopped short of this declaration.

The concept of mass was not used by Aristotle, and without calculus only average velocities could be considered, so any modern interpretation of

[1] Any similarity between Aristotle's exclusion of gravity forces and that of Einstein in his general relativity theory must be considered to be as coincidental as the similarities of the atomic theories of matter advanced by the Greek Democritus and those of modern physics.

Aristotelian mechanics is awkward.[1] It is only with deliberate oversimplification that we may say that Aristotle held that a given body would acquire a *velocity* proportional to the applied force, and in contrast later students of mechanics found *acceleration* to be force proportional. Whereas Aristotle envisaged "force-free" or "natural" motion generally as rest relative to the earth, later scholars (notably Galileo) considered constant speed translation to be the natural motion of a force-free body. There is a tremendous conceptual difference in these "laws," and yet Aristotle's arguments are not unreasonable. He is right in observing that a stone sliding on a large smooth table will eventually come to rest; only with the subtle introduction of *friction forces* can this evidence be made to support the modern position.

In order to penetrate to the truth in science, one must make careful experiments and precise measurements. The scholars of Greece had neither the inclination to conduct experiments nor the equipment to measure time with sufficient accuracy, although the spirit of inquiry was there. For nearly 2000 yr thereafter, even this spirit was somewhat subdued, and it was not until nearly 1600 that aristotelian mechanics was successfully overthrown.

1·1·3 Newtonian mechanics

It is not our purpose here to trace the historical evolution of mechanics, although the subject offers a fascinating study of man's struggle to understand and then to control his environment. We are concerned here only with the fact that in the course of the seventeenth century the discipline of mechanics evolved from the inadequate speculations of Aristotle to the incisive declarations of Newton, as reflected in his monumental work, the *Principia*,[2] in 1686. This period was spanned by the lives of Galileo and Newton (Newton was born in 1642, the year Galileo died), but the evolution of what is now called *newtonian mechanics* was served by other men as well. The *Principia* offers a brilliant distillation of the accumulated work of its time, and a generalization beyond the terrestrial confines of Galileo's very similar laws of mechanics.

1·2 NEWTON'S PRINCIPIA

1·2·1 Newton's definitions and laws

Newton sought to establish mechanics as an axiomatic discipline, based on a set of definitions and three axioms, or laws of motion. Among his many definitions

[1] See Aristotle, "Physics," vol. 2, pp. 257–259, Harvard University Press, Cambridge, Mass., 1934.

[2] Sir Isaac Newton, "Philosophiae Naturalis Principia Mathematica" (Mathematical Principles of Natural Philosophy), Cajori translation, University of California Press, Berkeley, 1962.

are the following (paraphrased from the *Principia*, with modern nomenclature inserted parenthetically):

1 The quantity of matter in a body (the *mass* of the body) is measured by the product of density and volume. Mass is proportional to weight.

2 The quantity of motion (the *linear momentum*) of a body is measured by the product of velocity and the quantity of matter (mass).

3 An impressed force is an action exerted on a body in order to change its state, either of rest or of uniform motion in a straight line.

Newton's three laws appear in the *Principia* as follows, with parenthetical insertions providing equivalent modern language:

1 Every body continues in its state of rest, or of uniform motion in a straight line, unless it is compelled to change that state by forces impressed on it.

2 The change of motion (the time derivative of the linear momentum) is proportional to the motive force impressed and is made in the direction of the straight line in which that force is impressed.

3 To every action there is always opposed an equal reaction; or, the mutual actions of two bodies on each other are always equal and directed to contrary parts.

1·2·2 Logical flaws and experimental contradictions

Every freshman physics student knows that Newton was not entirely successful in his attempt to proclaim the inviolate laws of physics. Upon close examination, his formulation fails also to meet the demands of logical completeness and internal consistency that are required of an axiomatic discipline. (Note that these are two separate and distinct failures: euclidean geometry, for example, may be imperfect in its representation of the universe, but it is nonetheless a flawless axiomatic discipline.) Despite these imperfections, the salient fact remains that Newton's laws are ubiquitous in twentieth-century technology, providing the fundamental basis not only for the design of civil structures, transportation vehicles, and the machines of industry, but also for such incredible technological achievements as lunar and planetary landings of space vehicles.

Awareness of the logical imperfections of newtonian mechanics developed slowly for 200 yr, but by the late nineteenth century evidence of the experimental inaccuracy of predictions based on Newton's laws began to accumulate.

(It was first noted in 1845 that observations of the motion of the planet Mercury were in conflict with the predictions of newtonian mechanics, and even more definitive conflicts arose after the famous Michelson-Morley experiments with light, performed in 1881 and again in 1887.) In beginning the serious study of mechanics, we are more concerned with possible ambiguity in the definitions and axioms of classical mechanics than with their limitations in predicting physical phenomena. First priority goes to any revision of Newton's exposition that may prove necessary to make the discipline of newtonian mechanics logically sound. Only when this is accomplished are we ready to confront the truth or falsity of its predictions in application to physical systems.

If we approach Newton's definitions and axioms critically, several questions arise:

1 What is a force? Definition 3 is no more than a restatement of Newton's first law.

2 If mass is the product of density and volume, what is density? It will of course not do to define density as the ratio of mass to volume, as is our practice.

3 What is meant by velocity? Velocity relative to what? Surely not the earth?

In addition, there are implied in Newton's words certain preconceptions concerning time and the relevance of geometry to physics. Newton imagined time and space as separate and distinct quantities, with time uniquely and identically defined for all points in physical space, which was assumed to be the space of euclidean geometry. We are not concerned right now with the important question of whether these concepts are physically correct; we wish only to be sure that all premises are explicitly included in our axiomatic formulation of the foundations of newtonian mechanics.

1·3 MODERN DEFINITIONS AND AXIOMS OF NEWTONIAN MECHANICS

1·3·1 Mach's approach to definitions and axioms

Many of the logical difficulties in the *Principia* can be accommodated without disturbing the basic structure of newtonian mechanics, following in a general way the line of argument advanced by Ernst Mach,[1] but with minor modernizations.

[1] Ernst Mach, "The Science of Mechanics," The Open Court Publishing Company, La Salle, Ill., 1893.

Definitions

1 The *velocity* of a point in a given frame of reference is the time derivative in that frame of a position vector[1] locating the point relative to any point fixed in the given frame; *acceleration* is the time derivative of velocity, in the given frame.

2 *Mass* is a fundamental attribute of matter, unchanging for a given quantity of material. The mass of a given material body can be defined only in terms of an arbitrary standard unit mass (therefore in terms of a standard body) and the equation of axiom 3 below.

3 A *particle* is a geometrical point possessing constant mass.

4 The *force* applied to a particle is the product of the scalar mass of the particle and its vector acceleration in an *inertial reference frame,* which is defined by axiom 2 below.

Axioms

1 Physical space is *euclidean,* and time is *newtonian,* so that the laws of euclidean geometry are experimentally verifiable for any accuracy of measurement, and time is uniquely and identically defined for all points in space.

2 There exists a frame of reference (called the *inertial frame*) such that an isolated particle (a particle completely free of interaction with other particles) maintains a constant velocity in that frame.

3 Any two particles P_1 and P_2, isolated from all other particles, will interact in such a way as to develop accelerations in the inertial frame that are vectors in opposite directions along the line joining P_1 and P_2, according to the rule

$$m_1\mathbf{A}_{12} = -m_2\mathbf{A}_{21}$$

where m_1 and m_2 are the masses of P_1 and P_2, respectively, and $\mathbf{A}_{12}$ and $\mathbf{A}_{21}$ are the inertial accelerations of P_1 (due to P_2) and P_2 (due to P_1).

4 Superposition applies, so that the acceleration $\mathbf{A}_1$ of P_1 in the presence of $P_2, P_3, \ldots, P_n$ is given by

$$\mathbf{A}_1 = \mathbf{A}_{12} + \mathbf{A}_{13} + \cdots + \mathbf{A}_{1n}$$

[1] The word *vector* in this text indicates a *gibbsian vector,* as defined in Appendix **B**; vectors are symbolized by boldface letters, e.g., **A**. Appendix B also provides alternative interpretations of the word vector.

The differences in these two alternative formulations of the foundations of newtonian mechanics are subtle, but worth considering carefully. A thoughtful comparison of the two sets of definitions and axioms reveals that Newton's first law is accommodated by definitions 1, 3, 4, and axiom 2 of the more modern formulation, while the second and third laws are together replaced by definitions 1 through 4 and axioms 3 and 4. Axiom 1 is a statement of assumptions implicit in Newton's original formulation. Thus the differences are significant only in terms of their logical structure, and not in terms of their physical implications. Either by examining Newton's second law with *body* replaced by *particle* so that mass is *explicitly* constant (as it was implicitly so to Newton), or by combining definitions 1 through 4 and axioms 3 and 4 of the more modern formulation, one obtains

$$\mathbf{F} = m\mathbf{A} \tag{1·1}$$

where $\mathbf{F}$ is the total force applied to a particle of mass m and acceleration $\mathbf{A}$ in an inertial reference frame. It is in this form that Newton's second law is most *useful*, and on this foundation we will build in future chapters. Before proceeding with such practical matters, however, it is instructive to pause briefly to reflect on the implications of some of the fundamental concepts of newtonian mechanics.

1·3·2 Velocities and accelerations

The jumble of words required for the definitions of *velocity* and *acceleration* can be confusing and, indeed, students of dynamics have traditionally found these concepts to be the most troublesome ones in the subject. However, until we get to Chap. 3 on kinematics, we need not worry about untangling the words used in definition 1. In that chapter we will adopt clear *mathematical* definitions and develop procedures reducing all relevant problems of velocity and acceleration determination to the very straightforward process of vector differentiation. You can expect to emerge from Chap. 3 with confidence in your individual ability to determine expressions for velocities and accelerations, however complex the problem (given sufficient time). (Unfortunately, similar confidence can never be expected in connection with other topics in dynamics.) Further examination of definition 1 can therefore safely be deferred until Chap. 3.

1·3·3 Particles

The carefully defined word *particle* has in the second formulation replaced Newton's more ambiguous word *body* in all of the laws. If body is instead understood to mean any material system, there arises the question of the meaning of *the* velocity or *the* acceleration of a body, since different points of the material system might have different motions. Furthermore, a particle is by definition of constant mass, so the new terminology removes any temptation into the error of applying Newton's second law directly to a "body" having time-varying mass (such as a rocket). Of course the word particle is a complete abstraction; strictly speaking there are no *point masses* in the universe. This should not be troublesome, as long as we remember that it remains the responsibility of the engineer to decide under what circumstances he can get adequate predictions of physical behavior on the basis of a particle model of a physical system. It may be quite acceptable to predict the trajectory of a large and extremely flexible space vehicle by treating it as a particle, and at the same time quite unacceptable to idealize a small lead bullet as a particle to find its trajectory. (The bullet is subjected to air drag, which depends on the attitude or orientation of the slightly elongated slug, whereas the influence of the space vehicle attitude on its trajectory may be negligible.)

1·3·4 The inertial reference frame

To the beginning student of mechanics, probably the most important difference in the two given formulations is in their points of view toward the inertial reference frame. Newton's representation in the *Principia* involves *the* quantity of motion; he considers it to be somehow self-evident that somewhere out there in space there is a frame of reference (the inertial frame) which is *absolutely fixed*, and he argues that all motions should be measured relative to this hypothetical reference frame. In the second formulation of Newton's laws offered here, the existence of an inertial frame of reference is an explicit and central *hypothesis* (axiom 2). When we reach the point of attempting application to physical situations, we'll find that this inertial reference frame is a will-o'-the-wisp we can never quite pin down precisely, but at the same time we'll see that for any given problem there is no difficulty in identifying a reference frame which may be assumed to be inertial for the problem at hand.

Experience teaches us that for most problems of dynamics in our solar system, we may adopt as the inertial reference frame a frame in which the center of mass of the solar system is fixed, and with respect to which certain distant stars have no discernible motion.[1] For problems of galactic dynamics, this assumption yields results in conflict with measurements, and some new hypothesis for the inertial frame must be found. For the dynamic analysis of artificial earth satellites with altitudes of practical interest, the indicated frame in which the sun's center and certain stars appear fixed (the *sun-star frame*) is quite adequate

[1] A more precise definition of this reference frame involves unfamiliar concepts of astronomy, and is foregone here. The phrase "center of mass of the solar system" describes a point which, by virtue of the sun's dominant contribution to the solar system mass, is nearly at the center of the sun; a more explicit definition will be introduced in Chap. 6.

as an inertial reference frame, but acceptable results are obtained also by adopting as the inertial frame a reference frame in which the center of the earth is fixed, and with respect to which those same distant stars have no discernible motion (the *earth-star frame*). (The stars are so far away that they can appear to be stationary with respect to both the sun-centered and the earth-centered frame, although these frames circle each other annually at a radius of about 93 million miles!) Evidently either the sun-star or the earth-star frame would suffice as an inertial reference frame for the solution of dynamics problems on earth, but either of these assumptions would complicate analysis unduly in most cases. For the great majority of dynamics problems on the earth and in the lower atmosphere, the spin of the earth in the earth-star frame is dynamically inconsequential, and it is quite adequate to treat the idealized rigid earth as establishing an inertial frame of reference.

The most important thing to understand at this point is that in the final analysis it is experience with physical observations that determines whether or not a particular reference frame can properly be assumed to be an inertial frame for the application of Newton's laws to a particular problem.

1·3·5 Force

The definition of force provided in the *Principia* is not very helpful to an engineer, nor is Mach's definition of force as the product of mass and inertial acceleration. This is particularly obvious when acceleration is zero, as in a *statics* problem. Yet, in practice we rarely find ourselves troubled by the abstract nature of these definitions; we feel that we know in any given physical situation just what a force is in terms of our experience with muscular or mechanical pushing or pulling. Such natural concepts of force are quite satisfactory for practical purposes, and we need not struggle with formal definitions and their philosophical advantages in order to solve engineering problems. However, it seems unwise to ignore the deep and persistent conceptual difficulties in the treatment of force that have plagued scientists and philosophers for centuries, and that remain less than fully resolved today.[1]

It is significant that Mach, while defining force as the product of mass and inertial acceleration, finds no need to use force in the axioms following his definitions. Force becomes a convenient word for communication, and perhaps in the intermediate stages of analysis; it is no longer a fundamental and necessary concept in mechanics. Modern scholars call force a "methodological intermediate." For most of us, however, the concept of force is the result of accumulated experience in observing the agencies of change of motion. We note that with a certain kind of muscular effort we can make a wagon accelerate, and by installing

[1] See Max Jammer, "Concepts of Force," Harvard University Press, Cambridge, Mass., 1957.

a jet engine on the wagon we can get the same result. When the wagon is on a steep incline, it again begins to move. All of these phenomena we accommodate with the idea that in each case a *force* is applied to the wagon, and we come to think of force without reference to the induced motion. This is particularly convenient in solving statics problems, which involve a balance of forces that together cause no motion at all. Thus, in what follows, we speak freely of the force of gravity, or the force applied by a spring or a man, without worrying about the fact that ultimately that force must be defined or measured in terms of the acceleration that it would impart to a given mass.

Whether we define force **F** as the product $m\mathbf{A}$ (following Mach), or accept Newton's interpretation of force as an action exerted on a body to change its velocity relative to an inertial reference frame, we are left with only two kinds of physical phenomena that can give rise to forces. A force on a body must be attributable to its interaction with another body, either by *direct contact* or by *remote interaction* (such as gravitational or electromagnetic interaction). We refer to these two classes of forces as *contact forces* and *field forces*, respectively. Because many analysts are accustomed to introducing concepts that are *not* forces under designations such as "effective force," "imaginary force," "inertial force," and so forth, it is important to understand at the outset that the only (real) forces are those that comprise **F** in the expression $\mathbf{F} = m\mathbf{A}$, where **A** is acceleration relative to inertial space.

Any useful definition of force must include its identification as a *bound vector* having *magnitude, direction* (including orientation in space and a positive sense), and a *line of action*. The last of these properties is implied by Mach's definition, or any definition identifying a force vector with a particle through which its line of action must pass. The identification of force as a vector implies a procedure for summing forces to obtain a *force resultant*. But the rules of vector addition supply no information concerning any point of application of the vector resultant, so it is necessary to introduce an *arbitrary* line of action before a force resultant becomes a bound vector. For example, in the study of the mechanics of a particle p, we generally encounter systems of *concurrent forces* each of which has a line of action passing through p; the resultant of such forces is usually conceived as a force having a line of action through p. In a more general context, other choices of a line of action can be made, as indicated in Chap. 7 where applications are first considered.

Note that the principle of "action and reaction," Newton's third law, is embodied in the third axiom of our alternative formulation (although we avoid the word force). The action-reaction principle causes little difficulty in problems of statics, but many students find it troublesome when motion is involved. If a boy pulls on his wagon with a force **F**, the wagon applies an equal and opposite force ($-\mathbf{F}$) to the boy's hand. This is true whether the boy is pulling his wagon down the road or trying unsuccessfully to get it out of a ditch. If it seems somehow troublesome that the boy may be moving forward while the force applied to him by the wagon is directed backward, just remember that the interaction forces between the boy and his wagon are not the only forces in the problem. The boy must also be pushing on the ground, and the ground on the boy. The friction force transmitted to the boy through his shoes must numerically exceed the horizontal force transmitted to his hand by the wagon if he is to accelerate forward on level ground. Remember that the force **F** appearing in Newton's second law is the *vector sum* of all forces or the *force resultant* applied to a given body.

Note the great liberty that has been taken here in applying postulated laws of *particle* dynamics to such obviously extended and nonrigid bodies as a boy and his wagon. There can be no formal justification for this identification, but at the same time there can be no illustration of particle dynamics in familiar physical terms without some such identification. Experience gained in later chapters will make this assumption appear more reasonable.

1·3·6 Mass

Newton's definition of mass is unsatisfactory unless it is accompanied by a definition of *density* (since he defines mass as the product of density and volume), but density is not defined in the *Principia*. The importance of Newton's definition is therefore neither in its conceptual contribution nor in its practical value, but instead in its distinction between mass and weight, the latter being essentially a measure of the gravitational force on a body. Although this differentiation was made also by some of Newton's predecessors (notably Galileo), it was not yet fully accepted in Newton's time.

Mach's definitions pose a different kind of problem. In order actually to determine the mass of a given body experimentally, one must, according to axiom 3, isolate that body together with some standard mass and measure certain accelerations. Such total isolation of two bodies from all other bodies in the universe is, of course, physically impossible. We might hope that with the help of the principle of superposition (axiom 4) we could measure the mass of an object even in the presence of many other bodies, simultaneously determining the mass of each in terms of all of the measured accelerations. Alas, the three-dimensional space in which we are obliged to perform our measurements provides only three equations, from which no more than three unknown masses can be determined simultaneously (see Prob. 1·5), so the proposed solution fails. We are left with the conclusion that Mach's scheme is *logically* superior to Newton's original proposal, but without advantage for the solution of physical problems.

In practice, the standard unit of mass is by international agreement the mass of a particular piece of metal called the International Prototype Kilogram

(IPK). The mass of any other object must theoretically be determined by performing an experiment involving the two bodies. If this is a dynamic experiment crudely simulating that proposed in axiom 3, the mass being determined is called the *inertial mass*. If, instead, the two bodies are simply placed on a balance scale to measure the relative gravitational attractions of the two bodies, the measured mass is called the *gravitational mass*. Although these two masses are conceptually different, the most painstaking experiments have failed to detect numerical differences between them, so the distinction is rarely made. (In general relativity theory, where these concepts must be treated separately, it is *assumed* that they are numerically identical.)

1·3·7 Units

Even when the principles of mechanics are agreed on, there remains the practical necessity of establishing, however arbitrarily, a body of nomenclature to be used in standard ways by all individuals regularly engaging in technical communication. Because of the limitations in global communication in centuries past, each technical community devised its own standard definitions for quantities of length, force, mass, and time, which are of central concern in the study of mechanics. The twentieth century is a period of transition from that multiplicity of unit systems to an international system of units. Although this imposes on the modern student the burden of working with several systems of units, at the same time it frees him from the distorted belief that there is something of fundamental significance in the particular unit system used in his technical community.

In mechanics, three basic units must be established, and a fourth unit may be defined in terms of the other three, using $\mathbf{F} = m\mathbf{A}$. Acceleration $\mathbf{A}$ requires units of length and time. In view of our *definition* of force as the product of mass and acceleration, it is most reasonable to establish a fundamental definition for the unit of mass and then establish a derived unit for force. This is indeed a widely adopted modern practice, but in many engineering contexts the basic unit remains that of force, with the mass unit derived.

It may be anticipated that in time a version of the so-called *mks* system will gain universal acceptance.[1] This system employs the *meter* (m) as the standard unit of length, the *kilogram* (kg) as the standard unit of mass, and the *second* (sec) as the standard time interval. A force of sufficient magnitude to impart 1 m/sec^2 acceleration to a mass of 1 kg is called a *newton*, providing the derived unit. Each of these definitions is related to specific physical objects or phenomena by international agreement (for example, the standard for the kilogram unit is the IPK mentioned previously). These definitions are periodically

[1] The International System of units is the particular mks system most universally accepted at present.

refined, so they are of transient character, and they have no great importance in the immediate context of the study of engineering mechanics.

The unit system still used most commonly in engineering work in the United States employs the *foot* (ft), *pound* (lb), and *second* (sec) as the standard units of length, force, and time. Official definitions of these units are usually expressed in terms of the official units of the mks system. A pound (more properly called a pound-force, in distinction to another unit called the pound-mass) is, for example, defined as that force that imparts an acceleration of 32.1740 ft/sec^2 to a mass that is 1/2.2046 of the IPK, and the foot is defined as the 1200/3937 part of the standard meter. Thus the primacy of the mks system is recognized even today. Engineering practice is rapidly changing, with increasing demands for cooperative work transcending national boundaries as well as the traditional boundaries between engineers and physicists. In newer fields (such as space technology) there is already substantial acceptance of metric systems of units. However, the unit of mass most commonly used in the United States is still the *slug,* defined as that quantity of mass to which a pound-force imparts an acceleration of 1 ft/sec^2.

It is inappropriate to emphasize the unit system in this introductory treatment of engineering mechanics, and for this reason numerical values given in problems and illustrations are generally expressed in terms of those units deemed most familiar to the American reader, namely, pounds, feet, and seconds. A table of conversion to the International System of units (SI units) appears on the endpapers of this book.

The major significance of units for the purposes of this text is in providing checks of dimensional homogeneity, or dimensional consistency, on problem solutions. It is a good practice to carry units through all numerical calculations, in order that those units that emerge from the calculations can be compared with those appropriate for the physical quantity being measured.

1·4 PERSPECTIVE

The differences in the definitions and axioms of the *Principia* and the more modern version recorded in Sec. 1·3 are purely *formal;* they are of no concern to the practicing engineer in his application of Newton's laws. Stripped of all the scholarly trappings, these formulations reduce to the same procedures for dynamic analysis. In application to a physical system that may safely be modeled as a particle of known mass m, we proceed as follows:

1 Assume a particular reference frame to be an inertial frame. (It may be necessary to confirm the acceptability of this assumption by comparing its implications to physical observations for a related problem.)

2 Derive an expression for the inertial acceleration vector $\mathbf{A}$ of the particle in terms of convenient variables, called *kinematic variables.* (For a statics problem, $\mathbf{A}$ is simply zero.)

3 Develop expressions for those vector forces applied to the particle by its environment, expressing these as functions of time, position, or velocity as necessary.

4 If the objective is the *analysis* of the motion of an object under *given forces*, record the force resultant $\mathbf{F}$, the vector equation $\mathbf{F} = m\mathbf{A}$, and three equivalent second-order scalar differential equations in the unknown kinematic variables. Then integrate these equations (in closed form or numerically, using a computer) to determine the motion.

5 If the objective is the *synthesis* of the motion of an object, we *choose* the forces that result in a desired motion. If the motion is fully given as an explicit function of time, we simply record $\mathbf{F} = m\mathbf{A}$ and determine those forces that must be added to the environmental forces to obtain the given acceleration. (A special case is the problem of *particle statics*, for which $\mathbf{A}$ is zero. This will be discussed in Chap. 2.)

6 In practice, the motion is rarely given in explicit detail, so the synthesis problem becomes more difficult. For example, if the goal is an unmanned lunar landing, we may be given an approximate destination and a maximum landing speed, and told to figure out how to get there in a reasonably efficient way. This problem admits a variety of approaches, and has more than one solution. For preliminary studies, we might break this mission into several stages: launching into earth orbit; traversing from earth orbit to the vicinity of the moon; and lunar landing. Newton's second law then serves as the basis for selecting rocket-thrust levels and directions for each stage of the mission. For more refined analysis, the several stages of the mission must be matched and combined, again using $\mathbf{F} = m\mathbf{A}$ to obtain equations of motion. If it should be required that the trajectory be *optimized* in some way (for example, for minimum fuel expenditure), then mathematical principles beyond those supplied by Newton must be applied (either classical variational calculus or more modern methods of optimization theory). In current practice the problem of accomplishing a lunar or planetary landing is solved by an integrated application of the principles of classical dynamics and modern control theory, with substantial reliance on digital computers for implementation.

Because of the possible complexity of the synthesis problem, emphasis in elementary mechanics focuses on the straightforward problem of analysis: Given a set of forces applied to an idealized body, determine its motion. This category still includes some very difficult problems, particularly when more than one particle is involved and the particles interact. Most of the problems of classical celestial mechanics are in this class, and these problems once occupied the attention of the most brilliant minds of the world.

It should be recognized at the outset, however, that the dynamics problems of tomorrow are for the most part not direct *analysis* problems. As technological developments give man increasing power to *control* his environment, and not simply to *predict* the natural course of events, dynamics is used more and more in problems of *synthesis*. We are concerned not solely with predicting the path of the spaceship, but also with *guiding* it; not only with predicting the behavior of a piece of machinery, but also with *controlling* it. Thus the basic elements of mechanics and control theory must be combined before the dynamics problems of greatest modern significance can be attempted.

PROBLEMS

1·1 Reread the section of your elementary physics text dealing with Newton's laws. What does it say about the hypothesized inertial reference frame? Was it quite clear to you as a freshman that $\mathbf{F} = m\mathbf{A}$ implies a special role for a particular reference frame in which acceleration is to be measured?

1·2 Compare the several definitions of force given here to the definition in your elementary physics text, or to the definition you carried into the present study of mechanics.

1·3 Try to explain Newton's laws to an intelligent friend with limited technical education. (Remember that we have all inadvertently learned more about physics than Aristotle ever knew.)

1·4 What kind of an experiment can you devise to demonstrate Newton's laws?

1·5 In a hypothetical (and physically impossible) experiment illustrated in Fig. P1·5, three particles of unknown masses m_1, m_2, and m_3 and a particle of known mass m_0 are isolated together as the only particles in the universe, and the accelerations of these particles relative to an inertial reference frame are measured. If $\mathbf{A}_i$ is the acceleration of the particle of mass $m_i (i = 0, 1, 2, 3)$, the measurements yield, in ft/sec²,

$$\mathbf{A}_0 = -10\hat{\mathbf{u}}_1 - 10\hat{\mathbf{u}}_2 - 15\hat{\mathbf{u}}_3$$
$$\mathbf{A}_1 = 2\hat{\mathbf{u}}_1 + 4\hat{\mathbf{u}}_2 + 2\hat{\mathbf{u}}_3$$
$$\mathbf{A}_2 = 4\hat{\mathbf{u}}_1 - 8\hat{\mathbf{u}}_2 + 11\hat{\mathbf{u}}_3$$
$$\mathbf{A}_3 = 6\hat{\mathbf{u}}_1 + 10\hat{\mathbf{u}}_2 + 10\hat{\mathbf{u}}_3$$

where $\hat{\mathbf{u}}_1$, $\hat{\mathbf{u}}_2$, and $\hat{\mathbf{u}}_3$ are orthogonal unit vectors. Determine the mass ratios m_1/m_0, m_2/m_0, and m_3/m_0. Would it be possible to solve this problem if there were more than three unknown masses?

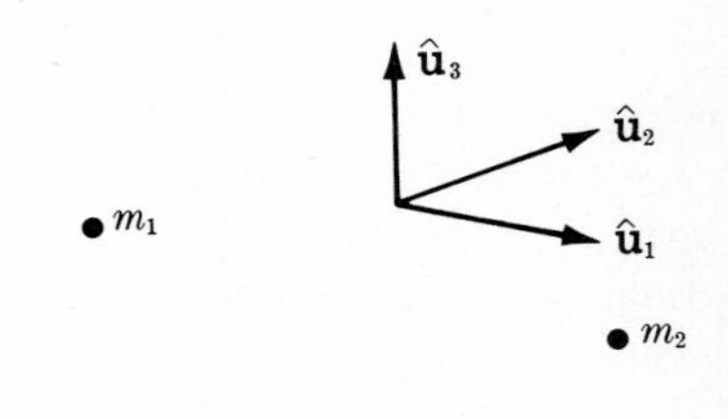

FIGURE P1·5

1·6 Evaluate, from external sources, the progress being made toward the universal adoption of an international system of units.

1·7 What is the mass of your body in slugs? in kilograms?

1·8 A differential equation of motion is found to be

$$-kx\hat{\mathbf{u}}_r - mg\hat{\mathbf{u}}_n + F_t\hat{\mathbf{u}}_t + F_n\hat{\mathbf{u}}_n = m(\ddot{x} - x\Omega^2)\hat{\mathbf{u}}_r + 2m\dot{x}\Omega\hat{\mathbf{u}}_t$$

where $\hat{\mathbf{u}}_r$, $\hat{\mathbf{u}}_t$, and $\hat{\mathbf{u}}_n$ are orthogonal unit vectors (dimensionless); x has units of length; m has units of mass; F_t and F_n have units of force; Ω has units of radians per unit time; k has units of force divided by length; and g has units of acceleration. A dot over a symbol indicates time differentiation, so that, for example, $\dot{x}$ has units of length per unit time.

Check the given equation for dimensional homogeneity or consistency. Would the units be consistent if Ω were given in terms of degrees per unit time?

CHAPTER TWO
STATICS: EQUILIBRIUM OF CONCURRENT FORCES

2·1 PARTICLE STATICS PROBLEMS

2·1·1 Classification

The simplest problem in mechanics occurs when a particle under applied forces remains at rest in inertial space; the particle is then said to be in *static equilibrium*, and the vector sum of the forces applied to the particle is zero. Since there is no inertial acceleration, the inertial mass of the particle is irrelevant, and the object in equilibrium could in most cases be called a *point* as well as a particle.[1] What follows is therefore applicable to any collection of forces passing through a common point; this class of problem is often called a *concurrent-force system*.

[1] In contrast the *gravitational mass* may be important since it determines the *gravitational force* exerted on the particle. The term *particle* (as opposed to *point*) should therefore be preserved whenever gravitational attraction is considered.

Problems in particle statics fall into two categories, depending on what is given and what is to be determined:

1 Given static-equilibrium positions of particles subjected to certain prescribed forces, determine the additional forces required to sustain static equilibrium.

2 Given all applied forces as functions of position, determine the position (or positions) of static equilibrium of a particle or system of particles.

Problems in the first category are conceptually relatively straightforward in solution since they reduce to the direct addition of known applied force vectors to obtain for each particle a resultant force that must be equilibrated by an equal and opposite reaction-force resultant. The problem may also involve distributing this resultant among several force-transmitting agencies, leading into a discipline sometimes called the *theory of structures;* for this task the equations of static equilibrium do not always suffice. Section 2·2 is concerned with applications, which illustrate the salient features of equilibrium-force-determination problems and their solution.

The second category of particle statics problems is somewhat more complex. In this case all forces are available as functions of position, so that from Newton's second law one may record for each particle an equation indicating that the sum of some vector functions equals zero. Solution of this vector algebraic equation (or equivalently three scalar algebraic equations) for the equilibrium position of a given particle (represented by a vector or three scalar coordinates) is a routine problem conceptually but, since the functions may be complex and nonlinear, the calculations may be difficult. When several particles are involved, there is a corresponding number of vector equations, which may introduce algebraic complexity by their coupling. The second category of problems is illustrated in Sec. 2·3.

2·1·2 Idealization of forces

Engineering problems are usually posed in physical terms, although in the classroom we often remove certain ambiguities by prescribing a mathematical idealization for a physical system. In mechanics this idealization involves not only the description of a material body as a particle, rigid body, or whatever, but also the idealized description of forces and force-transmission devices. Certain words and symbols have gained general acceptance for this purpose. For example, the word *spring* and its pictorial representation as some sort of wiggly line are generally understood to imply a device that transmits along its length a *restoring force* proportional to its deformation (elongation and contraction resulting, respectively, from tensile and compressive forces in the spring). If the elastic device offers linear resistance to rotation rather than translation, the term *rotary spring* is employed, and the pictorial representation is generally a helix. If the restoring force or torque is not directly proportional to deformation, but is instead some nonlinear function of deformation, the force- or torque-transmission device is called a *nonlinear spring*. Thus the unmodified word spring is generally understood to mean a *linear translational spring*. It is impossible, of course, to build a physical device that behaves in a precisely linear fashion; all physical structures are imperfect in their linearity even under small deformations, and under sufficiently large deformations they deviate grossly from linear response (they break!). Nonetheless, a great many physical systems can reasonably be modeled as springs over a limited range of deformation, and the analysis of a group of idealized particles connected by idealized springs has a surprising relevance to practical engineering problems. These remarks are offered in justification of the liberal use of such idealized concepts as springs, inextensible strings, pinned joints, inextensible rods, and rough surfaces in the following examples. All of these words have a place in our casual conversation, but they also have sharply defined technical meanings, as revealed in the applications following.

In addition to *contact forces* exerted on a particle by adjacent force-transmission devices, we are frequently concerned with *forces exerted at a distance*. Electromagnetic forces are in this category, but the force of gravitational attraction between two bodies is the most important example in newtonian mechanics of a force transmitted at a distance. Distinctions between the related concepts of *gravitational attraction* and *weight* have been blurred by conventional usages and thus deserve brief attention here. The weight of a body is generally understood to be a number (with units of force) corresponding to the recording on a set of scales supporting the body. When we fix a particular number for a particular object (as in "I weigh 160 lb"), we usually imply that the scales are located somewhere on the surface of the earth, but when we speak of the weightless state of an astronaut we imply that the scales are located in his satellite. In either case the scales are measuring not merely the force of gravitational attraction, but also certain properties of the motion of the object relative to an inertial reference frame. Since the earth is not an inertial frame of reference, a set of scales on the earth is influenced in its recording by the spin of the earth, as well as by the force of the earth's gravitational attraction. Because the weight of an object on the surface of the earth and the magnitude of its force of gravitational attraction are so nearly equal, the distinction is rarely made for terrestrial engineering problems. Furthermore, the variation of weight and gravity force with small changes in altitude is generally ignored for the study of the mechanics of bodies on or near the surface of the earth. In this text the term *weight* is therefore used to imply the magnitude of a constant force in the direction of the local vertical, as established by a plumb bob, equal to the product of the mass m and the gravitational acceleration constant g, given approximately by 32.2 ft/sec^2. When we wish to recognize the rotation of the earth in an inertial reference frame, or to accommodate the dependence of gravity force on the location of the object, we will avoid the word weight and work only with the terms *gravity force* and *gravitational attraction*.

The nature of the idealized force-transmission devices generally establishes the detailed character of the approach to a given problem, particularly when equilibrium forces are to be determined. This fact is perhaps best confirmed by considering the following specific examples.

2·2 EQUILIBRIUM–FORCE DETERMINATION

2·2·1 One string-supported particle

In attempting to illustrate particle statics problems, we are obliged to revert to schematically portrayed physical situations, such as Fig. 2·1a. We may interpret this figure as representing a body hung from the ceiling by a rope over a pulley, but we *idealize* the body as a particle, the pulley as a frictionless support, and the rope as an inextensible massless device capable of transmitting a force only when drawn taut and then only along its length. Such a force-transmission device is called a *massless inextensible string*, or simply a *string* (since an "elastic string"

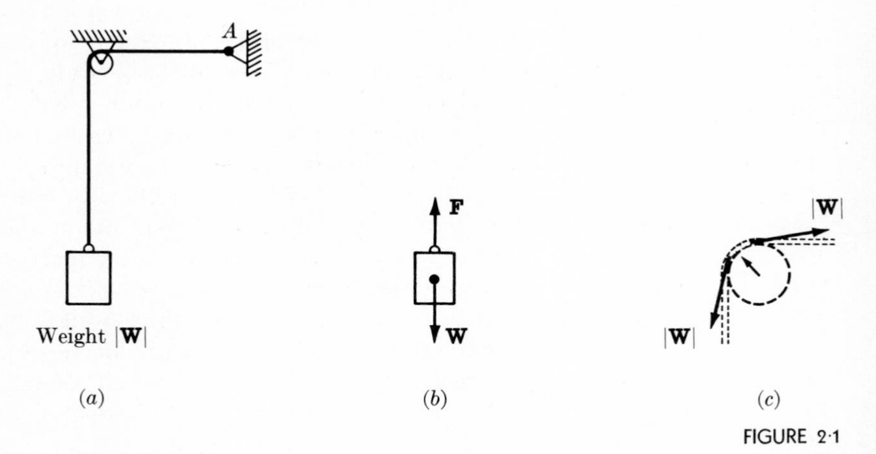

FIGURE 2·1

is really a spring according to previous definitions, and we shall not have occasion in this book to deal with continuous distributions of mass in deformable bodies). In this problem we are given the fact that the particle is in static equilibrium at the end of the string, and we are asked to find the force $\mathbf{F}$ applied to the particle by the string.

At this level of idealization the solution is obvious; static equilibrium requires that $\mathbf{F} + \mathbf{W} = 0$, or that the force $\mathbf{F}$ equal $-\mathbf{W}$, where $|\mathbf{W}|$ is the particle weight. Anticipating more difficult problems, we develop in this simple context a practice that proves invaluable in all of mechanics—we draw a *free-body diagram*, as shown in Fig. 2·1*b*. This involves simply isolating the particle of interest, and drawing it with arrowed line segments representing applied forces. Unknown forces (such as $\mathbf{F}$) then can often be determined by inspection, or by simple vector algebra.

It should be acknowledged in the context of this simple problem that the decision to focus attention on the particle at the end of the string was an arbitrary one. We could equally well concentrate on some point in the middle of the idealized string, draw our free-body diagram showing the equal and opposite forces applied to this point along the string line, and reach the obvious conclusion that at any point of the string there is a tensile or stretching force of magnitude $|\mathbf{W}|$. The frictionless support provided by the pulley lacks the capacity to change the tension in the string (see Fig. 2·1*c*) so that the force of magnitude $|\mathbf{W}|$ is also transmitted to the wall at support point A. It is the point of this digression to indicate once again that the methods of statics can be applied to idealized massless *points*, as well as to particles possessing mass. Although weight retains obvious significance in statics problems, mass in itself is irrelevant when there is no inertial acceleration.

Although by definition a string can carry only tensile load, it should be noted that if a string is under tension due to one given load it can under an *additional* load experience a reduction in tension, so that in terms of its response to the second load the string is effectively sustaining a compressive load. For example, if the bottom end of the string bearing the weight $|\mathbf{W}|$ in Fig. 2·1 is subjected to a small force $\mathbf{F}$ in a direction vertically upward, the magnitude of the tensile force in the string changes from $|\mathbf{W}|$ to $|\mathbf{W}| - |\mathbf{F}|$, so that the pretensioned string resists a compressive load. Of course the string is not able to sustain the total applied load unless $|\mathbf{F}| \leq |\mathbf{W}|$.

2·2·2 Two string-supported particles

Figure 2·2*a* represents a generalization of the previous problem in the sense that two particles are involved, but we shall see that equilibrium-force-determination problems are not made more difficult by the presence of more than one particle. In the figure, two identical particles of weight W are supported symmetrically on a single 15-ft-long string, and their equilibrium positions are given. In order to determine the forces applied to the particles at B and C by the string, we simply isolate one of the particles (say the particle at C) and sketch a free-body diagram (as in Fig. 2·2*b*). The string can sustain only tension, and the orientation of the string lengths CB and CD is given, so the forces applied to C are each of known orientation and sense, but of unknown magnitude. Identify the directions of these forces by the fully defined unit vectors[1] $\hat{\mathbf{u}}_D$ and $\hat{\mathbf{u}}_2$, and identify the unknown magnitudes of the forces by F_D and F_B. The force-equilibrium equation then becomes

$$W\hat{\mathbf{u}}_1 + F_B\hat{\mathbf{u}}_2 + F_D\hat{\mathbf{u}}_D = 0 \tag{2·1}$$

There remains a simple problem of vector algebra, which becomes a problem of scalar algebra with the substitution of the vector identity

$$\hat{\mathbf{u}}_D = -\frac{4}{5}\hat{\mathbf{u}}_1 - \frac{3}{5}\hat{\mathbf{u}}_2 \tag{2·2}$$

[1] Throughout this text, a vector with a caret (^) over it symbolizes a unit vector, which is a vector of length or magnitude equal to 1. By convention, two unit-vector symbols differing only in the numerical subscripts in the range 1, 2, 3 are orthogonal (so that in Fig. 2·2, $\hat{\mathbf{u}}_1 \cdot \hat{\mathbf{u}}_2 = 0$) unless noted otherwise; when the numerical subscripts 1, 2, and 3 are all used, the relationship $\hat{\mathbf{u}}_1 \times \hat{\mathbf{u}}_2 = \hat{\mathbf{u}}_3$ is implied.

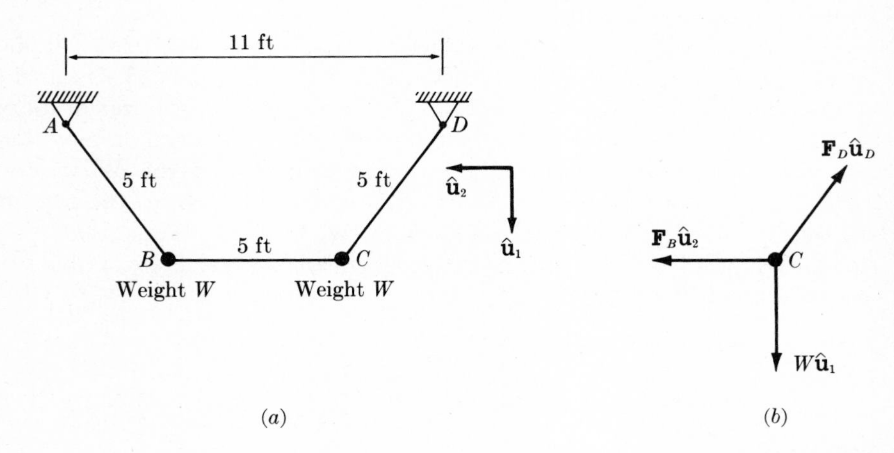

FIGURE 2·2

and the equating of the coefficients of $\hat{\mathbf{u}}_1$ and $\hat{\mathbf{u}}_2$ separately to zero. The result is

$$W - \frac{4}{5}F_D = 0$$

$$F_B - \frac{3}{5}F_D = 0$$

or

$$\begin{aligned} F_D &= 1.25W \\ F_B &= 0.75W \end{aligned} \tag{2·3}$$

Note that this result has been obtained without ever considering the particle at B. In order to complete the problem, we would in most cases find it necessary to proceed to the next particle, draw a new free-body diagram, and then write the new force-equilibrium equations and solve for any remaining unknowns. For the special case illustrated in Fig. 2·2*a*, however, it is apparent from symmetry considerations that the magnitudes of the tensile forces in strings AB and CD must be identical, so the problem is completely solved.

In the foregoing example, it was irrelevant which of the two particles B or C was isolated first for equilibrium-force determination. This would be the case even if symmetry were destroyed, as long as each particle was supported in the given plane by no more than two strings (or in three-dimensional space by no more than three strings). If, however, any of the particles is attached to more than three strings (or more than two for a planar problem), it becomes impossible to solve for the unknown forces from the equilibrium equation for that particle because there are more unknowns than there are scalar algebraic equations. In some cases this dilemma can be resolved by first obtaining the equilibrium equation for an adjacent particle with fewer strings attached, and solving for the force in the string connecting these adjacent particles. This possibility is illustrated by Fig. 2·3; for this planar configuration it is essential that the equilibrium equation of the lower particle be solved first, in order that the solution to the equilibrium equation for the upper particle be solvable. If we tried to write the force-equilibrium equation for the upper particle first, we would be stymied by the presence of three unknowns in just two equations.

It may happen, of course, that all particles in a given problem are connected to more than three strings (or more than two strings in a plane). Such a configuration is illustrated in Fig. 2·4, which portrays two particles supported in a plane by five lengths of string in such a way that each particle is connected to three strings in the plane. The forces in these strings cannot be determined by the equations of static equilibrium alone; such forces are called *statically indeterminate*. This concept will be discussed shortly in a wider context, but it may be remarked

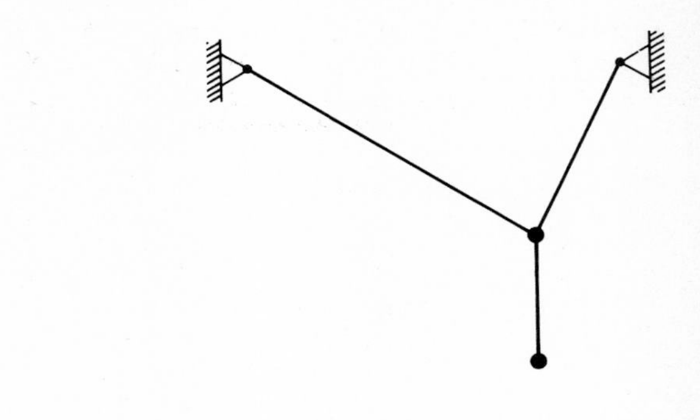

FIGURE 2·3

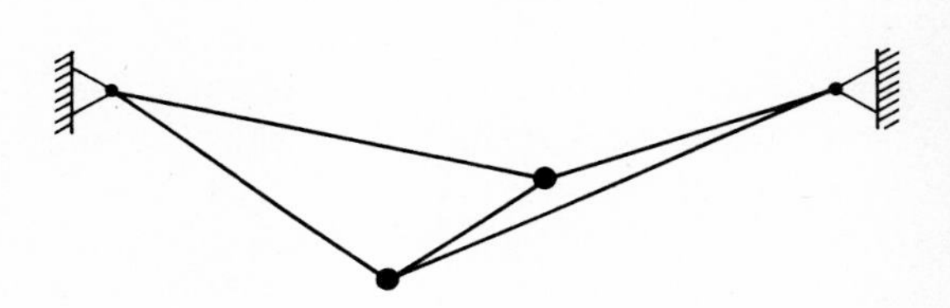

FIGURE 2·4

at this point that the difficulty stems from the idealization of the string as an *inextensible* force-transmission device. If in modeling the physical system we elect instead to represent the supporting ropes as springs rather than inextensible strings, we would find it possible to obtain sufficient additional information to solve this statically indeterminate problem. But then, of course, we would not know in advance exactly where the static-equilibrium position would lie, and we would be beyond the scope of the class of problem now being considered.

2·2·3 A truss-supported particle

In situations more complex than those illustrated in Figs. 2·1 to 2·4, the sense and even the orientation of the unknown force may be unknown. Sometimes (as in the preceding examples) the nature of the force-transmission device determines both sense and orientation (an idealized string can take only tension along its taut length). Often at least the orientation or line of action of the force is implied by the problem, although the sense is unknown. An example is illustrated in Fig. 2·5*a*. This schematic diagram shows two inextensible *rods* connected to each other and to a wall by idealized frictionless *pinned joints*. Such a structure is commonly called a *truss*. A *truss* as generally defined may include strings or springs as well as rods, but the members must be interconnected by frictionless hinges, and all loads must be idealized as concentrated forces applied at the joints. In §7·2·4 we'll see that this truss idealization *must* imply that the truss members can transmit forces only along their lengths, but for the present we

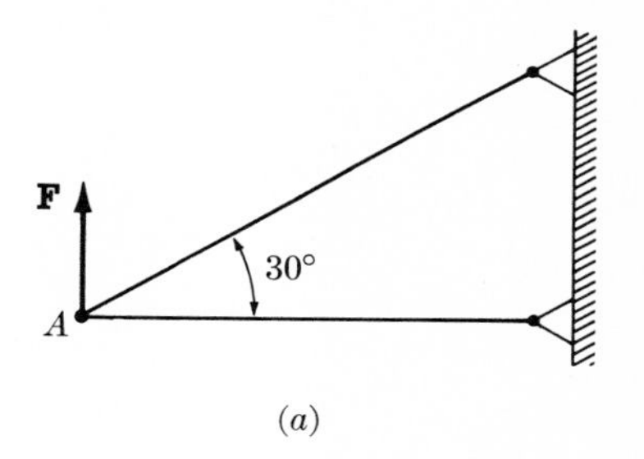

(*a*)

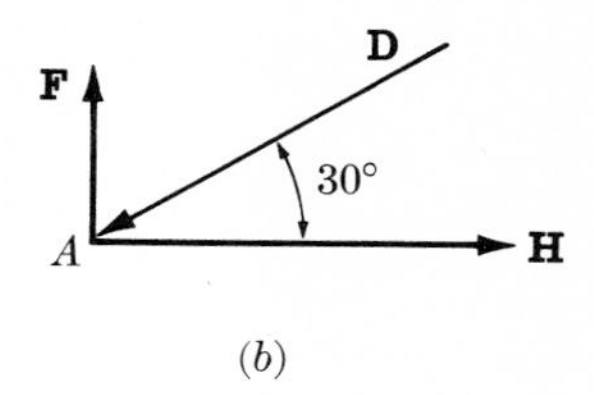

(*b*)

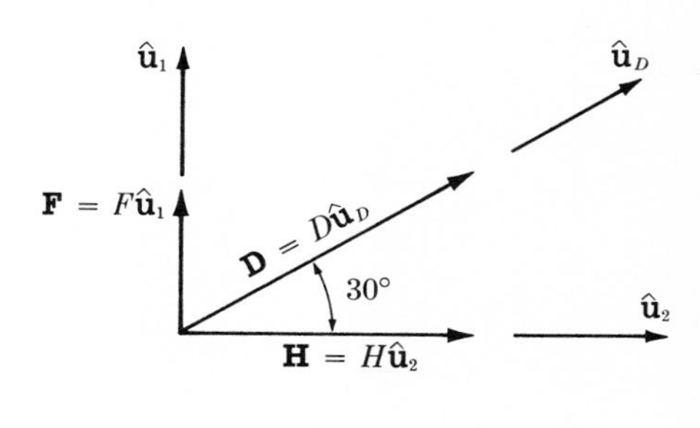

(*c*)

FIGURE 2·5

accept this declaration as an implication of the word truss. Unless a truss member is called a *string*, there is no implication as to whether the loads on truss members are tensile or compressive.

It is often possible to establish the sense of the force transmitted to the particle by inspection. In Fig. 2·5*a*, for example, the particle at A under load $\mathbf{F}$ must have a diagonal force $\mathbf{D}$ in the direction shown in Fig. 2·5*b* in order that its vertical component balance $\mathbf{F}$, and then the direction of the horizontal force $\mathbf{H}$ can be identified as that necessary to balance the horizontal component of $\mathbf{D}$.

In a more complex case, in which the *orientation* of the unknown force is implied by the force-transmission device but the *sense* of that force is not evident by inspection, we can always simply guess which way the force will go, and assign an arrow to the force line on the free-body diagram in accordance with this guess. Subsequent calculations will automatically show any error through the discovery of a minus sign in the answer.

Perhaps the easiest way to establish the sense of a force is by using unit vectors to establish a positive direction. If, for example, in Fig. 2·5*b* the free-body diagram had been drawn with the arrow on the other end of the diagonal line segment, as in Fig. 2·5*c*, the error of this guess of the direction of $\mathbf{D}$ can be automatically discovered as follows. Define a unit vector $\hat{\mathbf{u}}_D$ as in the figure and write the vector $\mathbf{D}$ as the product of the unknown scalar D and the unit vector $\hat{\mathbf{u}}_D$. The problem of finding $\mathbf{D}$ is now simply the task of finding the scalar D since the unit vector $\hat{\mathbf{u}}_D$ is fully defined. To find D in terms of the magnitude F of the applied force $\mathbf{F} = F\hat{\mathbf{u}}_1$ (see Fig. 2·5 for the vertical unit vector $\hat{\mathbf{u}}_1$), we observe from the required equilibrium-force components in the vertical direction that

$$D \sin 30° + F = 0$$

This result provides

$$D = -2F \tag{2·4}$$

The minus sign tells us the direction of the previously unknown force

$$\mathbf{D} = D\hat{\mathbf{u}}_D = -2F\hat{\mathbf{u}}_D \tag{2·5}$$

Thus $\mathbf{D}$ is of twice the magnitude of $\mathbf{F}$ and of direction *opposite* to that of $\hat{\mathbf{u}}_D$.

Similar operations may be performed with the unknown force $\mathbf{H}$, which we write as

$$\mathbf{H} = H\hat{\mathbf{u}}_2$$

where $\hat{\mathbf{u}}_2$ is a horizontal unit vector as shown in Fig. 2·5*c*. Now the equilibrium of horizontal forces provides

$$H + D \cos 30° = 0$$

or, with Eq. (2·4),

$$H = -D \cos 30° = -(-2F)\sqrt{\frac{3}{2}} = \sqrt{3}\, F \qquad (2{\cdot}6)$$

2·2·4 A systematic method

A more systematic approach to this problem should be sought since the previous wandering search first for D and then for H becomes unsatisfactory for more complex problems. Accordingly the problem is confronted again, this time with an approach that works easily for any particle statics problem of this class.

Step 1 Draw a free-body diagram of the particle, freely guessing the sense of any force for which orientation is fixed but sense is unknown.

Step 2 Define unit vectors (such as $\hat{\mathbf{u}}_1$, $\hat{\mathbf{u}}_D$, and $\hat{\mathbf{u}}_2$) anywhere on the free-body diagram you wish, so that you can conveniently write all vector forces in terms of scalars, known or unknown, and fully defined unit vectors.

Step 3 Record the given condition of force equilibrium as a vector equation. (In the example of Fig. 2·5, this step would yield $F\hat{\mathbf{u}}_1 + D\hat{\mathbf{u}}_D + H\hat{\mathbf{u}}_2 = 0$.) Now the problem is reduced completely to one of simple algebra.

Step 4 Solve the vector algebra problem. In practice this is often facilitated by selecting a set of three unit vectors that are mutually orthogonal (inventing new ones if necessary), and then writing all the unit vectors in terms of these three; these then become the *vector basis* for all vectors in the problem. (In many texts the three orthogonal unit vectors are always called **i**, **j**, and **k**. In this text they often appear with indices, as in $\hat{\mathbf{u}}_1$, $\hat{\mathbf{u}}_2$, and $\hat{\mathbf{u}}_3$ or $\hat{\mathbf{e}}_1$, $\hat{\mathbf{e}}_2$, and $\hat{\mathbf{e}}_3$, but we needn't think in such restricted terms. For the planar problem of Fig. 2·5, only two orthogonal vectors will actually be used, but we might for the sake of generality define a unit vector $\hat{\mathbf{u}}_3$ normal to the page, directed inward, and work with the orthogonal set $\hat{\mathbf{u}}_1$, $\hat{\mathbf{u}}_2$, and $\hat{\mathbf{u}}_3$.) By convention, the three orthogonal unit vectors are assigned a sequence such that the first vector cross-multiplied times the second gives the third, so that $\hat{\mathbf{u}}_1 \times \hat{\mathbf{u}}_2 = \hat{\mathbf{u}}_3$. Such a vector basis is called *dextral* or *right-handed* since it correlates with the *right-hand rule* of vector algebra.[1] Once the vector equation from step 3 is written wholly in terms of a set of three orthogonal unit vectors, the vector equation reduces to three scalar equations, some of which may provide the trivial identity $0 = 0$. Providing that the number of unknowns matches the number of independent nontrivial equations, the solution can readily be found.

In illustration of this procedure, the problem of Fig. 2·5*a* yields Fig. 2·5*c* in its first two steps, and step 3 provides

$$F\hat{\mathbf{u}}_1 + D\hat{\mathbf{u}}_D + H\hat{\mathbf{u}}_2 = 0 \tag{2·7}$$

In step 4, if we choose the orthogonal vectors $\hat{\mathbf{u}}_1$, $\hat{\mathbf{u}}_2$, and $\hat{\mathbf{u}}_3$ as the vector basis, we require the identity

$$\hat{\mathbf{u}}_D = \sin 30° \,\hat{\mathbf{u}}_1 + \cos 30° \,\hat{\mathbf{u}}_2 \tag{2·8}$$

which is available by inspection of Fig. 2·5*c*. Substituting Eq. (2·8) into Eq. (2·7) provides

$$F\hat{\mathbf{u}}_1 + D(\cos 30° \,\hat{\mathbf{u}}_2 + \sin 30° \,\hat{\mathbf{u}}_1) + H\hat{\mathbf{u}}_2 = 0 \tag{2·9}$$

or, since $\hat{\mathbf{u}}_1$ and $\hat{\mathbf{u}}_2$ are orthogonal vectors,

$$(F + D \sin 30°)\hat{\mathbf{u}}_1 = 0 \tag{2·10}$$

and

$$(D \cos 30° + H)\hat{\mathbf{u}}_2 = 0 \tag{2·11}$$

Since the unit vectors $\hat{\mathbf{u}}_1$ and $\hat{\mathbf{u}}_2$ are not zero vectors, it follows that the two scalar expressions in parentheses in Eqs. (2·10) and (2·11) must be zero. These equations provide the same results as Eqs. (2·1) and (2·5), although obtaining them has required less ingenuity and visualization and a little more routine mathematics.

In a formal sense there must always be three scalar equations emerging from a single vector equation such as Eq. (2·7). In this case, the third equation, which augments Eqs. (2·1) and (2·5), is the trivial *zero equals zero*, which comes from the components in the $\hat{\mathbf{u}}_3$ direction of the forces in Eq. (2·7), since all such components are zero.

2·2·5 Alternative methods

It should be acknowledged that the four-step analysis procedure just described is not the only systematic approach to a particle statics problem, nor even the

[1] If the fingers of the right hand are pointed in the direction of $\hat{\mathbf{u}}_1$ and then rotated into alignment with the direction of $\hat{\mathbf{u}}_2$, the thumb of the right hand points along $\hat{\mathbf{u}}_3$, in the direction in which a common screw with right-handed threads would advance if similarly turned.

most common approach. Probably most authors (and therefore most engineers) work *exclusively* with sets of orthogonal unit vectors (typically labeled $\hat{\mathbf{i}}$, $\hat{\mathbf{j}}$, and $\hat{\mathbf{k}}$). The unit vector $\hat{\mathbf{u}}_D$ does not fit easily within this framework, and many analysts would therefore avoid it. Instead, the unknown vector $\mathbf{D}$ in Fig. 2·5*c* would be written immediately in terms of the orthogonal unit vectors $\hat{\mathbf{u}}_1$ and $\hat{\mathbf{u}}_2$ (which might be labeled $\hat{\mathbf{i}}$ and $\hat{\mathbf{j}}$). This approach tends to run steps 2, 3, and 4 together, and blurs the distinction between the derivation of the vector equation [here Eq. (2·7)] and its solution. This traditional pattern of relying exclusively on sets of orthogonal unit vectors $\hat{\mathbf{i}}$, $\hat{\mathbf{j}}$, and $\hat{\mathbf{k}}$ seems to be a relic of the earlier practice of solving mechanics problems wholly in scalar terms, using cartesian coordinates. In modern texts we cast off this burden, and introduce unit vectors liberally and without the inhibition that comes from thinking always in terms of orthogonal triads, or sets of three.

2·2·6 Vector transformations

After obtaining the vector equation in step 3, the quest for its solution in step 4 does usually involve transformation of all unit vectors to a single vector basis (call the vectors $\hat{\mathbf{i}}$, $\hat{\mathbf{j}}$, and $\hat{\mathbf{k}}$ if you prefer). This transformation involves vector identities such as Eqs. (2·2) and (2·8). Such equations can be written after brief inspection of a sketch, just as Eq. (2·8) came from Fig. 2·5*c*. This operation sometimes seems difficult to the novice, but it is really very simple. Consider Fig. 2·6, which shows again the unit vectors $\hat{\mathbf{u}}_1$, $\hat{\mathbf{u}}_D$, and $\hat{\mathbf{u}}_2$. Remember that all of the vectors in the sketch are of length 1, as symbolized by the caret over the vector. In seeking an expression for $\hat{\mathbf{u}}_D$ in terms of the orthogonal vectors $\hat{\mathbf{u}}_1$ and $\hat{\mathbf{u}}_2$, we can resolve $\hat{\mathbf{u}}_D$ into components parallel to $\hat{\mathbf{u}}_1$ and $\hat{\mathbf{u}}_2$, and calculate the

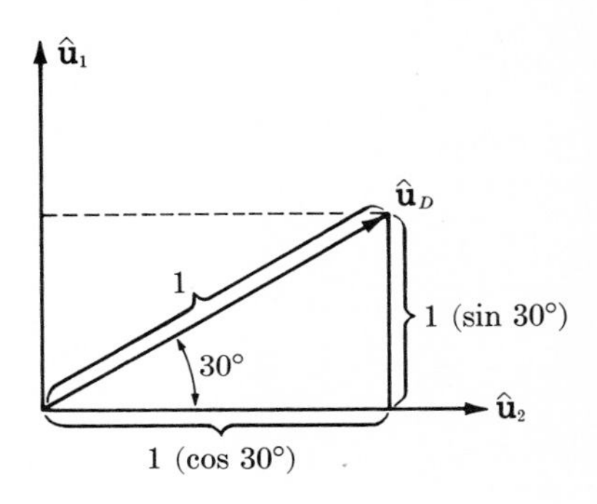

FIGURE 2·6

magnitude of these components. In this case the magnitudes are 1 (cos 30°) and 1 (sin 30°) in the horizontal and vertical directions, respectively (see Fig. 2·6). Now we can say verbally that the vector $\hat{\mathbf{u}}_D$ is equivalent to the sum of a vector of length 1 (cos 30°) in the direction of $\hat{\mathbf{u}}_2$ plus a vector of length 1 (sin 30°) in the direction of $\hat{\mathbf{u}}_1$. Symbolically this says

$$\hat{\mathbf{u}}_D = \sin 30° \,\hat{\mathbf{u}}_1 + \cos 30° \,\hat{\mathbf{u}}_2 \tag{2·12}$$

which is the result previously recorded as Eq. (2·8), with the assertion that it is "available by inspection" of Fig. 2·5*c*. With experience, this inspection requires no more than a glance, even when the previous planar example is generalized to three dimensions.

2·2·7 Statically indeterminate forces

Note that in the planar example of Fig. 2·5, the number of nontrivial scalar equations [two, from Eqs. (2·10) and (2·11)] equals the number of scalar unknowns (two, the scalars H and D). This must, of course, always be the case if the unknowns are to be determined uniquely, from basic algebraic principles. If there are more independent nontrivial equations to be satisfied than there are unknowns, it will be impossible to satisfy the equilibrium conditions. If, on the other hand, there are more unknowns than independent equations, there will be more than one set of values for the unknowns that satisfies the equations. Consider, for example, the result of adding another support member to the truss of Fig. 2·5*a*, as illustrated in Fig. 2·7. Now there are three forces of unknown mag-

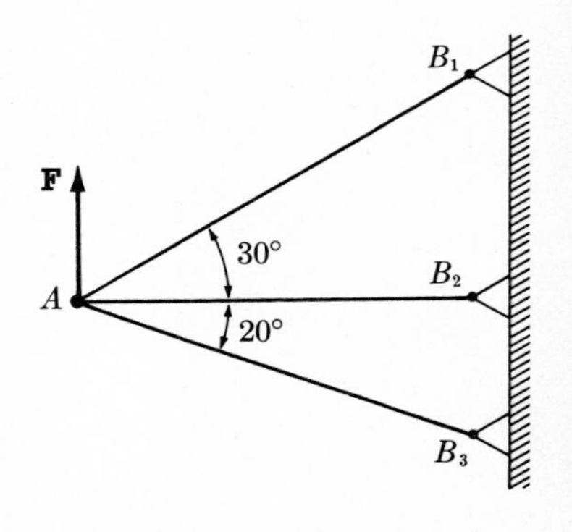

FIGURE 2·7

nitude and known direction applied at point A by members AB_1, AB_2, and AB_3; thus there are three scalar unknowns. But still there are only two nontrivial equilibrium equations for this planar problem. The difficulty now is not with finding values for these unknowns that satisfy the equations. Indeed, we know from our study of the truss in Fig. 2·5 that there exists one solution for which the loads in members AB_1, AB_2, and AB_3 are, respectively, $2F$, $\sqrt{3F}$, and 0. The difficulty lies in the fact that this is but one of an infinite number of solutions that satisfy the equilibrium equations. When the number of equations of static equilibrium is insufficient to determine the unknown forces uniquely, these forces are called *statically indeterminate*. (Recall that this phrase was used previously in the context of string-supported particle problems.)

We realize, of course, that there must be some way to determine all of the member forces in real structures such as those represented by Figs. 2·4 and 2·7; there must be other conditions to be satisfied besides those of static equilibrium. In the examples for which complete solutions have been obtained, it has been possible to determine the unknown forces by imagining the structural members to be inextensible force-transmission devices. In reality, however, *all* physical bodies are deformable. When we revise our mathematical model of the physical system in such a way as to permit deformations of the force-transmission devices (changing inextensible strings and rods to elastic springs), we will again find it possible to calculate the forces in these structural members. Because this step carries us into a class of problems in which we no longer know exactly where the equilibrium positions are, further discussion of statically indeterminate forces is deferred until Sec. 2·3.

2·2·8 Symmetry

In special situations it is possible to augment the equations of static equilibrium with *symmetry conditions*, and in this way to determine forces not available from equilibrium conditions alone. When there exists a *plane of symmetry* passing through a structure and dividing it into two identical substructures with identical loadings and identical projections on the plane, then the corresponding members of the two substructures must sustain the same load, by the logic of symmetry. Figure 2·8*a* illustrates this principle with a simple planar truss consisting of four members, and thus involving four unknown force magnitudes. The planar equilibrium equations for point A number only two, so the member forces are statically indeterminate. Yet the plane normal to the page and including the line of action of the force $\mathbf{F}$ is a plane of symmetry. We can cut the figure along this plane into two statically determinate problems, as in Fig. 2·8*b*, and then determine the member forces easily. (Because the support blocks are on frictionless

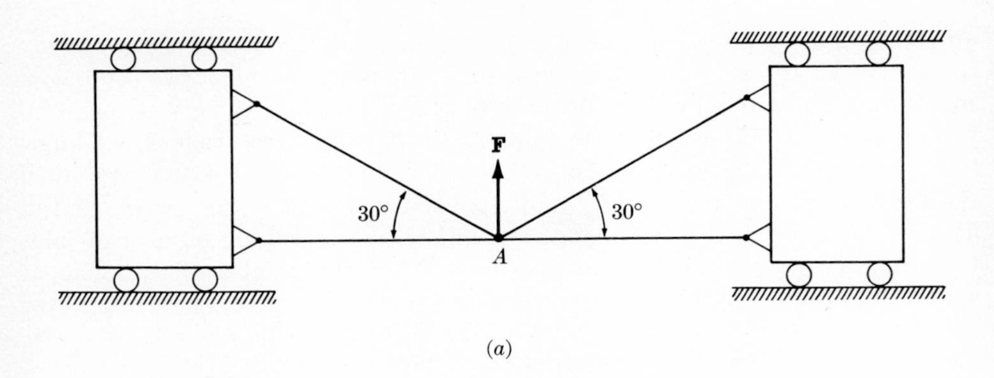

(a)

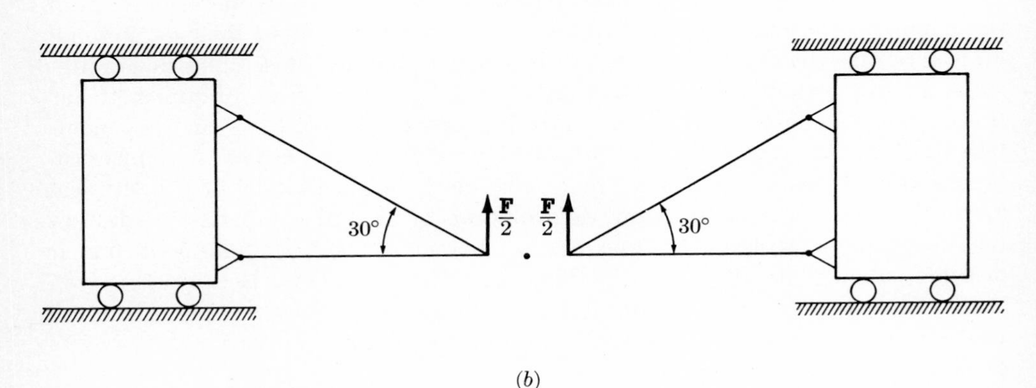

(b)

FIGURE 2·8

wheels there can be no horizontal interaction force at A between the two halves of the structure.)

2·2·9 A three-dimensional truss

The systematic procedure developed for the planar example of Fig. 2·5 is equally successful in application to three-dimensional problems, as illustrated by the following example.

A simple equilateral tripod truss is erected over the face of a clock that lies with its face in a horizontal plane (see Fig. 2·9a). The height of the tripod is a, the same as the radius. The feet of the tripod are supported at twelve o'clock, four o'clock, and eight o'clock. A horizontal force **F** is applied at the top of the tripod in a direction corresponding to one o'clock, 30 degrees out of the plane formed by points A, C_{12}, and O (see Fig. 2·9). Determine the forces applied to point A by the tripod legs.[1]

Step 1 Construct a free-body diagram, as shown in Fig. 2·9*b*, with assumed sense and known orientation for unknown forces.

[1] In what follows, nonorthogonal unit vectors are introduced with numerical indices corresponding to the numbers on the clock, so that we must temporarily set aside the convention whereby unit vectors such as $\hat{\mathbf{u}}_1$, $\hat{\mathbf{u}}_2$, and $\hat{\mathbf{u}}_3$ are orthogonal.

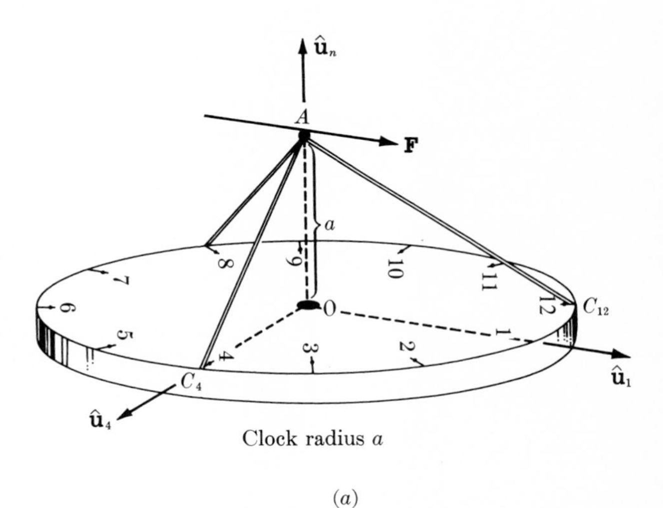

(*a*)

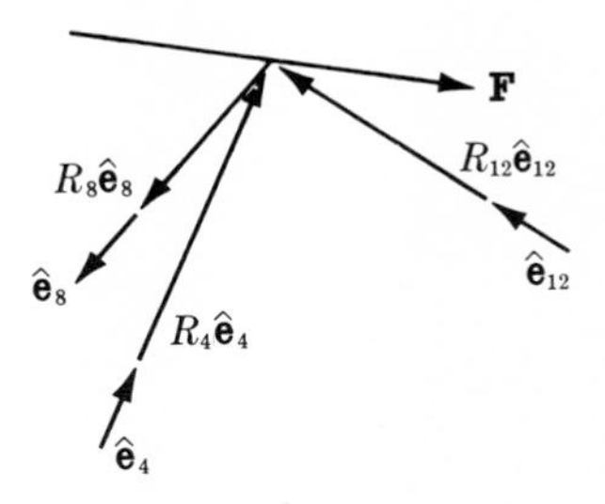

(*b*)

FIGURE 2·9

Step 2 Define unit vectors $\hat{\mathbf{e}}_{12}$, $\hat{\mathbf{e}}_4$, and $\hat{\mathbf{e}}_8$ to parallel the rods of the truss, as shown on the free-body diagram, and identify the unknown reactions as $R_{12}\hat{\mathbf{e}}_{12}$, $R_4\hat{\mathbf{e}}_4$, and $R_8\hat{\mathbf{e}}_8$.

Step 3 Record the equilibrium condition

$$\mathbf{F} + R_{12}\hat{\mathbf{e}}_{12} + R_4\hat{\mathbf{e}}_4 + R_8\hat{\mathbf{e}}_8 = 0 \tag{2·13}$$

This completes the derivation, and leaves only the task of rewriting the vector equation (2·13) as three scalar algebraic equations to be solved for the three unknowns R_{12}, R_4, and R_8.

Step 4 Solution might be facilitated by expressing all unit vectors in Eq. (2·13) in terms of a common orthogonal vector basis consisting of a unit vector $\hat{\mathbf{u}}_n$ normal to the clock and vertical upward and two unit vectors $\hat{\mathbf{u}}_1$ and $\hat{\mathbf{u}}_4$ paralleling lines from the clock center to the integers 1 and 4, respectively, directed radially outward (see Fig. 2·9*a*). Thus if F is the magnitude of the applied force, the substitution of $F\hat{\mathbf{u}}_1$ for $\mathbf{F}$ in Eq. (2·13) combines with the substitution of the purely geometrical relationships for $\hat{\mathbf{e}}_{12}$, $\hat{\mathbf{e}}_4$, and $\hat{\mathbf{e}}_8$ in terms of $\hat{\mathbf{u}}_n$, $\hat{\mathbf{u}}_1$, and $\hat{\mathbf{u}}_4$ to obtain the equilibrium equation in convenient form. It is helpful for the geometrical analysis to define unit vectors $\hat{\mathbf{u}}_2, \hat{\mathbf{u}}_3, \ldots, \hat{\mathbf{u}}_{12}$ so that one such vector points from the clock center to any number on the clock's face. Then geometry provides

$$\begin{aligned}
\hat{\mathbf{e}}_{12} &= \frac{\sqrt{2}}{2}\hat{\mathbf{u}}_n - \frac{\sqrt{2}}{2}\hat{\mathbf{u}}_{12} \\
\hat{\mathbf{e}}_4 &= \frac{\sqrt{2}}{2}\hat{\mathbf{u}}_n - \frac{\sqrt{2}}{2}\hat{\mathbf{u}}_4 \\
\hat{\mathbf{e}}_8 &= -\frac{\sqrt{2}}{2}\hat{\mathbf{u}}_n + \frac{\sqrt{2}}{2}\hat{\mathbf{u}}_8
\end{aligned} \tag{2·14}$$

Now the various unit vectors parallel to the plane of the clock face are written in terms of $\hat{\mathbf{u}}_1$ and $\hat{\mathbf{u}}_4$, to yield (noting that a 30-degree arc separates the numbers on a clock, and that $\cos 30° = \sqrt{3}/2$ and $\sin 30° = 1/2$)

$$\begin{aligned}
\hat{\mathbf{e}}_{12} &= \frac{\sqrt{2}}{2}\hat{\mathbf{u}}_n - \frac{\sqrt{2}}{2}\left(\frac{\sqrt{3}}{2}\hat{\mathbf{u}}_1 - \frac{1}{2}\hat{\mathbf{u}}_4\right) \\
\hat{\mathbf{e}}_4 &= \frac{\sqrt{2}}{2}\hat{\mathbf{u}}_n - \frac{\sqrt{2}}{2}\hat{\mathbf{u}}_4 \\
\hat{\mathbf{e}}_8 &= -\frac{\sqrt{2}}{2}\hat{\mathbf{u}}_n + \frac{\sqrt{2}}{2}\left(-\frac{\sqrt{3}}{2}\hat{\mathbf{u}}_1 - \frac{1}{2}\hat{\mathbf{u}}_4\right)
\end{aligned} \tag{2·15}$$

The indicated substitutions into Eq. (2·13) now give

$$F\hat{\mathbf{u}}_1 + R_{12}\left(\frac{\sqrt{2}}{2}\hat{\mathbf{u}}_n - \frac{\sqrt{6}}{4}\hat{\mathbf{u}}_1 + \frac{\sqrt{2}}{4}\hat{\mathbf{u}}_4\right) + R_4\left(\frac{\sqrt{2}}{2}\hat{\mathbf{u}}_n - \frac{\sqrt{2}}{2}\hat{\mathbf{u}}_4\right)$$
$$+ R_8\left(-\frac{\sqrt{2}}{2}\hat{\mathbf{u}}_n - \frac{\sqrt{6}}{4}\hat{\mathbf{u}}_1 - \frac{\sqrt{2}}{4}\hat{\mathbf{u}}_4\right) = 0 \qquad (2·16)$$

Coefficients of the independent unit vectors $\hat{\mathbf{u}}_1$, $\hat{\mathbf{u}}_n$, and $\hat{\mathbf{u}}_4$ can be collected and individually equated to 0 to obtain (after multiplying by 4 or 2 as appropriate)

$$\begin{aligned} \sqrt{6}\,R_{12} + \sqrt{6}\,R_8 &= 4F \\ \sqrt{2}\,R_{12} + \sqrt{2}\,R_4 - \sqrt{2}\,R_8 &= 0 \\ \sqrt{2}\,R_{12} - 2\sqrt{2}\,R_4 - \sqrt{2}\,R_8 &= 0 \end{aligned} \qquad (2·17)$$

By subtracting the third of Eqs. (2·17) from the second, we find $R_4 = 0$, so that $R_{12} = R_8 = 2F/\sqrt{6}$.

It may seem that such a simple answer should have been anticipated, and indeed an experienced analyst may have noted by inspection that equilibrium can be attained with no load in the four-o'clock leg, since the applied force **F** is in the plane established by the eight-o'clock and twelve-o'clock legs of the tripod. It should be noted that, although the tripod structure has three planes of symmetry, there is no plane of symmetry for this total system since the force does not lie in any of these three planes of structural symmetry.

The procedure applied here would work equally well with the force directed toward any other time of day, or if the tripod were not equilateral and the force not horizontal. The first three steps, resulting in vector equilibrium equations, are just as easy to find in the more general case. As long as the recommended procedure is followed, the difficulties introduced by increased generality are confined to the algebraic labors of solving a vector equation.

2·2·10 Force-transmission devices of unspecified direction

In the first example (Fig. 2·1*a*), the nature of the force-transmission mechanism (a string) is such that both orientation and sense of the unknown force are implied. A string (as idealized) can transmit a force only along the taut line to which it is stretched, and it can carry only a tensile load. When the nature of the force-transmission mechanism betrays both sense and orientation of the unknown force, the only unknown is the absolute value of the force. When the sense of the force is known in advance, it should always be possible to assign arrowed vectors to free-body diagrams properly, and thus no minus signs should appear in the results for force magnitudes.

In the third example (Fig. 2·5*a*), the structure is a pin-jointed truss, so the line of action of all forces is known, but the sense of the forces (tensile or compressive) is unknown. For this class of problem, the recommended approach involves the introduction of unit vectors along the known lines of action of the forces, in a direction corresponding to the analyst's guess regarding the sense of the unknown force. The only significant difference in this category of problem, as distinguished from those involving tension members only, is that here the analyst's guess might be wrong. As indicated previously in some detail, this little difficulty can always be accommodated by careful attention to algebraic signs since an erroneous guess results in a minus sign.

There remains a wide class of problems for which even the orientations of the unknown forces are not implied by the nature of the force-transmission mechanisms, and then a different procedure for setting up the equations is recommended. Perhaps the simplest example of this kind is the idealized *rough surface,* as illustrated by the classical problem of the block resting on an inclined plane, as in Fig. 2·10*a*.

2·2·11 A particle on an inclined plane

The plane in Fig. 2·10*a* is inclined an angle α from the horizontal line established by the uniform gravity field, but it is assumed that there is sufficient friction between block and plane to prevent sliding. The nature of the rough-surface idealization is such that the wall may apply to an adjacent body a force normal to the wall or inclined from the normal by a limited amount established by the *friction coefficient* for the two contact surfaces. This is therefore a problem for which the orientation of the force is *not* implied by the nature of the force-transmission mechanism.

For a specific problem as simple as that illustrated in Fig. 2·10*a*, the direction of the reaction force **R** is known immediately, as illustrated by the simple

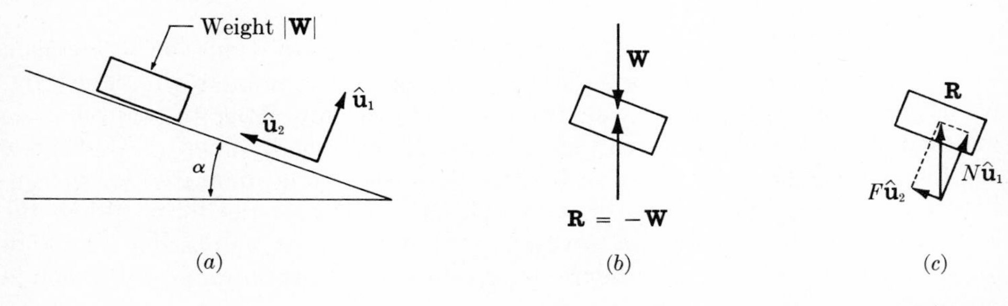

FIGURE 2·10

C. D. MICHALOPOULOS

free-body diagram in Fig. 2·10*b*. In a more complex problem, or even in this simple inclined plane problem when complicated by the presence of applied forces other than the force of gravity, the orientation of the reaction **R** may not be obvious. It then becomes advantageous to introduce an orthogonal set of unit vectors such as $\hat{\mathbf{u}}_1$ and $\hat{\mathbf{u}}_2$ of Fig. 2·10*a* (with an implied $\hat{\mathbf{u}}_3$ coming out of the page), and then to represent the unknown reaction in terms of its components in these directions. In Fig. 2·10*c*, the reaction **R**, unknown in magnitude and direction, is replaced by its equivalent, the two vectors $N\hat{\mathbf{u}}_1$ and $F\hat{\mathbf{u}}_2$, unknown in magnitude and sense but of well-defined orientation. The equilibrium equation then becomes

$$N\hat{\mathbf{u}}_1 + F\hat{\mathbf{u}}_2 + \mathbf{W} = 0 \tag{2·18}$$

or, with **W** resolved into components along $\hat{\mathbf{u}}_1$ and $\hat{\mathbf{u}}_2$,

$$N\hat{\mathbf{u}}_1 + F\hat{\mathbf{u}}_2 - W\cos\alpha\hat{\mathbf{u}}_1 - W\sin\alpha\hat{\mathbf{u}}_2 = 0 \tag{2·19}$$

where $W \triangleq |\mathbf{W}|$.† The nontrivial scalar equations

$$\begin{aligned} N - W\cos\alpha &= 0 \\ F - W\sin\alpha &= 0 \end{aligned} \tag{2·20}$$

follow from Eq. (2·19) (as well as the trivial $0 = 0$ from the $\hat{\mathbf{u}}_3$ components) since the scalar coefficients of the independent unit vectors $\hat{\mathbf{u}}_1$, $\hat{\mathbf{u}}_2$, and $\hat{\mathbf{u}}_3$ must all independently equal zero.

2·2·12 Friction forces

The force interaction between adjacent solid bodies is an extremely complex phenomenon when closely examined, but in a gross and macroscopic sense friction forces are commonly described in very simple terms. In the example of Fig. 2·10*a*, the force component $F\hat{\mathbf{u}}_2$ is generally described as a *friction force* since it is in the plane representing the idealized surface on which the block is resting. (Of course no physical surface is actually planar; in microscopic detail any surface is rough and irregular to some degree.) For a given pair of surfaces in contact with a given normal scalar component N of the interaction force, there is a limit to the magnitude of friction force that can develop. The ratio of the magnitude of the maximum friction force F_{max} to the normal force magnitude N is called the *coefficient of friction*, and is symbolized by μ. The relationship

$$F \leq \mu N \tag{2·21}$$

must therefore be satisfied in any problem involving friction between solid bodies

† In this text the symbol $\triangleq$ indicates equality by definition.

as traditionally idealized. The term *coulomb friction* is often applied to a force whose scalar components are related by Eq. (2·21).

In the problem of Fig. 2·10*a*, the solutions of Eqs. (2·20) provide

$$F = N \tan \alpha$$

We must therefore determine, before accepting this solution, that $\tan \alpha$ is less than or equal to the coefficient of friction μ for the given block and inclined surface.

It may seem somewhat puzzling that the block on the inclined plane could be treated as a *particle;* after all, friction depends on the presence of some contact surface, and a particle is a point, lacking dimensions. This concern must be allayed with the reminder that *friction forces* are abstract idealizations of a physical situation just as particles are abstract representations of material bodies. Once we define a *rough surface* in terms of an appropriate friction coefficient, it is quite consistent to analyze the motion (or rest) of a particle on this surface. In this context the term *smooth surface* implies that $\mu = 0$, so the surface is frictionless.

2·2·13 Summary of procedure for determining static-equilibrium forces

Conceptually, the problem of equilibrium-force determination for a particle statics problem appears trivial; we merely record the static-equilibrium equation from Newton's second law and then solve for the unknown forces. Yet the implementation of this simple advice is not entirely without difficulties, due largely to the possible complexity of determining the distribution of reaction forces within the structure supporting the particle.

In the context of the particular problem illustrated by Fig. 2·5*a*, an approach was developed that had the advantage of leading directly and yet systematically to the vector equilibrium equation, which could then be solved by the routine procedures of vector algebra. This method can be broadened somewhat to encompass the full range of statically determinate particle statics problems of equilibrium-force determination. The recommended procedure is as follows: First proceed by the most direct path to *derive* the equations of equilibrium, and *then solve* these equations by the well-established methods of vector algebra. In deriving equations of equilibrium, always begin with free-body diagrams in order to focus attention sharply on the particles in equilibrium, and then record the vector equations of equilibrium (the sum of all applied forces is zero), freely employing any convenient unit vectors in the process. Only *after* the vector equilibrium equations are written down should we worry about solving them, or about the representation of all of the forces in terms of a common dextral orthogonal vector basis. The practice of separating the derivation of the equilibrium equations from their solution is recommended even for the relatively simple problems appearing in a first course in mechanics. Realistic practical problems are frequently of such magnitude that the solution of the equations requires a computer. Then the practice recommended here becomes imperative.

2·3 EQUILIBRIUM–POSITION DETERMINATION

2·3·1 A particle on springs

Recall that for problems in this class, all applied forces are given as functions of position, and we are to find any static-equilibrium positions that exist. Since this class of problem is conceptually simple but potentially difficult in execution, it is again instructive to consider specific examples.

Consider the physical system illustrated in Fig. 2·11*a*. A particle of weight W is suspended from the center of a horizontal spring of undeformed length $2a$. Each half of the spring has the spring constant k, so that it develops a restoring force of magnitude k times its elongation. The problem is to determine the dis-

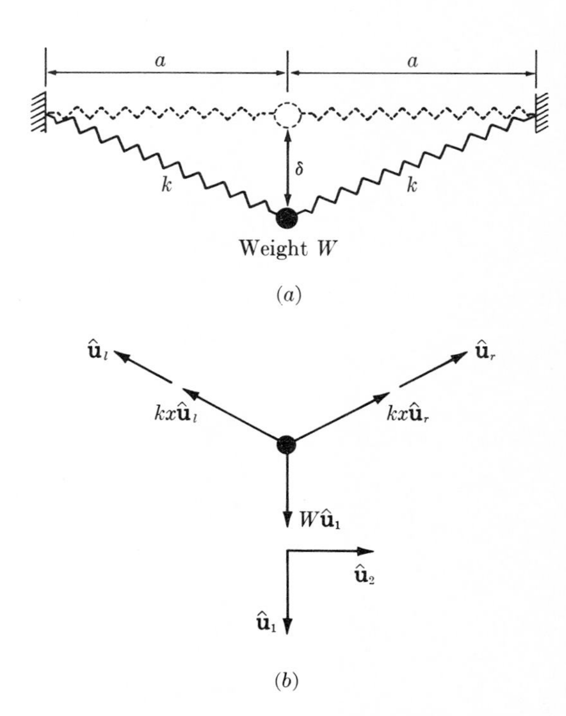

FIGURE 2·11

placement δ of the particle. The symmetry of this problem permits advance knowledge that the equilibrium displacement will be vertical.

Again, as in the equilibrium-force-determination problem, the recommended first step is the sketching of a free-body diagram, shown in Fig. 2·11*b*. (This practice may seem to be a waste of time in application to these simple problems, but the habit of drawing or at least imagining the free-body diagram must be instilled at an elementary level if it is to be useful when the problems get difficult.) Unit vectors $\hat{\mathbf{u}}_1$, $\hat{\mathbf{u}}_r$, and $\hat{\mathbf{u}}_l$ are freely assigned to the three lines of action shown on the free-body diagram, despite the fact that of these three only $\hat{\mathbf{u}}_1$ is of known direction. Now if x represents the elongation of each of the springs (which deform equally due to symmetry), the static-equilibrium equation can be written

$$W\hat{\mathbf{u}}_1 + kx\hat{\mathbf{u}}_l + kx\hat{\mathbf{u}}_r = 0 \qquad (2{\cdot}22)$$

The problem cannot yet be said to be reduced to vector algebra because the geometrical relationships between x and the orientations of $\hat{\mathbf{u}}_r$ and $\hat{\mathbf{u}}_l$ have yet to be obtained explicitly. It is convenient to employ the desired unknown δ in these relationships; for example, by the pythagorean theorem

$$(a + x)^2 = a^2 + \delta^2$$

or, equivalently,

$$x = \sqrt{a^2 + \delta^2} - a \qquad (2{\cdot}23)$$

Similarly, δ is used in combination with the dextral pair of unit vectors $\hat{\mathbf{u}}_1$ and $\hat{\mathbf{u}}_2$ to obtain

$$\begin{aligned} \hat{\mathbf{u}}_r &= -\left(\frac{\delta}{\sqrt{a^2 + \delta^2}}\right)\hat{\mathbf{u}}_1 + \left(\frac{a}{\sqrt{a^2 + \delta^2}}\right)\hat{\mathbf{u}}_2 \\ \hat{\mathbf{u}}_l &= -\left(\frac{\delta}{\sqrt{a^2 + \delta^2}}\right)\hat{\mathbf{u}}_1 - \left(\frac{a}{\sqrt{a^2 + \delta^2}}\right)\hat{\mathbf{u}}_2 \end{aligned} \qquad (2{\cdot}24)$$

Substitution of Eqs. (2·23) and (2·24) into (2·22) finally reduces the problem to vector algebra,

$$\begin{aligned} W\hat{\mathbf{u}}_1 &+ k(\sqrt{a^2 + \delta^2} - a)\left[-\left(\frac{\delta}{\sqrt{a^2 + \delta^2}}\right)\hat{\mathbf{u}}_1 - \left(\frac{a}{\sqrt{a^2 + \delta^2}}\right)\hat{\mathbf{u}}_2\right] \\ &+ k(\sqrt{a^2 + \delta^2} - a)\left[-\left(\frac{\delta}{\sqrt{a^2 + \delta^2}}\right)\hat{\mathbf{u}}_1 + \left(\frac{a}{\sqrt{a^2 + \delta^2}}\right)\hat{\mathbf{u}}_2\right] \\ &= 0 \end{aligned} \qquad (2{\cdot}25)$$

In actually solving this vector algebra problem for the single unknown scalar δ, we are pleased to note that we have but one nontrivial scalar equation implied by Eq. (2·25) since the sum of coefficients of $\hat{\mathbf{u}}_2$ in that equation is identically zero. Equating the sum of coefficients of $\hat{\mathbf{u}}_1$ to zero yields

$$W - 2k(\sqrt{a^2 + \delta^2} - a)\left(\frac{\delta}{\sqrt{a^2 + \delta^2}}\right) = 0 \tag{2·26}$$

which presents in its nonlinearity an obstacle to simple solution. It may be helpful to rewrite this equation in terms of the unknown ratio $\rho \triangleq \delta/a$, as

$$W - 2ka(\sqrt{1 + \rho^2} - 1)\left(\frac{\rho}{\sqrt{1 + \rho^2}}\right) = 0$$

or

$$W = 2ka\rho[1 - (1 + \rho^2)^{-1/2}] \tag{2·27}$$

The exponentiated term may be expanded into a binomial series, resulting in

$$W = 2ka\rho[1 - (1 - \frac{1}{2}\rho^2 + \frac{3}{8}\rho^4 - \frac{15}{48}\rho^6 + \cdots)]$$

or

$$W = ka(\rho^3 - \frac{3}{4}\rho^5 + \frac{15}{24}\rho^7 + \cdots) \tag{2·28}$$

If, for a given problem, physical considerations warrant the assumption $\delta \ll a$, so that $\rho \ll 1$, this series may be truncated to obtain a useful approximate solution for ρ (and therefore δ). In the limit as ρ approaches zero (small-deflection theory), the deflection δ approaches the value

$$\delta^* = \left(\frac{Wa^2}{k}\right)^{1/3} \tag{2·29}$$

Note that this equation has but one real solution for δ^*, and for this root $\delta^* > 0$.

It is important that we recognize once again that any difficulties encountered in the solution of this problem were algebraic difficulties. Strictly speaking, only Eq. (2·22) is an equation of mechanics; Eqs. (2·23) and (2·24) are equations of geometry. It is only after these three equations are combined in Eq. (2·25) that the difficulties of algebraic solution are confronted.

The problem of Fig. 2·11 may be generalized so as to destroy the symmetry and very substantially complicate the final solution. The spring supports may be three in number, or four, or more, and they may not be arranged in a plane. The distances labeled a in Fig. 2·11 may vary from one support to the next, and the spring constants may be different for the various spring supports. Each of the springs may be pretensioned so that, even when the vertical displacement is zero, there is a force in each spring. [If Δ_i is the initial elongation of the ith spring, and

k_i the spring constant, the force applied in the equilibrium position is, of course, $k_i(\Delta_i + x_i)$, where x_i is the added elongation induced by deflection under gravitational attraction.] Note carefully, however, that all of these added complexities serve only to complicate the algebra and the geometry; the equation of static equilibrium [Eq. (2·22)] can be recorded just as easily in the general case as in the special case developed here in detail (taking care to replace x_i by $\Delta_i + x_i$ when the ith spring is pretensioned).

In the general case there are three scalar unknowns required to determine the displacement of the particle under gravity, rather than the single unknown δ, so in general the vector algebraic equation (2·25) provides three nontrivial scalar equations, and not just one as in Eq. (2·26). This contributes materially to the algebraic complexity of the solution, but the mechanics per se remains very simple.

2·3·2 Application with more than one equilibrium solution

For an interesting variation of the first example, consider the system of Fig. 2·12 with each of the springs under compression an amount Δ_c when the vertical displacement δ of the particle is zero. The equation of equilibrium [Eq. (2·22)] then becomes

$$W\hat{\mathbf{u}}_1 + k(x - \Delta_c)\hat{\mathbf{u}}_l + k(x - \Delta_c)\hat{\mathbf{u}}_r = 0 \tag{2·30}$$

The equations of geometry [Eqs. (2·23) and (2·24)] are unchanged, and their substitution into Eq. (2·30) yields a vector equation with the scalar consequence

$$W - 2k(\sqrt{a^2 + \delta^2} - a - \Delta_c)\left(\frac{\delta}{\sqrt{a^2 + \delta^2}}\right) = 0 \tag{2·31}$$

in precise parallel to Eq. (2·26) as previously developed. There is, however, a major difference between Eqs. (2·31) and (2·26): When there is sufficient pre-

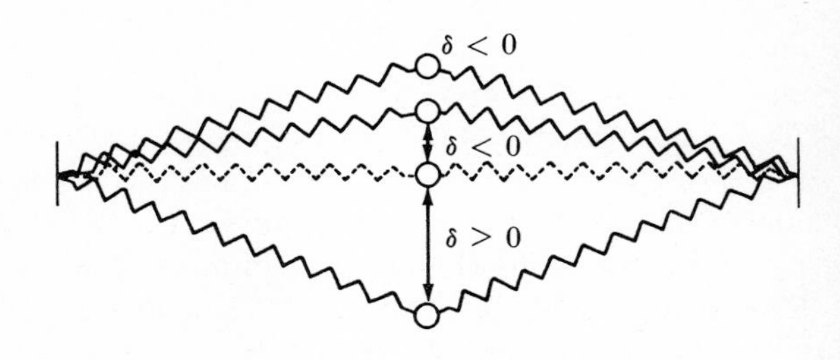

FIGURE 2·12

compression Δ_c so that the quantity $(\sqrt{a^2 + \delta^2} - a - \Delta_c)$ can be negative while δ is still numerically large enough to satisfy Eq. (2·31), then all three roots of the cubic equation are real, and there emerges from δ a pair of negative solutions as well as the expected positive solution. In other words, the physical system can for a certain range of values of W, a, k, and Δ_c adopt either of the three static-equilibrium positions shown in Fig. 2·12. This result would perhaps be readily anticipated only in the special case $W = 0$, when $\delta = 0$ and $\delta = \pm\sqrt{\Delta_c(2a + \Delta_c)}$ emerge as solutions of Eq. (2·31).

Explicit determination of the three equilibrium values of δ from the algebraic equation (2·31) is not a simple matter. The very existence of three solutions may escape the analyst if he does not look beyond Eq. (2·31) to the physical problem originally posed. The lesson of this example is not in conflict with the previously recommended practice of focusing attention on the simple principles of mechanics to obtain equilibrium equations before worrying about their algebraic solution. In generalizing from this and previous examples we may conclude that while concern about mathematical solution should not interfere with the simple task of deriving the equations of mechanics, the reverse is not true. It is often very helpful in solving algebraic equations of static equilibrium (or differential equations of motion) to take full advantage of your knowledge of mechanics and your powers of visualization of the behavior of physical systems.

2·3·3 Application with more than one particle in equilibrium

In the previous section, we noted that the determination of forces required to sustain a given equilibrium state for n particles is no more difficult than n separate determinations of the equilibrium forces required for a single particle. The only stipulation involves the judicious selection of the sequence in which to analyze the individual particles in those cases involving some particles supported by locally statically indeterminate forces.

In contrast, problems of equilibrium-position determination become substantially more difficult to solve with an increase in the number of particles in equilibrium. The added difficulties are, however, wholly algebraic; the derivation of the equations of equilibrium remains a simple matter.

Consider, for example, the problem of Fig. 2·13a. The system illustrated has only two particles, of weights W_1 and W_2, and these particles are suspended by three springs, all of which lie in a plane defined by unit vectors $\hat{\mathbf{e}}_1$ and $\hat{\mathbf{e}}_2$. When both particles lie on the horizontal line connecting suspension points A and B, the springs are of lengths a, b, and c and they are undeformed, so they transmit no forces. When stretched in amounts δ_a, δ_b, and δ_c, they transmit tensile forces of magnitudes $k_a\delta_a$, $k_b\delta_b$, and $k_c\delta_c$, respectively. The problem is simply stated: Find the equilibrium positions occupied by the two particles under uniform gravitational attraction. Since this is a planar problem, two scalars are required to locate each particle, so the scalar components x_1, y_1, x_2, and y_2 of the displacements are adopted as the four unknowns of the problem.

As in the previous examples, we approach this problem by separately considering three questions:

1. What are the vector equations of static equilibrium for the particles, written in terms of convenient symbols for forces?
2. What are the relationships among our force symbols and the system deformations or displacements?

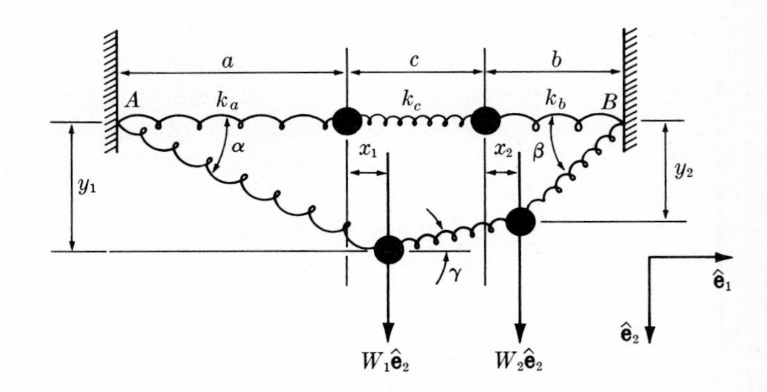

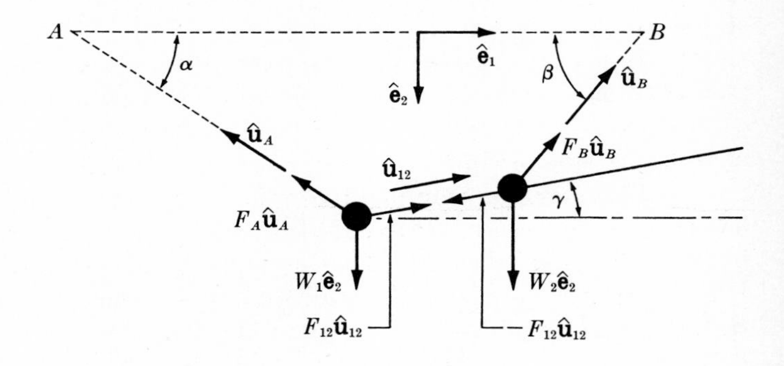

FIGURE 2·13

3 What are the geometrical relationships among the scalars chosen originally as unknowns and those symbols appearing in our answers to the first two questions?

In other words, we write *separately* (1) the *equations of static equilibrium,* (2) the *force-displacement relationships,* and (3) the *geometrical relationships* permitting (1) and (2) to be expressed in terms of the desired scalar unknowns. The first two steps involve principles of *mechanics,* and as we shall see they are never difficult. The third step involves only geometry and algebra, but in application it is this step that is sometimes tedious and difficult.

Writing the equations of static equilibrium is easy if you follow two rules: (1) Draw correct free-body diagrams and (2) freely introduce new symbols for any scalars or unit vectors that will facilitate the recording of equilibrium equations by inspection of the diagrams. Don't worry about expressing the equilibrium equations in terms of the final unknown scalars—there will be time to consider this requirement after a correct symbolic statement of equilibrium has been recorded.

For this two-particle problem, we require two free-body diagrams, as shown in Fig. 2·13*b*. Note that new unit vectors $\hat{\mathbf{u}}_A$, $\hat{\mathbf{u}}_B$, and $\hat{\mathbf{u}}_{12}$ have been introduced along the lines of action of the forces. The directions of these vectors are unknown since it is the displaced configuration we wish to determine, but we can still use the symbols without inhibition. Similarly, the symbols F_A, F_B, and F_{12} are unknown scalars that, when multiplied by appropriate unit vectors, define the unknown forces transmitted to the particles by the springs. Without digressing to express all of these unknowns in terms of the unknown displacement components x_1, y_1, x_2, and y_2 and the known quantities $\hat{\mathbf{e}}_1$, $\hat{\mathbf{e}}_2$, a, b, c, k_a, k_b, k_c, W_1, and W_2, we simply record the equilibrium equations

$$W_1\hat{\mathbf{e}}_2 + F_A\hat{\mathbf{u}}_A + F_{12}\hat{\mathbf{u}}_{12} = 0 \tag{2·32}$$
$$W_2\hat{\mathbf{e}}_2 + F_B\hat{\mathbf{u}}_B - F_{12}\hat{\mathbf{u}}_{12} = 0 \tag{2·33}$$

Note that Newton's third law (the action-reaction equivalence law) has been used in representing the force applied to body 2 by body 1 as the negative of the force $F_{12}\hat{\mathbf{u}}_{12}$ applied to body 1 by body 2.

The equilibrium equations (2·32) and (2·33) must be augmented by the *force-displacement* relationships

$$F_A = k_a\delta_a \tag{2·34}$$
$$F_B = k_b\delta_b \tag{2·35}$$
$$F_{12} = k_c\delta_c \tag{2·36}$$

which were given in the problem statement (recall that δ_a, δ_b, and δ_c are the elongations of the springs of unstretched lengths a, b, and c, respectively).

Equations (2·32) to (2·36) embody all the principles of mechanics required for this problem. There remains the problem of expressing the unknowns δ_a, δ_b, δ_c, $\hat{\mathbf{u}}_A$, $\hat{\mathbf{u}}_B$, and $\hat{\mathbf{u}}_{12}$ in terms of known quantities and the desired unknowns x_1, y_1, x_2, and y_2. This is a task of geometry and algebra. From the pythagorean theorem, we have

$$(a + \delta_a)^2 = (a + x_1)^2 + y_1^2 \tag{2·37}$$

$$(b + \delta_b)^2 = (b - x_2)^2 + y_2^2 \tag{2·38}$$

$$(c + \delta_c)^2 = (c + x_2 - x_1)^2 + (y_2 - y_1)^2 \tag{2·39}$$

Unit-vector geometry gives

$$\hat{\mathbf{u}}_A = -\hat{\mathbf{e}}_1 \cos \alpha - \hat{\mathbf{e}}_2 \sin \alpha \tag{2·40}$$

$$\hat{\mathbf{u}}_B = \hat{\mathbf{e}}_1 \cos \beta - \hat{\mathbf{e}}_2 \sin \beta \tag{2·41}$$

$$\hat{\mathbf{u}}_{12} = \hat{\mathbf{e}}_1 \cos \gamma - \hat{\mathbf{e}}_2 \sin \gamma \tag{2·42}$$

where

$$\alpha = \tan^{-1} \frac{y_1}{a + x_1} \tag{2·43}$$

$$\beta = \tan^{-1} \frac{y_2}{b - x_2} \tag{2·44}$$

$$\gamma = \tan^{-1} \frac{y_1 - y_2}{c + x_2 - x_1} \tag{2·45}$$

In order actually to obtain equations to be solved for the unknowns x_1, y_1, x_2, and y_2, we must combine Eqs. (2·32) to (2·45). This is a rather laborious task even for this planar problem involving only two particles, but it is a completely straightforward algebraic exercise. We must solve the quadratic equations (2·37) to (2·39) for δ_a, δ_b, and δ_c in terms of x_1, y_1, x_2, and y_2, and then substitute the results into Eqs. (2·34) to (2·36) to obtain expressions for the unknowns F_A, F_B, and F_{12} in the desired terms. Then Eqs. (2·40) to (2·45) must be used to obtain expressions for $\hat{\mathbf{u}}_A$, $\hat{\mathbf{u}}_B$, and $\hat{\mathbf{u}}_{12}$ in terms of x_1, y_1, x_2, and y_2. Finally, both results must be substituted into the equilibrium equations (2·32) and (2·33) to yield two vector equations, or four scalar equations, in the four unknowns x_1, y_1, x_2, and y_2. The static-equilibrium position can be found only by simultaneous solution of these highly nonlinear algebraic equations.

The system of Fig. 2·13*a* is very simple, and the equations of mechanics [Eqs. (2·32) to (2·36)] are easily found, but complications of geometry and algebra make even this little problem difficult. If a solution is required, we must turn to a digital computer for a numerical analysis. In many situations of practical interest, unrestricted solutions of the equilibrium equations are not required, and some form of approximate solution is sufficient. An illustration of such a situation has been given for the system shown in Fig. 2·11. For this system, Eq. (2·28) gives a useful approximation of Eq. (2·27) for small deflection $\delta = a\rho$. To obtain the equilibrium solution given by Eq. (2·29), only third-degree terms in ρ are retained, all higher-degree terms being ignored. Similar approximations might be used in Eqs. (2·40) to (2·45) of the example of Fig. 2·13*a*, but in this case the retention of third-degree terms in the unknowns x_1, y_1, x_2, and y_2 already leads to some complexity. This is the only meaningful approximation that can be considered for these two problems, however, because second-degree terms in the displacement variables do not appear in the equations, and there is also no linear approximation to the equilibrium solution. [Note that in Eq. (2·28) all linear terms in ρ disappear, as do all terms in ρ^2.]

In contrast to the examples considered thus far in this section, there is a wide class of structures for which linear approximations can be found for static-equilibrium solutions. Typical of such structures is the following example.

2·3·4 Loaded truss in equilibrium; small-deflection theory

Figure 2·14*a* illustrates a simple pin-jointed truss supporting a vertical load. The structural members of the truss are to be idealized as springs since inextensible rods would, of course, yield no deflection. This structure differs from others considered in this section (see Figs. 2·11 and 2·13) in one important respect: Prior to deformation the geometry of the truss is of a configuration that permits the load to be sustained, whereas the springs of Figs. 2·11 and 2·13 cannot bear the applied loads until they deform into a new geometry. This difference is important because it makes it possible to find an approximate equilibrium configuration of the truss

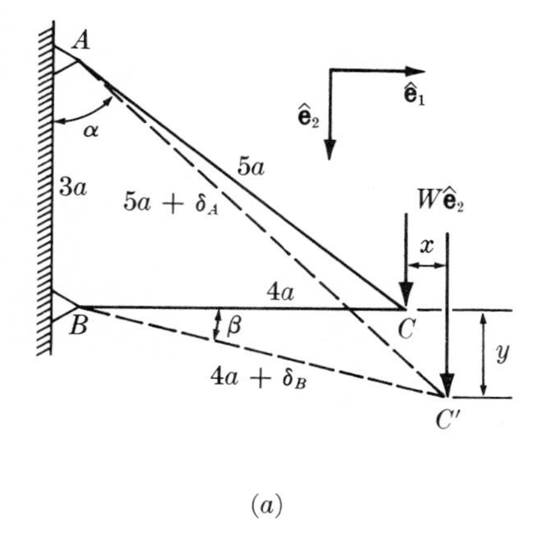

(*a*)

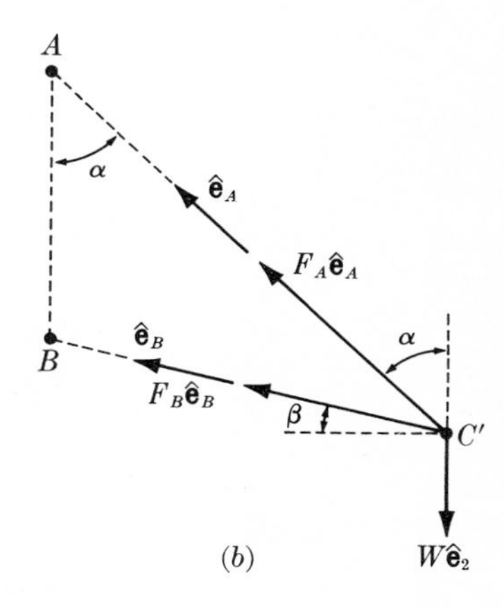

(*b*)

FIGURE 2·14

by assuming that deformations are so small that all terms above the first degree in the deformation variables are negligible. The equations to be solved for equilibrium are then linear equations. The previous examples have resulted in equilibrium equations, such as Eq. (2·26), that are nonlinear and have no linear approximation. The best that could be done in the case of Eq. (2·26) was to obtain a cubic equation, Eq. (2·28), ignoring all terms above the third degree.

When the objective is to find only the linear approximation of the equations for an equilibrium state, it is often possible to proceed directly to the linear equations, instead of writing first the full nonlinear equations and then specializing to the linear approximation. The difference in these procedures lies only in the third stage of the analysis, in which geometrical relationships are established. The equations of static equilibrium and the force-displacement relationships are easily written in symbolic form whether the deformation variables are assumed small or not.

To illustrate these procedures, all of the equations for the system of Fig. 2·14*a* are recorded. The truss in its undeformed geometry is a 3-4-5 triangle ABC; when deformed, it adopts the configuration ABC' shown dashed in Fig. 2·14*a*. (Note that in positioning C' in the figure no attempt has been made to estimate the true location; the figure is drawn as though both members of the truss stretched, in order to simplify the sign conventions by picturing positive displacement components x and y.)

As always, the first step is to draw a free-body diagram such as Fig. 2·14*b*, introducing unit vectors $\hat{\mathbf{e}}_A$ and $\hat{\mathbf{e}}_B$ paralleling lines $C'A$ and $C'B$, respectively, and introducing symbols F_A and F_B representing scalars that, when multiplied by $\hat{\mathbf{e}}_A$ and $\hat{\mathbf{e}}_B$, represent the forces applied to C' by the structural members connecting C' to A and to B. Without worrying for the moment that all of these new symbols are unknowns that must ultimately be expressed in terms of the unknown displacements x and y, we can record the static-equilibrium equation

$$F_A\hat{\mathbf{e}}_A + F_B\hat{\mathbf{e}}_B + W\hat{\mathbf{e}}_2 = 0 \tag{2·46}$$

The unknown scalars F_A and F_B are most easily expressed in terms of the elongations δ_A and δ_B sustained by the members $C'A$ and $C'B$. If again it is assumed that these structural members are *linearly elastic*, so that in essence they are *linear springs* (with spring constraints k_A and k_B), then the required force-displacement relationships are

$$F_A = k_A\delta_A \tag{2·47}$$
$$F_B = k_B\delta_B \tag{2·48}$$

The third stage of the analysis is the development of the geometrical relationships expressing δ_A, δ_B, $\hat{\mathbf{e}}_A$, and $\hat{\mathbf{e}}_B$ in terms of the unknowns x and y to be determined. Although we shall soon see that it is possible to record these relationships in terms that give us linear equations in the unknowns x and y directly, for this example we proceed first to the nonlinear equations obtained when the sizes of x and y are unrestricted. From the pythagorean theorem, we have

$$(4a + x)^2 + (3a + y)^2 = (5a + \delta_A)^2 \tag{2·49}$$

and

$$(4a + x)^2 + y^2 = (4a + \delta_B)^2 \tag{2·50}$$

The unit vectors $\hat{\mathbf{e}}_A$ and $\hat{\mathbf{e}}_B$ can be expressed in terms of the known unit vectors $\hat{\mathbf{e}}_1$ and $\hat{\mathbf{e}}_2$ in Fig. 2·14 as

$$\hat{\mathbf{e}}_A = -\hat{\mathbf{e}}_2 \cos\alpha - \hat{\mathbf{e}}_1 \sin\alpha \tag{2·51}$$
$$\hat{\mathbf{e}}_B = -\hat{\mathbf{e}}_2 \sin\beta - \hat{\mathbf{e}}_1 \cos\beta \tag{2·52}$$

where

$$\alpha = \tan^{-1}\frac{4a + x}{3a + y} = \tan^{-1}\frac{4 + (x/a)}{3 + (y/a)} \tag{2·53}$$

$$\beta = \tan^{-1}\frac{y}{4a + x} = \tan^{-1}\frac{y/a}{4 + (x/a)} \tag{2·54}$$

In order to obtain solutions for x and y without approximation, we must combine Eqs. (2·46) to (2·54) to obtain, after much labor, a pair of highly nonlinear algebraic equations in x and y. Since a literal (nonnumerical) solution of these equations is beyond our expectations, we would have to resort to a digital computer to obtain solutions. Our computer program might be designed to receive only the two nonlinear algebraic equations for solution, but it is more efficient to input the information in Eqs. (2·46) to (2·54) directly, so that we are spared much of the labor of combining these equations manually. We may therefore conclude that there is no point in showing the final pair of nonlinear equations in x and y.

We may, however, also consider the possibility of obtaining an approximate solution that is valid when x/a and y/a are sufficiently small in comparison with unity; this is, after all, the case for most problems of practical interest. If these variables are so small that terms above the first degree may be ignored, we can expect Eqs. (2·46) to (2·54) to reduce to a pair of linear algebraic equations for x and y, which we can solve without a computer.

Of course, the *static-equilibrium* equation (2·46) and the *force-displacement relationships*, Eqs. (2·47) and (2·48), are not explicitly influenced by the size of x and y; they appear in combination as

$$k_A\delta_A\hat{\mathbf{e}}_A + k_B\delta_B\hat{\mathbf{e}}_B + W\hat{\mathbf{e}}_2 = 0 \tag{2·55}$$

The scalars δ_A and δ_B may be written in terms of x and y by solving Eqs. (2·49) and (2·50) in the linear approximation, neglecting terms above the first degree in the small variables x, y, δ_A, and δ_B. These equations become

$$16a^2 + 8ax + 9a^2 + 6ay = 25a^2 + 10a\delta_A \tag{2·56}$$

and

$$16a^2 + 8ax = 16a^2 + 8a\delta_B \tag{2·57}$$

Eq. (2·56) provides

$$\delta_A = \frac{8x + 6y}{10} \tag{2·58}$$

and Eq. (2·57) becomes simply

$$\delta_B = x \tag{2·59}$$

We may now substitute δ_A and δ_B into Eq. (2·55) and write this vector equation in terms of the unit vectors $\hat{\mathbf{e}}_1$ and $\hat{\mathbf{e}}_2$ by using Eqs. (2·51) and (2·52) for $\hat{\mathbf{e}}_A$ and $\hat{\mathbf{e}}_B$, respectively. This substitution is greatly simplified by the decision to ignore nonlinear terms in x and y. Because $\hat{\mathbf{e}}_A$ and $\hat{\mathbf{e}}_B$ in Eq. (2·55) are multiplied by the small variables δ_A and δ_B, the appropriate substitutions for $\hat{\mathbf{e}}_A$ and $\hat{\mathbf{e}}_B$ from Eqs. (2·51) and (2·52) are the values adopted by these vectors when x and y are zero. Then Eqs. (2·53) and (2·54) become

$$\alpha = \tan^{-1}\frac{4}{3} \tag{2·60}$$

$$\beta = 0 \tag{2·61}$$

and $\hat{\mathbf{e}}_A$ and $\hat{\mathbf{e}}_B$ may be written as

$$\hat{\mathbf{e}}_A = -\frac{3}{5}\,\hat{\mathbf{e}}_2 - \frac{4}{5}\,\hat{\mathbf{e}}_1 \tag{2·62}$$

$$\hat{\mathbf{e}}_B = -\hat{\mathbf{e}}_1 \tag{2·63}$$

Finally, we may combine Eqs. (2·62), (2·63), (2·58), and (2·59) into Eq. (2·55) to obtain the vector equation

$$k_A\left(\frac{4x + 3y}{5}\right)\left(-\frac{3}{5}\,\hat{\mathbf{e}}_2 - \frac{4}{5}\,\hat{\mathbf{e}}_1\right) + k_Bx(-\hat{\mathbf{e}}_1) + W\hat{\mathbf{e}}_2 = 0 \tag{2·64}$$

or the equivalent scalar equations (after multiplying by 25)

$$-3k_A(4x + 3y) + 25W = 0 \tag{2·65}$$

$$-4k_A(4x + 3y) - 25k_Bx = 0 \tag{2·66}$$

The second of these equations provides

$$y = -\left(\frac{25k_B + 16k_A}{12k_A}\right)x \tag{2·67}$$

Combining Eqs. (2·67) and (2·65) and solving for x yields

$$-3k_A\left[4x - \left(\frac{3}{12k_A}\right)(25k_B + 16k_A)\,x\right] + 25W = 0$$

or

$$x = -\frac{4W}{3k_B} \tag{2·68}$$

Now Eqs. (2·68) and (2·67) provide the solution for y as

$$y = \left(\frac{25k_B + 16k_A}{9k_Ak_B}\right)W \tag{2·69}$$

It should by now be apparent that we have worked a lot harder than is required if all we want is a linear approximation for x and y. We recorded in Eqs. (2·51) to (2·54) a description of the deformed geometry, but *for the linear solution we needed only the geometry of the undeformed system.* In this respect *this example is typical of all structures that in their undeformed geometry have the capability of sustaining applied loads; it is always possible to develop an approximate solution for the displacements under load by calculating the directions of transmitted loads on the basis of the undeformed geometry.* The resulting equations are identical to those we would obtain if we recorded the full system of equations of equilibrium, force displacement, and geometry, combined them to obtain nonlinear algebraic equations in the displacement variables, and then formally linearized them, dropping terms above the first degree in these variables. These equations are sometimes referred to as the *small-deflection equations* or the equations of *first-order displacement theory.* The practice of analyzing the loads transmitted through the members of the truss by progressing systematically from joint to joint of the truss, applying the equations of force equilibrium at each location, is known as the *method of joints.* Alternative methods will be explored in Chaps. 6 and 7. We will also find in these later chapters that there are, for complex structures, more efficient ways of formulating and analyzing some of the equations examined in this chapter. Now, however, it is enough to understand the basic concepts of particle statics; facility in analysis will come later for those who need it.

2·3·5 Nonlinear force-displacement relationships

In each of the examples considered thus far, the force-displacement relationship has been that of the linear spring [see, for example, Eqs. (2·47) and (2·48)]. This assumption is quite commonly adopted in the description of engineering systems, although it is, of course, never realized precisely. A linear force-displacement relationship is clearly analytically advantageous, so it is adopted whenever it results in an adequate representation of the properties of the structural material.

Sometimes, however, other force-displacement relationships must be considered. Two alternatives are the following:

$$F_h = k_1\delta + k_3\delta^3 \tag{2·70}$$
$$F_s = k_1\delta - k_3\delta^3 \tag{2·71}$$

The first of these is the force-displacement relationship for the *hardening spring;* Fig. 2·15*a* is a plot of F_h versus δ for this case. Equation (2·71), illustrated in Fig. 2·15*b*, describes the *softening spring.*

Whatever the complexities of the force-displacement relationship, there arise no conceptual obstacles to recording these equations in combination with the vector equilibrium equations and those of geometry. The difficulties arise when we try to solve these equations. If the force-displacement relationship involves terms above the first degree in the displacements [as in Eqs. (2·70) and (2·71)], it becomes inconsistent to adopt the small-deflection theory and ignore terms above the first degree in the geometrical relationships. The equations then become nonlinear, and may be hopelessly complicated.

Even for the extremely simple system illustrated in Fig. 2·16, a nonlinear spring makes the static-equilibrium solution a nontrivial calculation. If the hardening spring of Eq. (2·70) is adopted, with δ the elongation of the spring in Fig. 2·16, the equilibrium equation

$$k_1\delta + k_3\delta^3 = W \tag{2·72}$$

is not as easily solved as we might wish. Nonetheless, there are many new and interesting phenomena associated with such nonlinear systems, and they will receive more attention as we develop our facilities for automatic computation.

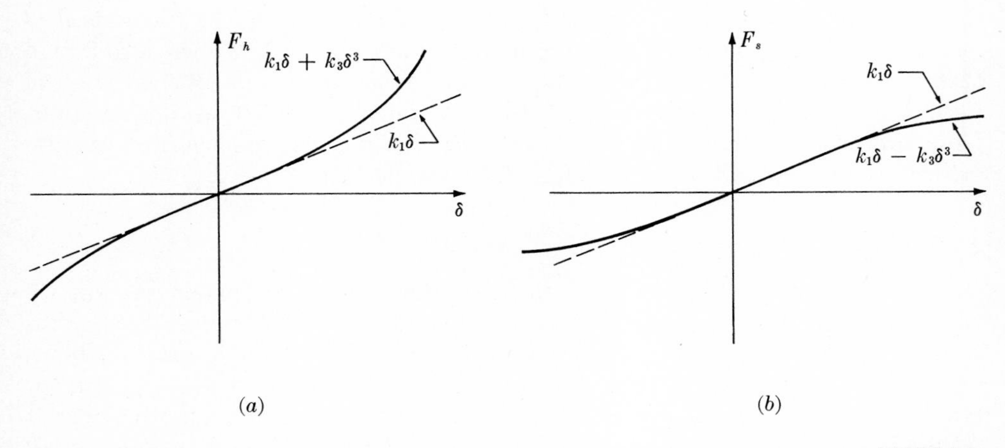

FIGURE 2·15

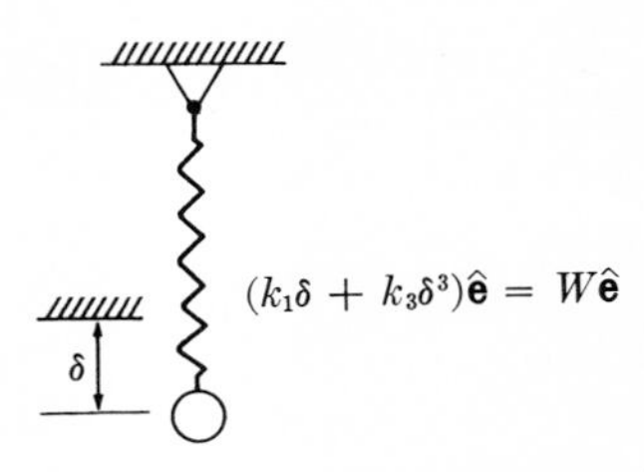

FIGURE 2·16

2·3·6 Summary of procedures for determining static-equilibrium positions

1 If no restrictions on the size of the displacement variables are permissible, proceed as follows:
   a. Draw free-body diagrams for each point of interest in the system, and record the static-equilibrium equations in symbolic form. These vector equations can be written in terms of symbols for unknown forces and unit vectors of unknown direction since in subsequent steps all such symbols will be related to the desired unknown displacements.
   b. Record the force-displacement relationships, expressing force symbols in terms of any convenient variables that describe displacements or deformations.
   c. Develop the geometrical relationships among the unit vectors and displacement or deformation variables used in (a) and (b).
   d. Solve the system of equations developed in (a), (b), and (c), either exactly or by some approximate procedure. (See §2·3·1 to 2·3·3.)

2 If the system can sustain the loads in its undeformed geometry, if small-displacement variables may be assumed, and if the number of displacement variables is not excessive, proceed as above, but in step three use the undeformed geometry in calculating lines of action of forces, and ignore nonlinear terms in displacement variables. (See §2·3·4 and Fig. 2·14.)

3 If the system is as described in case 2, except that a large number of displacement variables is involved, the methods of this chapter are impractical, and we should take advantage of the computational efficiency offered by the more refined methods developed in Chaps. 6 and 7.

Note that any of these methods for the determination of equilibrium position provides also convenient expressions for the forces in all force-transmission devices. Although in this section we have always been given, in the force-displacement relationships, enough information to express force as a function of position, we have not always been given all forces as explicit functions of the displacement variables of interest. In the process of finding these displacement variables, we have found explicit expressions for the forces.

In Sec. 2·2 we discovered that for many problems in which the geometry is fixed (by the assumption of inextensible members), forces can be determined by static-equilibrium equations alone. However, we also encountered a class of structures, called statically indeterminate, for which these equations would not suffice. Now we can see that if deformations of the force-transmission devices are permitted, we do not encounter the problem of indeterminacy because we are always augmenting the equations of static equilibrium with equations describing force-displacement relationships and geometrical relationships.

2·4 PERSPECTIVE

Conceptually, problems of particle statics are extremely simple. They involve the application of Newton's second law ($\mathbf{F} = m\mathbf{A}$) to particles without acceleration in an inertial frame of reference, so such problems reduce simply to the statement that the sum of forces applied to each particle in the system is zero ($\mathbf{F} = \mathbf{0}$). Because the inertial acceleration is zero, the particle inertial mass does not enter the problem, and we may as well imagine the application of Newton's laws to massless particles or *points*.

In practice, problems of particle statics can be difficult and time-consuming simply because of the possible complexities of geometry and the number of unknown variables that may have to be determined in order to establish the distribution of reaction forces among the elements of the supporting structure.

In the preceding sections, problems of particle statics are classified according to whether equilibrium *forces* or equilibrium *positions* are to be determined. This is a straightforward distinction in the case of statically determinate systems, but if the structure is statically indeterminate it must be idealized in such a way as to permit deformation so that displacements and forces can then be simultaneously determined.

In Sec. 2·2, on determining equilibrium forces, the equilibrium position was given, and forces were to be determined. Structural members sustaining loads were described as *inextensible* strings or rods, so the loaded structure had the same geometry that it had under zero load.

In Sec. 2·3, on methods of determining equilibrium positions, we found that useful approximations could often be obtained by assuming that displacements under load were sufficiently small to justify ignoring terms above the first degree in these variables. We found that, as an alternative to formally linearizing a set of nonlinear algebraic equations, we could proceed directly to the linear equations by assuming that the lines of action of the forces were established by the structure in its *undeformed geometry*.

Now we can see in retrospect that the method of force determination developed in Sec. 2·2 for structures with idealized inextensible members is also applicable to deformable systems, as long as we accept the assumption of sufficiently small deflections. This means that the method that gave exact results for forces in inextensible elements of structures in Sec. 2·2 also gives approximate results for structures with deformable (perhaps elastic) elements. These approximations are useful when the structure is expected to deform only slightly under applied loads, so that the geometrical configuration of the system is not greatly influenced by the deformation. This is, of course, the case for most structures of practical interest.

The application of Newton's second law to problems of particle *dynamics* involves many of the ingredients of the statics problem. In addition we must become proficient in calculating the inertial acceleration $\mathbf{A}$ appearing in $\mathbf{F} = m\mathbf{A}$ in terms of the time history of the particle position. This requirement leads us in the next chapter to the study of *kinematics*.

PROBLEMS

Note: If you rely upon your ability to find an answer by inspecting the figure, you will quickly answer most of these problems, but you will never master some of them (such as Probs. 2·30 and 2·18). Try instead to exercise the recommended systematic procedures on these simple problems, so that you'll be prepared for the more difficult ones.

2·1 Find the resultant $\mathbf{F}$ of the coplanar forces acting on the particle P shown in Fig. P2·1, expressing $\mathbf{F}$ in terms of the orthogonal unit vectors $\hat{\mathbf{u}}_1$ and $\hat{\mathbf{u}}_2$.

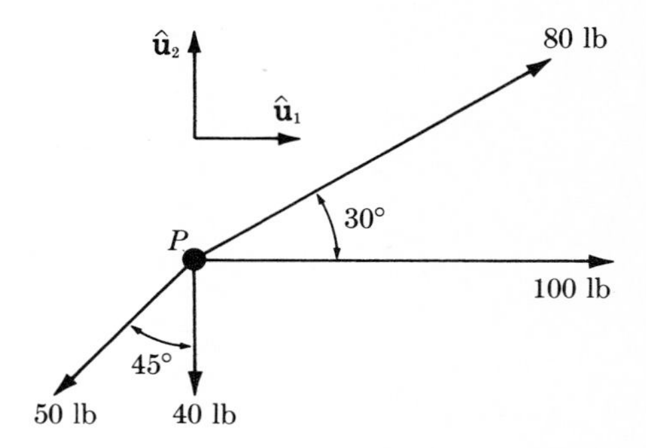

FIGURE P2·1

2·2 Find the resultant **F** of the forces acting on the particle P shown in Fig. P2·2, expressing **F** in terms of the orthogonal unit vectors $\hat{\mathbf{u}}_1$, $\hat{\mathbf{u}}_2$, and $\hat{\mathbf{u}}_3$.

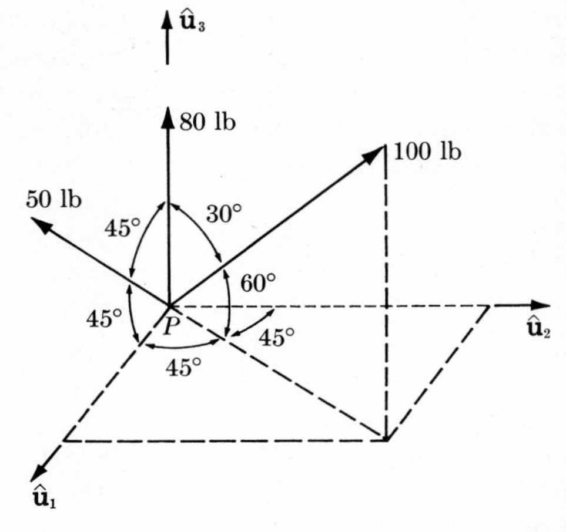

FIGURE P2·2

2·3 Express the unit vector $\hat{\mathbf{a}}$ in terms of the orthogonal unit vectors $\hat{\mathbf{u}}_1$, $\hat{\mathbf{u}}_2$, and $\hat{\mathbf{u}}_3$ shown in Fig. P2·3.

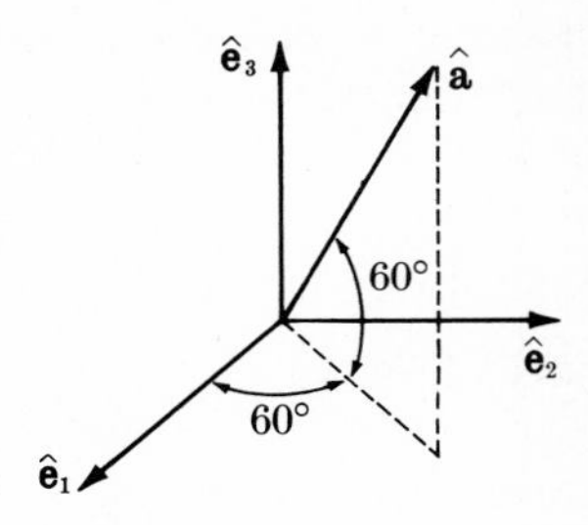

(a)

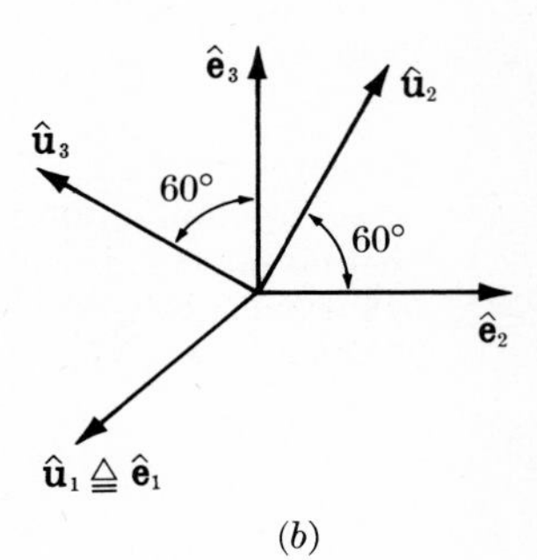

(b)

FIGURE P2·3

2·4 Find the forces applied to the particle P of weight W by the two supporting strings shown in Fig. P2·4.

2·5 Given the loaded system shown in Fig. P2·4, determine the change in string forces due to an additional horizontal force $H\hat{\mathbf{u}}_2$. Find the maximum ratio (H/W) for which the strings can sustain the loads.

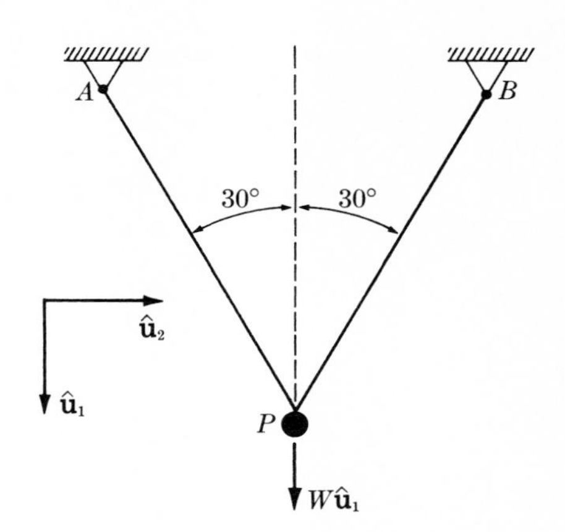

FIGURE P2·4

2·6 Find the forces applied by the two restraining strings to the point P under the horizontal applied force $\mathbf{F} = F\hat{\mathbf{u}}_1$, for the geometry shown in Fig. P2·6.

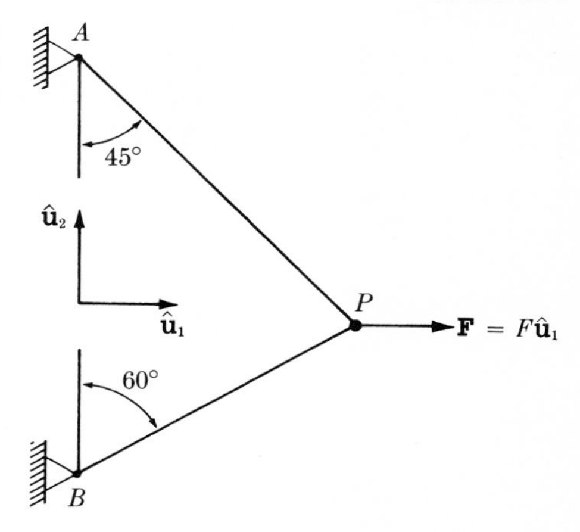

FIGURE P2·6

2·7 Find the tensile forces in the five restraining strings connecting points P_1, P_2, P_3, P_4, and P_5, when P_1 is a particle of weight W. (See Fig. P2·7, in which the strings are shown as solid lines and $\alpha = 30°$.)

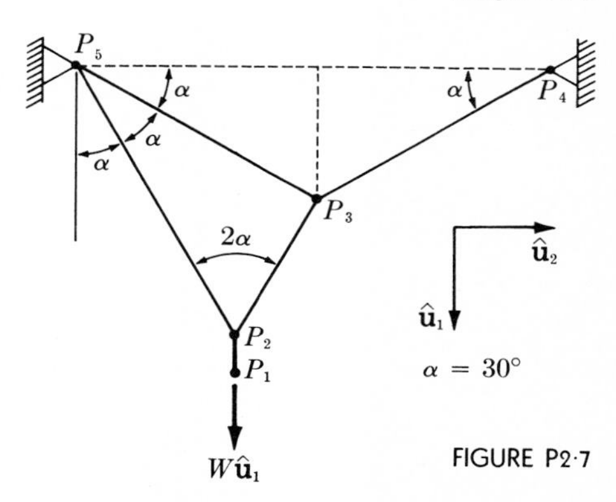

FIGURE P2·7

2·8 Two particles P_1 and P_2 having unequal weights W_1 and W_2 are attached to a string tied to the ends of a horizontal rod. Under the action of these weights, the string assumes the configuration shown in Fig. P2·8. Given that $W_1 = 10$ lb, find W_2.

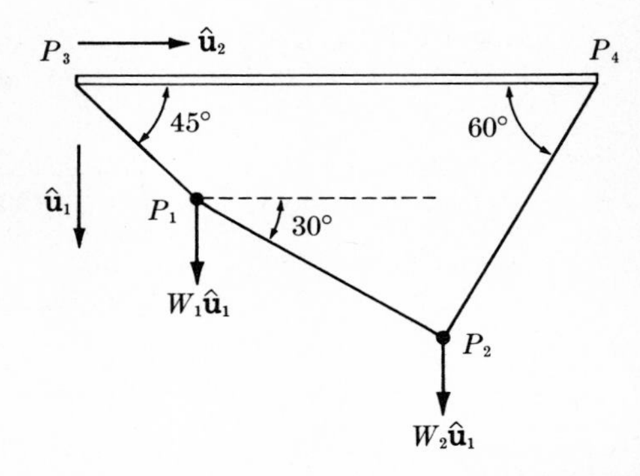

FIGURE P2·8

2·9 Find the forces applied to the particle P_0 of weight W by the three supporting strings of length a attached to a horizontal ceiling at the vertices P_1, P_2, and P_3 of the equilateral triangle of side dimension a shown in Fig. P2·9a, b.

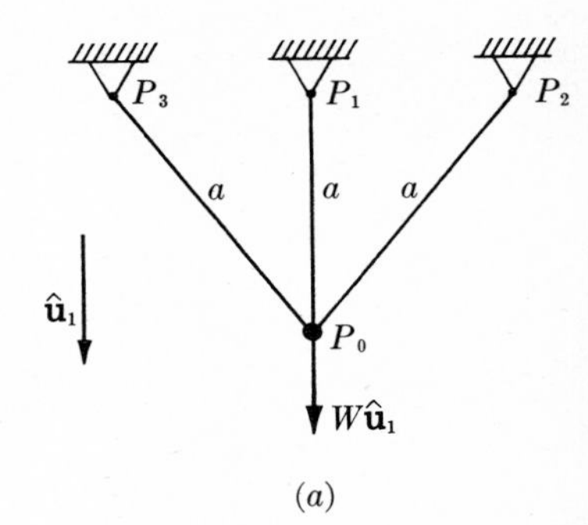

(a)

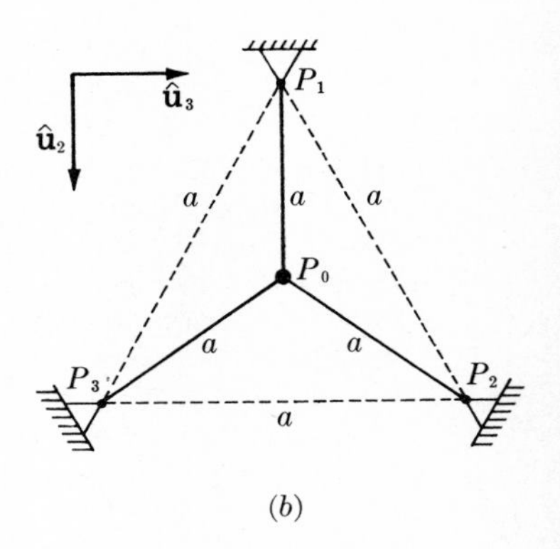

(b)

FIGURE P2·9

2·10 A particle P_1 of weight W is attached to a string passing over frictionless pulleys at points P_2 and P_3 while attached to a support at P_4 (see Fig. P2·10). Determine the forces applied by the ground support to points P_2, P_3, and P_4.

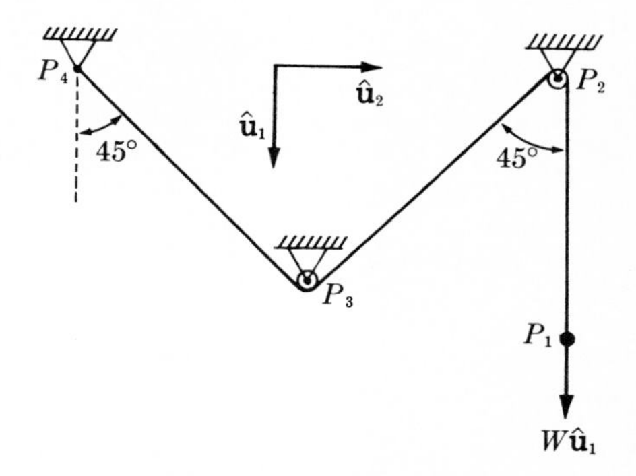

FIGURE P2·10

2·11 A particle P_0 weighing 200 lb is supported by a pair of frictionless pulleys as shown in Fig. P2·11, with a man applying a force of magnitude F to the end of the supporting string. What value of F maintains static equilibrium of P_0?

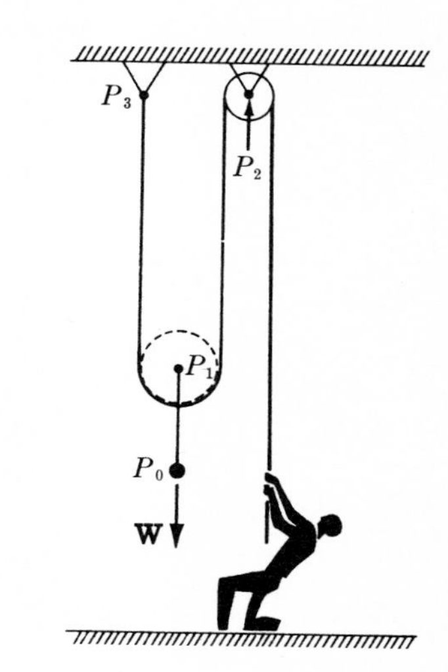

FIGURE P2·11

2·12 Determine the forces in the seven pin-jointed rods connecting points P_1 to P_5 in Fig. P2·12 when P_1 and P_2 are each particles of weight W. (Note that the nature of the support given P_4 by the wall is established by a symbol conventionally representing frictionless wheels, so that no vertical support is provided to P_4 by the wall.)

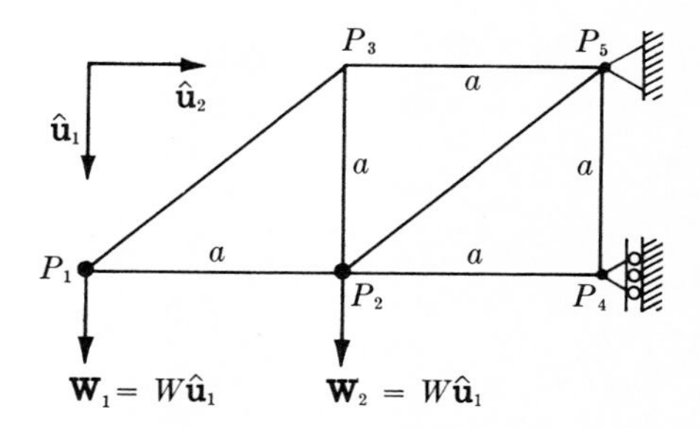

FIGURE P2·12

2·13 Determine the forces applied to points P_1 to P_5 by the seven pin-jointed rods in Fig. P2·13 when P_1 and P_2 are each particles of weight W and a horizontal force $\mathbf{F}$ of magnitude $2W$ is applied at P_3.

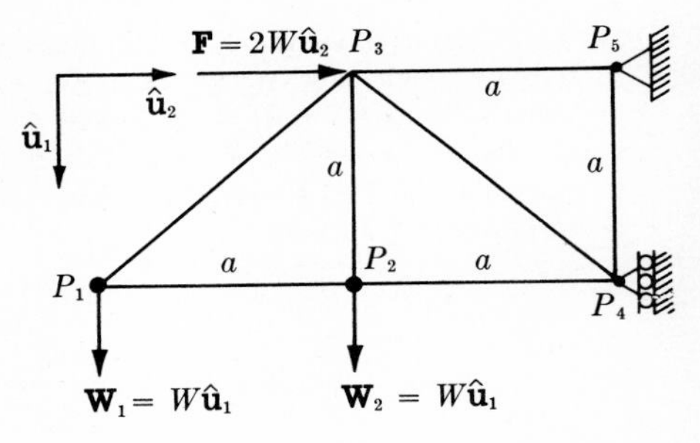

FIGURE P2·13

2·14 A particle P_0 of weight W is supported by a planar pin-jointed truss, as shown in Fig. P2·14. Determine the forces exerted on the five points P_1 to P_5 of the truss joints.

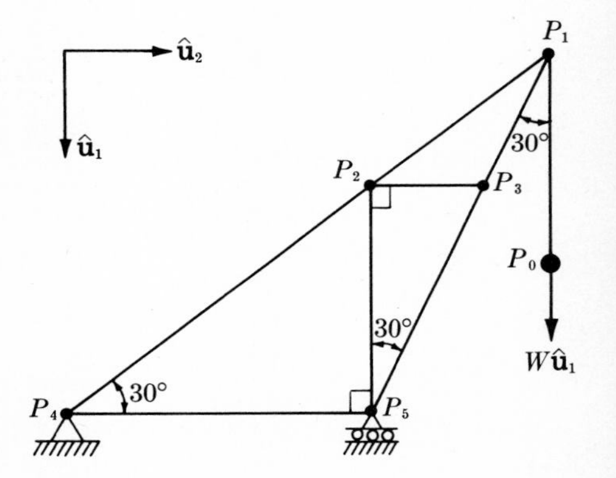

FIGURE P2·14

2·15 A particle P_0 of weight W is supported at point P_1 by a tripod truss whose members are rods anchored to a vertical wall at the vertices P_2, P_3, and P_4 of an equilateral triangle whose center is at C, as shown in Fig. P2·15. Determine the forces applied to P_1 by the members of the truss when the line from P_2 to C is

a. Vertical (as shown)
b. Horizontal

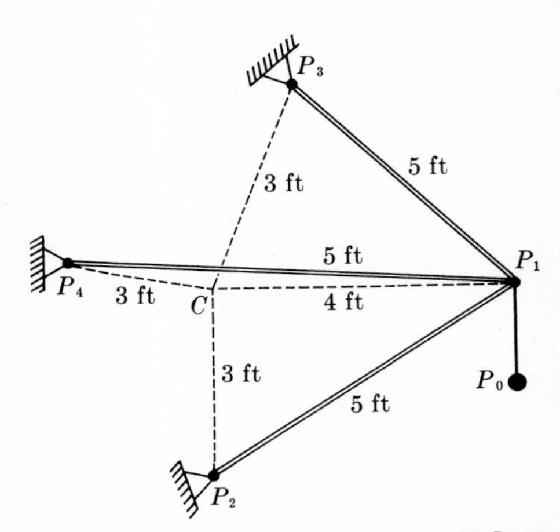

FIGURE P2·15

2·16 A particle P of weight W is supported a distance a above the ground by three pin-jointed rods connected to the vertices A, B, and C of an equilateral triangle of side dimension $2a$ on the ground, as shown in Fig. P2·16. P is in a vertical plane passing through A and B, and the shortest of the three rods is of length $2a$. Find the forces exerted on P by the rods.

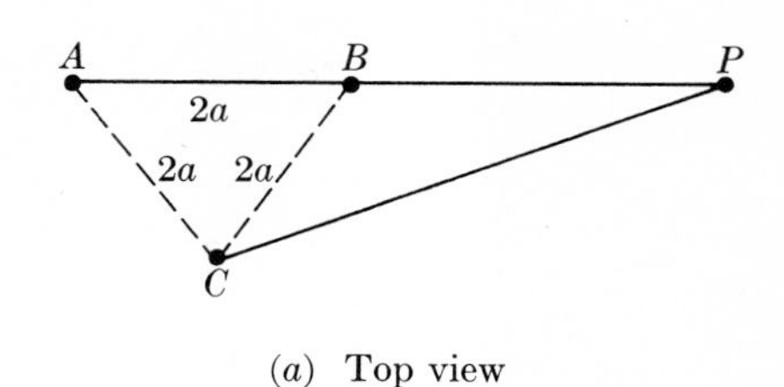

(a) Top view

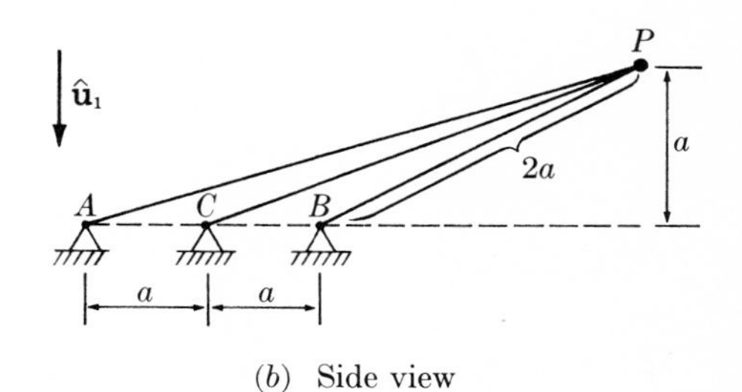

(b) Side view

FIGURE P2·16

2·17 If in the example treated in §2·2·9 the force **F** is replaced by a force **f** of magnitude f directed along the line from A to the two-o'clock position on the perimeter of the clock face, what would the forces in the members of the pin-jointed tripod become? (See Fig. 2·9.)

2·18 Point P is attached by means of pin-jointed rods to three corners A, B, and C of a rectangular parallelepipedal box, with dimensions as shown in Fig. P2·18. A force **F** of magnitude F has a line of action passing from P to the corner D of the box. Find the forces transmitted to P by the three rods, assuming that P is closer to the floor of the box than to its ceiling.

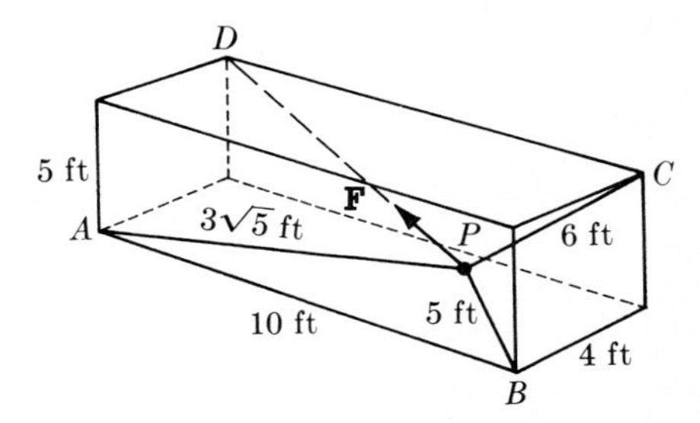

FIGURE P2·18

2·19 Determine for which of the systems in Fig. P2·19 the forces in the pin-jointed rods are statically determinate, noting those systems for which symmetry arguments must be used to establish determinacy.

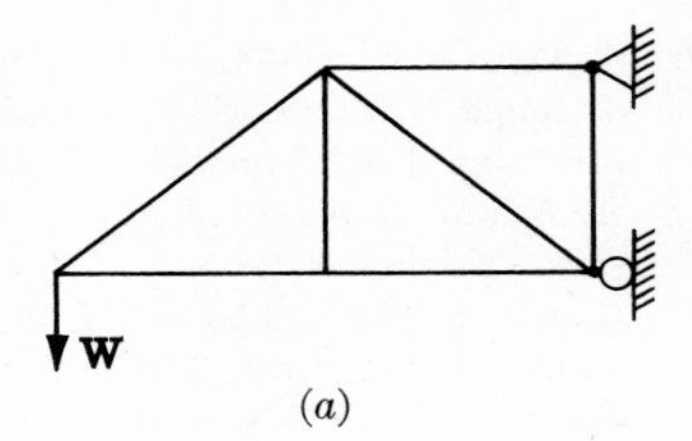

(*a*)

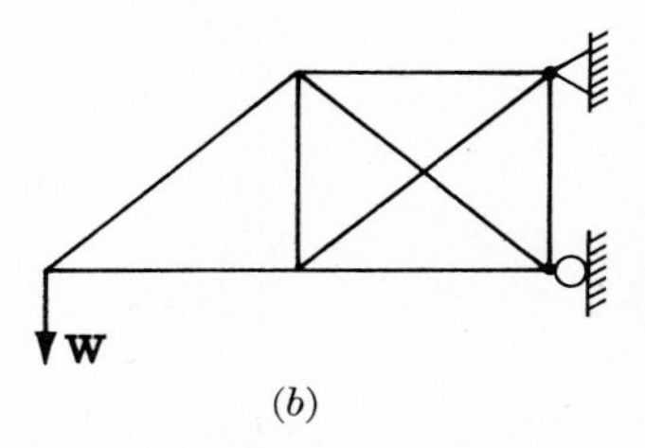

(*b*)

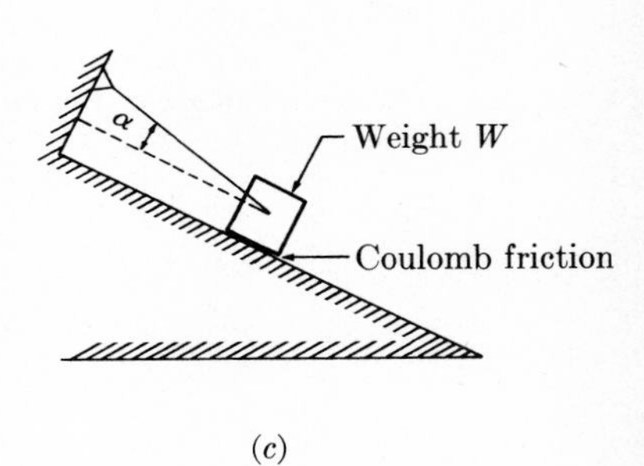

(*c*)

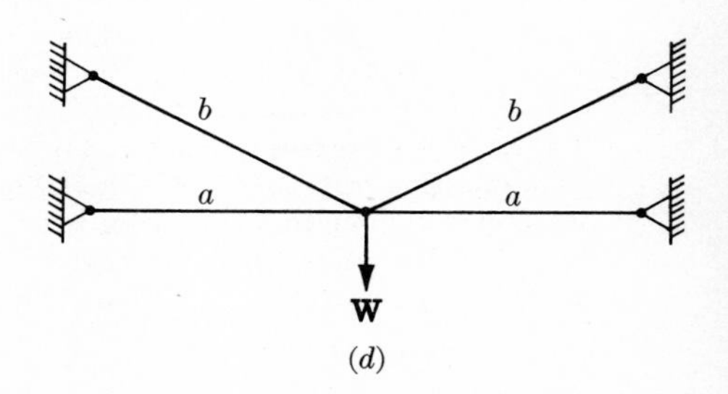

(*d*)

FIGURE P2·19

2·20 Determine for which of the following pin-jointed simple trusses (each shown in Fig. P2·20 in top and side views) the member forces are statically determinate, indicating when symmetry considerations must be used to establish determinacy.

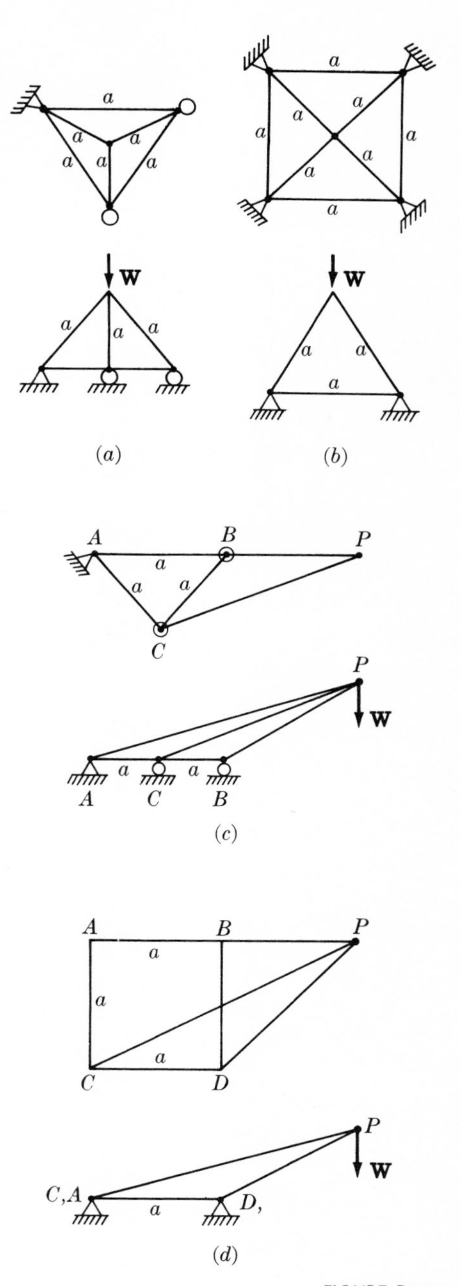

FIGURE P2·20

2·21 A particle P_1 of weight 200 lb and a particle P_2 of weight 100 lb are connected by a string passing over a frictionless pulley, as shown in Fig. P2·21. Each particle rests on a rough surface, with friction coefficient μ. Determine the minimum value of μ that permits static equilibrium. What happens if μ exceeds this value?

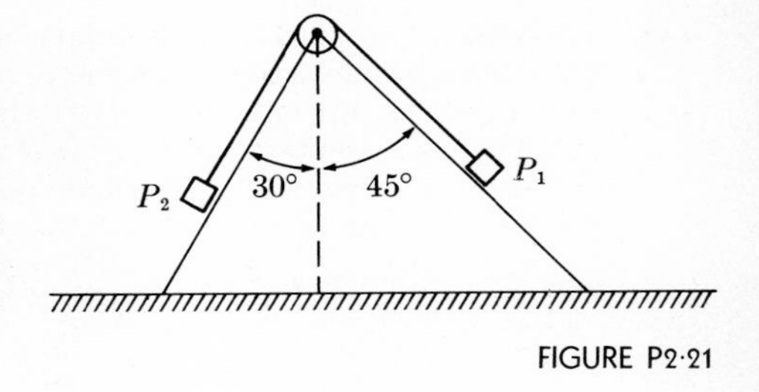

FIGURE P2·21

2·22 A particle P_1 of weight W is supported by a string passing over a frictionless pulley and then attached to a particle P_2 of weight 10 W, which rests on a rough horizontal surface having friction coefficient $\mu = 0.2$ (see Fig. P2·22). Is static equilibrium possible under these conditions? Express the force applied to P_2 by the floor in terms of orthogonal unit vectors $\hat{\mathbf{u}}_1$ and $\hat{\mathbf{u}}_2$.

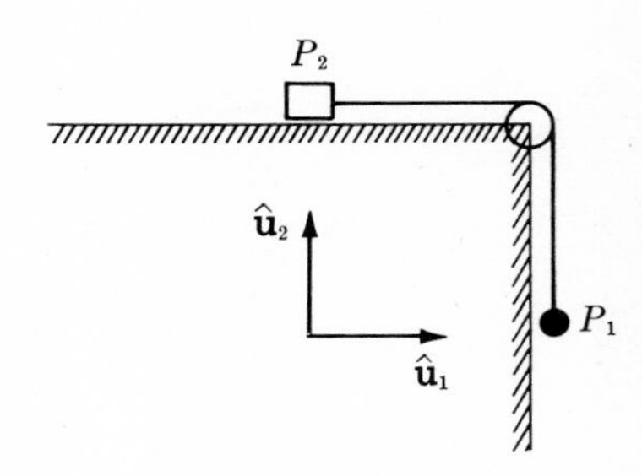

FIGURE P2·22

2·23 Three particles P_1, P_2, and P_3, weighing, respectively, $W_1 = 120$ lb, $W_2 = 20$ lb, and $W_3 = 20$ lb, are supported by frictionless pulleys as shown in Fig. P2·23, and in addition P_1 rests against a rough surface with friction coefficient $\mu = 0.2$, inclined at angle $\beta = 60°$. This system is *not* in static equilibrium until a man exerts a vertical force on the string at P_2 or P_3.

a. At which of these points can he apply the least force to equilibrate the system?

b. Does the answer change for any values of W_1, W_2, W_3, μ and β, with $0 < \beta < 90°$ and $W_2 < (W_1 - W_3) \sin \beta - \mu \cos \beta$?

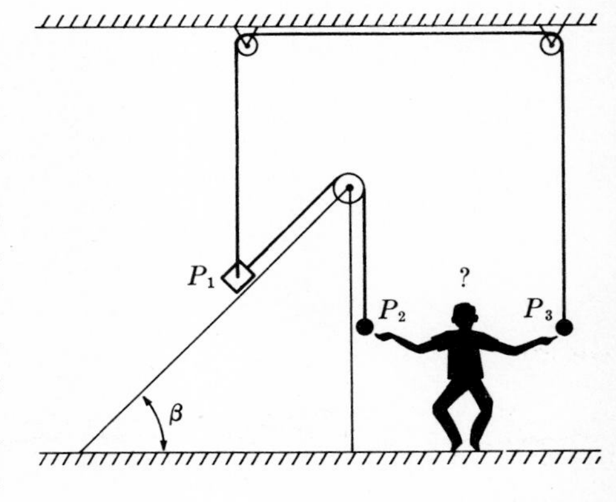

FIGURE P2·23

2·24 A 150-lb man displaces horizontally through a distance x a 1500-lb weight suspended by a 50-ft-long string from the ceiling, as shown in Fig. P2·24.

a. If the coefficient of friction between the man's shoes and the horizontal floor is $\mu = 0.8$, and if he applies only a horizontal force to the hanging weight, what is the maximum value of x for static equilibrium?

b. If the man applies a vertical force of magnitude V to the weight as he pushes it horizontally, he can displace the weight to an equilibrium distance x of 10 ft. Find V.

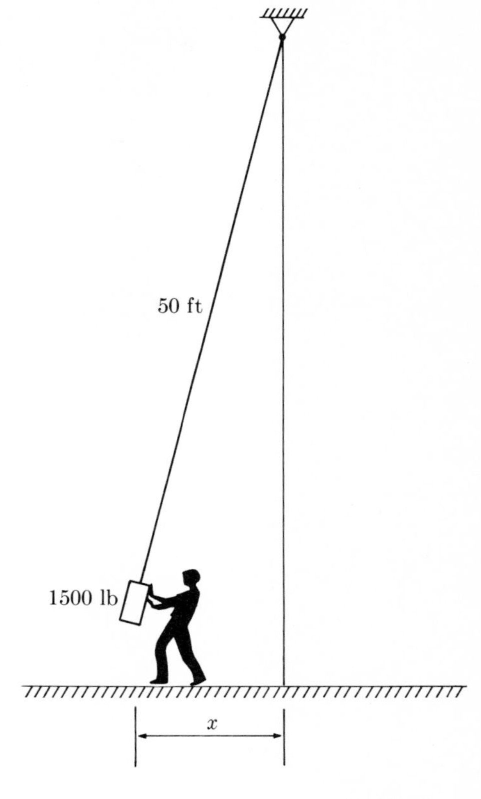

FIGURE P2·24

2·25 A particle P of weight W is supported vertically by two springs, as shown in Fig. P2·25, with each spring having unstretched length a, and the two spring constants being k_1 and k_2. Determine the displacement δ of P from the midpoint of the span between the supports at A and B.

2·26 Repeat Prob. 2·25, but change the unstretched lengths of the two springs from a to $2a$.

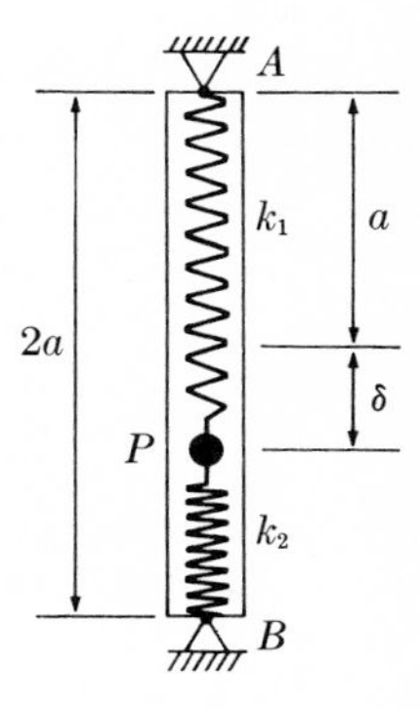

FIGURE P2·25

2·27 Find the equilibrium values of the distance r from the wall to the particle P of mass m in the two systems shown in Fig. P2·27. In each case P is confined to a tube that is free to rotate about a frictionless hinge at point A, and the displacement of P within the tube is resisted by a spring of unstretched length a and spring constant k, while P is subjected to an external force $\mathbf{F} = m\Omega^2 r\hat{\mathbf{e}}_1$, for constant Ω. Could your results be confirmed experimentally? Explain.

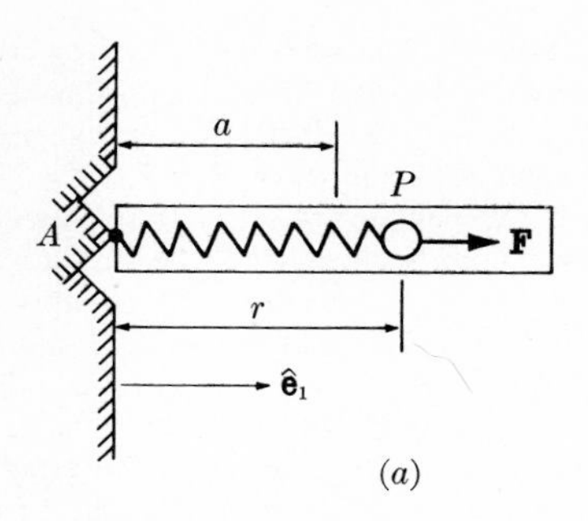

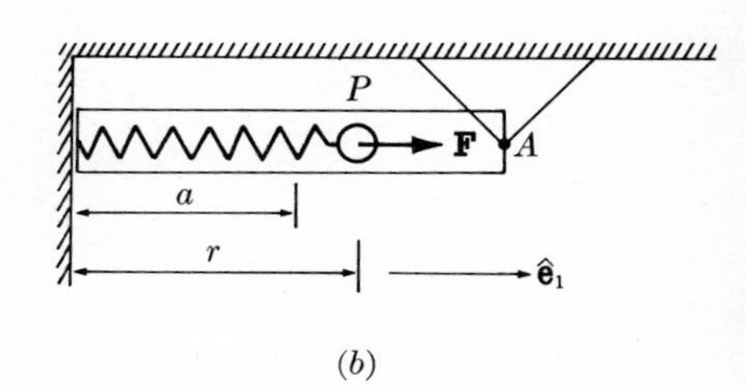

FIGURE P2·27

2·28 A particle P of weight W is suspended as a frictionless bead on a string of length $3\sqrt{2}\,L$, and the ends of the string are attached to supports at two points A and B which are a distance $3L$ apart horizontally and a distance L apart vertically (see Fig. P2·28). Find the horizontal distance x from A to P for static equilibrium.

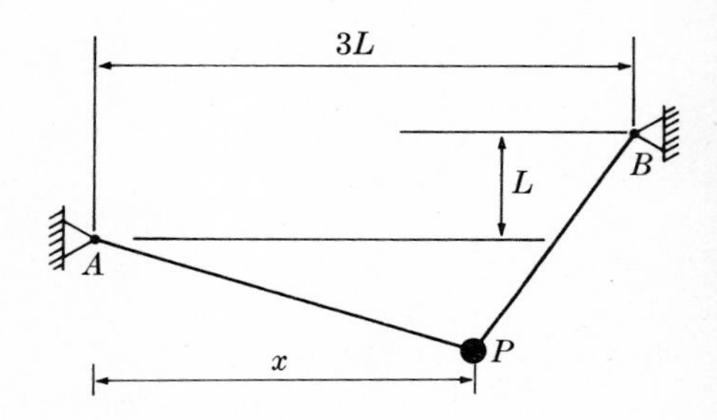

FIGURE P2·28

2·29 A particle P of weight $W = 10$ lb is placed in a smooth-walled hemispherical cup of radius $R = 2$ in.; a spring of undeformed length $2R$ and spring constant $k = W/R = 5$ lb/in. connects the particle to the rim of the cup, as shown in Fig. P2·29. Find the angle θ for which P is in static equilibrium.

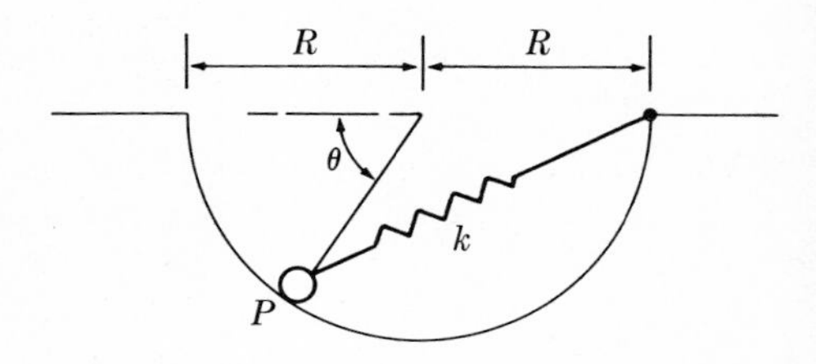

FIGURE P2·29

2·30 Particles P_1 and P_2 are attached as beads on a 25-ft-long frictionless string attached at its ends to points A and B and supported at point C by a frictionless pulley, as shown in Fig. P2·30. If P_1 weighs 100 lb and P_2 weighs 200 lb, what are the equations which must be solved for the equilibrium values of the deflections y_1 and y_2 of the particles below the horizontal line passing through A, B, and C?

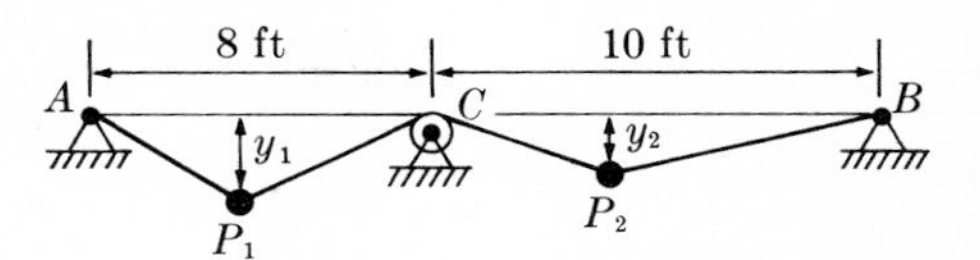

FIGURE P2·30

2·31 Determine the vertical displacement δ of particle P of weight W when supported as shown in Fig. P2·31, with the horizontal and diagonal members of the pin-jointed structure having, respectively, the axial spring constants k_1 and k_2.

a. Record the vector equations of equilibrium, the force-displacement relationships, and the exact geometrical relationships, and then combine these algebraically to obtain a single equation in the single unknown δ.

b. Define $\rho = \delta/a$, and obtain from part (a) a first-order approximation for ρ (neglecting terms above the first degree in ρ).

c. Check the answer to part (b) by using the equations of equilibrium and force displacement, from part (a), together with geometrical relationships suitable for the *undeformed* structure.

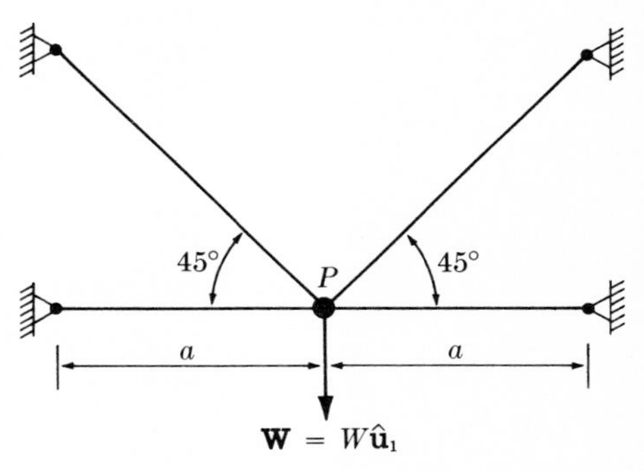

FIGURE P2·31

CHAPTER THREE
KINEMATICS

3·1 THE KINEMATIC PROPERTIES OF A POINT

3·1·1 Position of a point

The concept of *motion of a point in euclidean space* usually presents little difficulty in comprehension or visualization; we feel we have an "intuitive" idea of what this means because we have had the experience of watching small bodies move about in the space we occupy. Although a body, however small, is not a geometrical point, and the physical environment is a concept distinct from the space of euclidean geometry, the correspondence is good enough so that this natural appreciation of motion (and other aspects of physics) can greatly facilitate the acquisition of understanding of theoretical mechanics. Unless an engineer's feeling for kinematics and other aspects of dynamics is based on sound *analytical* foundations, however, it is unlikely that his skill will be sufficient for the complexities of modern engineering problems. You are therefore encouraged to work initially with the analytical formulations that follow, replacing them in part only gradually with subjective notions derived from physical interpretations.

The *position*, or location, of a point can be conceived only in reference to another point. Given a point a and a base point b, there remain numerous alternative ways to *describe* the location of a relative to b, but the mode of description is a less fundamental matter than the specification of what is to be described.

Consider the example of Fig. 3·1. Given merely a point a in (euclidean) space, "position" is meaningless. Given both a and b, relative position is fully established. This position may be *described* by indicating the values of three scalars, x, y, and z, once a cartesian coordinate system has been defined. Alternatively, the three scalars r, φ and λ fully characterize the position of a relative to b, with suitable definitions of these coordinates. The most satisfactory modes of description, however, employ *vectors*. A vector (see Appendix B) in three-dimensional space is a directed line segment with certain operational properties. It has a magnitude (length), an orientation, and a sense (positive direction). A vector is not *bound* to a point or line in space unless this is specified. Vectors add commutatively and associatively according to the *parallelogram rule*, and multiply according to the rules of scalar (*dot*) and vector (*cross*) multiplication.

The vector $\mathbf{r}$ from b to a, and indeed any vector, can be explicitly written in numberless ways. Although the strength of vector analysis lies in the independence of any coordinate system (which has to be specified for either of the scalar descriptions of position above), it may nonetheless sometimes be convenient to specify an arbitrarily selected cartesian coordinate system x, y, and z and define three unit vectors $\hat{\mathbf{i}}$, $\hat{\mathbf{j}}$, and $\hat{\mathbf{k}}$ parallel to the x, y, and z axes, respectively, so that the position vector $\mathbf{r}$ may be written as

$$\mathbf{r} = x\hat{\mathbf{i}} + y\hat{\mathbf{j}} + z\hat{\mathbf{k}} \tag{3·1}$$

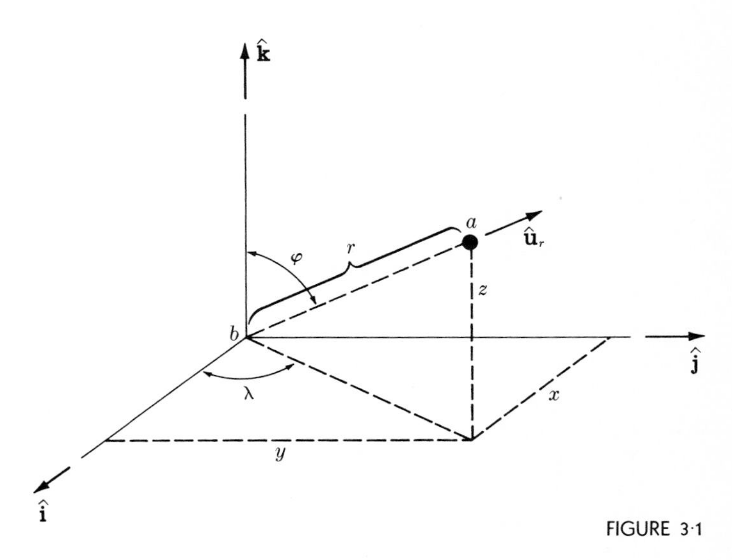

FIGURE 3·1

It is frequently more convenient, however, to define a single unit vector, such as $\hat{\mathbf{u}}_r$ in Fig. 3·1, parallel to the line from b to a. Then, since the length of the line segment from b to a is r, one may write

$$\mathbf{r} = r\hat{\mathbf{u}}_r \tag{3·2}$$

Remember that the unit vectors shown in Fig. 3·1 are *not bound.* They could each be drawn anywhere on the page without change of meaning, as long as magnitude, sense, and orientation are unchanged.

3·1·2 Velocity of a point

The velocity of a point has meaning only in reference to another point (expressed or implied) and only relative to a particular *frame of reference.* Figure 3·2 illustrates this fact with a physical example. If the ball labeled p remains fixed to a point a on the rotating disk d, there is no velocity of p or a relative to the disk center o in the frame of reference of d. (Some students find it helpful to imagine an observer seated on the disk, and to note that he would perceive no motion of a relative to o since both points are fixed in d.) In the base reference b, however, point a is moving, and point o at the center of the disk is stationary. Hence there *is* a velocity of a relative to o in reference frame b (an observer seated on b would perceive such a relative motion). Moreover, if p departs from a, a nonzero velocity of p relative

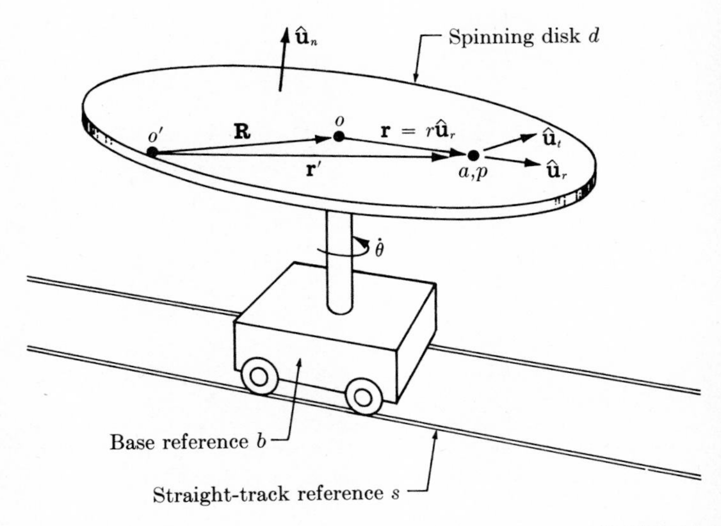

FIGURE 3·2

to a in frame d becomes possible, and this differs from the velocity of p relative to a in frame b.

In this example, the reference frames b and d are identified with idealized rigid bodies, the base block and the disk. Strictly speaking, we must draw a distinction between a rigid body (which has physical properties as well as size and shape) and the corresponding reference frame, which is a three-dimensional space of infinite dimensions in which the body is permanently embedded. It is quite possible, and often useful, to imagine a reference frame in which no idealized rigid body is embedded.

It is apparent that a fully descriptive symbol for the velocity of a relative to o in a reference frame f must convey the significance of each of these three symbols a, o, and f. On those rare occasions when such an elaborate symbol is needed, the velocity of a relative to o in f will be written $\mathbf{V}^{a_o f}$. This cumbersome symbol and the concept that it represents will be rarely used after this article. Such intricate symbology is useful in problem solving only when the complexity of the problem requires careful attention to "bookkeeping" details, although some form of rational symbology can also be an invaluable aid in the initial mastery of many of the basic concepts of kinematics.

The preceding remarks about velocity appeal to physical experience. The analytical counterpart may be expressed in the following *definition:* The velocity of point a relative to point o in reference frame f is the time derivative in frame f of the position vector $\mathbf{r}$ from o to a, which may be written (note the notational convention for *vector differentiation*)

$$\mathbf{V}^{a_o f} \triangleq \frac{{}^f d}{dt}\mathbf{r} \tag{3·3}$$

The magnitude of velocity is often referred to as *speed.*

As the role of the reference frame in vector differentiation may not be apparent, a very brief digression into vector calculus may be in order. The definition of the derivative in frame f of any vector, say $\mathbf{q}$, appears as a limit

$$\frac{{}^f d}{dt}\mathbf{q} \triangleq \text{limit in frame } f \text{ as } \Delta t \to 0 \text{ of } \frac{\mathbf{q}(t + \Delta t) - \mathbf{q}(t)}{\Delta t} \tag{3·4}$$

as for scalar derivatives, except that frame f must be specified. At this point, this definition has little operational value because the role of frame f in the limiting process has not been established. The frame designation is necessary because a vector may be rotating relative to one reference frame differently than it rotates in another, so that in the time interval Δt, the vector $\mathbf{q}$ may appear in different reference frames to undergo different changes of direction. We may anticipate from physical considerations [and later prove formally with Eq. (3·100)] that translation of a bound vector $\mathbf{q}$ (parallel translation of both its end points) and changes in the magnitude of $\mathbf{q}$ do not create differences in the $\mathbf{q}$ derivative in different reference frames. Thus if f and f' are two reference frames with no relative rotation, vector derivatives in these frames are identical, that is

$$\frac{{}^f d}{dt}\mathbf{q} = \frac{{}^{f'} d}{dt}\mathbf{q} \tag{3·5}$$

The symbol t and the use of words such as rotation and translation may seem to imply a restriction to time derivatives. This is not a necessary restriction, but this interpretation is useful in the present context.

Because of the correspondence between the mathematics of vector differentiation and the physics of velocity determination, it is convenient to draw examples illustrating the role of the reference frame in vector differentiation from physical experience. When we examined Fig. 3·2, we agreed that the velocity of a relative to o was different in the disk reference frame d and the base block frame b since the disk is rotating relative to the base block. We reached this conclusion without benefit of formal definitions or vector differentiations. Now let's re-examine the question in an analytical way, using the definition of velocity as a vector derivative.

The definition of Eq. (3·3) provides

$$\mathbf{V}^{a_o b} = \frac{{}^b d}{dt}\mathbf{r} \qquad \text{and} \qquad \mathbf{V}^{a_o d} = \frac{{}^d d}{dt}\mathbf{r}$$

As indicated in Eq. (3·2), $\mathbf{r}$ can be written $\mathbf{r} = r\hat{\mathbf{u}}_r$ so that by the product rule of differentiation (which applies to vectors as well as to scalars)

$$\mathbf{V}^{a_o b} = \frac{{}^b d}{dt}(r\hat{\mathbf{u}}_r) = \left(\frac{dr}{dt}\right)\hat{\mathbf{u}}_r + r\left(\frac{{}^b d}{dt}\hat{\mathbf{u}}_r\right) \tag{3·6}$$

and

$$\mathbf{V}^{a_o d} = \frac{{}^d d}{dt}(r\hat{\mathbf{u}}_r) = \left(\frac{dr}{dt}\right)\hat{\mathbf{u}}_r + r\left(\frac{{}^d d}{dt}\hat{\mathbf{u}}_r\right) \tag{3·7}$$

Note that no reference frame is designated for the scalar derivative dr/dt because a scalar derivative is the same for all reference frames; in other words, scalar differentiation is a concept that does not involve any frame of reference.

For the special case previously considered, for which both a and o are fixed in d, the scalar derivative $dr/dt = 0$, and the unit vector $\hat{\mathbf{u}}_r$ is fixed in d; thus the derivative $({}^d d/dt)\hat{\mathbf{u}}_r = 0$. But since, relative to the base b, the unit vector $\hat{\mathbf{u}}_r$ is constantly changing direction, we have $({}^b d/dt)\,\hat{\mathbf{u}}_r \neq 0$. Hence, as previously, the velocities differ and the need for reference frames is indicated.

If now the base b were placed on wheels and rolled down a straight track, and the reference frame established by the straight track were designated s, we would have

$$\mathbf{V}^{a_o s} = \frac{^s d}{dt}(r\hat{\mathbf{u}}_r) = \left(\frac{dr}{dt}\right)\hat{\mathbf{u}}_r + r\left(\frac{^s d}{dt}\hat{\mathbf{u}}_r\right) \tag{3·8}$$

The first term is the same as previously and, because $\hat{\mathbf{u}}_r$ is changing direction relative to s in precisely the same way that it changes direction in b,

$$\frac{^b d}{dt}\hat{\mathbf{u}}_r = \frac{^s d}{dt}\hat{\mathbf{u}}_r \tag{3·9}$$

Hence the velocities of a relative to o in b and s are identical,

$$\mathbf{V}^{a_o s} = \mathbf{V}^{a_o b} \tag{3·10}$$

Recall that in Eq. (3·5) this result was anticipated because as long as the track is straight there is no rotation of b relative to the frame s.

Now we'll allow the point p to move around on the disk, so that p no longer coincides with a. We'll define o' to be a point fixed on the perimeter of the disk (see Fig. 3·2), and we will then compare the velocities of p relative to o and o' in various reference frames.

By definition, if $\mathbf{r}$ and $\mathbf{r}'$ are respectively the position vectors to p from o and o', the relative velocities are

$$\mathbf{V}^{p_o d} = \frac{^d d}{dt}\mathbf{r} \tag{3·11}$$

and

$$\mathbf{V}^{p_{o'} d} = \frac{^d d}{dt}\mathbf{r}' \tag{3·12}$$

But, if $\mathbf{R}$ is the position vector from o' to o, we may substitute

$$\mathbf{r}' = \mathbf{R} + \mathbf{r} \tag{3·13}$$

into Eq. (3·12) to obtain

$$\mathbf{V}^{p_{o'} d} = \frac{^d d}{dt}(\mathbf{R} + \mathbf{r}) = \frac{^d d}{dt}\mathbf{R} + \mathbf{V}^{p_o d} \tag{3·14}$$

Since both o' and o are fixed in the disk d, the time derivative of $\mathbf{R}$ in frame d is zero, so we get the interesting conclusion

$$\mathbf{V}^{p_{o'} d} = \mathbf{V}^{p_o d} \tag{3·15}$$

This result is completely general. For a completely arbitrary point p and frame f, a unique vector describes the velocity of p in frame f relative to *any* point fixed in f; in other words, there is a single vector quantity that represents the velocity in frame f of point p with respect to a generic point or arbitrary typical point fixed in f. This vector may be called *the generic velocity*[1] *of* p *in* f, or simply *the velocity of* p *in* f, designated $\mathbf{V}^{pf}$. In practice this concept seems more useful than the velocity of one point relative to another in a reference frame in which neither is fixed, such as is defined by Eq. (3·3).

We can reformulate this idea with the following definition: The (generic) velocity of a point p in a reference frame f is

$$\mathbf{V}^{pf} \triangleq \frac{{}^f d}{dt} \mathbf{P} \tag{3·16}$$

where $\mathbf{P}$ is the position vector to point p from *any point fixed in* f.

When it becomes necessary to determine the *relative* velocity $\mathbf{V}^{p_q f}$ where both p and q are moving relative to f, it is always possible to return to the simpler concept of generic velocity. Let $\mathbf{P}$ and $\mathbf{Q}$ be position vectors to p and q, respectively, from any point fixed in f and let $\mathbf{r}$ be the vector from q to p; then write, from the definitions,

$$\mathbf{V}^{p_q f} = \frac{{}^f d}{dt} \mathbf{r} = \frac{{}^f d}{dt} (\mathbf{P} - \mathbf{Q}) = \frac{{}^f d}{dt} \mathbf{P} - \frac{{}^f d}{dt} \mathbf{Q} = \mathbf{V}^{pf} - \mathbf{V}^{qf} \tag{3·17}$$

Before proceeding, you may wish to pause and take stock of your understanding of the concepts of velocity introduced thus far. The notion of the velocity of one point p relative to another point q in a given reference frame f is *conceptually* the fundamental idea, but *in practice* we will often find it most convenient to work with generic velocities such as $\mathbf{V}^{pf}$ and $\mathbf{V}^{qf}$, for which the reference points may be any point fixed in f.

3·1·3 Vector differentiation

This chapter contains thus far only definitions and illustrations supporting the necessity of considering frames of reference. As actual determination of derivatives by evaluation of limits [see Eq. (3·4)] is impractical, no effective general method for determination of velocities has yet been presented.

If, however, any vector $\mathbf{Z}$ is written as

$$\mathbf{Z} = Z_1\hat{\mathbf{f}}_1 + Z_2\hat{\mathbf{f}}_2 + Z_3\hat{\mathbf{f}}_3 \tag{3·18}$$

where $\hat{\mathbf{f}}_1$, $\hat{\mathbf{f}}_2$, and $\hat{\mathbf{f}}_3$ are unit vectors fixed in reference frame f and not all parallel to the same plane, then the differentiation definition is easy to use *for differentiation in frame* f. The "product rule" for differentiation provides

$$\frac{{}^f d\mathbf{Z}}{dt} = \frac{dZ_1}{dt}\hat{\mathbf{f}}_1 + Z_1 \frac{{}^f d\hat{\mathbf{f}}_1}{dt} + \frac{dZ_2}{dt}\hat{\mathbf{f}}_2 + Z_2 \frac{{}^f d\hat{\mathbf{f}}_2}{dt} + \frac{dZ_3}{dt}\hat{\mathbf{f}}_3 + Z_3 \frac{{}^f d\hat{\mathbf{f}}_3}{dt} \tag{3·19}$$

[1] The term *absolute velocity* is often used, but in mechanics the phrase absolute velocity also has for many people the different meaning of velocity relative to an inertial reference frame, so this potentially confusing term will not be used at all in this text.

(Note again that derivative superscripts f don't appear for the scalar derivatives, where they would be meaningless.) Using limits for definitions now presents no handicap. The scalar derivatives in Eq. (3·19) are familiar, and, although originally defined in terms of limits, they are readily evaluated. The vector derivatives in Eq. (3·19) are all zero since $\hat{\mathbf{f}}_1$, $\hat{\mathbf{f}}_2$, and $\hat{\mathbf{f}}_3$ are all fixed in f. Hence, employing a dot ($\dot{}$) for $d(\ )/dt$ of a scalar, we write

$$\frac{{}^f d\mathbf{Z}}{dt} = \dot{Z}_1\hat{\mathbf{f}}_1 + \dot{Z}_2\hat{\mathbf{f}}_2 + \dot{Z}_3\hat{\mathbf{f}}_3 = \sum_{i=1}^{3} \dot{Z}_i\hat{\mathbf{f}}_i \tag{3·20}$$

Since any vector $\mathbf{Z}$ can be written in the form of Eq. (3·18) for unit vectors $\hat{\mathbf{f}}_1$, $\hat{\mathbf{f}}_2$, and $\hat{\mathbf{f}}_3$ fixed in any reference frame f, we now have a general method for differentiating vectors (and thereby determining velocities). Construction of Eq. (3·18) for a given problem may be rather difficult, however, and the subsequent differentiation of the scalars Z_1, Z_2, and Z_3 may be extremely tedious. If we follow this approach we have lost many of the advantages of vector analysis because we have again reverted to the use of a prescribed set of scalar coordinates (now called Z_1, Z_2, and Z_3 instead of x, y, and z). This is nonetheless one alternative procedure, and we cannot find a better way until we develop the foundation necessary for the introduction of the kinematical concept of *angular velocity*. The differentiation procedure in Eq. (3·20) may be familiar from elementary vector calculus, although the assumption that the unit vectors are fixed in the frame of differentiation may not have been noted explicitly. This procedure presents no conceptual difficulties, and it will suffice for the calculation of velocities in some examples to follow. When more efficient procedures for vector differentiation become available, we shall return to some of these examples for a comparison of methods.

3·1·4 Examples of linear velocity determination

First example Figure 3·3 shows a pair of pin-connected rigid bars, each of length a, connected at point p to a block free to slide in the direction of $\hat{\mathbf{e}}_x$. The angle θ is a specified function of time. We are to find the velocity of p in frame e.

By definition, the required velocity is

$$\mathbf{V}^{pe} = \frac{{}^e d}{dt}\mathbf{r} = \frac{{}^e d}{dt}(2a \sin \theta\hat{\mathbf{e}}_x) = 2a\dot{\theta} \cos \theta\hat{\mathbf{e}}_x \tag{3·21}$$

This calculation is very simple because the vector $\mathbf{r}$ to be differentiated in the frame e is readily expressed in terms of scalars and the unit vector $\hat{\mathbf{e}}_x$ fixed in e. We shall see in future examples that this is not always true.

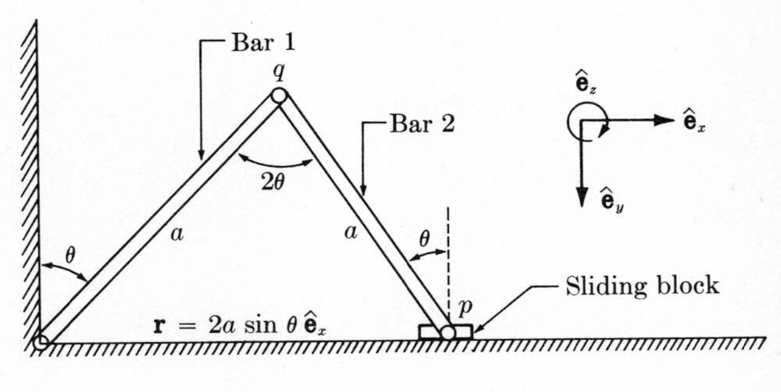

FIGURE 3·3

Second example An ant called a crawls along a meridian of a spinning cone c at a constant speed $v = \frac{1}{2}$ in./sec, as shown in Fig. 3·4. The cone is a right circular cone of height 4 in. and base radius 3 in., and it spins about its symmetry axis at the constant rate $\Omega = 2$ rad/sec relative to base b, with clockwise rotation as viewed from b. The ant began at the apex A at reference time $t = 0$, when unit vectors $\hat{\mathbf{c}}_1$, $\hat{\mathbf{c}}_2$, and $\hat{\mathbf{c}}_3$ fixed in c coincided respectively with $\hat{\mathbf{b}}_1$, $\hat{\mathbf{b}}_2$, and $\hat{\mathbf{b}}_3$ fixed in b.

We wish to find the velocity of a relative to the base b, and we require the answer in terms of unit vectors $\hat{\mathbf{c}}_1$, $\hat{\mathbf{c}}_2$, and $\hat{\mathbf{c}}_3$. The answer is to be recorded specifically for $t = 2$ sec.

By definition, the velocity $\mathbf{V}^{ab}$ is the time derivative of the position vector locating a from *any* point fixed in the reference frame of b. The most convenient reference point is the apex A since the vector from A to a is a straight line along the meridional path. (Note that point A qualifies because it is fixed in the reference frame of b; it is not essential that it be embedded in body b itself.) If we introduce the unit vector $\hat{\mathbf{u}}$ paralleling the path from A to a, and note that the distance between these points is vt, we have the expression

$$\mathbf{V}^{ab} = \frac{{}^b d}{dt}(vt\hat{\mathbf{u}}) \tag{3·22}$$

for the required velocity. This expression differs from Eq. (3·21) for the preceding example in one significant respect—the unit vector $\hat{\mathbf{u}}$ is not fixed in the reference frame b. Differentiation of the product in Eq. (3·22) yields

$$\mathbf{V}^{ab} = v\hat{\mathbf{u}} + vt\frac{{}^b d}{dt}\hat{\mathbf{u}} \tag{3·23}$$

However, to proceed further with our present limited methods of vector differentiation we will have to replace $\hat{\mathbf{u}}$ by some combination of vectors $\hat{\mathbf{b}}_1$, $\hat{\mathbf{b}}_2$, and $\hat{\mathbf{b}}_3$ fixed in b. From the geometry of the cone we have

$$\hat{\mathbf{u}} = \sin\alpha\,\hat{\mathbf{c}}_1 - \cos\alpha\,\hat{\mathbf{c}}_3 \tag{3·24}$$

where α is the half-angle of the cone. Since the base radius is 3 in. and the height is 4 in., the length of the meridian is 5 in., from the pythagorean theorem. Therefore the sine and cosine of α are available, and we have

$$\hat{\mathbf{u}} = \frac{3}{5}\hat{\mathbf{c}}_1 - \frac{4}{5}\hat{\mathbf{c}}_3$$

The next step requires the substitutions

$$\begin{aligned} \hat{\mathbf{c}}_1 &= \hat{\mathbf{b}}_1 \cos\Omega t + \hat{\mathbf{b}}_2 \sin\Omega t \\ \hat{\mathbf{c}}_2 &= \hat{\mathbf{b}}_3 \end{aligned} \tag{3·25}$$

and provides

$$\hat{\mathbf{u}} = \hat{\mathbf{b}}_1\left(\frac{3}{5}\cos\Omega t\right) + \hat{\mathbf{b}}_2\left(\frac{3}{5}\sin\Omega t\right) + \hat{\mathbf{b}}_3\left(-\frac{4}{5}\right) \tag{3·26}$$

Finally we substitute Eq. (3·26) into (3·23) and differentiate the scalars to obtain

$$\begin{aligned} \mathbf{V}^{ab} &= v\left[\hat{\mathbf{b}}_1\left(\frac{3}{5}\cos\Omega t\right) + \hat{\mathbf{b}}_2\left(\frac{3}{5}\sin\Omega t\right) + \hat{\mathbf{b}}_3\left(-\frac{4}{5}\right)\right] \\ &\qquad + vt\left[\hat{\mathbf{b}}_1\left(-\frac{3}{5}\Omega\sin\Omega t\right) + \hat{\mathbf{b}}_2\left(\frac{3}{5}\Omega\cos\Omega t\right)\right] \\ &= \hat{\mathbf{b}}_1\left[\frac{3}{5}(v\cos\Omega t - \Omega vt\sin\Omega t)\right] \\ &\qquad + \hat{\mathbf{b}}_2\left[\frac{3}{5}(v\sin\Omega t + \Omega vt\cos\Omega t)\right] + \hat{\mathbf{b}}_3\left(-\frac{4}{5}v\right) \end{aligned} \tag{3·27}$$

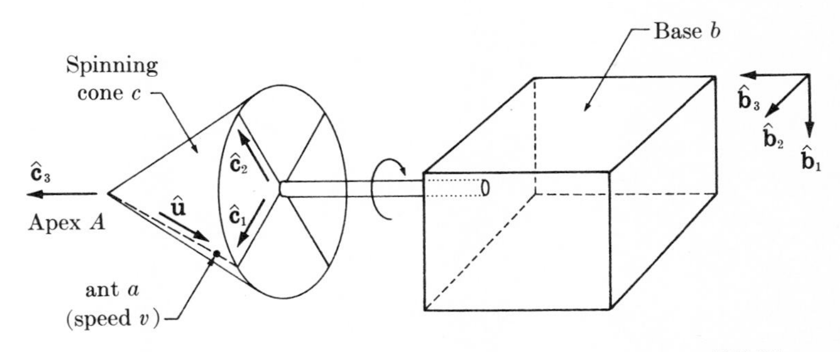

FIGURE 3·4

The problem statement requires the answer in terms of unit vectors $\hat{\mathbf{c}}_1$, $\hat{\mathbf{c}}_2$, and $\hat{\mathbf{c}}_3$, so we need the transformations

$$\begin{aligned}\hat{\mathbf{b}}_1 &= \hat{\mathbf{c}}_1 \cos \Omega t - \hat{\mathbf{c}}_2 \sin \Omega t \\ \hat{\mathbf{b}}_2 &= \hat{\mathbf{c}}_2 \cos \Omega t + \hat{\mathbf{c}}_1 \sin \Omega t \\ \hat{\mathbf{b}}_3 &= \hat{\mathbf{c}}_3\end{aligned} \tag{3·28}$$

Substituting Eqs. (3·28) into (3·27) and using the identity

$$\sin^2 \Omega t + \cos^2 \Omega t = 1$$

provides, after some algebraic manipulation,

$$\mathbf{V}^{ab} = \frac{3}{5} v\hat{\mathbf{c}}_1 + \frac{3}{5} \Omega v t\hat{\mathbf{c}}_2 - \frac{4}{5} v\hat{\mathbf{c}}_3 \tag{3·29}$$

For the given numerical values of v, Ω, and t, the answer is

$$\mathbf{V}^{ab} = 0.3\hat{\mathbf{c}}_1 + 1.2\hat{\mathbf{c}}_2 - 0.4\hat{\mathbf{c}}_3 \qquad \text{in./sec} \tag{3·30}$$

Figure 3·4 represents a very simple physical system, and the final expression in Eq. (3·29) is appropriately uncomplicated. A surprising amount of labor is required by this simple problem, and it would seem that there must be a better way. Indeed there is an alternative to our tedious procedure for obtaining $({}^b d/dt)\, \hat{\mathbf{u}}$ in Eq. (3·23), as we shall see later. However, to further sharpen our appetite for new and better methods of vector differentiation, we will consider what is involved in obtaining the velocity of a given point in the mechanism of the next example.

Third example Figure 3·5 shows a ring (center labeled p) free to slide along an arm a of total length L. One end of the arm can slide up and down the axis of the support block b, while the other end of arm a slides along the floor in such a way that the triangle formed by points c, o, and q remains normal to the floor. The support block b rotates relative to the floor about its centerline from o to q, which is normal to the floor. The figure indicates that the angle θ defines the rotation of block b relative to floor frame f, the angle φ defines the changing orientation of a relative to b, and the length z is the distance along the arm from q to p.

What would be required to obtain the velocity of p in reference frame f? If $\mathbf{P}$ is the position vector from o to p, the required velocity is given by

$$\mathbf{V}^{pf} = \frac{{}^f d}{dt} \mathbf{P} \tag{3·31}$$

from the definition of Eq. (3·16). The only differentiation procedure available thus far requires that vector $\mathbf{P}$ be written

$$\mathbf{P} = P_1\hat{\mathbf{f}}_1 + P_2\hat{\mathbf{f}}_2 + P_3\hat{\mathbf{f}}_2 \tag{3·32}$$

so that vector differentiation yields

$$\mathbf{V}^{pf} = \dot{P}_1\hat{\mathbf{f}}_1 + \dot{P}_2\hat{\mathbf{f}}_2 + \dot{P}_3\hat{\mathbf{f}}_3 \tag{3·33}$$

This is very easy to write down, but if we are interested in useful answers we must first find P_1, P_2, and P_3 in Eq. (3·32) and then differentiate these scalars.

By inspection of Fig. 3·5, we may write

$$\mathbf{P} = L \cos \varphi \hat{\mathbf{f}}_3 + z\hat{\mathbf{a}}_1 \tag{3·34}$$

and

$$\hat{\mathbf{a}}_1 = \hat{\mathbf{b}}_1 \sin \varphi - \hat{\mathbf{b}}_3 \cos \varphi = \hat{\mathbf{f}}_1(\sin \varphi \cos \theta) + \hat{\mathbf{f}}_2(\sin \varphi \sin \theta) - \hat{\mathbf{f}}_3 \cos \varphi \tag{3·35}$$

Substitution yields

$$\begin{aligned}\mathbf{P} &= \hat{\mathbf{f}}_1(z \sin \varphi \cos \theta) + \hat{\mathbf{f}}_2(z \sin \varphi \sin \theta) + \hat{\mathbf{f}}_3[(L - z) \cos \varphi] \\ &= P_1\hat{\mathbf{f}}_1 + P_2\hat{\mathbf{f}}_2 + P_3\hat{\mathbf{f}}_3\end{aligned} \tag{3·36}$$

Differentiation provides $\dot{P}_1$, $\dot{P}_2$, and $\dot{P}_3$ for substitution into Eq. (3·33), with the result

$$\begin{aligned}\mathbf{V}^{pf} = (\dot{z} \sin \varphi \cos \theta + z\dot{\varphi} \cos \varphi \cos \theta - z\dot{\theta} \sin \varphi \sin \theta)\hat{\mathbf{f}}_1 \\ + (\dot{z} \sin \varphi \sin \theta + z\dot{\varphi} \cos \varphi \sin \theta + z\dot{\theta} \sin \varphi \cos \theta)\hat{\mathbf{f}}_2 \\ + [-\dot{z} \cos \varphi - (L - z)\dot{\varphi} \sin \varphi]\hat{\mathbf{f}}_3\end{aligned} \tag{3·37}$$

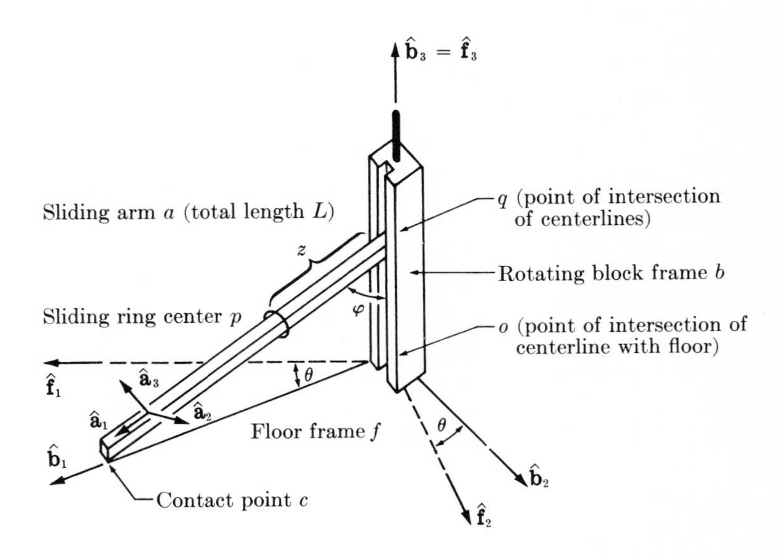

FIGURE 3·5

Perceptive examination of this seemingly complicated expression reveals that an expression for the velocity in terms of unit vectors $\hat{\mathbf{b}}_1$, $\hat{\mathbf{b}}_2$, and $\hat{\mathbf{b}}_3$ is somewhat simpler. The identities

$$\begin{aligned}\hat{\mathbf{f}}_1 &= \hat{\mathbf{b}}_1 \cos\theta - \hat{\mathbf{b}}_2 \sin\theta \\ \hat{\mathbf{f}}_2 &= \hat{\mathbf{b}}_2 \cos\theta + \hat{\mathbf{b}}_1 \sin\theta \\ \hat{\mathbf{f}}_3 &= \hat{\mathbf{b}}_3\end{aligned} \tag{3·38}$$

may be substituted into Eq. (3·37) to obtain

$$\begin{aligned}\mathbf{V}^{pf} = {} & \hat{\mathbf{b}}_1(\dot{z}\sin\varphi + z\dot{\varphi}\cos\varphi) + \hat{\mathbf{b}}_2(z\dot{\theta}\sin\varphi) \\ & + \hat{\mathbf{b}}_3[-\dot{z}\cos\varphi - (L - z)\dot{\varphi}\sin\varphi]\end{aligned} \tag{3·39}$$

Now it appears that the result may be even simpler in the vector basis $\hat{\mathbf{a}}_1$, $\hat{\mathbf{a}}_2$, $\hat{\mathbf{a}}_3$. Substituting

$$\begin{aligned}\hat{\mathbf{b}}_1 &= \hat{\mathbf{a}}_1 \sin\varphi + \hat{\mathbf{a}}_3 \cos\varphi \\ \hat{\mathbf{b}}_2 &= \hat{\mathbf{a}}_2 \\ \hat{\mathbf{b}}_3 &= \hat{\mathbf{a}}_3 \sin\varphi - \hat{\mathbf{a}}_1 \cos\varphi\end{aligned} \tag{3·40}$$

yields

$$\begin{aligned}\mathbf{V}^{pf} = {} & \hat{\mathbf{a}}_1(\dot{z} + L\dot{\varphi}\sin\varphi\cos\varphi) + \hat{\mathbf{a}}_2(z\dot{\theta}\sin\varphi) \\ & + \hat{\mathbf{a}}_3(z\dot{\varphi} - L\dot{\varphi}\sin^2\varphi)\end{aligned} \tag{3·41}$$

It is apparent once again that we are working very hard to get an answer which is relatively simple when expressed in the most advantageous vector basis. Often this vector basis is most convenient for physical interpretations of results, but it may not be the vector basis in which vector differentiations must be performed. As soon as we develop the concept of angular velocity, we will find methods of vector differentiation that avoid the cumbersome calculations of the last two examples and take us directly to the desired result in its simplest form. But before pursuing this path to more efficient vector differentiation procedures, we return to the fundamental business of establishing other definitions of kinematics.

3·1·5 Acceleration of a point

As might be anticipated, the acceleration of point p relative to point q in frame f is defined by

$$\mathbf{A}^{p_q f} \triangleq \frac{{}^f d}{dt}\mathbf{V}^{p_q f} \tag{3·42}$$

The designation of a reference frame is again required because $\mathbf{V}^{p_q f}$ is a vector, and a frame must be specified for the differentiation of any vector. The (generic) acceleration of p in f is, logically,

$$\mathbf{A}^{pf} = \frac{{}^f d}{dt} \mathbf{V}^{pf} \tag{3·43}$$

By substituting Eq. (3·17) into (3·42) and using the definition of generic acceleration in Eq. (3·43), we find that relative accelerations can be written as differences of generic accelerations, that is, that

$$\mathbf{A}^{p_q f} = \mathbf{A}^{pf} - \mathbf{A}^{qf} \tag{3·44}$$

In the use of Newton's second law, we must ascertain the generic acceleration of a particle in a particular reference frame called the *inertial frame* (designated as frame i in this text). Thus, in practice, one solves a particular problem by deciding at the outset on a reference frame that is "approximately inertial," labeling it i, and then using $\mathbf{F} = m\mathbf{A}^{pi}$ for a particle at point p. The objective of the study of kinematics is (in the present context) merely the development of methods for finding $\mathbf{A}^{pi}$ for a given point p and a given frame i. (When in Chap. 4 Lagrange's equations are used as fundamental, rather than Newton's laws, the objective changes to means of finding $\mathbf{V}^{pi}$.)

3·1·6 Examples of linear acceleration determination

No *conceptual* obstacles arise in using the definitions of Eqs. (3·42) and (3·43) to determine linear accelerations, but the task of performing the necessary differentiations is sometimes substantial. We can perhaps best appreciate the magnitude of this task by considering each of the three systems used to illustrate velocity calculations, now asking for corresponding accelerations.

First example If for the system of Fig. 3·3 we require the (generic) acceleration of point p in reference frame e, we combine the acceleration definition Eq. (3·43) with the velocity expression in Eq. (3·21). The result is

$$\begin{aligned}\mathbf{A}^{pe} &= \frac{{}^e d}{dt} \mathbf{V}^{pe} = \frac{{}^e d}{dt} (2a\dot{\theta} \cos \theta \hat{\mathbf{e}}_x) \\ &= \hat{\mathbf{e}}_x(2a\ddot{\theta} \cos \theta - 2a\dot{\theta}^2 \sin \theta)\end{aligned} \tag{3·45}$$

This is a simple calculation because the velocity expression in Eq. (3·21) involves only unit vectors fixed in the frame of differentiation, frame e.

Second example Consider next the ant on the spinning cone in Fig. 3·4, and calculate the ant's (generic) acceleration relative to the base b. By the definition of Eq. (3·43) this is simply

$$\mathbf{A}^{ab} = \frac{{}^b d}{dt} \mathbf{V}^{ab} \tag{3·46}$$

Three alternative forms for $\mathbf{V}^{ab}$ have already been calculated. Equation (3·27) is an expression for $\mathbf{V}^{ab}$ in the vector basis $\hat{\mathbf{b}}_1$, $\hat{\mathbf{b}}_2$, and $\hat{\mathbf{b}}_3$. Equation (3·29) is the equivalent but much simpler expression in the vector basis $\hat{\mathbf{c}}_1$, $\hat{\mathbf{c}}_2$, and $\hat{\mathbf{c}}_3$. Equation (3·30) expresses the velocity $\mathbf{V}^{ab}$ for the time $t = 2$ sec only, and this is certainly *not* the right expression to substitute into Eq. (3·46). Between the expressions in Eqs. (3·27) and (3·29), our preference would be to work with the latter because of its simpler form. Our limited procedures for vector differentiation force us, however, to work with $\mathbf{V}^{ab}$ expressed in terms of unit vectors fixed in the frame of differentiation (frame b), so we must substitute Eq. (3·27) into (3·46). The differentiation then becomes a straightforward but somewhat tedious task of scalar differentiation, leading to

$$\mathbf{A}^{ab} = \hat{\mathbf{b}}_1 \left[\frac{3}{5}(-2\Omega v \sin \Omega t - \Omega^2 vt \cos \Omega t)\right] + \hat{\mathbf{b}}_2 \left[\frac{3}{5}(2\Omega v \cos \Omega t - \Omega^2 vt \sin \Omega t)\right] \tag{3·47}$$

Once again we may recognize that our result takes a simpler form when written in terms of unit vectors $\hat{\mathbf{c}}_1$, $\hat{\mathbf{c}}_2$, and $\hat{\mathbf{c}}_3$. The basis transformation in Eq. (3·28) changes (3·47) to

$$\mathbf{A}^{ab} = -\frac{3}{5}\Omega^2 vt\, \hat{\mathbf{c}}_1 + \frac{6}{5}\Omega v\, \hat{\mathbf{c}}_2 \tag{3·48}$$

But we still find ourselves wondering if there is not some more efficient procedure for obtaining this simple result.

Third example Consider finally Fig. 3·5, and find the acceleration of the ring center p in the reference frame of the floor f. The definition gives

$$\mathbf{A}^{pf} = \frac{{}^f d}{dt} \mathbf{V}^{pf} \tag{3·49}$$

Expressions for $\mathbf{V}^{pf}$ have been written in various vector bases as Eqs. (3·37), (3·39), and (3·41). Equation (3·41) is much the simplest in form, and yet we are obliged by our present methods of differentiation to substitute Eq. (3·37). This task is so distasteful that we shall defer it until better methods of vector differentiation can be found. After some investment of time and energy in developing some new kinematic concepts, we will return to this problem and find the required acceleration quickly and efficiently.

3·2 THE KINEMATIC PROPERTIES OF RIGID BODIES AND REFERENCE FRAMES

3·2·1 Rigid bodies and reference frames

A *rigid body* is an idealized material system with physical dimensions such that a constant distance is maintained between any two points of the system. Although in definition a rigid body differs from a reference frame quite substantially, we imagine each rigid body to be embedded in a reference frame, so that kinematically there is no difference in the motion of the body and the motion of the associated frame of reference. Although a rigid body is a mathematical idealization, many of the physical bodies encountered in our environment are fairly rigid, i.e., any two points of these bodies remain essentially a constant distance apart, so again experience helps us conceive the mathematical abstraction of the moving rigid body. For this reason it is often more comfortable to think about rigid-body kinematics than reference frame kinematics, but the equivalence of these concepts should be recognized.

In order to characterize fully the position of every point of a rigid body, we need to know the position of any one such point relative to a base point and also the orientation or attitude of the body relative to some frame of reference or rigid body. As we have previously treated questions of position and changes of position of points, we now focus on orientation and changes of orientation with time.

3·2·2 Orientation in terms of direction cosines

We require a means of describing the orientation or attitude of one body or reference frame relative to another. It is convenient for this purpose to imagine either a set of cartesian axes or a solid cube embedded in each of the bodies or frames, so that the discussion reduces to the description of the orientation of one set of orthogonal axes relative to another.

Let us imagine, for example, that we are interested in the orientation of body (or reference frame) d relative to body (or frame) a. We begin by defining the dextral[1] orthogonal sets of unit vectors $\hat{\mathbf{d}}_1$, $\hat{\mathbf{d}}_2$, $\hat{\mathbf{d}}_3$ and $\hat{\mathbf{a}}_1$, $\hat{\mathbf{a}}_2$, $\hat{\mathbf{a}}_3$ paralleling axes fixed in d and a, respectively. As indicated in Fig. 3·6 it is clearly possible to describe the orientation in frame a of a line parallel to $\hat{\mathbf{d}}_1$ by the angles α_1, β_1, and γ_1 between that line and those parallel to $\hat{\mathbf{a}}_1$, $\hat{\mathbf{a}}_2$, and $\hat{\mathbf{a}}_3$, or by their cosines—the *direction cosines*. Two additional sets of angles, α_2, β_2, γ_2 and α_3, β_3, γ_3 may

[1] Recall that a dextral or right-handed set of unit orthogonal vectors $\hat{\mathbf{a}}_1$, $\hat{\mathbf{a}}_2$, and $\hat{\mathbf{a}}_3$ consists of orthogonal unit vectors with indices chosen to satisfy $\hat{\mathbf{a}}_1 \times \hat{\mathbf{a}}_2 = \hat{\mathbf{a}}_3$, $\hat{\mathbf{a}}_2 \times \hat{\mathbf{a}}_3 = \hat{\mathbf{a}}_1$, and $\hat{\mathbf{a}}_3 \times \hat{\mathbf{a}}_1 = \hat{\mathbf{a}}_2$.

be used to describe the orientations in a of $\hat{\mathbf{d}}_2$ and $\hat{\mathbf{d}}_3$. In terms of the cosines of these nine angles, we can relate the two sets of unit vectors by

$$\begin{aligned} \hat{\mathbf{d}}_1 &= \hat{\mathbf{a}}_1 \cos \alpha_1 + \hat{\mathbf{a}}_2 \cos \beta_1 + \hat{\mathbf{a}}_3 \cos \gamma_1 \\ \hat{\mathbf{d}}_2 &= \hat{\mathbf{a}}_1 \cos \alpha_2 + \hat{\mathbf{a}}_2 \cos \beta_2 + \hat{\mathbf{a}}_3 \cos \gamma_2 \\ \hat{\mathbf{d}}_3 &= \hat{\mathbf{a}}_1 \cos \alpha_3 + \hat{\mathbf{a}}_2 \cos \beta_3 + \hat{\mathbf{a}}_3 \cos \gamma_3 \end{aligned} \tag{3·50a}$$

as may be confirmed by dot multiplications such as $\hat{\mathbf{d}}_1 \cdot \hat{\mathbf{a}}_1 = \cos \alpha_1$, etc. If the direction cosines are labeled c_{ij} with $i, j, = 1, 2, 3$, Eqs. (3·50a) can be replaced by

$$\begin{aligned} \hat{\mathbf{d}}_1 &= c_{11}\hat{\mathbf{a}}_1 + c_{12}\hat{\mathbf{a}}_2 + c_{13}\hat{\mathbf{a}}_3 \\ \hat{\mathbf{d}}_2 &= c_{21}\hat{\mathbf{a}}_1 + c_{22}\hat{\mathbf{a}}_2 + c_{23}\hat{\mathbf{a}}_3 \\ \hat{\mathbf{d}}_3 &= c_{31}\hat{\mathbf{a}}_1 + c_{32}\hat{\mathbf{a}}_2 + c_{33}\hat{\mathbf{a}}_3 \end{aligned} \tag{3·50b}$$

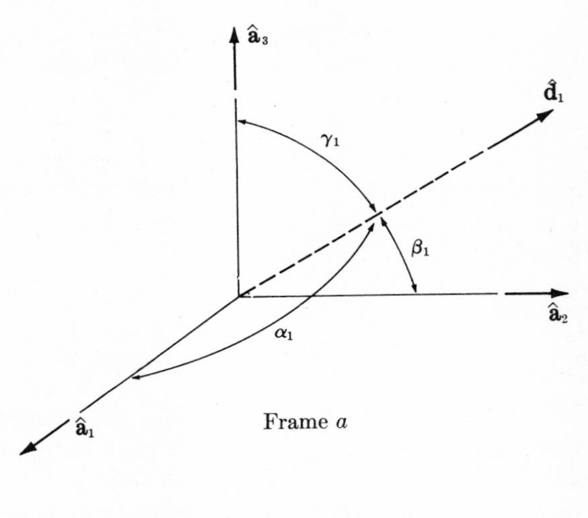

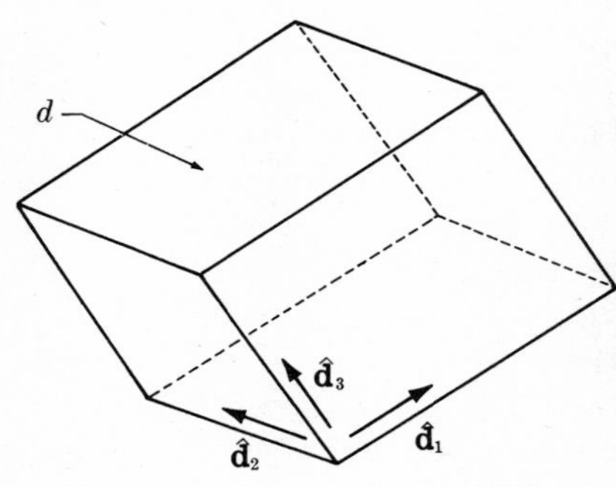

FIGURE 3·6

which can be written more compactly as

$$\hat{\mathbf{d}}_i = \sum_{k=1}^{3} c_{ik}\hat{\mathbf{a}}_k \qquad i = 1, 2, 3 \tag{3·50c}$$

Often the summation symbol is not written explicitly, summation over k being by convention implied by the repeated presence of the index k. (This summation convention is, however, *not* used in this text.)

It is generally convenient to group the vector equations (3·50*b*) in array form as

$$\begin{Bmatrix} \hat{\mathbf{d}}_1 \\ \hat{\mathbf{d}}_2 \\ \hat{\mathbf{d}}_3 \end{Bmatrix} = \begin{bmatrix} c_{11} & c_{12} & c_{13} \\ c_{21} & c_{22} & c_{23} \\ c_{31} & c_{32} & c_{33} \end{bmatrix} \begin{Bmatrix} \hat{\mathbf{a}}_1 \\ \hat{\mathbf{a}}_2 \\ \hat{\mathbf{a}}_3 \end{Bmatrix} \tag{3·50d}$$

By the rules of matrix multiplication (see Appendix B), Eqs. (3·50*d*) is equivalent to Eqs. (3·50*b*). In more compact form, Eqs. (3·50*d*) may be written as

$$\{\hat{\mathbf{d}}\} = [C]\{\hat{\mathbf{a}}\} \tag{3·50e}$$

The matrix $[C]$ is the *direction-cosine matrix.*

Equations (3·50*a*) to (3·50*d*) are all identical in content, although different in form and in amenability to mathematical manipulation.

3·2·3 Orthonormality

Because the sets $\hat{\mathbf{a}}_1$, $\hat{\mathbf{a}}_2$, $\hat{\mathbf{a}}_3$ and $\hat{\mathbf{d}}_1$, $\hat{\mathbf{d}}_2$, $\hat{\mathbf{d}}_3$ are each orthogonal triads of unit vectors, we have to preserve the relationships

$$\hat{\mathbf{a}}_i \cdot \hat{\mathbf{a}}_j = \begin{cases} 1 & \text{if } j = i \\ 0 & \text{if } j \neq i \end{cases} \tag{3·51}$$

and

$$\hat{\mathbf{d}}_i \cdot \hat{\mathbf{d}}_j = \begin{cases} 1 & \text{if } j = i \\ 0 & \text{if } j \neq i \end{cases} \tag{3·52}$$

If we now substitute Eqs. (3·50*b*) for $\hat{\mathbf{d}}_i$, and a similar expression for $\hat{\mathbf{d}}_j$ [changing i to j in Eqs. (3·50*b*)], we'll find that Eq. (3·52) becomes

$$\sum_{k=1}^{3} c_{ik}\hat{\mathbf{a}}_k \cdot \sum_{k=1}^{3} c_{jk}\hat{\mathbf{a}}_k = \begin{cases} 1 & \text{if } j = i \\ 0 & \text{if } j \neq i \end{cases} \tag{3·53}$$

If Eq. (3·51) is next used to simplify this result, we obtain a set of conditions to be imposed on the direction cosines, namely,

$$\sum_{k=1}^{3} c_{ik}c_{jk} = \begin{cases} 1 & \text{if } j = i = 1, 2, \text{ or } 3 \\ 0 & \text{if } j \neq i \text{ for } i \text{ and } j \text{ ranging over the values} \\ & \quad 1, 2, 3 \end{cases} \tag{3·54a}$$

It is convenient to use the single symbol δ_{ij} (called the *Kronecker delta*) for the right-hand side of Eq. (3·54*a*), permitting the equivalent representation

$$\sum_{k=1}^{3} c_{ik}c_{jk} = \delta_{ij} \tag{3·54b}$$

If we examine Eq. (3·54*a*) or (3·54*b*) carefully, considering all acceptable numerical combinations for i and j, we find that six independent scalar equations are represented. Since we began with nine direction cosines, and we now have six equations of constraint on the values to be taken by those nine scalars, we conclude that no more than three of the direction cosines can be specified independently. In other words, there are but three *degrees of freedom* in establishing the relative orientation of two bodies or reference frames.

In terms of the direction-cosine matrix $[C]$, Eqs. (3·54*a*) and (3·54*b*) are equivalent to

$$\begin{aligned}[C][C]^T &= \begin{bmatrix} c_{11} & c_{12} & c_{13} \\ c_{21} & c_{22} & c_{23} \\ c_{31} & c_{32} & c_{33} \end{bmatrix} \begin{bmatrix} c_{11} & c_{21} & c_{31} \\ c_{12} & c_{22} & c_{32} \\ c_{13} & c_{23} & c_{33} \end{bmatrix} \\ &= \begin{bmatrix} 1 & 0 & 0 \\ 0 & 1 & 0 \\ 0 & 0 & 1 \end{bmatrix} \triangleq [U] \end{aligned} \tag{3·54c}$$

where $[C]^T$ is called the *transpose* of $[C]$, obtained by interchanging like-numbered rows and columns of $[C]$, and $[U]$ is called the *unit matrix* or *identity matrix*. This result is easily confirmed by applying to Eq. (3·54*c*) the rules of matrix multiplication (see Appendix B), and comparing to Eq. (3·54*a*).

Having defined the direction cosines $c_{ij}(i, j = 1, 2, 3)$ by Eqs. (3·50*b*), we can display most explicitly the algebraic significance of these symbols by rewriting Eqs. (3·50*b*) in the form

$$\begin{aligned} \hat{\mathbf{d}}_1 &= (\hat{\mathbf{d}}_1 \cdot \hat{\mathbf{a}}_1)\hat{\mathbf{a}}_1 + (\hat{\mathbf{d}}_1 \cdot \hat{\mathbf{a}}_2)\hat{\mathbf{a}}_2 + (\hat{\mathbf{d}}_1 \cdot \hat{\mathbf{a}}_3)\hat{\mathbf{a}}_3 \\ \hat{\mathbf{d}}_2 &= (\hat{\mathbf{d}}_2 \cdot \hat{\mathbf{a}}_1)\hat{\mathbf{a}}_1 + (\hat{\mathbf{d}}_2 \cdot \hat{\mathbf{a}}_2)\hat{\mathbf{a}}_2 + (\hat{\mathbf{d}}_2 \cdot \hat{\mathbf{a}}_3)\hat{\mathbf{a}}_3 \\ \hat{\mathbf{d}}_3 &= (\hat{\mathbf{d}}_3 \cdot \hat{\mathbf{a}}_1)\hat{\mathbf{a}}_1 + (\hat{\mathbf{d}}_3 \cdot \hat{\mathbf{a}}_2)\hat{\mathbf{a}}_2 + (\hat{\mathbf{d}}_3 \cdot \hat{\mathbf{a}}_3)\hat{\mathbf{a}}_3 \end{aligned} \tag{3·55}$$

If we now construct the inverse relationships

$$\begin{aligned} \hat{\mathbf{a}}_1 &= (\hat{\mathbf{a}}_1 \cdot \hat{\mathbf{d}}_1)\hat{\mathbf{d}}_1 + (\hat{\mathbf{a}}_1 \cdot \hat{\mathbf{d}}_2)\hat{\mathbf{d}}_2 + (\hat{\mathbf{a}}_1 \cdot \hat{\mathbf{d}}_3)\hat{\mathbf{d}}_3 \\ \hat{\mathbf{a}}_2 &= (\hat{\mathbf{a}}_2 \cdot \hat{\mathbf{d}}_1)\hat{\mathbf{d}}_1 + (\hat{\mathbf{a}}_2 \cdot \hat{\mathbf{d}}_2)\hat{\mathbf{d}}_2 + (\hat{\mathbf{a}}_2 \cdot \hat{\mathbf{d}}_3)\hat{\mathbf{d}}_3 \\ \hat{\mathbf{a}}_3 &= (\hat{\mathbf{a}}_3 \cdot \hat{\mathbf{d}}_1)\hat{\mathbf{d}}_1 + (\hat{\mathbf{a}}_3 \cdot \hat{\mathbf{d}}_2)\hat{\mathbf{d}}_2 + (\hat{\mathbf{a}}_3 \cdot \hat{\mathbf{d}}_3)\hat{\mathbf{d}}_3 \end{aligned} \tag{3·56a}$$

and recognize the commutativity of vector dot multiplication (so $\hat{\mathbf{a}}_i \cdot \hat{\mathbf{d}}_j = \hat{\mathbf{d}}_j \cdot \hat{\mathbf{a}}_i$, where $i, j = 1, 2, 3$), we will see that the latter set of equations could as well be written

$$\begin{aligned} \hat{\mathbf{a}}_1 &= c_{11}\hat{\mathbf{d}}_1 + c_{21}\hat{\mathbf{d}}_2 + c_{31}\hat{\mathbf{d}}_3 \\ \hat{\mathbf{a}}_2 &= c_{12}\hat{\mathbf{d}}_1 + c_{22}\hat{\mathbf{d}}_2 + c_{32}\hat{\mathbf{d}}_3 \\ \hat{\mathbf{a}}_3 &= c_{13}\hat{\mathbf{d}}_1 + c_{23}\hat{\mathbf{d}}_2 + c_{33}\hat{\mathbf{d}}_3 \end{aligned} \tag{3·56b}$$

or

$$\hat{\mathbf{a}}_i = \sum_{k=1}^{3} c_{ki}\hat{\mathbf{d}}_k \qquad i = 1, 2, 3 \tag{3·56c}$$

or

$$\begin{Bmatrix} \hat{\mathbf{a}}_1 \\ \hat{\mathbf{a}}_2 \\ \hat{\mathbf{a}}_3 \end{Bmatrix} = \begin{bmatrix} c_{11} & c_{21} & c_{31} \\ c_{12} & c_{22} & c_{32} \\ c_{13} & c_{23} & c_{33} \end{bmatrix} \begin{Bmatrix} \hat{\mathbf{d}}_1 \\ \hat{\mathbf{d}}_2 \\ \hat{\mathbf{d}}_3 \end{Bmatrix} \tag{3·56d}$$

or simply as

$$\{\hat{\mathbf{a}}\} = [C]^T\{\hat{\mathbf{d}}\} \tag{3·56e}$$

If now we were to substitute Eqs. (3·56*b*) into (3·51), and utilize Eq. (3·52), proceeding in the same manner that led to Eqs. (3·54), we would find their counterparts to be

$$\sum_{k=1}^{3} c_{ki}c_{kj} = \delta_{ij} \tag{3·57a}$$

and

$$[C]^T[C] = [U] \tag{3·57b}$$

Comparing Eqs. (3·54*c*) and (3·57*b*), we see that when $[C]$ is either premultiplied or postmultiplied by $[C]^T$, the product is the unit matrix $[U]$. In general a matrix that produces $[U]$ when pre- or postmultiplied by a square matrix $[M]$ is called the *inverse* of that matrix, designated $[M]^{-1}$. Thus the transpose of the direction-cosine matrix $[C]$ is its inverse, that is,

$$[C]^T \equiv [C]^{-1} \tag{3·58}$$

A matrix with this property is called *orthonormal* (or, in some texts, *orthogonal*).

As you look back over the few pages preceding, you will realize that it was quite an arbitrary choice to use the symbols c_{ik} for the direction cosines in Eq. (3·50*c*), and c_{ki} for those in Eq. (3·56*c*); the symbols could as well have been reversed, with corresponding changes in each of Eqs. (3·50) and (3·56), exchanging $[C]$ and $[C]^T$, etc. Once you make a choice, however, you must remember it and stick with it. If this places an undue burden on the memory, or if there are many reference frames $a, b, c, d, \ldots$ involved in a problem, then it often proves convenient to adopt an explicit notation such as

$$\hat{\mathbf{a}}_i = \sum_{k=1}^{3} c_{ik}{}^{ab}\hat{\mathbf{b}}_k \qquad \{\hat{\mathbf{a}}\} = [C^{ab}]\{\hat{\mathbf{b}}\} \tag{3·59a}$$

and

$$\hat{\mathbf{b}}_i = \sum_{k=1}^{3} c_{ik}{}^{ba}\hat{\mathbf{a}}_k \qquad \{\hat{\mathbf{b}}\} = [C^{ba}]\{\hat{\mathbf{a}}\} \tag{3·59b}$$

With such a notational convention, we have only to remember the relationships typified by

$$[C^{ba}] = [C^{ab}]^T = [C^{ab}]^{-1} \tag{3·60}$$

3·2·4 Vector-basis transformations

The direction cosines and their matrices relate not only pairs of orthogonal triads of unit vectors, but also the corresponding scalar components of any vector **Z** in the two vector bases. If we write[1]

$$\mathbf{Z} = {}^dZ_1\hat{\mathbf{d}}_1 + {}^dZ_2\hat{\mathbf{d}}_2 + {}^dZ_3\hat{\mathbf{d}}_3 = \sum_{i=1}^{3} {}^dZ_i\hat{\mathbf{d}}_i \tag{3·61}$$

and

$$\mathbf{Z} = {}^aZ_1\hat{\mathbf{a}}_1 + {}^aZ_2\hat{\mathbf{a}}_2 + {}^aZ_3\hat{\mathbf{a}}_3 = \sum_{j=1}^{3} {}^aZ_j\hat{\mathbf{a}}_j \tag{3·62}$$

these expressions can be equated to provide [with Eq. (3·56c)]

$$\begin{aligned}\sum_{i=1}^{3} {}^dZ_i\hat{\mathbf{d}}_i &= \sum_{j=1}^{3} {}^aZ_j\hat{\mathbf{a}}_j = \sum_{j=1}^{3} {}^aZ_j \sum_{i=1}^{3} c_{ij}\hat{\mathbf{d}}_i \\ &= \sum_{i=1}^{3} \Big(\sum_{j=1}^{3} c_{ij}{}^aZ_j\Big)\,\hat{\mathbf{d}}_i\end{aligned} \tag{3·63}$$

Comparing the first and last of these expressions furnishes

$$^dZ_i = \sum_{j=1}^{3} c_{ij}{}^aZ_j \qquad i = 1, 2, 3 \tag{3·64a}$$

[1] In this text a presuperscript on a scalar, as in ${}^dZ_i$, or on a matrix, as in $[{}^dZ]$, indicates the vector basis, or choice of unit vectors, involved in obtaining the scalar components or matrix elements from a vector or dyadic (see Appendix B). This notational convention will be used only when the vector basis is not clear from the context.

In matrix terms, Eq. (3·64*a*) may be written as

$$\begin{bmatrix} {}^dZ_1 \\ {}^dZ_2 \\ {}^dZ_3 \end{bmatrix} = \begin{bmatrix} c_{11} & c_{12} & c_{13} \\ c_{21} & c_{22} & c_{23} \\ c_{31} & c_{32} & c_{33} \end{bmatrix} \begin{bmatrix} {}^aZ_1 \\ {}^aZ_2 \\ {}^aZ_3 \end{bmatrix} \tag{3·64b}$$

or, more compactly, as

$$[{}^dZ] = [C][{}^aZ] \tag{3·64c}$$

To obtain the inverse relationships, we can premultiply Eq. (3·64*c*) by $[C]^T \equiv [C]^{-1}$, to obtain

$$[{}^aZ] = [C]^T[{}^dZ] \tag{3·65a}$$

or, equivalently,

$$ {}^aZ_i = \sum_{j=1}^{3} c_{ji}\,{}^dZ_j \tag{3·65b}$$

3·2·5 Direction cosines in application

Suppose we are required to find the direction-cosine matrix $[C]$ in Eq. (3·50*e*) for unit vectors fixed in the blocks *a* and *d* of Fig. 3·7. This task requires the following calculations:

$$\begin{aligned} c_{11} &= \hat{\mathbf{d}}_1 \cdot \hat{\mathbf{a}}_1 & c_{12} &= \hat{\mathbf{d}}_1 \cdot \hat{\mathbf{a}}_2 & c_{13} &= \hat{\mathbf{d}}_1 \cdot \hat{\mathbf{a}}_3 \\ c_{21} &= \hat{\mathbf{d}}_2 \cdot \hat{\mathbf{a}}_1 & c_{22} &= \hat{\mathbf{d}}_2 \cdot \hat{\mathbf{a}}_2 & c_{23} &= \hat{\mathbf{d}}_2 \cdot \hat{\mathbf{a}}_3 \\ c_{31} &= \hat{\mathbf{d}}_3 \cdot \hat{\mathbf{a}}_1 & c_{32} &= \hat{\mathbf{d}}_3 \cdot \hat{\mathbf{a}}_2 & c_{33} &= \hat{\mathbf{d}}_3 \cdot \hat{\mathbf{a}}_3 \end{aligned}$$

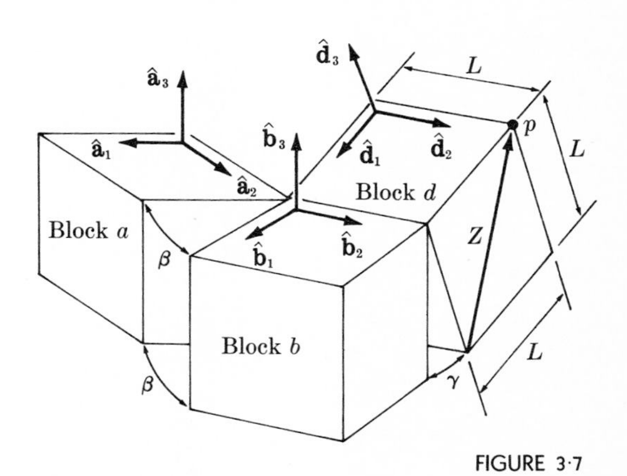

FIGURE 3·7

With the substitutions $\beta = 45°$ and $\gamma = 30°$, so that

$$\begin{aligned} \hat{\mathbf{d}}_1 &= \hat{\mathbf{b}}_1 \cos 30° - \hat{\mathbf{b}}_3 \sin 30° & \hat{\mathbf{a}}_1 &= \hat{\mathbf{b}}_1 \cos 45° - \hat{\mathbf{b}}_2 \sin 45° \\ \hat{\mathbf{d}}_2 &= \hat{\mathbf{b}}_2 & \hat{\mathbf{a}}_2 &= \hat{\mathbf{b}}_2 \cos 45° + \hat{\mathbf{b}}_1 \sin 45° \\ \hat{\mathbf{d}}_3 &= \hat{\mathbf{b}}_3 \cos 30° + \hat{\mathbf{b}}_1 \sin 30° & \hat{\mathbf{a}}_3 &= \hat{\mathbf{b}}_3 \end{aligned}$$

we obtain

$$\begin{aligned} c_{11} &= \cos 30° \cos 45° = \frac{\sqrt{6}}{4} \\ c_{12} &= \cos 30° \sin 45° = \frac{\sqrt{6}}{4} \\ c_{13} &= -\sin 30° \,(1) = -\frac{1}{2} \\ c_{21} &= -\sin 45° = -\frac{\sqrt{2}}{2} \\ c_{22} &= \cos 45° = \frac{\sqrt{2}}{2} \\ c_{23} &= 0 \\ c_{31} &= \sin 30° \cos 45° = \frac{\sqrt{2}}{4} \\ c_{32} &= \sin 30° \sin 45° = \frac{\sqrt{2}}{4} \\ c_{33} &= \cos 30° \,(1) = \frac{\sqrt{3}}{2} \end{aligned}$$

By arranging the nine scalars $c_{ij}(i, j = 1, 2, 3)$ as shown in Eqs. (3·50d), we obtain the direction-cosine matrix $[C]$ as

$$[C] = \begin{bmatrix} \frac{\sqrt{6}}{4} & \frac{\sqrt{6}}{4} & -\frac{1}{2} \\ -\frac{\sqrt{2}}{2} & \frac{\sqrt{2}}{2} & 0 \\ \frac{\sqrt{2}}{4} & \frac{\sqrt{2}}{4} & \frac{\sqrt{3}}{2} \end{bmatrix}$$

To check for orthonormality, we could multiply out $[C]^T[C]$ and $[C][C]^T$ to be sure that the result is $[U]$. It is equivalent, however, and more in the spirit of Eqs. (3·54b) and (3·57a), to be sure that the sum of the squares of the elements in each row of $[C]$ and each column of $[C]$ is unity, while the result of summing the products of like-numbered elements of any two *different* rows (or columns) is zero. Remember that the elements in the first row, for example, are simply the scalar components of $\hat{\mathbf{d}}_1$ in basis $\hat{\mathbf{a}}_1$, $\hat{\mathbf{a}}_2$, and $\hat{\mathbf{a}}_3$, and the length of this vector is unity; this requires $(\sqrt{6}/4)^2 + (\sqrt{6}/4)^2 + (-1/2)^2 = 1$ (checks). The second row represents $\hat{\mathbf{d}}_2$, requiring $(-\sqrt{2}/2)^2 + (\sqrt{2}/2)^2 = 1$ (checks), and since $\hat{\mathbf{d}}_1 \cdot \hat{\mathbf{d}}_2 = 0$ we must also have $(\sqrt{6}/4)(-\sqrt{2}/2) + (\sqrt{6}/4)(\sqrt{2}/2) = 0$ (checks). Similar interpretations may be applied successfully to each of the rows of $[C]$. Since the columns of $[C]$ are, respectively, the scalar components in basis $\hat{\mathbf{d}}_1$, $\hat{\mathbf{d}}_2$, $\hat{\mathbf{d}}_3$ of $\hat{\mathbf{a}}_1$, $\hat{\mathbf{a}}_2$, and $\hat{\mathbf{a}}_3$, the orthogonality of these unit vectors requires such checks as $(\sqrt{6}/4)^2 + (-\sqrt{2}/2)^2 + (\sqrt{2}/4)^2 = 1$, etc.

In order to illustrate the transformation of vector bases for the vector $\mathbf{Z}$ shown on block d of Fig. 3·7, we can first observe that

$$\mathbf{Z} = -L\hat{\mathbf{d}}_1 + L\hat{\mathbf{d}}_3$$

so that in Eq. (3·61) the scalar components in basis $\hat{\mathbf{d}}_1$, $\hat{\mathbf{d}}_2$, $\hat{\mathbf{d}}_3$ are

$$^dZ_1 = -L \qquad ^dZ_2 = 0 \qquad ^dZ_3 = L$$

If now we require the scalars $^aZ_1$, $^aZ_2$, and $^aZ_3$, representing the scalar components of $\mathbf{Z}$ in vector basis $\hat{\mathbf{a}}_1$, $\hat{\mathbf{a}}_2$, $\hat{\mathbf{a}}_3$, we can utilize Eq. (3·65a) and the direction-cosine matrix obtained previously to write

$$\begin{bmatrix} ^aZ_1 \\ ^aZ_2 \\ ^aZ_3 \end{bmatrix} = \begin{bmatrix} c_{11} & c_{21} & c_{31} \\ c_{12} & c_{22} & c_{32} \\ c_{13} & c_{23} & c_{33} \end{bmatrix} \begin{bmatrix} ^dZ_1 \\ ^dZ_2 \\ ^dZ_3 \end{bmatrix} = \begin{bmatrix} \frac{\sqrt{6}}{4} & -\frac{\sqrt{2}}{2} & \frac{\sqrt{2}}{4} \\ \frac{\sqrt{6}}{4} & \frac{\sqrt{2}}{2} & \frac{\sqrt{2}}{4} \\ -\frac{1}{2} & 0 & \frac{\sqrt{3}}{2} \end{bmatrix} \begin{bmatrix} -L \\ 0 \\ L \end{bmatrix}$$

providing, finally,

$$^aZ_1 = -\frac{\sqrt{6}}{4}L + \frac{\sqrt{2}}{4}L = -(\sqrt{6} - \sqrt{2})\frac{L}{4}$$

$$^aZ_2 = -\frac{\sqrt{6}}{4}L + \frac{\sqrt{2}}{4}L = -(\sqrt{6} - \sqrt{2})\frac{L}{4}$$

$$^aZ_3 = \frac{1}{2}L + \frac{\sqrt{3}}{2}L = (1 + \sqrt{3})\frac{L}{2}$$

3·2·6 Orientation in terms of attitude angles

Although the nine direction cosines can always be used to define relative orientation, this is not the only alternative, and it is not usually the most attractive in the practical description of simple physical systems. The difficulty with the direction cosines is that they are related by six constraint equations [for example, Eq. (3·54*b*)], so only three independent scalars may be specified. This is cumbersome, and it makes physical interpretation difficult. For digital-computer computations there may be certain advantages in using the nine direction cosines or various six-parameter or four-parameter systems for describing orientation, despite the fact that all these schemes require constraint equations relating the parameters.

For most simple engineering problems requiring physical interpretation of results, it is advantageous to use only the minimum number of orientation parameters, so some kind of three-parameter system is adopted. The most common set of three scalars used for attitude or orientation description consists of three angles called *Euler angles*.

The problem is to define the orientation of a rigid body (here called d) relative to a given reference frame (called a), using only three scalars. Remember that we have already treated the description of the position of one point of the body; now we consider only the *orientation* of the body. Let $\hat{\mathbf{a}}_1$, $\hat{\mathbf{a}}_2$, and $\hat{\mathbf{a}}_3$ be dextral orthogonal unit vectors paralleling cartesian axes with directions fixed in a, and let the unit vectors $\hat{\mathbf{d}}_1$, $\hat{\mathbf{d}}_2$, and $\hat{\mathbf{d}}_3$ parallel cartesian axes fixed in the body d (see Fig. 3·8). Note these sets of axes in the figure; it is necessary only that we find coordinates such as φ, θ, and ψ that uniquely describe the relative attitude of these two sets of axes. In order to visualize the meaning of the Euler angles, proceed as follows. Imagine that the orthogonal triads of unit vectors are bound

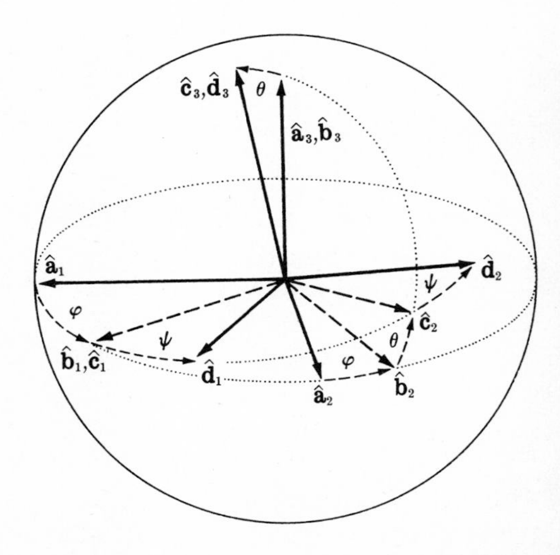

FIGURE 3·8

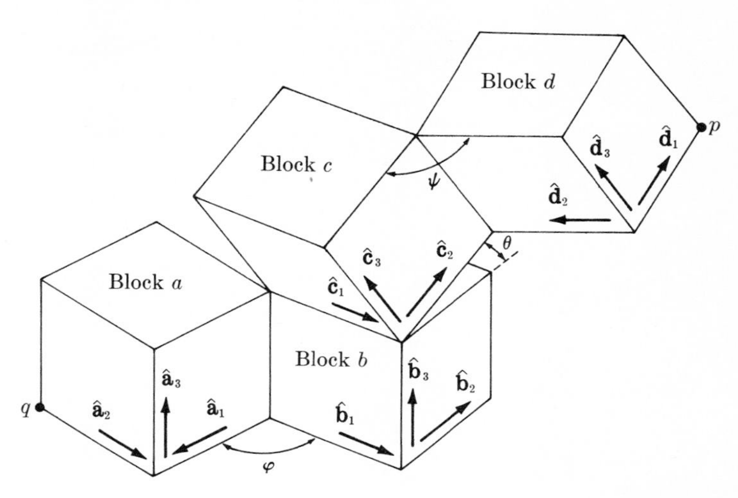

FIGURE 3·9

vectors emanating from a common point (as in the figure) and assume that initially $\hat{\mathbf{d}}_1$, $\hat{\mathbf{d}}_2$, and $\hat{\mathbf{d}}_3$ are, respectively, coincident with $\hat{\mathbf{a}}_1$, $\hat{\mathbf{a}}_2$, and $\hat{\mathbf{a}}_3$. Now perform a dextral rotation[1] of body d about the $\hat{\mathbf{a}}_3$ axis by an angle φ, and label these new (imagined) orientations of $\hat{\mathbf{d}}_1$, $\hat{\mathbf{d}}_2$, and $\hat{\mathbf{d}}_3$ with the symbols $\hat{\mathbf{b}}_1$, $\hat{\mathbf{b}}_2$, and $\hat{\mathbf{b}}_3$ (see Fig. 3·8). Then rotate the body d an angle θ about the line of $\hat{\mathbf{b}}_1$, defining the resulting unit vectors as $\hat{\mathbf{c}}_1$, $\hat{\mathbf{c}}_2$, and $\hat{\mathbf{c}}_3$. Finally, rotate d about $\hat{\mathbf{c}}_3$ an angle ψ to bring the body axes into the true orientations $\hat{\mathbf{d}}_1$, $\hat{\mathbf{d}}_2$, and $\hat{\mathbf{d}}_3$. Thus the three scalars φ, θ, and ψ describe a sequence of rotations uniquely establishing the attitude of d in a (and conversely, except that when $\theta = 0$, values of φ and ψ are not uniquely determined by the body attitude).[2] The figure shows the projections of these rotations on a sphere of radius unity, for convenience in visualization. We are not concerned at present with the fact that there might actually be translations of d relative to a in addition to the angular rotations described here.

The generation of these angles is also illustrated in Fig. 3·9, which has the advantage of representing geometrical units more readily visualized (cubes). Figure 3·9 also dispels any erroneous impression that a and d must have a common point. The interpretation of this figure closely parallels that of the preceding

[1] A dextral or right-handed rotation advances a familiar right-handed screw along the positive axis. It is always assumed here that positive rotations are right-handed.

[2] Ambiguity of parameter values for special orientations is characteristic of all three-parameter systems; the elimination of this ambiguity is one of the attractive features of systems for attitude description employing more than three parameters. For the Euler angles considered here, when θ is zero it becomes impossible to distinguish between φ and ψ, illustrating the noted ambiguity.

figure. Any reference frame a can be represented by a cube a as shown, and any rigid-body attitude is fully defined by the attitude of an embedded cube, such as d. It develops that any relative attitude of a and d can be defined by an appropriate selection of coordinates φ, θ, and ψ generated as follows. Imagine first that block d is so aligned with block a that unit vectors $\hat{\mathbf{d}}_1$, $\hat{\mathbf{d}}_2$, and $\hat{\mathbf{d}}_3$ parallel $\hat{\mathbf{a}}_1$, $\hat{\mathbf{a}}_2$, and $\hat{\mathbf{a}}_3$, respectively. It is not intended that the two blocks be coincident, merely that they adopt the specified relative attitude; in fact, it is convenient to imagine the relative position of the two blocks to be as shown in Fig. 3·10, in order that the blocks b and c can be introduced so that all unit vectors of corresponding subscript are aligned. (In studying these figures, remember that vectors are not bound to any particular point unless so restricted explicitly; the unit vectors of Fig. 3·10 may be differently located on their blocks than the corresponding vectors in Fig. 3·9, but their directions relative to their own blocks are unchanged.) Note that three of the lines in Fig. 3·10 are double lines (solid or dashed); these are *hinge* lines. They parallel $\hat{\mathbf{a}}_3$ (and $\hat{\mathbf{b}}_3$), $\hat{\mathbf{b}}_1$ (and $\hat{\mathbf{c}}_1$), and $\hat{\mathbf{c}}_3$ (and $\hat{\mathbf{d}}_3$). The succession of rotations φ, θ, and ψ about the hinges parallel to $\hat{\mathbf{a}}_3$, $\hat{\mathbf{b}}_1$, and $\hat{\mathbf{c}}_3$, respectively, provides Fig. 3·9, and again illustrates the generation of the Euler angles. These figures, in addition to providing this illustration, will later prove useful in relating unit vectors $\hat{\mathbf{d}}_1$, $\hat{\mathbf{d}}_2$, and $\hat{\mathbf{d}}_3$ to $\hat{\mathbf{a}}_1$, $\hat{\mathbf{a}}_2$, and $\hat{\mathbf{a}}_3$.

Although the Euler angles are the most widely used orientation parameters, it is important to realize that other alternatives are available. In the generation of the Euler angles from Fig. 3·8, successive rotations were performed about

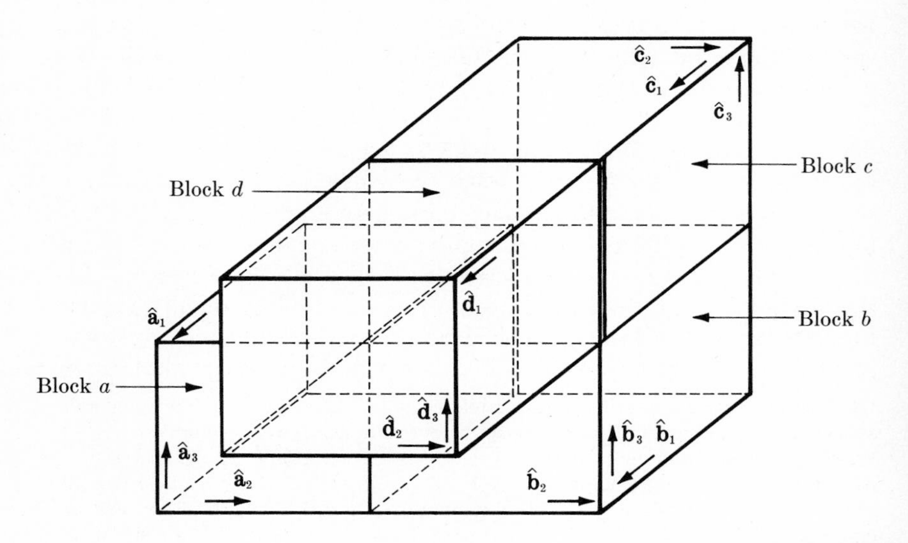

FIGURE 3·10

axes parallel to unit vectors $\hat{\mathbf{a}}_3$, $\hat{\mathbf{b}}_1$, and $\hat{\mathbf{c}}_3$, and in Figs. 3·9 and 3·10 these same rotations were about axes parallel to $\hat{\mathbf{a}}_3$, $\hat{\mathbf{b}}_1$, and $\hat{\mathbf{c}}_3$. In either illustration, there is a 3-1-3 sequence, involving only two of the three distinctly indexed axes. Alternatively, we could adopt a 1-2-3 sequence, involving all three distinctly indexed axes. Numerous others might be considered,[1] but the basic distinction is between *two-axis* and *three-axis systems of attitude angles.*

Since a general description of orientation can be generated as above by an ordered sequence of simple *planar rotations*, a general transformation matrix (previously described in terms of direction cosines) can be constructed as a matrix product representing a sequence of transformation matrices for simple rotations. Thus understanding of the general problem of description of relative orientation requires only a proper appreciation of the various modes of description of planar changes in orientation.

Specifically, if orthogonal unit vectors $\hat{\mathbf{a}}_1$ and $\hat{\mathbf{a}}_2$ are used as reference in describing orthogonal unit vectors $\hat{\mathbf{b}}_1$ and $\hat{\mathbf{b}}_2$, with all these vectors coplanar and related by an angle φ as in Fig. 3·11, we can write by inspection

$$\hat{\mathbf{b}}_1 = \hat{\mathbf{a}}_1 \cos\varphi + \hat{\mathbf{a}}_2 \sin\varphi$$
$$\hat{\mathbf{b}}_2 = -\hat{\mathbf{a}}_1 \sin\varphi + \hat{\mathbf{a}}_2 \cos\varphi$$

and if $\hat{\mathbf{a}}_1$, $\hat{\mathbf{a}}_2$, $\hat{\mathbf{a}}_3$ and $\hat{\mathbf{b}}_1$, $\hat{\mathbf{b}}_2$, $\hat{\mathbf{b}}_3$ are both dextral orthogonal sets,

$$\hat{\mathbf{b}}_3 = \hat{\mathbf{a}}_3$$

These relationships may be rewritten in convenient form using matrices and vector arrays as

$$\begin{Bmatrix} \hat{\mathbf{b}}_1 \\ \hat{\mathbf{b}}_2 \\ \hat{\mathbf{b}}_3 \end{Bmatrix} = \begin{bmatrix} \cos\varphi & \sin\varphi & 0 \\ -\sin\varphi & \cos\varphi & 0 \\ 0 & 0 & 1 \end{bmatrix} \begin{Bmatrix} \hat{\mathbf{a}}_1 \\ \hat{\mathbf{a}}_2 \\ \hat{\mathbf{a}}_3 \end{Bmatrix} \tag{3·66}$$

or in compact notation as

$$\{\hat{\mathbf{b}}\} = [C^3(\varphi)]\{\hat{\mathbf{a}}\} \tag{3·67}$$

Note that the superscript 3 identifies the axis about which the rotation takes place. If similar relationships are constructed for the rotations θ and ψ illustrated in Fig. 3·11, we have

$$\{\hat{\mathbf{c}}\} = [C^1(\theta)]\{\hat{\mathbf{b}}\} \tag{3·68}$$

and

$$\{\hat{\mathbf{d}}\} = [C^3(\psi)]\{\hat{\mathbf{c}}\} \tag{3·69}$$

[1] In fact, there is no unanimity among writers concerning even the classical Euler angles; see H. Goldstein, "Classical Mechanics," p. 108, Addison-Wesley Publishing Company, Inc., Reading, Mass., 1959. In his original work in 1760, Euler used a combination of right-handed and left-handed rotations, a convention unacceptable today.

where

$$[C^1(\theta)] = \begin{bmatrix} 1 & 0 & 0 \\ 0 & \cos\theta & \sin\theta \\ 0 & -\sin\theta & \cos\theta \end{bmatrix} \tag{3·70}$$

and from Eqs. (3·66),

$$[C^3(\psi)] = \begin{bmatrix} \cos\psi & \sin\psi & 0 \\ -\sin\psi & \cos\psi & 0 \\ 0 & 0 & 1 \end{bmatrix} \tag{3·71}$$

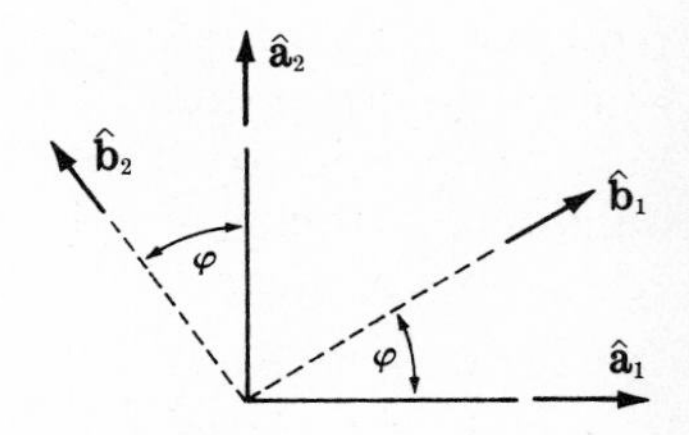

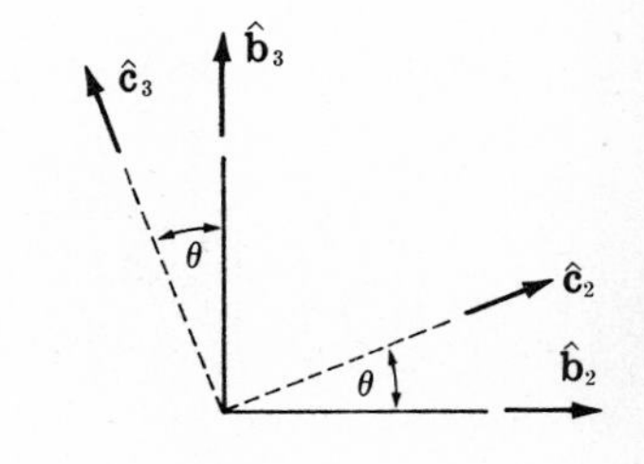

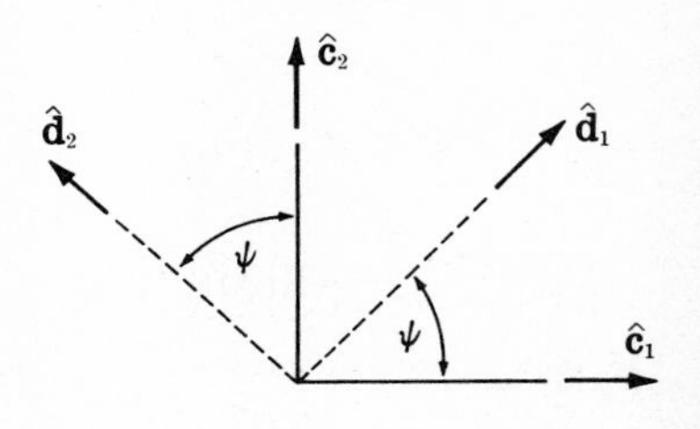

FIGURE 3·11

Now, if we require the unit vectors $\hat{\mathbf{d}}_1$, $\hat{\mathbf{d}}_2$, and $\hat{\mathbf{d}}_3$ in terms of the original vector basis $\hat{\mathbf{a}}_1$, $\hat{\mathbf{a}}_2$, $\hat{\mathbf{a}}_3$, we can write from Eqs. (3·67), (3·68), and (3·69) the matrix equation

$$\{\hat{\mathbf{d}}\} = [C^3(\psi)][C^1(\theta)][C^3(\varphi)]\{\hat{\mathbf{a}}\} \tag{3·72}$$

If the three planar-rotation matrices are multiplied together to obtain a single matrix $[C]$, this becomes

$$\{\hat{\mathbf{d}}\} = [C]\{\hat{\mathbf{a}}\} \tag{3·73}$$

The matrix $[C]$ is given in explicit detail after laborious matrix multiplication as

$$[C] = \begin{bmatrix} c\varphi c\psi - s\varphi c\theta s\psi & s\varphi c\psi + c\varphi c\theta s\psi & s\theta s\psi \\ -c\varphi s\psi - s\varphi c\theta c\psi & -s\varphi s\psi + c\varphi c\theta c\psi & s\theta c\psi \\ s\theta s\varphi & -c\varphi s\theta & c\theta \end{bmatrix} \tag{3·74}$$

with letters c and s symbolizing cosine and sine, respectively.

Equation (3·73) was written previously as (3·55), except that in that context the elements of the matrix $[C]$ were the direction cosines. Thus Eq. (3·74) provides explicit values of the nine direction cosines in terms of the 3-1-3 Euler angles φ, θ, and ψ. Having established in Eq. (3·58) the orthonormality of the matrix $[C]$, we can again write the inverse of Eq. (3·73) as simply

$$\{\hat{\mathbf{a}}\} = [C]^T\{\hat{\mathbf{d}}\} \tag{3·75}$$

Transformation matrices such as those generated here for the 3-1-3 Euler angles could equally easily be written for another sequence of three rotations; a 1-2-3 set is more convenient in some applications. If θ_1, θ_2, and θ_3 are a 1-2-3 set of attitude angles, the direction-cosine matrix $[C]$ in Eq. (3·73) becomes

$$[C] = [C^3(\theta_3)][C^2(\theta_2)][C^1(\theta_1)]$$

where $[C^3]$ and $[C^1]$ have the structure shown in Eqs. (3·66) and (3·70), while $[C^2(\theta_2)]$ is given by

$$[C^2(\theta_2)] = \begin{bmatrix} \cos\theta_2 & 0 & -\sin\theta_2 \\ 0 & 1 & 0 \\ \sin\theta_2 & 0 & \cos\theta_2 \end{bmatrix} \tag{3·76}$$

The indicated matrix multiplication then provides the direction-cosine matrix in the form

$$[C] = \begin{bmatrix} C_2C_3 & C_1S_3 + S_1S_2C_3 & S_1S_3 - C_1S_2C_3 \\ -C_2S_3 & C_1C_3 - S_1S_2S_3 & S_1C_3 + C_1S_2S_3 \\ S_2 & -S_1C_2 & C_1C_2 \end{bmatrix} \tag{3·77}$$

where $C_i \triangleq \cos\theta_i$ and $S_i \triangleq \sin\theta_i$ for $i = 1, 2, 3$. Whatever the details of the parametrization, the general transformation matrix is now easily constructed for three-angle systems. These matrices can always be used to obtain relationships among the unit vectors of the system, although it is often easier to obtain a particular relationship by inspection than by matrix multiplication. For example, if we need only to express the unit vector $\hat{\mathbf{a}}_3$ in Fig. 3·9 in terms of the vector basis $\hat{\mathbf{c}}_1$, $\hat{\mathbf{c}}_2$, and $\hat{\mathbf{c}}_3$, inspection of the figure provides

$$\hat{\mathbf{a}}_3 = \hat{\mathbf{b}}_3 = \hat{\mathbf{c}}_3 \cos\theta + \hat{\mathbf{c}}_2 \sin\theta$$

and the same information is more tediously extracted from the transformation matrices. The matrix formulation of the relationships among unit vectors is, in an elementary context, merely a shorthand notation that sometimes simplifies analysis. In a more advanced treatment of kinematics, the transformation matrix plays a more important role.

3·2·7 Attitude angles for a gyroscope

Figure 3·12*a* and *b* shows two portrayals of the same two-gimbal gyroscope; the difference is in the manner in which they are labeled. In Fig. 3·12*a* there is embedded in each of the rigid bodies *A*, *B*, *C*, and *D* a dextral, orthogonal triad of unit vectors with corresponding symbols (so that $\hat{\mathbf{A}}_1$, $\hat{\mathbf{A}}_2$, and $\hat{\mathbf{A}}_3$ are fixed in *A*, etc.). These vectors are so defined that if in some nominal state every vector with the same index is aligned, then any subsequent orientation of *D* relative to *A* can be established by the following succession of rotations: First rotate the rigid assembly *BCD* relative to *A* through an angle θ_1 about the gimbal axis parallel to $\hat{\mathbf{A}}_1$ and $\hat{\mathbf{B}}_1$; then rotate the rigid assembly *CD* relative to *B* through an angle θ_2 about the axis parallel to $\hat{\mathbf{B}}_2$ and $\hat{\mathbf{C}}_2$; and finally rotate body *D* relative to *C* through an angle θ_3 about the axis parallel to $\hat{\mathbf{C}}_3$ and $\hat{\mathbf{D}}_3$. Since this is a 1-2-3 set of attitude angles, the direction-cosine matrix relating the unit vectors in *A* and *D* is given by Eq. (3·77).

In Fig. 3·12*b* the four bodies are labeled with lowercase letters *a*, *b*, *c*, and *d*. Orthogonal triads of unit vectors with corresponding symbols are embedded in the bodies in such a way that if in some nominal state all unit vectors of the same index are aligned, then any subsequent orientation of *d* relative to *a* can be established by the following succession of rotations: First rotate the rigid assembly *bcd* relative to *a* through an angle φ about the gimbal axis parallel to $\hat{\mathbf{a}}_3$ and $\hat{\mathbf{b}}_3$; then rotate the rigid assembly *cd* relative to *b* through an angle θ about the axis parallel to $\hat{\mathbf{b}}_1$ and $\hat{\mathbf{c}}_1$; and finally rotate body *d* relative to *c* through an angle ψ about the axis parallel to $\hat{\mathbf{c}}_3$ and $\hat{\mathbf{d}}_3$. Since this is a 3-1-3 set of attitude angles, the direction-cosine matrix relating the unit vectors in *a* and *d* is given by Eq. (3·74).

The labeling schemes in Fig. 3·12*a* and *b* represent alternatives that will in general prove equally satisfactory (although the second option is prevalent), except that the angles θ_1, θ_2, and θ_3 prove inadequate for attitude description when $\theta_2 = \pi/2$, and the φ, θ, and ψ angles pose a similar problem when $\theta = 0$. These two troublesome cases describe the same physical situation (the rotor axis and the axis of rotation of the outer gimbal relative to the base frame are aligned), and introduce analytical ambiguity concerning the values of the remaining attitude angles. In application to gyroscopes, such a condition is known as *gimbal lock*

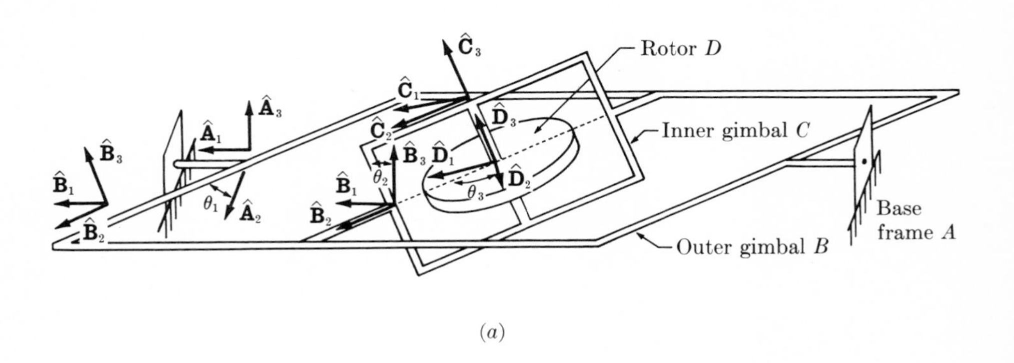

(*a*)

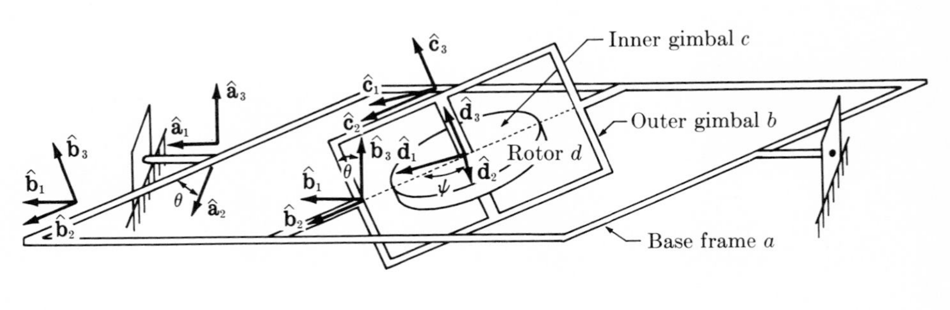

(*b*)

FIGURE 3·12

because in this singular state the rotor cannot be rotated about an axis perpendicular to the plane of the outer gimbal.

The primary difference in the two attitude coordinate schemes of Fig. 3·12*a* and 3·12*b* lies in the sequences adopted for the rotations; whereas the former is a 1-2-3 system, the latter is a 3-1-3 system. The sense in which these rotations are sequential should be noted carefully; this is more an analytical distinction than a physical one. To reach the states portrayed in the figures from the nominal states, we could physically accomplish in any sequence the *relative* rotations of *adjacent bodies*. If in Fig. 3·12*a* we first rotate D relative to C by the angle θ_3, and then accomplish the relative rotations of C relative to B and B relative to A (each time holding all other labeled angles fixed), we would reach the same state as with the reverse sequence. In practice, of course, all three of the angles change simultaneously. Nonetheless, in analytical work there remains a sequence of rotations in an important sense, as can be appreciated only by rereading the initial descriptions of the attitude angles and reconstructing the derivations of the direction-cosine matrices in Eqs. (3·74) and (3·77).

3·2·8 Infinitesimal rotations

In the preceding treatment of *planar* rotations, we might have been tempted to represent each planar rotation by a "vector" of magnitude corresponding to the magnitude of the angle of rotation, orientation established by the axis of rotation, and sense or sign defined by the right-hand rule. The first rotation illustrated in Fig. 3·8 would then be $\varphi\hat{\mathbf{a}}_3$, the second $\theta\hat{\mathbf{b}}_1$, and the last $\psi\hat{\mathbf{c}}_3$. It is important to recognize, however, that these three rotations *cannot* generally be combined as a single "orientation vector." In this respect a sequence of three planar rotations (e.g., three Euler angle rotations) differs in a very fundamental way from a sequence of linear translations (such as cartesian coordinates describe). The distinction is in the importance of *sequence* in performing successive rotations to obtain a given orientation. A succession of three linear displacements gives the same result regardless of the sequence of the steps, but this cannot be said for noninfinitesimal rotations. This is easily illustrated by performing two 90-degree rotations of a book about two orthogonal axes, and then repeating the process in opposite sequence. Since the sequence in which rotations are combined influences the final result, we say that noninfinitesimal rotations do not add *commutatively;* this is in violation of a fundamental property of a vector since it conflicts with the "parallelogram rule" for vector addition. Thus *noninfinitesimal angular displacements are not vectors*. This may be confirmed analytically by contrasting $[C^2(\theta_2)][C^1(\theta_1)]$ with $[C^1(\theta_1)][C^2(\theta_2)]$, even for the special case $\theta_1 = \theta_2 = \pi/2$ rad.

If, however, we restrict the discussion to *infinitesimal* rotations, the previous objections to a vectorial representation no longer apply. This is not surprising, as the distinguishing characteristic of a large-angle rotation of a body is that it carries any point of the body along a circular arc, and for infinitesimal rotations that arc becomes a straight-line segment. Thus, for a given point of the body, a sequence of *infinitesimal rotations* is indistinguishable from a sequence of *linear translations* of the body.

To establish this point analytically, we can reconstruct the direction-cosine matrix $[C] = [C^3(\theta_3)][C^2(\theta_2)][C^1(\theta_1)]$ originally presented as Eq. (3·77), but this time with the small-angle assumptions $\sin\theta_i \cong \theta_i$, $\cos\theta_i \cong 1$, where $i = 1, 2, 3$. The linearized approximation of $[C]$ is then given by

$$[C] \cong \begin{bmatrix} 1 & \theta_3 & 0 \\ -\theta_3 & 1 & 0 \\ 0 & 0 & 1 \end{bmatrix} \begin{bmatrix} 1 & 0 & -\theta_2 \\ 0 & 1 & 0 \\ \theta_2 & 0 & 1 \end{bmatrix} \begin{bmatrix} 1 & 0 & 0 \\ 0 & 1 & \theta_1 \\ 0 & -\theta_1 & 1 \end{bmatrix}$$

$$\cong \begin{bmatrix} 1 & 0 & 0 \\ 0 & 1 & 0 \\ 0 & 0 & 1 \end{bmatrix} - \begin{bmatrix} 0 & -\theta_3 & \theta_2 \\ \theta_3 & 0 & -\theta_1 \\ -\theta_2 & \theta_1 & 0 \end{bmatrix} \tag{3·78a}$$

where terms above the first degree in all angles have been ignored. If, *for this special case of small-angle rotations*, the rotational sequence were reversed, so that first the coincident reference frames b, c, and d were rotated relative to a by a small angle θ_3 on an axis parallel to $\hat{\mathbf{a}}_3$, and the frames c and d were together rotated with respect to b by a small angle θ_2 about an axis parallel to $\hat{\mathbf{b}}_2$, and finally d was rotated relative to c through a small angle θ_1 about an axis parallel to $\hat{\mathbf{c}}_1$, then the direction-cosine matrix relating the vector triads in a and d would be approximated by

$$[C] \cong \begin{bmatrix} 1 & 0 & 0 \\ 0 & 1 & \theta_1 \\ 0 & -\theta_1 & 1 \end{bmatrix} \begin{bmatrix} 1 & 0 & -\theta_2 \\ 0 & 1 & 0 \\ \theta_2 & 0 & 1 \end{bmatrix} \begin{bmatrix} 1 & \theta_3 & 0 \\ -\theta_3 & 1 & 0 \\ 0 & 0 & 1 \end{bmatrix}$$

$$\cong \begin{bmatrix} 1 & 0 & 0 \\ 0 & 1 & 0 \\ 0 & 0 & 1 \end{bmatrix} - \begin{bmatrix} 0 & -\theta_3 & \theta_2 \\ \theta_3 & 0 & -\theta_1 \\ -\theta_2 & \theta_1 & 0 \end{bmatrix} \tag{3·78b}$$

which is a linear approximation identical to that in (3·78*a*).

3·2·9 Angular velocity

Linear velocity (§3·1·2) is defined simply as a time derivative of a position vector. As we have noted, there exists no proper "orientation vector" which can be differentiated to define angular velocity. The simplest expression defining the orientation of one body or frame (say d) relative to another (say a) is the relationship

$$\{\hat{\mathbf{d}}\} = [C]\{\hat{\mathbf{a}}\} \tag{3·79}$$

of Eq. (3·73) or Eqs. (3·50). Here $[C]$ has been called the direction-cosine matrix, but its elements have been described variously as direction cosines [Eqs. (3·50)] or functions of attitude angles [Eqs. (3·74) and (3·77)]. The detailed description of the elements of $[C]$ is a less fundamental matter than $[C]$ itself, which has a significance beyond that of its component parts. Just as we have said that a position vector is a more fundamental concept than the particular coordinate system used to describe its elements (cartesian, polar, or whatever), so we can speak of the matrix $[C]$ without worrying about which system of parameters is used to describe its elements.

Thus far we have interpreted the matrix $[C]$ primarily as a *transformation matrix;* we have viewed it as a matrix that permits shifting or transforming from one vector basis to another, as in Eqs. (3·65) and (3·64). We can interpret $[C]$ also as a *rotation operator* since it describes the rotation of the body d or vector basis $\hat{\mathbf{d}}_1$, $\hat{\mathbf{d}}_2$, $\hat{\mathbf{d}}_3$ from coincidence with vector basis $\hat{\mathbf{a}}_1$, $\hat{\mathbf{a}}_2$, $\hat{\mathbf{a}}_3$ to its true orientation at a given time. In this sense the rotation matrix $[C]$ does the same job as the position vector, so it is reasonable to expect that angular velocity can somehow be defined in terms of the time derivative of the matrix $[C]$.

Equation (3·79) can be differentiated with respect to time only if we specify the reference frame of the differentiation, since the quantities in the vector arrays $\{\hat{\mathbf{d}}\}$ and $\{\hat{\mathbf{a}}\}$ are vectors. We adopt the symbol $\frac{{}^a d}{dt}\{\hat{\mathbf{d}}\}$ for the time derivative in frame a of the vector array $\{\hat{\mathbf{d}}\}$, which is the same thing as the array of vector derivatives, that is,

$$\frac{{}^a d}{dt}\{\hat{\mathbf{d}}\} \triangleq \begin{Bmatrix} \dfrac{{}^a d}{dt}\hat{\mathbf{d}}_1 \\ \dfrac{{}^a d}{dt}\hat{\mathbf{d}}_2 \\ \dfrac{{}^a d}{dt}\hat{\mathbf{d}}_3 \end{Bmatrix} \tag{3·80}$$

Because the time derivative of this vector array in frame d is zero, only the derivative in frame a is of interest here. If we simply differentiate Eq. (3·79) in frame a, we get

$$\frac{{}^a d}{dt}\{\hat{\mathbf{d}}\} = [\dot{C}]\{\hat{\mathbf{a}}\} + [C]\frac{{}^a d}{dt}\{\hat{\mathbf{a}}\} = [\dot{C}]\{\hat{\mathbf{a}}\} \tag{3·81}$$

where the time derivative of the matrix $[C]$ is the matrix of time derivatives of the individual elements. As noted in Eq. (3·56*e*), the orthonormality of $[C]$ permits the substitution

$$\{\hat{\mathbf{a}}\} = [C]^T\{\hat{\mathbf{d}}\} \tag{3·82}$$

in Eq. (3·81), with the result

$$\frac{^a d}{dt}\{\hat{\mathbf{d}}\} = [\dot{C}][C]^T\{\hat{\mathbf{d}}\} \tag{3·83}$$

This is an interesting result. It provides a matrix $[\dot{C}][C]^T$ that describes, in vector basis $\hat{\mathbf{d}}_1$, $\hat{\mathbf{d}}_2$, $\hat{\mathbf{d}}_3$, the vectors obtained by differentiating $\hat{\mathbf{d}}_1$, $\hat{\mathbf{d}}_2$, and $\hat{\mathbf{d}}_3$ in reference frame a. This matrix embodies all information concerning the rate of rotation of body (or frame) d with respect to body (or frame) a, so it contains the ingredients of the desired angular-velocity vector. The matrix $[\dot{C}][C]^T$ has the property of *skew-symmetry:* the transpose of this matrix is its negative. This may be confirmed by adding the matrix to its transpose, to obtain (using Appendix B to transpose the product)

$$\begin{aligned}[\dot{C}][C]^T + \{[\dot{C}][C]^T\}^T &= [\dot{C}][C]^T + [C][\dot{C}]^T \\ &= \frac{d}{dt}\{[C][C]^T\} = \frac{d}{dt}[U] = 0\end{aligned}$$

Note for future reference the consequence

$$[C][\dot{C}]^T = -[\dot{C}][C]^T \tag{3·84}$$

Skew-symmetry of $[\dot{C}][C]^T$ requires that all elements on the main diagonal be zero, and off-diagonal elements exist in pairs of opposite sign. Anticipating the intention to define angular velocity in terms of the elements of $[\dot{C}][C]^T$, we may introduce some suggestive new symbols and expand this matrix in the form (inserting a minus sign in adherence to convention)

$$[\dot{C}][C]^T \triangleq -\begin{bmatrix} 0 & -{}^d\omega_3{}^{da} & {}^d\omega_2{}^{da} \\ {}^d\omega_3{}^{da} & 0 & -{}^d\omega_1{}^{da} \\ -{}^d\omega_2{}^{da} & {}^d\omega_1{}^{da} & 0 \end{bmatrix} \tag{3·85}$$

and then *define* the angular velocity $\boldsymbol{\omega}^{da}$ by the expression[1]

$$\boldsymbol{\omega}^{da} \triangleq {}^d\omega_1{}^{da}\hat{\mathbf{d}}_1 + {}^d\omega_2{}^{da}\hat{\mathbf{d}}_2 + {}^d\omega_3{}^{da}\hat{\mathbf{d}}_3 \tag{3·86}$$

In other words, we are *defining* the scalar components of $\boldsymbol{\omega}^{da}$ in basis $\hat{\mathbf{d}}_1$, $\hat{\mathbf{d}}_2$, $\hat{\mathbf{d}}_3$ as those three scalars that appear in the skew-symmetric matrix $-[\dot{C}][C]^T$; ${}^d\omega_1{}^{da}$ is the element in the third row, second column of $-[\dot{C}][C]^T$, etc. Note that $\boldsymbol{\omega}^{da}$ is a *free vector,* not bound to any point or line in space.

This definition is rather elegant, but it has as yet no obvious connection with the phrase *angular velocity* as we have all used it in the past. We don't need a definition in terms of skew-symmetric matrices to determine, for example, the

[1] Recall our notational convention whereby a presuperscript identifies the vector basis of a set of scalar components of a vector.

angular velocity of a turntable rotating relative to its base. If we let Φ be the angle through which the turntable disk d rotates relative to its base a (as in Fig. 3·13), and define dextral orthogonal unit vectors $\hat{\mathbf{d}}_1$, $\hat{\mathbf{d}}_2$, and $\hat{\mathbf{d}}_3$ and $\hat{\mathbf{a}}_1$, $\hat{\mathbf{a}}_2$, and $\hat{\mathbf{a}}_3$ to be fixed in the bodies, with $\hat{\mathbf{d}}_3 \equiv \hat{\mathbf{a}}_3$ pointing along the common bearing axis in the direction the right-hand rule implies for positive rotation Φ, we expect the angular velocity of d relative to a to be $\dot{\Phi}\hat{\mathbf{d}}_3$. If the definition of angular velocity given by Eqs. (3·85) and (3·86) fails to support this expectation, we'll have to look for a more acceptable general definition.

To check this special case of planar motion, we must calculate $-[\dot{C}][C]^T$. Rotation of d through angle Φ relative to a produces

$$\begin{Bmatrix} \hat{\mathbf{d}}_1 \\ \hat{\mathbf{d}}_2 \\ \hat{\mathbf{d}}_3 \end{Bmatrix} = \begin{bmatrix} \cos\Phi & \sin\Phi & 0 \\ -\sin\Phi & \cos\Phi & 0 \\ 0 & 0 & 1 \end{bmatrix} \begin{Bmatrix} \hat{\mathbf{a}}_1 \\ \hat{\mathbf{a}}_2 \\ \hat{\mathbf{a}}_3 \end{Bmatrix} \tag{3·87}$$

Note that the matrix $[C]$ implied by Eq. (3·87) is the same as $[C^3(\Phi)]$ in Eq. (3·71), as it must be. The matrix $-[\dot{C}][C]^T$ is given by

$$\begin{aligned} -[\dot{C}][C]^T &= -\begin{bmatrix} -\dot{\Phi}\sin\Phi & \dot{\Phi}\cos\Phi & 0 \\ -\dot{\Phi}\cos\Phi & -\dot{\Phi}\sin\Phi & 0 \\ 0 & 0 & 0 \end{bmatrix} \begin{bmatrix} \cos\Phi & -\sin\Phi & 0 \\ \sin\Phi & \cos\Phi & 0 \\ 0 & 0 & 1 \end{bmatrix} \\ &= \begin{bmatrix} 0 & -\dot{\Phi} & 0 \\ \dot{\Phi} & 0 & 0 \\ 0 & 0 & 0 \end{bmatrix} \end{aligned} \tag{3·88}$$

Comparison of Eqs. (3·88), (3·85), and (3·86) indicates that the angular velocity of d relative to a is given by

$$\boldsymbol{\omega}^{da} = \dot{\Phi}\hat{\mathbf{d}}_3 \tag{3·89}$$

matching our expectations.

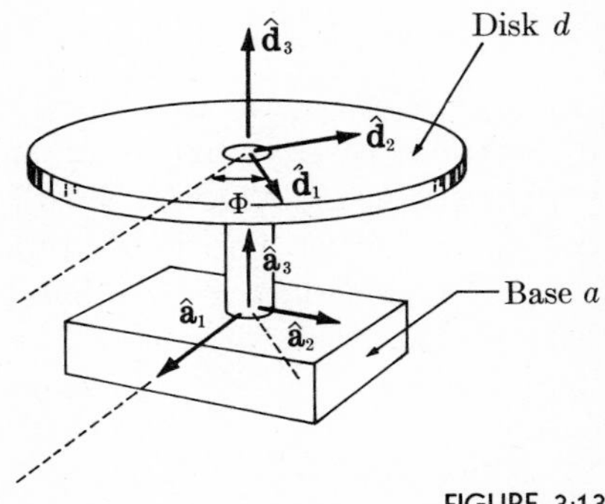

FIGURE 3·13

We now have in the matrix $-[\dot{C}][C]^T$ the basis for a formal definition of $\boldsymbol{\omega}^{da}$, and we have confirmed the correspondence of this definition to our expectation for the *planar motion* of the turntable. Of course it would be foolish to continue to use this definition in problems of planar motion; we can easily record angular velocities such as $\boldsymbol{\omega}^{da} = \dot{\Phi}\hat{\mathbf{d}}_3$ by inspection.

In §3·2·3 it was noted that a *general* description of orientation can be generated by an ordered sequence of planar rotations, as illustrated by the three Euler angle rotations. This raises the possibility of treating general rotational motions in time as a combination of simultaneous planar rotations, so that angular velocities would *always* be available by inspection. To pursue this question, we require new techniques of vector differentiation that can now be established with the help of the expression from which angular velocity is defined.

3·2·10 Vector differentiation using angular velocities

Vector differentiation has been an essential part of the preceding discussion of kinematics, but we have been required thus far to differentiate only vectors expressed in terms of unit vectors fixed in the reference frame of the differentiation (see §3·1·3). Equation (3·20) has offered the only path to vector differentiation. If we require, for example, the derivative of an arbitrary vector $\mathbf{Z}$ in frame a, we have been forced to write

$$\mathbf{Z} = {}^aZ_1\hat{\mathbf{a}}_1 + {}^aZ_2\hat{\mathbf{a}}_2 + {}^aZ_3\hat{\mathbf{a}}_3 \qquad (3\cdot90)$$

to obtain

$$\frac{{}^ad}{d}\mathbf{Z} = {}^a\dot{Z}_1\hat{\mathbf{a}}_1 + {}^a\dot{Z}_2\hat{\mathbf{a}}_2 + {}^a\dot{Z}_3\hat{\mathbf{a}}_3 \qquad (3\cdot91)$$

If we express this result in terms of matrices and vector arrays, we have

$$\mathbf{Z} = \{\hat{\mathbf{a}}_1\hat{\mathbf{a}}_2\hat{\mathbf{a}}_3\}\begin{bmatrix} {}^aZ_1 \\ {}^aZ_2 \\ {}^aZ_3 \end{bmatrix} \triangleq \{\hat{\mathbf{a}}\}^T[{}^aZ] \qquad (3\cdot92)$$

and

$$\frac{{}^ad}{dt}\mathbf{Z} = \frac{{}^ad}{dt}(\{\hat{\mathbf{a}}\}^T[{}^aZ]) = \{\hat{\mathbf{a}}\}^T[{}^a\dot{Z}] \qquad (3\cdot93)$$

We often find it convenient to write $\mathbf{Z}$ in some other vector basis, for example, to write

$$\mathbf{Z} = {}^dZ_1\hat{\mathbf{d}}_1 + {}^dZ_2\hat{\mathbf{d}}_2 + {}^dZ_3\hat{\mathbf{d}}_3 = \{\hat{\mathbf{d}}_1\hat{\mathbf{d}}_2\hat{\mathbf{d}}_3\}\begin{bmatrix} {}^dZ_1 \\ {}^dZ_2 \\ {}^dZ_3 \end{bmatrix} \triangleq \{\hat{\mathbf{d}}\}^T[{}^dZ] \qquad (3{\cdot}94)$$

Differentiation of this vector in frame d simply provides

$$\frac{{}^dd}{dt}\mathbf{Z} = \{\hat{\mathbf{d}}\}^T[{}^d\dot{Z}] \qquad (3{\cdot}95)$$

but differentiation in frame a of the expression in Eq. (3·94) has yet to be accomplished. We can surely substitute

$$\frac{{}^ad}{dt}\mathbf{Z} = \frac{{}^ad}{dt}(\{\hat{\mathbf{d}}\}^T[{}^dZ]) = \left(\frac{{}^ad}{dt}\{\hat{\mathbf{d}}\}^T\right)[{}^dZ] + \{\hat{\mathbf{d}}\}^T[{}^d\dot{Z}] \qquad (3{\cdot}96)$$

We obtain the satisfaction of recognizing the last term as that appearing in Eq. (3·95), but the first term is not so simple. In order to proceed, we must express the vector array $\{\hat{\mathbf{d}}\}^T$ in terms of vectors fixed in frame a. We have the relationships

$$\{\hat{\mathbf{d}}\} = [C]\{\hat{\mathbf{a}}\} \qquad \text{and} \qquad \{\hat{\mathbf{a}}\} = [C]^T\{\hat{\mathbf{d}}\} \qquad (3{\cdot}97a)$$

from Eqs. (3·79) and (3·82), and transposing these yields (see Appendix B)

$$\{\hat{\mathbf{d}}\}^T = \{\hat{\mathbf{a}}\}^T[C]^T \qquad \text{and} \qquad \{\hat{\mathbf{a}}\}^T = \{\hat{\mathbf{d}}\}^T[C] \qquad (3{\cdot}97b)$$

so we can effect a substitution to obtain

$$\frac{{}^ad}{dt}\{\hat{\mathbf{d}}\}^T = \frac{{}^ad}{dt}(\{\hat{\mathbf{a}}\}^T[C]^T) = \{\hat{\mathbf{a}}\}^T[\dot{C}]^T$$

Now Eq. (3·96) becomes

$$\frac{{}^ad}{dt}\mathbf{Z} = \{\hat{\mathbf{a}}\}^T[\dot{C}]^T[{}^dZ] + \frac{{}^dd}{dt}\mathbf{Z}$$

Equations (3·97b) can next be used to provide

$$\frac{{}^ad}{dt}\mathbf{Z} = \{\hat{\mathbf{d}}\}^T[C][\dot{C}]^T[{}^dZ] + \frac{{}^dd}{dt}\mathbf{Z}$$

and by virtue of Eq. (3·84) this is

$$\frac{{}^a d}{dt}\mathbf{Z} = \frac{{}^d d}{dt}\mathbf{Z} - \{\hat{\mathbf{d}}\}^T[\dot{C}][C]^T[{}^dZ] \tag{3·98}$$

In expanded form, the second term on the right can, with the help of Eq. (3·85), be written

$$\begin{aligned} -\{\hat{\mathbf{d}}\}^T[\dot{C}][C]^T[{}^dZ] &= \{\hat{\mathbf{d}}_1\hat{\mathbf{d}}_2\hat{\mathbf{d}}_3\}\begin{bmatrix} 0 & -{}^d\omega_3{}^{da} & {}^d\omega_2{}^{da} \\ {}^d\omega_3{}^{da} & 0 & -{}^d\omega_1{}^{da} \\ -{}^d\omega_2{}^{da} & {}^d\omega_1{}^{da} & 0 \end{bmatrix}\begin{bmatrix} {}^dZ_1 \\ {}^dZ_2 \\ {}^dZ_3 \end{bmatrix} \\ &= \{\hat{\mathbf{d}}_1\hat{\mathbf{d}}_2\hat{\mathbf{d}}_3\}\begin{bmatrix} {}^d\omega_2{}^{da}\,{}^dZ_3 - {}^d\omega_3{}^{da}\,{}^dZ_2 \\ {}^d\omega_3{}^{da}\,{}^dZ_1 - {}^d\omega_1{}^{da}\,{}^dZ_3 \\ {}^d\omega_1{}^{da}\,{}^dZ_2 - {}^d\omega_2{}^{da}\,{}^dZ_1 \end{bmatrix} \\ &= \hat{\mathbf{d}}_1({}^d\omega_2{}^{da}\,{}^dZ_3 - {}^d\omega_3{}^{da}\,{}^dZ_2) + \hat{\mathbf{d}}_2({}^d\omega_3{}^{da}\,{}^dZ_1 - {}^d\omega_1{}^{da}\,{}^dZ_3) \\ &\qquad + \hat{\mathbf{d}}_3({}^d\omega_1{}^{da}\,{}^dZ_2 - {}^d\omega_2{}^{da}\,{}^dZ_1) \end{aligned}$$

Now we can, with Eq. (3·86), make the remarkable observation that the expression just written is no more than the vector cross product of $\boldsymbol{\omega}^{da}$ and $\mathbf{Z}$, that is,

$$-\{\hat{\mathbf{d}}\}^T[\dot{C}][C]^T[{}^dZ] = \boldsymbol{\omega}^{da} \times \mathbf{Z} \tag{3·99}$$

This simplifies the vector derivative in Eq. (3·98) in a most satisfying way, providing the useful result

$$\frac{{}^a d}{dt}\mathbf{Z} = \frac{{}^d d}{dt}\mathbf{Z} + \boldsymbol{\omega}^{da} \times \mathbf{Z} \tag{3·100}$$

Equation (3·100) holds for any vector $\mathbf{Z}$, and it provides relationships between the vector derivatives in any two reference frames a and d.

This equation can be used to establish formally the reasonable-looking relationship

$$\boldsymbol{\omega}^{da} = -\boldsymbol{\omega}^{ad} \tag{3·101}$$

If we differentiate the arbitrary vector $\mathbf{Z}$ in frames a and d, we have

$$\frac{{}^a d}{dt}\mathbf{Z} = \frac{{}^d d}{dt}\mathbf{Z} + \boldsymbol{\omega}^{da} \times \mathbf{Z}$$

and

$$\frac{{}^d d}{dt}\mathbf{Z} = \frac{{}^a d}{dt}\mathbf{Z} + \boldsymbol{\omega}^{ad} \times \mathbf{Z}$$

from Eq. (3·100). These results combine as

$$\frac{{}^a d}{dt}\mathbf{Z} = \frac{{}^a d}{dt}\mathbf{Z} + \boldsymbol{\omega}^{ad} \times \mathbf{Z} + \boldsymbol{\omega}^{da} \times \mathbf{Z}$$

which requires

$$(\boldsymbol{\omega}^{ad} + \boldsymbol{\omega}^{da}) \times \mathbf{Z} = 0$$

Since **Z** is an arbitrary vector, the expression in parentheses is zero, proving Eq. (3·101).

Equation (3·100) has utility far beyond its use in proving in Eq. (3·101), a relationship that may appear self-evident. The skillful application of Eq. (3·100) permits substantial reduction in the labors of kinematic calculations, as later applications will demonstrate. This formula evidently requires the use of expressions for angular velocity, however, so before considering applications we should try to find more practical procedures for calculating angular velocities than are available at this point.

3·2·11 Angular Velocity determination

Although we have adopted a general definition of $\boldsymbol{\omega}^{da}$ for any frames a and d in terms of the direction-cosine matrix $[C]$ and its time derivative, this proves to be an unnecessarily cumbersome starting point for the determination of angular velocities in most cases of interest. For digital-computer simulations of complex dynamic systems, this definition might be used directly, but for manual calculations there are easier ways to find $\boldsymbol{\omega}^{da}$.

It has already been noted that it would be a waste of time to go back to this general definition in the case of planar motion; it is enough simply to inspect a figure (such as Fig. 3·13) and record an angular velocity (such as $\boldsymbol{\omega}^{da} = \dot{\Phi}\hat{\mathbf{d}}_3$).

We have already speculated that it might be possible to extend our knowledge that any *orientation* can be described as a sequence of planar rotations to a similar conclusion regarding *angular velocities*. This leads us to an important proposition, which we shall be able to prove with the assistance of the vector differentiation formula of Eq. (3·100).

Given n reference frames $f_1, f_2, \ldots, f_n$, the angular velocity of frame f_1 relative to frame f_n may be written in terms of the *chain rule*

$$\boldsymbol{\omega}^{f_1 f_n} = \boldsymbol{\omega}^{f_1 f_2} + \boldsymbol{\omega}^{f_2 f_3} + \boldsymbol{\omega}^{f_3 f_4} + \cdots + \boldsymbol{\omega}^{f_{n-1} f_n} \tag{3·102}$$

To prove this proposition, we let $\mathbf{Z}$ be a vector fixed in frame f_1, and repeatedly apply Eq. (3·100) $n - 1$ times to obtain

$$\frac{{}^{f_2}d}{dt}\mathbf{Z} = \frac{{}^{f_1}d}{dt}\mathbf{Z} + \boldsymbol{\omega}^{f_1f_2} \times \mathbf{Z} = \boldsymbol{\omega}^{f_1f_2} \times \mathbf{Z}$$

$$\frac{{}^{f_3}d}{dt}\mathbf{Z} = \frac{{}^{f_2}d}{dt}\mathbf{Z} + \boldsymbol{\omega}^{f_2f_3} \times \mathbf{Z} = (\boldsymbol{\omega}^{f_1f_2} + \boldsymbol{\omega}^{f_2f_3}) \times \mathbf{Z}$$

$$\cdots\cdots\cdots\cdots\cdots\cdots\cdots\cdots$$

$$\frac{{}^{f_n}d}{dt}\mathbf{Z} = \frac{{}^{f_{n-1}}d}{dt}\mathbf{Z} + \boldsymbol{\omega}^{f_{n-1}f_n} \times \mathbf{Z}$$

$$= (\boldsymbol{\omega}^{f_1f_2} + \boldsymbol{\omega}^{f_2f_3} + \cdots + \boldsymbol{\omega}^{f_{n-1}f_n}) \times \mathbf{Z} \tag{3·103}$$

If Eq. (3·100) is used once more to express

$$\frac{{}^{f_n}d}{dt}\mathbf{Z} = \boldsymbol{\omega}^{f_1f_n} \times \mathbf{Z} \tag{3·104}$$

and Eqs. (3·103) and (3·104) are compared, we may conclude that since $\mathbf{Z}$ is any vector fixed in f_1, the proposition (3·102) is established.

The utility of the chain rule for angular velocity determination is immediately apparent on consideration of a single example. Consider the gyroscope illustrated in Fig. 3·12*b*. The rotor d rotates relative to the gimbal frame c through angle ψ about an axis paralleling $\hat{\mathbf{d}}_3$ and $\hat{\mathbf{c}}_3$. Gimbal c rotates relative to gimbal b through the angle θ about an axis paralleling $\hat{\mathbf{c}}_1$ and $\hat{\mathbf{b}}_1$. Finally, the gimbal b rotates relative to the base frame a through the angle φ about an axis paralleling $\hat{\mathbf{b}}_3$ and $\hat{\mathbf{a}}_3$. Because there are three independent scalars ψ, θ, and φ defining the orientation of d relative to a, we recognize that a completely general orientation of d relative to a can be established. We note further that when $\psi = \theta = \varphi = 0$, all unit vectors of like index are aligned, and it takes a 3-1-3 sequence of rotations through angles φ, θ, and ψ to define the general orientation of d relative to a. This means that the angles φ, θ, and ψ are the Euler angles of §3·2·6. We have already established in Eq. (3·74) the expression for the direction-cosine matrix $[C]$ in terms of these angles, so we are prepared to calculate the angular velocity $\boldsymbol{\omega}^{da}$, or its negative $\boldsymbol{\omega}^{ad}$, in terms of the matrix $[\dot{C}][C]^T$. A glance at Eq. (3·74) is enough to discourage the direct use of this definition—differentiating $[C]$ and then postmultiplying by $[C]^T$ would be a dreary task.

As an alternative, consider the use of the chain rule in the form

$$\boldsymbol{\omega}^{da} = \boldsymbol{\omega}^{dc} + \boldsymbol{\omega}^{cb} + \boldsymbol{\omega}^{ba} \tag{3·105}$$

Each of these terms describes a planar rotation, which can be recorded by inspection of Fig. 3·12*b*. If we ignore everything but the planar rotation of gimbal frame *b* relative to base frame *a*, we readily recognize

$$\boldsymbol{\omega}^{ba} = \dot{\varphi}\hat{\mathbf{b}}_3 = \dot{\varphi}\hat{\mathbf{a}}_3 \tag{3·106}$$

Concentrating exclusively on the relative motion of gimbal frame *c* relative to frame *b* produces

$$\boldsymbol{\omega}^{cb} = \dot{\theta}\hat{\mathbf{c}}_1 = \dot{\theta}\hat{\mathbf{b}}_1 \tag{3·107}$$

Similarly we obtain

$$\boldsymbol{\omega}^{dc} = \dot{\psi}\hat{\mathbf{d}}_3 = \dot{\psi}\hat{\mathbf{c}}_3 \tag{3·108}$$

The last three equations may be substituted into Eq. (3·105) to find the desired angular velocity

$$\boldsymbol{\omega}^{da} = \dot{\psi}\hat{\mathbf{c}}_3 + \dot{\theta}\hat{\mathbf{c}}_1 + \dot{\varphi}\hat{\mathbf{b}}_3 \tag{3·109}$$

This is a complete result, but it may be in more useful form when expressed in terms of a single set of orthogonal unit vectors. By choosing the convenient set $\hat{\mathbf{c}}_1$, $\hat{\mathbf{c}}_2$, and $\hat{\mathbf{c}}_3$ and substituting

$$\hat{\mathbf{b}}_3 = \hat{\mathbf{c}}_3 \cos\theta + \hat{\mathbf{c}}_2 \sin\theta$$

we get

$$\boldsymbol{\omega}^{da} = \dot{\theta}\hat{\mathbf{c}}_1 + \dot{\varphi}\sin\theta\hat{\mathbf{c}}_2 + (\dot{\psi} + \dot{\varphi}\cos\theta)\,\hat{\mathbf{c}}_3 \tag{3·110}$$

It should be quite obvious that the determination of angular velocities by combining planar rotations with the chain rule is much easier in this case than obtaining the angular velocity in terms of the elements of the matrix $[\dot{C}][C]^T$. This conclusion can be generalized to apply to any motion described in terms of attitude angles. Even when there are no physical gimbal frames *b* and *c* as in Fig. 3·12*b*, there are nonphysical reference frames *b* and *c* as in Figs. 3·8 and 3·9, so this approach is not as restricted as it may appear to be initially. Examples in § 3·2·13 will reinforce this conclusion.

3·2·12 Angular acceleration

The angular acceleration of a body (or frame) *f* relative to a body (or frame) *f*′ is given by

$$\boldsymbol{\alpha}^{ff'} \triangleq \frac{{}^{f'}d}{dt}\boldsymbol{\omega}^{ff'} \tag{3·111}$$

Angular acceleration, like angular velocity, is a free vector, not bound to any point or line in space.

Note that the chain rule for angular velocity, Eq. (3·101), does *not* in general produce a chain rule for angular acceleration when substituted into Eq. (3·111), that is,

$$\boldsymbol{\alpha}^{f_1 f_n} \neq \boldsymbol{\alpha}^{f_1 f_2} + \boldsymbol{\alpha}^{f_2 f_3} + \cdots + \boldsymbol{\alpha}^{f_{n-1} f_n} \qquad (3\cdot112)$$

3·2·13 Examples of angular velocity and angular acceleration determination

First example Consider the two-bar linkage of Fig. 3·3, previously used to illustrate the calculation of a linear velocity. Now we are interested in the angular velocities and accelerations of the bars. This is a planar mechanism, so we should have no difficulty in recording angular velocities by inspection of the figure.

The angular velocity of bar 1 relative to frame e is given by

$$\boldsymbol{\omega}^{1e} = \dot{\theta}\hat{\mathbf{e}}_z \qquad (3\cdot113)$$

where $\hat{\mathbf{e}}_z$ is directed into the page. Note that we can speak of the angular velocity $\boldsymbol{\omega}^{e1}$ of frame e relative to bar 1 (although you may feel that somehow the frame e is more "fixed" than the bar) and write

$$\boldsymbol{\omega}^{e1} = -\dot{\theta}\hat{\mathbf{e}}_z \qquad (3\cdot114)$$

The only conceivable obstacle to these calculations is the determination of algebraic sign. To obtain the sign in Eq. (3·113), for example, we can imagine the angle θ to be zero, fix the frame e in mind, and advance bar 1 so as to increase θ. If the fingers of the right hand are aligned with bar 1 and, as it rotates, the fingers close on the palm, the right thumb will point along the angular velocity vector. This parallels the vector $\hat{\mathbf{e}}_z$ in Fig. 3·3, so $\boldsymbol{\omega}^{1e}$ is given by $\dot{\theta}$ times $\hat{\mathbf{e}}_z$. In the calculation of $\boldsymbol{\omega}^{e1}$, it is bar 1 that is fixed in the imagination, and frame e that rotates as θ increases. The right-hand rule then indicates a vector pointing out of the page, in direction $-\hat{\mathbf{e}}_z$. The angular velocity is thus given by $\dot{\theta}$ times $-\hat{\mathbf{e}}_z$, as in Eq. (3·114). Alternatively, of course, we could obtain $\boldsymbol{\omega}^{e1}$ as the negative of $\boldsymbol{\omega}^{1e}$, using Eq. (3·101).

The angular velocity of bar 2 relative to e is just as easily obtained as that of bar 1, despite the fact that no point of bar 2 is fixed in e. Without worrying that bar 2 is translating relative to e while it rotates, we can see that the angle θ between the bar and the vertical line fully defines the orientation of bar 2 relative to e. Thus the angular velocity is given by

$$\boldsymbol{\omega}^{2e} = -\dot{\theta}\hat{\mathbf{e}}_z \qquad (3\cdot115)$$

The algebraic sign can be established in the manner indicated (repeat the steps and be sure you see this).

Similarly, the angular velocity of bar 2 relative to bar 1 is, for the indicated geometry,

$$\boldsymbol{\omega}^{21} = -2\dot{\theta}\hat{\mathbf{e}}_z \tag{3·116}$$

For a simple confirmation of these results, we can check the chain rule, Eq. (3·102), which requires

$$\boldsymbol{\omega}^{2e} = \boldsymbol{\omega}^{21} + \boldsymbol{\omega}^{1e} \tag{3·117}$$

Angular accelerations for this planar problem are extremely easy to obtain by differentiation of angular velocities because the unit vector $\hat{\mathbf{e}}_z$ is fixed (has no time derivative) relative to reference frames e, 1, and 2. The definition of Eq. (3·111) provides in parallel with Eqs. (3·113) to (3·116)

$$\boldsymbol{\alpha}^{1e} \triangleq \frac{{}^e d}{dt}\boldsymbol{\omega}^{1e} = \ddot{\theta}\hat{\mathbf{e}}_z \tag{3·118}$$

$$\boldsymbol{\alpha}^{e1} \triangleq \frac{{}^1 d}{dt}\boldsymbol{\omega}^{e1} = -\ddot{\theta}\hat{\mathbf{e}}_z \tag{3·119}$$

$$\boldsymbol{\alpha}^{2e} \triangleq \frac{{}^e d}{dt}\boldsymbol{\omega}^{2e} = -\ddot{\theta}\hat{\mathbf{e}}_z \tag{3·120}$$

$$\boldsymbol{\alpha}^{21} \triangleq \frac{{}^1 d}{dt}\boldsymbol{\omega}^{21} = -2\ddot{\theta}\hat{\mathbf{e}}_z \tag{3·121}$$

Despite the fact that in Eq. (3·112) we specifically disavowed any "chain rule" for angular accelerations, we do find in this case the relationship

$$\boldsymbol{\alpha}^{2e} = \boldsymbol{\alpha}^{21} + \boldsymbol{\alpha}^{1e} \tag{3·122}$$

The validity of the chain rule in this special case is explained by the fact that the mechanism of Fig. 3·3 is limited to planar motion. In many respects planar motion is conceptually different from three-dimensional motion, and we must be very cautious about generalizing from this special case.

Second example Consider next the assembly used previously in Fig. 3·5 to illustrate the determination of linear velocity.

The angle θ describes the relative orientation of frames f and b, so that the angular velocity of b in f is

$$\boldsymbol{\omega}^{bf} = \dot{\theta}\hat{\mathbf{b}}_3 = \dot{\theta}\hat{\mathbf{f}}_3 \tag{3·123}$$

Although arm a is sliding down the slot in the block b, and rotating with b relative to the floor f, only the angle φ is required to establish the orientation of a relative to b, so the corresponding angular velocity is simply

$$\boldsymbol{\omega}^{ab} = -\dot{\varphi}\hat{\mathbf{b}}_2 = -\dot{\varphi}\hat{\mathbf{a}}_2 \tag{3·124}$$

If now we require the angular velocity of arm a relative to the floor, we use the chain rule

$$\boldsymbol{\omega}^{af} = \boldsymbol{\omega}^{ab} + \boldsymbol{\omega}^{bf} = -\dot{\varphi}\hat{\mathbf{b}}_2 + \dot{\theta}\hat{\mathbf{b}}_3 \tag{3·125}$$

Angular accelerations may be calculated from the definition, Eq. (3·111), as follows:

$$\boldsymbol{\alpha}^{bf} \triangleq \frac{{}^f d}{dt}\boldsymbol{\omega}^{bf} = \frac{{}^f d}{dt}(\dot{\theta}\hat{\mathbf{f}}_3) = \ddot{\theta}\hat{\mathbf{f}}_3$$

$$\boldsymbol{\alpha}^{ab} \triangleq \frac{{}^b d}{dt}\boldsymbol{\omega}^{ab} = \frac{{}^b d}{dt}(-\dot{\varphi}\hat{\mathbf{b}}_2) = -\ddot{\varphi}\hat{\mathbf{b}}_2$$

$$\boldsymbol{\alpha}^{af} \triangleq \frac{{}^f d}{dt}\boldsymbol{\omega}^{af} = \frac{{}^f d}{dt}(\dot{\theta}\hat{\mathbf{f}}_3 - \dot{\varphi}\hat{\mathbf{b}}_2) = \ddot{\theta}\hat{\mathbf{f}}_3 - \ddot{\varphi}\hat{\mathbf{b}}_2 - \dot{\varphi}\frac{{}^f d}{dt}\hat{\mathbf{b}}_2$$

To evaluate the last term, apply the differentiation formula in Eq. (3·100) to obtain [also using Eq. (3·123)]

$$\frac{{}^f d}{dt}\hat{\mathbf{b}}_2 = \frac{{}^b d}{dt}\hat{\mathbf{b}}_2 + \boldsymbol{\omega}^{bf} \times \hat{\mathbf{b}}_2 = \dot{\theta}\hat{\mathbf{b}}_3 \times \hat{\mathbf{b}}_2 = -\dot{\theta}\hat{\mathbf{b}}_1$$

Substitution now gives

$$\boldsymbol{\alpha}^{af} = \ddot{\theta}\hat{\mathbf{f}}_3 - \ddot{\varphi}\hat{\mathbf{b}}_2 + \dot{\varphi}\dot{\theta}\hat{\mathbf{b}}_1$$

Note that for this case there is no chain rule:

$$\boldsymbol{\alpha}^{af} \neq \boldsymbol{\alpha}^{ab} + \boldsymbol{\alpha}^{bf}$$

3.2.14 Examples of the use of angular velocity in linear velocity and linear acceleration determination

The problem of determining linear velocities and accelerations need not involve the concept of angular velocity at all. Articles 3·1·4 and 3·1·6 provide examples of such calculations, which preceded the introduction of the definition of angular velocity. We found there, however, that we were frustrated by the necessity of expressing all differentiated vectors in terms of a vector basis fixed in the reference frame of differentiation, as in Eq. (3·19). Now that we have introduced angular velocity and established the vector differentiation relationship of Eq. (3·100), the same problems solved in §3·1·4 and 3·1·6 will yield much more easily.

First example Consider again the system of Fig. 3·4. As in §3·1·4 and 3·1·6, the ant a crawls along a meridian of the cone c at constant speed v, while the cone rotates relative to its base b with angular speed Ω, with clockwise rotation as viewed from b. We require the linear velocity $\mathbf{V}^{ab}$ and linear acceleration $\mathbf{A}^{ab}$.

As calculated in §3·1·4, the required velocity is by definition

$$\mathbf{V}^{ab} = \frac{^b d}{dt} vt\hat{\mathbf{u}} = v\hat{\mathbf{u}} + vt \frac{^b d}{dt} \hat{\mathbf{u}} \tag{3·126}$$

Now we can use Eq. (3·100) to calculate

$$\begin{aligned} \frac{^b d}{dt} \hat{\mathbf{u}} &= \frac{^c d}{dt} \hat{\mathbf{u}} + \boldsymbol{\omega}^{cb} \times \hat{\mathbf{u}} = \Omega\hat{\mathbf{c}}_3 \times \hat{\mathbf{u}} \\ &= \Omega\hat{\mathbf{c}}_3 \times \left(\frac{3}{5}\hat{\mathbf{c}}_1 - \frac{4}{5}\hat{\mathbf{c}}_3\right) = \frac{3}{5}\Omega\hat{\mathbf{c}}_2 \end{aligned} \tag{3·127}$$

where $\hat{\mathbf{u}}$ has been expressed in the $\hat{\mathbf{c}}_1$, $\hat{\mathbf{c}}_2$, and $\hat{\mathbf{c}}_3$ basis as in Eq. (3·24), and the constancy of $\hat{\mathbf{u}}$ in the reference frame c has been recognized.

Combining Eqs. (3·126), (3·127), and (3·24) provides

$$\mathbf{V}^{ab} = \frac{3}{5} v\hat{\mathbf{c}}_1 + \frac{3}{5}\Omega vt\hat{\mathbf{c}}_2 - \frac{4}{5} v\hat{\mathbf{c}}_3 \tag{3·128}$$

This same result was obtained previously as Eq. (3·29), but with much more labor.

Next apply the definition of Eq. (3·43) to obtain

$$\mathbf{A}^{ab} \triangleq \frac{^b d}{dt} \mathbf{V}^{ab} \tag{3·129}$$

This definition was also used in §3·1·6, but then we could perform the differentiation only after writing $\mathbf{V}^{ab}$ in a vector basis fixed in b. Now we can employ Eq. (3·100) to differentiate $\mathbf{V}^{ab}$ in the simple form of Eq. (3·128). The result is

$$\begin{aligned} \mathbf{A}^{ab} &= \frac{^b d}{dt}\left(\frac{3}{5} v\hat{\mathbf{c}}_1 + \frac{3}{5}\Omega vt\hat{\mathbf{c}}_2 - \frac{4}{5} v\hat{\mathbf{c}}_3\right) \\ &= \frac{^c d}{dt}\left(\frac{3}{5} v\hat{\mathbf{c}}_1 + \frac{3}{5}\Omega vt\hat{\mathbf{c}}_2 - \frac{4}{5} v\hat{\mathbf{c}}_3\right) + \boldsymbol{\omega}^{cb} \times \left(\frac{3}{5} v\hat{\mathbf{c}}_1 + \frac{3}{5}\Omega vt\hat{\mathbf{c}}_2 - \frac{4}{5} v\hat{\mathbf{c}}_3\right) \\ &= \frac{3}{5}\Omega v\hat{\mathbf{c}}_2 + \Omega\hat{\mathbf{c}}_3 \times \left(\frac{3}{5} v\hat{\mathbf{c}}_1 + \frac{3}{5}\Omega vt\hat{\mathbf{c}}_2\right) \\ &= -\frac{3}{5}\Omega^2 vt\hat{\mathbf{c}}_1 + \frac{6}{5}\Omega v\hat{\mathbf{c}}_2 \end{aligned} \tag{3·130}$$

This result matches that obtained in Eq. (3·48), but again the procedure presently adopted is vastly more efficient.

Second example The mechanism shown in Fig. 3·5 has served to demonstrate the labors of calculating linear velocities by differentiation (see §3·1·4), and it has illustrated the chain rule for angular velocity determination in §3·2·8.

Recall that block b rotates relative to the floor f, arm a slides on the floor and in a slot in b, and the ring (center p) slides along arm a. To find the velocity of p in f, we can use the definition in Eq. (3·16)

$$\mathbf{V}^{pf} \triangleq \frac{^f d}{dt} \mathbf{P} \tag{3·131}$$

where the vector $\mathbf{P}$ locates p from some point (such as o) fixed in f. The vector $\mathbf{P}$ may therefore be expressed for differentiation as the sum of the vector from o to q and the vector from q to p, that is,

$$\mathbf{P} = L \cos \varphi \, \hat{\mathbf{f}}_3 + z\hat{\mathbf{a}}_1 \tag{3·132}$$

Substitution and differentiation yield

$$\mathbf{V}^{pf} = \frac{^f d}{dt} (L \cos \varphi \hat{\mathbf{f}}_3 + z\hat{\mathbf{a}}_1) = -L\dot{\varphi} \sin \varphi \hat{\mathbf{f}}_3 + \dot{z}\hat{\mathbf{a}}_1 + z \frac{^f d}{dt} \hat{\mathbf{a}}_1 \tag{3·133}$$

To accomplish the vector differentiation remaining in Eq. (3·133), we apply Eq. (3·100) in the form

$$\frac{^f d}{dt} \hat{\mathbf{a}}_1 = \frac{^a d}{dt} \hat{\mathbf{a}}_1 + \boldsymbol{\omega}^{af} \times \hat{\mathbf{a}}_1 = (\dot{\theta}\hat{\mathbf{b}}_3 - \dot{\varphi}\hat{\mathbf{b}}_2) \times \hat{\mathbf{a}}_1 \tag{3·134}$$

where the constancy of $\hat{\mathbf{a}}_1$ in frame a has been recognized, and $\boldsymbol{\omega}^{af}$ has been adopted from Eq. (3·125).

Combining Eqs. (3·133) and (3·134), performing the cross multiplication, and changing the vector basis, we get

$$\begin{aligned} \mathbf{V}^{pf} &= -L\dot{\varphi} \sin \varphi \hat{\mathbf{b}}_3 + \dot{z}\hat{\mathbf{a}}_1 + z(\dot{\theta} \sin \varphi \hat{\mathbf{a}}_2 + \dot{\varphi}\hat{\mathbf{a}}_3) \\ &= \hat{\mathbf{a}}_1(\dot{z} + L\dot{\varphi} \sin \varphi \cos \varphi) + \hat{\mathbf{a}}_2(z\dot{\theta} \sin \varphi) + \hat{\mathbf{a}}_3(z\dot{\varphi} - L\dot{\varphi} \sin^2 \varphi) \end{aligned} \tag{3·135}$$

Although this result is identical to that obtained as Eq. (3·41), this was much more easily found, as you should confirm by reexamining the derivation of Eq. (3·41).

In §3·1·6, the acceleration

$$\mathbf{A}^{pf} = \frac{^f d}{dt} \mathbf{V}^{pf}$$

was considered in the third example [see Eq. (3·49)], but the magnitude of the task of differentiating $\mathbf{V}^{pf}$ expressed in vector basis $\hat{\mathbf{f}}_1$, $\hat{\mathbf{f}}_2$, and $\hat{\mathbf{f}}_3$ seemed inordinate. Now that we have Eq. (3·100), we can retain $\mathbf{V}^{pf}$ in the vector basis used in Eq.

(3·135), and replace differentiation in f of unit vectors $\hat{\mathbf{a}}_1$, $\hat{\mathbf{a}}_2$, and $\hat{\mathbf{a}}_3$ by cross multiplication. The result is

$$
\begin{aligned}
\mathbf{A}^{pf} &= \frac{{}^a d}{dt}\mathbf{V}^{pf} + \boldsymbol{\omega}^{af} \times \mathbf{V}^{pf} \\
&= \hat{\mathbf{a}}_1[\ddot{z} + L\ddot{\varphi}\sin\varphi\cos\varphi + L\dot{\varphi}(\cos^2\varphi - \sin^2\varphi)] \\
&\qquad + \hat{\mathbf{a}}_2(\dot{z}\dot{\theta}\sin\varphi + z\ddot{\theta}\sin\varphi + z\dot{\theta}\dot{\varphi}\cos\varphi) \\
&\qquad + \hat{\mathbf{a}}_3(\dot{z}\dot{\varphi} + z\ddot{\varphi} - L\ddot{\varphi}\sin^2\varphi - 2L\dot{\varphi}^2\sin\varphi\cos\varphi) \\
&\qquad + (\dot{\theta}\hat{\mathbf{b}}_3 - \dot{\varphi}\hat{\mathbf{b}}_2) \times [\hat{\mathbf{a}}_1(\dot{z} + L\dot{\varphi}\sin\varphi\cos\varphi) \\
&\qquad + \hat{\mathbf{a}}_2(z\dot{\theta}\sin\varphi) + \hat{\mathbf{a}}_3(z\dot{\varphi} - L\dot{\varphi}\sin^2\varphi)] \\
&= \hat{\mathbf{a}}_1(\ddot{z} + L\ddot{\varphi}\sin\varphi\cos\varphi + L\dot{\varphi}^2\cos^2\varphi - z\dot{\varphi}^2 - z\dot{\theta}^2\sin^2\varphi) \\
&\qquad + \hat{\mathbf{a}}_2(z\ddot{\theta}\sin\varphi + 2\dot{z}\dot{\theta}\sin\varphi + 2z\dot{\theta}\dot{\varphi}\cos\varphi) \\
&\qquad + \hat{\mathbf{a}}_3(z\ddot{\varphi} - L\ddot{\varphi}\sin^2\varphi + 2\dot{z}\dot{\varphi} - z\dot{\theta}^2\sin\varphi\cos\varphi - L\dot{\varphi}^2\sin\varphi\cos\varphi)
\end{aligned} \tag{3·136}
$$

Although Eq. (3·136) is not a simple expression, it would be *vastly* more complicated if the vector $\mathbf{A}^{pf}$ were written in terms of the vector basis $\hat{\mathbf{f}}_1$, $\hat{\mathbf{f}}_2$, and $\hat{\mathbf{f}}_3$, as would be required by the differentiation method of §3·1·6. This example should illustrate forcefully the desirability and even the practical necessity of using Eq. (3·100) for vector differentiation. Remember that the mechanisms confronted by a practicing engineer today are often much more complex than the device considered in this example (Fig. 3·5).

3·2·15 Comment

The first two sections of this chapter are sufficient for the determination of any velocity or acceleration, whether linear or angular, and you should not proceed in this chapter unless these fundamental sections are understood. Section 3·3 contains information that is never necessary for the solution of a problem in dynamics, but this material has a well-established value in technical communication and it may serve to facilitate physical understanding and rapid analysis.

3·3 SPECIAL KINEMATIC EXPANSIONS AND INTERPRETATIONS

3·3·1 Transport velocity

When solving problems that involve the motion of a point relative to a body or frame that is itself moving relative to a base frame, it is sometimes convenient to introduce the concept of transport velocity. An example of such a problem is illustrated again by the turntable of Fig. 3·2, where ball p moves relative to disk

d that is in turn rotating relative to base b. Disk d is also both rotating and translating relative to the straight track s.

By definition, the velocity of p in frame s in Fig. 3·2 is the time derivative in frame s of a position vector **P** locating point p from any point q fixed in the track frame s, that is,

$$\mathbf{V}^{ps} = \frac{{}^s d}{dt}\mathbf{P} = \frac{{}^s d}{dt}(\mathbf{Q} + \mathbf{r}) = \mathbf{V}^{os} + \frac{{}^s d}{dt}(r\hat{\mathbf{e}}_r) \tag{3·137}$$

where **Q** is the vector from q to o. From formula (3·100), we have

$$\frac{{}^s d}{dt}(r\hat{\mathbf{e}}_r) = \frac{{}^d d}{dt}(r\hat{\mathbf{e}}_r) + \boldsymbol{\omega}^{ds} \times (r\hat{\mathbf{e}}_r) = \mathbf{V}^{pd} + \boldsymbol{\omega}^{ds} \times (r\hat{\mathbf{e}}_r) \tag{3·138}$$

Combining Eqs. (3·137) and (3·138) provides

$$\mathbf{V}^{ps} = \mathbf{V}^{pd} + \mathbf{V}^{os} + \boldsymbol{\omega}^{ds} \times (r\hat{\mathbf{e}}_r) \tag{3·139}$$

Thus the velocity of p in s is separated into the velocity of p relative to the turntable plus the *velocity relative to s that p would have if p were fixed on the turntable.* This second term is called the *transport velocity* and will be designated $\mathbf{V}^{p^*s}$ (so p^* is a point *fixed in d* under the moving ball p at any given instant).

Now the example will be generalized. The velocity of p relative to f is defined by Eq. (3·16) as $\mathbf{V}^{pf} \triangleq {}^f d\mathbf{P}/dt$, where **P** is a position vector to p from any point q fixed in f. If q' is any point fixed in another frame (or body) f', and the vector locating p from q' is called $\mathbf{P}'$, then we may freely substitute

$$\mathbf{P} = \mathbf{R} + \mathbf{P}' \tag{3·140}$$

where **R** is the vector from q to q' (see Fig. 3·14). Now the velocity of p in frame f may be written

$$\begin{aligned}\mathbf{V}^{pf} &= \frac{{}^f d}{dt}\mathbf{R} + \frac{{}^f d}{dt}\mathbf{P}' = \frac{{}^f d}{dt}\mathbf{R} + \frac{{}^{f'} d}{dt}\mathbf{P}' + \boldsymbol{\omega}^{f'f} \times \mathbf{P}' \\ &= \mathbf{V}^{pf'} + \mathbf{V}^{q'f} + \boldsymbol{\omega}^{f'f} \times \mathbf{P}'\end{aligned} \tag{3·141}$$

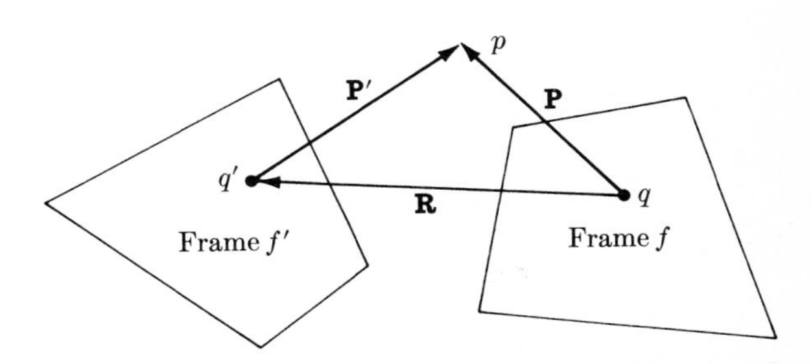

FIGURE 3·14

or

$$\mathbf{V}^{pf} = \mathbf{V}^{pf'} + \mathbf{V}^{p^*f} \tag{3·142}$$

where again $\mathbf{V}^{pf'}$ is the velocity of p relative to f' and $\mathbf{V}^{p^*f}$ is the velocity in f that p would have if p were fixed in f'. The vector $\mathbf{V}^{p^*f}$ is again the *transport velocity.*

A special case of Eq. (3·141) provides another useful relationship. If $p = p'$ and q' are *both* fixed in f', then this equation becomes

$$\mathbf{V}^{p'f} = \mathbf{V}^{q'f} + \boldsymbol{\omega}^{f'f} \times \mathbf{P}' \tag{3·143}$$

Thus if we know the angular velocity of a body f' in f and the velocity in f of any point q' fixed in f', we can readily obtain the velocity in f of any other point p' fixed in f'. This result can be used in the solution of the following problem.

3·3·2 Application to spacecraft docking

Consider the rendezvous and docking maneuver for the unmanned space vehicles illustrated in Fig. 3·15. Imagine that on-board sensors detect the inertial[1] angular velocities $\boldsymbol{\omega}^{bi}$ and $\boldsymbol{\omega}^{Bi}$ of spacecraft vehicle b and booster vehicle B, respectively,

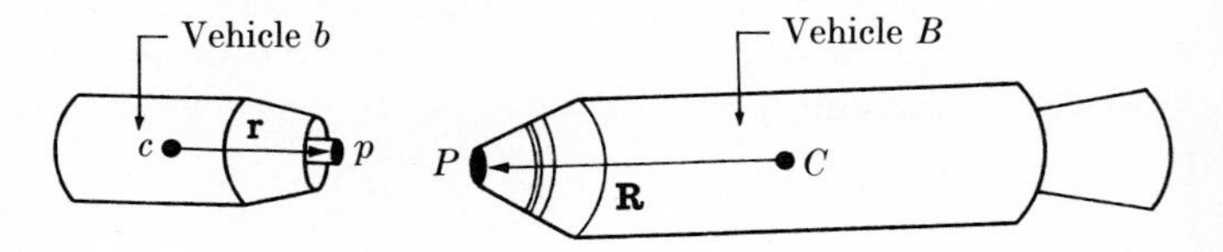

FIGURE 3·15

and the inertial linear velocities $\mathbf{V}^{ci}$ and $\mathbf{V}^{Ci}$ of the mass centers c and C are known from tracking data. In order to accomplish docking, the relative velocities of mating points p and P must be known, so we seek an expression for $\mathbf{V}^{pPi}$.

From Eq. (3·143), we have (changing f to i, etc.)

$$\mathbf{V}^{pi} = \mathbf{V}^{ci} + \boldsymbol{\omega}^{bi} \times \mathbf{r} \tag{3·144}$$

and

$$\mathbf{V}^{Pi} = \mathbf{V}^{Ci} + \boldsymbol{\omega}^{Bi} \times \mathbf{R} \tag{3·145}$$

where $\mathbf{R}$ and $\mathbf{r}$ are position vectors shown in Fig. 3·15. From Eq. (3·17), the relative velocity $\mathbf{V}^{pPi}$ is the difference in absolute velocities,

$$\mathbf{V}^{pPi} = \mathbf{V}^{ci} + \boldsymbol{\omega}^{bi} \times \mathbf{r} - \mathbf{V}^{Ci} - \boldsymbol{\omega}^{Bi} \times \mathbf{R} \tag{3·146}$$

[1] The inertial reference frame defined in Chap. 1 is designated as frame i.

3·3·3 Special acceleration terms

In §3·3·1 certain terms in an expansion of a velocity expression are collectively called the transport velocity. A parallel expansion of an expression for acceleration yields several terms or groups of terms bearing special names.

By definition in Eq. (3·43), the linear acceleration of p in f is

$$\mathbf{A}^{pf} \triangleq \frac{{}^f d}{dt} \mathbf{V}^{pf} \tag{3·147}$$

and the linear velocity of p in f is, by Eq. (3·16),

$$\mathbf{V}^{pf} \triangleq \frac{{}^f d}{dt} \mathbf{P} \tag{3·148}$$

with $\mathbf{P}$ a vector to p from some reference point fixed in f. As noted in the previous article [see Eq. (3·141)], $\mathbf{V}^{pf}$ can be written

$$\mathbf{V}^{pf} = \mathbf{V}^{pf'} + \mathbf{V}^{q'f} + \boldsymbol{\omega}^{f'f} \times \mathbf{P}' \tag{3·149}$$

where $\mathbf{P}'$ is a vector to p from any point q' fixed in an arbitrary reference frame f'. Substituting Eq. (3·149) into (3·147) gives

$$\begin{aligned}
\mathbf{A}^{pf} &= \frac{{}^f d}{dt} \mathbf{V}^{pf'} + \frac{{}^f d}{dt} \mathbf{V}^{q'f} + \left(\frac{{}^f d}{dt} \boldsymbol{\omega}^{f'f}\right) \times \mathbf{P}' + \boldsymbol{\omega}^{f'f} \times \frac{{}^f d}{dt} \mathbf{P}' \\
&= \frac{{}^{f'} d}{dt} \mathbf{V}^{pf'} + \boldsymbol{\omega}^{f'f} \times \mathbf{V}^{pf'} + \mathbf{A}^{q'f} + \boldsymbol{\alpha}^{f'f} \times \mathbf{P}' \\
&\qquad + \boldsymbol{\omega}^{f'f} \times \left(\frac{{}^{f'} d}{dt} \mathbf{P}' + \boldsymbol{\omega}^{f'f} \times \mathbf{P}'\right) \\
&= \mathbf{A}^{pf'} + \boldsymbol{\omega}^{f'f} \times \mathbf{V}^{pf'} + \mathbf{A}^{q'f} + \boldsymbol{\alpha}^{f'f} \times \mathbf{P}' \\
&\qquad + \boldsymbol{\omega}^{f'f} \times \mathbf{V}^{pf'} + \boldsymbol{\omega}^{f'f} \times (\boldsymbol{\omega}^{f'f} \times \mathbf{P}') \\
&= \mathbf{A}^{pf'} + \mathbf{A}^{q'f} + \boldsymbol{\alpha}^{f'f} \times \mathbf{P}' + \boldsymbol{\omega}^{f'f} \times (\boldsymbol{\omega}^{f'f} \times \mathbf{P}') + 2\boldsymbol{\omega}^{f'f} \times \mathbf{V}^{pf'}
\end{aligned} \tag{3·150}$$

Note that if in Eq. (3·150) frame f' is *translating* relative to f at a uniform rate, so that $\boldsymbol{\omega}^{f'f} \equiv 0$ and $\mathbf{A}^{q'f} \equiv 0$, then $\mathbf{A}^{pf} \equiv \mathbf{A}^{pf'}$. Hence, if f is an inertial frame, so that $\mathbf{F} = m\mathbf{A}^{pf}$, then also $\mathbf{F} = m\mathbf{A}^{pf'}$, and f' is also inertial. (It is obviously meaningless to say both are "fixed," so you should avoid the common practice of thinking of an inertial reference frame as somehow absolutely fixed or stationary.) Transformation from f to f' is called *galilean transformation* when $\boldsymbol{\omega}^{f'f} \equiv \mathbf{A}^{q'f} \equiv 0$.

In the calculation of Eq. (3·150) repeated use of the differentiation formula (3·100) and various definitions has separated the acceleration into parts that may permit some physical interpretation. Various descriptive names are used to identify each of the five terms finally appearing in Eq. (3·150), but conventions differ on all but the last two terms. The term $2\boldsymbol{\omega}^{f'f} \times \mathbf{V}^{pf'}$ is consistently called the *Coriolis acceleration*, and the term $\boldsymbol{\omega}^{f'f} \times (\boldsymbol{\omega}^{f'f} \times \mathbf{P}')$ is the *centripetal acceleration*. All terms in Eq. (3·150) that are independent of motion of p relative to frame f' are collectively called the *transport acceleration* because they provide the acceleration in f of a point p^* fixed in f' and coincident with p at any point in time, that is,

$$\mathbf{A}^{p^*f} = \mathbf{A}^{q'f} + (\boldsymbol{\alpha}^{f'f} \times \mathbf{P}') + [\boldsymbol{\omega}^{f'f} \times (\boldsymbol{\omega}^{f'f} \times \mathbf{P}')] \tag{3·151}$$

Note that Eq. (3·151) can be used directly to find the acceleration in f of any point fixed in body f', if the angular velocity and acceleration of f' in f are known, and the acceleration in f of any point q' fixed in f' is also known.

The physical significance of all these terms might best be grasped by considering once again the turntable mounted on a cart on wheels, as illustrated in Fig. 3·2.

The acceleration $\mathbf{A}^{ps}$ will be discussed for various cases. Consider first a situation in which the cart is stationary on the tracks, and the turntable is not rotating relative to the cart. The ball p is rolling about on the table in a given way, and $\mathbf{A}^{ps}$ is sought. We can either select a position vector locating p relative to some point fixed in s (vector $\mathbf{r}$ from o to p is convenient) and differentiate it twice, to obtain first $\mathbf{V}^{ps}$ and then $\mathbf{A}^{ps}$, or we can use Eq. (3·150) with $f \rightarrow s$, $f' \rightarrow d$, and $q' \rightarrow o$. Then, because $\boldsymbol{\omega}^{ds} = 0$ by assumption, the last three terms in Eq. (3·150) are zero, and since $\mathbf{A}^{os} = 0$ (cart stationary), the only term remaining is $\mathbf{A}^{pd}$. To evaluate this, twice differentiate a vector from any point fixed in d to p (vector $\mathbf{r}$ is convenient again). Clearly, for this simple problem, it is foolish to use Eq. (3·150); to get $\mathbf{A}^{ps}$, merely differentiate $\mathbf{r}$ twice.

Now assume a constant rotation ω of the table, holding other constraints as above. This provides two new terms in Eq. (3·150). A centripetal acceleration of magnitude $r\omega^2$ and radially inward direction appears, and a Coriolis acceleration develops of magnitude $2\omega V$ and direction normal to the path of the ball, but parallel to the plane of the table (V is the speed of p on the table).

If the rotation ω becomes variable, the term $\boldsymbol{\alpha}^{f'f} \times \mathbf{P}'$ takes some value, and finally, if the cart b begins to move down the track with changing velocity, the base-point acceleration term $\mathbf{A}^{q's} = \mathbf{A}^{os}$ makes a contribution to $\mathbf{A}^{ps}$.

3·3·4 Examples of the use of kinematic expansions

It is instructive to consider once again some of the mechanisms used previously to calculate linear velocities and accelerations, in order to see how the kinematic expansions of the preceding two sections can be applied.

First example Consider once more the ant a on the rotating cone c illustrated in Fig. 3·4. To obtain $\mathbf{V}^{ab}$ with the help of Eq. (3·141), write

$$\begin{aligned}\mathbf{V}^{ab} &= \mathbf{V}^{ac} + \mathbf{V}^{Ab} + \boldsymbol{\omega}^{cb} \times vt\hat{\mathbf{u}} \\ &= v\hat{\mathbf{u}} + 0 + \Omega\hat{\mathbf{c}}_3 \times vt\hat{\mathbf{u}} \\ &= v\left(\frac{3}{5}\hat{\mathbf{c}}_1 - \frac{4}{5}\hat{\mathbf{c}}_3\right) + \Omega vt\hat{\mathbf{c}}_3 \times \left(\frac{3}{5}\hat{\mathbf{c}}_1 - \frac{4}{5}\hat{\mathbf{c}}_3\right) \\ &= \frac{3}{5}v\hat{\mathbf{c}}_1 + \frac{3}{5}\Omega vt\hat{\mathbf{c}}_2 - \frac{4}{5}v\hat{\mathbf{c}}_3\end{aligned}$$

which checks Eq. (3·128). (Recall that v is the ant's speed along the cone, and Ω is the angular speed of cone c relative to base b.) We may interpret the term

$$\mathbf{V}^{a^*b} \triangleq \mathbf{V}^{Ab} + \boldsymbol{\omega}^{cb} \times vt\hat{\mathbf{u}} = \frac{3}{5}\Omega vt\hat{\mathbf{c}}_2$$

as the transport velocity since it is the value of $\mathbf{V}^{ab}$ when the velocity of a relative to c is eliminated.

To find $\mathbf{A}^{ab}$ with the expansion in Eq. (3·150), write

$$\begin{aligned}\mathbf{A}^{ab} &= \mathbf{A}^{ac} + \mathbf{A}^{Ab} + \boldsymbol{\alpha}^{cb} \times vt\hat{\mathbf{u}} + \boldsymbol{\omega}^{cb} \times (\boldsymbol{\omega}^{cb} \times vt\hat{\mathbf{u}}) + 2\boldsymbol{\omega}^{cb} \times \mathbf{V}^{ac} \\ &= 0 + 0 + 0 + \Omega\hat{\mathbf{c}}_3 \times (\Omega\hat{\mathbf{c}}_3 \times vt\hat{\mathbf{u}}) + 2\Omega\hat{\mathbf{c}}_3 \times v\hat{\mathbf{u}} \\ &= -\frac{3}{5}\Omega^2 vt\hat{\mathbf{c}}_1 + \frac{6}{5}\Omega v\hat{\mathbf{c}}_2\end{aligned}$$

This result agrees with Eq. (3·130), and may have been slightly easier to obtain. The two terms in the final expression may be identified in turn as the centripetal and Coriolis accelerations.

Second example In order to show that the kinematic expansions can be applied to a problem involving the relative motion of more than two reference frames, consider again the mechanism of Fig. 3·5. If the velocity of ring p relative to floor f is required, we may use Eq. (3·141) in more than one way, depending on the frame chosen as f'. If, for example, we choose frame a for f', we have,

$$\begin{aligned}\mathbf{V}^{pf} &= \mathbf{V}^{pa} + \mathbf{V}^{qf} + \boldsymbol{\omega}^{af} \times z\hat{\mathbf{a}}_1 \\ &= \dot{z}\hat{\mathbf{a}}_1 + \frac{d}{dt}(L\cos\varphi)\,\hat{\mathbf{b}}_3 + (\dot{\theta}\hat{\mathbf{b}}_3 - \dot{\varphi}\hat{\mathbf{a}}_2) \times z\hat{\mathbf{a}}_1 \\ &= \dot{z}\hat{\mathbf{a}}_1 - L\dot{\varphi}\sin\varphi\hat{\mathbf{b}}_3 + z\dot{\theta}\sin\varphi\hat{\mathbf{a}}_2 + z\dot{\varphi}\hat{\mathbf{a}}_3\end{aligned} \tag{3·152}$$

which agrees with Eq. (3·135). With this interpretation, the transport velocity is now every term in $\mathbf{V}^{pf}$ except $\dot{z}\hat{\mathbf{a}}_1$.

As an alternative, we could choose f' in Eq. (3·141) as b, and write

$$\begin{aligned}\mathbf{V}^{pf} &= \mathbf{V}^{pb} + \mathbf{V}^{of} + \boldsymbol{\omega}^{bf} \times (L\cos\varphi\hat{\mathbf{b}}_3 + z\hat{\mathbf{a}}_1) \\ &= \mathbf{V}^{pb} + \dot{\theta}\hat{\mathbf{b}}_3 \times z\hat{\mathbf{a}}_1 = \mathbf{V}^{pb} + z\dot{\theta}\sin\varphi\hat{\mathbf{a}}_2 \end{aligned} \tag{3·153}$$

since the term $\mathbf{V}^{of}$ is zero. The term $\mathbf{V}^{pb}$ can be obtained by a second application of Eq. (3·141) as

$$\begin{aligned}\mathbf{V}^{pb} &= \mathbf{V}^{pa} + \mathbf{V}^{qb} + \boldsymbol{\omega}^{ab} \times z\hat{\mathbf{a}}_1 \\ &= \dot{z}\hat{\mathbf{a}}_1 - L\dot{\varphi}\sin\varphi\hat{\mathbf{b}}_3 + z\dot{\varphi}\hat{\mathbf{a}}_3 \end{aligned} \tag{3·154}$$

Equations (3·153) and (3·154) combine to confirm the result of Eq. (3·152), but the second approach proceeds from a different point of view. In Eq. (3·153) only the term $z\dot{\theta}\sin\varphi\hat{\mathbf{a}}_2$ would be called the transport velocity. Thus we see that this phrase has no inherent physical meaning, and its usage depends on the perspective of the analyst.

Next we examine the possibility of applying the acceleration expansion in Eq. (3·150) to this mechanism (Fig. 3·5). Again we find at least two ways of using the expansion, depending on the choice of the frame to be used for f' in Eq. (3·150).

Choosing a for f' provides

$$\mathbf{A}^{pf} = \mathbf{A}^{pa} + \mathbf{A}^{qf} + \boldsymbol{\alpha}^{af} \times z\hat{\mathbf{a}}_1 + \boldsymbol{\omega}^{af} \times (\boldsymbol{\omega}^{af} \times z\hat{\mathbf{a}}_1) + 2\boldsymbol{\omega}^{af} \times \mathbf{V}^{pa} \tag{3·155}$$

The substitutions

$$\begin{aligned}\mathbf{V}^{pa} &= \dot{z}\hat{\mathbf{a}}_1 \\ \mathbf{A}^{pa} &= \ddot{z}\hat{\mathbf{a}}_1 \\ \mathbf{A}^{qf} &= \frac{{}^f d^2}{dt^2}(L\cos\varphi\hat{\mathbf{f}}_3) = (-L\ddot{\varphi}\sin\varphi - L\dot{\varphi}^2\cos\varphi)\hat{\mathbf{f}}_3 \\ \boldsymbol{\omega}^{af} &= \boldsymbol{\omega}^{ab} + \boldsymbol{\omega}^{bf} = \dot{\theta}\hat{\mathbf{b}}_3 - \dot{\varphi}\hat{\mathbf{a}}_2 = \dot{\theta}\hat{\mathbf{f}}_3 - \dot{\varphi}\hat{\mathbf{b}}_2 \\ \boldsymbol{\alpha}^{af} &= \frac{{}^f d}{dt}\boldsymbol{\omega}^{af} = \frac{{}^f d}{dt}(\dot{\theta}\hat{\mathbf{f}}_3 - \dot{\varphi}\hat{\mathbf{b}}_2) = \ddot{\theta}\hat{\mathbf{f}}_3 - \ddot{\varphi}\hat{\mathbf{b}}_2 - \dot{\varphi}\left(\frac{{}^b d}{dt}\hat{\mathbf{b}}_2 + \boldsymbol{\omega}^{bf} \times \hat{\mathbf{b}}_2\right) \\ &= \ddot{\theta}\hat{\mathbf{f}}_3 - \ddot{\varphi}\hat{\mathbf{b}}_2 + \dot{\varphi}\dot{\theta}\hat{\mathbf{b}}_1 \end{aligned}$$

provide

$$\begin{aligned}\mathbf{A}^{pf} &= \ddot{z}\hat{\mathbf{a}}_1 - L(\ddot{\varphi}\sin\varphi + \dot{\varphi}^2\cos\varphi)\hat{\mathbf{b}}_3 + (\ddot{\theta}\hat{\mathbf{b}}_3 - \ddot{\varphi}\hat{\mathbf{b}}_2 + \dot{\varphi}\dot{\theta}\hat{\mathbf{b}}_1) \times z\hat{\mathbf{a}}_1 \\ &\quad + (\dot{\theta}\hat{\mathbf{b}}_3 - \dot{\varphi}\hat{\mathbf{b}}_2) \times [(\dot{\theta}\hat{\mathbf{b}}_3 - \dot{\varphi}\hat{\mathbf{b}}_2) \times z\hat{\mathbf{a}}_1] + 2(\dot{\theta}\hat{\mathbf{b}}_3 - \dot{\varphi}\hat{\mathbf{b}}_2) \times \dot{z}\hat{\mathbf{a}}_1 \\ &= \ddot{z}\hat{\mathbf{a}}_1 - L(\ddot{\varphi}\sin\varphi + \dot{\varphi}^2\cos\varphi)(\hat{\mathbf{a}}_3\sin\varphi - \hat{\mathbf{a}}_1\cos\varphi) + z\ddot{\theta}\sin\varphi\hat{\mathbf{a}}_2 \\ &\quad + z\ddot{\varphi}\hat{\mathbf{a}}_3 + z\dot{\varphi}\dot{\theta}\cos\varphi\hat{\mathbf{a}}_2 + (\dot{\theta}\hat{\mathbf{b}}_3 - \dot{\varphi}\hat{\mathbf{b}}_2) \times (z\dot{\theta}\sin\varphi\hat{\mathbf{b}}_2 + z\dot{\varphi}\hat{\mathbf{a}}_3) \\ &\quad + 2\dot{z}\dot{\theta}\sin\varphi\hat{\mathbf{a}}_2 + 2\dot{z}\dot{\varphi}\hat{\mathbf{a}}_3 \\ &= \hat{\mathbf{a}}_1(\ddot{z} + L\ddot{\varphi}\sin\varphi\cos\varphi + L\dot{\varphi}^2\cos^2\varphi - z\dot{\theta}^2\sin^2\varphi - z\dot{\varphi}^2) \\ &\quad + \hat{\mathbf{a}}_2(z\ddot{\theta}\sin\varphi + 2z\dot{\varphi}\dot{\theta}\cos\varphi + 2\dot{z}\dot{\theta}\sin\varphi) \\ &\quad + \hat{\mathbf{a}}_3(-L\ddot{\varphi}\sin^2\varphi - L\dot{\varphi}^2\sin\varphi\cos\varphi + z\ddot{\varphi} - z\dot{\theta}^2\sin\varphi\cos\varphi + 2\dot{z}\dot{\varphi}) \end{aligned} \tag{3·156}$$

This result corresponds to Eq. (3·136), which was obtained without the use of the acceleration expansion of Eq. (3·150).

If, alternatively, we choose b for f' in Eq. (3·150), we obtain

$$\mathbf{A}^{pf} = \mathbf{A}^{pb} + \mathbf{A}^{of} + \boldsymbol{\alpha}^{bf} \times (L \cos \varphi \hat{\mathbf{b}}_3 + z\hat{\mathbf{a}}_1) + \boldsymbol{\omega}^{bf} \times [\boldsymbol{\omega}^{bf} \times (L \cos \varphi \hat{\mathbf{b}}_3 + z\hat{\mathbf{a}}_1)] + 2\boldsymbol{\omega}^{bf} \times \mathbf{V}^{pb} \qquad (3\cdot157)$$

Although the three middle terms in this expression are easier to evaluate than the corresponding terms in Eq. (3·155), the first and last terms must again be expanded to obtain

$$\mathbf{A}^{pb} = \mathbf{A}^{pa} + \mathbf{A}^{qa} + \boldsymbol{\alpha}^{ab} \times z\hat{\mathbf{a}}_1 + \boldsymbol{\omega}^{ab} \times (\boldsymbol{\omega}^{ab} \times z\hat{\mathbf{a}}_1) + 2\boldsymbol{\omega}^{ab} \times \mathbf{V}^{pa} \qquad (3\cdot158)$$

$$\mathbf{V}^{pb} = \mathbf{V}^{pa} + \mathbf{V}^{qb} + \boldsymbol{\omega}^{ab} \times z\hat{\mathbf{a}}_1 \qquad (3\cdot159)$$

by using Eqs. (3·150) and (3·141). Combination of Eqs. (3·157) to (3·159) should also produce the solution of Eq. (3·156); these approaches differ only in the manner in which the acceleration expansion of Eq. (3·150) has been used. It should be apparent that these two different perspectives yield different quantities for the individual terms of the expansion; thus, for example, the Coriolis acceleration $2\boldsymbol{\omega}^{af} \times \mathbf{V}^{pa}$ in Eq. (3·155) is not the same as the Coriolis acceleration $2\boldsymbol{\omega}^{bf} \times \mathbf{V}^{pb}$ in Eq. (3·157). It follows that the phrases "Coriolis acceleration" and "centripetal acceleration" do not have unambiguous physical significances. If such phrases are to be used, it must be quite clear which expansion is intended. Nonetheless, once this issue is clarified, the concepts represented by each of the five terms in the acceleration expansion and the three terms in the velocity expansion can be extremely useful aids to visualization and technical communication.

3·4 PERSPECTIVE

In practice many dynamicists have developed a strong preference for the use of Eq. (3·150) in obtaining accelerations for complex kinematical problems. Competent analysts acquire a feeling for the "physical significance" of each term, and with this understanding comes an ability to interpret the formula flexibly in application to a wide range of problems. Less competent engineers often rely too heavily on this formula and are as a consequence unable to handle dynamics problems more complex than the simplest "relative motion" problem of the sort illustrated by the turntable of Fig. 3·2. Similar observations may be made concerning the use of transport velocity [Eq. (3·141)].

In order to avoid these pitfalls, and to develop an ability to systematically determine velocities and accelerations for problems of any level of difficulty, students of mechanics should develop the habit of returning to basic definitions for the determination by vector differentiation of linear velocity, linear acceleration, and angular acceleration [see Eqs. (3·16), (3·43), and (3·111) for the critical definitions]. *Angular* velocity cannot be obtained by vector differentiation, but it is an extremely simple concept for planar rotations, and with the chain rule [Eq. (3·102)] all angular motions reduce to a succession of planar rotations. With these four basic relationships, and with some skill in the use of Eq. (3·100) for the necessary vector differentiation, virtually all kinematics problems are reduced to completely straightforward and systematic analysis. Kinematics, traditionally considered to be the most troublesome part of engineering dynamics, need pose no difficulty for the average undergraduate student of engineering.

PROBLEMS

3·1 Record the vector $\mathbf{r} = r\hat{\mathbf{u}}_r$ in Fig. 3·1 in terms of r, φ, λ, $\hat{\imath}$, $\hat{\jmath}$, and $\hat{\mathbf{k}}$.

3·2 Refer to Fig. 3·9, which portrays four cubes of side dimension L, and express the vector from q to p in terms of any unit vectors shown on the figure.

3·3 If, in Fig. P3·3, point p is traveling due north at 50 mi/hr and point q is traveling due east at 75 mi/hr, what is $\mathbf{V}^{pqe}$, where e is the reference frame established by the earth?

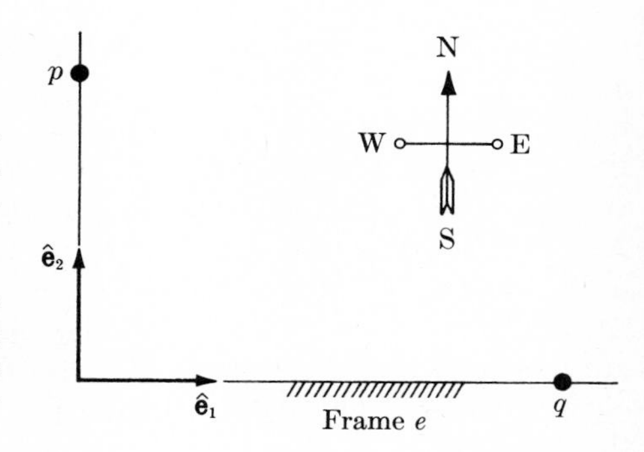

FIGURE P3·3

3·4 The mechanism shown in Fig. P3·4 is designed to advance point p horizontally a distance x when point A is constrained to move toward point B. Determine $\mathbf{V}^{pe}$ in terms of y, $\dot{y}$, and L, where the variable distance $2y$ between A and B is a known function of time. The distance from pin to pin along each arm of the linkage is L.

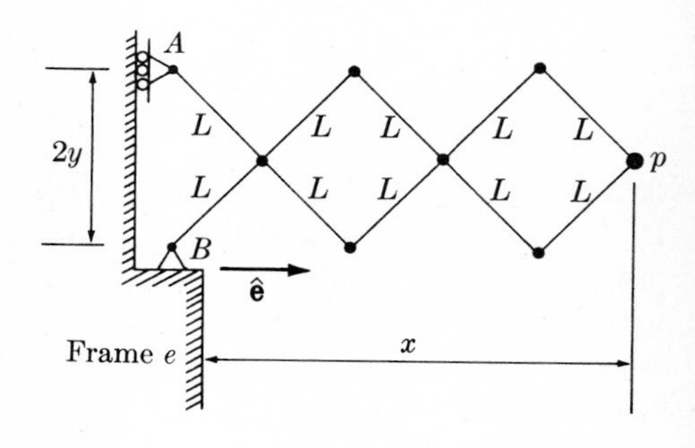

FIGURE P3·4

3·5 Refer to Fig. 3·7, and find $\mathbf{V}^{pa}$ in terms of $\hat{\mathbf{b}}_1$, $\hat{\mathbf{b}}_2$, and $\hat{\mathbf{b}}_3$ by differentiating a position vector expressed in terms of $\hat{\mathbf{a}}_1$, $\hat{\mathbf{a}}_2$, and $\hat{\mathbf{a}}_3$. (Note: Point p is fixed at the corner of block d as shown, but the angles β and γ are time-varying.)

3·6 For a television commercial, an electric grandfather clock is being tested as shown in Fig. P3·6 on a centrifuge spinning relative to the earth e with angular velocity $\Omega\hat{\mathbf{e}}_3$. (The test would fail for a traditional pendulum clock.) A mouse p

is running along the minute hand h with velocity $v\hat{\mathbf{u}}_1$, where $\hat{\mathbf{u}}_1$ is a unit vector fixed relative to h and pointing radially from its axis point q. If the clock functions properly, despite mouse and centrifuge, what is $\mathbf{V}^{pe}$ when the minute hand points to 9 and the mouse is a distance r from q, with $\Omega = 1$ rad/min; $v = \dot{r} = 5$ ft/min; $r = 6$ in.; and the distance $R = 10$ ft? Express your answer in terms of unit vectors $\hat{\mathbf{c}}_1$, $\hat{\mathbf{c}}_2$, and $\hat{\mathbf{c}}_3$ shown in the figure, and obtain $\mathbf{V}^{pe}$ by differentiating an appropriate vector expressed in terms of $\hat{\mathbf{e}}_1$, $\hat{\mathbf{e}}_2$, and $\hat{\mathbf{e}}_3$.

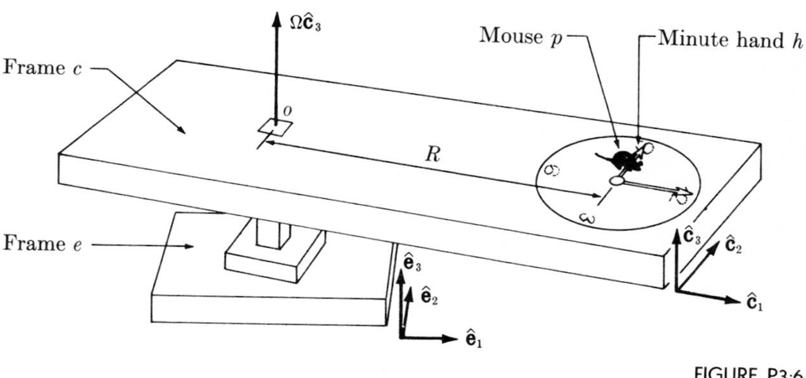

FIGURE P3·6

3·7 Determine $\mathbf{V}^{se}$, the velocity relative to the base e of the point s on the upper jaw of the power shovel shown in Fig. P3·7, by differentiating an appropriate position vector expressed in terms of $\hat{\mathbf{e}}_1$, $\hat{\mathbf{e}}_2$, and $\hat{\mathbf{e}}_3$ fixed in e. Substitute the dimensions 3 ft, 10 ft, 4 ft, and 2 ft, respectively, for the distances from o to p, p to q, q to r, and r to s (see Fig. P3·7), and consider only the special time for which $\varphi = \pi/2$ rad, $\theta = \pi/2$ rad, $\psi = \pi/4$ rad, $\dot{\varphi} = -2\sqrt{2}$ rad/min, $\dot{\theta} = 2\sqrt{2}$ rad/min, $\dot{\psi} = \sqrt{2}$ rad/min, and $\Omega = 1$ rad/min. Express your answer in terms of $\hat{\mathbf{d}}_1$, $\hat{\mathbf{d}}_2$, and $\hat{\mathbf{d}}_3$ fixed in the cab d. (Note that points o, p, q, r, and s are in a plane normal to $\hat{\mathbf{d}}_2$.)

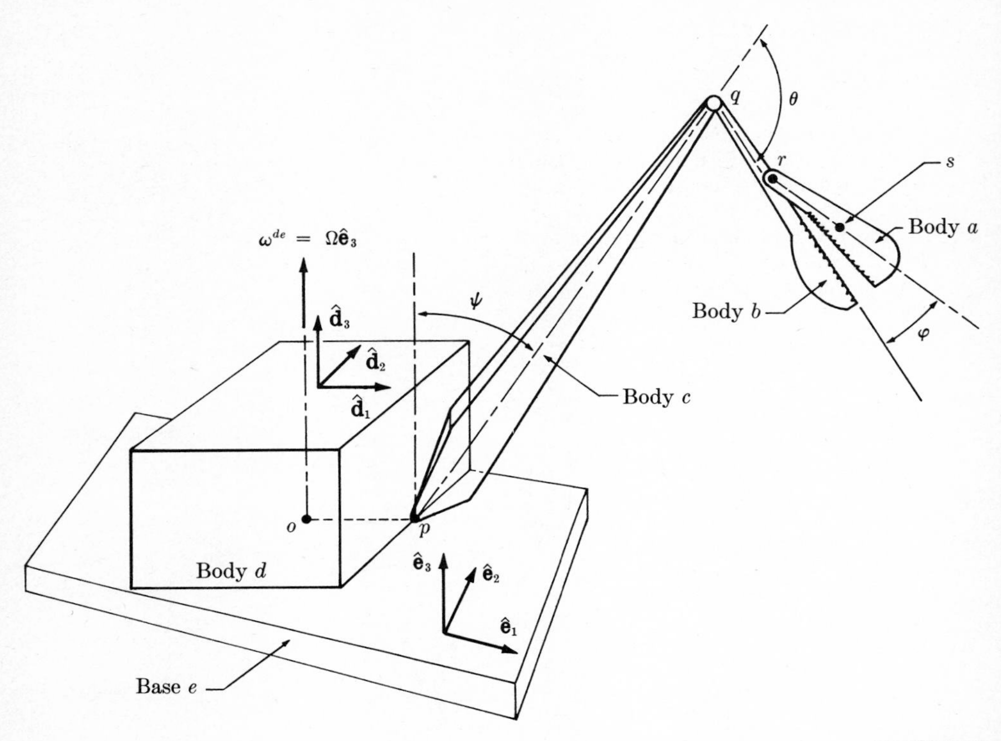

FIGURE P3·7

3·8 A projectile p of mass m is released from its point of attachment on a car c of a Ferris wheel d at time $t = 0$ (see Fig. P3·8a). The attachment point on the car is shown in Fig. P3·8b. Unit vectors $\hat{\mathbf{c}}_1$, $\hat{\mathbf{c}}_2$, and $\hat{\mathbf{c}}_3$ are fixed in c and rotate relative to $\hat{\mathbf{d}}_1$, $\hat{\mathbf{d}}_2$, and $\hat{\mathbf{d}}_3$ fixed in d by an angle θ, with $\hat{\mathbf{d}}_3 \equiv \hat{\mathbf{c}}_3$. Unit vectors $\hat{\mathbf{b}}_1$, $\hat{\mathbf{b}}_2$, and $\hat{\mathbf{b}}_3$ fixed in the base b rotate through an angle φ relative to d, with $\hat{\mathbf{b}}_3 \equiv \hat{\mathbf{d}}_3 \equiv \hat{\mathbf{c}}_3$. Finally, b rotates relative to unit vectors $\hat{\mathbf{e}}_1$, $\hat{\mathbf{e}}_2$, and $\hat{\mathbf{e}}_3$ fixed in the earth e, at the constant rate Ω, with $\hat{\mathbf{e}}_2 \equiv \hat{\mathbf{b}}_2$. Let h be the height of the axle at o above the ground and let R be the distance from o to the axle through the car at q.

If $\theta(t)$, $\varphi(t)$, Ω, R, h, and car dimensions A and B are known, and at $t = 0$ the

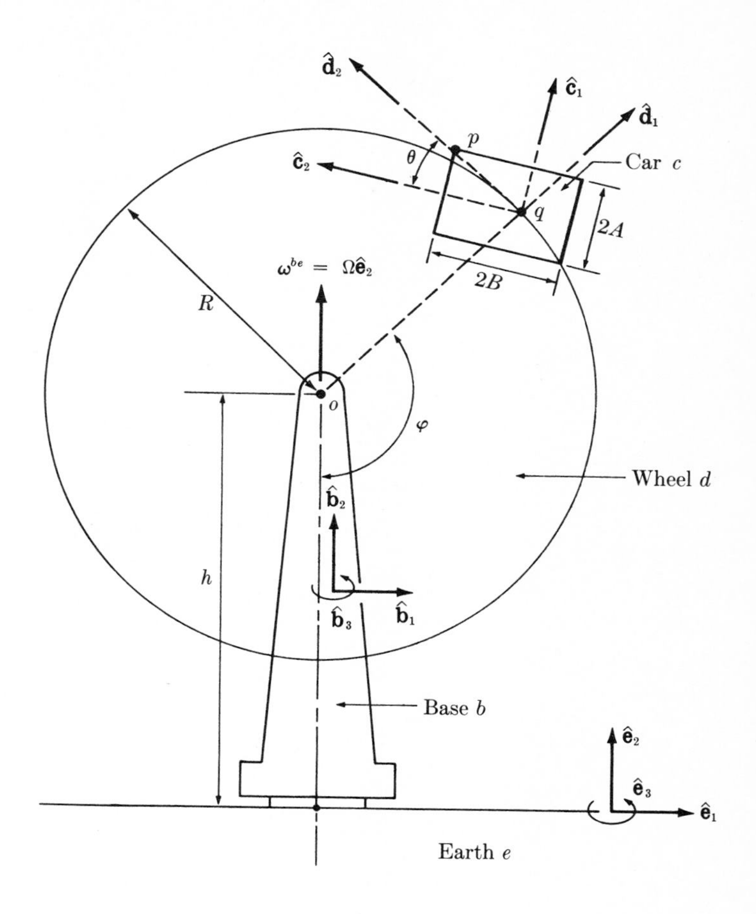

(a)

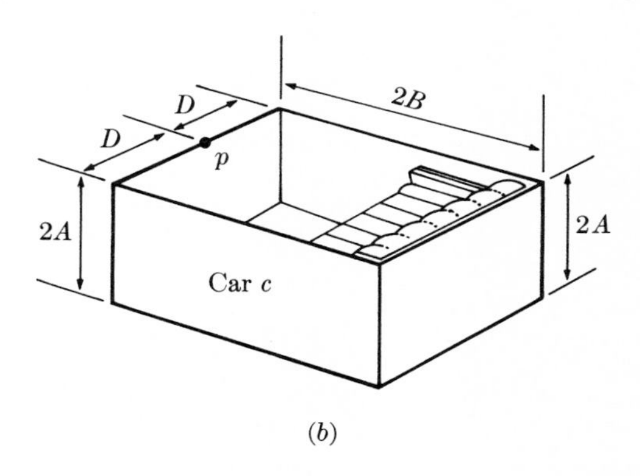

(b)

FIGURE P3·8

variables are $\theta(0) = 0$, $\varphi(0) = \pi/2$ rad, $\dot{\theta} = \dot{\theta}_0$, and $\dot{\varphi} = \dot{\varphi}_0$, what are the initial conditions for the particle trajectory? In other words, what are the inertial position and velocity vectors for p at $t = 0$, assuming e to be an inertial reference frame?

3·9 Find the acceleration $\mathbf{A}^{pe}$ for the mouse in Prob. 3·6 by differentiating $\mathbf{V}^{pe}$ written in terms of $\hat{\mathbf{e}}_1$, $\hat{\mathbf{e}}_2$, and $\hat{\mathbf{e}}_3$. Express your answer in terms of $\hat{\mathbf{c}}_1$, $\hat{\mathbf{c}}_2$, and $\hat{\mathbf{c}}_3$ for the same special case considered in Fig. P3·6, with the additional stipulation that Ω is constant and $\ddot{r} = 10$ ft/min².

3·10 Find the acceleration $\mathbf{A}^{se}$ of point s in Prob. 3·7 by differentiating $\mathbf{V}^{se}$ written in terms of $\hat{\mathbf{e}}_1$, $\hat{\mathbf{e}}_2$, and $\hat{\mathbf{e}}_3$, assuming that Ω is constant. Record your answer in terms of $\hat{\mathbf{d}}_1$, $\hat{\mathbf{d}}_2$, and $\hat{\mathbf{d}}_3$ only for the special time indicated in Prob. 3·7, when in addition to the previous data we have $\ddot{\varphi} = 2$ rad/min², $\ddot{\theta} = 3$ rad/min², and $\ddot{\psi} = 1$ rad/min².

3·11 Refer to Fig. P3·11 and find the direction-cosine matrices $[C^{ab}]$ and $[C^{bd}]$ such that

$$\{\hat{\mathbf{a}}\} = [C^{ab}]\{\hat{\mathbf{b}}\} \qquad \{\hat{\mathbf{b}}\} = [C^{bd}]\{\hat{\mathbf{d}}\}$$

and then multiply the matrices to obtain

$$[C^{ad}] = [C^{ab}][C^{bd}]$$

For the special case $\beta = 45°$ and $\gamma = 30°$, does your result check that obtained in the examples in §3·2·5, where $[C] \triangleq [C^{da}]$?

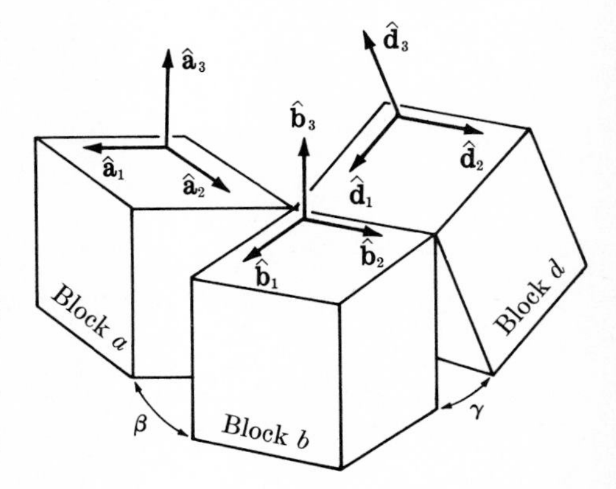

FIGURE P3·11

3·12 Refer to Fig. 3·9 and find the direction-cosine matrix $[C^{bd}]$ such that $\{\hat{\mathbf{b}}\} = [C^{bd}]\{\hat{\mathbf{d}}\}$; express your result in terms of θ and ψ.

3·13 Find the direction-cosine matrix $[C]$ for which $\{\hat{\mathbf{d}}\} = [C]\{\hat{\mathbf{a}}\}$ when the vector triad $\hat{\mathbf{d}}_1$, $\hat{\mathbf{d}}_2$, and $\hat{\mathbf{d}}_3$ is obtained from $\hat{\mathbf{a}}_1$, $\hat{\mathbf{a}}_2$, and $\hat{\mathbf{a}}_3$ by a 1-3-2 sequence of rotations θ_1, θ_3, and θ_2. [See Eqs. (3·74) and (3·77), where other sequences are considered.]

3·14 Find an approximation of $[C]$ as found in Prob. 3·13, retaining only linear terms in θ_1, θ_2, and θ_3. Compare your result to Eq. (3·78).

3·15 Use $[C^{da}] = [C^{ad}]^T$ from Prob. 3·11 and Eq. (3·85) to find the scalars ${}^d\omega_1{}^{da}$, ${}^d\omega_2{}^{da}$, and ${}^d\omega_3{}^{da}$, and record the angular velocity of d with respect to a.

3·16 Refer to Fig. P3·11 and use the chain rule of Eq. (3·102) to find ω^{da}, the angular velocity of d with respect to a. Compare the labors of finding $\boldsymbol{\omega}^{da}$ using Eqs. (3.102) and (3.85).

3·17 Find $\boldsymbol{\omega}^{ad}$ for the system of Fig. P3·17, which portrays a pair of hinged bars a and b mounted on a turntable c that rotates relative to a base d. The hinge line between a and b is perpendicular to that between b and c.

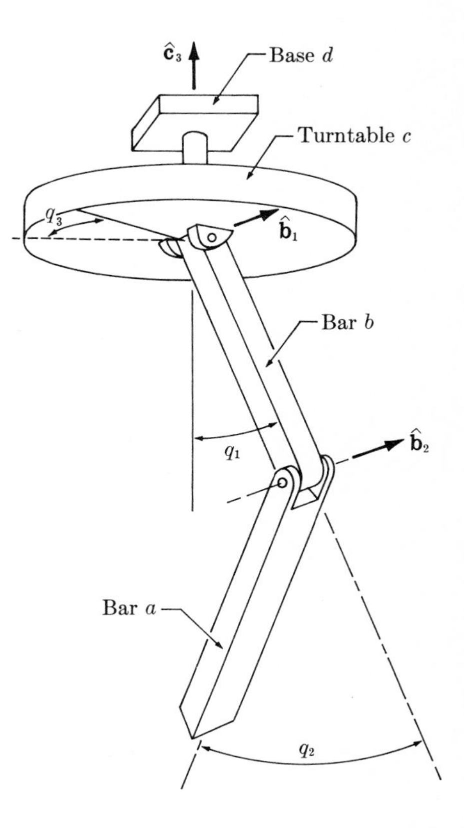

FIGURE P3·17

3·18 Directional control of the antenna a is accomplished by varying the two angles ψ and θ shown (positive) in Fig. P3·18. Structural analysis of the feed structure from q to p requires knowledge of the inertial acceleration of the small mass at p. Find the two intermediate quantities $\boldsymbol{\alpha}^{ab}$ and $\mathbf{V}^{pb}$, where b is the antenna base. Express the results in terms of unit vectors $\hat{\mathbf{s}}_1$, $\hat{\mathbf{s}}_2$, and $\hat{\mathbf{s}}_3$ shown in Fig. P3·18. (See Prob. 3·32 for the culmination of this task.)

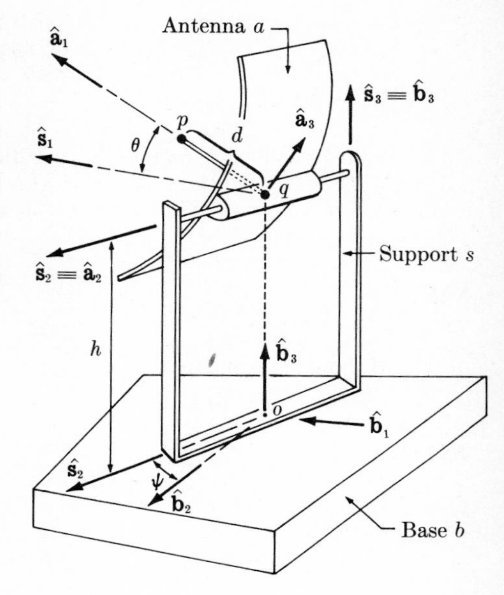

FIGURE P3·18

3·19 A dove p is perched on the end of a gun barrel b of a tank, which is moving forward in a straight line with variable velocity $V\hat{\mathbf{d}}_2$. The barrel-elevation angle β is changing, as is the gun-turret-pointing angle θ (see Fig. P3·19). Let q be the intersection point of the barrel axis (parallel to $\hat{\mathbf{b}}_2$), and the turret axis (parallel to $\hat{\mathbf{c}}_3$), and let L be the distance from q to p. The tank commander is understandably anxious to drive the dove away, but loathe to leave his tank or fire his weapon, so he tries to shake off the bird by accelerating its perch. As quantities intermediate to the calculation of $\mathbf{A}^{pe}$ (e being the frame of the earth), he requests (and you must provide) $\mathbf{V}^{pe}$ and $\boldsymbol{\alpha}^{be}$. (See Prob. 3·33 for the culmination of this calculation.)

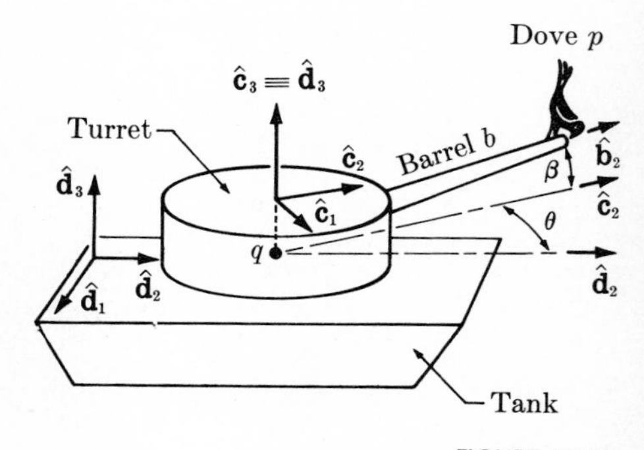

FIGURE P3·19

3·20 Determine the angular acceleration of disk d relative to base frame a for the gyroscope illustrated in Fig. 3·12b. Express your answer in terms of φ, θ, ψ, $\dot{\varphi}$, $\dot{\theta}$, and $\dot{\psi}$ and $\hat{\mathbf{c}}_1$, $\hat{\mathbf{c}}_2$, and $\hat{\mathbf{c}}_3$.

3·21 Determine the special conditions under which the inequality in Eq. (3·112) becomes an equality.

3·22 A wheel w with radius $R = 2$ ft rolls without slip on a belt moving to the right in Fig. P3·22 with a speed of 1 ft/sec. Relative to the earth e, the center q of the wheel moves to the left at 3 ft/sec.

a. Determine the rotation rate ω of the wheel w relative to e.

b. Let p be a point fixed on the rim of the wheel, and find $\mathbf{V}^{pe}$ for the instant illustrated in Fig. P3·22.

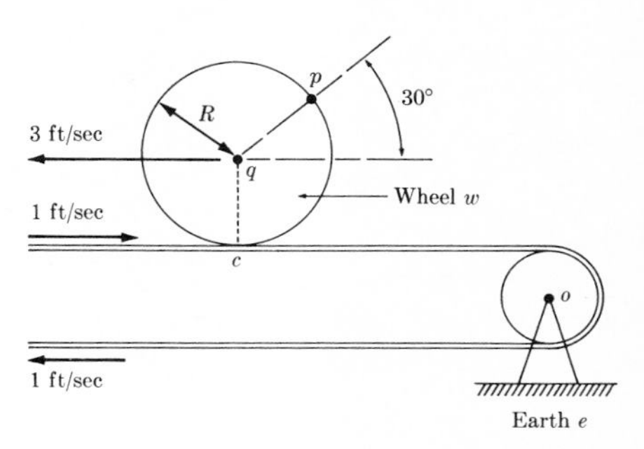

FIGURE P3·22

3·23 Find $\mathbf{V}^{pa}$ as defined in Prob. 3·5, making use of Eq. (3·100).

3·24 Find $\mathbf{V}^{pe}$ as defined in Prob. 3·6, making use of Eq. (3·100).

3·25 Find $\mathbf{V}^{se}$ as defined in Prob. 3·7, making use of Eq. (3·100).

3·26 Find $\mathbf{A}^{pe}$ as defined in Prob. 3·9, making use of Eq. (3·100).

3·27 Find $\mathbf{A}^{se}$ as defined in Prob. 3·10, making use of Eq. (3·100).

3·28 Use the results of Prob. 3·17 to find $\mathbf{V}^{pd}$, where p is a point on the free end of bar a.

3·29 Refer to Fig. 3·9, assume that each block shown is a cube of side dimension L, and find $\mathbf{V}^{pa}$ in terms of φ, θ, ψ, $\dot{\varphi}$, $\dot{\theta}$, $\dot{\psi}$, $\hat{\mathbf{d}}_1$, $\hat{\mathbf{d}}_2$, $\hat{\mathbf{d}}_3$, and L.

3·30 For the system of Fig. P3·30, find $\mathbf{V}^{pe}$ in terms of x, $\dot{x}$, θ, $\dot{\theta}$, φ, $\dot{\varphi}$, constant $\Omega = |\boldsymbol{\omega}^{de}|$, constant R, and $\hat{\mathbf{d}}_1$, $\hat{\mathbf{d}}_2$, and $\hat{\mathbf{d}}_3$.

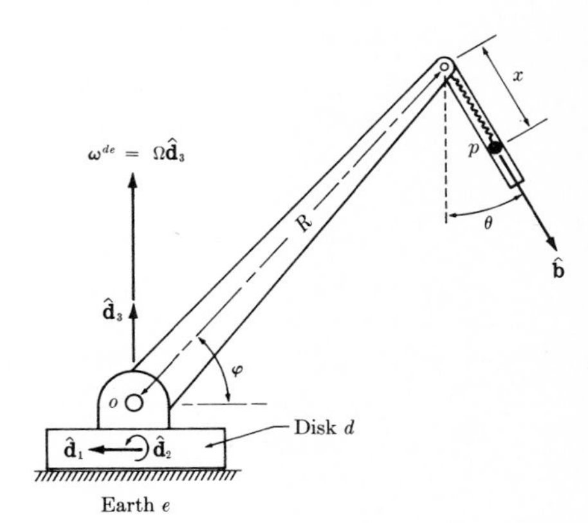

FIGURE P3·30

3·31 The centrifuge frame c in Fig. P3·31 revolves relative to the earth e at the constant rate Ω; the arm a rotates about an axle parallel to $\hat{\mathbf{c}}_2$ fixed in c; the ring p moves along the arm a in the direction of $\hat{\mathbf{a}}_1$, where $\hat{\mathbf{a}}_2 \equiv \hat{\mathbf{c}}_2$. The constant distance from o to q is R, and the variable distant from q to p is r. Determine $\mathbf{A}^{pe}$ in terms of $\hat{\mathbf{c}}_1$, $\hat{\mathbf{c}}_2$, and $\hat{\mathbf{c}}_3$, using Eq. (3·100) repeatedly.

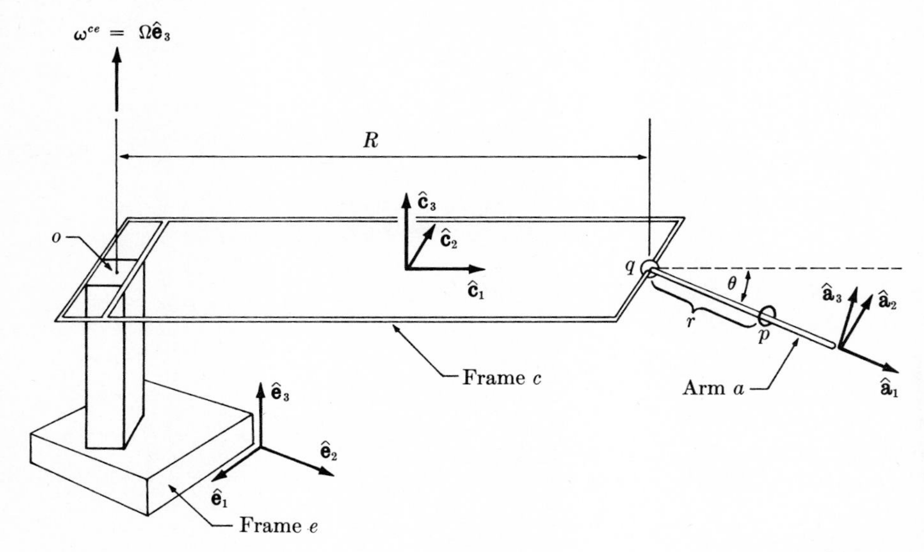

FIGURE P3·31

3·32 Refer to Fig. P3·18, and find $\mathbf{A}^{pb}$ in terms of $\hat{\mathbf{s}}_1$, $\hat{\mathbf{s}}_2$, and $\hat{\mathbf{s}}_3$, using Eq. (3·100) and any information from Prob. 3·18.

3·33 Refer to Fig. P3·19 and find $\mathbf{A}^{pe}$ in terms of $\hat{\mathbf{d}}_1$, $\hat{\mathbf{d}}_2$, and $\hat{\mathbf{d}}_3$, using Eq. (3·100) and any information from Prob. 3·19.

3·34 In application to the mouse p in Prob. 3·6, what would Eq. (3·142) indicate to be the transport velocity $\mathbf{V}^{p*f}$ if $f \triangleq e$ and $f' \triangleq h$? What if $f' \triangleq c$?

3·35 In Fig. P3·11, q' is a point of block d common also to blocks b and a, and p' is that corner of d shown farthest to the right. Find $\mathbf{V}^{p'a}$ by means of Eq. (3·143) and the results of Prob. 3·15 or 3·16.

3·36 Use Eq. (3·150) in combination with the results of Prob. 3·18 to find $\mathbf{A}^{pb}$, with p and b defined by Prob. 3·18.

3·37 Use Eq. (3·150) in combination with the results of Prob. 3·19 to find $\mathbf{A}^{pe}$, as defined in that problem.

3·38 The disk d in Fig. P3·38 rotates relative to the earth e at constant speed Ω, while an arm a attached to a point q a distance R from the center o of the disk rotates through an angle $\theta(t)$ in the vertical plane containing the line from o to q. A particle p oscillates along the axis of the arm a. Note that $\hat{\mathbf{e}}_3 \equiv \hat{\mathbf{d}}_3$ and $\hat{\mathbf{a}}_2 \equiv \hat{\mathbf{d}}_2$. Find $\mathbf{A}^{pe}$ by two means:

a. Use Eq. (3·150).
b. Use basic definitions [Eq. (3·43) and (3·16)], and vector differentiation [with Eq. (3·100)].

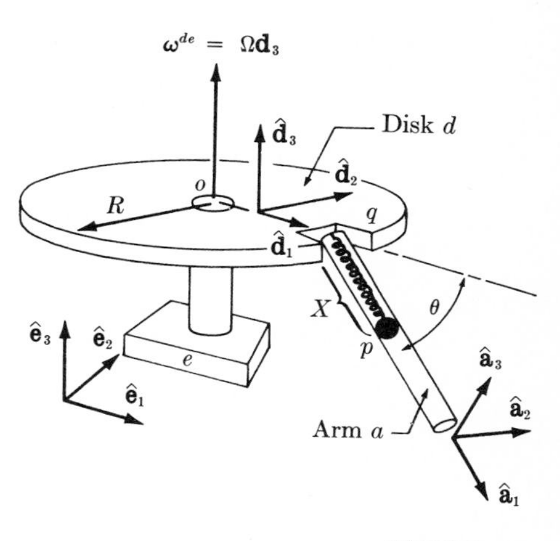

FIGURE P3·38

3·39 A tenacious ant a crawls out on the tail rotor of the helicopter shown in Fig.

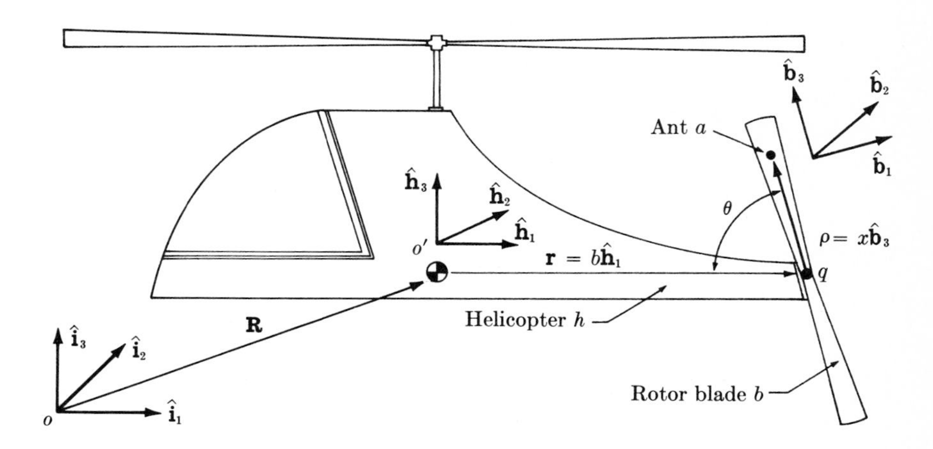

FIGURE P3·39

P3·39, while the rotor spins and the helicopter changes inertial position and orientation. To determine the force **F** applied to the ant, we require the ant's mass m and its inertial acceleration $\mathbf{A}^{ai}$.

a. Find $\mathbf{A}^{ai}$ in terms of the vectors $\mathbf{A}^{o'i}$, $\boldsymbol{\omega}^{hi}$, $\mathbf{r}$, $\hat{\mathbf{b}}_1$, $\hat{\mathbf{b}}_2$, and $\hat{\mathbf{b}}_3$ and the scalars x, $\dot{x}$, $\ddot{x}$, θ, $\dot{\theta}$, and $\ddot{\theta}$.

b. If you had explicit knowledge of $[C]$ in $\{\hat{\mathbf{h}}\} = [C]\{\hat{\mathbf{i}}\}$, how could you use it in writing $\mathbf{A}^{ai}$ from part (a) in matrix terms, for the vector basis $\hat{\mathbf{h}}_1$, $\hat{\mathbf{h}}_2$, and $\hat{\mathbf{h}}_3$?

3·40 A flea f clings to the perimeter of a disk d of radius R, while the disk rotates as shown in Fig. P3·40 with constant angular speed σ relative to its case c, and c keeps a point O fixed to the earth e and maintains a constant angular velocity relative to e given by $\boldsymbol{\omega}^{ce} = \Omega\hat{\mathbf{e}}_3$, so that the half-angle θ of the cone generated by the symmetry axis of c is constant.

A gnat g lands on the surface of c as shown, a distance $R + \epsilon$ from the symmetry axis along a line parallel to $\hat{\mathbf{c}}_1$

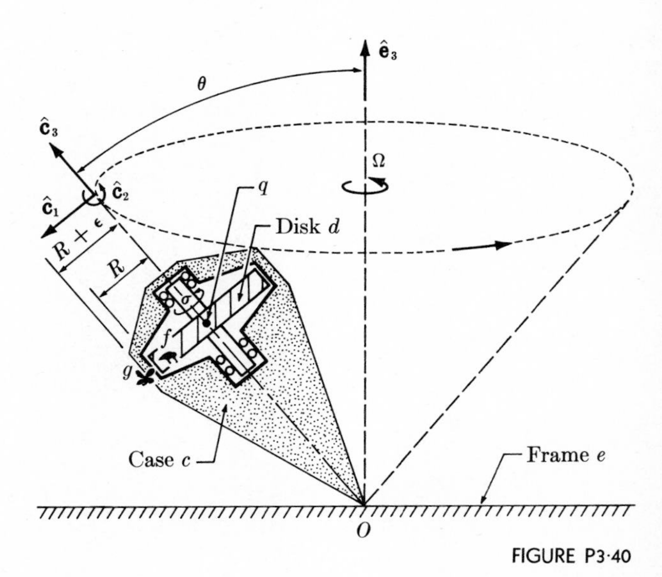

FIGURE P3·40

fixed in c. The flea f periodically passes beneath the gnat g.

a. Find $\mathbf{V}^{f_g e}$, the velocity of f relative to g in frame e, for a time $t = t^*$ when f and g are closest together. Neglect ϵ, and replace $R + \epsilon$ by R.

b. Find $\mathbf{A}^{f_g e}$, the acceleration of f relative to g in frame e at time t^*, neglecting ϵ.

3·41 An ant a marches steadily with constant velocity $v\hat{\mathbf{b}}_1$ relative to the blade of a fan, along a straight line intersecting the axis of the shaft at point p in Fig. P3·41. Let the symbols b, c, d, and e represent, respectively, the reference frames established by the blade, the motor case, the swiveling support, and the fixed base. Then the relative angular velocities are prescribed as

$$\boldsymbol{\omega}^{bc} = \sigma\hat{\mathbf{c}}_3$$
$$\boldsymbol{\omega}^{cd} = k \sin \omega t\, \hat{\mathbf{d}}_2$$
$$\boldsymbol{\omega}^{de} = K \sin \Omega t\, \hat{\mathbf{e}}_3$$

with σ, k, ω, K, and Ω constants, and $\hat{\mathbf{c}}_3$, $\hat{\mathbf{d}}_2$, and $\hat{\mathbf{e}}_3$ unit vectors as shown. Point q is the common intersection point of blade shaft, motor-case hinge axis, and vertical axis of rotation of d with respect to e. The distance from q to p is L.

a. Find $\mathbf{V}^{ae}$, the velocity of the ant relative to the earth.

b. Find $\boldsymbol{\alpha}^{be}$, the angular acceleration of the blade relative to the earth.

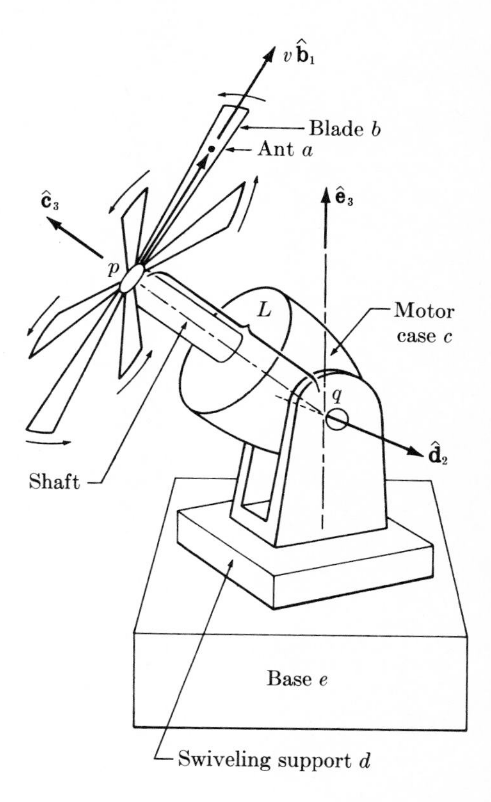

FIGURE P3·41

CHAPTER FOUR PARTICLE DYNAMICS: FORMULATIONS

4·1 APPLICATION OF NEWTON'S SECOND LAW

4·1·1 Introductory remarks

Kinetics is the branch of dynamics that treats the effects of *forces* on the motions of material systems. We are concerned in this chapter with procedures for predicting the motions of a particle by direct application of Newton's second law, $\mathbf{F} = m\mathbf{A}$. Here $\mathbf{A}$ is the acceleration of the particle p relative to some frame of reference i postulated as an inertial frame, so we can write $\mathbf{A}$ in explicit form† as $\mathbf{A}^{pi}$. By definition [see Eqs. (3·43) and (3·16)], $\mathbf{A}^{pi}$ is ${}^i d^2\mathbf{R}/dt^2$, where $\mathbf{R}$ is the position vector to p from some point o fixed in i. Thus Newton's second law provides a second-order vector differential equation. Because any vector in three-

† In this and subsequent chapters, the superscript i indicating an inertial reference frame will often be deleted in order to simplify expressions for velocities and accelerations. When the reference frame is not explicitly identified in the symbol, it is to be understood to be the inertial frame i.

dimensional space may be represented by some ordered sequence of three scalars, depending on the vector basis chosen, this vector differential equation is equivalent to three scalar second-order ordinary differential equations. One or more of these equations may reduce to *zero equals zero*, as for rectilinear and planar motions, but in every case dynamic analysis of particle motion via Newton's second law reduces finally to the formulation of second-order scalar differential equations and their solution or partial solution.

The task of deriving the equations of motion of a particle by application of $\mathbf{F} = m\mathbf{A}$ is straightforward if definitions of force, mass, and inertial acceleration are clearly understood. As noted in Chap. 1, the quantities *force* and *mass* are not lacking in conceptual difficulty, but through repeated usage we have grown familiar with these ideas, and they rarely cause practical difficulties in engineering applications. The inertial acceleration $\mathbf{A}$ is conceptually very simple since it is defined by the straightforward mathematical process of repeated vector differentiation, but this is a major analytical obstacle for many practicing engineers and engineering students. The primary objective of the preceding chapter on kinematics is to systematize the calculation of linear acceleration $\mathbf{A}$ and to reduce this task to a succession of routine vector differentiations. Unless the first two sections of Chap. 3 are well understood, you should not proceed with the discussion of dynamics in the present chapter.

You may recall from Sec. 1·4 that the first thing you must do in approaching any dynamics problem is to decide on a reference frame that can be designated an inertial reference frame for the problem at hand. Because we face so many problems in which the earth as a rigid body establishes an adequate inertial reference frame, we often take this choice for granted, but we should think about it carefully. Because the final decision on the acceptability of assuming a particular reference frame to be inertial must rest on the comparison of prediction and observation, no purely mathematical criterion can be provided. You can, however, as in the following example, consider more than one option as the inertial frame and compare your predictions. If you have a preconception, based on experience, of the acceptability of one of your candidates for the inertial frame, then you can make a quantitative assessment of the adequacy of the second candidate.

4·1·2 Inertial reference frame for lunar orbit

A good approximation of the orbital motion of the earth around the sun can be obtained from the premise that an inertial reference frame is established by the center of the sun, the unit vector $\hat{\mathbf{s}}_3$ normal to the plane of the earth's orbit around the sun, and a unit vector $\hat{\mathbf{s}}_1$ in that orbital plane and parallel to the line of intersection of the earth's orbital plane and its equatorial plane (see Fig. 4·1). We shall designate this sun-centered reference frame as s, and accept on the authority of others its adequacy as an inertial frame for predicting the earth's orbit.

The experience of others also tells us that to obtain a good approximation of the motion of a low-altitude, artificial earth-satellite, it is sufficient to adopt as the inertial frame a reference frame established by the center of the earth, a unit vector $\hat{\mathbf{e}}_3$ along the earth's polar axis, and a unit vector $\hat{\mathbf{e}}_1$ identical to $\hat{\mathbf{s}}_1$. We shall let e designate the earth-centered frame.

Given this information, what can we adopt as an inertial frame in order to obtain a preliminary determination of the moon's orbit? Surely it is acceptable to call s an inertial frame, but what about e?

If the moon can be idealized as a particle p of mass m under force $\mathbf{F}$, then we accept

$$\mathbf{F} = m\mathbf{A}^{ps}$$

by postulate, having decided that s will serve as an inertial frame. If e is also to qualify as an inertial frame of reference, then this equation must give nearly the same result when s is replaced by e; in other words, $\mathbf{A}^{ps} - \mathbf{A}^{pe}$ must be small in

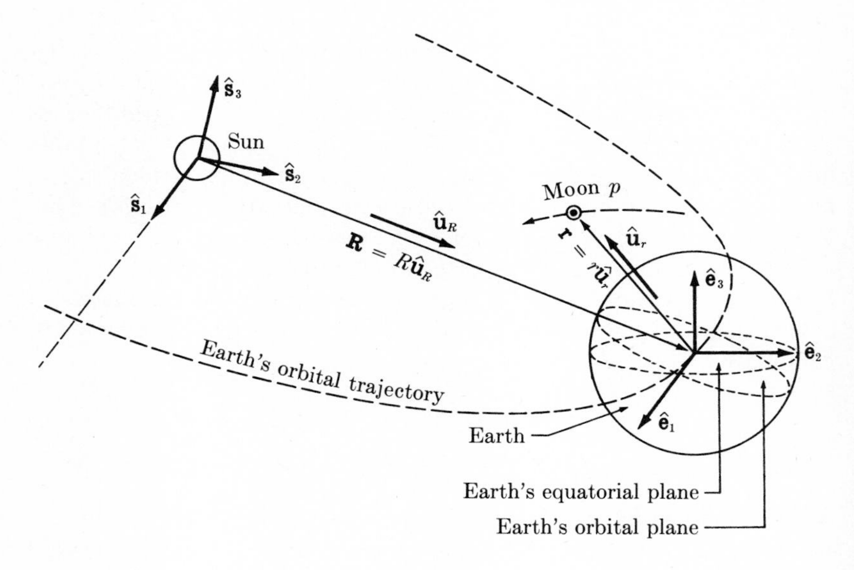

FIGURE 4·1

comparison with the inertial acceleration. From Fig. 4·1 and the kinematics of Chap. 3, we have

$$\mathbf{A}^{ps} \triangleq \frac{{}^s d^2}{dt^2} (R\hat{\mathbf{u}}_R + r\hat{\mathbf{u}}_r)$$

where $R\hat{\mathbf{u}}_R$ is the position vector of the earth with respect to the sun, and $r\hat{\mathbf{u}}_r$ is that of the moon relative to the earth. For comparison, we have

$$\mathbf{A}^{pe} \triangleq \frac{{}^e d^2}{dt^2} (r\hat{\mathbf{u}}_r)$$

But for *any* vector $\mathbf{V}$ we have from Eq. (3·100)

$$\frac{{}^s d}{dt}\mathbf{V} = \frac{{}^e d}{dt}\mathbf{V} + \boldsymbol{\omega}^{es} \times \mathbf{V} = \frac{{}^e d}{dt}\mathbf{V}$$

since $\boldsymbol{\omega}^{es} \equiv 0$. Thus we conclude that

$$\mathbf{A}^{ps} = \frac{{}^s d^2}{dt^2} (R\hat{\mathbf{u}}_R) + \mathbf{A}^{pe}$$

For preliminary calculations we now imagine the earth to be in circular orbit of radius $R = 93 \times 10^6$ mi and orbital period 1 yr, while the moon orbits the earth in a circle of radius $r = 240{,}000$ mi and orbital period 1 mo. For these uniform circular motions, the accelerations are simply

$$\begin{aligned} \mathbf{A}^{pe} &= -r\omega^2\hat{\mathbf{u}}_r \\ \mathbf{A}^{ps} &= -R\Omega^2\hat{\mathbf{u}}_R - r\omega^2\hat{\mathbf{u}}_r \end{aligned}$$

where $\Omega = 2\pi$ rad/1 yr and $\omega = 2\pi$ rad/($\frac{1}{12}$) yr. These approximations provide

$$\begin{aligned} \mathbf{A}^{pe} &= -(0.24 \times 10^6 \times 4\pi^2 \times 144) \text{ mi/yr}^2\, \hat{\mathbf{u}}_r \\ &\cong -1.37 \times 10^9 \text{ mi/yr}^2\, \hat{\mathbf{u}}_r \end{aligned}$$

and

$$\begin{aligned} \mathbf{A}^{ps} &= -(93 \times 10^6 \times 4\pi^2) \text{ mi/yr}^2\, \hat{\mathbf{u}}_R - 1.37 \times 10^9 \text{ mi/yr}^2\, \hat{\mathbf{u}}_r \\ &\cong -3.68 \times 10^9 \text{ mi/yr}^2\, \hat{\mathbf{u}}_R - 1.37 \times 10^9 \text{ mi/yr}^2\, \hat{\mathbf{u}}_r \end{aligned}$$

indicating quite clearly that if s is acceptable as an inertial reference frame in determining the lunar orbit, then the assumption that e is an inertial reference frame will lead to a very poor estimate of the moon's motion.

4·1·3 Inertial "forces"

A common practice in engineering analysis is to write Newton's second law in the form

$$\mathbf{F} - m\mathbf{A} = 0 \tag{4·1}$$

This representation is often referred to in modern engineering textbooks as *d'Alembert's principle,* although in classical terminology this phrase more frequently describes a result we will cite in §6·1·1. Those dynamicists who favor the practice of working with Newton's second law in the form of Eq. (4·1) often refer to the term $-m\mathbf{A}$ as an *inertial "force."*

It is quite obvious that no substantial change in analytical complexity results from the decision to write $\mathbf{F} = m\mathbf{A}$ in the form $\mathbf{F} - m\mathbf{A} = 0$. Any advantages in this approach must therefore be purely subjective. Some engineers find it more comfortable to think about equations of motion as arising from a kind of dynamic equilibrium, for which all "forces" (real and inertial) balance. Others find this practice confusing, and prefer to reserve the concept of force to describe physical interactions with other bodies, due principally to physical contact or gravitational attraction. Confusion is particularly common when the concept of inertial "forces" is combined with the acceleration expansion of Eq. (3·150). This expansion transforms Newton's second law into the form

$$\mathbf{F} = m\mathbf{A}^{pi} = m[\mathbf{A}^{pf} + \mathbf{A}^{qi} + \boldsymbol{\alpha}^{fi} \times \mathbf{r} + \boldsymbol{\omega}^{fi} \times (\boldsymbol{\omega}^{fi} \times \mathbf{r}) + 2\boldsymbol{\omega}^{fi} \times \mathbf{V}^{pf}] \tag{4·2}$$

where f = any reference frame
q = a point fixed in f
$\mathbf{r}$ = a position vector from q to particle at p.

We may, of course, rewrite Eq. (4·2) in the form of (4·1), to obtain

$$\mathbf{F} - m\mathbf{A}^{pf} - m\mathbf{A}^{qi} - m\boldsymbol{\alpha}^{fi} \times \mathbf{r} - m\boldsymbol{\omega}^{fi} \times (\boldsymbol{\omega}^{fi} \times \mathbf{r}) - 2m\boldsymbol{\omega}^{fi} \times \mathbf{V}^{pf} = 0 \tag{4·3}$$

Two of the inertial "force" terms in this expression have widely accepted names; the term $-2m\boldsymbol{\omega}^{fi} \times \mathbf{V}^{pf}$ is often called a *Coriolis "force,"* while the term $-m\boldsymbol{\omega}^{fi} \times (\boldsymbol{\omega}^{fi} \times \mathbf{r})$ is called a *centrifugal "force."* Recall that in Chap. 3 the terms $2\boldsymbol{\omega}^{fi} \times \mathbf{V}^{pf}$ and $\boldsymbol{\omega}^{fi} \times (\boldsymbol{\omega}^{fi} \times \mathbf{r})$ were called *Coriolis accelerations* and *centripetal accelerations,* respectively, even when the inertial frame i is replaced by an arbitrary reference frame. These labels can be convenient in communicating ideas, but they can also cause confusion. You must be careful not to include such terms as centrifugal "force" both in the force term symbolized by $\mathbf{F}$ and in the term $-m\mathbf{A}$ in Eq. (4·1). You must also realize that Eq. (4·3) is valid for any arbitrary reference frame f, and precisely what is meant by the terms centrifugal "force" and Coriolis "force" depends on the choice for f. In the turntable example of Fig. 3·2 and the spinning cone example of Fig. 3·4, little ambiguity in these terms arises since it is quite natural (although not necessary) to select the reference frame f for first the turntable disk d and then the cone c. In the example of Fig. 3·5, however, either the rotating block frame b or the sliding arm frame a may be chosen to replace f in Eq. (4·3). In the former case, the Coriolis "force" is, if the floor is an inertial frame,

$$\begin{aligned} -2m\boldsymbol{\omega}^{bi} \times \mathbf{V}^{pb} &= -2m\dot{\theta}\hat{\mathbf{b}}_3 \times \left[\frac{{}^b d}{dt}(L\cos\varphi\hat{\mathbf{b}}_3 + z\hat{\mathbf{a}}_1)\right] \\ &= -2m\dot{\theta}\hat{\mathbf{b}}_3 \times \left[-L\dot{\varphi}\sin\varphi\hat{\mathbf{b}}_3 + \dot{z}\hat{\mathbf{a}}_1 + z\left(\frac{{}^a d}{dt}\hat{\mathbf{a}}_1 + \boldsymbol{\omega}^{ab} \times \hat{\mathbf{a}}_1\right)\right] \\ &= -2m\dot{\theta}\hat{\mathbf{b}}_3 \times [\dot{z}(\hat{\mathbf{b}}_1 \sin\varphi - \hat{\mathbf{b}}_3\cos\varphi) + z(-\dot{\varphi}\hat{\mathbf{a}}_2 \times \hat{\mathbf{a}}_1)] \\ &= -2m\dot{\theta}\hat{\mathbf{b}}_3 \times [\dot{z}\sin\varphi\hat{\mathbf{b}}_1 + z\dot{\varphi}\,(\hat{\mathbf{b}}_3\sin\varphi + \hat{\mathbf{b}}_1\cos\varphi)] \\ &= -2m\dot{\theta}(\dot{z}\sin\varphi + z\dot{\varphi}\cos\varphi)\hat{\mathbf{b}}_2 \end{aligned} \tag{4·4}$$

In contrast, if a is selected as the frame f in Eq. (4·3), the Coriolis "force" becomes

$$\begin{aligned} -2m\boldsymbol{\omega}^{ai} \times \mathbf{V}^{pa} &= -2m(\boldsymbol{\omega}^{ab} + \boldsymbol{\omega}^{bi}) \times \dot{z}\hat{\mathbf{a}}_1 \\ &= -2m(\dot{\theta}\hat{\mathbf{b}}_3 - \dot{\varphi}\hat{\mathbf{a}}_2) \times \dot{z}\hat{\mathbf{a}}_1 \\ &= -2m(\dot{\theta}\dot{z}\sin\varphi\hat{\mathbf{a}}_2 + \dot{z}\dot{\varphi}\hat{\mathbf{a}}_3) \end{aligned} \tag{4·5}$$

The expressions in Eqs. (4·5) and (4·4) clearly differ, illustrating that there is no single quantity that may be called *the* Coriolis "force" (or *the* centrifugal "force") for a given problem. Either of these expressions might be employed in the process of writing the equations of motion of the ring p in Fig. 3·5.

Although we can certainly write $\mathbf{F} = m\mathbf{A}^{pi}$ directly, obtaining $\mathbf{A}^{pi}$ by a repeated process of vector differentiation as in §3·2·9, it is also possible to begin with Eq. (4·3), with f replaced by either a or b. Then each term in Eq. (4·3) must be determined from the definitions and vector differentiation.

Every dynamicist must be familiar with the concept of inertial "forces," and specifically with centrifugal and Coriolis "forces." But, in my opinion, you will do well to emphasize the mastery of the method of direct application of Newton's law in the form $\mathbf{F} = m\mathbf{A}^{pi}$, at least in your basic course in engineering mechanics. This will accordingly be the emphasis in this book, with only an occasional reference to the inertial "force" interpretation.

4·1·4 Oscillating particle on turntable

Consider the cart of mass m shown in Fig. 4·2 to be moving on a frictionless track mounted on a diametral line of a turntable rotating inertially with constant angular velocity $\Omega\hat{\mathbf{e}}_n$. The cart is idealized as a particle p, and it is connected to each end of the track with a spring of spring constant $k/2$. Both springs are unstretched when the cart is at the center of the turntable, so a displacement $R\hat{\mathbf{e}}_r$ causes a force $-kR\hat{\mathbf{e}}_r$ to be applied to the cart.

Imagine first that the values of R and $\dot{R}$ are known at some point in time, which we shall call $t = 0$, and that subsequent values of R are desired. We will also be interested in determining the forces applied by the track to the cart, and vice versa. These objectives can be met by writing $\mathbf{F} = m\mathbf{A}^{pi}$ and solving the resulting differential equations.

A systematic approach to this problem might first involve the careful calculation of the acceleration

$$\mathbf{A}^{pi} \triangleq \frac{{}^i d^2}{dt^2} \mathbf{R} \tag{4·6}$$

(where $\mathbf{R} = R\hat{\mathbf{e}}_r$ is the vector from O to p in Fig. 4·2), and then the identification of all contact and gravitational forces that combine in $\mathbf{F}$. The first task might be written out in detail as follows:

$$\mathbf{A}^{pi} = \frac{{}^i d^2}{dt^2} R\hat{\mathbf{e}}_r = \frac{{}^i d}{dt}\left(\dot{R}\hat{\mathbf{e}}_r + R\,\frac{{}^i d}{dt}\,\hat{\mathbf{e}}_r\right) \tag{4·7}$$

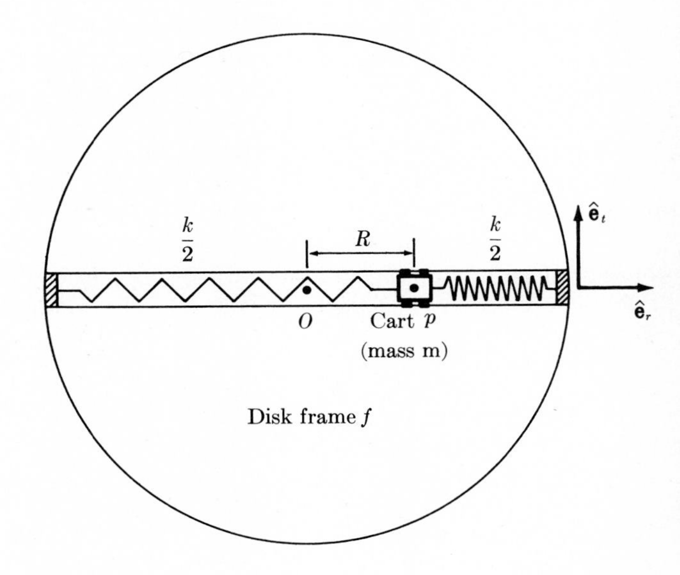

FIGURE 4·2

But from Eq. (3·100) we have

$$\frac{{}^i d}{dt}\,\hat{\mathbf{e}}_r = \frac{{}^f d}{dt}\,\hat{\mathbf{e}}_r + \boldsymbol{\omega}^{fi} \times \hat{\mathbf{e}}_r = \boldsymbol{\omega}^{fi} \times \hat{\mathbf{e}}_r = \Omega\hat{\mathbf{e}}_n \times \hat{\mathbf{e}}_r = \Omega\hat{\mathbf{e}}_t \tag{4·8}$$

so $\mathbf{A}^{pi}$ becomes

$$\mathbf{A}^{pi} = \frac{{}^i d}{dt}\,(\dot{R}\hat{\mathbf{e}}_r + R\Omega\hat{\mathbf{e}}_t) = \ddot{R}\hat{\mathbf{e}}_r + \dot{R}\,\frac{{}^i d}{dt}\,\hat{\mathbf{e}}_r + \dot{R}\Omega\hat{\mathbf{e}}_t + R\Omega\,\frac{{}^i d}{dt}\,\hat{\mathbf{e}}_t \tag{4·9}$$

Into Eq. (4·9) we substitute Eq. (4·8) and its counterpart

$$\frac{{}^i d}{dt}\,\hat{\mathbf{e}}_t = \frac{{}^f d}{dt}\,\hat{\mathbf{e}}_t + \boldsymbol{\omega}^{fi} \times \hat{\mathbf{e}}_t = \boldsymbol{\omega}^{fi} \times \hat{\mathbf{e}}_t = \Omega\hat{\mathbf{e}}_n \times \hat{\mathbf{e}}_t = -\Omega\hat{\mathbf{e}}_r \tag{4·10}$$

to obtain

$$\begin{aligned}\mathbf{A}^{pi} &= \ddot{R}\hat{\mathbf{e}}_r + \dot{R}\Omega\hat{\mathbf{e}}_t + \dot{R}\Omega\hat{\mathbf{e}}_t + R\Omega(-\Omega\hat{\mathbf{e}}_r) \\ &= \ddot{R}\hat{\mathbf{e}}_r + 2\dot{R}\Omega\hat{\mathbf{e}}_t - R\Omega^2\hat{\mathbf{e}}_r\end{aligned} \tag{4·11}$$

Newton's second law next requires the force $\mathbf{F}$, which in this problem is the combination of forces applied to the particle (cart) p by the spring, the tracks, and gravitational attraction of the earth. The spring force has already been described as $-kR\hat{\mathbf{e}}_r$, and the gravitational attraction is $-mg\hat{\mathbf{e}}_n$. The track force on the cart is unknown, but since the track is frictionless this force can have components only in the directions of $\hat{\mathbf{e}}_t$ and $\hat{\mathbf{e}}_n$. Call the track force $F_t\hat{\mathbf{e}}_t + F_n\hat{\mathbf{e}}_n$. Now we can record Newton's second law in the form

$$-kR\hat{\mathbf{e}}_r - mg\hat{\mathbf{e}}_n + F_t\hat{\mathbf{e}}_t + F_n\hat{\mathbf{e}}_n = m(\ddot{R} - R\Omega^2)\hat{\mathbf{e}}_r + 2m\dot{R}\Omega\hat{\mathbf{e}}_t \tag{4·12}$$

This vector second-order differential equation of motion provides three scalar equations when components of the orthogonal unit vectors $\hat{\mathbf{e}}_r$, $\hat{\mathbf{e}}_t$, and $\hat{\mathbf{e}}_n$ are considered. The result is

$$m\ddot{R} + (k - m\Omega^2)R = 0 \tag{4·13}$$

$$F_t = 2m\dot{R}\Omega \tag{4·14}$$

$$-mg + F_n = 0 \tag{4·15}$$

Recall that the original objective is to solve for the unknown R, given certain initial conditions. This requires the solution of the differential equation (4·13). We also want to know the force applied to the cart by the track. The constant component $F_n = mg$ is no surprise since the track must clearly support the weight of the cart. But the "sideways" force $F_t = 2m\Omega\dot{R}$ is a consequence of the motion that may have been unanticipated.

Our problem is of course not solved until we confront the differential equation (4·13), but we'll postpone this important subject until Chap. 5. For the present we'll concentrate on the business of *deriving* equations of motion for particles. We may not always be able to solve these equations of motion analytically; indeed, in any practical situation the digital computer usually handles the differential equation numerically. But even in the framework of an elementary course, we can develop full confidence in our ability to obtain equations of motion for any well-defined particle dynamics problem, given sufficient time.

It may be instructive to return to our example and view it from another side, obtaining the acceleration from the expansion of Eq. (3·150), namely,

$$\mathbf{A}^{pi} = \mathbf{A}^{oi} + \mathbf{A}^{pf} + \boldsymbol{\alpha}^{fi} \times (R\hat{\mathbf{e}}_r) + 2\boldsymbol{\omega}^{fi} \times \mathbf{V}^{pf} + \boldsymbol{\omega}^{fi} \times (\boldsymbol{\omega}^{fi} \times R\hat{\mathbf{e}}_r) \qquad (4\cdot16)$$

In the application of Fig. 4·2, $\mathbf{A}^{oi} = 0$, $\boldsymbol{\omega}^{fi} = \Omega\hat{\mathbf{e}}_n$, and

$$\boldsymbol{\alpha}^{fi} \triangleq \frac{{}^i d}{dt}\boldsymbol{\omega}^{fi} = \frac{{}^i d}{dt}\Omega\hat{\mathbf{e}}_n = 0$$

$$\mathbf{V}^{pf} \triangleq \frac{{}^f d}{dt}R\hat{\mathbf{e}}_r = \dot{R}\hat{\mathbf{e}}_r$$

$$\mathbf{A}^{pf} \triangleq \frac{{}^f d}{dt}\mathbf{V}^{pf} = \frac{{}^f d}{dt}\dot{R}\hat{\mathbf{e}}_r = \ddot{R}\hat{\mathbf{e}}_r$$

so Eq. (4·16) becomes

$$\begin{aligned}\mathbf{A}^{pi} &= \ddot{R}\hat{\mathbf{e}}_r + 2\Omega\hat{\mathbf{e}}_n \times \dot{R}\hat{\mathbf{e}}_r + \Omega\hat{\mathbf{e}}_n \times (\Omega\hat{\mathbf{e}}_n \times R\hat{\mathbf{e}}_r)\\ &= (\ddot{R} - R\Omega^2)\hat{\mathbf{e}}_r + 2\Omega\dot{R}\hat{\mathbf{e}}_t \end{aligned} \qquad (4\cdot17)$$

This is the same result as found in Eq. (4·11), but some may find it easier or faster to obtain. The necessity of looking up Eq. (3·150) or remembering it is a drawback to this approach, although with experience the terms in this equation acquire a physical significance that makes them easily remembered and readily checked against expectations based on similar problems. Just a little experience with related problems teaches us, for example, to expect that the Coriolis acceleration $2\boldsymbol{\omega}^{fi} \times \mathbf{V}^{pf}$ will require a "sideways" force such as that applied to the cart by the track. This may not be anticipated if you're obtaining $\mathbf{A}^{pi}$ directly by repeated vector differentiation. A useful practice is to return always to basic definitions and vector differentiations to find $\mathbf{A}^{pi}$ initially, but then to employ the expansion of Eq. (3·150) to check your answer, if only mentally.

What advantage can we find for this problem in the method of inertial "forces" based on the relationship $\mathbf{F} - m\mathbf{A}^{pi} = 0$? If we combine this approach with the kinematic expansion, as in Eq. (4·3), we obtain (ignoring terms equaling zero)

$$-kR\hat{\mathbf{e}}_r - mg\hat{\mathbf{e}}_n + F_t\hat{\mathbf{e}}_t + F_n\hat{\mathbf{e}}_n - m\ddot{R}\hat{\mathbf{e}}_r + m\Omega^2R\hat{\mathbf{e}}_r - 2m\Omega\dot{R}\hat{\mathbf{e}}_t = 0 \qquad (4\cdot18)$$

We may write this result as

$$-kR\hat{\mathbf{e}}_r - m\ddot{R}\hat{\mathbf{e}}_r + m\Omega^2 R\hat{\mathbf{e}}_r = 0 \tag{4·19}$$
$$F_t\hat{\mathbf{e}}_t - 2m\Omega\dot{R}\hat{\mathbf{e}}_t = 0 \tag{4·20}$$
$$-mg\hat{\mathbf{e}}_n + F_n\hat{\mathbf{e}}_n = 0 \tag{4·21}$$

and describe each of these terms verbally. Equation (4·19) says that the sum of the spring force, the relative-acceleration inertial "force," and the centrifugal "force" is zero. Note that the last of these is directed *away from* the turntable center o (*centri-fugal* suggests this if you associate the *fugal* root with *fugitive: one who flees*). The spring force, on the other hand, is directed *toward* the center in this problem. The term $-m\ddot{R}\hat{\mathbf{e}}_r$ is of either sign, but if this term is zero the centrifugal "force" must be counterbalanced by some inwardly directed force such as that supplied by the spring. Imagine for a moment that the spring is not present, so that the cart moves to the perimeter of the disk and is supported by the little wall or "stop" shown in Fig. 4·2. Dynamicists who favor the inertial "force" approach like to say the wall applies a force to the cart that balances the centrifugal "force." Others prefer to keep the contact and gravity forces in **F** separated by an equals sign from the mass-acceleration product, and they would say that the wall force on the cart is directed radially inward because it is providing the centripetal-acceleration term $m\boldsymbol{\omega}^{fi} \times (\boldsymbol{\omega}^{fi} \times R\hat{\mathbf{e}}_r)$ (which is opposite in sign to the centrifugal "force"). But what if the wall suddenly breaks, and there is no spring or other contact force to push in the radial direction? Equation (4.19) then becomes

$$-m\ddot{R}\hat{\mathbf{e}}_r + m\Omega^2 R\hat{\mathbf{e}}_r = 0$$

so we see that the term $-m\ddot{R}\hat{\mathbf{e}}_r$ takes up the slack to preserve the equality. This requires $\ddot{R} = \Omega^2 R$ so, of course, the cart accelerates away from the turntable center.

Now consider Eq. (4·20), which is equivalent verbally to the statement that the track force on the cart plus the Coriolis "force" is zero. Don't get these two confused. The force $F_t\hat{\mathbf{e}}_t$ applied to the cart by the track equals the product of mass m and the Coriolis acceleration. This contact force is the *opposite* of the Coriolis "force."

What would happen if the track failed to generate the force $F_t\hat{\mathbf{e}}_t = 2m\Omega\dot{R}\hat{\mathbf{e}}_t$, and the cart jumped the track? If this occurs when the cart is moving radially outward with $\dot{R} > 0$, which way will the cart leave the tracks, to the right or left of its forward path?

4·2 IMPULSE AND MOMENTUM

4·2·1 Linear momentum and impulse

In a later chapter we will examine methods of solving the second-order scalar ordinary differential equations that arise from the application of Newton's second law to specific problems. We might inquire first, however, about the possibility of confronting the vector second-order differential equation $\mathbf{F} = m\, {}^i d^2\mathbf{R}/dt^2$ directly, looking for ways to perform general integrations. Since time t is the independent variable, our first thought is to seek a time integral. Consider then the definite integral over an interval from t_1 to t_2

$$\int_{t_1}^{t_2} \mathbf{F}\, dt \Big|_i = \int_{t_1}^{t_2} \left(m \frac{{}^i d^2}{dt^2} \mathbf{R} \right) dt \Big|_i = m \int_{t_1}^{t_2} \frac{{}^i d}{dt} \left(\frac{{}^i d\mathbf{R}}{dt} \right) dt \Big|_i$$
$$= m \left[\frac{{}^i d\mathbf{R}}{dt}(t_2) - \frac{{}^i d\mathbf{R}}{dt}(t_1) \right] \tag{4·22}$$

where the reference frame for the integration is the inertial frame i. Just as we recognized the importance of designating a reference frame in defining any vector differentiation, so also we must identify the frame for integration, and this is the significance of the notation $\Big|_i$ following the differential dt in Eq. (4·22). The importance of this distinction is later illustrated by example.

The quantity

$$\mathbf{I} \triangleq \int_{t_1}^{t_2} \mathbf{F}\, dt \Big|_i \tag{4·23}$$

is called the (inertial) *impulse* exerted by force $\mathbf{F}$ in time interval $t_2 - t_1$, and the quantity

$$m\mathbf{V}^{pi} \triangleq m \frac{{}^i d}{dt} \mathbf{R} \triangleq m\dot{\mathbf{R}} \tag{4·24}$$

is called the (inertial) *linear momentum*. In the limiting case for which $t_2 - t_1$ approaches zero, $\mathbf{I}$ is called an *ideal impulse*. Thus Eq. (4·22) says that the (inertial) *impulse* in a time interval $t_2 - t_1$ equals the change in (inertial) *linear momentum* during that time interval, that is,

$$\mathbf{I} = m\mathbf{V}^{pi}(t_2) - m\mathbf{V}^{pi}(t_1) \tag{4·25}$$

This relationship is called the *impulse-momentum equation*, or often the impulse-momentum "principle." Since it is derived as a first time integral of Newton's second law, however, it is not a "principle" in any fundamental sense, but merely a consequence of $\mathbf{F} = m\mathbf{A}$. It would, of course, be possible to define linear momentum more generally than in Eq. (4·24), to permit an arbitrary frame f instead of the inertial frame i. Similarly, a more general concept of impulse could be defined, permitting integration in any reference frame f. But Newton's second law applies only for acceleration relative to an inertial frame i, so the impulse-momentum equation (4·25) is restricted to *inertial* impulse and momentum. For this reason, the more general concepts find little utility, and they are ignored here. Henceforth the restrictive modifier *inertial* is omitted but implied in the terms *impulse* and *linear momentum.*

4·2·2 Application to spacecraft rendezvous

A spacecraft s/c of mass $m = 20$ lb sec²/ft coasting through free space with velocity $\mathbf{V}_0 = 25{,}000$ ft/sec $\hat{\mathbf{i}}_1$, where $\hat{\mathbf{i}}_1$ is an inertially fixed unit vector, requires a trajectory correction if it is to accomplish a scheduled rendezvous 24 hr hence. On its uncorrected course it will, at the scheduled rendezvous time $t^* = 24$ hr, be a distance $\epsilon = 864{,}000$ ft from the planned rendezvous location at P^*, and this will be the time of closest approach (see Fig. 4·3). The spacecraft carries a small trajectory-correction motor with constant thrust magnitude $T = 10$ lb and controlled thrust direction. We must decide on an appropriate time interval Δt for the motor burn. Since the trajectory adjustment consists entirely of a correction in the $\hat{\mathbf{i}}_2$ direction, it is clear that the midcourse motor should fire in this direction. (This would not be the case if the time of uncorrected closest approach were not t^*.) If you simply apply $\mathbf{F} = m\mathbf{A}$ to the spacecraft and integrate $\mathbf{A}$ twice with respect to time, you find that the acceleration $\mathbf{A}$ of 0.5 (ft/sec²)$\hat{\mathbf{i}}_2$ during the time interval Δt produces a velocity increment given by

$$\Delta v\,\hat{\mathbf{i}}_2 = \int_{t=0}^{t=\Delta t} \mathbf{A}\,dt = 0.5t \text{ ft/sec}^2\,\hat{\mathbf{i}}_2 \Big|_0^{\Delta t} = 0.5\,\Delta t \text{ ft/sec}^2\,\hat{\mathbf{i}}_2$$

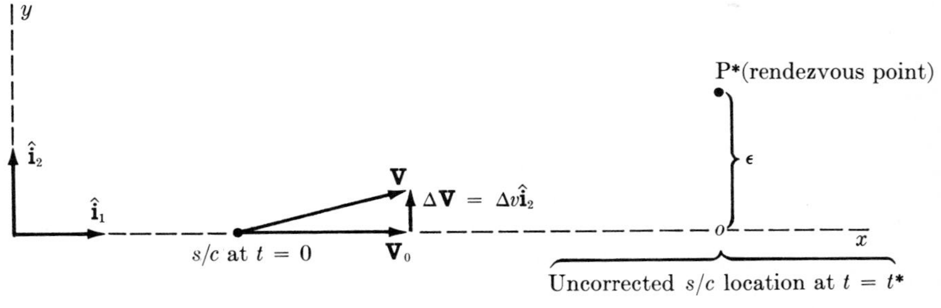

FIGURE 4·3

and a change in position with respect to the uncorrected trajectory given by

$$y\ (\Delta t)\ \hat{\mathbf{i}}_2 = \int_{t=0}^{t=\Delta t} 0.5t\ dt\ \text{ft/sec}^2\ \hat{\mathbf{i}}_2 = 0.25t^2 \Big|_0^{\Delta t} \text{ft/sec}^2\ \hat{\mathbf{i}}_2$$
$$= 0.25\ (\Delta t)^2\ \text{ft/sec}^2\ \hat{\mathbf{i}}_2.$$

Subsequent to the motor firing, $\mathbf{A} = 0$, so the velocity remains constant, and the additional translation in the y direction is directly proportional to Δv; thus, at $t = t^*$, y is given by

$$\begin{aligned} y(t^*) &= y\ (\Delta t) + (\Delta v)\ (t^* - \Delta t) \\ &= [0.25\ (\Delta t)^2 + (0.5\ \Delta t)\ (t^* - \Delta t)] \qquad \text{ft/sec}^2 \\ &= [0.5\ (\Delta t)\ t^* - 0.25\ (\Delta t)^2] \qquad \text{ft/sec}^2 \end{aligned}$$

Since we require $y(t^*) = \epsilon = 864{,}000$ ft, we could substitute ϵ and $t^* = 24$ hr $=$ 86,400 sec into $y(t^*)$ to obtain

$$864{,}000 - 0.5(86{,}400)\ \Delta t + 0.25\ (\Delta t)^2 = 0$$

and then solve this quadratic equation for Δt.

The solution is

$$\begin{aligned} \Delta t &= 2\{43{,}200 \pm [(43{,}200)^2 - 864{,}000]^{1/2}\} \\ &\cong 2[43{,}200 \pm (43{,}190 - 1.15 \times 10^{-3})] \end{aligned}$$

Choosing the minus sign, we have

$$\Delta t \cong 19.9977\ \text{sec}$$

The quantity Δt is so much smaller than t^* that the term $0.25\ (\Delta t)^2$ in the quadratic equation has little importance; if we ignore this term we find

$$\Delta t = \frac{864{,}000}{0.5(86{,}400)} = 20\ \text{sec}$$

which is an excellent approximation (much better than some others we are making in this problem). If we explore the physical implications of this assumption, we see that we are in effect assuming that the motor applies to the spacecraft an ideal impulse, i.e., a force of infinite magnitude and infinitesimal duration giving rise to an impulse

$$\mathbf{I} = (10\ \text{lb})(20\ \text{sec})\ \hat{\mathbf{i}}_2 = 200\ \text{lb sec}\ \hat{\mathbf{i}}_2$$

With the motor applying an ideal impulse, the velocity change $\Delta v\ \hat{\mathbf{i}}_2 = \mathbf{I}/m$ occurs instantaneously, and $y(t^*)$ is simply $\Delta v\ t^*$, matching the result previously obtained by suppressing the term $0.25\ (\Delta t)^2$ in the quadratic equation for Δt.

Having now gained some experience in the application of $\mathbf{F} = m\mathbf{A}$ to this problem, we can better appreciate the advantages of instead applying the impulse-momentum equation, assuming an ideal impulse. Equation (4·25) would then provide

$$\mathbf{I} = m\,\Delta v\,\hat{\mathbf{i}}_2$$

for our problem, and Fig. 4·3 would furnish

$$\epsilon = \Delta v\,t^*$$

and hence

$$\begin{aligned}\mathbf{I} &= \frac{m\epsilon}{t^*}\hat{\mathbf{i}}_2 = \frac{(20\text{ lb sec}^2/\text{ft})(864{,}000\text{ ft})\,\hat{\mathbf{i}}_2}{86{,}400\text{ sec}}\\ &= 200\text{ lb sec}\,\hat{\mathbf{i}}_2\end{aligned}$$

Only after this result is obtained does it become necessary to abandon the ideal impulse idealization and recognize that the 10 lb motor thrust limitation requires

$$(10\text{ lb})(\Delta t\text{ sec})\,\hat{\mathbf{i}}_2 = \mathbf{I} = 200\text{ lb sec}\,\hat{\mathbf{i}}_2$$

and hence

$$\Delta t = 20\text{ sec}$$

In this simple problem, the impulse-momentum equation has provided a quick and easy way to impose a physically meaningful approximation of a result available alternatively from $\mathbf{F} = m\mathbf{A}$. In more complex problems, an exact solution of $\mathbf{F} = m\mathbf{A}$ in closed form may be impossible.

4·2·3 Linear momentum conservation

In many applications it is misleading to consider Eq. (4·25) as a partial solution to $\mathbf{F} = m\mathbf{A}$; although formally we have found a first integral, this is no help unless the integral $\mathbf{I}$ in Eq. (4·23) can be obtained explicitly. This is generally impossible unless $\mathbf{F}$ is a given function of time, and most frequently it is not. Although textbooks are laden with exercises in which $\mathbf{I}$ is given, practicing engineers are often denied this luxury. Even when the impulse $\mathbf{I}$ applied to a particle is unknown, however, the impulse-momentum equation can be very useful if for the system of particles of interest the resultant external impulse is zero, so that the total linear momentum of the system of particles is conserved, even though the particles may be bumping into one another and exerting equal and opposite impulses on one another. The zero impulse case is trivial when the system is a single particle, as in the present chapter, but it becomes interesting when we begin to deal with systems of particles in Chap. 6.

The concept of linear momentum conservation is useful when dealing with a single particle only when certain components of the impulse $\mathbf{I}$ are known to be zero, since then the corresponding linear momentum components are unchanged. Consider, for example, the problem of a ball bouncing obliquely off of a *frictionless wall*. The absence of friction precludes the existence of any forces on the ball except for the force vector *normal* to the wall. This force may be an unknown and highly complex function during the interval $t_2 - t_1$ of the collision, so the impulse **I** is not known in detail. It is known, however, that the impulse vector **I** is normal to the wall, so there can be no change in the linear momentum of the ball except for its normal component. Thus the components of the ball's velocity in the plane of the wall are unchanged by the collision. This frictionless wall is, of course, an abstraction that can never be realized physically, but it may be a useful approximation. With this idealization we have quickly reduced the three-dimensional problem of an oblique collision to the same level of difficulty as a one-dimensional normal collision problem. We still cannot solve that one-dimensional problem unless someone either gives us the impulse **I** or shows us a way to avoid **I** entirely. The latter approach requires the concept of energy, which is developed in Sec. 4·3.

4·2·4 Application to a jet-propelled slot car

Consider the slot car shown in Fig. 4·4 to be running around a horizontal slot track consisting of two circular curves of radius a separated by two straight stretches of track of length L. Imagine first that the car has no propulsion capability, but that the track is frictionless (later we'll put a jet engine on the slot car). The car is given an initial speed v_0 when positioned at point A, and we are to apply the impulse-momentum equation to see what we can learn about the subsequent motion.

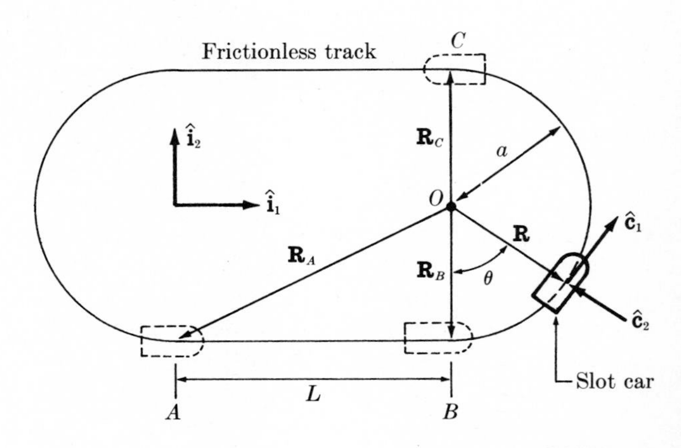

FIGURE 4·4

Consider the motion of the car on the straight track from A to B, for the time interval from t_A to t_B. The impulse-momentum equation

$$\int_{t_A}^{t_B} \mathbf{F}\,dt\bigg|_i = m\mathbf{V}_B - m\mathbf{V}_A \tag{4·26}$$

[where $\mathbf{V}_B \triangleq \mathbf{V}(t_B)$ and $\mathbf{V}_A \triangleq \mathbf{V}(t_A)$] is trivial since no force is applied to the car in this time interval. Since $\mathbf{V}_A$ is $v_0\hat{\mathbf{i}}_1$ from the initial conditions, $\mathbf{V}_B$ has this value also. (As shown in Fig. 4·4, $\hat{\mathbf{i}}_1$ and $\hat{\mathbf{i}}_2$ are orthogonal unit vectors fixed relative to the track, which is assumed to be an inertial reference frame.)

Next examine the impulse-momentum equation for the time interval from t_B to t_C, when the car rounds the curve from B to C. There must be an impulse for this period since $\mathbf{V}_C$ (the velocity at time t_C) is opposite in direction from $\mathbf{V}_B$ (the velocity at t_B). This impulse is given by

$$\int_{t_B}^{t_C} \mathbf{F}\,dt\bigg|_i = m\mathbf{V}_C - m\mathbf{V}_B \tag{4·27}$$

Although the track is frictionless and we have no propulsion device on our car, there does exist a force $\mathbf{F}$ to substitute into Eq. (4·27), namely, the force applied by the track to the car in the radial direction ($\hat{\mathbf{c}}_2$ in the figure). We can obtain this force as a function of the unknown speed v of the car from $\mathbf{F} = m\mathbf{A}$,

$$\mathbf{F} = m\left(\frac{{}^id^2}{dt^2}\mathbf{R}\right) = m\frac{{}^id^2}{dt^2}(-a\hat{\mathbf{c}}_2) = -ma\frac{{}^id^2}{dt^2}\hat{\mathbf{c}}_2 \tag{4·28}$$

But we have

$$\frac{{}^id}{dt}\hat{\mathbf{c}}_2 = \frac{{}^cd}{dt}\hat{\mathbf{c}}_2 + \boldsymbol{\omega}^{ci} \times \hat{\mathbf{c}}_2$$

where c is the reference frame of the car, in which $\hat{\mathbf{c}}_1$ and $\hat{\mathbf{c}}_2$ are fixed. Furthermore, in Fig. 4·4 the angular velocity $\boldsymbol{\omega}^{ci}$ is given by $\dot{\theta}\hat{\mathbf{c}}_3$, and since the product of θ with the radius a is the distance along the arc of the track, we have $\dot{\theta} = v/a$. Thus the previous equation becomes

$$\frac{{}^id}{dt}\hat{\mathbf{c}}_2 = -\frac{v}{a}\hat{\mathbf{c}}_1$$

and consequently

$$\begin{aligned}\frac{{}^id^2}{dt^2}\hat{\mathbf{c}}_2 &= -\frac{\dot{v}}{a}\hat{\mathbf{c}}_1 - \frac{v}{a}\frac{{}^id}{dt}\hat{\mathbf{c}}_1 \\ &= -\frac{\dot{v}}{a}\hat{\mathbf{c}}_1 - \frac{v}{a}\left(\frac{v}{a}\hat{\mathbf{c}}_3 \times \hat{\mathbf{c}}_1\right) \\ &= -\frac{\dot{v}}{a}\hat{\mathbf{c}}_1 - \frac{v^2}{a^2}\hat{\mathbf{c}}_2\end{aligned}$$

Equation (4·28) then becomes

$$\mathbf{F} = m\dot{v}\hat{\mathbf{c}}_1 + \frac{mv^2}{a}\hat{\mathbf{c}}_2$$

Since the track is frictionless, there can be no component of $\mathbf{F}$ along $\hat{\mathbf{c}}_1$, so $\dot{v}$ must be zero. Therefore v remains the constant v_0, and the force applied to the cart by the track is

$$\mathbf{F} = \frac{mv_0^{\,2}}{a}\hat{\mathbf{c}}_2 \tag{4·29}$$

We have already solved the problem of determining the car's motion by using Newton's second law directly. Since v is always v_0, the change in linear momentum must be

$$m\mathbf{V}_C - m\mathbf{V}_B = -mv_0\hat{\mathbf{i}}_1 - mv_0\hat{\mathbf{i}}_1 = -2mv_0\hat{\mathbf{i}}_1 \tag{4·30}$$

It may be instructive to examine the impulse-momentum equation (4·27) even with the solution known, since there is an opportunity here to illustrate the importance played by reference frames in vector integration. If Eqs. (4·29) and (4·30) are substituted into Eq. (4·27), the result is

$$\int_{t_B}^{t_C} \frac{mv_0^{\,2}}{a}\hat{\mathbf{c}}_2\, dt\Big|_i = -2mv_0\hat{\mathbf{i}}_1 \tag{4·31}$$

The symbol $dt\,|_i$ reminds us that the integration in Eq. (4·31) is accomplished in an inertial reference frame, and only inertially fixed vectors within the integral can be taken outside of the integrand. Specifically, we must substitute (from Fig. 4·4)

$$\hat{\mathbf{c}}_2 = \hat{\mathbf{i}}_2 \cos\theta - \hat{\mathbf{i}}_1 \sin\theta$$

into Eq. (4·31) to obtain

$$\int_{t_B}^{t_C} \frac{mv_0^{\,2}}{a}(\hat{\mathbf{i}}_2\cos\theta - \hat{\mathbf{i}}_1\sin\theta)\, dt\Big|_i = \hat{\mathbf{i}}_2 \int_{t_B}^{t_C} \frac{mv_0^{\,2}}{a}\cos\theta\, dt - \hat{\mathbf{i}}_1 \int_{t_B}^{t_C} \frac{mv_0^{\,2}}{a}\sin\theta\, dt$$

Now we change the integration variable by substituting for dt from

$$\frac{d\theta}{dt} = \frac{v_0}{a}$$

and change integration limits to obtain for the impulse

$$\int_{t_B}^{t_C} \mathbf{F}\, dt\Big|_i = \hat{\mathbf{i}}_2 \int_0^{\pi} mv_0 \cos\theta\, d\theta - \hat{\mathbf{i}}_1 \int_0^{\pi} mv_0 \sin\theta\, d\theta$$
$$= \hat{\mathbf{i}}_1 mv_0 \cos\theta\Big|_0^{\pi} = -2mv_0\hat{\mathbf{i}}_1$$

which checks Eq. (4·31).

If we had failed to interpret the integral correctly as an integration to be performed in an inertial frame of reference, we might have failed to replace $\hat{\mathbf{c}}_2$ by inertially fixed unit vectors before attempting the integration. If, for example, we had replaced the inertial impulse by the integral in the cart frame c, we would have obtained

$$\int_{t_B}^{t_C} \mathbf{F}\, dt \bigg|_c = \int_{t_B}^{t_C} \frac{mv_0^2}{a} \hat{\mathbf{c}}_2\, dt \bigg|_c = \int_{t_B}^{t_C} \frac{mv_0^2}{a}\, dt \hat{\mathbf{c}}_2$$

$$= \frac{mv_0^2}{a}(t_C - t_B)\hat{\mathbf{c}}_2 = \frac{mv_0^2}{a} \frac{a\pi}{v_0} \hat{\mathbf{c}}_2$$

$$= \pi m v_0\, \hat{\mathbf{c}}_2$$

This is clearly not the same thing as the impulse $-2mv_0\hat{\mathbf{i}}_1$ calculated previously.

The primary objective of this example is to illustrate the role played by reference frames in vector integration. Having made this point, we now abandon the practice of identifying inertial frame integration with the symbol $dt\,|_i$, reserving this explicit notation for the rare instance of a noninertial integration, such as $\int \mathbf{F}\, dt\,|_c$ in the preceding paragraph.

Now we'll put a jet engine on the car, and take a new look at the impulse-momentum relationships. We represent the jet force by $F_1(t)\hat{\mathbf{c}}_1$, and we define the symbol

$$I \triangleq \int_0^{\Delta\tau} F_1(\tau)\, d\tau \tag{4·32}$$

to represent the magnitude of the impulse applied to the car by the engine in time interval $\Delta\tau$, while the car is held stationary in a test fixture.

If we are constrained by fuel supply to a given value of the impulse magnitude I, we may as slot-car racers want to know if there is some optimum location on the track to accomplish our motor burn for maximum speed increment. Are we better off to turn on the jet on the straight section between A and B in Fig. 4·4, or will we impart more additional speed to the car by firing the motor as we round the turn from B to C? Does it make any difference when we apply the motor thrust?

These are not simple questions. We have not fully defined the force $\mathbf{F}$ to be substituted into $\mathbf{F} = m\mathbf{A}$, requiring only that Eq. (4·32) be satisfied for a given I. Even the duration $\Delta\tau$ of the thrust is unspecified. Here we have another example of a problem for which a form of the impulse-momentum equation better suits our needs than does Newton's second law.

Consider first the option of firing during the journey from A to B. The impulse then becomes

$$\int_{t_A}^{t_B} \mathbf{F}\, dt = \int_{t_A}^{t_B} F_1\hat{\mathbf{i}}_1\, dt = \int_{t_A}^{t_B} F_1\, dt\, \hat{\mathbf{i}}_1 = I\hat{\mathbf{i}}_1$$

where we have selected $\Delta\tau = t_B - t_A$ and taken advantage of the opportunity to remove the inertially fixed unit vector $\hat{\mathbf{i}}_1$ from the inertial integral defining the impulse. The impulse-momentum equation then provides

$$I\hat{\mathbf{i}}_1 = m\mathbf{V}_B - m\mathbf{V}_A = m(v_B - v_A)\hat{\mathbf{i}}_1 = m\,\Delta v\,\hat{\mathbf{i}}_1$$

where symbols $v_A \triangleq |\mathbf{V}_A|$ and $v_B \triangleq |\mathbf{V}_B|$ are employed, with $\Delta v = v_B - v_A$ representing the increase in speed of the car. Thus we conclude that, whatever the detailed behavior of the thrust function $F_1(\tau)$ in Eq. (4·32), the increment in speed available from the impulse of magnitude I is given by

$$\Delta v = \frac{I}{m} \tag{4·33}$$

Next consider the option of firing the jet engine as the car rounds the curve from B to C, with the thrust magnitude F_1 still obeying Eq. (4·32) for the time interval $\Delta\tau = t_C - t_B$ of this part of the race. We want to know how the new increment in speed $\Delta v'$ compares to Δv in Eq. (4·33).

Direct application of the impulse-momentum equation of the previous section provides

$$\mathbf{I} = \int_{t_B}^{t_C} \mathbf{F}\,dt = m\mathbf{V}_C - m\mathbf{V}_B$$

where $\mathbf{V}_C$ and $\mathbf{V}_B$ are the car velocities at times t_C and t_B, respectively. Since we know in advance the directions of $\mathbf{V}_C$ and $\mathbf{V}_B$, we can introduce $v_B \triangleq |\mathbf{V}_B| = v_0$ and $v_C \triangleq |\mathbf{V}_C| = v_0 + \Delta v'$ and record the linear momentum in the form

$$m\mathbf{V}_C - m\mathbf{V}_B = [-m(v_0 + \Delta v') - mv_0]\hat{\mathbf{i}}_1 = -m(2v_0 + \Delta v')\hat{\mathbf{i}}_1$$

The impulse $\mathbf{I}$ must be calculated as the time integral of the force vector

$$\mathbf{I} = \int_{t_B}^{t_C} \left(F_1\hat{\mathbf{c}}_1 + \frac{mv^2}{a}\,\hat{\mathbf{c}}_2\right) dt \tag{4·34}$$

[where Eq. (4·29) has been employed to represent the force $(mv^2/a)\hat{\mathbf{c}}_2$ applied to the car by the track]. It is very tempting at this point to replace $\int_{t_B}^{t_C} F_1\hat{\mathbf{c}}_1\,dt$ by $\int_{t_B}^{t_C} F_1\,dt\,\hat{\mathbf{c}}_1 = I\hat{\mathbf{c}}_1$. But we discovered in the calculation of the impulse imparted by the track forces, Eq. (4·31), that we can get into trouble by removing from the integrand a vector that is not fixed in the reference frame of integration, which is here the inertial frame. In order properly to evaluate the integral in Eq. (4·34), we must substitute

$$\hat{\mathbf{c}}_1 = \hat{\mathbf{i}}_1 \cos\theta + \hat{\mathbf{i}}_2 \sin\theta$$
$$\hat{\mathbf{c}}_2 = \hat{\mathbf{i}}_2 \cos\theta - \hat{\mathbf{i}}_1 \sin\theta$$

where θ is given by

$$\theta = \int_{t_B}^{t} \frac{v}{a}\, dt$$

with v an unknown function of time. The calculation of the impulse vector is getting quite complicated, and the objective of finding $\Delta v'$ by using the impulse-momentum principle seems to be quite unrealizable. Perhaps there is an easier way to find $\Delta v'$.

Our attempt to use the impulse-momentum equation became discouragingly difficult when we realized that we could not remove the car-fixed unit vectors $\hat{\mathbf{c}}_1$ and $\hat{\mathbf{c}}_2$ from the integral in Eq. (4·34). We might therefore try to eliminate this problem by going back to the starting point, $\mathbf{F} = m\mathbf{A}$, and dot-multiplying both sides by $\hat{\mathbf{c}}_1$ *before* we perform the integrations leading to the vector impulse-momentum equation. We would then obtain

$$\int_{t_B}^{t_C} \mathbf{F} \cdot \hat{\mathbf{c}}_1\, dt = \int_{t_B}^{t_C} m\mathbf{A} \cdot \hat{\mathbf{c}}_1\, dt \tag{4·35}$$

By substituting the contact force

$$\mathbf{F} = F_1\hat{\mathbf{c}}_1 + \frac{mv^2}{a}\hat{\mathbf{c}}_2$$

and the inertial acceleration

$$\mathbf{A} \triangleq \frac{{}^i d^2}{dt^2}\mathbf{R} = \frac{{}^i d}{dt}(v\hat{\mathbf{c}}_1) = \dot{v}\hat{\mathbf{c}}_1 + v(\dot{\theta}\hat{\mathbf{c}}_3 \times \hat{\mathbf{c}}_1) = \dot{v}\hat{\mathbf{c}}_1 + \dot{\theta}v\hat{\mathbf{c}}_2$$

into Eq. (4·35), we obtain

$$\int_{t_B}^{t_C} F_1\, dt = \int_{t_B}^{t_C} m\dot{v}\, dt$$

Now the definition of the scalar I from Eq. (4·32) can be substituted, and the integration accomplished as

$$I = \int_{t_B}^{t_C} m\frac{dv}{dt}\, dt = \int_{v_0}^{v_0+\Delta v'} m\, dv = m\, \Delta v'$$

By comparing with Eq. (4·33), we see that

$$\Delta v' = \frac{I}{m} = \Delta v$$

In other words, the increment in speed imparted by the jet engine does not depend on the location or duration of the firing, provided that the magnitude F_1 of the thrust integrated over $\Delta\tau$ maintains a fixed value I.

Note that in exploring this example we have *not* succeeded in illustrating the usefulness of the vector impulse-momentum equation derived as Eq. (4·25)—indeed, we tried that approach and got stuck after writing Eq. (4·34)! Even the direct application of Newton's second law proved unsatisfactory for this difficult problem since **F** was not available in sufficiently explicit form. Our eventual success came from Eq. (4·35)—which is *not* the vector impulse-momentum equation but an equation obtained by operating on $\mathbf{F} = m\mathbf{A}$ in an ad hoc fashion, dot-multiplying both sides by a unit vector pointing in the direction of forward motion, as this particular problem required. Although we can readily generalize from this example a method applicable to all problems in which a body is confined to a track, thus having but one degree of freedom of motion, yet there are no completely general instructions to be offered in trying to teach all the subtle ways in which the basic equation $\mathbf{F} = m\mathbf{A}$ can be used.

4·2·5 Angular momentum/moment of momentum

The vector equation $\mathbf{F} = m\mathbf{A}$ can of course provide the starting point for an indefinite number of valid relationships in mechanics since we can multiply both sides of this equation by any function or operator and find correct results. We must expect, however, that most multipliers won't help us in solving the equation $\mathbf{F} = m\mathbf{A}$, and we must understand that there is nothing we can do with this equation that invests it with any information not already contained in its original form. We do find that occasionally some light is shed by a judicious choice of multiplier, as in §4·2·4.

For example, if we precross-multiply Newton's second law by **R**, the position vector to particle p from a point o fixed in inertial space i, we have

$$\mathbf{R} \times \mathbf{F} = \mathbf{R} \times m\mathbf{A}^{pi} = \mathbf{R} \times m \frac{{}^i d^2}{dt^2}\mathbf{R} = \frac{{}^i d}{dt}(m\mathbf{R} \times \dot{\mathbf{R}}) \tag{4·36}$$

where $\dot{\mathbf{R}}$ is again the inertial derivative ${}^i d\mathbf{R}/dt$. The quantity $m\mathbf{R} \times \dot{\mathbf{R}}$ is an example of a function variously called the *angular momentum* or the *moment of momentum*. Since $\mathbf{R} \times \mathbf{F}$ is known as the *moment*† of the force **F** about the inertially fixed point o with respect to which **R** is measured, Eq. (4·36) may be expressed verbally by saying that a certain moment equals the inertial time derivative of a certain angular momentum (or moment of momentum).

By integrating Eq. (4·36) over a time interval $t_2 - t_1$, we obtain an impulse-momentum equation analogous to Eq. (4·25), but for *angular impulse* and *angular momentum*. If we allow $\mathbf{H}^o$ to denote the angular momentum about point o, namely, $m\mathbf{R} \times \dot{\mathbf{R}}$, this equation becomes

$$\int_{t_1}^{t_2} \mathbf{R} \times \mathbf{F}\, dt = \mathbf{H}^o(t_2) - \mathbf{H}^o(t_1) \tag{4·37}$$

† The *moment* of a bound vector is defined formally in Chap. 7.

If the angular impulse is *zero* (as, for example, when $\mathbf{R}$ and $\mathbf{F}$ are parallel), angular momentum $\mathbf{H}^o$ is constant, and Eq. (4·37) becomes a statement of the conservation of angular momentum.

Although the idea of angular momentum or moment of momentum is occasionally convenient in the dynamic analysis of individual particles, we can get along quite well without it. A much more fundamental role is played by this concept in the treatment of systems of particles and rigid bodies. In these contexts it becomes convenient to generalize the definition $\mathbf{H}^o \triangleq m\mathbf{R} \times \dot{\mathbf{R}}$ in a variety of ways. One common extension of this definition is obtained by replacing the vector $\mathbf{R}$ from inertially fixed o to particle p by an arbitrary vector $\mathbf{r}$ locating p from any reference point q, thus obtaining

$$\mathbf{h}^q \triangleq m\mathbf{r} \times \dot{\mathbf{R}} \tag{4·38}$$

A more useful generalization frequently encountered in modern practice is given by

$$\mathbf{H}^q \triangleq m\mathbf{r} \times \dot{\mathbf{r}} \tag{4·39}$$

Note that $\mathbf{h}^q$ and $\mathbf{H}^q$ are quite fundamentally different quantities, although both are sometimes called angular momentum and/or moment of momentum in the literature. Because both definitions are commonplace, but few texts acknowledge more than one definition, it is very easy to get confused in passing from one book or technical paper to another. In an effort to maintain some measure of precision, the term *moment of momentum* is applied in this text to $\mathbf{h}^q$, as defined by Eq. (4·38), and the term *angular momentum* is reserved for $\mathbf{H}^q$, as defined by Eq. (4·39). The original concept $\mathbf{H}^o \triangleq m\mathbf{R} \times \dot{\mathbf{R}}$ may be interpreted as a special case of either definition, so either name or symbol could be applied to it. In fact, in most (but not all) practical applications, the reference point is selected in such a way that these definitions become equivalent. These concepts are explored in detail in Part Two, "Mechanics of Rigid Bodies."

4·2·6 Application to a particle orbiting an attracting center

To obtain a preliminary approximation of the orbital motion of an artificial satellite encircling the earth, or of a planet orbiting the sun, it is often satisfactory to assume that the relatively massive body of the unequal pair in the problem is uninfluenced in its motions by the lesser body, and even to assume that the more massive body is fixed in inertial space. Then the problem reduces to the task of determining the motion of a single particle p (the lesser body) under the influence of the gravitational attraction of a mass idealized as a particle located at an inertially fixed point, which is then called an attracting center. If this fixed point is designated o and the vector from o to p is called $\mathbf{R}$, we can write the vector differential equation of motion

$$\mathbf{F} = m \frac{{}^i d^2}{dt^2} \mathbf{R}$$

where m is the mass of p, and then determine the orbital motion of p by solving this equation for $\mathbf{R}(t)$. The vector $\mathbf{F}$, of course, must represent the force of gravity, which we know to be inversely proportional to the square of the distance from o to p and directed from p toward o; for present purposes it will suffice to write

$$\mathbf{F} = -F\hat{\mathbf{u}}_r$$

where $\hat{\mathbf{u}}_r$ is the unit vector shown in Fig. 4·5.

We will later have occasion to write F in more explicit form and try to solve the equation of motion (see §5·2·3). This is not a trivial task (even for this greatly simplified problem), but it becomes slightly easier if you recognize the usefulness of the concepts of moment of momentum and angular momentum in this problem.

If we substitute $\mathbf{F} = -F\hat{\mathbf{u}}_r$ and $\mathbf{R} = r\hat{\mathbf{u}}_r$ into Eq. (4·37) (see Fig. 4·5), and note that $\hat{\mathbf{u}}_r \times \hat{\mathbf{u}}_r = 0$, we determine that the angular momentum (or moment of momentum) $\mathbf{H}^o \triangleq m\mathbf{R} \times \dot{\mathbf{R}}$ is a constant during the orbital motion. This information is most valuable in facilitating the solution of the scalar equations of motion when expressed in terms of the scalars r and φ, as defined in Fig. 4·5, where p' is some inertially fixed reference point such that the plane established by points o, p, and p' remains fixed in inertial space. (Such a point p' always exists because the central force $\mathbf{F}$ does not induce p ever to depart from the inertially fixed plane of motion established by point o and the velocity $\dot{\mathbf{R}}$ at any given time.) In these terms, we have

$$\dot{\mathbf{R}} \triangleq \frac{{}^i d}{dt} \mathbf{R} = \frac{{}^i d}{dt} (r\hat{\mathbf{u}}_r) = \dot{r}\hat{\mathbf{u}}_r + r \frac{{}^i d}{dt} \hat{\mathbf{u}}_r$$

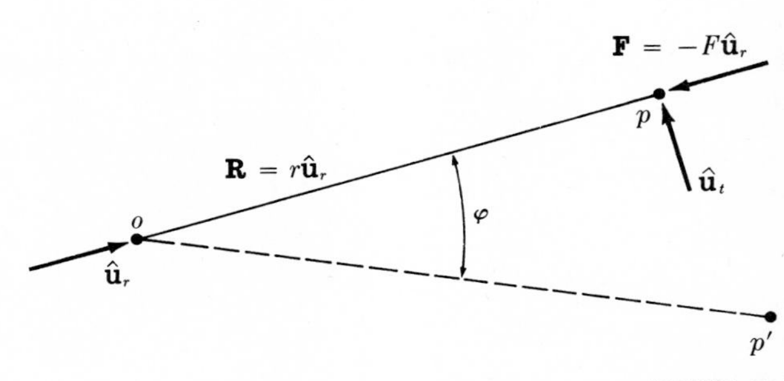

FIGURE 4·5

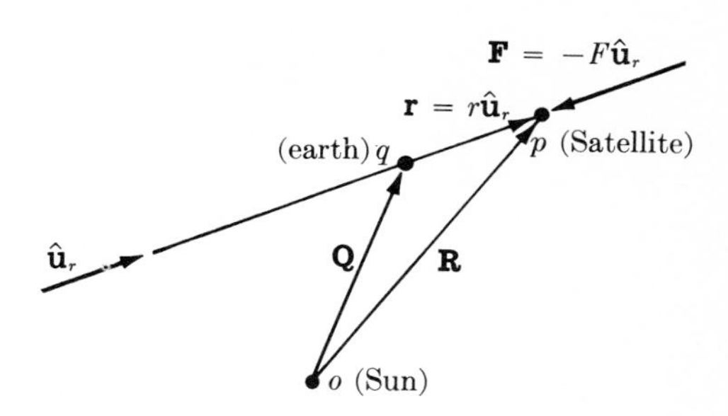

FIGURE 4·6

If we now introduce a reference frame u in which we fix unit vector $\hat{\mathbf{u}}_r$ and another unit vector $\hat{\mathbf{u}}_n$, which remains normal to the orbital plane defined by 0, p, and p', then we have [see Eq. (3·100)]

$$\begin{aligned}\dot{\mathbf{R}} &= \dot{r}\hat{\mathbf{u}}_r + r\left(\frac{^u d}{dt}\hat{\mathbf{u}}_r + \boldsymbol{\omega}^{ui} \times \hat{\mathbf{u}}_r\right) = \dot{r}\hat{\mathbf{u}}_r + r(0 + \dot{\varphi}\hat{\mathbf{u}}_n \times \hat{\mathbf{u}}_r)\\ &= \dot{r}\hat{\mathbf{u}}_r + r\dot{\varphi}\hat{\mathbf{u}}_t\end{aligned}$$

where $\hat{\mathbf{u}}_t \triangleq \hat{\mathbf{u}}_n \times \hat{\mathbf{u}}_r$. Thus the constant angular momentum is

$$\mathbf{H}^o = m(r\hat{\mathbf{u}}_r) \times (\dot{r}\hat{\mathbf{u}}_r + r\dot{\varphi}\hat{\mathbf{u}}_t) = mr^2\dot{\varphi}\hat{\mathbf{u}}_n$$

The particle p may therefore be said to orbit the attracting center o in such a way that the product $mr^2\dot{\varphi}$ remains constant, although both r and $\dot{\varphi}$ change with time. Although even without the concept of angular momentum this constant of the motion can be found mathematically as a first integral of the equations of motion (see §5·2·3), with the idea of angular momentum conservation you can make this discovery most easily. Having found it, you are still well short of determining the orbital motion, but you have taken a constructive step in that direction. We shall take a few more such steps in §5·2·3.

Note that in this problem it has not been necessary to distinguish between the concept of angular momentum $\mathbf{H}^q$ as defined in Eq. (4·39) and moment of momentum $\mathbf{h}^q$ as defined by Eq. (4·38) since heretofore q has been point o, assumed fixed in inertial space. By generalizing a bit from this example, however, we can illustrate the differences in these two concepts.

Let's imagine that p is a particle representing a satellite orbiting the earth, as previously, but now we'll idealize the earth as a particle q orbiting the sun, which we shall treat as an inertially fixed particle o. In terms of the symbols shown on Fig. 4·6, we can, from Eqs. (4·39) and (4·38), write

$$\mathbf{H}^q \triangleq m\mathbf{r} \times \dot{\mathbf{r}}$$

and

$$\mathbf{h}^q \triangleq m\mathbf{r} \times \dot{\mathbf{R}} = m\mathbf{r} \times (\dot{\mathbf{Q}} + \dot{\mathbf{r}}) = m\mathbf{r} \times \dot{\mathbf{Q}} + \mathbf{H}^q$$

Except under the extraordinary circumstances when $\mathbf{r}$ and $\dot{\mathbf{Q}}$ are parallel, we have $\mathbf{h}^q \neq \mathbf{H}^q$.

Note that neither $\mathbf{h}^q$ nor $\mathbf{H}^q$ fits directly into the impulse-momentum equation as presented in Eq. (4·37) since the reference point q is not inertially fixed. If we attempt to parallel the derivation of Eq. (4·37), cross-multiplying $\mathbf{F}$ and $m\mathbf{A}$ by $\mathbf{r}$ instead of $\mathbf{R}$, we have

$$\begin{aligned}\mathbf{r} \times \mathbf{F} &= \mathbf{r} \times m\ddot{\mathbf{R}} = \mathbf{r} \times m(\ddot{\mathbf{Q}} + \ddot{\mathbf{r}}) \\ &= \mathbf{r} \times m\ddot{\mathbf{Q}} + \frac{{}^i d}{dt}(m\mathbf{r} \times \dot{\mathbf{r}}) = \mathbf{r} \times m\ddot{\mathbf{Q}} + \dot{\mathbf{H}}^q\end{aligned}$$

or else

$$\begin{aligned}\mathbf{r} \times \mathbf{F} &= \mathbf{r} \times m\ddot{\mathbf{R}} = \frac{{}^i d}{dt}(m\mathbf{r} \times \dot{\mathbf{R}}) - m\dot{\mathbf{r}} \times \dot{\mathbf{R}} \\ &= \frac{{}^i d}{dt}\mathbf{h}^q - m\dot{\mathbf{r}} \times (\dot{\mathbf{Q}} + \dot{\mathbf{r}}) = \dot{\mathbf{h}}^q - \dot{\mathbf{r}} \times m\dot{\mathbf{Q}}\end{aligned}$$

In either case, if the force exerted on p by o is ignored, the moment $\mathbf{r} \times \mathbf{F}$ is zero, and we find

$$\dot{\mathbf{H}}^q = -m\mathbf{r} \times \ddot{\mathbf{Q}}$$

and

$$\dot{\mathbf{h}}^q = m\dot{\mathbf{r}} \times \dot{\mathbf{Q}}$$

Neither the angular momentum $\mathbf{H}^q$ nor the moment of momentum $\mathbf{h}^q$ is a constant of the motion. Only by evaluating $\dot{\mathbf{H}}^q$ or $\dot{\mathbf{h}}^q$ numerically for a given problem can we evaluate the magnitude of the error introduced in the preceding analysis by assuming the earth (particle q) to be inertially stationary (see §4·1·1).

4·3 WORK AND ENERGY

4·3·1 Work, activity, and kinetic energy

In §4·2·1 we found that it was possible to obtain a first integral of the vector equation of particle motion, $\mathbf{F} = m\mathbf{A}$, simply by integrating over time t. The resulting vector equation, Eq. (4·25), equated the linear impulse $\mathbf{I}$ to the change in linear momentum $m\mathbf{V}$ over a given time interval. In §4·2·3 to 4·2·5 we found that it is sometimes most useful to perform some operation on both sides of the equation $\mathbf{F} = m\mathbf{A}$ *before* accomplishing the time integration. In §4·2·3 we simply dot-multiplied both sides by a convenient unit vector, and in §4·2·4 and 4·2·5 we cross-multiplied both sides by a position vector before integrating. In the present section, we discover that if we dot-multiply both sides of $\mathbf{F} = m\mathbf{A}^{pi}$ by $\mathbf{V}^{pi}$, the inertial velocity of the particle, subsequent time integration provides a very useful scalar equation called the *work-energy* equation (often called the *work-energy principle*).

The suggested dot multiplication provides immediately

$$\mathbf{F} \cdot \mathbf{V}^{pi} = m\mathbf{A}^{pi} \cdot \mathbf{V}^{pi} = m\left[\frac{{}^i d}{dt}(\mathbf{V}^{pi})\right] \cdot \mathbf{V}^{pi}$$

$$= \frac{{}^i d}{dt}[\tfrac{1}{2} m\mathbf{V}^{pi} \cdot \mathbf{V}^{pi}] = \frac{d}{dt}(\tfrac{1}{2}m\mathbf{V}^{pi} \cdot \mathbf{V}^{pi})$$

With the definition of the *kinetic energy* T of a particle p

$$T \triangleq \tfrac{1}{2}m\mathbf{V}^{pi} \cdot \mathbf{V}^{pi} \tag{4·40a}$$

the preceding equation becomes

$$\mathbf{F} \cdot \mathbf{V}^{pi} = \dot{T} \tag{4·41a}$$

It is, however, occasionally convenient to introduce the quantity

$$T^f \triangleq \tfrac{1}{2}m\mathbf{V}^{pf} \cdot \mathbf{V}^{pf} \tag{4·40b}$$

and then to write $\mathbf{F} = m\mathbf{A}$ in the expanded form shown in Eq. (4·2) or, equivalently, as

$$\mathbf{F} = m(\mathbf{A}^{pf} + \mathbf{A}^{p^*i} + 2\boldsymbol{\omega}^{fi} \times \mathbf{V}^{pf})$$

where the *transport acceleration*

$$\mathbf{A}^{p^*i} \triangleq \mathbf{A}^{qi} + \boldsymbol{\alpha}^{fi} \times \mathbf{r} + \boldsymbol{\omega}^{fi} \times (\boldsymbol{\omega}^{fi} \times \mathbf{r}) \tag{4·42}$$

is the same as that introduced following Eq. (3·150). Dot-multiplying by $\mathbf{V}^{pf}$ then produces

$$\mathbf{F} \cdot \mathbf{V}^{pf} = m\frac{{}^f d}{dt}(\mathbf{V}^{pf}) \cdot \mathbf{V}^{pf} + m\mathbf{A}^{p^*i} \cdot \mathbf{V}^{pf} + 2m(\boldsymbol{\omega}^{fi} \times \mathbf{V}^{pf}) \cdot \mathbf{V}^{pf}$$

$$= \frac{{}^f d}{dt}(\tfrac{1}{2}m\mathbf{V}^{pf} \cdot \mathbf{V}^{pf}) + m\mathbf{A}^{p^*i} \cdot \mathbf{V}^{pf}$$

or

$$(\mathbf{F} - m\mathbf{A}^{p^*i}) \cdot \mathbf{V}^{pf} = \frac{d}{dt}(\tfrac{1}{2}m\mathbf{V}^{pf} \cdot \mathbf{V}^{pf})$$

If now in the spirit of §4·1·2 we introduce the "effective force for frame f,"

$$\mathbf{F}^f \triangleq \mathbf{F} - m\mathbf{A}^{p^*i} \tag{4·43}$$

we can write a generalized version of Eq. (4·41a) as

$$\mathbf{F}^f \cdot \mathbf{V}^{pf} = \dot{T}^f \tag{4·41b}$$

While it may seem offhand that Eq. (4·41b) can offer no advantage over Eq. (4·41a), the elimination of the term $2m(\boldsymbol{\omega}^{fi} \times \mathbf{V}^{pf}) \cdot \mathbf{V}^{pf}$ may be helpful. See, for example, Prob. 4·17.

The quantity $\mathbf{F} \cdot \mathbf{V}^{pi}$ is sometimes called the *activity*, and Eq. (4·41a) is then the *activity-energy equation*. Equation (4·41b) might be called a *generalized-activity-energy equation*. These equations have some immediate utility, but because they are single scalar second-order differential equations in the position coordinates, they do not offer complete equations of motion unless there is only one single unknown. (Recall that an unconstrained particle has *three* unknown coordinates.) Examples in §4·3·2 illustrate the class of problem for which this equation is most useful.

A more important relationship can be obtained by taking a time integral of the activity-energy equations to obtain the *work-energy equation*

$$\int_{t_1}^{t_2} \mathbf{F} \cdot \mathbf{V}^{pi}\, dt = \int_{t_1}^{t_2} \frac{dT}{dt}\, dt = T_2 - T_1 \tag{4·44a}$$

where $T_2 \triangleq T(t_2)$ and $T_1 \triangleq T(t_1)$, or its generalized form

$$\int_{t_1}^{t_2} \mathbf{F}^f \cdot \mathbf{V}^{pf}\, dt = T_2{}^f - T_1{}^f \tag{4·44b}$$

The integral surviving in Eq. (4·44a) is known as the *work* W done by the force $\mathbf{F}$ on particle p in the time interval from t_1 to t_2, while the integral in Eq. (4·44b) is the *effective work* W^f *for frame* f. Note carefully that the velocity appearing in (4·44a) is the *inertial* velocity $\mathbf{V}^{pi}$, while that in (4·44b) is the velocity relative to frame f. In terms of the position vector $\mathbf{R}$ locating particle p relative to a point *fixed in inertial space*, the *work definition* may be written

$$W \triangleq \int_{t_1}^{t_2} \mathbf{F} \cdot \mathbf{V}^{pi}\, dt = \int_{t_1}^{t_2} \mathbf{F} \cdot \frac{{}^i d}{dt} \mathbf{R}\, dt = \int_{P_1}^{P_2} \mathbf{F} \cdot d\mathbf{R} \tag{4·45a}$$

where $d\mathbf{R}$ is a differential element along a path in inertial space traversed by particle p in going from position P_1 at time t_1 to position P_2 at time t_2. In similar terms, if $\mathbf{r}$ is the position vector of p relative to a point q fixed in frame f, then

$$W^f \triangleq \int_{t_1}^{t_2} \mathbf{F}^f \cdot \mathbf{V}^{pf}\, dt = \int_{t_1}^{t_2} \mathbf{F}^f \cdot \frac{{}^f d}{dt} \mathbf{r}\, dt = \int_{P_1}^{P_2} \mathbf{F}^f \cdot d\mathbf{r} \tag{4·45b}$$

where $d\mathbf{r}$ is a differential element along a path in frame f traversed by particle p in going from position P_1 at time t_1 to position P_2 at time t_2.

Only rarely is there any advantage in working with the abstract generalized concepts represented by Eqs. (4·44*b*) and (4·45*b*), so we shall concentrate on the special cases represented by Eqs. (4·44*a*) and (4·45*a*).

Although the expression in Eq. (4·45*a*) for work W in terms of the *line integral* or *path integral* is frequently taken as the *definition* of work, this can be a dangerous practice. In the present context we are restricted to the dynamics of a single particle, so it hardly need be emphasized that in the path-integral expression $\mathbf{F}$ is the force applied to the particle and $\mathbf{dR}$ is the differential inertial change in the position of the particle. It then becomes equivalent to say that $\mathbf{dR}$ is the differential inertial change in the point of application of the force. But this is no longer true when we widen the scope of our applications to embrace systems of particles and rigid bodies, since then the force may shift in time from one moving particle of the material system to another. It then becomes essential to recognize that $\mathbf{dR}$ is *not* in general the differential inertial change in the point of application of the force, although it *is* the differential inertial change in position of that particle to which $\mathbf{F}$ is applied at a given instant. Thus it is both conceptually and operationally easier to stick with the work definition in terms of the time integral appearing first in Eq. (4·45*a*), using for $\mathbf{V}^{pi}$ the inertial velocity of that particle instantaneously subjected to force $\mathbf{F}$. Because this important distinction can be elusive, we will briefly abandon the single-particle restriction to consider the multiparticle or rigid-body system of Fig. 4·7.

Imagine that a massive grinding wheel is rotating freely with inertial

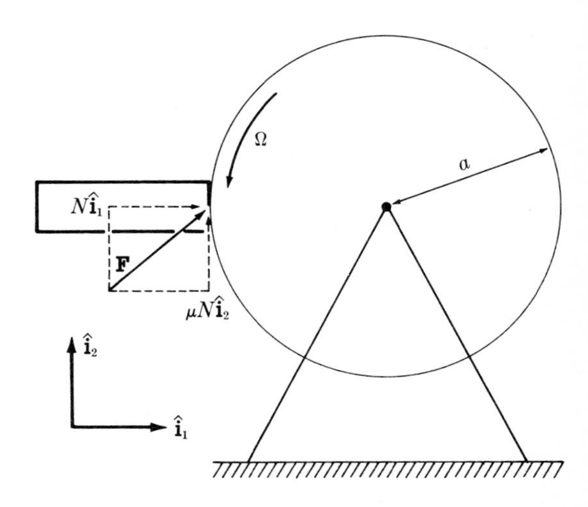

FIGURE 4·7

angular velocity $\Omega\hat{\mathbf{i}}_3$, when we apply a tool to be sharpened. If we hold the tool steadily with a constant radial force $N\hat{\mathbf{i}}_1$ applied to the wheel, the friction force $\mu N\hat{\mathbf{i}}_2$ must also be transmitted to the wheel, and we know from experience that this will gradually reduce the angular speed Ω of the wheel. Although we have not formally defined the kinetic energy of a rigid body or system of particles yet, you will surely agree that this thing we are going to call the kinetic energy of the wheel must diminish in time, as the "particles" in the wheel all lose velocity. Consequently there must be some work W done on the wheel by the force $\mathbf{F} = N\hat{\mathbf{i}}_1 + \mu N\hat{\mathbf{i}}_2$. But the point in inertial space at which the force $\mathbf{F}$ is applied is stationary, so if $\mathbf{dR}$ in Eq. (4·45a) represents the motion of this point, the work integral would be zero. This interpretation of $\mathbf{dR}$ represents the differential displacement of the point of the wheel that instantaneously is in contact with the tool, that is,

$$\mathbf{dR} = -a\Omega\, dt\, \hat{\mathbf{i}}_2$$

While this expression for $\mathbf{dR}$ may correctly be substituted into the path integral in Eq. (4·45a), it is equivalent and much simpler to substitute

$$\mathbf{V}^{pi} = \Omega\hat{\mathbf{i}}_3 \times (-a\hat{\mathbf{i}}_1) = -a\Omega\hat{\mathbf{i}}_2$$

into the original work-integral definition. In either case we get

$$W = \int_{t_1}^{t_2} (N\hat{\mathbf{i}}_1 + \mu N\hat{\mathbf{i}}_2) \cdot (-a\Omega\hat{\mathbf{i}}_2)\, dt = \int_{t_1}^{t_2} (-\mu N a\Omega)\, dt$$
$$= -\mu N a \int_{t_1}^{t_2} \Omega\, dt \neq 0$$

It should be noted explicitly that, since the tool in Fig. 4·7 is stationary, no work is being done *on* the tool, either by the wheel or by the workman holding the tool.

4·3·2 Application of the activity-energy equation

Figure 4·8 shows a stationary tubular ring of radius a containing a block (particle) p of mass m. The plane of the ring is vertical in a gravity field, so the block is subjected to the gravity force $mg\hat{\mathbf{i}}_1$ as well as those forces applied by the tube walls. The contact surface between block and wall is frictionless. The earth is assumed an inertial frame.

We can, of course, apply $\mathbf{F} = m\mathbf{A}$ directly to the block without difficulty. We must recognize that there are unknown wall forces on the block, and if we seek a single differential equation in θ that does not involve these unknown and unwanted forces, we must record $\mathbf{F} = m\mathbf{A}$ in the vector basis $\hat{\mathbf{b}}_1$, $\hat{\mathbf{b}}_2$, and $\hat{\mathbf{b}}_3$ fixed in the block rather than the inertially fixed vector basis $\hat{\mathbf{i}}_1$, $\hat{\mathbf{i}}_2$, and $\hat{\mathbf{i}}_3$. The successful application of $\mathbf{F} = m\mathbf{A}$ therefore requires a bit of judgment.

Alternatively, we can apply the activity-energy equation, and write Eq. (4·41a) as

$$\mathbf{F} \cdot \mathbf{V}^{pi} = \dot{T}$$

where

$$T = \tfrac{1}{2} m \mathbf{V}^{pi} \cdot \mathbf{V}^{pi}$$

The velocity $\mathbf{V}^{pi}$ is simply

$$\mathbf{V}^{pi} = \frac{{}^i d}{dt} \mathbf{R} = \frac{{}^i d}{dt} (a\hat{\mathbf{b}}_1) = a \frac{{}^i d}{dt} \hat{\mathbf{b}}_1$$

The vector differentiation relationship, Eq. (3·100), provides

$$\frac{{}^i d}{dt} \hat{\mathbf{b}}_1 = \frac{{}^b d}{dt} \hat{\mathbf{b}}_1 + \boldsymbol{\omega}^{bi} \times \hat{\mathbf{b}}_1 = 0 + \dot{\theta}\hat{\mathbf{b}}_3 \times \hat{\mathbf{b}}_1 = \dot{\theta}\hat{\mathbf{b}}_2$$

where b is the reference frame established by the block, with respect to which $\hat{\mathbf{b}}_1$ is fixed. Thus the velocity is

$$\mathbf{V}^{pi} = a\dot{\theta}\hat{\mathbf{b}}_2$$

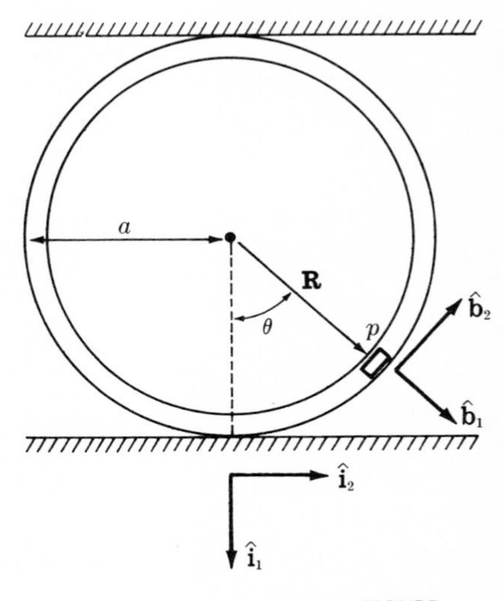

FIGURE 4·8

(as might have been ascertained by inspection of this simple problem). The kinetic energy is simply

$$T = \tfrac{1}{2}m(a\dot{\theta})^2$$

and differentiation provides

$$\frac{dT}{dt} = ma^2\dot{\theta}\ddot{\theta}$$

The left-hand side of Eq. (4·41a) is determined more simply than $\mathbf{F}$ itself, since only the component of $\mathbf{F}$ in the direction of $\mathbf{V}^{pi} = a\dot{\theta}\hat{\mathbf{b}}_3$ survives the dot multiplication. (The whole point in this approach is to avoid the unknown wall forces by dot multiplication.) The result is

$$\mathbf{F} \cdot \mathbf{V}^{pi} = (-mg \sin \theta)(a\dot{\theta})$$

Equating the last two equations gives us

$$-mag\dot{\theta} \sin \theta = ma^2\dot{\theta}\ddot{\theta}$$

and if the common factor $ma\dot{\theta}$ is nonzero, we have

$$a\ddot{\theta} = -g \sin \theta \tag{4·46}$$

We may have trouble *solving* this nonlinear differential equation, but we can leave that question for another day, confident that we have *derived* an equation of motion that fully describes the movement of the block in the tube.

The activity-energy equation has perhaps provided Eq. (4·46) slightly more efficiently than would be possible with $\mathbf{F} = m\mathbf{A}$, but the former method is useful in only a special class of problems. Not only is it required that the problem involve only a single unknown coordinate (as θ above), but further restrictions must be imposed to keep $\mathbf{F} \cdot \mathbf{V}^{pi}$ free of any unknown forces. You might consider this method, for example, in application to the problem of Fig. 4·8 modified so that the tubular ring is spinning about its vertical diametral line at a constant rate, say Ω. Then the velocity $\mathbf{V}^{pi}$ can be shown to be

$$\mathbf{V}^{pi} = a\dot{\theta}\hat{\mathbf{b}}_2 + a\Omega \sin \theta\hat{\mathbf{b}}_3$$

The dot product $\mathbf{F} \cdot \mathbf{V}^{pi}$ now preserves the unknown force applied by the tube wall to the block in the direction of $\hat{\mathbf{b}}_3$, so the scalar equation emerging from Eq. (4·41a) contains an unknown in addition to θ. This equation is therefore insufficient to solve the problem, and we would be better off using $\mathbf{F} = m\mathbf{A}$ directly, and eliminating the wall forces mathematically. We could apply the activity-energy equation to this problem only if we also use $\mathbf{F} = m\mathbf{A}$ to obtain an expression for the wall force (see Prob. 4·14).

As a second illustration of the activity-energy equation, we might return to the jet-propelled slot car of Fig. 4·4, moving on frictionless tracks. In §4·2·4 we demonstrated, using a scalar variation of the impulse-momentum equation, that a given impulsive thrust from the car's engine must impart to it the same increment in speed, regardless of the position of the car on the track. This result can also be obtained quite easily with the activity-energy equation.

Recall that $\hat{\mathbf{c}}_1$ is a unit vector fixed in the car, pointing forward, and $\mathbf{F} = F_1\hat{\mathbf{c}}_1$ is the force applied to the car by the jet. If s is the distance traveled by the car around the track, then $\dot{s}\hat{\mathbf{c}}_1$ is the car's velocity relative to the inertially fixed track, and the kinetic energy of the car (as a particle) is $(m/2)\dot{s}\hat{\mathbf{c}}_1 \cdot \dot{s}\hat{\mathbf{c}}_1 = (m/2)\dot{s}^2$. The activity-energy equation (4·41a) then provides

$$(F_1\hat{\mathbf{c}}_1 + F_2\hat{\mathbf{c}}_2) \cdot \dot{s}\hat{\mathbf{c}}_1 = \frac{d}{dt}\left(\frac{m}{2}\,\dot{s}^2\right)$$

where $F_2\hat{\mathbf{c}}_2$ is the unknown force applied to the car by the track, or

$$F_1\dot{s} = m\dot{s}\ddot{s}$$

which for nonzero $\dot{s}$ implies

$$F_1 = m\ddot{s} \tag{4·47}$$

Since this result holds everywhere on the track, it follows that the motor burn can be accomplished anywhere with equal increment in speed resulting, thus confirming our previous conclusion (see §4·2·4). This result has now been obtained somewhat more directly, and in a form that permits the following generalization: *If a particle p moves through a distance $s(t)$ on an inertially fixed path, and if F_1 is the component of the force exerted on p in the direction of the tangent to the path at p, then Eq. (4·47) holds.* This result appears as an obvious consequence of $\mathbf{F} = m\mathbf{A}$ when the path is a straight line, and it can always be obtained from $\mathbf{F} = m\mathbf{A}$ when this vector equation is dot-multiplied by a unit vector that remains tangential to the fixed trajectory of p at the instantaneous location of p.

4·3·3 Application of the work-energy equation

A straight, smooth-walled tube a is rotating in a horizontal plane at constant angular velocity $\Omega\hat{\mathbf{i}}_3$ relative to its inertially fixed base i as shown in Fig. 4·9. A smooth sphere (particle) of mass m is at time $t = 0$ placed in the tube through an opening directly above the supporting shaft, but a tiny initial speed v_0 is imparted to the ball along the tube. The sphere slides along the tube without frictional restraint, while a motor maintains the rotation rate of the tube at the constant value Ω. What is the *speed* v_L of the sphere relative to the base i when it reaches the end of the tube of length L?

The work-energy equation offers an attractive method for this problem since the desired result is speed v (and not velocity) and this we can find easily once we determine the kinetic energy

$$T = \tfrac{1}{2}m\mathbf{V}^{pi} \cdot \mathbf{V}^{pi} = \tfrac{1}{2}mv^2 \tag{4·48}$$

If t_L is the time at which the ball leaves the tube, the work-energy equation may be written

$$\begin{aligned} W &\triangleq \int_0^{t_L} \mathbf{F} \cdot \mathbf{V}^{pi}\, dt = T(t_L) - T(0) \\ &= \tfrac{1}{2}mv_L{}^2 - \tfrac{1}{2}mv_0{}^2 \cong \tfrac{1}{2}mv_L{}^2 \end{aligned} \tag{4·49}$$

since the initial speed v_0 is negligibly small. The unknown scalar v_L is thus available directly from the work integral W. As noted in Eq. (4·45*a*), we can write the work integral as a path integral

$$W \triangleq \int_0^{t_L} \mathbf{F} \cdot \mathbf{V}^{pi}\, dt = \int_{P_1}^{P_2} \mathbf{F} \cdot d\mathbf{R} \tag{4·50}$$

(with the integration limits P_1 and P_2 denoting, respectively, the original and final position of the sphere), as long as we are careful in our interpretation of $d\mathbf{R}$. Here (as always in this book) $\mathbf{R}$ is the position vector to p from an inertially fixed point, so if R is the distance from the center to p at any time, then $\mathbf{R} = R\hat{\mathbf{a}}_1$ (see Fig. 4·9). If we were so foolish as to believe that $d\mathbf{R} = dR\hat{\mathbf{a}}_1$, we would find that, in the absence of friction, the product $\mathbf{F} \cdot d\mathbf{R}$ would be zero. This would tell us that no work is done on the particle, which would force us to the absurd conclusion that it has no change in kinetic energy and thus exit speed would be v_0! We must guard against this error in interpreting $d\mathbf{R}$, and this is perhaps most

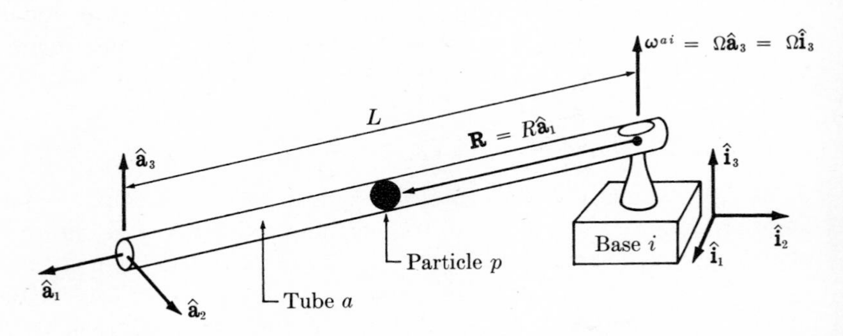

FIGURE 4·9

easily accomplished by returning to the original expression for W in terms of a time integral. Then to find W we need the velocity

$$\begin{aligned}
\mathbf{V}^{pi} &= \frac{{}^id}{dt}\mathbf{R} = \frac{{}^id}{dt}(R\hat{\mathbf{a}}_1) = \dot{R}\hat{\mathbf{a}}_1 + R\frac{{}^id}{dt}\hat{\mathbf{a}}_1 \\
&= \dot{R}\hat{\mathbf{a}}_1 + R\left(\frac{{}^ad}{dt}\hat{\mathbf{a}}_1 + \boldsymbol{\omega}^{ai} \times \hat{\mathbf{a}}_1\right) = \dot{R}\hat{\mathbf{a}}_1 + R(0 + \Omega\hat{\mathbf{a}}_3 \times \hat{\mathbf{a}}_1) \\
&= \dot{R}\hat{\mathbf{a}}_1 + R\Omega\hat{\mathbf{a}}_2
\end{aligned} \tag{4·51}$$

Since $\mathbf{F}$ has no component along $\hat{\mathbf{a}}_1$ (no friction force), only the component of $\mathbf{F}$ along $\hat{\mathbf{a}}_2$ does work. This is an unknown force, applied to the sphere by the tube wall, and the only way we can find an expression for this force is by using $\mathbf{F} = m\mathbf{A}$.

To obtain $\mathbf{A}$ (or $\mathbf{A}^{pi}$, in explicit notation), we may use the kinematic expansion of Eq. (3·150), or we can simply differentiate $\mathbf{V}^{pi}$ from Eq. (4·51). The latter option gives

$$\begin{aligned}
\mathbf{A}^{pi} &= \frac{{}^id}{dt}\mathbf{V}^{pi} = \frac{{}^id}{dt}(\dot{R}\hat{\mathbf{a}}_1 + R\Omega\hat{\mathbf{a}}_2) \\
&= \ddot{R}\hat{\mathbf{a}}_1 + \dot{R}\frac{{}^id}{dt}\hat{\mathbf{a}}_1 + \dot{R}\Omega\hat{\mathbf{a}}_2 + R\Omega\frac{{}^id}{dt}\hat{\mathbf{a}}_2 \\
&= \ddot{R}\hat{\mathbf{a}}_1 + \dot{R}\boldsymbol{\omega}^{ai} \times \hat{\mathbf{a}}_1 + \dot{R}\Omega\hat{\mathbf{a}}_2 + R\Omega\boldsymbol{\omega}^{ai} \times \hat{\mathbf{a}}_2 \\
&= \ddot{R}\hat{\mathbf{a}}_1 + \dot{R}\Omega\hat{\mathbf{a}}_3 \times \hat{\mathbf{a}}_1 + \dot{R}\Omega\hat{\mathbf{a}}_2 + R\Omega^2\hat{\mathbf{a}}_3 \times \hat{\mathbf{a}}_2 \\
&= (\ddot{R} - R\Omega^2)\hat{\mathbf{a}}_1 + 2\Omega\dot{R}\hat{\mathbf{a}}_2
\end{aligned} \tag{4·52}$$

Now the required component of $\mathbf{F}$ in the direction $\hat{\mathbf{a}}_2$ is available as $2m\Omega\dot{R}\hat{\mathbf{a}}_2$, and the work integral becomes

$$\begin{aligned}
W &\triangleq \int_0^{t_L} \mathbf{F} \cdot \mathbf{V}^{pi}\, dt = \int_0^{t_L} \left(2m\Omega\frac{dR}{dt}\right)(R\Omega)\, dt \\
&= \int_0^L 2m\Omega^2 R\, dR = 2m\Omega^2 \frac{R^2}{2}\Big|_0^L = m\Omega^2 L^2
\end{aligned} \tag{4·53}$$

The work-energy equation, Eq. (4·49), now provides

$$m\Omega^2 L^2 = \tfrac{1}{2} m v_L{}^2$$

so the exit speed is given by

$$v_L = \Omega L\sqrt{2} \tag{4·54}$$

Having solved this problem by the use of the work-energy equation in the basic form found in Eq. (4·44a), we now consider its solution by means of the generalized work-energy equation in Eq. (4·44b). Choosing f as the tube frame a, we have simply

$$T^a = \tfrac{1}{2}m\mathbf{V}^{pa} \cdot \mathbf{V}^{pa} = \tfrac{1}{2}m\dot{R}^2$$

and the "effective force" is

$$\mathbf{F}^a = \mathbf{F} - m\mathbf{A}^{p^*i} = \mathbf{F} + m\Omega^2 R\hat{\mathbf{a}}_1$$

The "effective work" for frame a in the given interval is

$$W^a \triangleq \int_0^{t_L} \mathbf{F}^a \cdot \mathbf{V}^{pa}\,dt = \int_0^{t_L} (\mathbf{F} + m\Omega^2 R\hat{\mathbf{a}}_1) \cdot \dot{R}\hat{\mathbf{a}}_1$$

$$= m\Omega^2 \int_0^{t_L} R\dot{R}\,dt = m\Omega^2 \int_0^{L} R\,dR = \frac{m\Omega^2 L^2}{2}$$

since, for the frictionless tube, $\mathbf{F} \cdot \hat{\mathbf{a}}_1 = 0$. Thus Eq. (4·44b) furnishes

$$\tfrac{1}{2}m\Omega^2 L^2 = \tfrac{1}{2}m\dot{R}^2(t_L)$$

or

$$\dot{R}(t_L) = \Omega L$$

If we seek the exit speed v_L of the sphere relative to the base i (as previously), we must recognize that $v_L = |\mathbf{V}^{pi}| = |\mathbf{V}^{pf} + \boldsymbol{\omega}^{fi} \times L\hat{\mathbf{a}}_1| = |\Omega L\hat{\mathbf{a}}_1 + \Omega L\hat{\mathbf{a}}_2| = \sqrt{2}\,\Omega L$ in order to check the conclusion found in Eq. (4·54).

We elected in this problem to use work-energy equations to find the exit speed but, as always, we could instead have relied directly on the solution of the scalar differential equations available from $\mathbf{F} = m\mathbf{A}$. Since we have already calculated $\mathbf{A}$ in Eq. (4·52), we can easily record these equations and compare methods. In this problem there is no force in direction $\hat{\mathbf{a}}_1$ (no friction) and no resultant force in direction $\hat{\mathbf{a}}_3$ (the wall-contact force must balance the gravity force since there is no acceleration in this direction). The only resultant force on the sphere is the wall force in direction $\hat{\mathbf{a}}_2$, and we'll designate this force as $F_2\hat{\mathbf{a}}_2$. Equating this force with the product of mass m and $\mathbf{A}$ from Eq. (4·52) gives two scalar equations

$$0 = m(\ddot{R} - R\Omega^2) \tag{4·55}$$

$$F_2 = 2m\Omega\dot{R} \tag{4·56}$$

Equation (4·55) alone is sufficient to solve for $R(t)$, and since it is a linear constant-coefficient homogeneous differential equation, its solution by the methods of Chap. 5 is conceptually routine. Nonetheless, it is not a trivial task to determine v_L from this solution, and this approach turns out to be somewhat more difficult than the preceding solution employing the work-energy equation (See Prob. 5·15).

4·3·4 Conservative forces and potential energy

In general, the integral

$$W \triangleq \int_{t_1}^{t_2} \mathbf{F} \cdot \mathbf{V}^{pi}\, dt = \int_{P_1}^{P_2} \mathbf{F} \cdot d\mathbf{R}$$

which represents the work done by a given force function $\mathbf{F}$ on the particle p as it moves from position P_1 to position P_2, is dependent not only on the end points P_1 and P_2, but also on the path taken between those points, on the manner in which the path is traversed, and on the time at which the journey is begun. For example, a speedboat engine will use less fuel (and the propulsive force will do less work) in propelling the boat at constant speed directly across a calm lake than in reaching the opposite side by following the shoreline at that same speed. Thus even when the force $\mathbf{F}$ has a constant magnitude, the work it does on a particle can depend on the path taken. Even if the direct crossing path is selected, the speedboat engine will clearly use more fuel if it operates at maximum speed than it would for a slightly submaximum speed since the resistance offered by the water increases with the speed of the boat. In this case $\mathbf{F}$ is a given function of velocity, but it varies in magnitude from one case to the next. Thus, even for a given path, the work done by $\mathbf{F}$ can vary from case to case if $\mathbf{F}$ depends not only on the path itself but also on the speed of the particle on the path. Finally, more fuel will be required if the speedboat crosses the lake on another day when head winds offer added resistance to forward motion. Thus the work integral of the propulsive force may vary even for a given path traversed at a given rate if $\mathbf{F}$ is a function of time t.

We wish now to explore the possibility that there may exist certain forces $\mathbf{F}$ for which the work integral $\int_{P_1}^{P_2} \mathbf{F} \cdot d\mathbf{R}$ depends *only* on positions P_1 and P_2. This is clearly impossible if $\mathbf{F}$ depends explicitly on the time t or the velocity $\mathbf{V}^{pi}$ of the particle, but some success might be anticipated when $\mathbf{F}$ depends only on the position $\mathbf{R}$ of the particle. This is surely not a *sufficient* restriction on $\mathbf{F}$ (see the seventh example in §4·3·5), but it is clearly a *necessary* restriction.

Examination of the integral $\int_{P_1}^{P_2} \mathbf{F} \cdot d\mathbf{R}$ reveals that if this scalar is to depend only on P_1 and P_2, the dot product $\mathbf{F} \cdot d\mathbf{R}$ must be an exact scalar differential, which we shall call dU. In this special case the work integral becomes

$$W \triangleq \int_{t_1}^{t_2} \mathbf{F} \cdot \mathbf{V}^{pi}\, dt = \int_{P_1}^{P_2} \mathbf{F} \cdot d\mathbf{R} = \int_{U_1}^{U_2} dU = U_2 - U_1 \tag{4·57}$$

where U is a scalar depending only on position $\mathbf{R}$, and U_1 and U_2 are the values of U at positions P_1 and P_2, respectively. The scalar U, called the *force potential*, is an example of a *potential function*, such as we find appearing throughout mathematical physics. We can obtain an explicit expression for $\mathbf{F}$ in terms of U most easily by first introducing a set of cartesian coordinates R_1, R_2, and R_3 establishing the particle position in terms of inertial vectors $\hat{\mathbf{i}}_1$, $\hat{\mathbf{i}}_2$, and $\hat{\mathbf{i}}_3$. Then we have

$$\mathbf{R} = R_1\hat{\mathbf{i}}_1 + R_2\hat{\mathbf{i}}_2 + R_3\hat{\mathbf{i}}_3$$
$$\mathbf{F} = F_1\hat{\mathbf{i}}_1 + F_2\hat{\mathbf{i}}_2 + F_3\hat{\mathbf{i}}_3$$

and

$$\mathbf{F} \cdot d\mathbf{R} = F_1\,dR_1 + F_2\,dR_2 + F_3\,dR_3 \tag{4·58}$$

The differential dU is given by

$$dU = \frac{\partial U}{\partial R_1}\,dR_1 + \frac{\partial U}{\partial R_2}\,dR_2 + \frac{\partial U}{\partial R_3}\,dR_3 \tag{4·59}$$

Thus the equality $\mathbf{F} \cdot d\mathbf{R} = dU$ requires

$$F_1 = \frac{\partial U}{\partial R_1} \qquad F_2 = \frac{\partial U}{\partial R_2} \qquad F_3 = \frac{\partial U}{\partial R_3} \tag{4·60}$$

or, in vector terms,

$$\mathbf{F} = \frac{\partial U}{\partial R_1}\hat{\mathbf{i}}_1 + \frac{\partial U}{\partial R_2}\hat{\mathbf{i}}_2 + \frac{\partial U}{\partial R_3}\hat{\mathbf{i}}_3 \triangleq \boldsymbol{\nabla} U \tag{4·61}$$

where the del operator $\boldsymbol{\nabla}$ may be written in terms of cartesian coordinates as

$$\boldsymbol{\nabla} \triangleq \frac{\partial}{\partial R_1}\hat{\mathbf{i}}_1 + \frac{\partial}{\partial R_2}\hat{\mathbf{i}}_2 + \frac{\partial}{\partial R_3}\hat{\mathbf{i}}_3 \tag{4·62}$$

or in terms of spherical polar coordinates r, λ, φ in Fig. 4·10 as

$$\boldsymbol{\nabla} = \frac{\partial}{\partial r}\hat{\mathbf{e}}_r + \frac{1}{r}\frac{\partial}{\partial \varphi}\hat{\mathbf{e}}_\varphi + \frac{1}{r\sin\varphi}\frac{\partial}{\partial \lambda}\hat{\mathbf{e}}_\lambda \tag{4·63}$$

The symbol $\boldsymbol{\nabla} U$ (often called del U) signifies the *gradient* of U, sometimes written **grad U**. A force $\mathbf{F}$ can be written as the gradient of a potential function U whenever the measure numbers F_1, F_2, and F_3 of $\mathbf{F}$ meet the test $\boldsymbol{\nabla} \times \mathbf{F} = 0$ (the "curl of F'' is zero), or equivalently, in terms of inertial cartesian coordinates,

$$\frac{\partial F_1}{\partial R_2} = \frac{\partial F_2}{\partial R_1} \qquad \frac{\partial F_2}{\partial R_3} = \frac{\partial F_3}{\partial R_2} \qquad \frac{\partial F_3}{\partial R_1} = \frac{\partial F_1}{\partial R_3} \tag{4·64}$$

(The validity of this test follows from the identity $\nabla \times \nabla U = 0$.) Such forces are called *conservative*. When *only* conservative forces are applied to a particle, the work-energy equation, Eq. (4·42), becomes

$$\begin{aligned} W &= \int_{t_1}^{t_2} \mathbf{F} \cdot \mathbf{V}^{pi}\, dt = \int_{P_1}^{P_2} \mathbf{F} \cdot d\mathbf{R} = \int_{U_1}^{U_2} dU \\ &= U_2 - U_1 = T_2 - T_1 \end{aligned} \tag{4·65}$$

The force potential U evidently has units of energy, such as foot-pounds. It is conventional to define the *potential energy* V as the negative of the force potential U plus an arbitrary constant, that is, by definition

$$V \triangleq -U + C \tag{4·66}$$

The free constant C may be chosen so as to make the potential energy equal to zero for some arbitrarily chosen datum point. Then the potential energy of a particle p at some point P_1 is the work done by the conservative force $\mathbf{F}$ as the particle moves from P_1 to the datum point. Equivalently, it is the energy that must be expended in order to move the particle *from* the datum *to* the point P_1.

In terms of potential energy, Eq. (4·65) may be written

$$(C - V_2) - (C - V_1) = T_2 - T_1$$

or

$$T_1 + V_1 = T_2 + V_2 \tag{4·67a}$$

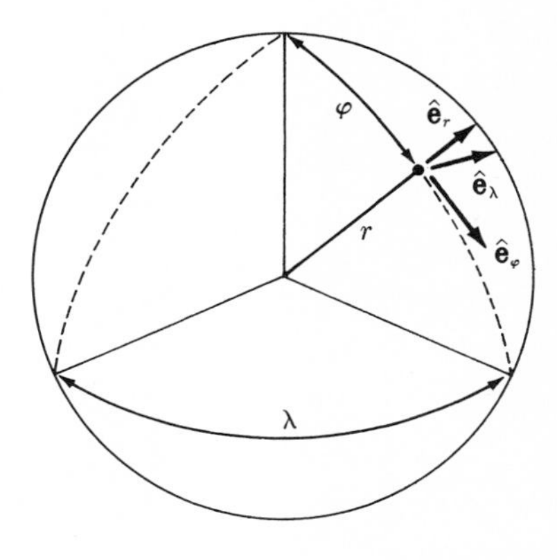

FIGURE 4·10

The sum of the kinetic energy T and the potential energy V is called the *total mechanical energy*, and from Eq. (4·67*a*) this sum is a constant for particles subjected *only* to conservative forces. (In mechanics the practice is to speak of energy being conserved only when *mechanical* energy is constant.) The identification of a force as conservative or not depends on the inertial reference frame selected (see Prob. 4·15. Thus the constancy of $T + V$ depends on the coordinate selection.

In parallel with the development of the force potential U and the potential energy V from the work W, beginning with Eq. (4·57), we can construct from the effective work W^f for frame f [see Eq. (4·45*b*)] the concepts of *effective force potential* U^f and *effective potential energy* V^f for frame f. Equations analogous to the series from Eq. (4·57) to (4·67) can be written by inspection of these equations, replacing i by f and adding a superscript f to the symbols $\mathbf{F}$, W, U, V, and T. In order for $\mathbf{F}^f$ to qualify as a *conservative effective force*, it is necessary that $\nabla \times \mathbf{F}^f = 0$, so that if in terms of some vector basis $\hat{\mathbf{e}}_1$, $\hat{\mathbf{e}}_2$, and $\hat{\mathbf{e}}_3$ we have

$$\mathbf{r} = r_1\hat{\mathbf{e}}_1 + r_2\hat{\mathbf{e}}_2 + r_3\hat{\mathbf{e}}_3$$

and

$$\mathbf{F}^f = F^f{}_1\hat{\mathbf{e}}_1 + F^f{}_2\hat{\mathbf{e}}_2 + F^f{}_3\hat{\mathbf{e}}_3$$

then in parallel with Eq. (4·64) we must satisfy

$$\frac{\partial F^f{}_1}{\partial r_2} = \frac{\partial F^f{}_2}{\partial r_1} \qquad \frac{\partial F^f{}_2}{\partial r_3} = \frac{\partial F^f{}_3}{\partial r_2} \qquad \frac{\partial F^f{}_3}{\partial r_1} = \frac{\partial F^f{}_1}{\partial r_3}$$

The scalars $F^f{}_j$, where $j = 1, 2, 3$, are available from Eqs. (4·42) and (4·43) as

$$F^f{}_j = \hat{\mathbf{e}}_j \cdot [F - m\mathbf{A}^{qi} - m\boldsymbol{\alpha}^{fi} \times \mathbf{r} - m\boldsymbol{\omega}^{fi} \times (\boldsymbol{\omega}^{fi} \times \mathbf{r})]$$

Substituting

$$\begin{aligned}
\mathbf{F} &= F_1\hat{\mathbf{e}}_1 + F_2\hat{\mathbf{e}}_2 + F_3\hat{\mathbf{e}}_3 \\
\mathbf{A}^{qi} &= A_1\hat{\mathbf{e}}_1 + A_2\hat{\mathbf{e}}_2 + A_3\hat{\mathbf{e}}_3 \\
\boldsymbol{\alpha}^{fi} &= \alpha_1\hat{\mathbf{e}}_1 + \alpha_2\hat{\mathbf{e}}_2 + \alpha_3\hat{\mathbf{e}}_3 \\
\boldsymbol{\omega}^{fi} &= \omega_1\hat{\mathbf{e}}_1 + \omega_2\hat{\mathbf{e}}_2 + \omega_3\hat{\mathbf{e}}_3
\end{aligned}$$

and performing the required cross multiplication yields

$$\begin{aligned}
F^f{}_1 &= F_1 - mA_1 - m(\alpha_2 r_3 - \alpha_3 r_2) - m[\omega_1\omega_2 r_2 \\
&\qquad + \omega_1\omega_3 r_3 - (\omega_2{}^2 + \omega_3{}^2)r_1] \\
F^f{}_2 &= F_2 - mA_2 - m(\alpha_3 r_1 - \alpha_1 r_3) - m[\omega_2\omega_3 r_3 \\
&\qquad + \omega_2\omega_1 r_1 - (\omega_1{}^2 + \omega_3{}^2)r_2] \\
F^f{}_3 &= F_3 - mA_3 - m(\alpha_1 r_2 - \alpha_2 r_1) - m[\omega_3\omega_1 r_1 \\
&\qquad + \omega_3\omega_2 r_2 - (\omega_1{}^2 + \omega_2{}^2)r_3]
\end{aligned}$$

with the resulting partial derivatives

$$\frac{\partial F^f{}_1}{\partial r_2} = \frac{\partial F_1}{\partial r_2} - m(\omega_1\omega_2 - \alpha_3) - m\frac{\partial A_1}{\partial r_2}$$
$$\frac{\partial F^f{}_2}{\partial r_1} = \frac{\partial F_2}{\partial r_1} - m(\omega_1\omega_2 + \alpha_3) - m\frac{\partial A_2}{\partial r_1}$$
$$\frac{\partial F^f{}_2}{\partial r_3} = \frac{\partial F_2}{\partial r_3} - m(\omega_2\omega_3 - \alpha_1) - m\frac{\partial A_2}{\partial r_3}$$
$$\frac{\partial F^f{}_3}{\partial r_2} = \frac{\partial F_3}{\partial r_2} - m(\omega_2\omega_3 + \alpha_1) - m\frac{\partial A_3}{\partial r_2}$$
$$\frac{\partial F^f{}_3}{\partial r_1} = \frac{\partial F_3}{\partial r_1} - m(\omega_1\omega_3 - \alpha_2) - m\frac{\partial A_3}{\partial r_1}$$
$$\frac{\partial F^f{}_1}{\partial r_3} = \frac{\partial F_1}{\partial r_3} - m(\omega_1\omega_3 + \alpha_2) - m\frac{\partial A_1}{\partial r_3}$$

Thus $\mathbf{F}^f$ is a conservative effective force if and only if it meets the following criteria: (1) $\mathbf{F}$ is conservative, (2) $\boldsymbol{\alpha}^{fi} \equiv 0$, (3) $\nabla \times \mathbf{A}^{qi} = 0$, and (4) $\mathbf{F}^f$ does not depend on time t or velocity $\mathbf{V}^{pf}$.

When $\mathbf{F}^f$ is conservative, and V^f is defined as an effective potential energy, then in parallel with Eq. (4·67a) we have

$$T^f{}_1 + V^f{}_1 = T^f{}_2 + V^f{}_2 \tag{4·67b}$$

where T^f is the kinetic energy for frame f [see Eq. (4·40b)], and subscripts 1 and 2 indicate conditions at times t_1 and t_2, respectively.

4·3·5 Examples of conservative forces and potential energy functions

In this article a number of vector force functions are considered, and potential energy functions are obtained whenever the forces are conservative.

First example $\mathbf{F} = -kR_1\hat{\mathbf{i}}_1$ is the force in a linear spring with spring constant k and extension R_1. The test for a conservative force [Eq. (4·64)] is satisfied since

$$\frac{\partial F_1}{\partial R_2} = \frac{\partial F_1}{\partial R_3} = 0 \qquad \text{and} \qquad F_2 = F_3 = 0$$

A force potential U is given by

$$U = \int F_1\, dR_1 + C_0 = -\tfrac{1}{2}kR_1{}^2$$

choosing C_0 as zero. The potential energy becomes

$$V = \tfrac{1}{2}kR_1^2 \tag{4·68}$$

if the constant C in Eq. (4·66) is chosen as zero.

Second example $\mathbf{F} = -mg\hat{\mathbf{i}}_3$ is the force on a particle of mass m in a uniform gravity field with $\hat{\mathbf{i}}_3$ directed "upward," where g is the gravity field constant. (For the earth, g is usually taken as 32.2 ft/sec².) The test for a conservative force is satisfied since

$$\frac{\partial F_3}{\partial R_1} = \frac{\partial F_3}{\partial R_2} = 0 \qquad \text{and} \qquad F_1 = F_2 = 0$$

A force potential U is given by

$$U = \int F_3\, dR_3 + C_0 = -mgR_3$$

choosing C_0 as zero. The potential energy of a particle in the gravity field becomes

$$V = mgR_3 \tag{4·69}$$

if the free constant in the definition is chosen so as to make $V = 0$ when $R_3 = 0$. (By convention, a point on the local surface of the earth is usually taken as the datum, where V and R_3 are zero.)

Third example $\mathbf{F} = -(GMm/r^2)\hat{\mathbf{e}}_r$ is the force on a particle or sphere of mass m in an inverse-square gravity field of a sphere of mass M, where r is the distance separating the centers of the two bodies, $\hat{\mathbf{e}}_r$ is a unit vector paralleling the line from the center of mass M to the body of mass m, and G is the "universal gravitational constant." The test for a conservative force is satisfied since from the polar coordinate expansion for ∇ in Eq. (4·63)

$$\nabla \times \mathbf{F} = \left(\frac{\partial}{\partial r}\hat{\mathbf{e}}_r + \frac{1}{r}\frac{\partial}{\partial \varphi}\hat{\mathbf{e}}_\varphi + \frac{1}{r \sin\varphi}\frac{\partial}{\partial \lambda}\hat{\mathbf{e}}_\lambda\right) \times \left(-\frac{GMm}{r^2}\hat{\mathbf{e}}_r\right) = 0$$

A force potential is given by

$$U = \int -\frac{GMm}{r^2}\, dr + C_0 = \frac{GMm}{r}$$

choosing C_0 as zero.

The potential energy of a particle in an inverse-square gravity field may now be written

$$V = -\frac{GmM}{r} \tag{4·70}$$

where the free constant in the definition has been selected so as to make the potential energy zero on a datum surface an infinite distance from the center of the gravity field. Due to this conventional choice of a datum surface, the potential energy of a particle in an inverse-square gravity field is always negative.

Fourth example

$$\mathbf{F} = 100(R_2R_3\hat{\mathbf{i}}_1 + R_1R_3\hat{\mathbf{i}}_2 + R_1R_2\hat{\mathbf{i}}_3)$$

In accordance with Eq. (4·64) we have

$$\frac{\partial F_1}{\partial R_2} = 100R_3 = \frac{\partial F_2}{\partial R_1}$$

$$\frac{\partial F_2}{\partial R_3} = 100R_1 = \frac{\partial F_3}{\partial R_2}$$

$$\frac{\partial F_3}{\partial R_1} = 100R_2 = \frac{\partial F_1}{\partial R_3}$$

so the given force is conservative. To obtain the corresponding force potential U, we note the following:

$$\frac{\partial U}{\partial R_1} = F_1$$

so

$$U = \int F_1\,dR_1 + G_1(R_2,R_3) = 100R_1R_2R_3 + G_1(R_2,R_3)$$

where $G_1(R_2,R_3)$ is an arbitrary function of R_2 and R_3.

$$\frac{\partial U}{\partial R_2} = F_2$$

so

$$U = \int F_2\,dR_2 + G_2(R_1,R_3) = 100R_1R_2R_3 + G_2(R_1,R_3)$$

where $G_2(R_1,R_3)$ is an arbitrary function of R_1 and R_3.

$$\frac{\partial U}{\partial R_3} = F_3$$

so

$$U = \int F_3\,dR_3 + G_3(R_1,R_2) = 100R_1R_2R_3 + G_3(R_1,R_2)$$

where $G_3(R_1,R_2)$ is an arbitrary function of R_1 and R_2. Evidently all of these relationships are satisfied if

$$G_1 \equiv G_2 \equiv G_3 \equiv 0$$

so U is simply

$$U = 100R_1R_2R_3$$

The corresponding potential energy then becomes

$$V = -100R_1R_2R_3$$

if the free constant is chosen so that $V = 0$ at the origin.

Fifth example $\mathbf{F} = -c\dot{R}_1\hat{\mathbf{i}}_1$ is the force transmitted by a viscous dashpot. Because $\mathbf{F}$ depends explicitly on velocity, it cannot be a conservative force. This matches our expectations since we recognize a dashpot as a device for transforming mechanical energy into thermal energy through viscous friction in the fluid.

Sixth example See Fig. 4·2, which shows a spring-mounted particle oscillating on tracks that span the diameter of a turntable rotating at a constant rate Ω. The force applied to the particle by the spring is, of course, conservative, as noted in the first example. But what about the transverse force $F_t\hat{\mathbf{e}}_t$ applied by the frictionless track?

If we recall that this is the force that provides the Coriolis acceleration, so that

$$F_t = 2m\Omega\dot{R}$$

it becomes apparent that $F_t\hat{\mathbf{e}}_t$ cannot be a conservative force since it depends explicitly on $\dot{R}$.

In contrast to the dashpot force, the force transmitted by the frictionless track may *seem* to be conservative since it is less obvious that mechanical energy is being changed into some other form. In this case it is the motor driving the turntable that dissipates or introduces mechanical energy to the particle as it oscillates. It is the nonconservative *constraint forces* such as this one that are most often overlooked. Because of the nonconservative constraint force applied by the frictionless track in this problem, the total mechanical energy $T + V$ is not constant.

Seventh example

$$\mathbf{F} = \mathbf{f} \times \mathbf{R} = \hat{\mathbf{i}}_1(f_2R_3 - f_3R_2) + \hat{\mathbf{i}}_2(f_3R_1 - f_1R_3) + \hat{\mathbf{i}}_3(f_1R_2 - f_2R_1)$$

Note that

$$\frac{\partial F_1}{\partial R_2} = -f_3 \qquad \frac{\partial F_2}{\partial R_1} = f_3$$

$$\frac{\partial F_2}{\partial R_3} = -f_1 \qquad \frac{\partial F_3}{\partial R_2} = f_1$$

$$\frac{\partial F_3}{\partial R_1} = -f_2 \qquad \frac{\partial F_1}{\partial R_3} = f_2$$

so that Eq. (4·64) is not satisfied and $\mathbf{F}$ is not a conservative force.

4·3·6 Virtual work

The direct application of Newton's second law in the form $\mathbf{F} = m\mathbf{A}$ is always possible for any particle dynamics problem, and no alternative approach within the framework of newtonian mechanics can provide any information not available from $\mathbf{F} = m\mathbf{A}$. Nonetheless, we have seen in the preceding sections that it is often convenient to operate on both sides of $\mathbf{F} = m\mathbf{A}$ in one way or another, dot- or cross-multiplying by any quantity we choose so as to produce results such as the impulse-momentum equation, the work-energy equation, etc. Now we consider another example of a quantity that sometimes gives useful results when dot-multiplied by $\mathbf{F}$ and $m\mathbf{A}$.

Let $\mathbf{R}$ be a position vector locating particle p in an inertial reference frame so that, in the course of the particle motion, $\mathbf{R}$ varies with time. At any instant of time, $\mathbf{R}$ has a certain value, but we are free to imagine the particle p to have instead a different position, and to let $\delta\mathbf{R}$ represent the vector from p to its imagined position. We might wish to consider p as having experienced a displacement $\delta\mathbf{R}$ from its true position to its imagined position, but this displacement occurs only in the imagination, and without the elapse of time. The quantity $\delta\mathbf{R}$ is called a *virtual displacement*. Whatever the value of $\delta\mathbf{R}$, it can clearly be used as a dot multiplier of $\mathbf{F} = m\mathbf{A}$, to obtain

$$\mathbf{F} \cdot \delta\mathbf{R} = m\mathbf{A} \cdot \delta\mathbf{R} \tag{4·71}$$

The quantity $\mathbf{F} \cdot \delta\mathbf{R}$ is called *virtual work* [note the correspondence with work as expressed in Eq. (4·45*a*)]. In application to a single particle, as is the case here, the virtual-work concept is of limited usefulness; we can always find some other quantity (such as a convenient unit vector) to multiply by $\mathbf{F}$ and $m\mathbf{A}$ to obtain the desired results without introducing imaginary displacements, as will be shown in the following example. Appreciation of the possibilities inherent in the virtual-work concept must be deferred until Chap. 6.

The particle p in Fig. 4·11 is constrained to remain on the axis of a straight tube of length L that undergoes a prescribed constant angular velocity $\Omega\hat{\mathbf{n}}_n$. A massless spring with spring constant k and undeformed length L connects the particle to the outer end of the tube. The equations of motion of p can, of course, be obtained from $\mathbf{F} = m\mathbf{A}$. But this vector equation is equivalent to *three* scalar equations, and only *one* scalar equation is required to define the motions of this particle with a single degree of freedom. To obtain a scalar equation that does not involve the unknown forces of constraint applied to p by the tube, we can apply Eq. (4·71) with $\delta\mathbf{R}$ chosen as $\delta R \hat{\mathbf{u}}_r$. The result (after obtaining $\mathbf{A}$ by the methods of Chap. 3) is

$$-kR\,\delta R = m(\ddot{R} - \Omega^2 R)\,\delta R$$

The arbitrary magnitude δR of the virtual displacement $\delta\mathbf{R}$ can now be canceled from both sides to provide the equation of motion of the particle in terms of the single unknown R, namely,

$$m\ddot{R} + (k - m\Omega^2)R = 0$$

In this problem we introduced a virtual displacement $\delta\mathbf{R}$ *compatible with the constraints* imposed on the particle by the tube; this is a general procedure that will be employed in a more significant way in Chap. 6. For this little problem we could have obtained the final equation of motion even more easily simply by dot-multiplying $\mathbf{F}$ and $m\mathbf{A}$ by the unit vector $\hat{\mathbf{u}}_r$.

4·3·7 Lagrange's equations

Because the virtual displacement $\delta\mathbf{R}$ dot-multiplied by $\mathbf{F} = m\mathbf{A}$ in the scalar equation (4·71) is completely arbitrary, we can obtain as many different scalar equations of this kind as we wish. Not all such equations will, however, be independent of one another. Since an individual particle can have no more than three degrees of freedom (moving in a three-dimensional space), no more than three equations such as Eq. (4·71) can be independent, and it is pointless to record

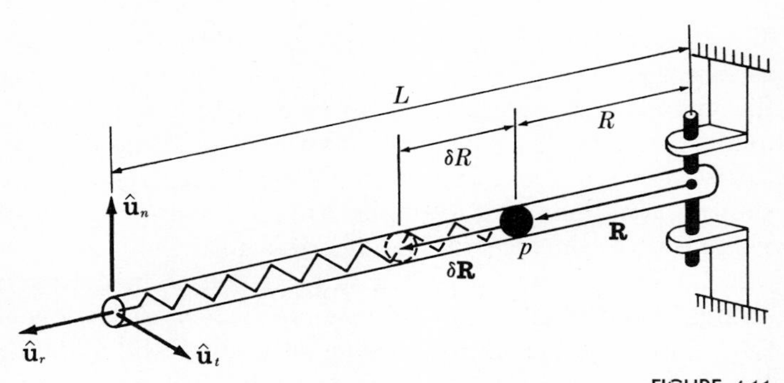

FIGURE 4·11

more than three. In the event that the particle is constrained (as in the system of Fig. 4·11), it may be unproductive to write more such scalar equations than we have independent variables (degrees of freedom), since equations involving unknown constraint forces are not helpful in determining the motion of the particle.

It should not be taken for granted that it will always be possible to select virtual displacements matching in number the degrees of freedom of the particle and sufficient to provide the corresponding number of scalar equations without involving unknown constraint forces. We will find, however, that this is possible whenever we can choose a set of *independent* scalar coordinates the values of which at any given time fully establish the position of the particle.

For a particle moving freely through space, there are three degrees of freedom and therefore three independent coordinates; we might choose some set of cartesian coordinates x, y, and z, or a set of spherical polar coordinates r, φ, and λ as shown in Fig. 4·10. In either case it is customary in the context of *lagrangian mechanics* to adopt the labels q_1, q_2, and q_3. In the general case requiring n coordinates, the symbols $q_1, q_2, \ldots, q_n$ are called *generalized coordinates* since they often represent mathematical abstractions related to such simple geometrical coordinates as noted above only by coordinate transformations.

For a particle moving on the surface of a sphere (perhaps a ship at sea, on the surface of the earth), we could choose sphere-centered cartesian coordinates x, y, and z, but it would be necessary to impose the *constraint equation*

$$x^2 + y^2 + z^2 - R_0^2 = 0$$

where R_0 is the radius of the sphere. The spherical polar coordinates of Fig. 4·10 present a far more attractive choice, however, since then the constraint equation is simply $r = R_0$. It would in fact suffice to define the generalized coordinates q_1 and q_2 of a ship at sea as the latitude and longitude, ignoring the third (constrained) coordinate entirely.

If for some reason we are so obtuse as to insist on using the cartesian coordinates x, y, and z to locate a ship at sea, then we can always solve the indicated constraint equation for $z = \pm(R_0^2 - x^2 - y^2)^{1/2}$ and work subsequently with only the two unknowns x and y.

To generalize the preceding observation, we note that if we have $n + m$ coordinates $q_1, q_2, \ldots, q_{n+m}$ related by m constraint equations of the form

$$\mathfrak{F}_i(q_1, q_2, \ldots, q_{n+m}, t) = 0 \qquad i = 1, \ldots, m \tag{4·72}$$

then we can (at least in theory) solve for m of the coordinates in terms of the remaining n coordinates, and in this way obtain a set of n *independent* generalized coordinates. This process of obtaining independent coordinates is possible only if any constraints among the original coordinates can be represented by equations of the special form of Eq. (4·72); such constraints are called *holonomic*, and all others are called *nonholonomic*.

For an example of a nonholonomic constraint, consider the constraint on the cartesian coordinates of a satellite, which must satisfy

$$x^2 + y^2 + z^2 \geq R_0^2$$

in order to remain on or above the surface of the spherical earth.

A second example of a nonholonomic constraint is presented by the particle constrained to move in free space at the constant speed V. If $\mathbf{R} = R_1\hat{\mathbf{i}}_1 + R_2\hat{\mathbf{i}}_2 + R_3\hat{\mathbf{i}}_3$ is the inertial position vector of the particle in terms of inertially fixed unit vectors $\hat{\mathbf{i}}_1$, $\hat{\mathbf{i}}_2$, and $\hat{\mathbf{i}}_3$, then the nonholonomic constraint equation becomes

$$\dot{R}_1^2 + \dot{R}_2^2 + \dot{R}_3^2 = V^2$$

It is assumed in the versions of Lagrange's equations considered here that the generalized coordinates are independent. This implies that any constraint equations that may have related other coordinates of the system represent holonomic constraints. The dynamical system is then called holonomic.

Lagrange's equations provide, for a single particle or any other system of particles and rigid bodies, a systematic procedure for obtaining the minimum number of independent scalar equations. These equations will be derived here for a single particle, first by building on Eq. (4·71) in the traditional manner, and then by a more modern procedure springing directly from $\mathbf{F} = m\mathbf{A}$.

For a particle with n degrees of freedom (n being either 1, 2, or 3), let $q_1(t), \ldots, q_n(t)$ be independent scalar generalized coordinates that fully characterize the position of the particle for any time t. In general, the inertial position vector $\mathbf{R}$ of the particle is available in terms of $q_1, \ldots, q_n$, and possibly also time t (as in the example of Fig. 4·11), so that

$$\mathbf{R} = \mathbf{R}(q_1, \ldots, q_n, t) \tag{4·73}$$

If the virtual displacement $\delta\mathbf{R}$ in Eq. (4·71) is restricted so as to make it compatible with any constraints imposed on the problem, then $\delta\mathbf{R}$ may be written (noting that time is not involved) as

$$\delta\mathbf{R} = \delta\mathbf{R}(q_1, \ldots, q_n) = \sum_{\alpha=1}^{n} \frac{\partial \mathbf{R}}{\partial q_\alpha}\,\delta q_\alpha$$

In this development, all vector derivatives, total or partial, are taken relative to an inertial reference frame. The virtual work in Eq. (4·71) then becomes

$$\mathbf{F} \cdot \delta\mathbf{R} = \mathbf{F} \cdot \sum_{\alpha=1}^{n} \frac{\partial \mathbf{R}}{\partial q_\alpha} \delta q_\alpha$$

where $\delta q_1, \ldots, \delta q_n$ are to be considered as generalized virtual displacements. In terms of the scalars,

$$Q_\alpha \triangleq \mathbf{F} \cdot \frac{\partial \mathbf{R}}{\partial q_\alpha} \qquad \alpha = 1, \ldots, n \tag{4·74}$$

called the *generalized forces*, the virtual work can be written

$$\mathbf{F} \cdot \delta\mathbf{R} = \sum_{\alpha=1}^{n} Q_\alpha \, \delta q_\alpha \tag{4·75}$$

The right-hand side of Eq. (4·71) then becomes

$$\begin{aligned} m\mathbf{A} \cdot \delta\mathbf{R} &= m\ddot{\mathbf{R}} \cdot \delta\mathbf{R} = m\left[\frac{d}{dt}(\dot{\mathbf{R}} \cdot \delta\mathbf{R}) - \dot{\mathbf{R}} \cdot \frac{d}{dt}(\delta\mathbf{R})\right] \\ &= m\left[\frac{d}{dt}\left(\dot{\mathbf{R}} \cdot \sum_{\alpha=1}^{n} \frac{\partial \mathbf{R}}{\partial q_\alpha} \delta q_\alpha\right) - \dot{\mathbf{R}} \cdot \frac{d}{dt}\left(\sum_{\alpha=1}^{n} \frac{\partial \mathbf{R}}{\partial q_\alpha} \delta q_\alpha\right)\right] \\ &= \sum_{\alpha=1}^{n} \frac{d}{dt}\left(m\dot{\mathbf{R}} \cdot \frac{\partial \mathbf{R}}{\partial q_\alpha}\right) \delta q_\alpha - \sum_{\alpha=1}^{n} \left(m\dot{\mathbf{R}} \cdot \frac{d}{dt}\frac{\partial \mathbf{R}}{\partial q_\alpha}\right) \delta q_\alpha \end{aligned} \tag{4·76}$$

where the time-independence of the generalized virtual displacements δq_α has been recognized. The expansion

$$\dot{\mathbf{R}} = \sum_{\alpha=1}^{n} \frac{\partial \mathbf{R}}{\partial q_\alpha} \dot{q}_\alpha + \frac{\partial \mathbf{R}}{\partial t}$$

permits the substitution

$$\frac{\partial \mathbf{R}}{\partial q_\alpha} = \frac{\partial \dot{\mathbf{R}}}{\partial \dot{q}_\alpha} \tag{4·77}$$

in the first sum in the final expression for $m\mathbf{A} \cdot \delta\mathbf{R}$, and in the second sum in this expression the expansion

$$\frac{d}{dt}\frac{\partial \mathbf{R}}{\partial q_\alpha} = \sum_{\beta=1}^{3} \frac{\partial^2 \mathbf{R}}{\partial q_\beta \partial q_\alpha} \dot{q}_\beta + \frac{\partial^2 \mathbf{R}}{\partial t \partial q_\alpha} = \frac{\partial}{\partial q_\alpha}\left(\sum_{\beta=1}^{3} \frac{\partial \mathbf{R}}{\partial q_\beta} \dot{q}_\beta + \frac{\partial \mathbf{R}}{\partial t}\right) = \frac{\partial \dot{\mathbf{R}}}{\partial q_\alpha}$$

is substituted, to obtain

$$m\mathbf{A}\cdot\delta\mathbf{R} = \sum_{\alpha=1}^{n}\left[\frac{d}{dt}\left(m\dot{\mathbf{R}}\cdot\frac{\partial\dot{\mathbf{R}}}{\partial\dot{q}_\alpha}\right) - m\dot{\mathbf{R}}\cdot\frac{\partial\dot{\mathbf{R}}}{\partial q_\alpha}\right]\delta q_\alpha$$

$$= \sum_{\alpha=1}^{n}\left\{\frac{d}{dt}\left[\frac{\partial}{\partial\dot{q}_\alpha}\left(\frac{1}{2}m\dot{\mathbf{R}}\cdot\dot{\mathbf{R}}\right)\right] - \frac{\partial}{\partial q_\alpha}\left(\frac{1}{2}m\dot{\mathbf{R}}\cdot\dot{\mathbf{R}}\right)\right\}\delta q_\alpha \tag{4·78}$$

When this expression is written in terms of the kinetic energy

$$T \triangleq \tfrac{1}{2}m\dot{\mathbf{R}}\cdot\dot{\mathbf{R}} \tag{4·79}$$

and combined in Eq. (4·71) with the virtual work as expressed in Eq. (4·75), we have

$$\sum_{\alpha=1}^{n} Q_\alpha\delta q_\alpha = \sum_{\alpha=1}^{n}\left[\frac{d}{dt}\left(\frac{\partial T}{\partial\dot{q}_\alpha}\right) - \frac{\partial T}{\partial q_\alpha}\right]\delta q_\alpha$$

or

$$\sum_{\alpha=1}^{n}\left[\frac{d}{dt}\left(\frac{\partial T}{\partial\dot{q}_\alpha}\right) - \frac{\partial T}{\partial q_\alpha} - Q_\alpha\right]\delta q_\alpha = 0$$

Because each of the generalized virtual displacements $\delta q_1, \ldots, \delta q_n$ is independent, the expressions in the brackets must vanish individually for the n values of α, that is,

$$\frac{d}{dt}\frac{\partial T}{\partial\dot{q}_\alpha} - \frac{\partial T}{\partial q_\alpha} = Q_\alpha \qquad \alpha = 1, \ldots, n \tag{4·80}$$

for a holonomically constrained or unconstrained particle. This result is a form of *Lagrange's equation*, derived here for the very special case of a single particle with n independent coordinates.

In the special case for which all components of the resultant force $\mathbf{F}$ which contribute to $Q_1, \ldots, Q_n$ are *conservative* in the sense introduced in §4·3·4, we can express the generalized force Q_α in terms of the potential energy V. Equations (4·61) and (4·66) then show that $\mathbf{F} = -\nabla V$, and hence

$$Q_\alpha \triangleq \mathbf{F}\cdot\frac{\partial\mathbf{R}}{\partial q_\alpha} = -\nabla V\cdot\frac{\partial\mathbf{R}}{\partial q_\alpha} \qquad \alpha = 1, \ldots, n$$

Writing both $\mathbf{R}$ and the gradient operator ∇ in terms of cartesian coordinates R_1, R_2, and R_3 and corresponding unit vectors $\hat{\mathbf{u}}_1$, $\hat{\mathbf{u}}_2$, and $\hat{\mathbf{u}}_3$ provides

$$Q_\alpha = -\left(\frac{\partial V}{\partial R_1}\hat{\mathbf{u}}_1 + \frac{\partial V}{\partial R_2}\hat{\mathbf{u}}_2 + \frac{\partial V}{\partial R_3}\hat{\mathbf{u}}_3\right)\cdot\left[\frac{\partial}{\partial q_\alpha}(R_1\hat{\mathbf{u}}_1 + R_2\hat{\mathbf{u}}_2 + R_3\hat{\mathbf{u}}_3)\right]$$

$$= -\left(\frac{\partial V}{\partial R_1}\frac{\partial R_1}{\partial q_\alpha} + \frac{\partial V}{\partial R_2}\frac{\partial R_2}{\partial q_\alpha} + \frac{\partial V}{\partial R_3}\frac{\partial R_3}{\partial q_\alpha}\right)$$

or

$$Q_\alpha = -\frac{\partial V}{\partial q_\alpha} \qquad \alpha = 1, \ldots, n \tag{4·81}$$

Combining Eqs. (4·80) and (4·81) thus provides, for holonomic conservative systems, Lagrange's equation in the form

$$\frac{d}{dt}\left(\frac{\partial T}{\partial \dot{q}_\alpha}\right) - \frac{\partial T}{\partial q_\alpha} + \frac{\partial V}{\partial q_\alpha} = 0 \qquad \alpha = 1, \ldots, n \tag{4·82}$$

In terms of the quantity $L \triangleq T - V$ (called the *lagrangian*), this equation can be written

$$\frac{d}{dt}\left(\frac{\partial L}{\partial \dot{q}_\alpha}\right) - \frac{\partial L}{\partial q_\alpha} = 0 \qquad \alpha = 1, \ldots, n \tag{4·83}$$

since V depends only on position, so $\partial V/\partial \dot{q}_\alpha \equiv 0$ always.

Having obtained Lagrange's equations in the traditional way by means of the concepts of virtual work and virtual displacement, it is a simple matter to rederive these equations by operating on $\mathbf{F} = m\mathbf{A}$ in a more direct manner. If again we accept $q_1, \ldots, q_n$ as a complete and independent set of generalized coordinates, then $\mathbf{R} = \mathbf{R}(q_1, \ldots, q_n, t)$ and the vectors $\partial\dot{\mathbf{R}}/\partial\dot{q}_1, \ldots, \partial\dot{\mathbf{R}}/\partial\dot{q}_n$ are independent. The equation $\mathbf{F} = m\ddot{\mathbf{R}}$ can then be dot-multiplied by each of these vectors, to provide a complete set of equations of motion in the form

$$\mathbf{F} \cdot \frac{\partial \dot{\mathbf{R}}}{\partial \dot{q}_\alpha} = m\ddot{\mathbf{R}} \cdot \frac{\partial \dot{\mathbf{R}}}{\partial \dot{q}_\alpha} \qquad \alpha = 1, \ldots, n \tag{4·84}$$

From Eqs. (4·74) and (4·77), we find that the left side of this equation is again Q_α, and the right side can be manipulated as previously to obtain

$$\begin{aligned}
Q_\alpha &= \frac{d}{dt}\left(m\dot{\mathbf{R}} \cdot \frac{\partial \dot{\mathbf{R}}}{\partial \dot{q}_\alpha}\right) - m\dot{\mathbf{R}} \cdot \frac{d}{dt}\frac{\partial \dot{\mathbf{R}}}{\partial \dot{q}_\alpha} \\
&= \frac{d}{dt}\left[\frac{\partial}{\partial \dot{q}_\alpha}\left(\frac{1}{2}m\dot{\mathbf{R}} \cdot \dot{\mathbf{R}}\right)\right] - m\dot{\mathbf{R}} \cdot \frac{d}{dt}\left(\frac{\partial \mathbf{R}}{\partial q_\alpha}\right) \\
&= \frac{d}{dt}\left[\frac{\partial}{\partial \dot{q}_\alpha}\left(\frac{1}{2}m\dot{\mathbf{R}} \cdot \dot{\mathbf{R}}\right)\right] - m\dot{\mathbf{R}} \cdot \frac{\partial \dot{\mathbf{R}}}{\partial q_\alpha} \\
&= \frac{d}{dt}\left[\frac{\partial}{\partial \dot{q}_\alpha}\left(\frac{1}{2}m\dot{\mathbf{R}} \cdot \dot{\mathbf{R}}\right)\right] - \frac{\partial}{\partial q_\alpha}\left(\frac{1}{2}m\dot{\mathbf{R}} \cdot \dot{\mathbf{R}}\right)
\end{aligned}$$

which with the substitution of T from Eq. (4·79) provides Lagrange's equations in the form of Eq. (4·80).

A primary advantage of Lagrange's equations in comparison with the direct application of $\mathbf{F} = m\mathbf{A}$ is that the latter may contain in $\mathbf{F}$ certain unknown forces of constraint that automatically disappear from the generalized forces [see Eqs. (4·74) and (4·77)]

$$Q_\alpha = \mathbf{F} \cdot \frac{\partial \dot{\mathbf{R}}}{\partial \dot{q}_\alpha} \tag{4·85}$$

Any of the forces comprising $\mathbf{F}$ that do not contribute to Q_α are called *nonworking constraint forces.*

4·3·8 Applications of Lagrange's equations

Although the potency of Lagrange's equations as a method of obtaining equations of motion cannot be appreciated by considering the dynamics of a single particle, in this chapter a pair of sample problems will be solved in order to illustrate in a basic way some of the features of the lagrangian approach. More substantial examples will be found in Chap. 6, which deals with systems containing many particles.

In §4·1·3 the equations of motion of the particle p illustrated in Fig. 4·2 were obtained by writing $\mathbf{F} = m\mathbf{A}$; Eqs. (4·13) to (4·15) were the resulting set of three scalar equations. Now we can apply Lagrange's equations to this problem, and compare the features of the two methods.

At the outset we consider the special case of the problem in which the turntable is not rotating in inertial space ($\Omega = 0$). Then the problem is simply one of rectilinear motion of a particle on a spring, with $R\hat{\mathbf{e}}_r$ representing the displacement of p from a point o fixed in inertial space, and $-kR\hat{\mathbf{e}}_r$ representing the force exerted on p by the spring. The gravitational force on p is $-mg\hat{\mathbf{e}}_n$ (the turntable is horizontal so $\hat{\mathbf{e}}_n$ comes out of the paper). In addition we would admit the possibility that the track applies a force to p; since the track is frictionless, it will suffice to let $F_t\hat{\mathbf{e}}_t + F_n\hat{\mathbf{e}}_n$ represent the force exerted on p by the track. (You may know from $\mathbf{F} = m\mathbf{A}$ that F_t is zero when $\Omega = 0$, but you won't need this information to apply Lagrange's equations.) We choose R as the single generalized coordinate, previously called $q_1 (n = 1)$.

To apply Eq. (4·80), we need [from Eq. (4·79)]

$$T = \tfrac{1}{2}m\mathbf{V}^{pi} \cdot \mathbf{V}^{pi} = \tfrac{1}{2}m(\dot{R}\hat{\mathbf{e}}_r) \cdot (\dot{R}\hat{\mathbf{e}}_r) = \tfrac{1}{2}m\dot{R}^2$$

and [from Eq. (4·85)], with $\dot{q}_1 = \dot{R}$

$$\begin{aligned} Q_1 &= (-kR\hat{\mathbf{e}}_r - mg\hat{\mathbf{e}}_n + F_t\hat{\mathbf{e}}_t + F_n\hat{\mathbf{e}}_n) \cdot \frac{\partial}{\partial \dot{R}}(\dot{R}\hat{\mathbf{e}}_r) \\ &= -kR \end{aligned}$$

Thus Eq. (4·80) provides

$$\frac{d}{dt}\left[\frac{\partial}{\partial \dot{R}}\left(\frac{1}{2}m\dot{R}^2\right)\right] - \frac{\partial}{\partial R}\left(\frac{1}{2}m\dot{R}^2\right) = -kR$$

or

$$\frac{d}{dt}(m\dot{R}) - 0 = -kR$$

or

$$m\ddot{R} + kR = 0$$

This matches the result of Eq. (4·13) when $\Omega = 0$. This is perhaps the simplest possible application of Lagrange's equations, and it holds no surprises. The force terms $-mg\hat{\mathbf{e}}_n + F_t\hat{\mathbf{e}}_t + F_n\hat{\mathbf{e}}_n$ do not contribute to Q_1; in fact we know from our application of $\mathbf{F} = m\mathbf{A}$ in §4·1·3 that the sum of these forces is zero when $\Omega = 0$.

But what if the turntable is rotating at a prescribed constant rate ($\Omega \neq 0$)? Then our experience with $\mathbf{F} = m\mathbf{A}$ tells us that $F_t \neq 0$ [see Eq. (4·14)], and we might wonder if $F_t\hat{\mathbf{e}}_t$ still qualifies as a "nonworking constraint force," with no contribution to Q_1. Now we have velocity

$$\mathbf{V}^{pi} = \dot{R}\hat{\mathbf{e}}_r + R\Omega\hat{\mathbf{e}}_t$$

so that

$$T = \tfrac{1}{2}m(\dot{R}^2 + R^2\Omega^2)$$

and

$$Q_1 = (-kR\hat{\mathbf{e}}_r - mg\hat{\mathbf{e}}_n + F_t\hat{\mathbf{e}}_t + F_n\hat{\mathbf{e}}_n) \cdot \frac{\partial}{\partial \dot{R}}(\dot{R}\hat{\mathbf{e}}_r + R\Omega\hat{\mathbf{e}}_t)$$
$$= -kR$$

The generalized force Q_1 is unchanged, and the equation of motion from Eq. (4·80) is

$$\frac{d}{dt}\left[\frac{\partial}{\partial \dot{R}}\left(\frac{1}{2}m\dot{R}^2 + \frac{1}{2}mR^2\Omega^2\right)\right] - \frac{\partial}{\partial R}\left(\frac{1}{2}m\dot{R}^2 + \frac{1}{2}mR^2\Omega^2\right) = -kR$$

or

$$m\ddot{R} - mR\Omega^2 = -kR$$

which matches Eq. (4·13).

Thus in the traditional language adopted here the force $F_t\hat{\mathbf{e}}_t$ remains a "nonworking constraint force," even when $\Omega \neq 0$. This language might be particularly disturbing when considered in the context of §4·3·3, in which the work-energy equation is applied to a system (Fig. 4·9) conceptually identical to the one treated here (Fig. 4·11), except for the spring. In §4·3·3 it was clearly indicated that the force here called $F_t\hat{\mathbf{e}}_t$ contributes to the integral

$$W = \int_{t_1}^{t_2} \mathbf{F} \cdot \mathbf{V}^{pi}\, dt$$

and hence certainly does do work on the particle. The term *nonworking constraint force* is derived from the fact that such a force does no work in the course of virtual displacements δq_α of the generalized coordinates. Rather than attempt to classify a force by trying to figure out when it does work, you should simply determine from Eq. (4·85) whether it contributes to Q_α.

An example of a force that *would* contribute to Q_1 in the problem of Fig. 4·11 is a coulomb friction force applied by the track to p in the direction of relative motion, and represented analytically by

$$F_f\hat{\mathbf{e}}_r = -\mu F_t \operatorname{sgn}(\dot{R})\hat{\mathbf{e}}_r$$

Here sgn $(\dot{R})$, to be read signum $\dot{R}$, is $+1$ if $\dot{R} > 0$ and -1 if $\dot{R} < 0$, so the friction force always opposes the motion. The symbol μ is the coefficient of friction, as first introduced in §2·2·12. (In reality two material surfaces in contact may have one value of μ when at relative rest and a different value of μ when sliding relative to one another, but we ignore this complication.)

Because $(F_f\hat{\mathbf{e}}_r) \cdot \hat{\mathbf{e}}_r = F_f = -\mu F_t \operatorname{sgn}(\dot{R})$ is obtained from Eq. (4·85) as an addition to Q_1 previously calculated for this problem, it cannot be ignored as a "nonworking constraint force." Yet because it involves the unknown F_t, it cannot simply be tacked onto the equation of motion without further investigation. In this case we would be obliged to use $\mathbf{F} = m\mathbf{A}$ to determine F_t [see Eq. (4·14)]. Lagrange's equations in the form of Eq. (4·80) are simply insufficient to solve the problem when Coulomb friction is introduced.

It should be noted that in the *absence* of coulomb friction, the problem illustrated in Fig. 4·11 yields not only to the formulation of Lagrange's equation in Eq. (4·80) but also to the special case in Eq. (4·83). The generalized force $Q_1 = -kR$ can be obtained as in Eq. (4·81) from the potential energy function $V = \frac{1}{2}kR^2$ (compare with the first example in §4·3·5). Thus for $\Omega \neq 0$, the lagrangian is

$$L \triangleq T - V = \tfrac{1}{2}m(\dot{R}^2 + R^2\Omega^2) - \tfrac{1}{2}kR^2$$

and Eq. (4·83) provides once again

$$\frac{d}{dt}(m\dot{R}) - mR\Omega^2 + kR = 0$$

4·4 PERSPECTIVE

This chapter began with a discussion of the procedures used in obtaining the equations of motion of a particle from Newton's second law, $\mathbf{F} = m\mathbf{A}$, and concluded with exploration of the alternative of using Lagrange's equations to obtain these equations. Most students find that the difficulties in this task of deriving equations of motion disappear as understanding of the kinematical concepts of the previous chapter develops. The end product of the application of $\mathbf{F} = m\mathbf{A}$ to a particle is a vector second-order differential equation, or a set of three generally coupled scalar second-order differential equations, while Lagrange's equations lead directly to as many scalar second-order differential equations as the particle has degrees of freedom, provided that any constraints among the coordinates are holonomic.

In most cases of genuine modern interest, the equations of motion, even for a single particle, cannot easily be solved. We defer discussion of this topic to the next chapter, and even there solutions will be found only in special cases. Before confronting the task of solving equations of motion simply as mathematicians confronting equations, we have tried in Chap. 4 to explore the possibilities of using physically meaningful concepts like *momentum* and *energy* to facilitate the task of determining the motion. These are invaluable aids to engineers and physicists, and their importance deserves emphasis. At the same time it must be understood that these concepts are helpful only in certain classes of problems, and in many applications we must solve the scalar equations of motion directly. In the next chapter we will begin to explore this challenging assignment.

PROBLEMS

4·1 If a shot-putter can put a 16-lb shot a distance of 60 ft when standing on level ground, how far could he put the shot if he were performing on a special enclosed car of a train traveling on a straight, level track at a uniform speed of 100 ft/sec?

4·2 In order to provide astronauts with a sense of weight such as they experience when standing on the earth, a space station of toroidal configuration (see Fig. P4·2) is designed to spin about its axis of symmetry. If the radial distance to the astronaut's heart (the critical organ) is 64.4 ft, what spin rate Ω is required to establish for the heart the same weight that it has on earth? (Recall the definition of weight in §2·1·2.) Is the dynamic environment of the heart made *identical* to that on earth by this spin? Will it change

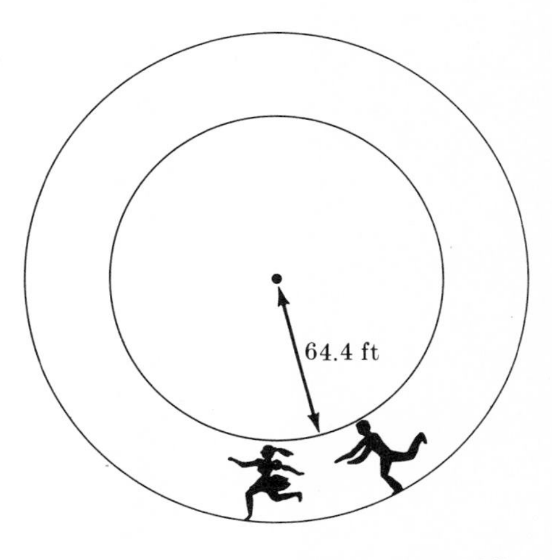

FIGURE P4·2

the weight of the heart if the astronaut runs around the perimeter of his craft?

4·3 While balancing a small ball on his nose as shown in Fig. P4·3, a circus clown rides a scooter on a circle of radius R painted concentrically on a horizontal, circular platform spinning about an axis through its center at the constant rate of Ω rad/sec relative to the earth, here assumed an inertial reference.

The clown is anxious to determine the scooter motion that will simplify his act to the level of difficulty experienced on a stationary platform. As an engineering graduate of the early 1970s, he knows about Eqs. (4·2) and (4·3), and he reasons that in order to counterbalance the centrifugal "forces" (directed radially outward), he must develop through his motion relative to the turntable an equal and opposite Coriolis "force." If he rides on the circle in opposition to the direction in which he is transported by the table, traveling at speed $v = R\Omega/2$, he concludes that his balancing act will not be complicated by the turntable motion, since then in Eq. (4·3), with f the turntable frame, m the ball mass, and $\mathbf{r}$ the ball position vector with respect to the table center,

$$-m\boldsymbol{\omega}^{fi} \times (\boldsymbol{\omega}^{fi} \times \mathbf{r}) - 2m\boldsymbol{\omega}^{fi} \times \mathbf{V}^{pf} = 0$$

Is the clown correct in his reasoning? Have you an alternative choice for v? In your response, speak first as an engineer, using Eq. (4·2) or (4·3), and then speak as a clown, using your head.

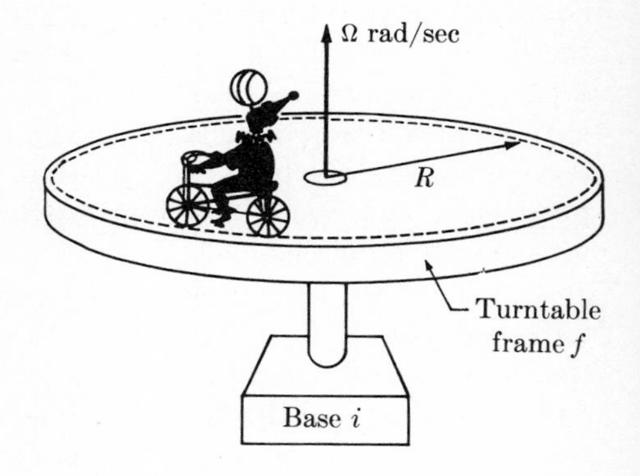

FIGURE P4·3

4·4 A fireman named p of weight W hangs onto the end of a ladder a that is telescoping out of a ladder b at constant linear speed $\mathbf{V}^{ab}$ as shown in Fig. P4·4. Simultaneously, b is rotating relative to ladder c at constant angular speed $\dot{\gamma}$, c is rotating relative to turntable (disk) d at a constant angular speed $\dot{\beta}$ (a negative number—note that β in Fig. P4·4 is shown negative), and d rotates on the

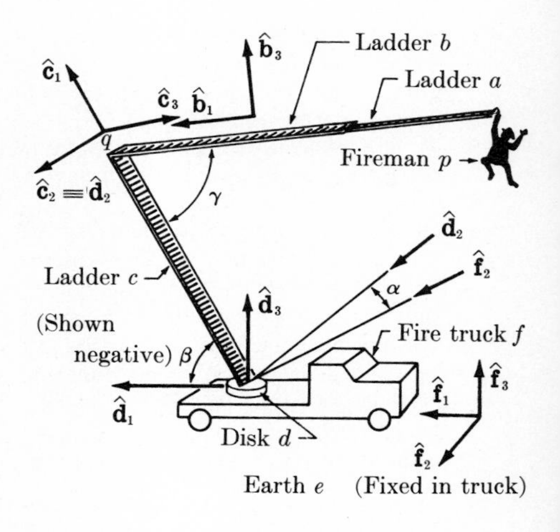

FIGURE P4·4

stationary fire engine f with constant angular speed $\dot{\alpha}$. Ladders a, b, and c are of length L, and at time $t = 0$, ladder a is telescoped wholly within b.

a. Find $\mathbf{V}^{pe}$, the velocity of p relative to the earth, in the vector basis $\hat{\mathbf{c}}_1$, $\hat{\mathbf{c}}_2$, and $\hat{\mathbf{c}}_3$ of Fig. P4·4.

b. Develop an expression for the force p transmits to a as he hangs on, assuming that e is inertial and p is a particle. Do not trouble to differentiate $\mathbf{V}^{pe}$ explicitly; you may instead retain $\mathbf{V}^{pe}$ as a symbol in your answer.

4·5 A particle p of mass m is moving on frictionless tracks that traverse a diameter of a horizontal disk a, as shown in Fig. P4·5. Linear springs attached to p and to the ends of the track apply a force $-kr\hat{\mathbf{a}}_1$ to p when displaced a distance r in direction $\hat{\mathbf{a}}_1$ along the track. The center q of a is attached a distance R from the center of a second disk b, while the center of b is fixed in an inertial frame i. The angular velocities of b and a are prescribed constants such that $\boldsymbol{\omega}^{bi} = \Omega\hat{\mathbf{n}}$ and $\boldsymbol{\omega}^{ab} = \omega\hat{\mathbf{n}}$, with $\dot{\Omega} = \dot{\omega} = 0$ and $\hat{\mathbf{n}}$ normal to both disks.

You are to apply $\mathbf{F} = m\mathbf{A}^{pi}$ to p, obtaining scalar equations of motion in a vector basis fixed in a.

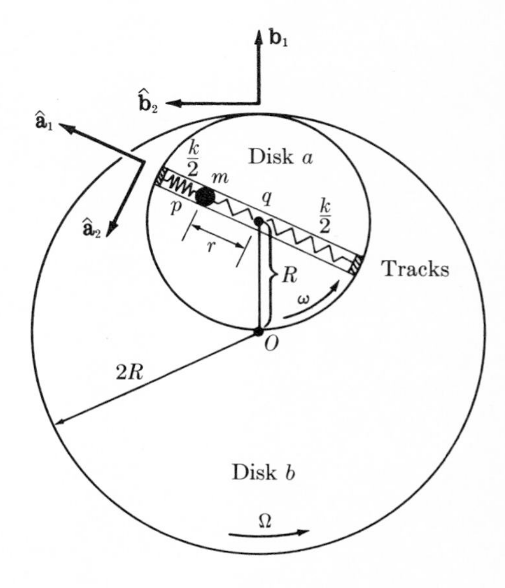

FIGURE P4·5

4·6 Detonation of the charge of a bullet in the firing chamber of a revolver sends the slug, of mass m, on a straight line out of the barrel with inertial velocity $v\hat{\mathbf{u}}$. What is the impulse $\mathbf{I}$ imparted to the slug? What is the impulse $\mathbf{I}'$ imparted by the gun to the hand that holds it? (Note that the simplicity of this problem belies the complexity of the physical situation; although it would be virtually impossible to determine with any accuracy the time history and distribution of forces within the gun, we *can* readily determine the impulse it imparts to the slug and to the hand.)

4·7 A particle p of mass m leaves point A shown in Fig. P4·7 with initial speed v, collides at time $t = t^*$ with the wall at point C, and rebounds to reach point B at $t = 2t^*$. What is the impulse $\mathbf{I}$ imparted to p by the wall during the collision at C?

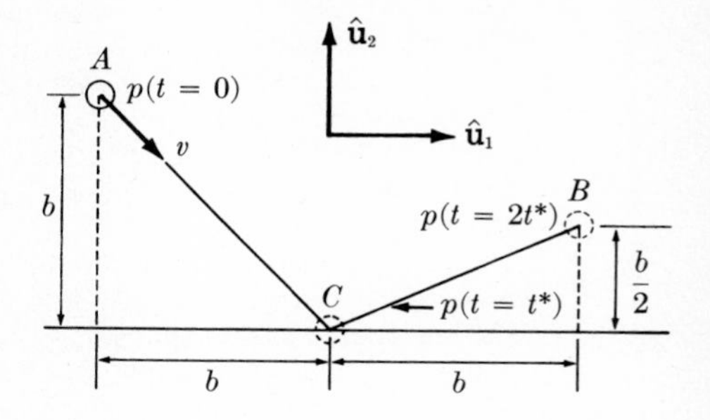

FIGURE P4·7

4·8 Consider the example of the spacecraft illustrated in Fig. 4·3, and repeat the calculations in the text for a new initial position for $t = 0$. On its uncorrected course, the s/c will pass within the same distance $\epsilon = 864{,}000$ ft of the planned rendezvous point P^* as shown in Fig. 4·3, but it will reach this point of closest approach to P^* at time $t = 23$ hr, 57 min, 07 sec, which is 173 sec too early for the planned rendezvous. Making the assumption that the midcourse motor imparts an ideal (instantaneous) impulse to the s/c, find the desired correction, including burning time and motor thrust direction.

4·9 According to the example of the jet-propelled slot car in §4·2·4 (see Fig. 4·4), it is immaterial to the car's speed on completing a lap whether I impart to it an impulse while on the straightaway (say at point A) or as I enter a turn (say at point B). Yet when I test this theory on my son's model racer (which imparts the impulse mechanically rather than with a jet), I find that it is better to impart the impulse at point A. What could you offer as a speculative explanation?

4·10 A slot car on a frictionless spiral track on a circular table is shown in Fig. P4·10. If at the perimeter of the table the car is given an impulse $\mathbf{I} = m\,\Delta v\,\hat{\mathbf{c}}$, where m is the car mass and $\hat{\mathbf{c}}$ is a unit vector fixed in the car and pointing forward, what will be the speed of the car when it crosses the dotted line at the table center? Justify your answer analytically.

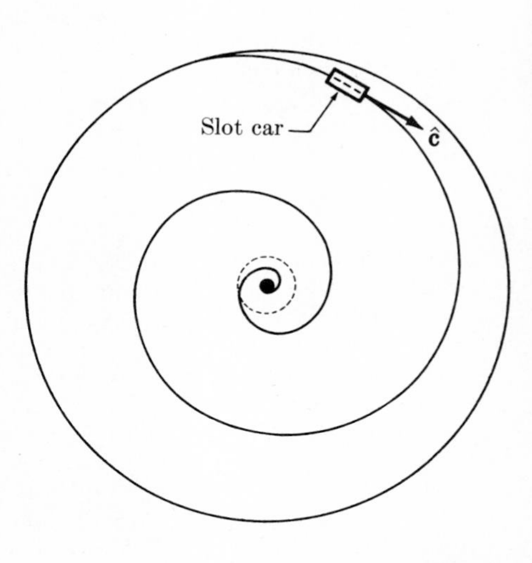

FIGURE P4·10

4·11 A particle p of mass $m = 0.01$ slug is attached to a string that passes without frictional restraint through a hole in the center of a circular table and terminates in a spool that reels the string in at the rate $V_1 = 0.5$ ft/sec. At time $t = 0$, p is located at the perimeter of the table, a distance $R = 12$ ft from the center, and p has a velocity $V_0\hat{\mathbf{u}}_t$, where $V_0 = 5$ ft/sec and $\hat{\mathbf{u}}_t$ is tangential to the table's perimeter (see Fig. P4·11). Find an expression for $\varphi(t)$, where φ is the angle shown in Fig. P4·11. What is the numerical value of φ when $t = 12$ sec?

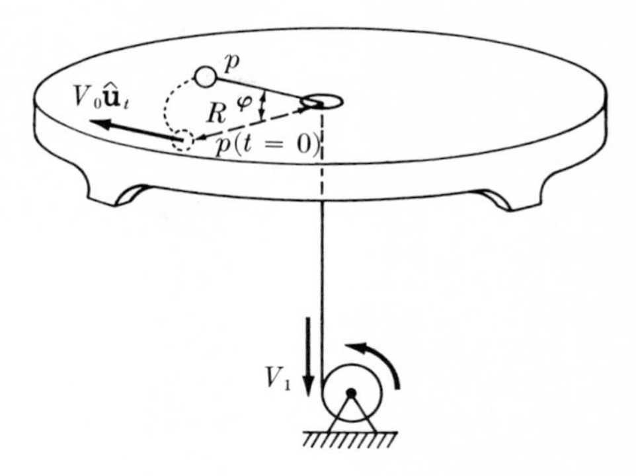

FIGURE P4·11

4·12 A particle p of mass m is traveling horizontally with inertial velocity $V_1\hat{\mathbf{u}}_1$ on a line normal to the wing of an aircraft diving vertically with velocity $V_2\hat{\mathbf{u}}_2$; p is about to strike the wing at point O, a distance L from the mass center C of the airplane, as shown in Fig. P4·12. Use Eqs. (4·39) and (4·38) to find expressions for the angular momentum $\mathbf{H}^C$ and the moment of momentum $\mathbf{h}^C$ of p when collision is impending. Substitute the numerical values $V_1 = 1200$ ft/sec; $V_2 = 500$ ft/sec; $L = 8$ ft, and $m = 0.001$ slug.

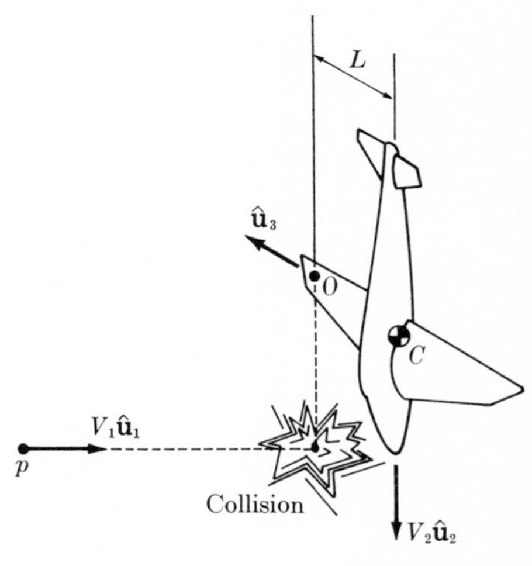

FIGURE P4·12

4·13 Two laborers, A and B, both have the capacity to exert a force of magnitude P on a cart with frictionless wheel bearings and weight W that each is obliged to push to the top of a loading platform, a height h above the ground level. Mr. A takes the direct approach, pushing his cart up a straight ramp of inclination α with the horizontal, while Mr. B pushes his cart up the spiral ramp, which has at all points the inclination β from the horizontal (see Fig. P4·13). Both men release their carts at the top, where they crash into a wall, and both men come down firemen's poles.

If each man applies a constant force $P = 0.6W$ (although this quickly produces unrealistically rapid motions), and the angles are $\alpha = 30°$ and $\beta = 11°32'$, who finishes the job first? Which of the workmen does more work on the cart in accomplishing his task? If you can do the

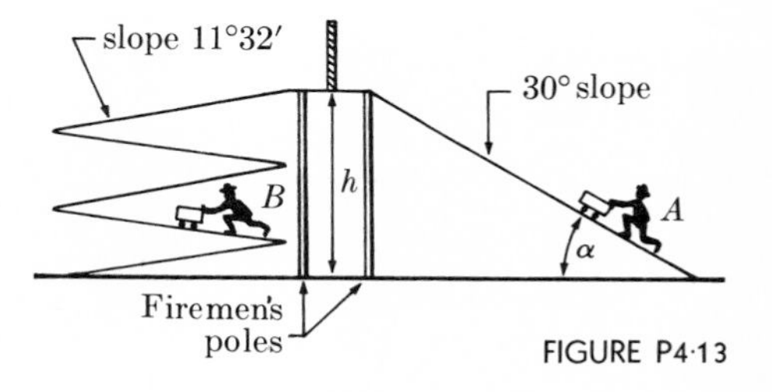

FIGURE P4·13

job by pushing with constant force $P^* = 0.75W$, which of the two paths would you choose?

4·14 A smooth-walled circular tube of radius R in which a slug p of mass m slides without friction, under a uniform gravitational attraction of magnitude mg, is shown in Fig. P4·14. The tube is constrained to rotate about a vertical diametral axis at the constant angular velocity $\Omega\hat{\mathbf{u}}_v$. The position of p in the tube is established by the angle θ.

Obtain a scalar equation of motion for θ in two ways:

a. Use $\mathbf{F} = m\mathbf{A}$ exclusively.
b. Use $\mathbf{F} \cdot \mathbf{V}^{pi} = \dot{T}$ [Eq. (4·41*a*)] and $\mathbf{F} = m\mathbf{A}$ as required.

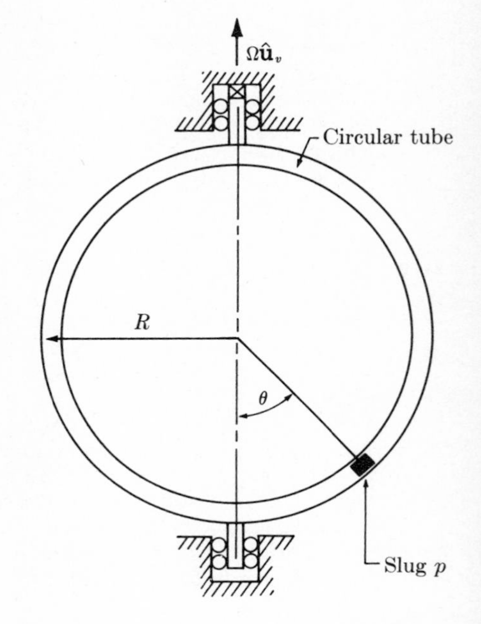

FIGURE P4·14

4·15 The cart shown in Fig. P4·15 contains a linear oscillator consisting of a particle p of mass m attached to springs, which apply to p a force of $-kx\hat{\mathbf{u}}_x$ when p is displaced relative to the cart through a distance x as shown.

a. Imagine that the cart is fixed to the tracks, which may be considered to define an inertial reference frame i, and that p is released from an initial displacement $x(0) = -\delta$, with $\dot{x}(0) = 0$ and $\delta > 0$. Use the work-energy equation, Eq. (4·44*a*), to find the maximum value reached by $\dot{x}(t)$ subsequent to release. Also calculate the potential energy V and kinetic energy T. Is $T + V$ constant?
b. Imagine now that the cart moves along the straight track with constant speed v_0. If the track is an inertial frame, is not the cart now *also* an inertial frame? If so, then the initial conditions $x(0) = -\delta$ and $\dot{x}(0) = 0$ must result in the same $x(t)$ as in part (a). (If the equations of motion were to be solved by the methods of Chap. 5, the solution would be $x(t) = -\delta \cos \sqrt{k/m}\, t$ in both cases.) Equation (4·44*a*) must now apply whether frame i designates

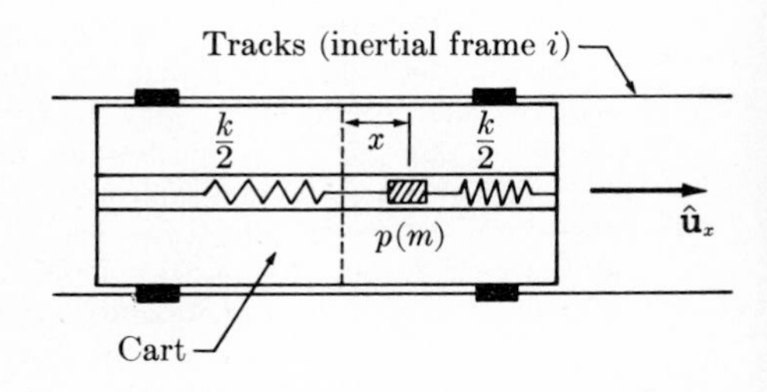

FIGURE P4·15

the track or the cart. Knowing the solution for $x(t)$, you are now to demonstrate the applicability of Eq. (4·44a) when i represents the frame of the track. How must you now interpret $d\mathbf{R}$ if you choose to define W as in Eq. (4·45)? Is the total mechanical energy $T + V$ still constant?

4·16 A length of straight track is supported at its center C and constrained to rotate in a horizontal plane at constant rate Ω relative to its inertially fixed base, as shown in Fig. P4·16. A cart (particle) p of mass m is connected to each end of the track by a spring with spring constant $k/2$. The length of each spring when undeformed is R, which is half the track length. Let $r\hat{\mathbf{u}}_r$ be the position vector of p relative to C, and let $r(0) = R$ and $\dot{r}(0) = -v_0$ (so, for $v_0 > 0$, p is initially moving inward). Let $k \equiv m\Omega^2$; ignore friction.

a. Apply $\mathbf{F} = m\mathbf{A}$ to p and determine $r(t)$.

b. Show that the work done by forces exerted on p equals the change in kinetic energy of p as r changes from R to zero.

c. Determine the work done on p *by the spring*. Compare this work to the loss of potential energy (strain energy) in the spring.

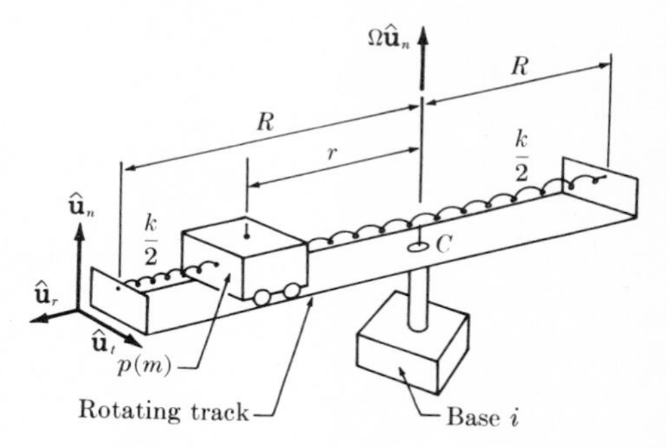

FIGURE P4·16

4·17 For the system of Fig. P4·16, with k no longer restricted to the value $k = m\Omega^2$ and $v_0 = 0$, determine the minimum value that k must have in order for the spring to push the cart all the way back to center C, by each of the methods following:

a. $\mathbf{F} = m\mathbf{A}$

b. Equation (4·44a)

c. Equation (4·44b)

4·18 A frictionless track which joins two circles, as shown in Fig. P4·18, is mounted on a turntable with the constant angular velocity $\Omega\hat{\mathbf{u}}_n$. A slot car, initially on the track at the outer circle, which has radius R, is accelerated by a massless spring

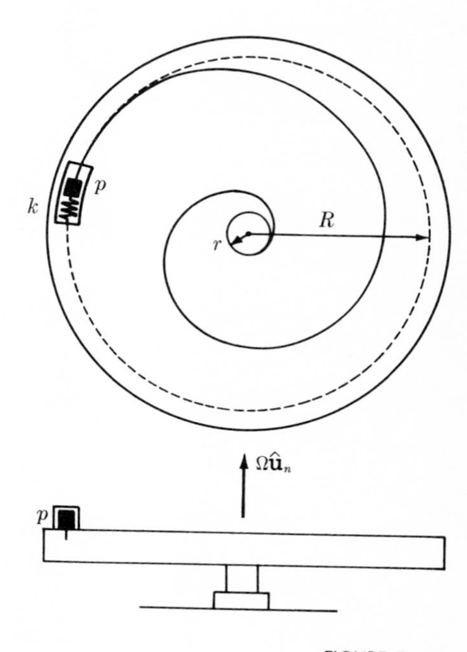

FIGURE P4·18

with spring constant k, initially compressed an amount δ and then released. Find the minimum value of δ required to get the car onto the inner circular track, which has radius r.

4·19 Determine the potential energy function V for any of the following forces $\mathbf{F}$ or effective forces $\mathbf{F}^f$ which you find to be conservative.

a. $\mathbf{F} = -kx^3\hat{\mathbf{u}}_x$
b. $\mathbf{F}^f = mr\Omega^2\hat{\mathbf{e}}_r - \mu mr^{-2}\hat{\mathbf{e}}_r$
c. $\mathbf{F} = -k\mathbf{r}(\mathbf{r}\cdot\mathbf{r})^{-3/2}$
d. $\mathbf{F} = x_1x_2x_3\hat{\mathbf{u}}_1 + x_1{}^2x_3\hat{\mathbf{u}}_2 + x_1{}^2x_2\hat{\mathbf{u}}_3$
e. $\mathbf{F} = x_1x_2{}^2x_3\hat{\mathbf{u}}_1 + x_1{}^2x_2x_3{}^2\hat{\mathbf{u}}_2 + x_1{}^2x_2{}^2x_3\hat{\mathbf{u}}_3$

4·20 A lunar-landing vehicle p weighing 1000 lb on earth is ejected from its parent spacecraft above the moon in such a way that when $t = t_0$, it is 5×10^6 ft above the moon's surface and at rest relative to an assumed inertial reference frame i that includes the moon's center C. Assuming that the only force on p is $\mathbf{F} = -GMmr^{-2}\hat{\mathbf{u}}_r$, where G is the universal gravitational constant, M is the lunar mass, m is the vehicle mass, r is the distance from p to C, and $\hat{\mathbf{u}}_r$ is a unit vector pointing from C toward p, find the vehicle's impact speed with the lunar surface. (Note: $GM = 1.735 \times 10^{14}$ ft^3/sec^2 and the lunar radius is $R = 11.4 \times 10^6$ ft.)

4·21 Reconsider Prob. 4·20, changing the conditions at $t = t_0$ in such a way that $\mathbf{V}^{pi}(t_0) = 300$ ft/sec $\hat{\mathbf{u}}_t$, where $\hat{\mathbf{u}}_t$ is normal to $\hat{\mathbf{u}}_r$, and the position is as given previously. Find the vehicle's velocity at the time of impact with the lunar surface.

4·22 A section of track designed for an amusement park is shown in Fig. P4·22. The track lies in a vertical plane, and the indicated radii are $R = 50$ ft and $r = 40$ ft. Will a loaded vehicle weighing 322 lb finish the course without leaving the track if at the top of the first rise it has virtually no speed and no energy is dissipated?

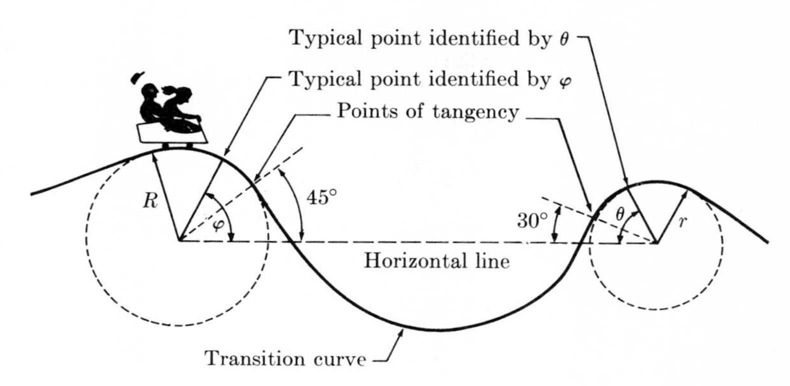

FIGURE P4·22

Does the weight of the occupants make a difference?

4·23 Apply Lagrange's equations in the form of Eq. (4·80) and also in the form of Eq. (4·83) to a particle p of mass m attached to an inertially fixed point s by a rigid massless rod of length L, as shown in Fig. P4·23, and constrained to a planar motion as defined by θ.

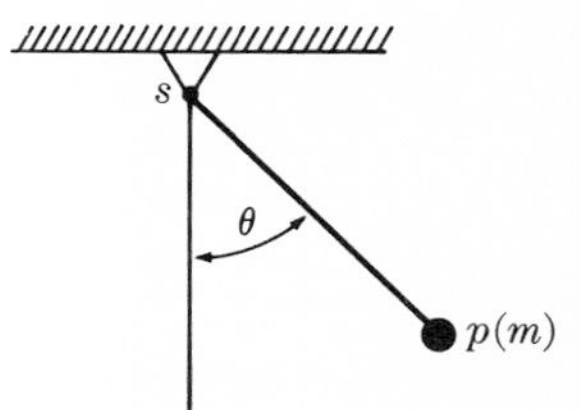

FIGURE P4·23

4·24 Apply Lagrange's equations in appropriate form to the particle p of mass m shown in Fig. P4·24 to be attached by a rigid massless rod of length L to a point s experiencing a translation in inertial space as prescribed by $\mathbf{S} = C\hat{\mathbf{u}}_1 + A \sin \omega t \hat{\mathbf{u}}_1$ where $\mathbf{S}$ is the position vector of s with respect to an inertially fixed point o, and A, C, and ω are constants. The planar motion of p is defined by θ.

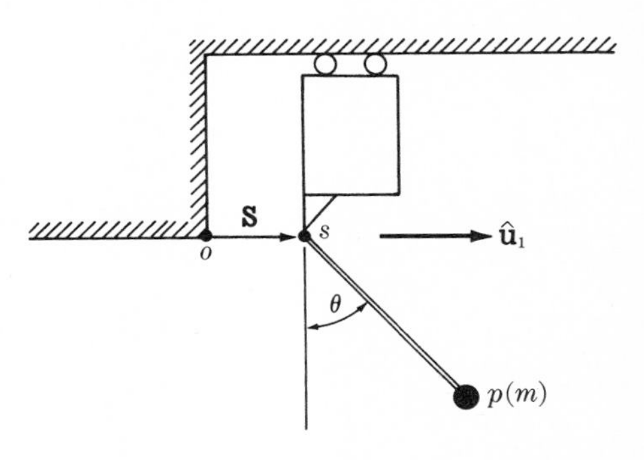

FIGURE P4·24

4·25 A block of ice b slides without friction down a straight slot in a conical frustrum c, which rotates with constant angular velocity $\Omega\hat{\mathbf{c}}_3$ relative to an inertially fixed base i, as shown in Fig. P4·25. The block, of mass m, begins at the top of the slope at rest relative to the cone, and the combination of gravity force $-mg\hat{\mathbf{c}}_3$ and spinning motion makes b slide; the symbol z denotes the distance b moves along the slot.

In terms of the unit vectors in Fig. P4·25 and the symbols m, g, φ, Ω, r, z, and $\dot{z}$ defined above and in the figure, express the following:

a. The inertial angular velocity of b
b. The force f *applied to b by c*
c. The speed of the center of b in inertial space (neglecting dimensions of b relative to R and r), for time t^* when b reaches the bottom of the slope, in terms of the symbols g, h, R, r, and Ω

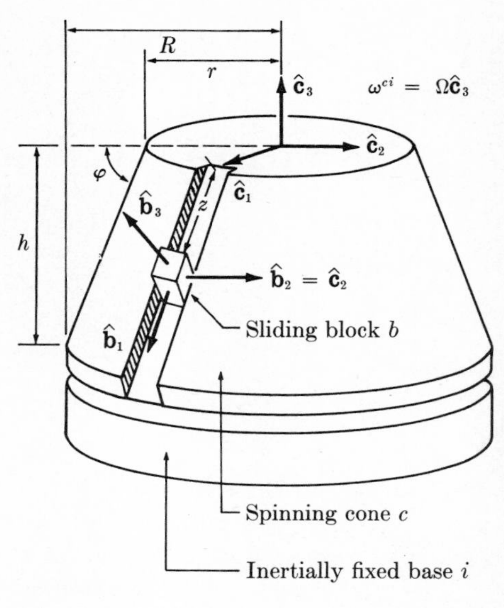

FIGURE P4·25

4·26 A particle p of mass m is suspended from a massless linear spring with spring constant k and undeformed length L, as shown in Fig. P4·26. Use the indicated generalized coordinates q_1 and q_2 to find the following:

a. The kinetic energy T
b. The generalized forces Q_1 and Q_2
c. The potential energy V
d. The equations of motion

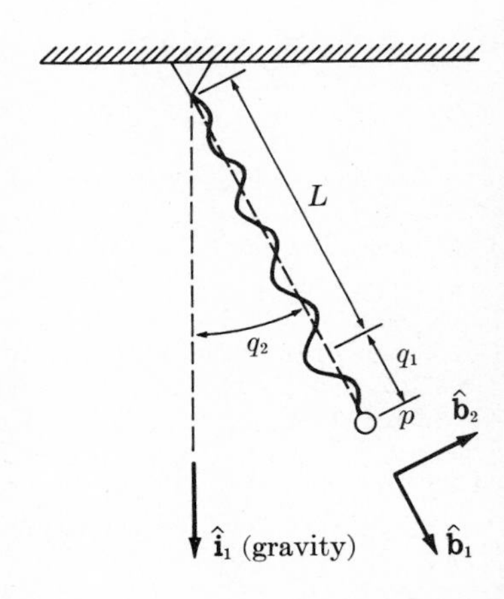

FIGURE P4·26

CHAPTER FIVE
PARTICLE DYNAMICS: SOLUTIONS

5·1 SOLUTIONS FOR MOTION OF A PARTICLE WITH ONE DEGREE OF FREEDOM

5·1·1 Differential equations of motion in one coordinate

A dynamic system is said to have n degrees of freedom if a complete specification of the motion of the system requires n independent scalar functions of time. A particle in three-dimensional space can have at most three degrees of freedom, which might, for example, be represented by three scalar cartesian coordinates called x, y, and z.

If somehow the particle is constrained in its motion, it must have fewer than three degrees of freedom. If, for example, the particle is constrained to remain on a track (but free to move *along* the track), it has only one degree of freedom. If the track forms a straight line fixed in an inertial reference frame, then we can fix the origin of a cartesian coordinate system in that frame, fix the x axis along the track, and completely define the motion with the single scalar $x(t)$. The acceleration of the particle in this inertial frame is clearly $\ddot{x}\hat{\mathbf{i}}_x$, where $\hat{\mathbf{i}}_x$

is a unit vector paralleling the track. Newton's second law then easily provides the vector equation of motion

$$\mathbf{F} = F\hat{\mathbf{i}}_x = m\ddot{x}\hat{\mathbf{i}}_x$$

from which the scalar equation of motion

$$F = m\ddot{x} \tag{5·1}$$

is obtained. Here F is the force scalar component in the $\hat{\mathbf{i}}_x$ direction.

In general F may be a function of time, and it may also depend on x and $\dot{x}$. The scalar equation of motion might therefore be written in the explicit general form

$$m\ddot{x} = F(x, \dot{x}, t) \tag{5·2}$$

If the track is fixed in inertial space, but no longer straight, the formulation of the equations of motion is somewhat more complicated but the result is the same as Eq. (5·1), with x representing the distance traveled along the track. This result was established as Eq. (4·47), and it can be obtained as a special case of the most general result developed in what follows.

Next consider the form of the equations of motion when the particle is constrained to remain on a straight track that *moves* in inertial space in a prescribed way. Figure 4·2 shows an example of this kind: a particle on a straight track fixed to a rotating turntable. The particle still has only one degree of freedom, represented by R in the example of Fig. 4·2, but its motion is no longer confined to an inertially fixed line. Newton's second law is now most conveniently written for a vector basis fixed in the track, with one unit vector paralleling the track (as $\hat{\mathbf{e}}_r$ does in Fig. 4·2). In the general case of the straight moving track, we can still fix the origin of a cartesian coordinate system in the track, and align one axis (say the x axis) with the track. If unit vectors $\hat{\mathbf{e}}_x$, $\hat{\mathbf{e}}_y$, and $\hat{\mathbf{e}}_z$ parallel axes x, y, and z, Newton's second law may be written

$$\mathbf{F} = F_x\hat{\mathbf{e}}_x + F_y\hat{\mathbf{e}}_y + F_z\hat{\mathbf{e}}_z = m(A_x\hat{\mathbf{e}}_x + A_y\hat{\mathbf{e}}_y + A_z\hat{\mathbf{e}}_z)$$

The corresponding scalar equations then become

$$F_x = mA_x \tag{5·3}$$
$$F_y = mA_y \tag{5·4}$$
$$F_z = mA_z \tag{5·5}$$

Now consider the five-term kinematic expansion for acceleration originally shown as Eq. (3·150), namely,

$$\mathbf{A}^{pf} = \mathbf{A}^{pf'} + \mathbf{A}^{q'f} + \boldsymbol{\alpha}^{f'f} \times \mathbf{P}' + \boldsymbol{\omega}^{f'f} \times (\boldsymbol{\omega}^{f'f} \times \mathbf{P}') + 2\boldsymbol{\omega}^{f'f} \times \mathbf{V}^{pf'} \tag{5·6}$$

If we let f represent the inertial frame, and f' represent the moving frame of the track, then the first term in the expansion becomes

$$\mathbf{A}^{pf'} = \ddot{x}\hat{\mathbf{e}}_x \tag{5·7}$$

indicating that $\ddot{x}$ is an ingredient of A_x in Eq. (5·3). In general, A_x is not identically $\ddot{x}$ since A_x may have contributions as well from other terms in the expansion. Without reservation, however, we can be sure that Eq. (5·3) has the general form of Eq. (5·2), provided that we understand that now $F(x, \dot{x}, t)$ in Eq. (5·2) stands not always for a *force*, but in general for a *forcing function* embracing the force $F_x(x, \dot{x}, t)$ from Eq. (5·3) and any terms from $-mA_x$ that remain after $-m\ddot{x}$ is removed. For the specific example of the particle on the turntable track in Fig. 4·2, the scalar equation of motion corresponding to Eq. (5·3) is (after exchanging x for R)

$$-kx = m(\ddot{x} - \Omega^2 x)$$

as shown in Eq. (4·13). In the standard form adopted as Eq. (5·2), this equation becomes

$$m\ddot{x} = (m\Omega^2 - k)x$$

Note that in this case $F(x, \dot{x}, t)$ includes both the force term $-kx$ and the inertial term $m\Omega^2 x$ sometimes called the centrifugal "force." In this special example, the general term $F(x, \dot{x}, t)$ depends only on x, and may be written as simply $F(x)$.

The scalar equations of motion represented by Eqs. (5·4) and (5·5) need not be used to determine the motion unless F_x in Eq. (5·3) somehow depends on F_y or F_z. This may occur, for example, when F_x includes a coulomb friction force proportional to F_y. In such a case, it becomes necessary to substitute mA_y for F_y in the expression for F_x, with A_y properly expanded as a function of x and $\dot{x}$. The final result is still a single differential equation of the structure of Eq. (5·2).

Consider finally the most general case of the motion of a particle with only one degree of freedom. We might imagine a cart rolling down the tracks of a roller coaster that is moving and constantly changing the geometry of its tracks in an explicitly prescribed manner. We can still let x be the distance the cart has traveled along the track from some starting point, and let $\hat{\mathbf{e}}_x$, $\hat{\mathbf{e}}_y$, and $\hat{\mathbf{e}}_z$ be a vector basis *fixed in the cart* with $\hat{\mathbf{e}}_x$ pointing forward. Newton's second law still yields the scalar equations of motion [Eqs. (5·3) to (5·5)], and A_x still contains $\ddot{x}$ (among other things). To appreciate this fact, consider again the five-term expansion for acceleration, Eq. (5·5), and this time let the reference frame f' be established at any point in time by that portion of the track lying under the cart at the given time. With this definition of f', the validity of Eq. (5·7) is preserved, so we are faced again with a differential equation of the form of Eq. (5·2). The derivation of the function $F(x, \dot{x}, t)$ may for the most general case be a substantial task, calling for all your skill in kinematics developed from Chap. 3, but when the laborious vector differentiations are accomplished, the result must have the form of Eq. (5·2).

In many applications the most attractive selection for the variable to play the role of x in Eq. (5·2) is not a linear displacement, but an angular displacement. The pendulum shown in Fig. 5·1 provides an illustration of such an application. The particle p of mass m at the tip of a massless rigid rod of length r is constrained by the rod to remain a fixed distance r from the support point O. Such a device is sometimes called a *simple pendulum*, or a *mathematical pendulum*. The hinge at O keeps the rod in an inertially fixed plane, so the particle must remain on the perimeter of an inertially fixed circle centered at O. The scalar coordinate that most readily describes the particle location is the angle marked x.

If we wish to use the five-term expansion in Eq. (5·6) to determine the inertial acceleration of the particle p, we can select frame f' as the reference frame established by the rigid rod, with f representing an inertial frame. Then $\mathbf{A}^{pf'}$ and $\mathbf{V}^{pf'}$ are zero, with $\boldsymbol{\omega}^{f'f} = \dot{x}\hat{\mathbf{e}}_n$ and $\boldsymbol{\alpha}^{f'f} = \ddot{x}\hat{\mathbf{e}}_n$, where $\hat{\mathbf{e}}_n$ is a unit vector coming out of the page. Substitution of $\mathbf{P}' = r\hat{\mathbf{e}}_r$ (where $\hat{\mathbf{e}}_r$ parallels the line from O to p) then provides

$$\begin{aligned}\mathbf{A}^{pf} &= \ddot{x}\hat{\mathbf{e}}_n \times r\hat{\mathbf{e}}_r + \dot{x}\hat{\mathbf{e}}_n \times (\dot{x}\hat{\mathbf{e}}_n \times r\hat{\mathbf{e}}_r) \\ &= r\ddot{x}\hat{\mathbf{e}}_t - r\dot{x}^2\hat{\mathbf{e}}_r\end{aligned}$$

where $\hat{\mathbf{e}}_t = \hat{\mathbf{e}}_n \times \hat{\mathbf{e}}_r$. If Eq. (5·6) is not at hand, and you don't happen to remember the five-term expansion, it is no more difficult to obtain the acceleration of p

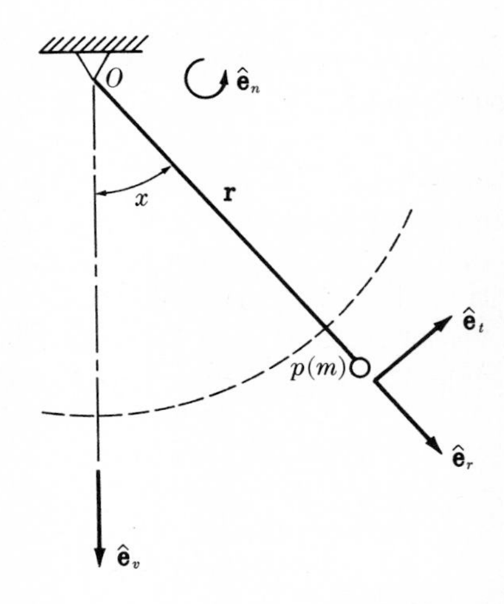

FIGURE 5·1

simply by differentiating the vector $r\hat{\mathbf{e}}_r$ twice relative to an inertial reference frame.

Now if the force is resolved into the same vector basis, and written $\mathbf{F} = F_r\hat{\mathbf{e}}_r + F_t\hat{\mathbf{e}}_t + F_n\hat{\mathbf{e}}_n$, then the three scalar equations analogous to Eqs. (5·3) to (5·5) become

$$F_r = -mr\dot{x}^2 \tag{5·8a}$$
$$F_t = mr\ddot{x} \tag{5·8b}$$
$$F_n = 0 \tag{5·8c}$$

The second of these may be recognized as having the form of Eq. (5·2), since F_t cannot depend on $\ddot{x}$, and might depend on x, $\dot{x}$, or t.

In the special case for which the only force applied to the particle is that applied by the rod, we might hope to solve the problem. If the hinge at O is a frictionless pinned joint, then the rod can transmit force only in the direction of $\hat{\mathbf{e}}_r$, so F_t is zero. This makes $\ddot{x}$ zero, so $\dot{x}$ is a constant, say Ω. The particle then goes around the point O at a constant rate, and the radial force in the rod is simply $F_r = -mr\Omega^2$. Attempts to solve the pendulum problem for other cases of interest, with F_t nonzero, will be deferred until we discuss in more general terms the challenges of differential equations of the form of Eq. (5·2).

There remains in general for the single-degree-of-freedom system the mathematical problem of solving a differential equation of the form

$$m\ddot{x} = F(x, \dot{x}, t) \tag{5·2}$$

and here the difficulty really begins. In the entire realm of mathematics there do not exist procedures for solving exactly and explicitly the full range of differential equations implied by Eq. (5·2). If we insist on analytical procedures leading to closed-form solutions (ignoring approximate methods and computer simulations), we must be satisfied with an understanding of solutions for special cases of Eq. (5·2). Those special cases of greatest practical interest are explored in the following articles.

5·1·2 Forcing function depending on time only

Substituting $F(t)$ for $F(x, \dot{x}, t)$ in Eq. (5·2) yields the differential equation

$$m\ddot{x} = F(t) \tag{5·9}$$

The equivalent expression

$$m\frac{d\dot{x}}{dt} = F(t)$$

suggests the separation of the variables $\dot{x}$ and t. The derivative $d\dot{x}/dt$ is the ratio of the differentials $d\dot{x}$ and dt, so $d\dot{x}/dt$ may for present purposes be treated as a single symbol representing the second time derivative of x or as the ratio of two symbols representing differentials. The latter interpretation permits us to multiply both sides by dt, in order to obtain an expression with $\dot{x}$ and t separated, namely,

$$m\,d\dot{x} = F(t)\,dt$$

This expression is readily integrated. If symbols v and τ are used as dummy variables for $\dot{x}$ and t, respectively, the integrated equation becomes

$$\int_{v_0}^{\dot{x}} m\,dv = m\dot{x} - mv_0 = \int_0^t F(\tau)\,d\tau \tag{5·10}$$

where v_0 is the value of $\dot{x}$ when $t = 0$. (Note that we have arbitrarily assigned the value zero to the "starting time"; we could as well call it t_0.) [Equation (5·10) should be compared to the vector impulse-momentum equation (4·25), of which it represents a special case.] With the definition

$$G(t) \triangleq \int_0^t F(\tau)\,d\tau \tag{5·11}$$

the solution has the form

$$\dot{x} = v_0 + \frac{G(t)}{m} \tag{5·12}$$

The indefinite integral $G(t)$ in Eq. (5·11) may not in a given application be easily evaluated, but we have managed to define it in terms of an explicit function of t, and this permits us to claim in Eq. (5·12) a closed-form solution for $\dot{x}$. More precisely, we say that "we have reduced the solution to quadrature," that is, to the evaluation of an integral. In the same sense, we can now obtain from Eq. (5·12) a closed-form solution for x, writing

$$\dot{x} \triangleq \frac{dx}{dt} = v_0 + \frac{G(t)}{m}$$

and then, with the introduction of the dummy variables ξ for x and τ for t in the integrands,

$$\int_{x_0}^{x} d\xi = x - x_0 = \int_0^t v_0\,d\tau + \frac{1}{m}\int_0^t G(\tau)\,d\tau$$

where x_0 is the value of x when $t = 0$. The final solution for x in quadrature form then becomes

$$x = x_0 + v_0 t + \frac{1}{m}\int_0^t G(\tau)\,d\tau \tag{5·13}$$

with $G(t)$ as defined in Eq. (5·11), and initial conditions on x and $\dot{x}$ defined by x_0 and v_0, respectively.

5·1·3 Variable thrust, constant-mass space vehicle

As a specific example, consider the equation

$$m\ddot{x} = F_0(1 - e^{-\alpha t})$$

In an approximate sense this might represent the equation of motion of a vehicle in space, subjected only to the propulsion system thrust that begins at zero and rises exponentially to the constant value F_0. The mass m of the vehicle is assumed constant, so expended propellant must have a negligible influence. (Perhaps the propulsion system is an ion engine, which expels small amounts of mass at great speed.) The inertial direction of the thrust is assumed constant, and x is the distance traveled along the line of thrust.

The solution (5·12) involves $G(t)$, given by Eq. (5·11) as

$$G(t) \triangleq \int_0^t F(\tau)\,d\tau = \int_0^t F_0(1 - e^{-\alpha\tau})\,d\tau$$

or

$$G(t) = \left(F_0\tau + \frac{F_0}{\alpha}e^{-\alpha\tau}\right)\Big|_0^t = F_0 t - \frac{F_0}{\alpha}(1 - e^{-\alpha t})$$

With this expression for $G(t)$, Eq. (5·13) provides

$$x = x_0 + v_0 t + \frac{1}{m}\left(\int_0^t F_0\tau\,d\tau - \frac{F_0}{\alpha}\int_0^t d\tau + \frac{F_0}{\alpha}\int_0^t e^{-\alpha\tau}\,d\tau\right)$$

or

$$x = x_0 + v_0 t + \frac{F_0}{m}\left[\frac{t^2}{2} - \frac{t}{\alpha} + \frac{1}{\alpha^2}(1 - e^{-\alpha t})\right]$$

In more convenient form, the solution becomes

$$x = x_0 + \frac{F_0}{m\alpha^2}(1 - e^{-\alpha t}) + \left(v_0 - \frac{F_0}{m\alpha}\right)t + \frac{F_0 t^2}{2m}$$

After a sufficiently long time elapses, this expression is dominated by the last term, which is the result that would be predicted if the thrust were represented simply by the constant F_0.

5·1·4 Forcing function depending on position only

A second special case of Eq. (5·2) is given by

$$m\ddot{x} = F(x) \tag{5·14}$$

or

$$m \frac{d\dot{x}}{dt} = F(x)$$

In terms of the differentials $d\dot{x}$ and dt, we have

$$m\, d\dot{x} = F(x)\, dt$$

In order to separate variables, we must multiply this result by $\dot{x} = dx/dt$. Interpreting $\dot{x}$ both as a derivative function and as a ratio of differentials provides the separated equation

$$m\dot{x}\, d\dot{x} = F(x)\, dt \left(\frac{dx}{dt}\right) = F(x)\, dx$$

Integration then yields

$$\int_{v_0}^{\dot{x}} mv\, dv = \tfrac{1}{2}m\dot{x}^2 - \tfrac{1}{2}mv_0^2 = \int_{x_0}^{x} F(\xi)\, d\xi \tag{5·15}$$

where ξ is the dummy variable for x in the integrand. [Equation (5·15) should be compared with the work-energy equation (4·44*a*), of which it represents a special case.] Solving Eq. (5·15) for $\dot{x}$ gives an expression for the speed

$$\dot{x} = \pm \left(v_0^2 + \frac{2}{m}\int_{x_0}^{x} F(\xi)\, d\xi\right)^{1/2} \tag{5·16}$$

Equation (5·16) raises questions of algebraic sign. Confusion is usually avoided by choosing the sign initially to coincide with that established by v_0, but this practice can be troublesome if the motion is oscillatory, with frequent sign reversals.

A second integration is more conveniently recorded with the definition

$$G(x) \triangleq \int_{x_0}^{x} F(\xi)\, d\xi \tag{5·17}$$

Replacing $\dot{x}$ in Eq. (5·16) by the ratio dx/dt and rearranging then provides the separated form

$$dt = \frac{dx}{\pm[v_0^2 + (2/m)\, G(x)]^{1/2}}$$

Integration then yields

$$\int_0^t d\tau = t = \pm \int_{x_0}^{x} \left(v_0^2 + \frac{2}{m} G(\xi)\right)^{-1/2} d\xi \tag{5·18}$$

We have succeeded in reducing to quadrature the expression for t as a function of x and the initial conditions x_0 and v_0. This is not always the most convenient form of the answer—we usually want x as an explicit function of t and initial conditions—but it is the best we can do in general terms. In specific cases of interest, the inversion is straightforward. In other cases, Eq. (5·18) is of little value, and more useful information is obtained from Eq. (5·16), which does not involve time at all.

5·1·5 Particle in space under gravitational attraction

Consider the specific equation of motion

$$m\ddot{x} = \frac{-\mu m}{x^2}$$

A particle in an inverse-square gravity field of an inertially fixed attracting body is represented in its motion by this equation if x is the radial distance to the particle from the center of the attracting body, and there is initial motion only along the radial line. The constant μ then represents the product of the mass of the attracting body and a constant called the *universal gravitational constant.*

A first integration of the equation of motion yields, from Eq. (5·16), with $F(\xi) = -\mu m \xi^{-2}$,

$$\dot{x} = \pm \left(v_0^2 - 2\mu \int_{x_0}^{x} \xi^{-2}\, d\xi \right)^{1/2} = \pm [v_0^2 + 2\mu(x^{-1} - x_0^{-1})]^{1/2}$$

If, for example, we want to know the impact speed $\dot{x}_i$ of an object released from rest at a distance h from the surface of an atmosphereless moon of radius R, this expression provides

$$\dot{x}_i = -\left[2\mu\left(\frac{1}{R} - \frac{1}{R+h}\right)\right]^{1/2}$$
$$= \frac{-(2\mu h)^{1/2}}{[R(R+h)]^{1/2}}$$

If the weight of the object on the moon's surface defines a gravitational constant g by

$$mg = \frac{m\mu}{R^2}$$

then the impact speed may be written as

$$\dot{x}_i = \frac{-(2gh)^{1/2}}{[1 + (h/R)]^{1/2}}$$

When h is much less than R, this expression reduces to the familiar $-\sqrt{2gh}$ for the impact speed of a body falling under constant gravity force.

If speed is not of interest, and the relationship between x and t is required, we may use Eq. (5·18), with Eq. (5·17) providing

$$G(x) = -\int_{x_0}^{x} \mu m \xi^{-2}\, d\xi = \mu m(x^{-1} - x_0^{-1})$$

The resulting expression is

$$t = \pm \int_{x_0}^{x} (v_0^2 - 2\mu x_0^{-1} + 2\mu \xi^{-1})^{-1/2}\, d\xi$$

Although with the transformation to the new variable $u \triangleq \xi^{-1}$, we could write the integral for t in terms of an integral usually found in integral tables, the result involves an inverse tangent function of an awkward expression in ξ (as you will discover if you complete Prob. 5·2*e*), and little would be gained by writing it out here. Evidently there is a good deal of labor ahead for the man who requires detailed knowledge of x as a function of t. Fortunately, the digital computer can be employed to facilitate this task.

5·1·6 Simple pendulum under uniform gravitational attraction

Equations (5·8) describe the motion of a simple mathematical pendulum subjected to arbitrary force $\mathbf{F} = F_r\hat{\mathbf{e}}_r + F_t\hat{\mathbf{e}}_t + F_n\hat{\mathbf{e}}_n$. (See Fig. 5·1.) We consider now a special case with the pendulum in a uniform gravitational force field. Then the particle is subjected to a gravity force $mg\hat{\mathbf{e}}_v$ (where $\hat{\mathbf{e}}_v$ points "down" in Fig. 5·1), as well as the force transmitted by the rod. We might anticipate that the rod will remain in tension, and represent the force applied to the particle by the rod as $-T\hat{\mathbf{e}}_r$. Then the vector equation $F = m\mathbf{A}$ provides

$$mg\hat{\mathbf{e}}_v - T\hat{\mathbf{e}}_r = -mr\dot{x}^2\hat{\mathbf{e}}_r + mr\ddot{x}\hat{\mathbf{e}}_t$$

The vector identity

$$\hat{\mathbf{e}}_v = \hat{\mathbf{e}}_r \cos x - \hat{\mathbf{e}}_t \sin x$$

may be substituted to yield

$$(-T + mg \cos x)\hat{\mathbf{e}}_r - mg \sin x\hat{\mathbf{e}}_t = -mr\dot{x}^2\hat{\mathbf{e}}_r + mr\ddot{x}\hat{\mathbf{e}}_t$$

The scalar equation

$$mr\ddot{x} = -mg \sin x$$

must be solved if we are to determine the motion of the pendulum. After both sides are divided by r, this equation is of the form of Eq. (5·2), with $F = -(mg/r) \sin x$. Because F depends only on x, the indicated solutions for the special case represented by Eq. (5·14) are applicable. The general expressions for $\dot{x}$ and

t, represented by Eqs. (5·16) and (5·18), both involve

$$G(x) \triangleq \int_{x_0}^{x} F(\xi)\, d\xi = \int_{x_0}^{x} -\frac{mg}{r} \sin \xi \, d\xi$$

$$= \frac{mg}{r} \cos \xi \Big|_{x_0} = \frac{mg}{r} (\cos x - \cos x_0)$$

Substituting this result into Eq. (5·16) yields the angular speed

$$\dot{x} = \pm \left[v_0^2 + \frac{2}{m}\left(\frac{mg}{r}\right)(\cos x - \cos x_0)\right]^{1/2}$$

where v_0 is the initial angular speed and x_0 the initial angular displacement. This result can be used directly to obtain information about the pendulum motion without actually solving explicitly for x as a function of t. We can determine the value of x when $\dot{x} = 0$, for example, by

$$\cos x = \cos x_0 - \frac{rv_0^2}{2g}$$

Comparison with a general expression for $\cos x$ reveals that the preceding expression also provides the maximum value of the amplitude of oscillation x.

If we need more information, we can turn to the general expression for t in Eq. (5·18). For the pendulum, the result is

$$t = \int_{x_0}^{x} \left(v_0^2 + \frac{2}{m} G(\xi)\right)^{-1/2} d\xi$$

$$= \int_{x_0}^{x} \left(v_0^2 - \frac{2g}{r} \cos x_0 + \frac{2g}{r} \cos \xi \right)^{-1/2} d\xi$$

Again we encounter an expression for t in terms of a hostile-looking integral. This time we can evaluate it only if we recognize that substitution of the identity

$$\cos \xi = 1 - 2 \sin^2 \frac{\xi}{2}$$

puts the integral into a classical form called an *elliptic integral*, for which results are tabulated (see Prob. 5·3).

The message of the last two examples should be clear; even for particles with a single degree of freedom subjected to idealized representations of physically interesting forces depending on the particle coordinate x, an explicit solution for x is often very difficult to obtain.

5·1·7 Forcing function depending on speed only

In the two preceding articles, the differential equation $m\ddot{x} = F(x, \dot{x}, t)$ has been reduced to quadrature for the simpler cases $m\ddot{x} = F(t)$ and $m\ddot{x} = F(x)$. The third special case of this kind

$$m\ddot{x} = F(\dot{x}) \tag{5·19}$$

can also be solved formally.

Rewrite Eq. (5·19) as

$$\frac{d\dot{x}}{dt} = \frac{1}{m} F(\dot{x})$$

or

$$dt = \frac{m\, d\dot{x}}{F(\dot{x})} \tag{5·20}$$

Integration then yields a relationship between t and $\dot{x}$:

$$\int_0^t d\tau = t = m \int_{v_0}^{\dot{x}} F^{-1}(\xi)\, d\xi \tag{5·21}$$

If an expression for x in terms of $\dot{x}$ is more useful, the definition $\dot{x} = dx/dt$ may be used to provide in conjunction with Eq. (5·20) to yield

$$dx = \dot{x}\, dt = m\dot{x}F^{-1}(\dot{x})\, d\dot{x}$$

Integration produces the result

$$\int_{x_0}^{x} d\xi = x - x_0 = m \int_{v_0}^{\dot{x}} \xi F^{-1}(\xi)\, d\xi \tag{5·22}$$

Equations (5·21) and (5·22) may seem less than wholly satisfactory as solutions to Eq. (5·19); an expression for x in terms of t is often preferable, and this requires first the integration in Eq. (5·22) and then substitution for $\dot{x}$ of the results of the integration in Eq. (5·21). For specific functions $F(\dot{x})$, this can be accomplished explicitly, as shown in the example following.

5·1·8 Particle subjected to constant force plus aerodynamic drag

Imagine a particle to be subject to a constant force of magnitude F_0, producing an acceleration that is resisted by the surrounding medium. Perhaps the particle is falling under a gravitational force approximated by the constant $F_0 = mg$, or perhaps the particle represents a rocket under thrust F_0. In either case, the surrounding medium provides a drag force that depends on $\dot{x}$; call it $-Df(\dot{x})$. The equation of motion then becomes

$$m\ddot{x} = F_0 - Df(\dot{x})$$

In application, the constant D is established by the medium, the projected area of the projectile, and its shape. The function $f(\dot{x})$ depends on the fluid and the projectile speed; for slow speeds in viscous fluids, the linear law $f(\dot{x}) = \dot{x}$ may suffice, and for a wide class of aerodynamics problems, $f(\dot{x}) = \dot{x}^2$ will provide a useful approximation. In either case, as long as $f(\dot{x})$ increases with $\dot{x}$, there must be a *terminal speed* V that the particle approaches as time goes to infinity. This is, of course, the speed at which the net applied force is zero, so it is given by

$$f(V) = \frac{F_0}{D}$$

For the case of linear resistance or *viscous drag*, the equation of motion is simply

$$m\ddot{x} = F_0 - D\dot{x}$$

Because this is a linear, ordinary differential equation with constant coefficients, solution by standard methods is straightforward. If Eq. (5·21) is used instead, the immediate result is

$$t = m \int_{v_0}^{\dot{x}} \frac{d\xi}{F_0 - D\xi} = -\frac{D}{m} \ln \frac{F_0 - D\dot{x}}{F_0 - Dv_0}$$

Let $\beta \triangleq D/m$, and note that the terminal speed is $V = F_0/D$. With these substitutions, exponentiation of the expression above yields

$$e^{-\beta t} = \frac{V - \dot{x}}{V - v_0}$$

Solving for $\dot{x}$, we find

$$\dot{x} = V - (V - v_0)e^{-\beta t}$$

In this case we need not use Eq. (5·22) to complete the solution, but can integrate the $\dot{x}$ result to get

$$x = Vt + \frac{1}{\beta}(V - v_0)e^{-\beta t} + C_1$$

where C_1 is a constant of integration. If x_0 is the value of x when $t = 0$, then $C_1 = x_0 - (1/\beta)(V - v_0)$, and the solution becomes

$$x = x_0 + Vt - \frac{1}{\beta}(V - v_0)(1 - e^{-\beta t})$$

The representation of fluid drag by quadratic approximation also permits ready solution. The equation of motion

$$m\ddot{x} = F_0 - D\dot{x}^2$$

is now nonlinear in $\dot{x}$, and may seem more difficult. Equations (5·21) and (5·22) still apply, however, and they can be used to provide

$$t = m \int_{v_0}^{\dot{x}} \frac{d\dot{\xi}}{F_0 - D\dot{\xi}^2}$$

and

$$x = x_0 + m \int_{v_0}^{\dot{x}} \frac{\dot{\xi}\, d\dot{\xi}}{F_0 - D\dot{\xi}^2}$$

The terminal speed is now given by

$$V^2 = \frac{F_0}{D}$$

and in terms of this useful concept, the integration results are

$$t = \frac{mV^2}{F_0} \int_{v_0}^{\dot{x}} \frac{d\dot{\xi}}{V^2 - \dot{\xi}^2} = \frac{mV}{F_0} \left(\tanh^{-1} \frac{\dot{x}}{V} - \tanh^{-1} \frac{v_0}{V} \right)$$

and

$$x = x_0 + \frac{mV^2}{F_0} \int_{v_0}^{\dot{x}} \frac{\dot{\xi}\, d\dot{\xi}}{V^2 - \dot{\xi}^2} = x_0 + \frac{mV^2}{2F_0} \ln \frac{V^2 - v_0^2}{V^2 - \dot{x}^2}$$

Proper interpretation of these results in terms of the time behavior of x is not as simple as for the viscous drag, but it is remarkable that we can get even this far with the given nonlinear differential equation.

5·1·9 Damped linear oscillator equation

Figure 5·2 illustrates schematically an object of mass m constrained to remain on a straight track, and subjected to an applied force $F(t)\hat{\mathbf{e}}_x$ as well as a restraining force applied by the mechanism shown. The track is fixed in inertial space, X is the scalar coordinate measuring the cart's distance along the track from the support point, and the constant X_0 is the distance to the cart location for which the spring is unstretched. The force applied by the spring to the cart is therefore $-k(X - X_0)$, and the dashpot force is $-c\dot{X}$, where k and c are constants of the support mechanism. With the definition

$$x \triangleq X - X_0 \tag{5·23}$$

providing a direct measure of the spring distortion, the scalar equation of motion in the $\hat{\mathbf{e}}_x$ direction becomes

$$m\ddot{x} = -kx - c\dot{x} + F(t) \tag{5·24}$$

In anticipation of the form of the solution, we define the symbols

$$p \triangleq \sqrt{\frac{k}{m}} \tag{5·25}$$

$$\zeta \triangleq \frac{c}{2mp} = \frac{c}{2\sqrt{km}} \tag{5·26}$$

and

$$\mathfrak{F}(t) \triangleq \frac{F(t)}{m} \tag{5·27}$$

The equation of motion then takes the form[1]

$$\ddot{x} + 2\zeta p\dot{x} + p^2x = \mathfrak{F}(t) \tag{5·28}$$

Equation (5·28) may be classified as a *linear, inhomogeneous, second-order ordinary scalar differential equation with constant coefficients.* Because it is linear in the dependent variable x and its derivatives $\dot{x}$ and $\ddot{x}$, the *principle of superposition* may be applied. Therefore we can solve the homogeneous equation

$$\ddot{x}_h + 2\zeta p\dot{x}_h + p^2x_h = 0 \tag{5·29}$$

and to the homogeneous solution x_h satisfying Eq. (5·29) we can add any *particular solution* of the inhomogeneous equation (5·28) that can be found. The search for a particular solution can be difficult for certain classes of functions $F(t)$, but the homogeneous solution is always readily available. We therefore direct our attention first to the solution of Eq. (5·29).

[1] This standard form clearly becomes inappropriate when in Eq. (5·24) $k = 0$ and $c \neq 0$, so at this point $k > 0$ is assumed.

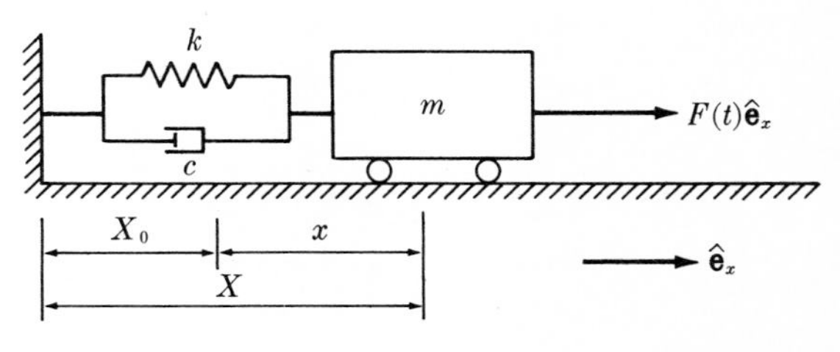

FIGURE 5·2

5·1·10 Homogeneous solution for the oscillator equation

A linear, constant coefficient differential equation can be shown by substitution to have a solution of the form

$$x_h = A_s e^{\lambda_s t} \tag{5·30}$$

To establish those scalars λ_s for which Eq. (5·30) provides a solution for Eq. (5·29), we can simply substitute the trial solution, to obtain

$$\lambda_s{}^2 A_s e^{\lambda_s t} + \lambda_s 2\zeta p A_s e^{\lambda_s t} + p^2 A_s e^{\lambda_s t} = 0$$

Since the common factor $A_s e^{\lambda_s t}$ is nonzero for any but the trivial solution $x_h \equiv 0$, this term may be canceled, to provide the *characteristic equation* for λ_s:

$$\lambda_s{}^2 + 2\zeta p \lambda_s + p^2 = 0 \tag{5·31}$$

This quadratic equation has the solutions

$$\lambda_s = -\zeta p \pm ip\sqrt{1 - \zeta^2} \tag{5·32}$$

where $i \triangleq \sqrt{-1}$.

Since we have begun with a second-order ordinary differential equation, we require in our general solution the possibility of specifying independently the initial values of x and $\dot{x}$; this means that we need two free constants in our solution, to be determined from initial conditions. We have confirmed by substitution the existence of a solution of the form $x_h = A_s e^{\lambda_s t}$, with λ_s given by Eq. (5·32). If this equation gives two distinct roots λ_1 and λ_2 (as it will unless $\zeta = 1$), we may combine the two associated solutions to obtain

$$x_h = A_1 e^{\lambda_1 t} + A_2 e^{\lambda_2 t} \tag{5·33}$$

as the general solution for Eq. (5·29). In the event $\zeta = 1$, however, the two roots from Eq. (5·32) are identical, and a solution of the form of Eq. (5·33) would have in truth only one free constant, the sum $A_1 + A_2$. We need a general solution with two free constants. It will suffice simply to observe that when $\zeta = 1$, the homogeneous equation is satisfied by a solution of the form

$$x_h = tA_s e^{\lambda_s t} \tag{5·34}$$

as well as by $x_h = A_s e^{\lambda_s t}$. This can be confirmed in this case by substitution, and can further be shown to be generally valid whenever the characteristic equation yields repeated roots, as Eq. (5·31) does when $\zeta = 1$.

We are now prepared to record and interpret the homogeneous solution x_h for each of three classes of system: (1) $\zeta < 1$; (2) $\zeta > 1$; and (3) $\zeta = 1$.

Equation (5·32) for λ_s is already in the form most suitable for $\zeta < 1$. The homogeneous solution in Eq. (5·33) then becomes

$$x_h = A_1 \exp[(-\zeta p + ip\sqrt{1-\zeta^2})\,t] + A_2 \exp[(-\zeta p - ip\sqrt{1-\zeta^2})\,t] \tag{5·35}$$

With the identity

$$e^{a+b} = e^a e^b \tag{5·36}$$

this solution becomes

$$x_h = e^{-\zeta pt}[A_1 \exp ip\sqrt{1-\zeta^2}\,t) + A_2 \exp(-ip\sqrt{1-\zeta^2}\,t)] \tag{5·37}$$

The relationships (for any α)

$$\sin\alpha = \frac{1}{2i}(e^{i\alpha} - e^{-i\alpha}) \tag{5·38}$$

$$\cos\alpha = \tfrac{1}{2}(e^{i\alpha} + e^{-i\alpha}) \tag{5·39}$$

combine to provide

$$e^{i\alpha} = \cos\alpha + i\sin\alpha \tag{5·40}$$

permitting the homogeneous solution to be written as

$$x_h = e^{-\zeta pt}(C_1 \sin p\sqrt{1-\zeta^2}\,t + C_2 \cos p\sqrt{1-\zeta^2}\,t) \tag{5·41}$$

where the constants $C_2 \triangleq (A_1 + A_2)$ and $C_1 \triangleq i(A_1 - A_2)$ are to be determined by the initial conditions. Equation (5·41) is the most convenient form for the homogeneous solution as long as $\zeta < 1$, since then the result is expressed wholly in terms of real quantities. [Note that A_1 and A_2 in Eq. (5·35) must in general be complex since x_h is real.]

Remember that x_h represents the general solution for Eq. (5·29), for $\zeta \neq 1$, but that this is only a *part* of the solution of the original equation of motion (5·28). When we evaluate the constants C_1 and C_2 from initial conditions, we must apply these conditions to the *total* solution. For the present we shall assume that the oscillator is free of the driving force $F(t)$, and that the initial conditions are given by

$$x(0) = x_0 \tag{5·42}$$

$$\dot{x}(0) = v_0 \tag{5·43}$$

with

$$\mathfrak{F}(t) \equiv 0 \tag{5·44}$$

The total solution for $x(t)$ is then $x_h(t)$ from Eq. (5·41), and the initial-condition equations are

$$x_0 = C_2 \tag{5·45}$$

$$v_0 = -\zeta p C_2 + p\sqrt{1-\zeta^2}\,C_1 \tag{5·46}$$

Solving these equations for C_1 and C_2 and substituting into Eq. (5·41) provides the free-vibration solution

$$x(t) = e^{-\zeta pt}\left(\frac{v_0 + \zeta p x_0}{p\sqrt{1-\zeta^2}}\sin p\sqrt{1-\zeta^2}\,t + x_0 \cos p\sqrt{1-\zeta^2}\,t\right) \tag{5·47}$$

A plot of this solution is shown as Fig. 5·3*a*, which portrays a harmonic oscillation of exponentially decaying amplitude. Such a response is termed *subcritically damped.*

The second case of interest is said to have *supercritical damping,* with $\zeta > 1$. The solution for the homogeneous solution x_h in Eq. (5·41) is still correct, as is the free-vibration response in Eq. (5·47), but the imaginary arguments of the trigonometric functions suggest a more convenient formulation of these results. With the identities (for any α)

$$\cos i\alpha = \cosh \alpha \tag{5·48}$$

$$\sin i\alpha = i \sinh \alpha \tag{5·49}$$

Eq. (5·41) may be written in terms of hyperbolic functions as

$$x_h(t) = e^{-\zeta pt}(B_1 \sinh p\sqrt{\zeta^2 - 1}\,t + B_2 \cosh p\sqrt{\zeta^2 - 1}\,t) \tag{5·50}$$

where, in terms of C_1 and C_2 from Eq. (5·41), $B_1 = iC_1$ and $B_2 = C_2$. The free-vibration response of Eq. (5·47) becomes

$$x(t) = e^{-\zeta pt}\left(\frac{v_0 + \zeta p x_0}{p\sqrt{\zeta^2 - 1}}\sinh p\sqrt{\zeta^2 - 1}\,t + x_0 \cosh p\sqrt{\zeta^2 - 1}\,t\right) \tag{5·51}$$

Although this is a useful form of the solution for $\zeta > 1$, it does not provide a clear picture of the asymptotic behavior of $x(t)$ as t goes to infinity. For this interpretation it is convenient to obtain the solution in exponential form. This may be accomplished by substituting into Eq. (5·51) the relationships

$$\sinh \alpha = \tfrac{1}{2}(e^{\alpha} - e^{-\alpha}) \tag{5·52}$$

$$\cosh \alpha = \tfrac{1}{2}(e^{\alpha} + e^{-\alpha}) \tag{5·53}$$

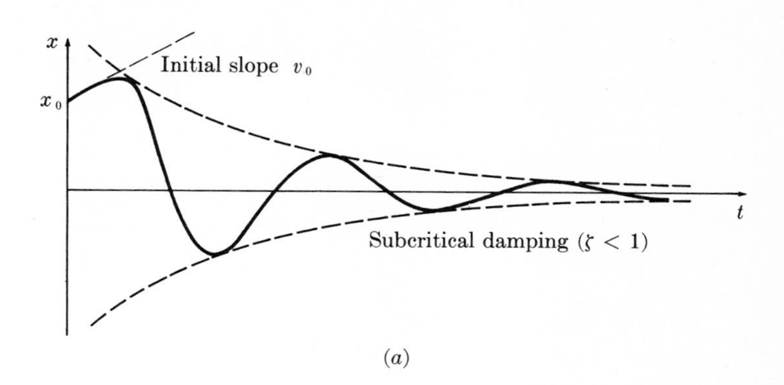

(a)

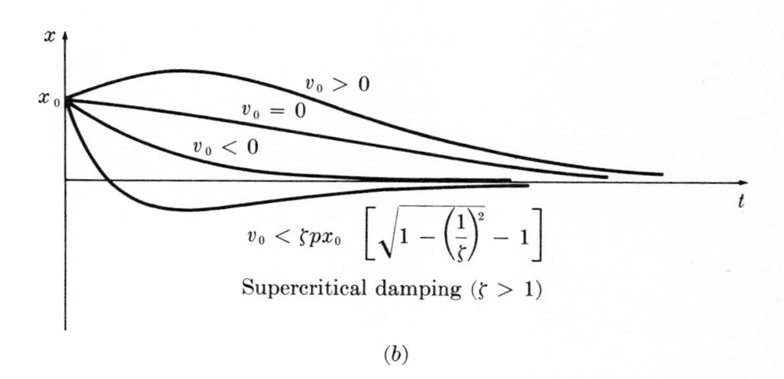

(b)

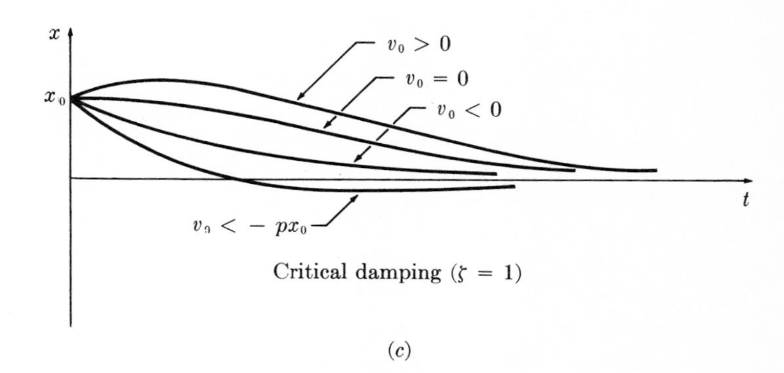

(c)

FIGURE 5·3

or by substituting directly into Eq. (5·33) the roots of Eq. (5·32). The result is

$$x_h(t) = A_1 \exp\left[\left(-\zeta p + p\sqrt{\zeta^2 - 1}\right) t\right] + A_2 \exp\left[\left(-\zeta p - p\sqrt{\zeta^2 - 1}\right) t\right] \qquad (5{\cdot}54)$$

which can be recognized as the sum of two decaying exponentials. The free response to initial conditions x_0 and v_0 then becomes

$$x(t) = \frac{v_0 + \zeta p x_0}{2p\sqrt{\zeta^2 - 1}} \left\{\exp\left[\left(p\sqrt{\zeta^2 - 1} - \zeta p\right)t\right] - \exp\left[\left(-p\sqrt{\zeta^2 - 1} - \zeta p\right)t\right]\right\} + \frac{x_0}{2}\left\{\exp\left[\left(p\sqrt{\zeta^2 - 1} - \zeta p\right)t\right] + \exp\left[\left(-p\sqrt{\zeta^2 - 1} - \zeta p\right)t\right]\right\} \qquad (5{\cdot}55)$$

as an equivalent alternative to Eq. (5·51).

By equating $x(t)$ to zero in Eq. (5·51), and noting that the hyperbolic tangent $\tanh p\sqrt{\zeta^2 - 1}\, t$ approaches unity as time goes to infinity, we can deduce that a zero crossing occurs when $v_0 < \zeta p x_0[\sqrt{1 - (1/\zeta)^2} - 1]$, as shown in Fig. 5·3*b*.

The final possibility is the critical case $\zeta = 1$. Now the roots of Eq. (5·32) are no longer distinct, and the solution takes the form

$$x_h = A_1 e^{-\zeta p t} + A_2 t e^{-\zeta p t} = (A_1 + A_2 t) e^{-pt} \qquad (5{\cdot}56)$$

as suggested by Eq. (5·34).

The free response to the initial conditions of Eqs. (5·42) to (5·44) then becomes

$$x(t) = x_0 e^{-pt} + (v_0 + p x_0) t e^{-pt} \qquad (5{\cdot}57)$$

The term $x_0 e^{-pt}$ is a simple exponential decay, but the asymptotic behavior of the second term in the solution requires L'Hospital's rule for evaluation. Since the limit as $t \to \infty$ of t/e^{pt} is the same as the limit of $1/pe^{pt}$, namely zero, the solution decays, as shown in Fig. 5·3*c*. By setting $x(t) = 0$ in Eq. (5·57), we see that there exists a zero crossing whenever $v_0 < -p x_0$.

In Eqs. (5·41), (5·50), and (5·56), we have three forms of the general homogeneous solution to the original inhomogeneous equation (5·28). The last of these is the required expression when $\zeta = 1$; the first two are both correct when $\zeta \neq 1$, but they are presented in forms convenient for $\zeta < 1$ and $\zeta > 1$, respectively. Each of these solutions contains two scalar constants, to be evaluated from initial conditions. Before evaluating these constants, however, we must find some expression for $x(t)$ that satisfies the original inhomogeneous equation (5·28); that is, we must find a *particular solution*.

5·1·11 Particular solutions by the method of undetermined coefficients

We have seen the variety of forms that the solution for a linear, constant-coefficient *homogeneous* differential equation may adopt. Basically they all stem from $e^{\lambda t}$ or $t^n e^{\lambda t}$ for some integer n, but noted identities permit expressions in terms of trigonometric functions, and hyperbolic trigonometric functions as well, to appear. As a general statement, we could say that any expression of the form

$$\mathfrak{F}(t) = \sum_{s=1}^{n} P_s(t) e^{\alpha_s t} \tag{5·58}$$

may occur, where $P_s(t)$ is a polynomial in t, and α_s is a constant, which may be complex. Now we will find that whenever the inhomogeneous part (right-hand side) of a linear, constant-coefficient ordinary differential equation has the structure of Eq. (5·58), we can solve for a particular solution in closed form by the straightforward *method of undetermined coefficients*. In other words, when our forcing function $\mathfrak{F}(t)$ is itself a solution to some other linear, constant-coefficient ordinary differential equation, we'll be able to find a particular solution easily. The method applicable in this common but restricted case is not a general method, however, and for forcing functions that cannot be written as in Eq. (5·58) an alternative must be found. But, for the present, we will concentrate on the function $\mathfrak{F}(t)$ of the special kind indicated.

When the inhomogeneous equation

$$\ddot{x} + 2\zeta p\dot{x} + p^2 x = \mathfrak{F}(t) \tag{5·59}$$

has a forcing function $\mathfrak{F}(t)$ of the special kind indicated by Eq. (5·58), and when in that expression none of the constants α_s is a root of the characteristic equation

$$\lambda_s^2 + 2\zeta p\lambda_s + p^2 = 0 \tag{5·60}$$

[as in Eq. (5·31)], then a particular solution of Eq. (5·59) exists in the form

$$x_p(t) = D_1\mathfrak{F}(t) + D_2\dot{\mathfrak{F}}(t) + D_3\ddot{\mathfrak{F}}(t) + \cdots \tag{5·61}$$

continuing as long as differentiation produces linearly independent functions. The coefficients D_1, D_2, . . . are determined by substituting Eq. (5·61) into (5·59) and requiring satisfaction of the equation for all time t.

When $\mathfrak{F}(t)$ is of the form in Eq. (5·58), and some constant α_s does appear as a root λ_s of Eq. (5·60), a second look is required. When a term in $\mathfrak{F}(t)$ or any of its derivatives is of the same functional structure as a term in the homogeneous solution, then the offending term or terms in Eq. (5·61) must be multiplied by t^k, choosing k no higher than necessary to avoid repetition of the form of the corresponding term in the homogeneous equation.

5·1·12 Oscillator response to harmonic excitation

The general procedures just described will first be illustrated by the example of a damped linear oscillator under harmonic excitation. The equation of motion is

$$m\ddot{x} + c\dot{x} + kx = F_0 \sin \omega t$$

or

$$\ddot{x} + 2\zeta p\dot{x} + p^2 x = \mathfrak{F}_0 \sin \omega t \tag{5·62}$$

where $\mathfrak{F}_0 \triangleq F_0/m$, and other symbols are as previously defined.

For subcritical damping, with $\zeta < 1$, the homogeneous solution is most conveniently written as in Eq. (5·41)

$$x_h = e^{-\zeta pt}(C_1 \sin p\sqrt{1-\zeta^2}\,t + C_2 \cos p\sqrt{1-\zeta^2}\,t) \tag{5·63}$$

To apply the method of undetermined coefficients, we first compare this homogeneous solution to the forcing function $\mathfrak{F}(t) = \mathfrak{F}_0 \sin \omega t$. If the latter is linearly independent of the functions in the homogeneous solution, so that we cannot generate $\mathfrak{F}(t)$ by any linear combination of the functions that are added together to comprise the homogeneous solution, then the particular solution may be expressed in terms of undetermined coefficients as

$$x_p = D_1 \sin \omega t + D_2 \cos \omega t \tag{5·64}$$

Note that for the first term we adopt simply $D_1 \sin \omega t$, not troubling to include the multiplier $\mathfrak{F}_0$, as a literal interpretation of Eq. (5·61) would imply. Similarly, the constant $\omega\mathfrak{F}_0$ in $\dot{\mathfrak{F}}(t)$ is absorbed in the undetermined coefficient D_2. The proposed solution does not contain terms in D_3, etc., because further differentiation would only yield more terms in $\sin \omega t$ and $\cos \omega t$, and these functions are already included in the proposal.

Substituting Eq. (5·64) into (5·62) provides

$$-\omega^2(D_1 \sin \omega t + D_2 \cos \omega t) + 2\zeta p\omega(D_1 \cos \omega t - D_2 \sin \omega t) + p^2(D_1 \sin \omega t + D_2 \cos \omega t) = \mathfrak{F}_0 \sin \omega t \tag{5·65}$$

From this single equation, the two scalar coefficients D_1 and D_2 must be determined. This is accomplished by collecting terms multiplied by each of the linearly independent functions $\sin \omega t$ and $\cos \omega t$, setting each set of coefficients independently to zero. This gives the two algebraic equations

$$(p^2 - \omega^2)D_1 - 2\zeta p\omega D_2 = \mathfrak{F}_0 \tag{5·66}$$

$$2\zeta p\omega D_1 + (p^2 - \omega^2)D_2 = 0 \tag{5·67}$$

to be solved simultaneously for D_1 and D_2. Either by direct substitution or by

Cramer's rule we may find the solutions

$$D_1 = \frac{\begin{vmatrix} \mathfrak{F}_0 & -2\zeta p\omega \\ 0 & p^2 - \omega^2 \end{vmatrix}}{\begin{vmatrix} p^2 - \omega^2 & -2\zeta p\omega \\ 2\zeta p\omega & p^2 - \omega^2 \end{vmatrix}} = \frac{\mathfrak{F}_0(p^2 - \omega^2)}{(p^2 - \omega^2)^2 + 4\zeta^2 p^2\omega^2} \tag{5·68}$$

$$D_2 = \frac{\begin{vmatrix} p^2 - \omega^2 & \mathfrak{F}_0 \\ 2\zeta p\omega & 0 \end{vmatrix}}{\begin{vmatrix} p^2 - \omega^2 & -2\zeta p\omega \\ 2\zeta p\omega & p^2 - \omega^2 \end{vmatrix}} = \frac{-2\zeta p\omega\mathfrak{F}_0}{(p^2 - \omega^2)^2 + 4\zeta^2 p^2\omega^2} \tag{5·69}$$

Interpretation of the solution is facilitated by the identity

$$D_1 \sin \omega t + D_2 \cos \omega t = D \sin (\omega t - \varphi) \tag{5·70}$$

where

$$D^2 = D_1{}^2 + D_2{}^2 \tag{5·71}$$

and

$$\tan \varphi = -\frac{D_2}{D_1} \tag{5·72}$$

The validity of this identity can be seen in Fig. 5·4, where $D_1 \sin \omega t$ and $D_2 \cos \omega t$ represent horizontal projections of orthogonal vectors of lengths D_1 and D_2. With this identity, the particular solution becomes

$$x_p = \frac{\mathfrak{F}_0 \sin (\omega t - \varphi)}{[(p^2 - \omega^2)^2 + 4\zeta^2 p^2\omega^2]^{1/2}}$$

$$= \frac{(\mathfrak{F}_0/p^2) \sin (\omega t - \varphi)}{\{[1 - (\omega^2/p^2)]^2 + 4\zeta^2(\omega^2/p^2)\}^{1/2}} \tag{5·73}$$

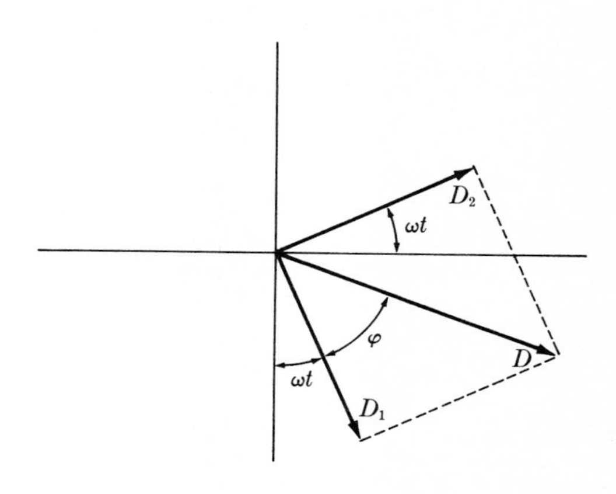

FIGURE 5·4

where

$$\tan \varphi = \frac{2\zeta p\omega}{p^2 - \omega^2} = \frac{2\zeta(\omega/p)}{1 - (\omega^2/p^2)} \tag{5·74}$$

When we return to the original definitions, we find

$$\frac{\mathfrak{F}_0}{p^2} = \frac{F_0/m}{k/m} = \frac{F_0}{k} \triangleq \Delta \tag{5·75}$$

where Δ represents the value x would adopt under the static load F_0. Thus an expression for x_p normalized by the static deflection Δ may be written as

$$\frac{x_p}{\Delta} = Q \sin(\omega t - \varphi) \tag{5·76}$$

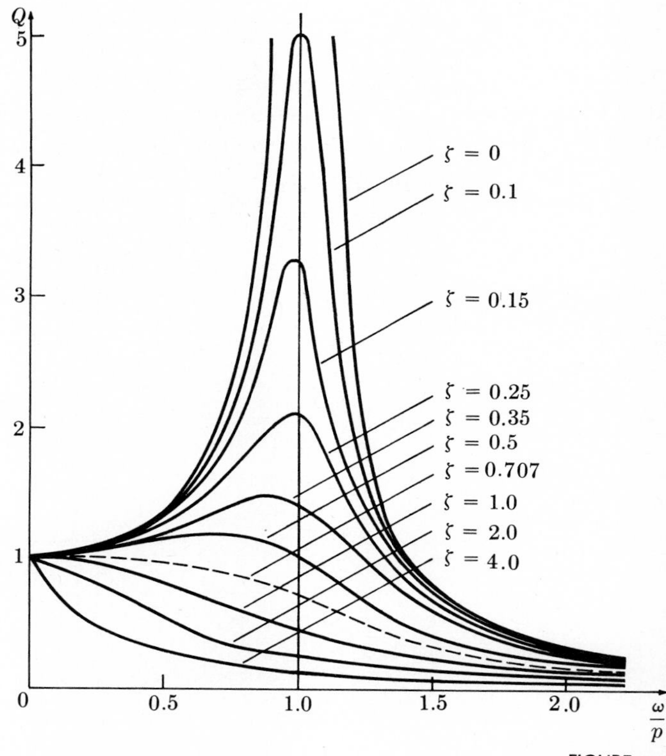

FIGURE 5·5

where the *gain* or *magnification factor* Q is given by

$$Q = \left[\left(1 - \frac{\omega^2}{p^2}\right)^2 + 4\zeta^2\frac{\omega^2}{p^2}\right]^{-1/2} \tag{5·77}$$

Figs. 5·5 and 5·6 show how gain Q and phase lag φ depend on the damping ratio ζ and the frequency ratio ω/p. When the frequency ratio *approaches* unity, the gain Q reaches a maximum value which, from Eq. (5·77), is approximated by

$$Q_{\max} \cong \frac{1}{2\zeta} \tag{5·78}$$

The term *resonance* is applied when $\omega = p$.

It would appear from Eqs. (5·76) and (5·77) that when $\zeta = 0$, the resonance condition results in unbounded response since in the absence of damping, Q goes to infinity as ω/p approaches unity. This would be a hasty conclusion, however, since we will find on examination that when $\zeta = 0$ and $\omega = p$, the premises on which we based the solution of Eq. (5·64) are no longer valid. Note that when $\zeta = 0$, the homogeneous solution of Eq. (5·63) becomes

$$x_h = C_1 \sin pt + C_2 \cos pt \tag{5·79}$$

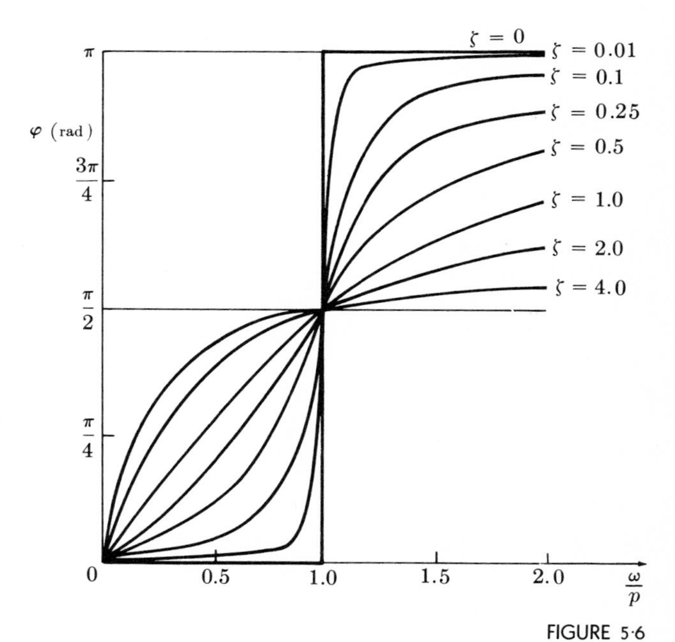

FIGURE 5·6

When $\omega = p$, the forcing function becomes

$$\mathfrak{F}(t) = \mathfrak{F}_0 \sin pt \tag{5·80}$$

We can no longer adopt the solution proposed in Eq. (5·64) since, when $\omega = p$, the terms in that expression are not linearly independent of the homogeneous solution in Eq. (5·79). Following the indicated procedure, we must now advance a solution of the form

$$x_p = D_1 t \sin pt + D_2 t \cos pt \tag{5·81}$$

Substitution into the differential equation

$$\ddot{x} + p^2 x = \mathfrak{F}_0 \sin pt \tag{5·82}$$

yields

$$(-pD_2 - p^2 D_1 t - pD_2) \sin pt + (pD_1 + pD_1 - p^2 D_2 t) \cos pt + p^2(D_1 t \sin pt + D_2 t \cos pt) = \mathfrak{F}_0 \sin pt$$

Cancellation provides the convenient result

$$(-2pD_2 - \mathfrak{F}_0) \sin pt + 2pD_1 \cos pt = 0$$

The independence of sin pt and cos pt permits the conclusions

$$D_1 = 0$$

$$D_2 = -\frac{\mathfrak{F}_0}{2p}$$

The particular solution sought therefore becomes

$$x_p = -\frac{\mathfrak{F}_0}{2p} t \cos pt \tag{5·83}$$

It should be acknowledged at this point that Eq. (5·83) is of little engineering significance since we may be sure that every system of physical origin has some degree of energy dissipation (although it may not be properly represented by viscous damping). While the difference in the solutions represented by Eqs. (5·83) and (5·76) is very real mathematically, limitations in the mathematical model make this distinction relatively inconsequential. All we can say for sure about the motion of an essentially undamped physical system when the linear vibration equations indicate resonance is that the amplitude of its oscillations becomes so large that the equation becomes an unreliable indicator of behavior. The distinction between solution (5·83) and solution (5·76) for undamped resonance serves here primarily to illustrate the alternative approaches required by the method of undetermined coefficients for finding particular solutions of linear, constant-coefficient ordinary differential equations.

5·1·13 General solution for harmonically forced oscillator

In our quest for specific methods of generating particular and homogeneous solutions, we must not lose sight of the fact that the general solution to the original equation for forced vibration of a damped linear oscillator is the *combination* of homogeneous and particular solutions.

We began with an inhomogeneous equation of motion which we wrote in the form [see Eq. (5·28)]

$$\ddot{x} + 2\zeta p\dot{x} + p^2x = \mathfrak{F}(t)$$

Homogeneous solutions were written as Eqs. (5·41), (5·50), and (5·56), and particular solutions for the special case $\mathfrak{F}(t) = \mathfrak{F}_0 \sin \omega t$ were recorded as Eqs. (5·76) and (5·83). To illustrate the combination process and the evaluation of free constants from initial conditions, we focus on the class of problem with $\zeta = 0$ and $\mathfrak{F}(t) = \mathfrak{F}_0 \sin \omega t$, where $\omega \neq p$. The combination of Eqs. (5·41) and (5·76) then provides

$$x = C_1 \sin pt + C_2 \cos pt + \frac{\Delta \sin \omega t}{(1 - \omega^2/p^2)} \tag{5·84}$$

The complete solution must satisfy the initial conditions $x(0) = x_0$, $\dot{x}(0) = v_0$, so we have

$$x_0 = C_2$$

$$v_0 = C_1p + \frac{\omega\Delta}{(1 - \omega^2/p^2)}$$

Finally, the complete solution may be written

$$x = \frac{1}{p}\left[v_0 - \frac{\omega\Delta}{(1 - \omega^2/p^2)}\right] \sin pt + x_0 \cos pt + \frac{\Delta \sin \omega t}{(1 - \omega^2/p^2)} \tag{5·85}$$

Because we have chosen for simplicity to examine a case with $\zeta = 0$, both the homogeneous and the particular parts of our solution in Eq. (5·85) persist for all time. When viscous damping is incorporated, the homogeneous solutions of Eqs. (5·41), (5·50), and (5·56) decay asymptotically with time, so this part of the solution is called *transient*. In contrast, the *steady-state* particular solution is maintained as long as the system is excited. It is for this reason that the particular solution, described by Eqs. (5·76) and (5·77), and by Figs. 5·5 and 5·6, receives greatest attention; if you wait long enough this represents the total solution for *any* physical system since damping in some form is always present, and the homogeneous solution will become attenuated in time.

5·1·14 Response to step-function force by method of undetermined coefficients

The determination of the dynamic response of the damped oscillator to the force illustrated in Fig. 5·7 is both simple and useful. The equation of motion (5·24) applicable for $t \geq 0$ is simply

$$m\ddot{x} + c\dot{x} + kx = F_0 \tag{5·86}$$

With the usual definition of the force per unit mass $\mathfrak{F}_0 = F_0/m$, frequency $p = \sqrt{k/m}$, and the damping ratio $\zeta = c/(2mp)$, this equation becomes

$$\ddot{x} + 2\zeta p\dot{x} + p^2 x = \mathfrak{F}_0 \tag{5·87}$$

In order to focus attention on a specific case, we shall assume $\zeta < 1$, although what follows is equally straightforward in other cases. The homogeneous solution (5·41) applies, and we need only find a new particular solution. It may not immediately be apparent that the constant $\mathfrak{F}_0$ qualifies as a special case of the expression (5·58), $\mathfrak{F}(t) = \sum_{s=1}^{n} P_s(t)e^{\alpha_s t}$, where $P_s(t)$ is a polynomial in t. We must consider the trivial case with $n = 1$, $\alpha_1 = 0$, and $P_1(t) = \mathfrak{F}_0$ to recognize this qualification. (The first term in a polynomial in t is generally a coefficient multiplied by $t^0 = 1$, providing a constant we call $\mathfrak{F}_0$.) Thus the method of undetermined coefficients does apply to the step function $\mathfrak{F}(t) = \mathfrak{F}_0$, as it does to the ramp function $\mathfrak{F}(t) = at$, the parabolic function $\mathfrak{F}(t) = at^2$, etc.

The general form of the particular solution proposed as Eq. (5·61) is easy to apply here since the time derivatives of our $\mathfrak{F}(t) = \mathfrak{F}_0$ are all zero. Thus the particular solution is merely

$$x_p = D_1$$

Substitution into Eq. (5·87) provides

$$D_1 = \frac{\mathfrak{F}_0}{p^2} = \frac{F_0}{mp^2} = \frac{F_0}{k} \triangleq \Delta$$

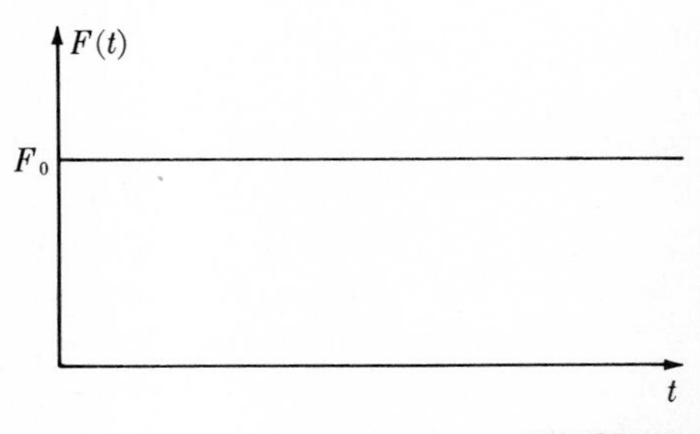

FIGURE 5·7

adopting the notation of Eq. (5·75) for static deflection Δ. Thus the total solution for Eq. (5·87) becomes

$$x = \Delta + e^{-\zeta pt}(C_1 \sin p\sqrt{1-\zeta^2}\,t + C_2 \cos p\sqrt{1-\zeta^2}\,t) \tag{5·88}$$

Coefficients C_1 and C_2 are calculated from initial conditions $x(0) = x_0$ and $\dot{x}(0) = v_0$. The result is

$$x = \Delta + e^{-\zeta pt}\left[\frac{v_0 + \zeta p(x_0 - \Delta)}{p\sqrt{1-\zeta^2}} \sin p\sqrt{1-\zeta^2}\,t + (x_0 - \Delta)\cos p\sqrt{1-\zeta^2}\,t\right] \tag{5·89}$$

For the special case $x_0 = v_0 = 0$, the step-function response becomes

$$x = \Delta\left[1 - e^{-\zeta pt}\left(\cos p\sqrt{1-\zeta^2}\,t + \frac{\zeta}{\sqrt{1-\zeta^2}} \sin p\sqrt{1-\zeta^2}\,t\right)\right] \tag{5·90}$$

The identity of Eq. (5·70) has the counterpart

$$D_1 \sin \omega t + D_2 \cos \omega t = D \cos(\omega t - \psi) \tag{5·91}$$

where

$$D^2 = D_1^2 + D_2^2$$

and

$$\tan \psi = \frac{D_1}{D_2}$$

The application of this identity to Eq. (5·90) provides the step-function response in the useful form

$$x = \Delta\left[1 - \frac{e^{-\zeta pt}\cos(p\sqrt{1-\zeta^2}\,t - \psi)}{\sqrt{1-\zeta^2}}\right] \triangleq \Delta\Gamma(t) \tag{5·92}$$

The term *indicial admittance* is sometimes applied to the unit step-function[1] response, represented by the symbol $\Gamma(t)$. From Fig. 5·8, we see that $\Gamma(t)$ can never exceed the value 2.

[1] The unit step function is often called the Heaviside function.

5·1·15 Superposition integrals

Thus far we have examined only the method of undetermined coefficients in our quest for particular solutions, and this method is restricted to forcing functions of the special kind described in Eq. (5·58). We must be prepared to cope with more general forcing functions. In engineering practice we are particularly concerned with the response to shock loadings of brief duration, and with forcing functions for which only experimental data is available, without functional description. In the *superposition integral* we find an approach ideally suited even to these difficult problems.

We consider first the response to a force of constant magnitude and brief duration, as represented by Fig. 5·9. In the limit as the duration ϵ goes to zero, subject to the constraint that the area under the curve in Fig. 5·9 remains a constant [represented by $I \triangleq \int_{t_0}^{t_0+\epsilon} F(t)\, dt$], this forcing function is called an *ideal impulse*.

Recall that in the general definition of Eq. (4·23), the term impulse describes a *vector* integral of unspecified duration. In that more general context, we developed in Eq. (4·25) an impulse-momentum relationship that, in the special case at hand, provides

$$\mathbf{I} = \int_{t_0}^{t_0+\epsilon} \mathbf{F}(t)\, dt = m[\dot{x}(t_0 + \epsilon) - v_0]\hat{\mathbf{e}}_x$$

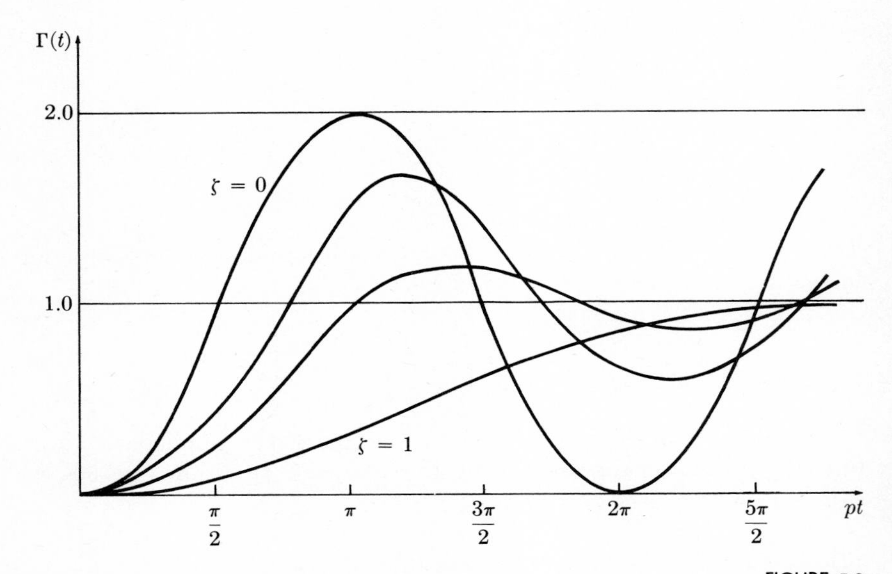

FIGURE 5·8

where $v_0 \triangleq \dot{x}(t_0)$. If we now assume that the duration ϵ is so brief that in that time interval there is no change in the coordinate x, then we conclude that

$$x(t_0 + \epsilon) = x(t_0) \triangleq x_0 \tag{5·93a}$$

and we are prepared in the interval of application of the impulsive external force to ignore the relatively smaller forces applied to the mass by the spring and the dashpot. Thus the response to an ideal impulse of magnitude I is an instantaneous change of speed to the value

$$\dot{x}(t_0 + \epsilon) = v_0 + \frac{I}{m} \triangleq v_0 + \mathcal{I} \tag{5·93b}$$

where v_0 is the speed prior to the impulsive loading at time t_0, and $\mathcal{I}$ is the impulse magnitude per unit mass. After t_0, when there is no longer any force applied, the free-vibration solution of Eq. (5·41) applies (assuming $\zeta \neq 1$), with initial conditions given by Eq. (5·93). The result is a variation of Eq. (5·47), given by

$$x(t - t_0) = e^{-\zeta p(t-t_0)} \left\{ \left(\frac{v_0 + \mathcal{I} + \zeta p x_0}{p\sqrt{1-\zeta^2}} \right) \sin \left[p\sqrt{1-\zeta^2}\,(t - t_0)\right] + x_0 \cos \left[p\sqrt{1-\zeta^2}\,(t - t_0)\right] \right\}$$

or more conveniently by

$$x(t - t_0) = e^{-\zeta p(t-t_0)} \left\{ \left(\frac{v_0 + \zeta p x_0}{p\sqrt{1-\zeta^2}} \right) \sin \left[p\sqrt{1-\zeta^2}\,(t - t_0)\right] + x_0 \cos \left[p\sqrt{1-\zeta^2}\,(t - t_0)\right] + \mathcal{I} g(t - t_0) \right\} \tag{5·94}$$

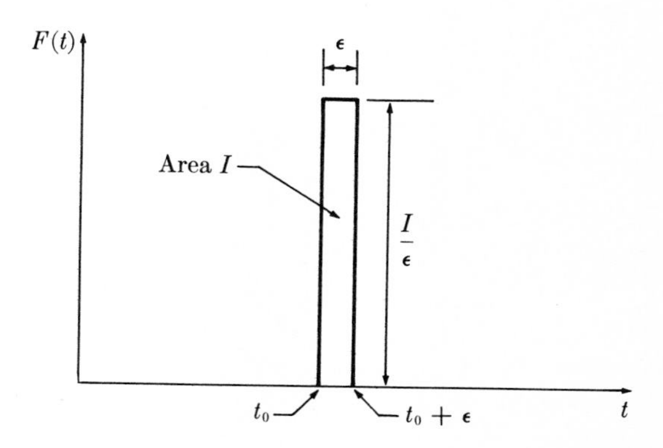

FIGURE 5·9

where $g(t - t_0)$, called the *unit impulse response* for Eq. (5·28), is

$$g(t - t_0) = e^{-\zeta p(t-t_0)} \sin \frac{[p\sqrt{1 - \zeta^2}\,(t - t_0)]}{p\sqrt{1 - \zeta^2}} \tag{5·95}$$

Note that in this case we have escaped from the task of actually solving an inhomogeneous equation describing an impulsive forcing function; we have instead solved a homogeneous initial-value problem, using the impulse-momentum relationship to calculate the appropriate initial values. The solution (5·94) does *not* represent the combination of homogeneous and particular solutions in the manner previously described [illustrated, for example, by Eqs. (5·88) and (5·89)]. We have, in fact, not yet even agreed on an analytical formulation for the ideal impulse; we have not even written out equations of motion.

The force represented by Fig. 5·9 is a perfectly sensible concept, approximately representative of many physical situations. We have described an ideal impulse as a limiting case of this force, requiring ϵ to approach zero; the result is a force of infinite magnitude over zero duration with integral or impulse given by the finite constant I. In this limiting sense, the concept becomes nonphysical, and even difficult to justify mathematically. Because the concept is a useful one, however, it has been widely adopted in engineering analysis, and a class of functions called *singularity functions* has been devised to provide analytical representation of such concepts as the ideal impulse. We choose for example to represent an ideal impulse of magnitude I applied at time t_0 by the expression

$$F(t) = I\delta(t - t_0) \tag{5·96}$$

where the *Dirac delta function* $\delta(t - t_0)$ is defined by

$$\delta(t - t_0) = 0 \qquad t \neq t_0 \tag{5·97a}$$

$$\int_0^\infty \delta(t - t_0)\,dt = 1 \qquad 0 < t_0 < \infty \tag{5·97b}$$

so that

$$\int_0^\infty F(t)\,dt = \int_0^\infty I\,\delta(t - t_0)\,dt = I\int_0^\infty \delta(t - t_0)\,dt = I$$

As consideration of Fig. 5·10 suggests, the delta function has the property

$$\int_0^\infty F(t)\,\delta(t - \tau)\,dt = F(\tau) \tag{5·98}$$

Not only can we represent the ideal impulse analytically with the Dirac delta function, but we can also use the unit impulse response $g(t)$ to a delta-function excitation as the basis for representing the response to arbitrary excitation.

Examine Fig. 5·10 once again, and imagine that it represents the forcing function in the equation

$$m\ddot{x} + c\dot{x} + kx = F(t)$$

or

$$\ddot{x} + 2\zeta p\dot{x} + p^2x = \mathfrak{F}(t) \tag{5·99}$$

We might be willing to accept the approximation of $F(t)$ represented by the staircase curve in Fig. 5·10 even for noninfinitesimal values of pulse widths $\Delta\tau$, and satisfaction with this approximation increases as $\Delta\tau$ diminishes. As $\Delta\tau$ approaches zero, each of these rectangular pulses approaches an ideal impulse of magnitude $I = F(\tau)\,\Delta\tau$. If initial conditions are ignored, the response to the single impulse at $t = \tau$ is, from Eq. (5·94),

$$x(t - \tau) = \mathfrak{I}(\tau)g(t - \tau) = \frac{I(\tau)}{m}g(t - \tau)$$
$$= \frac{F(\tau)\,\Delta\tau}{m}g(t - \tau) = \mathfrak{F}(\tau)\,\Delta\tau\,g(t - \tau)$$

The total response to the excitation $F(t)$ (or $m\mathfrak{F}(t)$) then becomes the sum of these many individual responses since the homogeneous equation under consideration is linear. In the limit as $\Delta\tau$ vanishes, we can replace the summation informally by an integration, to obtain the total response

$$x(t) = \int_0^t x(t - \tau)\,d\tau = \int_0^t \mathfrak{F}(\tau)g(t - \tau)\,d\tau \tag{5·100}$$

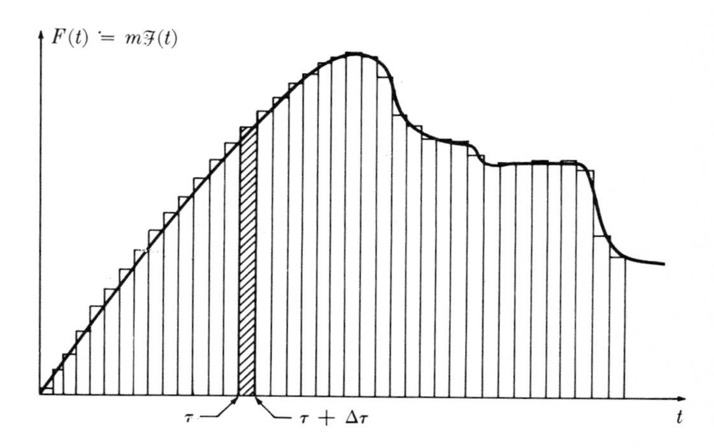

FIGURE 5·10

If at time zero there are initial conditions x_0 and v_0, we can see from Eq. (5·94) that the total response is

$$x(t) = e^{-\zeta pt}\left(\frac{v_0 + \zeta p x_0}{p\sqrt{1-\zeta^2}} \sin p\sqrt{1-\zeta^2}\,t + x_0 \cos p\sqrt{1-\zeta^2}\,t\right) + \int_0^t \mathfrak{F}(\tau) g(t-\tau)\,d\tau \qquad (5\cdot101)$$

for initial $t_0 = 0$.

The integrals in Eqs. (5·100) and (5·101) are examples of *superposition integrals* (sometimes called *convolution integrals* or *Duhamél integrals*), and the function $g(t - \tau)$ is *Green's function* for Eq. (5·99). Both of these concepts prove valuable in a more general framework than is revealed here.

5·1·16 Application of superposition integral to step-function excitation

We have noted previously the solution to the problem of determining the dynamic response of a damped oscillator subjected to a step-function force, as portrayed in Fig. 5·7. The differential equation of motion is given by Eqs. (5·86) and (5·87), and the solution is obtained by the method of undetermined coefficients as in Eq. (5·89). We now seek to illustrate the superposition-integral approach by attacking the same problem.

By direct substitution into Eq. (5·101) of $\mathfrak{F}(\tau) = \mathfrak{F}_0$, and $g(t - \tau)$ from Eq. (5·95), we obtain

$$\begin{aligned} x(t) &= e^{-\zeta pt}\left(\frac{v_0 + \zeta p x_0}{p\sqrt{1-\zeta^2}} \sin p\sqrt{1-\zeta^2}\,t + x_0 \cos p\sqrt{1-\zeta^2}\,t\right) \\ &\qquad + \int_0^t \left\{\mathfrak{F}_0 e^{-\zeta p(t-\tau)} \frac{\sin [p\sqrt{1-\zeta^2}\,(t-\tau)]}{p\sqrt{1-\zeta^2}}\right\} d\tau \\ &= e^{-\zeta pt}\left(\frac{v_0 + \zeta p x_0}{p\sqrt{1-\zeta^2}} \sin p\sqrt{1-\zeta^2}\,t + x_0 \cos p\sqrt{1-\zeta^2}\,t\right) \\ &\qquad + \frac{p\,\Delta e^{-\zeta pt}}{\sqrt{1-\zeta^2}} \int_0^t e^{\zeta p\tau}(\sin p\sqrt{1-\zeta^2}\,t \cos p\sqrt{1-\zeta^2}\,\tau \\ &\qquad - \cos p\sqrt{1-\zeta^2}\,t \sin p\sqrt{1-\zeta^2}\,\tau)\,d\tau \end{aligned} \qquad (5\cdot102)$$

since by definition $\Delta \triangleq \mathfrak{F}_0/p^2$.

From tables of integrals we find

$$\int e^{a\xi} \sin b\xi\,d\xi = \frac{e^{a\xi}(a \sin b\xi - b \cos b\xi)}{a^2 + b^2} \qquad (5\cdot103a)$$

and

$$\int e^{a\xi} \cos b\xi \, d\xi = \frac{e^{a\xi}(a \cos b\xi + b \sin b\xi)}{a^2 + b^2} \tag{5·103b}$$

From these tabulated integrals we obtain

$$\sin p \sqrt{1 - \zeta^2}\, t \int_0^t e^{\zeta p \tau} \cos p \sqrt{1 - \zeta^2}\, \tau \, d\tau - \cos p \sqrt{1 - \zeta^2}\, t \int_0^t e^{\zeta p \tau} \sin p \sqrt{1 - \zeta^2}\, \tau \, d\tau$$

$$= \frac{e^{\zeta p \tau}}{p^2} (p \sqrt{1 - \zeta^2}) - \frac{\zeta p}{p^2} \sin p \sqrt{1 - \zeta^2}\, t - \frac{p \sqrt{1 - \zeta^2}}{p^2} \cos p \sqrt{1 - \zeta^2}\, t \tag{5·104}$$

Combining Eqs. (5·104) and (5·102) produces

$$x(t) = e^{-\zeta p t} \left(\frac{v_0 + \zeta p x_0}{p \sqrt{1 - \zeta^2}} \sin p \sqrt{1 - \zeta^2}\, t + x_0 \cos p \sqrt{1 - \zeta^2}\, t \right) + \Delta - \Delta e^{-\zeta p t} \left(\frac{\zeta}{\sqrt{1 - \zeta^2}} \sin p \sqrt{1 - \zeta^2}\, t + \cos p \sqrt{1 - \zeta^2}\, t \right)$$

This result matches that obtained previously as Eq. (5·89).

It should be evident from this example that in execution the method depends on an integration, and in practice we cannot expect to find conveniently tabulated very many of the integrals we encounter. We have nonetheless in the development of the general solution in Eq. (5·101) reduced to quadrature the task of obtaining a general solution for the equation

$$\ddot{x} + 2\zeta p \dot{x} + p^2 x = \mathfrak{F}(t)$$

5·1·17 Comments on the general case

Section 5·1 on the motion of a particle with one degree of freedom began with the derivation of

$$m\ddot{x} = F(x, \dot{x}, t)$$

as the general equation of motion to be solved for this class of problem. The balance of this section has been devoted to the solution of the following special cases:

$$\begin{aligned} m\ddot{x} &= F(t) \\ m\ddot{x} &= F(x) \\ m\ddot{x} &= F(\dot{x}) \\ m\ddot{x} + c\dot{x} + kx &= F(t) \end{aligned}$$

Although investigation of these special cases has required time and energy, we have just begun to explore the difficulties of the general problem. To continue this exploration is an adventure in itself, to be deferred for other books and other courses. It is important, however, for every student of dynamics to realize that the theory of ordinary differential equations plays a dominant role in his work, and he must meet the substantial challenges of this subject if he is to become proficient in dynamics. The emphasis in the study of dynamics is on the *derivation* of equations of motion, and in the classroom we often restrict our attention to those equations we can readily solve. You should recognize, however, that we are at this point considering only the simplest imaginable problem in dynamics—the motion of a single particle with a single degree of freedom. Even for this problem we are unable to cope with the general differential equations, and must take recourse to more advanced analytical methods, procedures for approximate or asymptotic solution, or numerical methods for computer simulation. It is for this reason that we choose to separate the relatively straightforward task of deriving equations of motion from the often difficult business of solving them.

5·2 MULTIPLE-COORDINATE SOLUTIONS FOR MOTION OF A PARTICLE

5·2·1 Differential equations of motion

With the specification of three scalar functions, to be called $x_1(t)$, $x_2(t)$, and $x_3(t)$, it is always possible to define the motion of a particle fully. These scalars might be the cartesian coordinates of the particle for some chosen set of axes, but this is frequently not the most attractive alternative. If, for example, the particle is constrained to remain on the surface of a sphere of radius r, the spherical polar coordinates of Fig. 5·11 will offer a more convenient alternative. With this choice, x_1 and x_2 are angular measures, while x_3 is a measure of linear distance, constrained to maintain the constant value r.

When the particle has fewer than three degrees of freedom, an arbitrarily chosen set of three scalar coordinates will be related by one or more *constraint equations*. In the example of the particle on the sphere of radius r, the sum of the squares of each of three cartesian coordinates must be r^2 if the origin of the coordinate system is at the center of the sphere; with alternative choice of the spherical polar coordinates shown in Fig. 5·11, the constraint equation becomes simply $x_3 = r$.

Whether or not there exist any constraint equations, there are always equations of motion available from the vector equation $\mathbf{F} = m\mathbf{A}$. Whatever the physical significance of the coordinates, we know that the vectors $\mathbf{F}$ and $\mathbf{A}$ can depend on them and on certain of their time derivatives, as well as on time itself. We can therefore represent Newton's second law for a single particle in the explicit form

$$\mathbf{F}(x_1, x_2, x_3, \dot{x}_1, \dot{x}_2, \dot{x}_3, t) = m\mathbf{A}(x_1, x_2, x_3, \dot{x}_1, \dot{x}_2, \dot{x}_3, \ddot{x}_1, \ddot{x}_3, \ddot{x}_3, t) \tag{5·105a}$$

or in the equivalent symbolic form

$$\mathbf{F}(x, \dot{x}, t) = m\mathbf{A}(x, \dot{x}, \ddot{x}, t) \tag{5·105b}$$

From the discussion of kinematics in Chap. 3 it may be noted that the acceleration $\mathbf{A}$ can include only linear terms in the second derivatives $\ddot{x}_1$, $\ddot{x}_2$, and $\ddot{x}_3$; these scalars are never found in $\mathbf{A}$ multiplied by each other or raised to any but the first power.

The vector equations (5·105) can, for any choice of vector basis, be written as three scalar equations, which must also be linear in $\ddot{x}_1$, $\ddot{x}_2$, and $\ddot{x}_3$, that is, the scalar equations must have the form

$$m_{11}\ddot{x}_1 + m_{12}\ddot{x}_2 + m_{13}\ddot{x}_3 = G_1(x_1, x_2, x_3, \dot{x}_1, \dot{x}_2, \dot{x}_3, t) \tag{5·106a}$$
$$m_{21}\ddot{x}_1 + m_{22}\ddot{x}_2 + m_{23}\ddot{x}_3 = G_2(x_1, x_2, x_3, \dot{x}_1, \dot{x}_2, \dot{x}_3, t) \tag{5·106b}$$
$$m_{31}\ddot{x}_1 + m_{32}\ddot{x}_2 + m_{33}\ddot{x}_3 = G_3(x_1, x_2, x_3, \dot{x}_1, \dot{x}_2, \dot{x}_3, t) \tag{5·106c}$$

for some unspecified functions G_1, G_2, and G_3.

When the particle has fewer than three degrees of freedom, either a constraint equation must be introduced to augment the equations of motion in Eqs. (5·106), or each constraint equation must be used to eliminate one of the

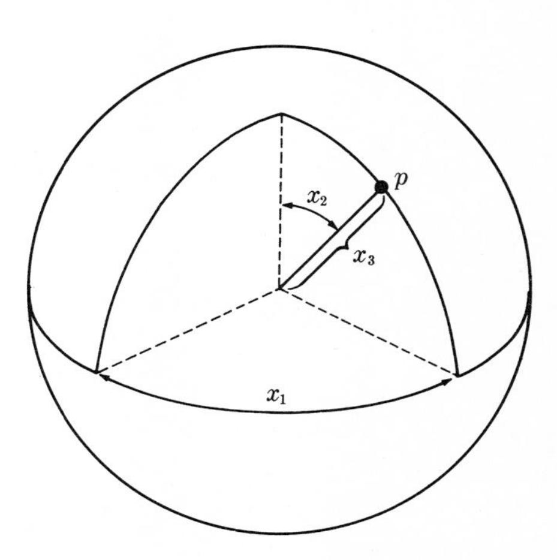

FIGURE 5·11

variables. If, for example, a constraint equation is used to eliminate x_3 from Eqs. (5·106), the resulting three differential equations in the two unknowns $x_1(t)$ and $x_2(t)$ cannot be independent. It must then be possible to manipulate the equations algebraically to obtain two independent second-order equations in x_1 and x_2, of the form

$$a_{11}\ddot{x}_1 + a_{12}\ddot{x}_2 = g_1(x_1, x_2, \dot{x}_1, \dot{x}_2, t) \tag{5·107a}$$
$$a_{21}\ddot{x}_1 + a_{22}\ddot{x}_2 = g_2(x_1, x_2, \dot{x}_1, \dot{x}_2, t) \tag{5·107b}$$

By judicious choice of vector basis, it is usually not difficult to obtain at the outset equations of motion of two-degree-of-freedom systems in the form of Eqs. (5·107).

Because Eqs. (5·106) and (5·107) are linear in the quantities $\ddot{x}_1$, $\ddot{x}_2$, and $\ddot{x}_3$, it is always possible to manipulate them algebraically to obtain equations in the form (for the more general case)

$$\ddot{x}_1 = f_1(x, \dot{x}, t) \tag{5·108a}$$
$$\ddot{x}_2 = f_2(x, \dot{x}, t) \tag{5·108b}$$
$$\ddot{x}_3 = f_3(x, \dot{x}, t) \tag{5·108c}$$

where x symbolizes x_1, x_2, and x_3, and f_1, f_2, and f_3 are unspecified functions.

Now that we have established the form of the scalar second-order equations of motion of a single particle with more than one degree of freedom, we might well wonder what to do with these equations. We know from our experience with particles with one degree of freedom (Sec. 5·1) that we cannot hope to establish general procedures for solution, and will have to resort to a computer simulation for all but a relatively small number of special cases.

5·2·2 A particle subject to a constant gravity force

The simplest type of motion of a particle is a motion described by *uncoupled* equations, that is, by equations such as Eqs. (5·108) but, for the special case

$$\ddot{x}_1 = f_1(x_1, \dot{x}_1, t) \tag{5·109a}$$
$$\ddot{x}_2 = f_2(x_2, \dot{x}_2, t) \tag{5·109b}$$
$$\ddot{x}_3 = f_3(x_3, \dot{x}_3, t) \tag{5·109c}$$

Now the problem is reduced to three uncoupled problems, each of which is of the same form as Eq. (5·2), the general equation of motion of a particle with a single degree of freedom. As we discovered in Sec. 5·1, the simplification afforded by uncoupling the equations still leaves equations of motion that we can solve only in special cases. Among the simplest of the special problems in this category is the idealized problem of determining the trajectory of a particle in a vacuum, subject to uniform gravitational attraction. This is an idealization with little immediate utility beyond the classroom, but it provides an appropriate starting point for more precise calculations.

If a particle p of mass m is given an initial velocity $\mathbf{V}_0$, and the particle is influenced in its flight only by a constant gravitational attraction of magnitude mg, its vector equation of motion might be written in terms of cartesian coordinates as

$$-mg\hat{\mathbf{e}}_3 = m(\ddot{x}_1\hat{\mathbf{e}}_1 + \ddot{x}_2\hat{\mathbf{e}}_2 + \ddot{x}_3\hat{\mathbf{e}}_3)$$

Here $\hat{\mathbf{e}}_3$ is a unit vector directed vertically upward, and $\hat{\mathbf{e}}_1$ and $\hat{\mathbf{e}}_2$ complete an orthogonal triad. If $\hat{\mathbf{e}}_1$ is selected so that the plane formed by $\hat{\mathbf{e}}_1$ and $\hat{\mathbf{e}}_3$ is the plane of the initial velocity vector $\mathbf{V}_0$, and if α is the angle of inclination of the initial velocity from the horizontal, then the initial velocity may be written

$$\mathbf{V}_0 = V_0 \cos \alpha\hat{\mathbf{e}}_1 + V_0 \sin \alpha\hat{\mathbf{e}}_3 \tag{5·110}$$

where $V_0 \triangleq |\mathbf{V}_0|$.

Scalar equations of motion are obtained from the vector equation as

$$\ddot{x}_1 = 0 \tag{5·111a}$$
$$\ddot{x}_2 = 0 \tag{5·111b}$$
$$\ddot{x}_3 = -g \tag{5·111c}$$

If the origin of the coordinate system is selected so as to coincide with the position when $t = 0$, the initial conditions become

$$\begin{aligned} x_1(0) &= x_2(0) = x_3(0) = 0 \\ \dot{x}_1(0) &= V_0 \cos \alpha \\ \dot{x}_2(0) &= 0 \\ \dot{x}_3(0) &= V_0 \sin \alpha \end{aligned}$$

These equations have the simple solutions

$$\begin{aligned} x_1 &= A_1 + B_1t \\ x_2 &= A_2 + B_2t \\ x_3 &= A_3 + B_3t - \frac{gt^2}{2} \end{aligned}$$

Initial conditions provide

$$\begin{aligned} A_1 &= A_2 = A_3 = 0 \\ B_1 &= \dot{x}_1(0) = V_0 \cos \alpha \\ B_2 &= 0 \\ B_3 &= \dot{x}_3(0) = V_0 \sin \alpha \end{aligned}$$

thus the final solution is

$$x_1 = (V_0 \cos \alpha)t \tag{5·112a}$$
$$x_2 = 0 \tag{5·112b}$$
$$x_3 = -\tfrac{1}{2}gt^2 + (V_0 \sin \alpha)t \tag{5·112c}$$

In this case there has been no obstacle to finding the coordinates x_1, x_2, and x_3 as functions of time t. If we want to know the shape of the trajectory in space, we can easily substitute $x_1/(V_0 \cos \alpha)$ for t in the expression for x_3, to obtain

$$x_3 = -\frac{1}{2}\frac{gx_1^2}{(V_0 \cos \alpha)^2} + x_1 \tan \alpha$$

This equation has the form $x_3 = Ax_1^2 + Bx_1$, which describes a parabola passing through the origin, as shown in Fig. 5·12. It is a simple matter to obtain the range

$$R = \frac{2V_0^2}{g} \sin \alpha \cos \alpha = \frac{V_0^2}{g} \sin 2\alpha \tag{5·113}$$

by solving for x_1 when x_3 returns to zero. Evidently the value of the angle α which maximizes R is $\pi/4$.

The peak altitude $(x_3)_{\max}$ must be reached when the slope dx_3/dx_1 is zero, which occurs at midrange. Substitution of $x_1 = (V_0^2/g) \sin \alpha \cos \alpha$ into the expression for x_3 provides

$$(x_3)_{\max} = \frac{V_0^2}{2g} \sin^2 \alpha \tag{5·114}$$

5·2·3 A particle subject to an inverse-square gravity force

The problem of the motion of a particle in an inverse-square central force field provides an introduction to the field of celestial mechanics, and a second illus-

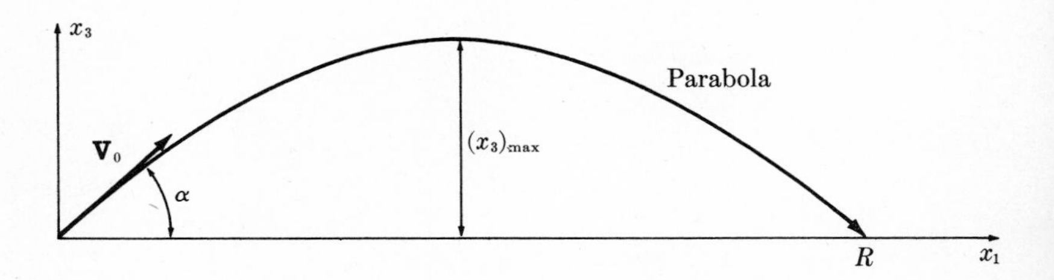

FIGURE 5·12

tration of a special case of Eqs. (5·108). The motions of celestial bodies have always fascinated man, and it was the attempt to explain these motions that gave primary impetus to the classical scholars in their formulation of the principles and procedures of dynamics. The advent of the "space age" in the late 1950s injected new spirit into this field, which has come to be called *astrodynamics*. Yet it must be recognized that much of the specialized methodology and nomenclature employed in determining the orbits of bodies in space is unique to that field, and tangential to the objectives of an introductory textbook in mechanics. We will therefore examine the problem of the motion of a particle p subject to forces proportional to the mass m of p and inversely proportional to the square of its distance from an attracting center, but only insofar as this exploration provides a useful example of methods and results of general applicability.

As illustrated in Fig. 5·13, the idealized system is a free particle p, subjected to a force directed toward a center o fixed in inertial space. The magnitude of the force is inversely proportional to the distance r from o to p. We assume that we know the position $\mathbf{R}_0$ and velocity $\mathbf{V}_0$ at some point in time we call *zero*, and we want to know the subsequent motion of the particle.

As always, we need only record Newton's second law to obtain the vector equation of motion. In order to record the gravitational force conveniently, we define the unit vector $\hat{\mathbf{u}}_r$ shown in Fig. 5·13 to parallel the line from o to p (so $\hat{\mathbf{u}}_r$ changes orientation in inertial space as the particle p moves). If m is the mass of particle p, and μ is a proportionality constant, the inverse-square gravity force may be written as

$$\mathbf{F} = -\frac{\mu m}{r^2}\hat{\mathbf{u}}_r \tag{5·115}$$

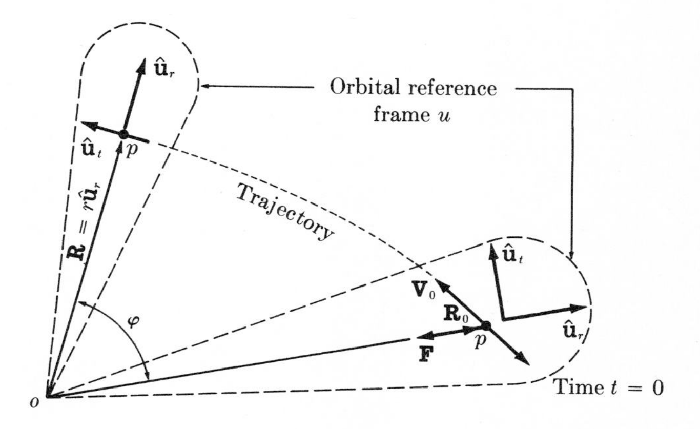

FIGURE 5·13

Newton's second law then provides the vector equation

$$\mathbf{F} = -\frac{\mu m}{r^2}\,\hat{\mathbf{u}}_r = m\mathbf{A}^p = m\,\frac{{}^i d^2}{dt^2}\,\mathbf{R} \tag{5·116}$$

There remains the task of vector differentiation.

Convenience is served by introducing a new unit vector $\hat{\mathbf{u}}_t$ orthogonal to $\hat{\mathbf{u}}_r$ and in the plane formed by $\mathbf{V}_0$ and $\mathbf{R}_0$. The unit vector $\hat{\mathbf{u}}_n$ completes the triad, with $\hat{\mathbf{u}}_n \triangleq \hat{\mathbf{u}}_r \times \hat{\mathbf{u}}_t$. Since the initial motion is in the plane established by the initial values of $\hat{\mathbf{u}}_r$ and $\hat{\mathbf{u}}_t$, and the force (and hence the inertial acceleration) are also in this plane, we may be assured at the outset that this is indeed a problem of motion confined to a plane remaining fixed in inertial space. Differentiation of vector $\mathbf{R}$ is facilitated by the introduction of an *orbital reference frame* u in which point o and unit vectors $\hat{\mathbf{u}}_r$, $\hat{\mathbf{u}}_t$ and $\hat{\mathbf{u}}_n$ are fixed. Frame u does not correspond to anything physical; although it is portrayed in Fig. 5·13 by a silhouette of dashed lines, this is just an artifice for representing the abstraction of the rotating frame.

The inertial velocity of p is now readily calculated as

$$\begin{aligned}\mathbf{V}^p \triangleq \frac{{}^i d}{dt}\,\mathbf{R} &= \frac{{}^i d}{dt}\,(r\hat{\mathbf{u}}_r) = \dot{r}\hat{\mathbf{u}}_r + r\,\frac{{}^i d}{dt}\,\hat{\mathbf{u}}_r \\ &= \dot{r}\hat{\mathbf{u}}_r + r\left(\frac{{}^u d}{dt}\,\hat{\mathbf{u}}_r + \boldsymbol{\omega}^u \times \hat{\mathbf{u}}_r\right)\end{aligned}$$

By definition of u, the derivative $({}^u d/dt)\,\hat{\mathbf{u}}_r$ is zero. Since the particle p moves in a plane fixed in inertial space, the inertial angular velocity $\boldsymbol{\omega}^u$ may be written as

$$\boldsymbol{\omega}^u = \dot{\varphi}\hat{\mathbf{u}}_n$$

where φ is the angle between $\mathbf{R}$ and $\mathbf{R}_0$. Thus the inertial velocity is

$$\mathbf{V}^p = \dot{r}\hat{\mathbf{u}}_r + r\dot{\varphi}\hat{\mathbf{u}}_t \tag{5·117}$$

The inertial acceleration of the particle is then

$$\begin{aligned}\mathbf{A}^p &= \frac{{}^i d}{dt}\,(\mathbf{V}^p) = \frac{{}^i d}{dt}\,(\dot{r}\hat{\mathbf{u}}_r + r\dot{\varphi}\hat{\mathbf{u}}_t) \\ &= \ddot{r}\hat{\mathbf{u}}_r + \dot{r}\,\frac{{}^i d}{dt}\,\hat{\mathbf{u}}_r + \dot{r}\dot{\varphi}\hat{\mathbf{u}}_t + r\ddot{\varphi}\hat{\mathbf{u}}_t + r\dot{\varphi}\,\frac{{}^i d}{dt}\,\hat{\mathbf{u}}_t \\ &= \ddot{r}\hat{\mathbf{u}}_r + \dot{r}\left(\frac{{}^u d}{dt}\,\hat{\mathbf{u}}_r + \dot{\varphi}\hat{\mathbf{u}}_n \times \hat{\mathbf{u}}_r\right) + \dot{r}\dot{\varphi}\hat{\mathbf{u}}_t + r\ddot{\varphi}\hat{\mathbf{u}}_t \\ &\qquad + r\dot{\varphi}\left(\frac{{}^u d}{dt}\,\hat{\mathbf{u}}_t + \dot{\varphi}\hat{\mathbf{u}}_n \times \hat{\mathbf{u}}_t\right) \\ &= (\ddot{r} - r\dot{\varphi}^2)\,\hat{\mathbf{u}}_r + (r\ddot{\varphi} + 2\dot{r}\dot{\varphi})\,\hat{\mathbf{u}}_t\end{aligned} \tag{5·118}$$

We could, of course, equally well obtain Eq. (5·118) by using the five-term kinematic expansion of Eq. (3·150), choosing the inertial frame i for frame f and the orbiting frame u for frame f'.

Newton's second law provides the vector equation of motion

$$-\frac{\mu m}{r^2}\,\hat{\mathbf{u}}_r = m(\ddot{r} - r\dot{\varphi}^2)\,\hat{\mathbf{u}}_r + m(r\ddot{\varphi} + 2\dot{r}\dot{\varphi})\,\hat{\mathbf{u}}_t \tag{5·119}$$

By equating coefficients of like unit vectors and dividing by m, we find the scalar differential equations

$$\ddot{r} - r\dot{\varphi}^2 = -\frac{\mu}{r^2} \tag{5·120a}$$

$$r\ddot{\varphi} + 2\dot{r}\dot{\varphi} = 0 \tag{5·120b}$$

Thus far we have merely derived the equations of planar motion of a particle under an inverse-square central force. As we anticipated, they readily adopt the form of Eqs. (5·108). There remains the major task of solving this pair of coupled nonlinear second-order ordinary differential equations. This time we don't have any general procedures or formulas to adopt; we simply have to examine this particular problem in search for solutions. If we examine Eq. (5·120*b*) thoughtfully, we may discover a first integral merely by inspection. This equation is equivalent to

$$\frac{1}{r}\frac{d}{dt}(r^2\dot{\varphi}) = 0 \tag{5·121}$$

Thus the quantity $r^2\dot{\varphi}$ must be a constant, which we designate as

$$k \triangleq r^2\dot{\varphi} \tag{5·122}$$

The result obtained as Eq. (5·122) is available directly from (5·120*b*), without any consideration of the physical problem described by the equation. Yet this result is not so obvious that you can see it at a glance; you might stare at Eq. (5·120*b*) for a long time without recognizing its equivalence to the following equation. When a differential equation describes the motion of a physical system, as Eq. (5·120) does, the search for a first integral is often much easier if you keep the physics of the problem in mind. The physical quantities most frequently helpful in the search for first integrals are forms of momentum (linear or angular) and energy quantities. For the particle in a central-force field, the colinearity of the force vector $\mathbf{F} = (-\mu m/r^2)\,\hat{\mathbf{u}}_r$ and the position vector $\mathbf{R} = r\hat{\mathbf{u}}_r$ guarantees that with respect to the fixed attracting center o, the particle is subjected to no external moment. If we go back to the definition of Eq. (4·36), we see that this means that the angular momentum or moment of momentum $\mathbf{H}^o \triangleq m\mathbf{R} \times \dot{\mathbf{R}}$ is constant.

In fact, we have in §4·2·6 already made this discovery, and recorded the constancy of the angular momentum $\mathbf{H}^o = mr^2\dot{\varphi}\hat{\mathbf{u}}_n$ for this problem. This conclusion was reached without even recording the scalar equations of motion, much less solving them. Thus we find that the constancy of the quantity $r^2\dot{\varphi}$ could have been discovered from either physical or mathematical argument.

To complete the solution of the original equations of motion, we must confront Eq. (5·120a), using the constancy of $r^2\dot{\varphi} = k$ wherever we can. It is a simple matter to substitute $\dot{\varphi} = k/r^2$ into Eq. (5·120a) to obtain

$$\ddot{r} - \frac{k^2}{r^3} = -\frac{\mu}{r^2} \tag{5·123}$$

This is a second-order differential equation in $r(t)$ alone, with all reference to φ removed, and it is quite natural to try to complete the solution of this problem by concentrating on this equation. It is, however, a nonlinear ordinary differential equation, and there do not exist general procedures for such equations; we may not be able to solve it.

After struggling in vain to solve Eq. (5·123), we might ask ourselves if there might be some way to get useful information concerning the solution without actually finding r as an explicit function of t by integrating this difficult equation. The compromise objective of greatest value is to find r as a function of φ, with time t removed from consideration.

The differentiation rule

$$\frac{dr}{dt} = \frac{dr}{d\varphi}\frac{d\varphi}{dt} = \frac{dr}{d\varphi}\left(\frac{k}{r^2}\right)$$

permits the elimination of time from Eq. (5·123). Substituting

$$\ddot{r} = \frac{d}{d\varphi}(\dot{r})\frac{d\varphi}{dt} = \frac{d}{d\varphi}\left(\frac{dr}{d\varphi}\frac{k}{r^2}\right)\frac{k}{r^2}$$
$$= \left[\frac{d^2r}{d\varphi^2}\left(\frac{k}{r^2}\right) - 2\left(\frac{dr}{d\varphi}\right)^2\frac{k}{r^3}\right]\frac{k}{r^2}$$

into Eq. (5·123) and multiplying by r^5/k^2 provides

$$r\frac{d^2r}{d\varphi^2} - 2\left(\frac{dr}{d\varphi}\right)^2 - r^2 = -\frac{\mu r^3}{k^2} \tag{5·124}$$

At first appearance, this equation is even more formidable than Eq. (5·123). You might examine it for a long time before you discover that the transformation to the new variable

$$w \triangleq r^{-1} \tag{5·125}$$

simplifies the equation remarkably. The substitutions

$$r = w^{-1}$$
$$\frac{dr}{d\varphi} = -w^{-2}\frac{dw}{d\varphi}$$
$$\frac{d^2r}{d\varphi^2} = 2w^{-3}\left(\frac{dw}{d\varphi}\right)^2 - w^{-2}\frac{d^2w}{d\varphi^2}$$

into Eq. (5·124) produce the gratifying result

$$2w^{-4}\left(\frac{dw}{d\varphi}\right)^2 - w^{-3}\frac{d^2w}{d\varphi^2} - 2w^{-4}\left(\frac{dw}{d\varphi}\right)^2 - w^{-2} = \frac{\mu w^{-3}}{k^2}$$

After canceling the first and third terms and multiplying by w^3, we find the amazingly simple equation

$$\frac{d^2w}{d\varphi^2} + w = \frac{\mu}{k^2} \tag{5·126}$$

This is a linear constant-coefficient equation with a constant on the right-hand side. This class of equation we know we can solve completely and easily.

The homogeneous solution of Eq. (5·126) must be of the form

$$w_h = C_1e^{\lambda_1\varphi} + C_2e^{\lambda_2\varphi}$$

where λ_1 and λ_2 are found by substituting $e^{\lambda\varphi}$ for w in the homogeneous counterpart of Eq. (5·126). The result is

$$\lambda^2e^{\lambda\varphi} + e^{\lambda\varphi} = 0$$

providing

$$\lambda = \pm i$$

where $i \triangleq \sqrt{-1}$. The homogeneous solution is therefore

$$\begin{aligned} w_h &= C_1e^{i\varphi} + C_2e^{-i\varphi} \\ &= C_1(\cos\varphi + i\sin\varphi) + C_2(\cos\varphi - i\sin\varphi) \\ &= A_1\cos\varphi + A_2\sin\varphi \end{aligned} \tag{5·127}$$

[by using the identity of Eq. (5·40)]. Quantities A_1 and A_2 are to be established by initial conditions, after the total solution to Eq. (5·126) is found. This requires the addition of a particular solution. Since the right-hand side of Eq. (5·126) is a constant, the method of undetermined coefficients readily provides the particular solution

$$w_p = \frac{\mu}{k^2} \tag{5·128}$$

The general solution of Eq. (5·126) is thus

$$w = A_1 \cos \varphi + A_2 \sin \varphi + \frac{\mu}{k^2} \tag{5·129}$$

Substituting the definition of w as r^{-1} does provide a result in which r is expressed as a function of φ, but interpretation is awkward until the form of this result is changed. For this purpose it is helpful to multiply all terms by $k^2 r/\mu$. The identity written as Eq. (5·91) is also useful in facilitating interpretation. With these alterations, and with the introduction of the symbol P for k^2/μ, Eq. (5·129) becomes

$$\frac{k^2}{\mu} \triangleq P = r[1 + e \cos (\varphi - \psi)] \tag{5·130}$$

where e and ψ are free constants to be established from initial conditions.

In the form of Eq. (5·130), the trajectory may be identified as a *conic section*, that is, either an *ellipse*, a *parabola*, or an *hyperbola*. The choice among these three alternatives is established by the quantity e, and hence by the initial conditions. Evidently, if $e < 1$, constancy of P is preserved if r varies harmonically from a minimum value of

$$r_{\min} = \frac{P}{1 + e} \tag{5·131}$$

occurring when $\varphi = \psi$, or $\psi + 2\pi n$, for any integer n, to a maximum value of

$$r_{\max} = \frac{P}{1 - e} \tag{5·132}$$

occurring when $\varphi = \psi + n\pi$, for integral n. (For an earth satellite, the position of minimum r is called the *perigee*, and the position of maximum r is called the *apogee*.) Thus the condition $e < 1$ corresponds to the elliptical orbit, as illustrated in Fig. 5·14. When initial conditions make $e = 1$, the trajectory is parabolic, and $r_{\max}$ goes to infinity. Hyperbolic trajectories occur when $e > 1$.

The trajectory represented by Eq. (5·130) does not represent in all respects a complete solution to the problem originally posed by the differential equations (5·120). Not unless we obtain r and φ as functions of time t can we claim a complete solution. Although it is possible to make progress in this direction by introducing more new variables and attacking the problem indirectly, this procedure is of interest primarily in the specialized field of orbital mechanics, and will be foregone here. You may be interested, however, in the early history of the problem of motion of a satellite in a gravity field.

In the preceding paragraphs we have examined the problem of determining from Newton's second law the motion of a particle in an inverse-square gravitational force field. Since the planets are small relative to the sun, and the influence of secondary celestial bodies on planetary motion is much less than that of the sun, we can with considerable accuracy predict the motions of the planets with dynamical theory only a little more complex than that preceding. Historically, however, the painstaking observations and exhaustive data analysis of the early astronomers provided orbital information before Newton's laws were available as the basis for analytical predictions. Indeed, when Newton devised his formulation of the laws of mechanics, he was aware of the empirically based conclusions of Johannes Kepler regarding planetary motions. Newton found that by hypothesizing a gravitational attraction between two bodies proportional to their mass product and inversely proportional to the distance between them, he could predict their motions by making use of his newly formulated laws of mechanics.

Kepler's principal conclusions from celestial observations are usually presented in the form of three laws, as follows:

1 Planetary orbits are ellipses, with the sun at a focus.

2 Equal areas are swept out by the position vector from sun to planet in equal time intervals.

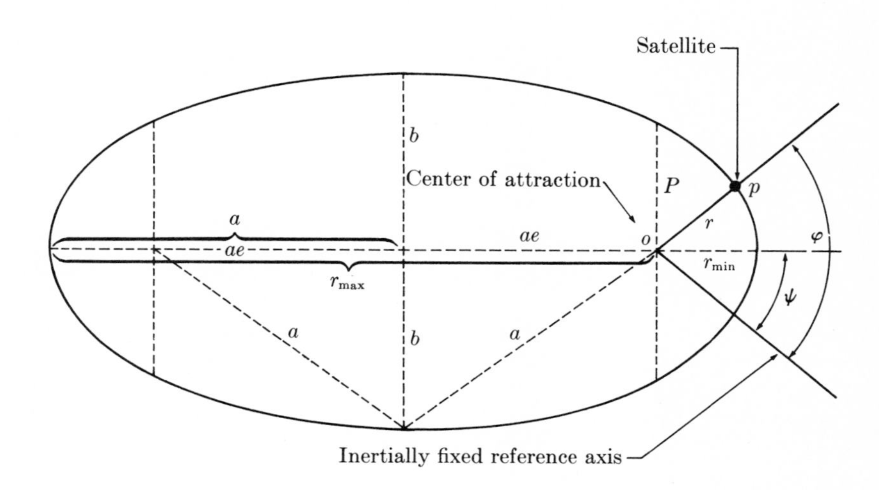

FIGURE 5·14

3 The square of an orbital period is proportional to the cube of the semimajor axis of the orbital ellipse.

Although Kepler's laws were deduced from observations, now that Newton's laws are available, the laws of Kepler can all be obtained from simple calculations. We have already shown, in Eq. (5·130), the compatibility of Kepler's first law and the newtonian laws. The second keplerian law follows directly from Eq. (5·122), which expresses the constancy of angular momentum per unit mass, k. In the time interval Δt in which the angle φ increases by amount $\Delta\varphi$, the area approximated by $\Delta A = \frac{1}{2}r(r\,\Delta\varphi)$ is swept out. In the limit as $\Delta A/\Delta t$ becomes dA/dt and $\Delta\varphi/\Delta t$ becomes $d\varphi/dt$, the expression for area swept out per unit time becomes, from Eq. (5·122),

$$\frac{dA}{dt} = \frac{1}{2}r^2\frac{d\varphi}{dt} = \frac{1}{2}k \qquad \text{constant} \tag{5·133}$$

Finally, consider what Newton's laws would predict for the period T of a planet's orbit around the sun. Since the rate dA/dt at which the area of the ellipse is swept out by the position vector is constant, the total area of the ellipse must equal the product of T and dA/dt. If a and b are, respectively, the semimajor axis and semiminor axis of the ellipse, as shown in Fig. 5·14, the total area is πab, and we have

$$\pi ab = T\frac{dA}{dt} = T(\tfrac{1}{2}r^2\dot{\varphi}) = \frac{Tk}{2}$$

Thus the period may be expressed as

$$T = \frac{2\pi ab}{k}$$

From the geometry of the ellipse we find [using Eqs. (5·131) and (5·132)]

$$2a = r_{\max} + r_{\min} = \frac{P}{1-e} + \frac{P}{1+e} = \frac{2P}{1-e^2}$$

Since $a - r_{\min} = ae$, the distance from the center of the ellipse to its focus is ae (explaining why e is called the *eccentricity* of the orbit).

To relate b to a and e, recall that the sum of the distances to any point on an ellipse from its two foci is a constant, given by $2a$. Thus from Fig. 5·14 and the pythagorean theorem we have

$$b^2 = a^2 - (ae)^2 = a^2(1 - e^2)$$

The orbital period may therefore be written

$$T = \frac{2\pi a^2 \sqrt{1 - e^2}}{k}$$

Since k may be written as $\sqrt{\mu P}$, and P in turn is equal to $a(1 - e^2)$, we have

$$T = \frac{2\pi a^2(1 - e^2)^{1/2}}{[\mu a(1 - e^2)]^{1/2}} = 2\pi \sqrt{\frac{a^3}{\mu}} \tag{5·134}$$

Eq. (5·134) tells us that T^2 is proportional to a^3, confirming Kepler's third law.

To establish the geometrical interpretation of P indicated by Fig. 5·14, note that

$$P + [P^2 + (2ae)^2]^{1/2} = 2a$$

since $P = a(1 - e^2)$.

5·2·4 Small oscillations of a particle

In Sec. 5·1, dealing with solutions for motion of a particle with *one* degree of freedom, particular success was achieved in solving the linear equation

$$m\ddot{x} + c\dot{x} + kx = F(t) \tag{5·24}$$

Quite analogous progress can be made in the present context of more general motions of a particle, when described by the linear constant-coefficient equations

$$m_{11}\ddot{x}_1 + m_{12}\ddot{x}_2 + m_{13}\ddot{x}_3 + c_{11}\dot{x}_1 + c_{12}\dot{x}_2 + c_{13}\dot{x}_3 + k_{11}x_1 + k_{12}x_2 + k_{13}x_3 = F_1(t) \tag{5·135a}$$

$$m_{21}\ddot{x}_1 + m_{22}\ddot{x}_2 + m_{23}\ddot{x}_3 + c_{21}\dot{x}_1 + c_{22}\dot{x}_2 + c_{23}\dot{x}_3 + k_{21}x_1 + k_{22}x_2 + k_{23}x_3 = F_2(t) \tag{5·135b}$$

$$m_{31}\ddot{x}_1 + m_{32}\ddot{x}_2 + m_{33}\ddot{x}_3 + c_{31}\dot{x}_1 + c_{32}\dot{x}_2 + c_{33}\dot{x}_3 + k_{31}x_1 + k_{32}x_2 + k_{33}x_3 = F_3(t) \tag{5·135c}$$

which represent a special case of Eqs. (5·106).

Example An example of a physical system providing such equations is afforded by the elastic simple pendulum of Fig. 5·15, restricted to arbitrarily small values of the angle x_1 and the linear displacement x_2. For this system, the scalar x_3, representing displacements of the particle in the direction of the unit vector $\mathbf{\hat{u}}_n$ (normal to the page) is constrained to be zero. The coordinate x_2 measures the elongation of the spring beyond the elongation $\Delta \triangleq mg/k$ that it would experience due to gravitational attraction when $x_1 \equiv 0$. To obtain these equations from

$\mathbf{F} = m\mathbf{A}$, we simply equate

$$\mathbf{F} = mg\hat{\mathbf{e}}_v - k\,(\Delta + x_2)\hat{\mathbf{u}}_r$$

to $m\mathbf{A}$ as obtained from Eq. (4·2), with f the frame of the massless tube in Fig. 5·15, namely,

$$m\mathbf{A} = m\{\ddot{x}_2\hat{\mathbf{u}}_r + \ddot{x}_1\hat{\mathbf{u}}_n \times (L + \Delta + x_2)\hat{\mathbf{u}}_r + \dot{x}_1\hat{\mathbf{u}}_n \times [\dot{x}_1\hat{\mathbf{u}}_n \times (L + \Delta + x_2)\hat{\mathbf{u}}_r] + 2\dot{x}_1\hat{\mathbf{u}}_n \times \dot{x}_2\hat{\mathbf{u}}_r\}$$

Because of the restriction to arbitrarily small values of x_1 and x_2, nonlinear terms in these variables can be dropped from $m\mathbf{A}$, and the vector

$$\hat{\mathbf{e}}_v = \hat{\mathbf{u}}_r \cos x_1 - \hat{\mathbf{u}}_t \sin x_1$$

can be replaced by $\hat{\mathbf{u}}_r - x_1\hat{\mathbf{u}}_t$. With these substitutions and linearizations, the vector equation $\mathbf{F} = m\mathbf{A}$ becomes

$$-mgx_1\hat{\mathbf{u}}_t - kx_2\hat{\mathbf{u}}_r = m\ddot{x}_2\hat{\mathbf{u}}_r + m(L + \Delta)\,\ddot{x}_1\hat{\mathbf{u}}_t$$

Note that the terms $mg\hat{\mathbf{u}}_r$ and $-k\Delta\,\hat{\mathbf{u}}_r$ have canceled each other. The equivalent scalar equations

$$m(L + \Delta)\,\ddot{x}_1 + mgx_1 = 0 \qquad (5\cdot136a)$$

$$m\ddot{x}_2 + kx_2 = 0 \qquad (5\cdot136b)$$

are particularly simple special cases of Eqs. (5·135).

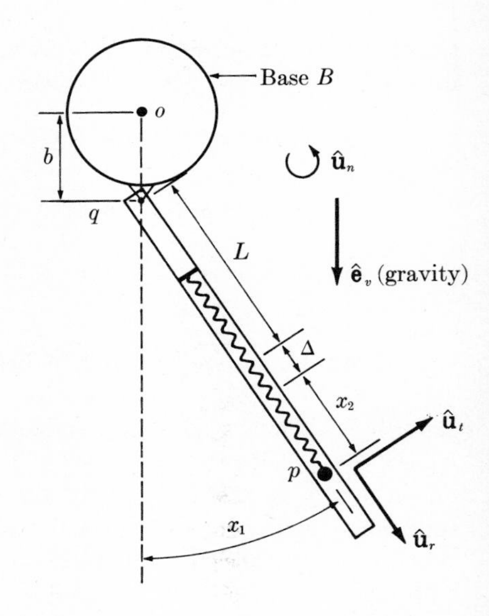

FIGURE 5·15

If viscous damping is incorporated in the mathematical model of the helical spring in Fig. 5·15, these equations would be modified slightly to yield equations of the form

$$m(L + \Delta)\, \ddot{x}_1 + mgx_1 = 0 \qquad (5{\cdot}137a)$$
$$m\ddot{x}_2 + c\dot{x}_2 + kx_2 = 0 \qquad (5{\cdot}137b)$$

If, in addition, the base B of the pendulum in Fig. 5·15 were given the constant angular velocity

$$\boldsymbol{\omega}^B = \Omega \hat{\mathbf{u}}_n$$

with a point o a distance b above the support point q in Fig. 5·15 remaining inertially fixed, and x_1 interpreted as the angle between the lines oq and qp, the vector equation of motion $\mathbf{F} = m\mathbf{A}$ would become in linearized approximation

$$\begin{aligned} mg[\hat{\mathbf{u}}_r(\cos \Omega t - x_1 \sin \Omega t) - \hat{\mathbf{u}}_t(\sin \Omega t + x_1 \cos \Omega t)] - k\,(\Delta + x_2)\hat{\mathbf{u}}_r - c\dot{x}_2\hat{\mathbf{u}}_r \\ = m[\ddot{x}_2\hat{\mathbf{u}}_r - \Omega^2 b(\hat{\mathbf{u}}_r - x_1\hat{\mathbf{u}}_t) + \ddot{x}_1(L + \Delta)\,\hat{\mathbf{u}}_t - \Omega^2(L + \Delta + x_2)\hat{\mathbf{u}}_r \\ + 2\Omega\dot{x}_1(L + \Delta)\,\hat{\mathbf{u}}_t + 2\Omega\dot{x}_2\hat{\mathbf{u}}_t] \end{aligned}$$

where now the steady-state elongation of the spring is

$$\Delta = m(b + L + \Delta)\,\frac{\Omega^2}{k}$$

or

$$\Delta = \left[m(b + L)\,\frac{\Omega^2}{k}\right]\left(1 - m\,\frac{\Omega^2}{k}\right)^{-1}$$

The resulting scalar equations of motion

$$\begin{aligned} m(L + \Delta)\,\ddot{x}_1 + mgx_1 \cos \Omega t + 2m\Omega(L + \Delta)\,\dot{x}_1 + 2m\Omega\dot{x}_2 + mb\Omega^2 x_1 &= -mg \sin \Omega t \\ m\ddot{x}_2 + c\dot{x}_2 + (k - m\Omega^2)x_2 + mgx_1 \sin \Omega t &= mg \cos \Omega t \end{aligned}$$

do *not* satisfy the constant-coefficient restriction on Eqs. (5·135) unless gravity is ignored ($g = 0$), with the result

$$m(L + \Delta)\,\ddot{x}_1 + 2m\Omega(L + \Delta)\,\dot{x}_1 + 2m\Omega\dot{x}_2 + mb\Omega^2 x_1 = 0 \qquad (5{\cdot}138a)$$
$$m\ddot{x}_2 + c\dot{x}_2 + (k - m\Omega^2)x_2 = 0 \qquad (5{\cdot}138b)$$

These equations are again a special case of Eqs. (5·135).

In each of the several variations of the elastic pendulum problem considered here, the particle is permitted only small oscillations on a surface. When that surface is inertially fixed (as in the simpler examples), equations of the form of Eqs. (5·135) result, but this is not always the case when the surface undergoes a prescribed motion (as when the base rotates). In proceeding with Eqs. (5·135) we are therefore treating only a subclass of the total group of problems concerning linear oscillations of a particle.

For the simplest class of differential equations of the form of Eqs. (5·135), the two equations are uncoupled, and the results of §5·1·9 may be applied directly to each of the equations individually. Equations (5·136) and (5·137) are in this category, and Eq. (5·138*b*) is uncoupled from Eq. (5·138*a*), but we must be prepared to face the more general case of coupled equations.

5·2·5 Homogeneous solutions of particle oscillation problems

In previous dealings with Eq. (5·24), we noted that it must admit a homogeneous solution of the form [see Eq. (5·30)]

$$x_h = A_s e^{\lambda_s t}$$

Analogously, it can be shown by substitution that the homogeneous form of Eqs. (5·135) admits the solution

$$x_{1h} = A_1^{(s)} e^{\lambda_s t} \tag{5·139a}$$
$$x_{2h} = A_2^{(s)} e^{\lambda_s t} \tag{5·139b}$$
$$x_{3h} = A_3^{(s)} e^{\lambda_s t} \tag{5·139c}$$

This substitution will again provide a characteristic equation to be solved for λ_s [see Eq. (5·31)], except in this instance the equation is of either fourth or sixth degree in λ_s, depending on whether the particle has two or three degrees of freedom. When the roots of the characteristic equation are distinct (nonrepeated), we can write the general homogeneous solution to Eqs. (5·135) as

$$x_{1h} = \sum_{s=1}^{6} A_1^{(s)} e^{\lambda_s t} \tag{5·140a}$$
$$x_{2h} = \sum_{s=1}^{6} A_2^{(s)} e^{\lambda_s t} \tag{5·140b}$$
$$x_{3h} = \sum_{s=1}^{6} A_3^{(s)} e^{\lambda_s t} \tag{5·140c}$$

Conceptual differences between the homogeneous solution of the single second-order equation (5·24) and the pair of second-order equations (5·135) arise only when the characteristic equation has repeated roots. To improve our perspective on this problem, we shall adopt a new mode of presentation of Eqs. (5·135).

Equations (5·135) may be written in combination as the single matrix equation

$$\begin{bmatrix} m_{11} & m_{12} & m_{13} \\ m_{21} & m_{22} & m_{23} \\ m_{31} & m_{32} & m_{33} \end{bmatrix} \begin{bmatrix} \ddot{x}_1 \\ \ddot{x}_2 \\ \ddot{x}_3 \end{bmatrix} + \begin{bmatrix} c_{11} & c_{12} & c_{13} \\ c_{21} & c_{22} & c_{23} \\ c_{31} & c_{32} & c_{33} \end{bmatrix} \begin{bmatrix} \dot{x}_1 \\ \dot{x}_2 \\ \dot{x}_3 \end{bmatrix} + \begin{bmatrix} k_{11} & k_{12} & k_{13} \\ k_{21} & k_{22} & k_{23} \\ k_{31} & k_{32} & k_{33} \end{bmatrix} \begin{bmatrix} x_1 \\ x_2 \\ x_3 \end{bmatrix} = \begin{bmatrix} F_1(t) \\ F_2(t) \\ F_3(t) \end{bmatrix} \qquad (5·141)$$

With the introduction of symbols for the matrices, Eq. (5·141) becomes

$$[m][\ddot{x}] + [c][\dot{x}] + [k][x] = [F(t)] \qquad (5·142)$$

presenting an alternative representation of the scalar equations (5·135). Except for the brackets (which are used in this basic text to signify matrices printed unadorned in more advanced work), Eq. (5·142) is identical in form to the single scalar equation (5·24). It is instructive to consider both similarities and differences in the solutions of these equations.

It has been noted that Eqs. (5·135) have a homogeneous solution given by Eqs. (5·139). In matrix terms this result is the statement that Eq. (5·142) has a homogeneous solution

$$[x] = [A^{(s)}]e^{\lambda_s t} \qquad (5·143)$$

where $[A^{(s)}]$ is now a column matrix with elements $A_1^{(s)}$, $A_2^{(s)}$, and $A_3^{(s)}$. When this proposed solution is substituted into Eq. (5·142), the result is

$$\{[m]\lambda_s^2 + [c]\lambda_s + [k]\}[A^{(s)}]e^{\lambda_s t} = 0 \qquad (5·144)$$

The scalar $e^{\lambda_s t}$ can, as previously, be ignored, being nonzero. The result is a system of three homogeneous algebraic equations in the four unknowns λ_s, $A_1^{(s)}$, $A_2^{(s)}$, and $A_3^{(s)}$. These equations admit a nonzero solution for $[A^{(s)}]$ only if the determinant of the matrix in braces in Eq. (5·144) is zero, that is, if

$$|[m]\lambda_s^2 + [c]\lambda_s + [k]| = 0 \qquad (5·145)$$

as may be noted, for example, by the application of Cramer's rule to the solution of the scalar algebraic equations embodied in Eq. (5·144). Equation (5·145) provides the sixth-degree characteristic equation previously anticipated for λ_s. As indicated by Eqs. (5·140), when the roots of this equation are distinct, the total homogeneous solution may be written

$$[x_h] = \sum_{s=1}^{6} [A^{(s)}]e^{\lambda_s t} \qquad (5·146)$$

Equation (5·144) provides more than just the characteristic equation following it, however, since once λ_s is found from Eq. (5·145) it becomes possible to return to Eq. (5·144) to establish the *relative* values of $A_1^{(s)}$, $A_2^{(s)}$, and $A_3^{(s)}$. Because we began in Eq. (5·144) with three algebraic equations in four unknowns, we cannot expect to solve for $A_1^{(s)}$, $A_2^{(s)}$, and $A_3^{(s)}$ from this equation. We can, however, learn their relative magnitudes, thereby permitting representation of the previously unknown matrix $[A^{(s)}]$ as the product of an unknown scalar (say α_s) and a known column matrix (say $[V]^{(s)}$), or

$$[A^{(s)}] = \alpha_s[V^{(s)}] \tag{5·147}$$

The introduction of the column matrix $[V^{(s)}]$ corresponding to the scalar λ_s marks a conceptual departure of the solution of the matrix equation (5·147) from the pattern established for the scalar equation (5·24), and this difference influences the form of the solution when the characteristic equation has repeated roots. It should be noted that the scalars λ_s are conventionally called the *eigenvalues* of the differential equation (5·142), and the column matrices $[V^{(s)}]$ are called *normalized eigenvectors* of this second-order system.

If Eqs. (5·146) and (5·147) are combined in the form

$$[x_h] = \sum_{s=1}^{6} \alpha_s[V^{(s)}]e^{\lambda_s t} \tag{5·148}$$

then the six free and independent constants $\alpha_1, \ldots, \alpha_6$, which are to be determined from the initial conditions, are set apart from the matrices $[V^{(s)}]$ and the scalars $e^{\lambda_s t}$, both of which are known from Eqs. (5·144) and (5·145).

As noted in the discussion of the solution of the scalar equation

$$m\ddot{x} + c\dot{x} + kx = F(t) \tag{5·24}$$

the proposed homogeneous solution

$$x_h = A_1e^{\lambda_1 t} + A_2e^{\lambda_2 t} \tag{5·33}$$

becomes unacceptable when $\lambda_2 = \lambda_1$, because it then has only one constant, the sum $A_1 + A_2$. It becomes necessary then to seek out the new solution

$$x_h = A_1e^{\lambda_1 t} + A_2te^{\lambda_1 t}$$

The introduction of the troublesome factor t may be unnecessary when two roots of the *matrix* characteristic equation (5·145) coalesce. If, for example, $\lambda_1 = \lambda_2$ in the proposed homogeneous solution in Eq. (5·148), it would appear in detail as

$$[x_h] = \alpha_1[V^{(1)}]e^{\lambda_1 t} + \alpha_2[V^{(2)}]e^{\lambda_1 t} + \alpha_3[V^{(3)}]e^{\lambda_3 t} + \cdots + \alpha_6[V^{(6)}]e^{\lambda_6 t} \tag{5·149}$$

If the normalized eigenvectors $[V^{(1)}]$ and $[V^{(2)}]$ are independent, then Eq. (5·149) continues to provide six independent scalars $\alpha_1, \ldots, \alpha_6$ as required in order to satisfy arbitrary initial conditions. Only if $[V^{(2)}]$ is some scalar multiple of $[V^{(1)}]$ does it become necessary (for the preservation of six free constants in the final homogeneous solution) to insert a factor t, as in

$$[x_h] = \alpha_1[V^{(1)}]e^{\lambda_1 t} + \alpha_2 t[V^{(1)}]e^{\lambda_1 t} + \alpha_3[V^{(3)}]e^{\lambda_3 t} + \cdots + \alpha_6[V^{(6)}]e^{\lambda_6 t} \tag{5·150}$$

when $\lambda_2 = \lambda_1$ and $[V^{(2)}]$ is proportional to $[V^{(1)}]$.

Initial conditions If the forcing function $F(t)$ in the given differential equation is zero [see Eq. (5·142)], then the initial conditions $[x(t_0)] \triangleq [x_0]$ and $[\dot{x}(t_0)] \triangleq [V_0]$ can be used directly with Eq. (5·149) [or Eq. (5·150) if appropriate] to find $\alpha_1, \ldots, \alpha_6$ in terms of the six elements of $[x_0]$ and $[v_0]$. To be specific, we'll consider a formal procedure for the incorporation of initial conditions into the homogeneous solution in Eq. (5·149), which applies when the eigenvectors $[V^{(1)}], \ldots, [V^{(n)}]$ are independent.

We begin by rewriting Eq. (5·149) in the form

$$[x(t)] = [V^{(1)} \vdots \cdots \vdots V^{(6)}] \begin{bmatrix} e^{\lambda_1 t} & 0 & \cdots & & 0 \\ 0 & e^{\lambda_2 t} & & & \vdots \\ \vdots & & \ddots & & \vdots \\ & & & e^{\lambda_5 t} & 0 \\ 0 & \cdots & & 0 & e^{\lambda_6 t} \end{bmatrix} \begin{bmatrix} \alpha_1 \\ \vdots \\ \vdots \\ \alpha_6 \end{bmatrix} \tag{5·151}$$

where the 3×1 matrices $V^{(1)}, \ldots, V^{(6)}$ are assembled as the partitioned columns of a 3×6 matrix, and where $x_h(t)$ is replaced by $x(t)$ because in the absence of $[F(t)]$ this is the total solution. When $t = t_0$, Eq. (5·151) furnishes

$$[x_0] = [V^{(1)} \vdots \cdots \vdots V^{(6)}] \begin{bmatrix} e^{\lambda_1 t_0} & & \\ & \ddots & \\ & & e^{\lambda_6 t_0} \end{bmatrix} \begin{bmatrix} \alpha_1 \\ \vdots \\ \alpha_6 \end{bmatrix} \tag{5·152}$$

Differentiating Eq. (5·151) and substituting $t = t_0$ yields

$$[v_0] = [\lambda_1 V^{(1)} \vdots \cdots \vdots \lambda_6 V^{(6)}] \begin{bmatrix} e^{\lambda_1 t_0} & & \\ & \ddots & \\ & & e^{\lambda_6 t_0} \end{bmatrix} \begin{bmatrix} \alpha_1 \\ \vdots \\ \alpha_6 \end{bmatrix} \tag{5·153}$$

Eqs. (5·152) and (5·153) provide six scalar algebraic equations that can be solved for the six unknowns $\alpha_1, \ldots, \alpha_6$. This can be accomplished formally by defining the 6×6 matrix

$$B \triangleq \left[\begin{array}{c|c|c|c} V^{(1)} & V^{(2)} & \cdots & V^{(6)} \\ \hline \lambda_1 V^{(1)} & \lambda_2 V^{(2)} & \cdots & \lambda_6 V^{(6)} \end{array}\right] \tag{5·154}$$

and writing Eqs. (5·152) and (5·153) in combination as

$$\begin{bmatrix} x_0 \\ \hline v_0 \end{bmatrix} = [B] \begin{bmatrix} e^{\lambda_1 t_0} & & & \\ & \cdot & & \\ & & \cdot & \\ & & & \cdot \\ & & & & e^{\lambda_6 t_0} \end{bmatrix} \begin{bmatrix} \alpha_1 \\ \cdot \\ \cdot \\ \cdot \\ \alpha_6 \end{bmatrix} \tag{5·155}$$

Inverting this equation yields

$$\begin{bmatrix} \alpha_1 \\ \cdot \\ \cdot \\ \cdot \\ \alpha_6 \end{bmatrix} = \begin{bmatrix} e^{-\lambda_1 t_0} & & & \\ & \cdot & & \\ & & \cdot & \\ & & & \cdot \\ & & & & e^{-\lambda_6 t_0} \end{bmatrix} [B]^{-1} \begin{bmatrix} x_0 \\ \hline v_0 \end{bmatrix} \tag{5·156}$$

which, on substitution into Eq. (5·151), provides

$$[x(t)] = [V^{(1)} | V^{(2)} | \cdots | V^{(6)}] \begin{bmatrix} e^{\lambda_1 (t-t_0)} & & & \\ & \cdot & & \\ & & \cdot & \\ & & & \cdot \\ & & & & e^{\lambda_6 (t-t_0)} \end{bmatrix} [B]^{-1} \begin{bmatrix} x_0 \\ \hline v_0 \end{bmatrix} \tag{5·157}$$

5·2·6 Free oscillations of a particle

A particle of mass m is attached to the top of an axisymmetric massless elastic beam, rising vertically from a fixed base, as shown in Fig. 5·16. The beam resists the translation of the particle in any direction as does an elastic spring, with spring constants k_1, k_2, and k_3 characterizing the spring force resisting motions in directions parallel to $\hat{\mathbf{u}}_1$, $\hat{\mathbf{u}}_2$, and $\hat{\mathbf{u}}_3$. The homogeneous solution of the equations of small oscillations is sought.

If x_1, x_2, and x_3 are translations of the particle in the directions of unit vectors $\hat{\mathbf{u}}_1$, $\hat{\mathbf{u}}_2$, and $\hat{\mathbf{u}}_3$, the equations of motion

$$\begin{aligned} m\ddot{x}_1 + k_1 x_1 &= 0 \\ m\ddot{x}_2 + k_2 x_2 &= 0 \\ m\ddot{x}_3 + k_3 x_3 &= 0 \end{aligned}$$

may be written by inspection. Although in this degenerate case it is simplest to solve these equations separately, we prefer here to write them as the matrix equations

$$\begin{bmatrix} m & 0 & 0 \\ 0 & m & 0 \\ 0 & 0 & m \end{bmatrix} \begin{bmatrix} \ddot{x}_1 \\ \ddot{x}_2 \\ \ddot{x}_3 \end{bmatrix} + \begin{bmatrix} k_1 & 0 & 0 \\ 0 & k_2 & 0 \\ 0 & 0 & k_3 \end{bmatrix} \begin{bmatrix} x_1 \\ x_2 \\ x_3 \end{bmatrix} = \begin{bmatrix} 0 \\ 0 \\ 0 \end{bmatrix}$$

Substitution of the homogeneous solution [see Eq. (5·143)] then provides

$$\left\{ \begin{bmatrix} m & 0 & 0 \\ 0 & m & 0 \\ 0 & 0 & m \end{bmatrix} \lambda_s^2 + \begin{bmatrix} k_1 & 0 & 0 \\ 0 & k_2 & 0 \\ 0 & 0 & k_3 \end{bmatrix} \right\} \begin{bmatrix} A_1^{(s)} \\ A_2^{(s)} \\ A_3^{(s)} \end{bmatrix} e^{\lambda_s t} = 0$$

in parallel with Eq. (5·144). The characteristic equation is [see Eq. (5·145)]

$$\left| \begin{bmatrix} m & 0 & 0 \\ 0 & m & 0 \\ 0 & 0 & m \end{bmatrix} \lambda_s^2 + \begin{bmatrix} k_1 & 0 & 0 \\ 0 & k_2 & 0 \\ 0 & 0 & k_3 \end{bmatrix} \right| = 0$$

or

$$(m\lambda_s^2 + k_1)(m\lambda_s^2 + k_2)(m\lambda_s^2 + k_3) = 0$$

Define $p_i^2 \triangleq k_i/m$ for $i = 1, 2, 3$, so that the above becomes

$$(\lambda_s^2 + p_1^2)(\lambda_s^2 + p_2^2)(\lambda_s^2 + p_3^2) = 0$$

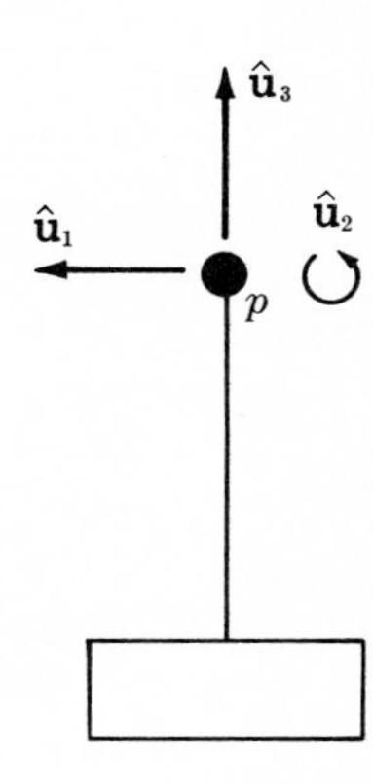

FIGURE 5·16

The roots thus become (with arbitrary numbering)

$$\lambda_1 = ip_1 \qquad \lambda_2 = -ip_1 \qquad \lambda_3 = ip_2 \qquad \lambda_4 = -ip_2 \qquad \lambda_5 = ip_3 \qquad \lambda_6 = -ip_3$$

The corresponding eigenvectors are then obtained from Eq. (5·144), which in this case leads to

$$\left\{\begin{bmatrix} m & 0 & 0 \\ 0 & m & 0 \\ 0 & 0 & m \end{bmatrix} \lambda_s^2 + \begin{bmatrix} k_1 & 0 & 0 \\ 0 & k_2 & 0 \\ 0 & 0 & k_3 \end{bmatrix}\right\} \begin{bmatrix} A_1^{(s)} \\ A_2^{(s)} \\ A_3^{(s)} \end{bmatrix} = 0$$

Substituting 1 for s (and noting that $\lambda_1^2 = -p_1^2 = -k_1/m$) leads to

$$\begin{bmatrix} -k_1 + k_1 & 0 & 0 \\ 0 & -k_1 + k_2 & 0 \\ 0 & 0 & -k_1 + k_3 \end{bmatrix} \begin{bmatrix} A_1^{(1)} \\ A_2^{(1)} \\ A_3^{(1)} \end{bmatrix} = 0$$

while the substitution of 2 for s furnishes a result that differs only in the replacement of the parenthetical superscript 1 by 2. The substitution of $s = 3$ furnishes

$$\begin{bmatrix} -k_2 + k_1 & 0 & 0 \\ 0 & -k_2 + k_2 & 0 \\ 0 & 0 & -k_2 + k_3 \end{bmatrix} \begin{bmatrix} A_1^{(3)} \\ A_2^{(3)} \\ A_3^{(3)} \end{bmatrix} = 0$$

with results for $s = 4$ differing only in the parenthetical superscripts. The substitutions of $s = 5$ and $s = 6$ lead instead to equations typified by

$$\begin{bmatrix} -k_3 + k_1 & 0 & 0 \\ 0 & -k_3 + k_2 & 0 \\ 0 & 0 & -k_3 + k_3 \end{bmatrix} \begin{bmatrix} A_1^{(5)} \\ A_2^{(5)} \\ A_3^{(5)} \end{bmatrix} = 0$$

It is now apparent that many of the elements in the column matrices $[A^{(1)}], \ldots, [A^{(6)}]$ must be zero, while other elements can be anything at all. For example, in the matrix $[A^{(1)}]$, the terms $A_2^{(1)}$ and $A_3^{(1)}$ must be zero, while $A_1^{(1)}$ can have any value. As suggested by a comparison of Eqs. (5·146) and (5·148), we can replace $[A^{(s)}]$ by $\alpha_s[V^{(s)}]$, and on the basis of the preceding equations write

$$\begin{bmatrix} A_1^{(1)} \\ A_2^{(1)} \\ A_3^{(1)} \end{bmatrix} = \alpha_1 \begin{bmatrix} 1 \\ 0 \\ 0 \end{bmatrix} \qquad \begin{bmatrix} A_1^{(2)} \\ A_2^{(2)} \\ A_3^{(2)} \end{bmatrix} = \alpha_2 \begin{bmatrix} 1 \\ 0 \\ 0 \end{bmatrix} \qquad \begin{bmatrix} A_1^{(3)} \\ A_2^{(3)} \\ A_3^{(3)} \end{bmatrix} = \alpha_3 \begin{bmatrix} 0 \\ 1 \\ 0 \end{bmatrix}$$

$$\begin{bmatrix} A_1^{(4)} \\ A_2^{(4)} \\ A_3^{(4)} \end{bmatrix} = \alpha_4 \begin{bmatrix} 0 \\ 1 \\ 0 \end{bmatrix} \qquad \begin{bmatrix} A_1^{(5)} \\ A_2^{(5)} \\ A_3^{(5)} \end{bmatrix} = \alpha_5 \begin{bmatrix} 0 \\ 0 \\ 1 \end{bmatrix} \qquad \begin{bmatrix} A_1^{(6)} \\ A_2^{(6)} \\ A_3^{(6)} \end{bmatrix} = \alpha_6 \begin{bmatrix} 0 \\ 0 \\ 1 \end{bmatrix}$$

From Eq. (5·149), the total homogeneous solution is available as

$$[x_h] = \alpha_1 \begin{bmatrix}1\\0\\0\end{bmatrix} e^{ip_1t} + \alpha_2 \begin{bmatrix}1\\0\\0\end{bmatrix} e^{-ip_1t} + \alpha_3 \begin{bmatrix}0\\1\\0\end{bmatrix} e^{ip_2t} + \alpha_4 \begin{bmatrix}0\\1\\0\end{bmatrix} e^{-ip_2t} + \alpha_5 \begin{bmatrix}0\\0\\1\end{bmatrix} e^{ip_3t} + \alpha_6 \begin{bmatrix}0\\0\\1\end{bmatrix} e^{-ip_3t}$$

Note that in this example the eigenvectors $[V^{(s)}]$ have been normalized to establish a uniform length of unity.

Physical interpretation of this result is facilitated by the identities

$$\sin p_st = \frac{1}{2i}(e^{ip_st} - e^{-ip_st})$$
$$\cos p_st = \tfrac{1}{2}(e^{ip_st} + e^{-ip_st})$$

These permit the solution to be written in the form

$$[x_h] \triangleq \begin{bmatrix}x_{1h}\\x_{2h}\\x_{3h}\end{bmatrix} = C_1 \begin{bmatrix}1\\0\\0\end{bmatrix} \cos p_1t + C_2 \begin{bmatrix}1\\0\\0\end{bmatrix} \sin p_1t + C_3 \begin{bmatrix}0\\1\\0\end{bmatrix} \cos p_2t + C_4 \begin{bmatrix}0\\1\\0\end{bmatrix} \sin p_2t + C_5 \begin{bmatrix}0\\0\\1\end{bmatrix} \cos p_3t + C_6 \begin{bmatrix}0\\0\\1\end{bmatrix} \sin p_3t$$

Evidently this merely says

$$x_{1h} = C_1 \cos p_1t + C_2 \sin p_1t$$
$$x_{2h} = C_3 \cos p_2t + C_4 \sin p_2t$$
$$x_{3h} = C_5 \cos p_3t + C_6 \sin p_3t$$

with $C_1, \ldots, C_6$ real constants to be determined from initial conditions. Now it is clear that the particle in Fig. 5·16 is simply vibrating freely and independently in directions indicated by $\hat{\mathbf{u}}_1$, $\hat{\mathbf{u}}_2$, and $\hat{\mathbf{u}}_3$.

Even if the beam were axisymmetric, with $k_2 = k_1$, so that $\lambda_3 = \lambda_1$ and $\lambda_4 = \lambda_2$, this solution would stand; the fact that symmetry of the beam makes the independent vibrations in the directions of $\hat{\mathbf{u}}_1$ and $\hat{\mathbf{u}}_2$ have the same frequency presents no problems, and surely does not introduce a factor t into any term of the solution.

Formal application of Eq. (5·157) to obtain the solution to this problem provides

$$\begin{bmatrix} x_1(t) \\ x_2(t) \\ x_3(t) \end{bmatrix} = \begin{bmatrix} 1 & 1 & 0 & 0 & 0 & 0 \\ 0 & 0 & 1 & 1 & 0 & 0 \\ 0 & 0 & 0 & 0 & 1 & 1 \end{bmatrix} \begin{bmatrix} e^{ip_1(t-t_0)} & & & \\ & \cdot & & \\ & & \cdot & \\ & & & \cdot \\ & & & & e^{-ip_3(t-t_0)} \end{bmatrix} [B]^{-1} \begin{bmatrix} x_{10} \\ x_{20} \\ x_{30} \\ v_{10} \\ v_{20} \\ v_{30} \end{bmatrix}$$

where $[B]^{-1}$ can be shown to be given by

$$[B]^{-1} = \frac{1}{2} \begin{bmatrix} 1 & 0 & 0 & -\dfrac{i}{p_1} & 0 & 0 \\ 1 & 0 & 0 & \dfrac{i}{p_1} & 0 & 0 \\ 0 & 1 & 0 & 0 & -\dfrac{i}{p_2} & 0 \\ 0 & 1 & 0 & 0 & \dfrac{i}{p_2} & 0 \\ 0 & 0 & 1 & 0 & 0 & -\dfrac{i}{p_3} \\ 0 & 0 & 1 & 0 & 0 & \dfrac{i}{p_3} \end{bmatrix}$$

For a problem as simple as this example, it is obviously simpler just to use the initial values of x_1 and $\dot{x}_1$ to solve for C_1 and C_2 in the expressions

$$\begin{aligned} x_1 &= C_1 \cos p_1 t + C_2 \sin p_1 t \\ \dot{x}_1 &= -p_1 C_1 \sin p_1 t + p_1 C_2 \cos p_1 t \end{aligned}$$

and to proceed similarly for x_2 and x_3. Equation (5·157) may be appropriate for a more highly coupled problem, and it is often useful in finding the solutions to inhomogeneous equations.

5·2·7 Matrix superposition integral solutions

Although special procedures can be adopted to obtain inhomogeneous solutions for special kinds of forcing functions $[F(t)]$ in the matrix differential equation [see Eq. (5·142)]

$$[m][\ddot{x}] + [c][\dot{x}] + [k][x] = [F(t)]$$

attention is here directed immediately to the solution obtained by the matrix superposition integral, the matrix counterpart to the solution for the scalar $x(t)$ found in Eq. (5·101).

Recall that in the derivation of Eq. (5·101) the rationale was to superimpose on the unforced solution to the linear system of differential equations an integral that accommodated the arbitrary forcing function by treating it as a collection of ideal impulses multiplied by the unit impulse response $g(t - t_0)$. A parallel derivation guided by the same rationale leads to the parallel expression

$$[x(t)] = [V^{(1)} \vdots \cdots \vdots V^{(6)}] \begin{bmatrix} e^{\lambda_1(t-t_0)} & & & \\ & \cdot & & \\ & & \cdot & \\ & & & \cdot \\ & & & & e^{\lambda_6(t-t_0)} \end{bmatrix} \begin{bmatrix} V^{(1)} & \vdots & \cdots & \vdots & V^{(6)} \\ \hline \lambda_1 V^{(1)} & \vdots & \cdots & \vdots & \lambda_6 V^{(6)} \end{bmatrix}^{-1} \begin{bmatrix} x_0 \\ \hline v_0 \end{bmatrix} + \int_{t_0}^{t} [g(t - \tau)][m]^{-1}[F(\tau)]\, d\tau \tag{5·158}$$

where the first term is the unforced solution, from Eq. (5·157), and the second term involves the matrix unit impulse response $[g(t - \tau)]$, whose columns characterize the response of the system to unitary impulsive changes in each of the three coordinates. Specifically, if Eq. (5·157) is used to provide the response to $x_0 = 0$, $v_0 = [1 \quad 0 \quad 0]^T$, we have the 3×1 matrix

$$[g^{(1)}(t - t_0)] = [V^{(1)} \vdots \cdots \vdots V^{(6)}] \begin{bmatrix} e^{\lambda_1(t-t_0)} & & & \\ & \cdot & & \\ & & \cdot & \\ & & & \cdot \\ & & & & e^{\lambda_6(t-t_0)} \end{bmatrix} \begin{bmatrix} V^{(1)} & \vdots & \cdots & \vdots & V^{(6)} \\ \hline \lambda_1 V^{(1)} & \vdots & \cdots & \vdots & \lambda_6 V^{(6)} \end{bmatrix}^{-1} \begin{bmatrix} 0 \\ 0 \\ 0 \\ 1 \\ 0 \\ 0 \end{bmatrix} \tag{5·159}$$

If $[g^{(2)}(t - t_0)]$ and $[g^{(3)}(t - t_0)]$ are similarly calculated for $[v_0] = [0 \quad 1 \quad 0]^T$ and $[v_0] = [0 \quad 0 \quad 1]^T$, respectively, then the 3×3 matrix $[g(t - \tau)]$ in Eq. (5·158) is

$$[g(t - \tau)] = [g^{(1)}(t - \tau) \vdots g^{(2)}(t - \tau) \vdots g^{(3)}(t - \tau)] \tag{5·160}$$

5·2·8 Application of the matrix superposition integral

Suppose that the beam-particle system of Fig. 5·16 is subjected to a vertical force of magnitude

$$F_3(t) = \sin \omega t$$

and given initial conditions $[x_0] = [v_0] = 0$. For such a simple problem it is, of course, wise to examine the (uncoupled) scalar equations of motion directly, but for illustrative purposes the general solution represented by Eq. (5·158) will be used. For the given problem this equation becomes

$$\begin{bmatrix} x_1(t) \\ x_2(t) \\ x_3(t) \end{bmatrix} = \int_{t_0}^{t} [g^{(1)}(t-\tau) \mid g^{(2)}(t-\tau) \mid g^{(3)}(t-\tau)] \begin{bmatrix} m & 0 & 0 \\ 0 & m & 0 \\ 0 & 0 & m \end{bmatrix}^{-1} \begin{bmatrix} 0 \\ 0 \\ \sin \omega\tau \end{bmatrix} d\tau$$

where from Eq. (5·159) and the previous example

$$[g^{(1)}(t-\tau)] = \frac{1}{2} \begin{bmatrix} 1 & 1 & 0 & 0 & 0 & 0 \\ 0 & 0 & 1 & 1 & 0 & 0 \\ 0 & 0 & 0 & 0 & 1 & 1 \end{bmatrix} \begin{bmatrix} e^{ip_1(t-\tau)} & & & \\ & \cdot & & \\ & & \cdot & \\ & & & \cdot \\ & & & & e^{-ip_3(t-\tau)} \end{bmatrix}$$

$$\begin{bmatrix} 1 & 0 & 0 & -\frac{i}{p_1} & 0 & 0 \\ 1 & 0 & 0 & \frac{i}{p_1} & 0 & 0 \\ 0 & 1 & 0 & 0 & -\frac{i}{p_2} & 0 \\ 0 & 1 & 0 & 0 & \frac{i}{p_2} & 0 \\ 0 & 0 & 1 & 0 & 0 & -\frac{i}{p_3} \\ 0 & 0 & 1 & 0 & 0 & \frac{i}{p_3} \end{bmatrix} \begin{bmatrix} 0 \\ 0 \\ 0 \\ 1 \\ 0 \\ 0 \end{bmatrix} = \frac{1}{p_1} \sin p_1(t-\tau) \begin{bmatrix} 1 \\ 0 \\ 0 \end{bmatrix}$$

in which the identity

$$\sin p_1 t = \frac{1}{2i} (e^{ip_1 t} - e^{-ip_1 t})$$

has been employed. Similar calculations provide

$$[g^{(2)}(t-\tau)] = \frac{1}{p_2}\sin p_2(t-\tau)\begin{bmatrix}0\\1\\0\end{bmatrix}$$

and

$$[g^{(3)}(t-\tau)] = \frac{1}{p_3}\sin p_3(t-\tau)\begin{bmatrix}0\\0\\1\end{bmatrix}$$

Returning to the integral expression for $[x(t)]$, we find

$$\begin{bmatrix}x_1(t)\\x_2(t)\\x_3(t)\end{bmatrix} = \int_{t_0}^{t}\begin{bmatrix}\frac{1}{p_1}\sin p_1(t-\tau) & 0 & 0\\ 0 & \frac{1}{p_2}\sin p_2(t-\tau) & 0\\ 0 & 0 & \frac{1}{p_3}\sin p_3(t-\tau)\end{bmatrix}$$

$$\begin{bmatrix}\frac{1}{m} & 0 & 0\\ 0 & \frac{1}{m} & 0\\ 0 & 0 & \frac{1}{m}\end{bmatrix}\begin{bmatrix}0\\0\\\sin\omega\tau\end{bmatrix}d\tau = \int_{t_0}^{t}\begin{bmatrix}0\\0\\\frac{1}{mp_3}\sin p_3(t-\tau)\sin\omega\tau\end{bmatrix}d\tau$$

so that

$$\begin{aligned}
x_1(t) &= 0\\
x_2(t) &= 0\\
x_3(t) &= \frac{1}{mp_3}\int_{t_0}^{t}\sin p_3(t-\tau)\sin\omega\tau\,d\tau\\
&= \frac{1}{mp_3}\sin p_3t\int_{t_0}^{t}\cos p_3\tau\sin\omega\tau\,d\tau - \frac{1}{mp_3}\cos p_3t\int_{t_0}^{t}\sin p_3\tau\sin\omega\tau\,d\tau
\end{aligned}$$

Tables of integrals are required for a final expression for $x_3(t)$; the result is recorded only for the special case $t_0 = 0$:

$$\begin{aligned}
x_3(t) = \frac{1}{mp_3}\sin p_3t\left[-\frac{\cos(\omega-p_3)\tau}{2(\omega-p_3)} - \frac{\cos(\omega+p_3)\tau}{2(\omega+p_3)}\right]\Bigg|_0^t\\
- \frac{1}{mp_3}\cos p_3t\left[\frac{\sin(\omega-p_3)\tau}{2(\omega-p_3)} - \frac{\sin(\omega+p_3)\tau}{2(\omega+p_3)}\right]\Bigg|_0^t
\end{aligned}$$

$$= \frac{1}{mp_3} \sin p_3 t \left[\frac{1}{2(\omega - p_3)} + \frac{1}{2(\omega + p_3)} \right]$$

$$+ \frac{1}{2mp_3(\omega - p_3)} [-\sin p_3 t(\cos \omega t \cos p_3 t + \sin \omega t \sin p_3 t)$$

$$- \cos p_3 t(\sin \omega t \cos p_3 t - \cos \omega t \sin p_3 t)]$$

$$+ \frac{1}{2mp_3(\omega + p_3)} [-\sin p_3 t(\cos \omega t \cos p_3 t - \sin \omega t \sin p_3 t)$$

$$+ \cos p_3 t(\sin \omega t \cos p_3 t + \cos \omega t \sin p_3 t)]$$

$$= \frac{\omega \sin p_3 t}{mp_3(\omega^2 - p_3^2)} - \frac{\sin \omega t}{2mp_3(\omega - p_3)} + \frac{\sin \omega t}{2mp_3(\omega + p_3)}$$

$$= \frac{1}{m(\omega^2 - p_3^2)} \left(-\sin \omega t + \frac{\omega}{p_3} \sin p_3 t \right)$$

5·2·9 The state equation

Although it should be abundantly clear that the differential equations of motion of dynamic systems of typical engineering complexity will rarely yield to closed-form solution, yet it is necessary somehow to squeeze information from these equations. This may require the numerical integration of the equations on a digital computer, generally using standard programs designed to receive the differential equations in standard form. Alternatively, we might acquire useful qualitative information from the equations by applying certain techniques for stability analysis or approximate solution; often these methods are formally structured to apply to differential equations in standard form.

Although the differential equations of motion of mechanical systems have been presented in this text as *second-order* differential equations, the standard form generally adopted for numerical and analytical studies requires that differential equations appear in the form

$$\begin{aligned} \dot{y}_1 &= Y_1(y_1, y_2, \ldots, y_n, t) \\ \dot{y}_2 &= Y_2(y_1, y_2, \ldots, y_n, t) \\ &\cdots\cdots\cdots\cdots \\ \dot{y}_n &= Y_n(y_1, y_2, \ldots, y_n, t) \end{aligned} \tag{5·161}$$

or, equivalently, as the matrix equation

$$\begin{bmatrix} \dot{y}_1 \\ \cdot \\ \cdot \\ \cdot \\ \dot{y}_n \end{bmatrix} = \begin{bmatrix} Y_1(y_1, \ldots, y_n, t) \\ \cdot \\ \cdot \\ \cdot \\ Y_n(y_1, \ldots, y_n, t) \end{bmatrix} \tag{5·162a}$$

which is written symbolically as

$$[\dot{y}] = [Y(y, t)] \tag{5·162b}$$

If, for numerical or analytical treatment by formally structured procedures, it is required that equations of motion of a particle be in the form of Eq. (5·162) rather than in the second-order form [see Eq. (5·106)]

$$\begin{aligned} m_{11}\ddot{x}_1 + m_{12}\ddot{x}_2 + m_{13}\ddot{x}_3 &= F_1(x_1, x_2, x_3, \dot{x}_1, \dot{x}_2, \dot{x}_3, t) \\ m_{21}\ddot{x}_1 + m_{22}\ddot{x}_2 + m_{23}\ddot{x}_3 &= F_2(x_1, x_2, x_3, \dot{x}_1, \dot{x}_2, \dot{x}_3, t) \\ m_{31}\ddot{x}_1 + m_{32}\ddot{x}_2 + m_{33}\ddot{x}_3 &= F_3(x_1, x_2, x_3, \dot{x}_1, \dot{x}_2, \dot{x}_3, t) \end{aligned} \tag{5·163a}$$

or in the symbolic matrix form

$$[m][\ddot{x}] = [F(x, \dot{x}, t)] \tag{5·163b}$$

this transformation can be accomplished with the definitions

$$\begin{aligned} y_1 &\triangleq x_1 \qquad & y_4 &\triangleq \dot{x}_1 \\ y_2 &\triangleq x_2 & y_5 &\triangleq \dot{x}_2 \\ y_3 &\triangleq x_3 & y_6 &\triangleq \dot{x}_3 \end{aligned} \tag{5·164}$$

Assuming that matrix $[m]$ in Eq. (5·163b) is nonsingular, so that its inverse exists, we can first write this equation in the form

$$[\ddot{x}] = [m]^{-1}[F(x, \dot{x}, t)] \tag{5·165}$$

and then rewrite this result in combination with the definitions of Eqs. (5·164), to obtain in partitioned matrix form

$$\begin{bmatrix} \dot{y}_1 \\ \dot{y}_2 \\ \dot{y}_3 \\ \hdashline \dot{y}_4 \\ \dot{y}_5 \\ \dot{y}_6 \end{bmatrix} = \begin{bmatrix} y_4 \\ y_5 \\ y_6 \\ \hdashline \\ [m]^{-1}[F(y, t)] \\ \\ \end{bmatrix} \tag{5·166}$$

which has the structure of Eq. (5·162).

When differential equations of motion are written in the standard first-order matrix form represented by Eq. (5·162), they are called *equations of state* or *state equations*. The quantities $y_1, \ldots, y_n$ are called the *state variables;* in this example they represent the position coordinates x_1, x_2, and x_3 and their first time derivatives.

The state equation concept is a much broader one than is indicated by the present context; the ordinary differential equations characterizing the behavior in time of a wide variety of systems (whether mechanical, electrical, economic, or whatever) can be cast in this standard form. We are in this introductory discussion of dynamics not prepared to discuss the numerical and qualitative or approximate analytical procedures to which the state equation is particularly well adapted; it is enough at this point simply to recognize that the second-order differential equations treated explicitly in this text can easily be transformed into (first-order) state equations. Alternative methods of formulating the equations of motion of mechanical systems (via Hamilton's equations, as described in any intermediate or advanced dynamics text) lead directly to first-order equations.

5·3 PERSPECTIVE

It is particularly important that you pause after examining Chap. 5 to reestablish your perspective on the study of dynamics. Among the characteristics of this book is the degree to which there have been emphasized the conceptual and operational differences in the three basic tasks of the dynamicist: (1) devising mathematical models for physical systems, (2) deriving equations of motion, and (3) solving or qualitatively analyzing these equations. This emphasis is suggested by the trends in engineering practice, dictated primarily by the digital computer.

The number of dynamics problems that give rise to differential equations amenable to closed-form solution is a very small percentage of those of potential interest to man. Before the development of high-speed digital computers, it was natural for students of dynamics to concentrate their attention on those problems that could be solved in closed form, and within this limited framework it was often advantageous to combine the tasks of idealization, derivation, and solution of equations of motion into a single project. The digital computer has, however, made it possible to extract useful information from the equations of motion of even the most complex dynamic systems; and although the computer has had a pervasive influence on each of the three tasks of idealization, derivation, and solution of equations, only in the last of these areas has its influence been direct and overwhelming. Because the engineer knows the computer can process highly complex equations of motion, he is inclined toward more complex (and more accurate) mathematical idealizations of physical systems, and he may in consequence select more general or more highly structured procedures for deriving the equations of motion. (As computer capacity for symbolic, algebraic manipulation is developed, the computer will exert an increasingly direct influence on the selection of methods of *deriving* equations of motion also.)

Rather than confine attention to a few classic examples of dynamics problems for which solutions are known, the presentation in this text is intended to examine the broader question of differential equation solution. In order to keep the discussion from becoming too abstract, this confrontation with a general set of equations has been injected into the treatment of the dynamics of a single particle. The message of this presentation is simply stated: *The closed-form solution of equations of motion is possible only in very special cases, even when the physical system is idealized as a single particle.* Obviously the difficulties grow greater as the complexity of the idealization increases. The task of seeking solutions for equations of motion provides a sharp contrast with the business of deriving those equations; the latter job is always straightforward, if sometimes intricate and time-consuming. With Chaps. 3 and 4 understood, you should feel confident of your ability to formulate equations of motion of a single particle, but it is not intended that Chap. 5 should generate the same confidence in your ability to solve these equations.

PROBLEMS

5·1 Solve the following equations of motion for $x(t)$, in terms of $x_0 \triangleq x(0)$ and $v_0 \triangleq \dot{x}(0)$.

a. $m\ddot{x} = F_0 \cos \omega t$

b. $m\ddot{x} = F_0 \cosh \omega t$

c. $m\ddot{x} = \alpha t + \beta t^2$

d. $m\ddot{x} = F_0 t(a^2 - t^2)^{-1/2}$

e. $m\ddot{x} = F_0 \sin^2 t$

5·2 Solve the following equations of motion for $\dot{x}(x)$ and $t(x)$ in terms of $x_0 \triangleq x(0)$ and $v_0 \triangleq \dot{x}(0)$

a. $m\ddot{x} = F_0(\alpha + \beta x)$ with $x_0 = v_0 = 0$

b. $m\ddot{x} = -kx^3$ with $x_0 = v_0 = 0$

c. $m\ddot{x} = -kx$

d. $m\ddot{x} = -\frac{1}{2} m \sin x$ $\quad x_0 = 1$, $v_0 = 1$

e. $m\ddot{x} = \dfrac{-\mu m}{x^2}$

5·3 Show that the formula for the period of oscillation of a simple pendulum of length l through an angle θ under uniform gravitational attraction of magnitude mg is

$$T = 4\sqrt{\frac{l}{g}} \int_0^{\pi/2} \frac{d\varphi}{\left(1 - \sin^2 \dfrac{\theta_0}{2} \sin^2 \varphi\right)^{1/2}}$$

where θ_0 is the amplitude of the oscillation. (This integrand is called the complete elliptic integral of the first kind, with argument $\theta_0/2$, and its numerical values can be found in mathematical tables.)

5·4 A sphere of mass m is dropped from a helicopter hovering above a fixed point of the earth on a windless day. If the gravitational force is idealized as a constant, of magnitude mg, and quadratic aerodynamic drag forces have coefficient $D = 0.0001\ mg$, what is the terminal speed of descent, and how many seconds will it take for the speed to go from zero to half of its terminal value?

5·5 A projectile of mass m traveling on a straight line through a resisting medium is subjected to a force of magnitude $F = \alpha m - \beta\dot{x} - \gamma\dot{x}^2$. Find its terminal speed.

5·6 A projectile of mass m traveling on a straight line through a resisting medium is subjected to a force of magnitude $F = \alpha\dot{x}(a + b\dot{x})$. Find $\dot{x}(t)$, and indicate a procedure for obtaining $x(t)$.

5·7 Find the solution of $\ddot{x} + 4\dot{x} + 4x = 0$, with initial conditions $x(0) = 3$ ft and $\dot{x}(0) = 4$ ft/sec.

5·8 A force $F(t)\hat{\mathbf{u}}_x = F_0 \sin \omega t \hat{\mathbf{u}}_x$ is applied to a mass m for which displacement x in the $\hat{\mathbf{u}}_x$ direction is resisted by a spring with spring constant k, with $k \neq m\omega^2$ (see Fig. P5·8). Find the values of $x_0 \triangleq x(0)$ and $v_0 \triangleq \dot{x}(0)$ such that $x(t)$ is a sinusoidal oscillation.

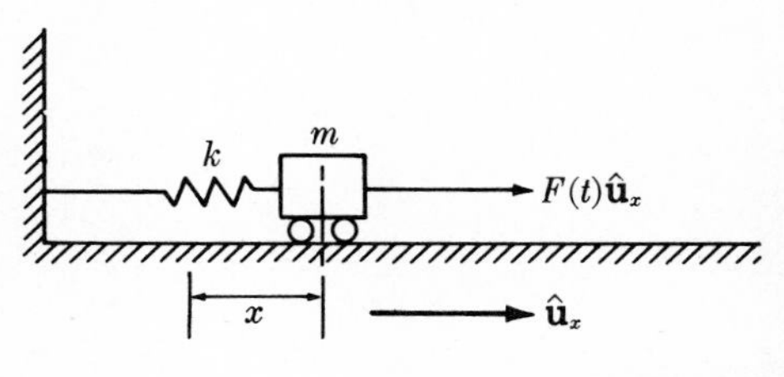

FIGURE P5·8

5·9 A particle of mass m is attached to a wall by a spring with constant k in parallel with a dashpot with constant c (see Fig. P5·9). When released from rest with the spring stretched 10 in., the particle first compresses the spring and then stretches it again, and the extension on the first rebound is 9 in. Find $\zeta \triangleq c/(2\sqrt{km})$, the percentage of critical damping.

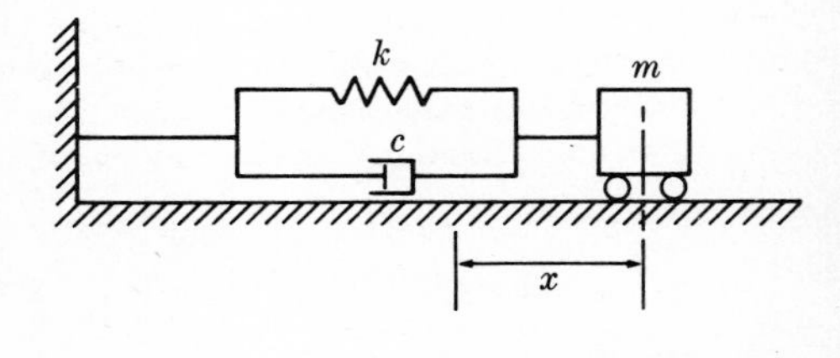

FIGURE P5·9

5·10 The *time constant* τ of a damped free oscillation is the time interval required for its amplitude to decay to $1/e$ times its original value. Find an expression for τ for subcritical damping, in terms of any appropriate basic quantities among m, c, and k. Does this expression apply also for supercritical damping? For critical damping, is the time constant greater than, equal to, or less than the value given by your expression applicable for subcritical damping?

5·11 A particle of mass m is attached to one end of a massless spring of unstretched length L and spring constant k, while the other end of the spring is attached to the roof of a massive box (see Fig. P5·11). Determine the time history of the length of the spring if it is released from rest in a stretched length $1.1L$ when placed on the moon, and compare your result to that obtained on earth, where the weight is 6 times greater.

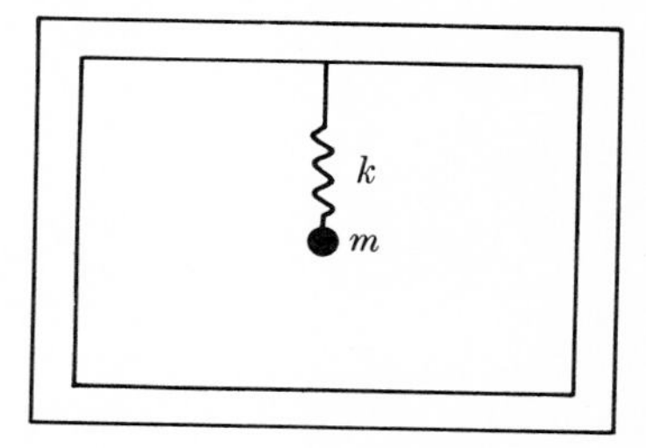

FIGURE P5·11

5·12 A particle of mass m is attached by a massless spring with unstretched length L and spring constant k to a foot that follows a track the elevation of which is defined by

$$\xi(y) = A \sin \omega y$$

The particle remains directly above the foot (see Fig. P5·12) as the system moves along the track at constant horizontal speed V. Derive and solve the equation of motion of the particle in its vertical oscillations in a uniform gravity field with acceleration constant g.

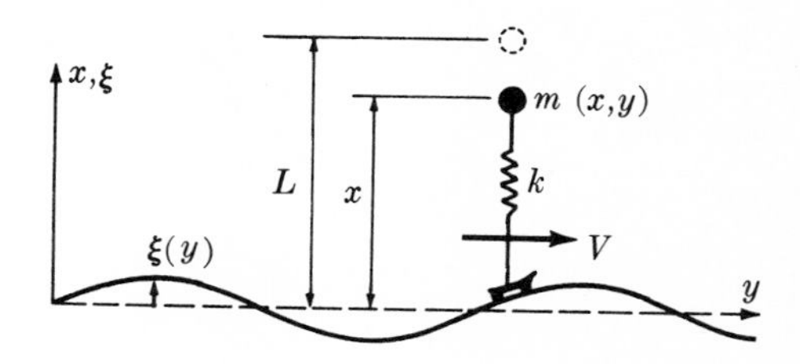

FIGURE P5·12

5·13 Solve the differential equation

$$\ddot{x} + 4x = \cos 2t + 2e^t$$

for initial values $x(0) = x_0$ and $\dot{x}(0) = v_0$.

5·14 A spring-mounted inverted pendulum under uniform gravitational attraction is shown in Fig. P5·14. Determine the condition that must be met by k in order that small oscillations about the vertical equilibrium can occur. If this condition is met, what is the natural frequency of the oscillation?

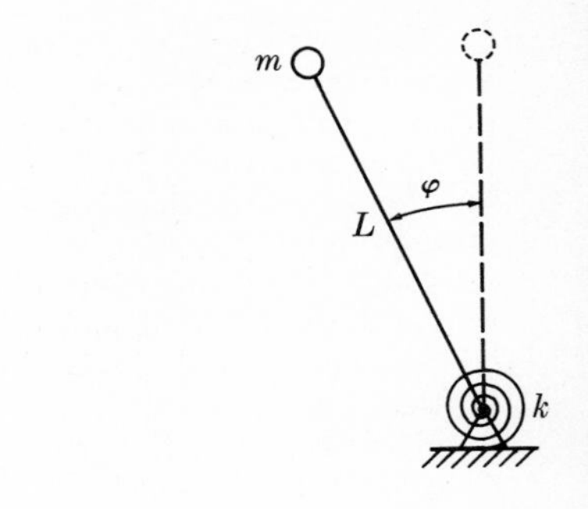

FIGURE P5·14

5·15 Is unbounded response (infinite x) possible for a damped, self-excited system with the following equation of motion?

$$m\ddot{x} + c\dot{x} + kx = Kx$$

If so, under what conditions? If not, why not?

5·16 Preflight check-out of small spacecraft often involves a shake-table test during which the vehicle support undergoes, as shown in Fig. P5·16a, vibratory motion $\xi \cong a \sin \omega t$ of specified amplitude $a(\omega)$, with frequency ω varying continuously over a wide specified range. Use the single degree of freedom, undamped, mathematical model shown in Fig. P5·16b to explain the importance of the rate at which ω sweeps through any value corresponding to a natural frequency of the spacecraft.

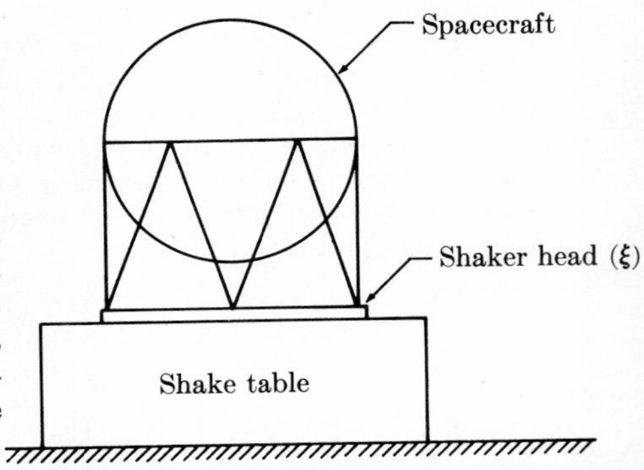

(a)

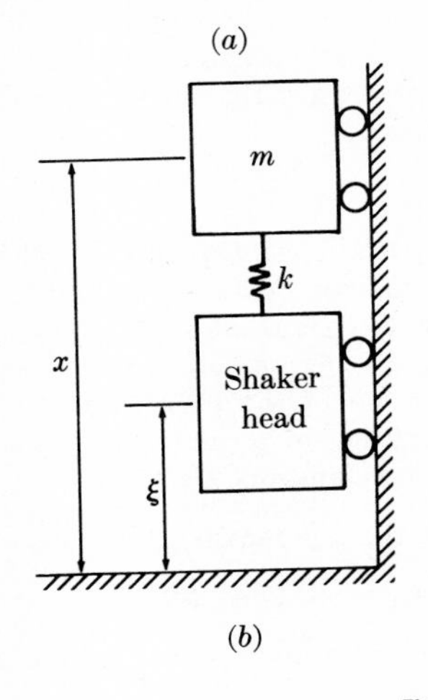

(b)

FIGURE P5·16

5·17 The force $F(t)$ shown in Fig. P5·17a is a very rough approximation of that applied to the spring-mounted cart in Fig. P5·17b when subjected to a blast loading. Determine the response of the initially stationary cart to the blast.

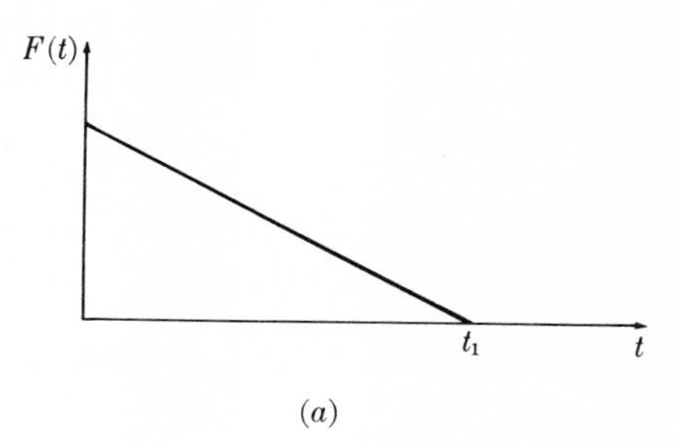

(a)

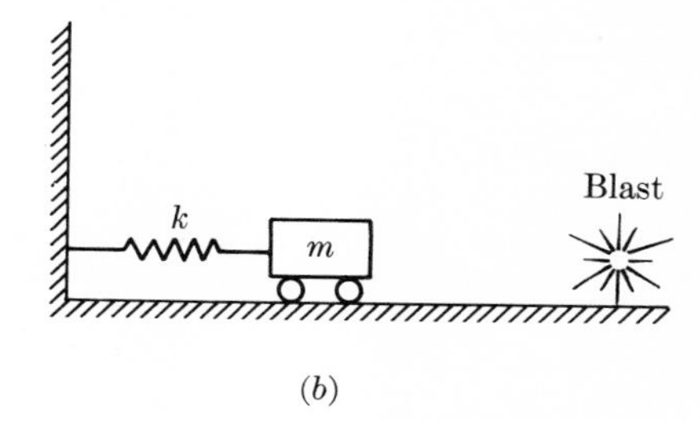

(b)

FIGURE P5·17

5·18 An initially stationary spring-mounted mass is subjected to the force $F(t)$ shown in Fig. P5·18a.

a. Find the response by superimposing the homogeneous and particular solutions, obtaining the latter by the method of undetermined coefficients.

b. Find the response by using the convolution integral.

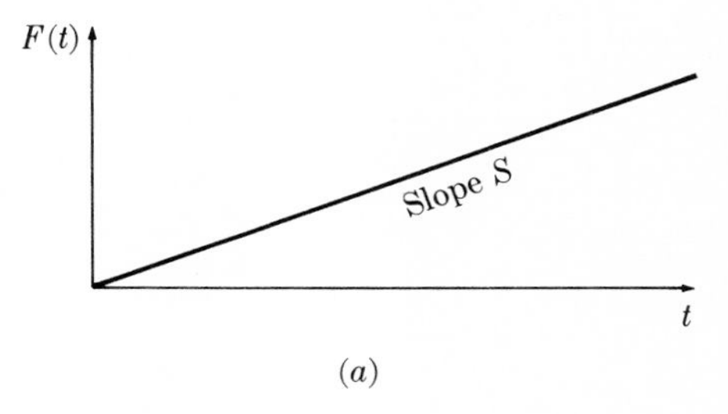

(a)

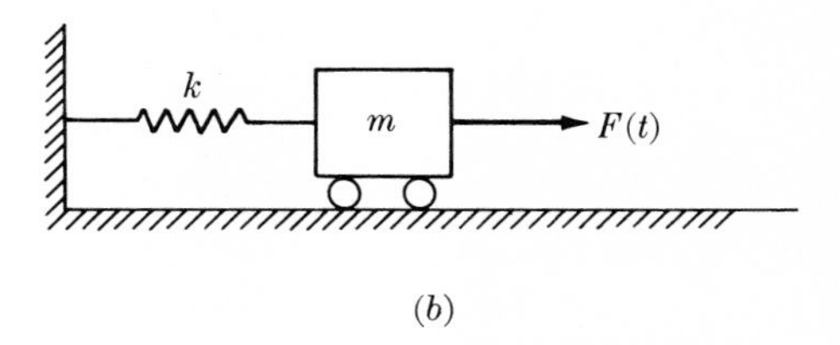

(b)

FIGURE P5·18

5·19 A mass m is attached to a wall by a spring k and dashpot c in parallel (see Fig. P5·19), and subjected to a force $F(t)$. When $F(t) = F_0$, the spring extends an amount Δ. If only m, k, c, and F_0 are known, and you can measure the amplitude of steady-state response of the mass to an excitation $F(t) = F_0 \sin \omega t$, can you determine indirectly the driving frequency ω? If not, what minimal additional information do you need? Provide an expression for ω in support of your answer.

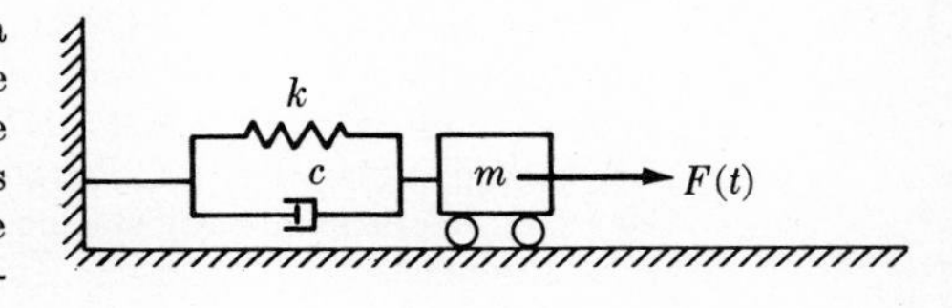

FIGURE P5·19

5·20 A block of mass m slides on frictionless bearings down a vertical tube a distance h under uniform gravity force, and collides without rebound with a massless table supported by a spring k and dashpot c in parallel (see Fig. P5·20). Find an expression for the subsequent change $x(t)$ in the length of the spring. What is the relationship between your result and the indicial admittance plotted in Fig. P5·20?

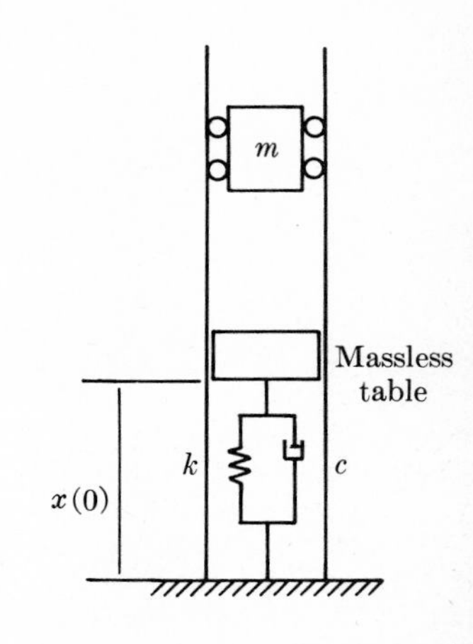

FIGURE P5·20

5·21 When the fan in Fig. P5·21 is going at full speed, the plate to which it is attached deflects a steady-state amount Δ against its springs. How much space ϵ must be provided between the plate and its support to accommodate the transient deflection of the plate induced by turning on the fan? (Assuming that the fan does not at any time provide greater flow of air than in its steady-state operation.)

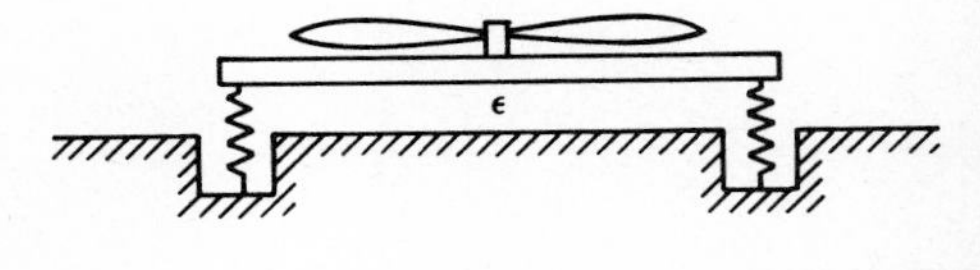

FIGURE P5·21

5·22 A projectile of mass m is given a speed v by releasing a compressed spring housed in a cylinder attached to a table of mass M; the table in turn is attached to the floor by a spring-dashpot system with effective spring constant k and damping constant c, as shown in Fig. P5·22. What must be the minimum clearance ϵ between the table and its support to avoid collision, for the following parameters:

$$m = 0.1 \text{ slug} \qquad v = 5000 \text{ ft/sec}$$
$$M = 10 \text{ slugs} \qquad k = 500 \text{ lb/ft}$$
$$c = 70 \text{ lb sec/ft}$$

Relate your result to the unit impulse response found in Eq. (5·95).

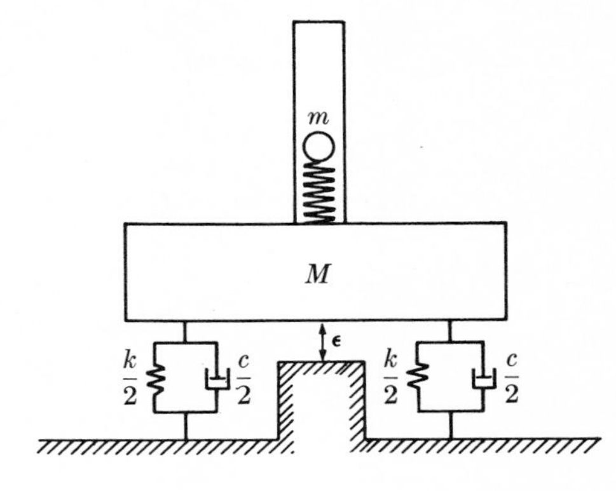

FIGURE P5·22

5·23 In Chap. 4, the example of Fig. 4·9 was treated as an illustration of the application of the work-energy equation. Show that this problem can also be solved by the direct solution of the differential equations arising from $\mathbf{F} = m\mathbf{A}$.

5·24 Albert and Harold are enjoying a teeter-totter ride in the park. Somehow, they manage to keep the board oscillating steadily, with inclination $\theta(t) = \theta_0 \sin \omega t$ where $\theta_0 = 0.5$ rad and $\omega = 0.2$ rad/sec. At time $t = 0$, Bill brings a 10-lb sack containing Albert's lunch, and places it on the teeter-totter board at the fulcrum F. The sack slides on the board without friction, so after $t = 0$ it begins to slide toward Albert (see Fig. P5·24). But in order for Albert to reach the sack it must slide 4 ft, and before this occurs Albert fears that the sack will begin sliding back toward Harold.

Derive the equation of motion of the sack. You won't be able to solve it, but by making what appear to be conservative approximations and considering limiting cases, you are to try to determine whether or not Albert gets his lunch.

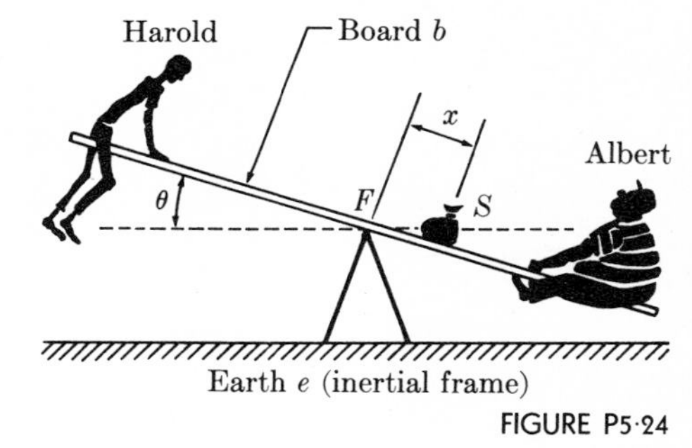

FIGURE P5·24

5·25 In what the controversial Norman Mailer called "the most monstrous exhibition of ego by a brave man in many a year," an astronaut hit a golf ball on the moon, claiming that he "knocked it a mile." If the man could hit a golf ball 200 yd on earth, and if you estimate that with no air resistance he might hit it 300 yd on earth, what might you estimate to be his maximum range on the moon, which has roughly 1 percent of the earth's mass and 25 percent of the earth's diameter?

5·26 A catapult of length $L = 5$ ft and rotary spring constant $k = 10{,}400/\pi^2$ ft lb/rad has an arm of negligible mass with a cup weighing 5 lb at the end (see Fig. P5·26). The spring is unstressed when the arm is vertical ($\theta = 0$). When a 15-lb shot is used for the projectile, what range would the catapult develop on level ground if aerodynamic drag could be neglected?

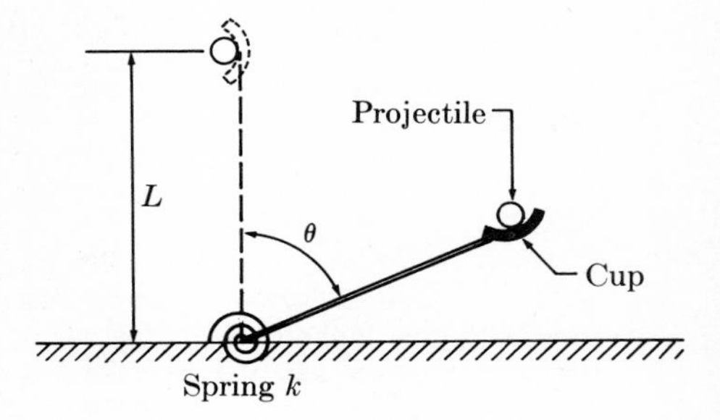

FIGURE P5·26

5·27 A variation of a pinball game is shown in Fig. P5·27. The idea is to compress the (massless) spring (having spring constant k) by just the right amount δ so that when the spring is released it will impart to the slug (mass m) just enough velocity to make it slide to the top of the hill and then fall through vertical distance d, dropping exactly into a hole located a horizontal distance d from the edge of the cliff. Neglecting air resistance and friction, find δ in terms of m, k, d, and g.

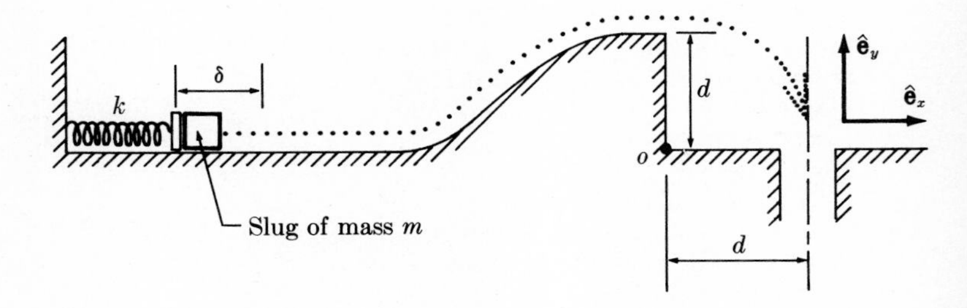

FIGURE P5·27

5·28 A projectile p is attached to the rim of a disk d of radius R (see Fig. P5·28) while d rotates at angular speed ω about its horizontal axis through a point q, and the base b in which d is mounted rotates at rate Ω relative to the earth e. When the projectile is detached from d, it flies off, landing a distance L from the vertical shaft through q. Find an equation (perhaps transcendental) from which the angle θ characterizing the release point can be determined assuming the following special conditions:

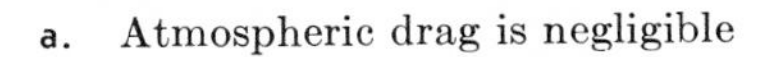

a. Atmospheric drag is negligible

b. $\dfrac{L}{R} \gg 1$

c. $\omega \equiv \Omega$

Hint: The required equation involves only θ and the quantity $N \triangleq R\Omega^2/g$.

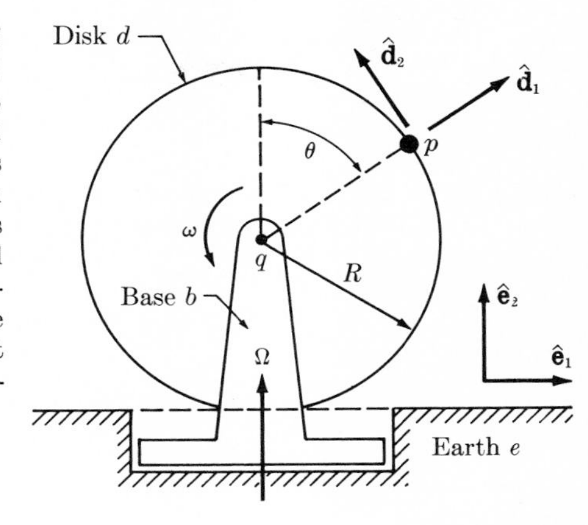

FIGURE P5·28

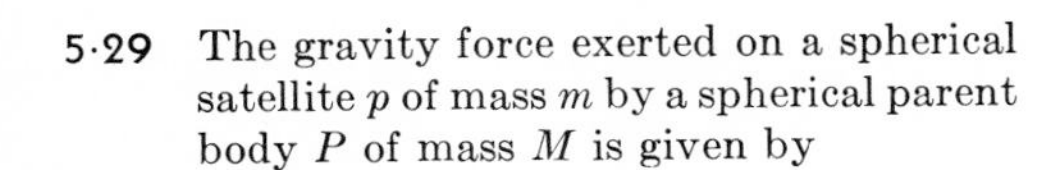

5·29 The gravity force exerted on a spherical satellite p of mass m by a spherical parent body P of mass M is given by

$$\mathbf{F} = -GMmr^{-2}\hat{\mathbf{u}}_r$$

where $r\hat{\mathbf{u}}_r$ is the position vector from P to p, and where G is the "universal gravitational constant." [Compare with Eq. (5·115), in which μ replaces GM.]

The earth satellite *Tiros I* had an orbital period $T = 99.2$ min, a maximum (apogee) height above the earth's surface of 7.54×10^5 m, and a minimum (perigee) height of 6.93×10^5 m. The earth may be approximated as a sphere of radius 6.38×10^6 m and mass $M = 6 \times 10^{24}$ kg.

a. Determine a numerical value for G from the above orbital data without using the known constant g for the gravitational acceleration at the earth's surface. Obtain your result in units of newtons, meters, and kilograms.

b. Obtain G by using $g = 9.81$ m/sec^2.

5·30 An earth satellite in an elliptical orbit of eccentricity $e = 0.36$ is subjected by a brief rocket motor burn to a velocity increment $\mathbf{\Delta V}$ when at apogee (see Fig. P5·30). The increment $\mathbf{\Delta V}$ is given by $\lambda \mathbf{V}_a$, where $\mathbf{V}_a$ is the inertial velocity at apogee. What must be the value of λ to change the orbit from elliptical to circular?

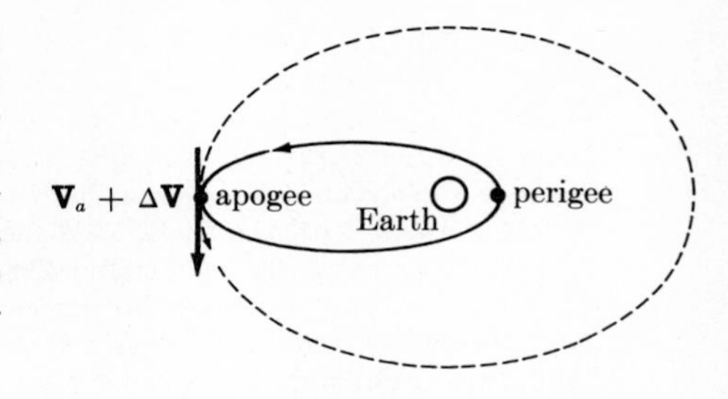

FIGURE P5·30

5·31 An earth satellite s is injected into orbit by a booster that cuts off at time $t = 0$, when its distance $r(0)$ from the center of the earth is twice the earth's radius (see Fig. P5·31), so that $r(0) = 2R_E$. At booster cutoff, the inertial velocity of s is $\mathbf{V}^s = V_{ro}\hat{\mathbf{u}}_r + V_{to}\hat{\mathbf{u}}_t$, where $V_{ro}/V_{to} = 0.1$, and $V_{ro} > 0$; and the centripetal acceleration $-r(0)\dot{\theta}^2(0)$ has magnitude $0.3\ g$ (where $g = 9.81$ m/sec²). Given these initial conditions, find the angle $\theta(0)$ which locates the perigee of the orbit with respect to the satellite position at $t = 0$.

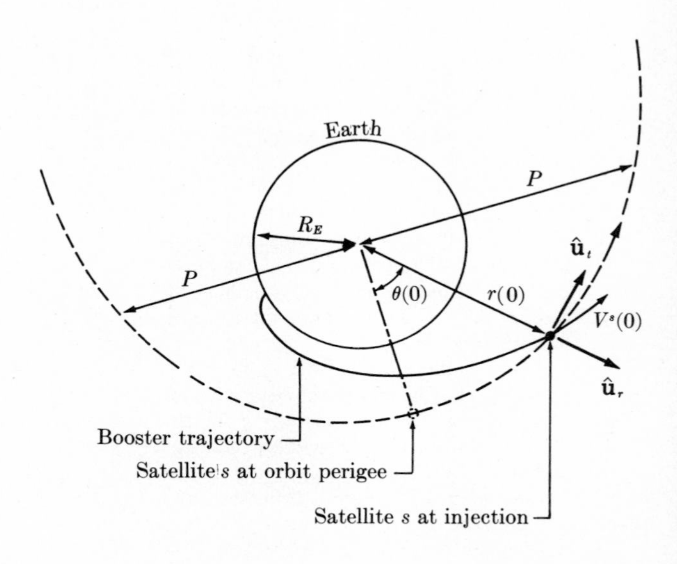

FIGURE P5·31

5·32 A particle p of mass m on a vertical massless elastic beam identical to that treated in the example illustrated by Fig. 5·16 is shown in Fig. P5·32, except that here the beam base is attached at a radial distance R to a turntable spinning at constant rate Ω, and the spring constants characterizing the beam stiffnesses in the directions of $\hat{\mathbf{u}}_1$, $\hat{\mathbf{u}}_2$, and $\hat{\mathbf{u}}_3$ are $k_1 = k_2 = 4m\Omega^2$ and k_3 infinite. Even when the beam is not vibrating, point p sustains a steady-state deflection $\mathbf{\Delta} = R\Omega^2/(k/m - \Omega^2)\hat{\mathbf{u}}_1$. Write the equations of motion in terms of the deviation $\mathbf{x} = x_1\hat{\mathbf{u}}_1 + x_2\hat{\mathbf{u}}_2 + x_3\hat{\mathbf{u}}_3$ from this steady-state position, in both scalar and matrix form.

a. Find the system eigenvalues λ_j and eigenvectors $[V^{(j)}]$ for $j = 1, 2, 3, 4$, with $V_2^{(j)}$ set equal to unity for normalization.

b. Find the solution for the equations of motion, as indicated by Eq. (5·157).

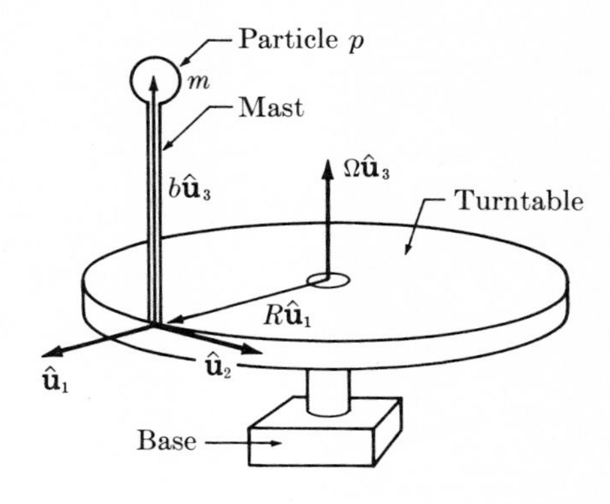

FIGURE P5·32

5·33 The particle p atop the massless elastic beam in Fig. P5·32 is subjected to the force

$$\mathbf{F} = F_1 \sin \omega t \hat{\mathbf{u}}_1 + F_2 \cos \omega t \hat{\mathbf{u}}_2$$

Find the response, using Eq. (5·158).

CHAPTER SIX
MECHANICS OF SYSTEMS OF PARTICLES

6·1 D'ALEMBERT'S PRINCIPLE AND MASS-CENTER MOTION

6·1·1 D'Alembert's principle

In previous chapters we have focused on individual particles. In Chap. 1 we postulated laws of motion for a particle, and in Chaps. 2 and 4 we applied those laws to problems of statics and dynamics of individual particles. Even when more than one particle was considered in Chap. 2, we wrote static-equilibrium equations for the particles *one at a time*.

Now we will begin to consider the possibility of developing new principles we can apply to collections of particles considered together as a *system* of particles. We will build the new concepts firmly on the foundation established in earlier chapters. Specifically, if the N particles of our system are called $p_1, p_2, \ldots, p_N$, and we let m_j be the mass of p_j, $\mathbf{R}_j$ be the inertial position vector of p_j, and $\mathbf{F}_j$ be the vector sum of forces applied to p_j, for $j = 1, \ldots, N$, then from Newton's second law we can write

$$\mathbf{F}_j = m_j\ddot{\mathbf{R}}_j \qquad j = 1, \ldots, N \tag{6·1}$$

with each dot over a vector indicating time differentiation in an inertial reference frame. If we can obtain explicit expressions for $\mathbf{F}_j$ then, as in Chap. 4, we can obtain a set of differential equations containing all motion information provided by newtonian mechanics. The word *if* is an important one, however, since often we have incomplete knowledge of the forces applied to the individual particles of a given collection or system of particles. In particular, we are often quite ignorant of the *interaction forces among particles* of a system. Whereas these unknown forces might be of primary interest in a statics problem, they may be of no concern at all in a dynamics problem. In the latter case we would like to be able to predict certain aspects of the system motion without bothering with internal interaction forces at all.

Consider, for example, the problem of determining the motion of a balloon in the atmosphere. We might have a fairly accurate knowledge of the forces applied to the balloon by the surrounding air and the nearby earth, but we have little knowledge of the forces exerted between individual molecules of gas within the balloon. Of course, if we simply idealized the whole balloon as a particle, we could predict its gross motion by a single application of $\mathbf{F} = m\mathbf{A}$. It does seem therefore that even if we let each molecule of gas be idealized as a particle, it should be possible to obtain some information about the gross motion of the balloon, even in ignorance of particle-interaction forces.

In pursuit of this general objective, we separate the force $\mathbf{F}_j$ applied to p_j into the two parts $\mathbf{f}_j$ and $\mathbf{f}'_j$, with the latter representing the resultant of all forces exerted on p_j by other particles within the system. Thus Eq. (6·1) becomes

$$\mathbf{f}_j + \mathbf{f}'_j = m_j\ddot{\mathbf{R}}_j \qquad j = 1, \ldots, N$$

Summing these N vector equations then furnishes

$$\sum_{j=1}^{N} \mathbf{f}_j + \sum_{j=1}^{N} \mathbf{f}'_j = \sum_{j=1}^{N} m_j\ddot{\mathbf{R}}_j \tag{6·2}$$

If there are two particles in the system ($N = 2$), then from Newton's third law (the action-reaction law), we know that $\mathbf{f}'_1 = -\mathbf{f}'_2$, so that the sum $\sum_{j=1}^{N} \mathbf{f}'_j$ vanishes. The proposition that this sum of resultant internal forces vanishes in general, so that

$$\sum_{j=1}^{N} \mathbf{f}'_j = 0 \tag{6·3}$$

is due to d'Alembert, and is sometimes referred to as *d'Alembert's principle.*

Actually we could deduce Eq. (6·3) from Newton's third law simply by considering the particle-interaction forces as pairs of equal and opposite forces between pairs of particles. Since d'Alembert's principle does not indicate in detail just *how* the internal forces sum to zero, it may be considered to be a generalization of Newton's third law. But since Eq. (6·3) can be written and used without the logical requirement of any new principles beyond those offered by Newton, d'Alembert's principle is most noteworthy here simply because it provides the first result applicable uniquely to *systems* of particles.

Note that the combination of Eqs. (6·2) and (6·3) may be written

$$\sum_{j=1}^{N} (\mathbf{f}_j - m_j\ddot{\mathbf{R}}_j) = 0 \tag{6·4}$$

Equation (6·4) [rather than Eq. (6·3)] is often referred to in modern literature as d'Alembert's principle, and many texts use this term to describe Eq. (6·4) even in the special case of a single particle. This seems to represent a substantial departure from d'Alembert's intent, since then there are no "internal" forces to be considered, but this usage has become quite conventional. Equation (6·4) then becomes Eq. (4·1).

6·1·2 Mass-center definition

A system consisting of N particles of masses $m_1, m_2, \ldots, m_N$, respectively located with respect to an arbitrary point q by position vectors $\mathbf{r}_1, \mathbf{r}_2, \ldots, \mathbf{r}_N$, has a *mass center* or *center of mass*† c whose position vector with respect to point q is $\mathbf{r}_c$, as given by

$$\mathbf{r}_c \sum_{j=1}^{N} m_j = \sum_{j=1}^{N} m_j\mathbf{r}_j \tag{6·5}$$

As we shall see, the mass center is the point at which we can concentrate the total mass of a material system in order to achieve some form of equivalence between the material system idealized as a single particle and that same material system idealized as a collection of particles. The nature of this equivalence remains to be established.

6·1·3 Equations of mass-center motion

Equation (6·5) defines the position vector $\mathbf{r}_c$ which locates the mass center of a system of particles from an arbitrary point q. If we choose the reference point q

† To be distinguished from the *center of gravity*, with which it coincides for a typical material system only when the gravitational force on a particle of the system is idealized as being independent of the position of the particle. (See §7·1·3 for the definition of center of gravity.)

as a point fixed in an inertial reference frame, then we should for consistency with previous notation use the symbol $\mathbf{R}_c$ rather than $\mathbf{r}_c$, and then as a special case of Eq. (6·5) we have

$$\mathbf{R}_c \sum_{j=1}^{N} m_j = \sum_{j=1}^{N} m_j \mathbf{R}_j$$

If now we let the symbol $\mathfrak{M}$ denote the total mass of the system of particles, and differentiate this equation with respect to time twice in an inertial reference frame, we have

$$\mathfrak{M}\ddot{\mathbf{R}}_c = \sum_{j=1}^{N} m_j \ddot{\mathbf{R}}_j$$

This result can be substituted into Eq. (6·4) to obtain

$$\sum_{j=1}^{N} \mathbf{f}_j = \mathfrak{M}\ddot{\mathbf{R}}_c \tag{6·6}$$

Equation (6·6) is the result we have been seeking since it permits us to formulate an equation of motion to be solved for the gross behavior of the system as represented by its mass-center motion, even if we have no knowledge of the interaction forces among the particles.

If we introduce the symbol $\mathbf{F}$ to represent the vector sum or resultant of the external forces $\mathbf{f}_1, \ldots, \mathbf{f}_N$ applied to the N particles of the system, then we have from Eq. (6·6)

$$\mathbf{F} = \mathfrak{M}\ddot{\mathbf{R}}_c \tag{6·7a}$$

or in the more explicit notation of Chap. 3,

$$\mathbf{F} = \mathfrak{M}\mathbf{A}^{ci} \tag{6·7b}$$

Thus we see that Newton's second law applies not only to a particle but also to the mass center c of a system of N particles, provided that $\mathfrak{M}$ is the total system mass and $\mathbf{F}$ is the resultant of all external forces exerted on the system. Now perhaps you can better understand why we have such great success with the idealization of complex mechanical objects as simple particles. An enormous space vehicle, with its many flexible components, its tanks of fuel, and its astronauts moving about inside, can be idealized as a particle of mass $\mathfrak{M}$ located at the spacecraft mass center c, and Eq. (6·7b) can be used to predict its trajectory quite accurately because the resultant force $\mathbf{F}$ applied to the vehicle is essentially the same as would be applied to a particle of mass $\mathfrak{M}$ located at point c.

6·1·4 A falling sack of sand

Among the earliest of the "payload delivery systems" devised for civil or military use is the catapult, a portable version of which is illustrated in Fig. 6·1. In order to minimize the weight of the system for increased mobility, it may have proven advantageous to actuate the catapult with sacks of sand or other locally collected material. To the technologists of those days, the question of catapult performance had to begin with the question of the dynamics of a falling sack of sand. We will deal with this preliminary problem here, and return to the catapult dynamics problem later when we have developed a wider range of capabilities.

Even with Newton's laws available to us, we find it necessary to make certain "engineering decisions" before we can attack the problem analytically. If we can treat the sack of sand as a single particle of mass $\mathfrak{M}$, and assume that the only force applied to the sack during its fall is the constant $-\mathfrak{M}g\hat{\mathbf{e}}_3$, where $\hat{\mathbf{e}}_3$ is a unit vector directed vertically upward and g is an empirically generated scalar such that $\mathfrak{M}g$ is the weight of the sand, then if s is the distance the sack falls, we have the single scalar equation of motion

$$-\mathfrak{M}g = -\mathfrak{M}\ddot{s}$$

which by virtue of the initial conditions presented by springing an ideal trapdoor

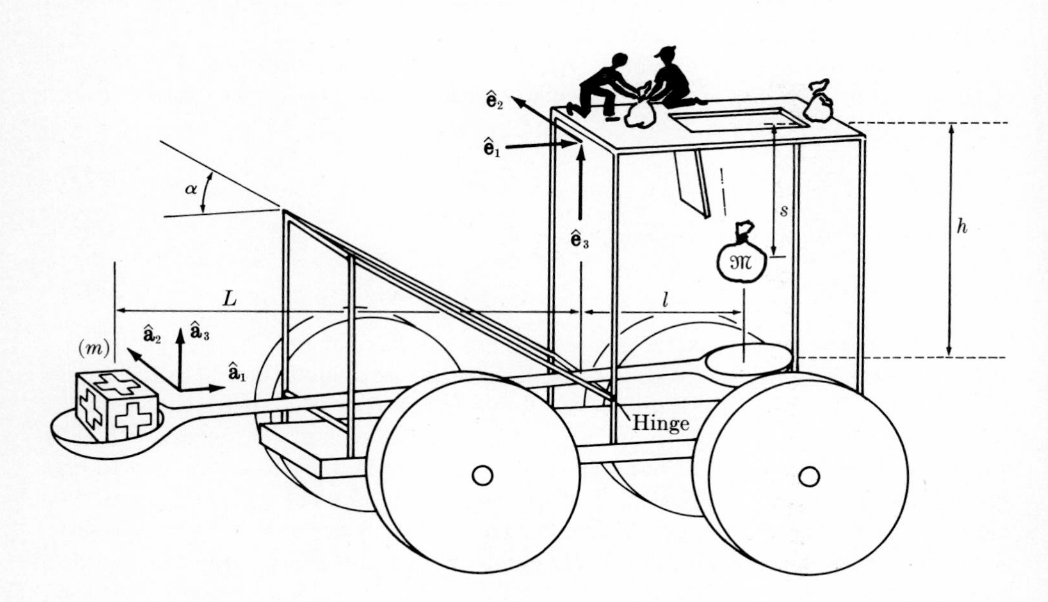

FIGURE 6·1

leads (see §5·2·2) to

$$s = \tfrac{1}{2}gt^2 \tag{6·8}$$

applicable until $s = h$, and the sack strikes the catapult.

Before we accept this simple solution, we must reexamine our analytical model. In order to assess the validity of our assumptions regarding **F**, it would be necessary to repeat the analysis under different assumptions, perhaps with some representation of air resistance or the inverse-square nature of the force of gravity. Difficulties thereby introduced would require the methods of Chap. 5. In order to assess the validity of our idealization of the sack of sand as a single particle, it would be necessary to idealize it differently, perhaps treating each *grain* of sand as a particle, and then compare the results of the two analyses.

If the number of particles N in our mathematical model equals the number of grains in a sack of sand, then it is of little comfort to us that we can write N vector equations of motion for the particles, particularly since we have no knowledge of the contact forces of interaction among the particles. With Eq. (6·7*a*) or its equivalent Eq. (6·7*b*), we can see immediately that the *mass center* of the sack does move as though the sack were a single particle located at that point, and possessing the total system mass $\mathfrak{M}$. Thus the solution in Eq. (6·8) is unchanged by the new model of the sack of sand, as long as s defines the motion of the mass center of the sack.

6·2 MOMENTUM CONCEPTS

6·2·1 Linear momentum

When in Chap. 4 the vector equation $\mathbf{F} = m\ddot{\mathbf{R}}$ was examined as it applies to a particle, it was noted that certain first integrals can always be extracted from this equation. The first of these led to the *impulse-momentum equation* [Eq. (4·25)], which stated the equality of the (inertial) impulse

$$\mathbf{I} \triangleq \int_{t_1}^{t_2} \mathbf{F}\, dt \bigg|_i \tag{6·9}$$

[Eq. (4·23)] and the inertial time derivative of the (inertial) linear momentum $m\dot{\mathbf{R}}$.

Equation (6·7*a*) has the same form as $\mathbf{F} = m\ddot{\mathbf{R}}$, and although the symbols now denote properties of a system of particles rather than a single particle, the mathematical arguments used to find two first integrals must apply to Eq. (6·7*a*) without change. Nothing new can be added but physical interpretation and the conventions of new language.

A system of N particles $p_1, \ldots, p_N$, having masses $m_1, \ldots, m_N$, respectively, and being located with respect to an inertially fixed point o by position vectors $\mathbf{R}_1, \ldots, \mathbf{R}_N$, is said to have a *linear momentum* equaling the vector sum of the linear momenta of its constituent particles, and by Eq. (6·5) this momentum can be written in terms of the total system mass $\mathfrak{M}$ and the inertial velocity $\dot{\mathbf{R}}_c$ of the system mass center in the form

$$\mathfrak{M}\dot{\mathbf{R}}_c = \sum_{j=1}^{N} m_j \dot{\mathbf{R}}_j \tag{6·10}$$

Thus the equation

$$\mathbf{I} = \mathfrak{M}\dot{\mathbf{R}}_c(t_2) - \mathfrak{M}\dot{\mathbf{R}}_c(t_1) \tag{6·11}$$

derived from $\mathbf{F} = \mathfrak{M}\ddot{\mathbf{R}}_c$ by arguments paralleling those leading to Eq. (4·25), continues to be interpretable as the equality of (resultant) impulse and the inertial time derivative of (system) linear momentum.

It is important to this result that all interaction forces among the particles of a system cancel each other in their contributions to the resultant impulse in Eq. (6·9); thus we could obtain Eq. (6·11) alternatively by adding up the N impulse-momentum equations that apply to the N particles of the system. In the event that the interaction forces among particles are unknown (as when the particles collide with one another), the impulse-momentum equation for the system may provide useful information when no information at all is available from the equations of motion of individual particles.

6·2·2 Angular momentum/moment of momentum

The concepts of *angular momentum* and *moment of momentum* of a particle with respect to a point, as introduced in §4·2·5, lead quite naturally to the much more useful application of these concepts to systems of particles.

The angular momentum of a system of particles with respect to a point q is the sum of the contributions of its constituent parts, that is, for a system of N particles $p_1, \ldots, p_N$ of masses $m_1, \ldots, m_N$ and position vectors $\mathbf{r}_1, \ldots, \mathbf{r}_N$ relative to an arbitrary point q, the angular momentum with respect to q is

$$\mathbf{H}^q = \sum_{j=1}^{N} m_j \mathbf{r}_j \times \dot{\mathbf{r}}_j \tag{6·12}$$

Similarly, the moment of momentum of the same system of particles $p_1, \ldots, p_N$ with respect to point q is given by

$$\mathbf{h}^q = \sum_{j=1}^{N} m_j \mathbf{r}_j \times \dot{\mathbf{R}}_j \tag{6·13}$$

where $\mathbf{R}_j$ is the position vector of p_j with respect to an inertially fixed point o.

(Recall from §4·2·5 that the terms *angular momentum* and *moment of momentum* are applied without consistency in the literature of dynamics, with either or both terms applied to both $\mathbf{H}^q$ and $\mathbf{h}^q$ as defined and separately labeled in this text.)

We might well wonder what new information can be obtained by the use of $\mathbf{H}^q$ or $\mathbf{h}^q$, since for a system of N particles all dynamical information is contained in N vector equations of the form

$$\mathbf{F}_j = m_j\ddot{\mathbf{R}}_j \qquad j = 1, \ldots, N \tag{6·14}$$

where $\mathbf{R}_j$ and m_j are as defined previously, and $\mathbf{F}_j$ is the force applied to particle p_j. We noted when dealing with a single particle that the concepts of angular momentum and moment of momentum were auxiliary and never essential to the solution of a problem. Although the same may be said when dealing with particle systems, it often improves the efficiency of a dynamic analysis and lends physical insight to consider the time history of the angular momentum or the moment of momentum.

By forming the time derivative of the angular momentum $\mathbf{H}^q$ in an inertial frame of reference, we find

$$\dot{\mathbf{H}}^q = \frac{{}^i d}{dt}\left(\sum_{j=1}^{N} m_j\mathbf{r}_j \times \dot{\mathbf{r}}_j\right) = \sum_{j=1}^{N} m_j\mathbf{r}_j \times \ddot{\mathbf{r}}_j$$

and by replacing $\ddot{\mathbf{r}}_j$ by $\ddot{\mathbf{R}}_j - \ddot{\mathbf{R}}_q$ (see Fig. 6·2) and employing Eq. (6·14) we obtain the interpretation

$$\begin{aligned}\dot{\mathbf{H}}^q &= \sum_{j=1}^{N} \mathbf{r}_j \times m_j\ddot{\mathbf{R}}_j - \sum_{j=1}^{N} m_j\mathbf{r}_j \times \ddot{\mathbf{R}}_q \\ &= \sum_{j=1}^{N} \mathbf{r}_j \times \mathbf{F}_j + \ddot{\mathbf{R}}_q \times \sum_{j=1}^{N} m_j\mathbf{r}_j\end{aligned}$$

The product $\mathbf{r}_j \times \mathbf{F}_j$ is called the *moment* of the force $\mathbf{F}_j$ with respect to the point q, and the sum of such moments is the total moment $\mathbf{M}^q$ of the forces of the system for point q. Further simplification of the $\dot{\mathbf{H}}^q$ expression is effected by the substitution (see Fig. 6·2)

$$\mathbf{r}_j = \mathbf{r}_c + \boldsymbol{\rho}_j$$

since by mass-center definition the sum $\sum_{j=1}^{N} m_j\boldsymbol{\rho}_j$ is *zero*. With this substitution, $\dot{\mathbf{H}}^q$ becomes

$$\dot{\mathbf{H}}^q = \mathbf{M}^q + \ddot{\mathbf{R}}_q \times \sum_{j=1}^{N} m_j\mathbf{r}_c$$

or, with $\mathfrak{M}$ representing the system mass,

$$\dot{\mathbf{H}}^q = \mathbf{M}^q + \mathfrak{M}\ddot{\mathbf{R}}_q \times \mathbf{r}_c \tag{6·15}$$

Differentiating the moment of momentum $\mathbf{h}^q$ in inertial space produces an expression somewhat different from Eq. (6·15). In the derivative

$$\dot{\mathbf{h}}^q = \frac{{}^i d}{dt}\left(\sum_{j=1}^{N} m_j \mathbf{r}_j \times \dot{\mathbf{R}}_j\right) = \sum_{j=1}^{N} m_j \mathbf{r}_j \times \ddot{\mathbf{R}}_j + \sum_{j=1}^{N} m_j \dot{\mathbf{r}}_j \times \dot{\mathbf{R}}_j$$

the substitutions $\dot{\mathbf{R}}_j = \dot{\mathbf{R}}_q + \dot{\mathbf{r}}_j$ and $\mathbf{F}_j = m_j \ddot{\mathbf{R}}_j$ produce

$$\dot{\mathbf{h}}^q = \sum_{j=1}^{N} \mathbf{r}_j \times \mathbf{F}_j + \sum_{j=1}^{N} m_j \dot{\mathbf{r}}_j \times \dot{\mathbf{R}}_q$$

With the introduction of $\mathbf{M}^q$, the moment of the applied forces with respect to point q, and the substitution $\dot{\mathbf{r}}_j = \dot{\mathbf{r}}_c + \dot{\boldsymbol{\rho}}_j$ (see Fig. 6·2), we obtain

$$\dot{\mathbf{h}}^q = \mathbf{M}^q + \dot{\mathbf{r}}_c \times \mathfrak{M}\dot{\mathbf{R}}_q \tag{6·16}$$

By comparing Eqs. (6·15) and (6·16) we can recognize the rather subtle differences in the concepts we have chosen to distinguish as angular momentum $\mathbf{H}^q$ and moment of momentum $\mathbf{h}^q$. If the reference point q is fixed in inertial space, so that $\dot{\mathbf{R}}_q = \ddot{\mathbf{R}}_q = 0$, these equations say $\dot{\mathbf{H}}^q = \mathbf{M}^q = \dot{\mathbf{h}}^q$, and there is no practical need to distinguish between $\mathbf{H}^q$ and $\mathbf{h}^q$. Similarly, if q is the mass center of the particle system, the vector $\mathbf{r}_c \equiv 0$, and again $\dot{\mathbf{H}}^q = \mathbf{M}^q = \dot{\mathbf{h}}^q$. Since in practical applications it is almost always convenient to choose the point q as either inertially fixed or mass-center coincident, it is rarely necessary to distinguish between $\mathbf{H}^q$ and $\mathbf{h}^q$. The occasion does arise, however, when point q is most wisely selected so as to require the general forms of Eqs. (6·15) and (6·16), and when that time comes, we must make a clear choice between $\mathbf{H}^q$ and $\mathbf{h}^q$ and stick

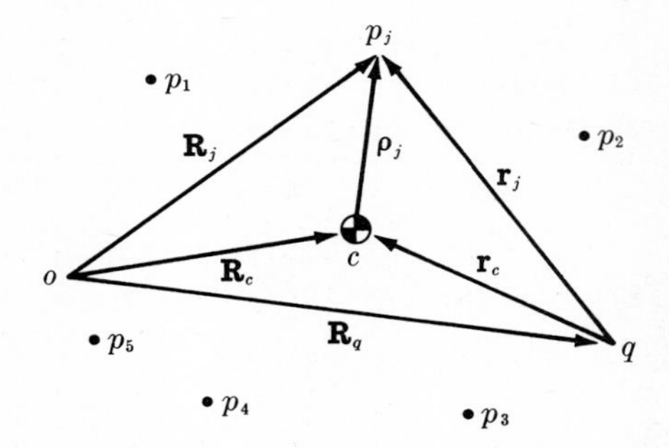

FIGURE 6·2

with it carefully. In this book, the angular momentum $\mathbf{H}^q$ is given the preferred position, for reasons that will become clear only in Chap. 8.

In practical application, it would be very cumbersome if for the calculation of the moment $\mathbf{M}^q$ the definition

$$\mathbf{M}^q \triangleq \sum_{j=1}^{N} \mathbf{r}_j \times \mathbf{F}_j \tag{6·17}$$

were required, since this would often involve an exacting task of calculation of the resultant force $\mathbf{F}_j$ applied to each particle, and subsequent summation. If each such force $\mathbf{F}_j$ is written as the sum of the resultant force $\mathbf{f}'_j$ exerted on the jth particle by other particles internal to the system, and the resultant force $\mathbf{f}_j$ exerted on the jth particle by external influences, so that Eq. (6·17) becomes

$$\mathbf{M}^q = \sum_{j=1}^{N} \mathbf{r}_j \times \mathbf{f}_j + \sum_{j=1}^{N} \mathbf{r}_j \times \mathbf{f}'_j$$

an important simplification is accomplished. Newton's third law (see §1·2·1) indicates that "the mutual actions of two bodies on each other are always equal, and directed to contrary parts." This means that the force exerted by the ith particle on the kth particle (say $\mathbf{f}'_{ik}$) is the negative of that exerted by the kth particle on the ith particle (say $\mathbf{f}'_{ki}$). If, as in Fig. 6·3, these forces are directed along the line joining the two particles, then $\mathbf{r}_i \times \mathbf{f}'_{ki} + \mathbf{r}_k \times \mathbf{f}'_{ik} = 0$, and since all forces labeled $\mathbf{f}'_j$ appear in such pairs, the moment $\mathbf{M}^q$ becomes

$$\mathbf{M}^q = \sum_{j=1}^{N} \mathbf{r}_j \times \mathbf{f}_j \tag{6·18}$$

Although in almost all physical systems of immediate interest the inter-

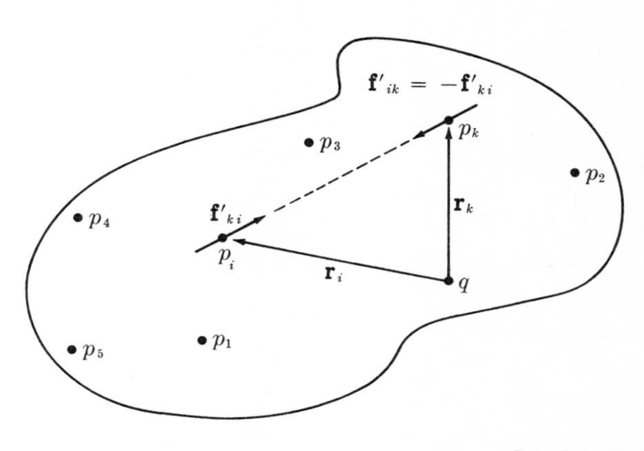

FIGURE 6·3

action forces among particles are directed along the line joining the particles, there are exceptions for which Eq. (6·18) is invalid. The electromagnetic interaction forces between pairs of moving electrically charged particles are equal and opposite, as required by Newton's third law, but in direction these forces are orthogonal to their relative inertial velocity vector. When such forces are included in the system, we must bypass Eq. (6·18) for its more complex predecessor to obtain $\mathbf{M}^q$, but there is no further obstacle to applying the equations of motion as found in Eqs. (6·15) and (6·16).

6·2·3 Spacecraft rendezvous and docking

In §4·2·6 we established that for a spacecraft (a particle p) orbiting the earth (an inertially fixed attracting center o), the angular momentum $\mathbf{H}^o$ for point o and the moment of momentum $\mathbf{h}^o$ for point o are identical constants, given by $mr\dot{\varphi}^2\hat{\mathbf{u}}_n$, where m is the mass of p, r is the distance from o to p, $\hat{\mathbf{u}}_n$ is a unit vector perpendicular to the orbital plane, and φ is an angle in the orbital plane between the ray from o to p and a reference line (or ray) emanating from o (see Fig. 4·5). In §5·2·3 we actually confronted the equations of motion of p, and learned more about the orbital motion. Now we wish to consider two space vehicles, p_1 and p_2, having masses m_1 and m_2, respectively, and see what we can determine about their dynamic behavior as they approach one another and are linked together in orbit.

If we neglect any interaction forces between the two vehicles other than actual contact forces (ignoring such tiny forces as those due to mutual gravitational attraction), then prior to linkup the space vehicles each move independently, and their motions can be predicted by the methods of Chaps. 4 and 5. In order to be sure that the vehicles will come together for docking, we must equip one or the other with a propulsive capability so that its orbit can be changed to accomplish a rendezvous with the second vehicle. Let us imagine that p_1 has propulsive capability, while p_2 has none.

Prior to linkup, particle p_2 moves in an elliptic orbit about the earth, as indicated in §5·2·3, while p_1 experiences deviations from such an orbit as the result of its trajectory-correction motor. If we imagine these corrections to result from brief periods of accelerated flight due to motor thrust, followed by long periods with the motor turned off (as in the example in §4·2·2), then when p_1 and p_2 approach each other for docking they might both be in elliptic orbits resulting from "free fall" in the gravity field. Although these orbits must intersect at the point of rendezvous, they will not be identical, so that after docking neither spacecraft will be permitted to persist in its original orbit, being deflected slightly from that course by the small interaction forces involved in the docking maneuver (see Fig. 6·4). We want to learn whatever we can about the motion that ensues after docking.

During the (presumably brief) interval of the minor collision during docking, the two vehicles are subjected to equal and opposite impulses **I**, as defined by Eq. (6·9). We can't obtain **I** by integrating the force **F** applied to one of the vehicles over time, as Eq. (6·9) suggests, because we lack detailed knowledge of the function $\mathbf{F}(t)$. Instead, we focus attention on the *system* of particles including both p_1 and p_2, and then try to find the impulse **I** applied to the system by external forces. The only forces external to the system are those of gravity, and over the vanishingly small interval of docking these forces contribute nothing to **I** in Eq. (6·9). Hence the linear momentum of the system is conserved (although p_1 and p_2 both change momentum), and Eq. (6·11) provides

$$\mathfrak{M}\dot{\mathbf{R}}_c(t_2) = \mathfrak{M}\dot{\mathbf{R}}_c(t_1)$$

where $\mathbf{R}_c(t)$ is the position vector of the system mass center c at time (t), and $\mathfrak{M} = m_1 + m_2$. Since the linear momentum of the system at t_1 (just prior to

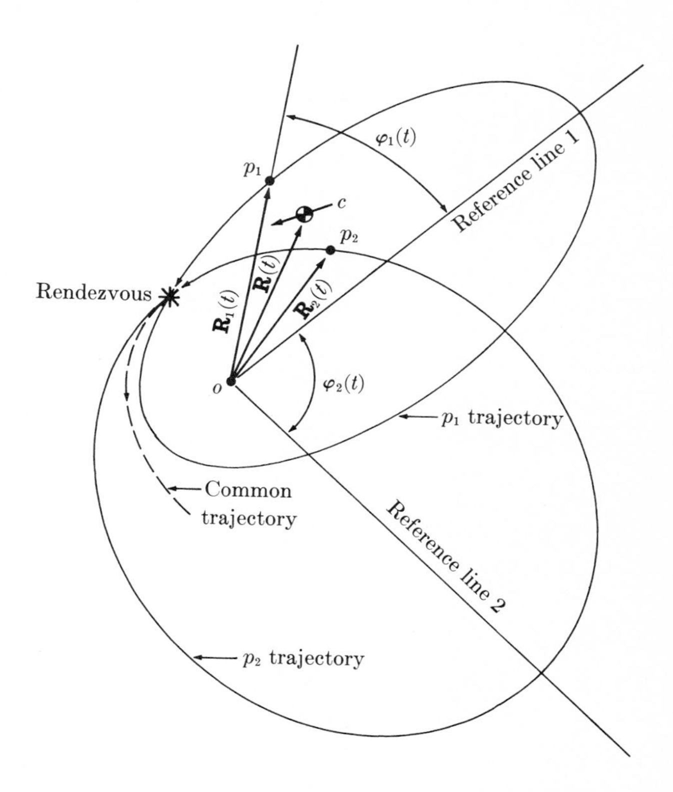

FIGURE 6·4

docking) is given by

$$\mathfrak{M}\dot{\mathbf{R}}_c(t_1) = m_1\dot{\mathbf{R}}_1(t_1) + m_2\dot{\mathbf{R}}_2(t_1)$$

where $\mathbf{R}_j(t)$ is the position vector of p_j with respect to o at time t, the velocity of the combined system just after docking is

$$\dot{\mathbf{R}}_c(t_2) = \frac{m_1\dot{\mathbf{R}}_1(t_1) + m_2\dot{\mathbf{R}}_2(t_1)}{m_1 + m_2}$$

After the docking has been accomplished, the two particles p_1 and p_2 become for present purposes a single particle p of mass $\mathfrak{M} = m_1 + m_2$, and p then follows a new elliptical orbit as prescribed by §5·2·3, with "initial" conditions at rendezvous time t_2 given by $\mathbf{R}_c(t_2)$ and $\dot{\mathbf{R}}_c(t_2)$.

We know from §4·2·6 that the angular momentum of p_1 relative to o prior to docking is a constant, which we may call $\mathbf{H}_1{}^o = m_1 r_1 \dot{\varphi}_1{}^2 \mathbf{u}_{n1}$, while that for p_2 is another constant, which we'll call $\mathbf{H}_2{}^o = m_2 r_2 \dot{\varphi}_2{}^2 \mathbf{u}_{n2}$. After docking, the system is equivalent to the single particle p, so once again the angular momentum (call it $\mathbf{H}^o$) remains constant. From Eq. (6·12), we can write

$$\mathbf{H}^o = \sum_{j=1}^{N} m_j \mathbf{R}_j \times \dot{\mathbf{R}}_j$$

We could substitute $\mathbf{R}_1 = \mathbf{R}_2 = \mathbf{R}_c(t_2)$ and $\dot{\mathbf{R}}_1 = \dot{\mathbf{R}}_2 = \dot{\mathbf{R}}_c(t_2)$ into this expression, using the value for $\dot{\mathbf{R}}_c(t_2)$ obtained from system linear momentum conservation. It is surely much easier, however, simply to observe from Eq. (6·15) that

$$\dot{\mathbf{H}}^o = \mathbf{M}^o - \mathbf{R}_c \times \mathfrak{M}\ddot{\mathbf{R}}_0 = 0$$

so that $\mathbf{H}^o$ is constant in time, even over the time interval of the docking maneuver. Thus we can write

$$\mathbf{H}^o(t_2) = \mathbf{H}^o(t_1) = \mathbf{H}_1{}^o + \mathbf{H}_2{}^o = m_1 r_1 \dot{\varphi}_1{}^2 \mathbf{u}_{n1} + m_2 r_2 \dot{\varphi}_2{}^2 \mathbf{u}_{n2}$$

to obtain the constant angular momentum $\mathbf{H}^o = \mathbf{H}^o(t_2)$.

6·3 WORK AND ENERGY

6·3·1 Work and kinetic energy

Another first integral obtained from $\mathbf{F} = m\ddot{\mathbf{R}}$ for a single particle in Chap. 4 is the *work-energy* equation, Eq. (4·42). Mathematical arguments parallel to those of §4·3·1 produce when applied to Eq. (6·7*a*) a result parallel to Eq. (4·42), namely,

$$\int_{t_1}^{t_2} \mathbf{F} \cdot \dot{\mathbf{R}}_c \, dt = \frac{\mathfrak{M}}{2} \dot{\mathbf{R}}_c(t_2) \cdot \dot{\mathbf{R}}_c(t_2) - \frac{\mathfrak{M}}{2} \dot{\mathbf{R}}_c(t_1) \cdot \dot{\mathbf{R}}_c(t_1) \tag{6·19}$$

Although the integral in Eq. (6·19) may be interpreted as the work done by the resultant force acting on the system mass center, and this integral is equated to the change in the kinetic energy of a particle having the total system mass $\mathfrak{M}$ and the location of the system mass center, still Eq. (6·19) cannot be said to relate the work done by system forces to the change in system kinetic energy. This is evident if we consider the "dumbbell" of Fig. 6·5, which portrays two particles separated by the fixed distance $2a$ and subjected to the equal and opposite forces $\mathbf{F}_1$ and $\mathbf{F}_2 = -\mathbf{F}_1$. The mass center c of this two-particle system is fixed in inertial space, so the resultant $\mathbf{F}$ of external forces is zero, and both sides of Eq. (6·19) disappear. Yet we know that the dumbbell of Fig. 6·5 revolves faster and faster under the forces $\mathbf{F}_1$ and $\mathbf{F}_2$, with consequent increase in the kinetic energy of each of the particles of the system. If we were to apply the particle work-energy equation [Eq. (4·42)] to each of the particles in Fig. 6.5 individually and add the results, we would obtain a work-energy equation for particle systems that would be quite different from Eq. (6·19).

Given a system of particles $p_1, \ldots, p_N$ having masses $m_1, \ldots, m_N$ and applied forces $\mathbf{F}_1, \ldots, \mathbf{F}_N$, we can write Eq. (4·42) for each particle and then sum the results, to provide

$$\sum_{j=1}^{N} \int_{t_1}^{t_2} \mathbf{F}_j \cdot \dot{\mathbf{R}}_j \, dt = \sum_{j=1}^{N} [\tfrac{1}{2} m_j \dot{\mathbf{R}}_j(t_2) \cdot \dot{\mathbf{R}}_j(t_2) - \tfrac{1}{2} m_j \dot{\mathbf{R}}_j(t_1) \cdot \dot{\mathbf{R}}_j(t_1)] \tag{6·20}$$

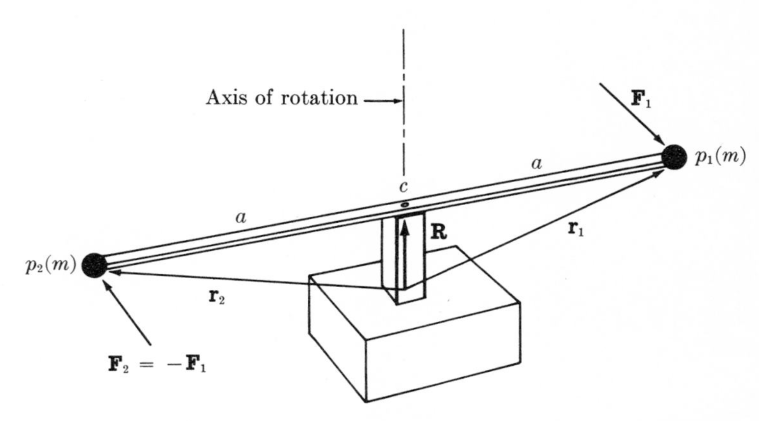

FIGURE 6·5

where $\mathbf{R}_j$ is the position vector of p_j relative to an inertially fixed point o. In order to extract from this equation a useful version of the work-energy equation for the system of particles, we separate each force $\mathbf{F}_j$ into two parts $\mathbf{f}_j$ and $\mathbf{f}'_j$, with the latter including all forces applied to particle p_j by other particles of the system, and also replace $\mathbf{R}_j$ by the vector sum $\mathbf{R}_c + \boldsymbol{\rho}_j$, where $\mathbf{R}_c$ is the position vector of the mass center c relative to o (see Fig. 6·6). With these substitutions, and reversal of the sequence of summation and integration, Eq. (6·20) becomes

$$\int_{t_1}^{t_2} \sum_{j=1}^{N} (\mathbf{f}_j + \mathbf{f}'_j) \cdot (\dot{\mathbf{R}}_c + \dot{\boldsymbol{\rho}}_j)\, dt = \frac{1}{2} \sum_{j=1}^{N} m_j (\dot{\mathbf{R}}_c + \dot{\boldsymbol{\rho}}_j) \cdot (\dot{\mathbf{R}}_c + \dot{\boldsymbol{\rho}}_j) \Big|_{t_1}^{t_2}$$

Since by Newton's third law the sum $\sum_{j=1}^{N} \mathbf{f}'_j$ is zero, and the vector $\dot{\mathbf{R}}_c$ can be removed from the summation, this equation becomes

$$\int_{t_1}^{t_2} \dot{\mathbf{R}}_c \cdot \sum_{j=1}^{N} \mathbf{f}_j\, dt + \int_{t_1}^{t_2} \sum_{j=1}^{N} (\mathbf{f}_j + \mathbf{f}'_j) \cdot \dot{\boldsymbol{\rho}}_j\, dt$$
$$= \left[\frac{1}{2} \dot{\mathbf{R}}_c \cdot \dot{\mathbf{R}}_c \sum_{j=1}^{N} m_j + \dot{\mathbf{R}}_c \cdot \sum_{j=1}^{N} m_j \dot{\boldsymbol{\rho}}_j + \frac{1}{2} \sum_{j=1}^{N} m_j \dot{\boldsymbol{\rho}}_j \cdot \dot{\boldsymbol{\rho}}_j \right] \Bigg|_{t_1}^{t_2}$$

The mass-center definition provides $\sum_{j=1}^{N} m_j \boldsymbol{\rho}_j = 0$, eliminating one term in this equation. With the total mass $\sum_{j=1}^{N} m_j$ represented by $\mathfrak{M}$, and the resultant force

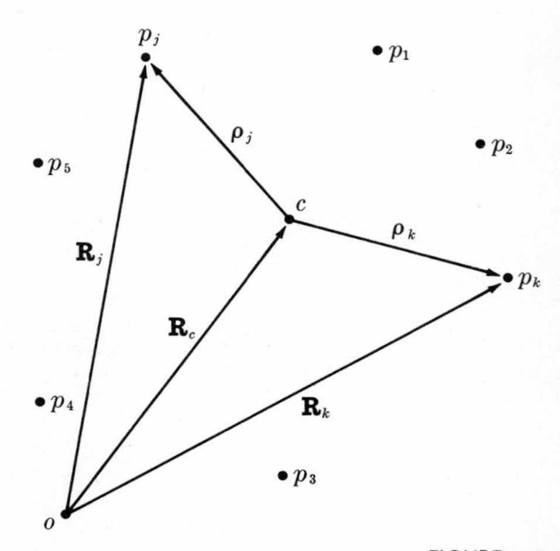

FIGURE 6·6

on the system represented by **F**, the work-energy equation for the system takes the form

$$\int_{t_1}^{t_2} \mathbf{F} \cdot \dot{\mathbf{R}}_c \, dt + \int_{t_1}^{t_2} \sum_{j=1}^{N} (\mathbf{f}_j + \mathbf{f}_j') \cdot \dot{\boldsymbol{\rho}}_j \, dt = \frac{1}{2} \mathfrak{M} \dot{\mathbf{R}}_c(t_2) \cdot \dot{\mathbf{R}}_c(t_2) - \frac{1}{2} \mathfrak{M} \dot{\mathbf{R}}_c(t_1) \cdot \dot{\mathbf{R}}_c(t_1) + \frac{1}{2} \sum_{j=1}^{N} m_j \dot{\boldsymbol{\rho}}_j(t_2) \cdot \dot{\boldsymbol{\rho}}_j(t_2) - \frac{1}{2} \sum_{j=1}^{N} m_j \dot{\boldsymbol{\rho}}_j(t_1) \cdot \dot{\boldsymbol{\rho}}_j(t_1) \qquad (6{\cdot}21)$$

The left side of Eq. (6·21) represents the work done by all forces in the system in the time interval $t_2 - t_1$, while the right side is the change in system kinetic energy in that interval.

When we compare Eqs. (6·19) and (6·21), we can see that all of the terms in the former are repeated in the latter; subtracting Eq. (6·19) from (6·21) thus yields a third form of the work-energy equation, specifically

$$\int_{t_1}^{t_2} \sum_{j=1}^{N} (\mathbf{f}_j + \mathbf{f}_j') \cdot \dot{\boldsymbol{\rho}}_j \, dt = \frac{1}{2} \sum_{j=1}^{N} m_j \dot{\boldsymbol{\rho}}_j(t_2) \cdot \dot{\boldsymbol{\rho}}_j(t_2) - \frac{1}{2} \sum_{j=1}^{N} m_j \dot{\boldsymbol{\rho}}_j(t_1) \cdot \dot{\boldsymbol{\rho}}_j(t_1) \qquad (6{\cdot}22)$$

While Eq. (6·21) offers the basic statement: "Work done by all forces equals change in system kinetic energy," Eqs. (6·19) and (6·22) provide the useful observations that both work and kinetic energy can be separated into two distinct parts, and a form of work-energy equation can be written for each part.

6·3·2 Potential energy

In §4·3·4, on the dynamics of a single particle, it was noted that certain forces called *conservative forces* can be expressed as the gradient of some potential function (written $\mathbf{F} = \nabla U$). The *potential energy* V of a particle subjected to a conservative force **F** was defined in terms of the potential function U by Eq. (4·66), such that

$$\mathbf{F} = -\nabla V \qquad (6{\cdot}23)$$

Generalization for a single particle to a system of N particles is completely straightforward when each of the particles is subjected to a force that is independent of the position of all other particles (as when the system of particles moves under the attraction of the earth's gravitational field). When conservative forces act between pairs of particles (as for the three spring-connected particles in Fig. 6·7), we face a slightly different aspect of potential energy than we find for one-particle systems. Whereas previously the conservative force depended only on the position of the single particle in the system, now it may depend on the *relative* position of one or more particles. For example, if the unstressed lengths of the three springs in Fig. 6·7 are each L, and Δ_{12}, Δ_{23}, and Δ_{31} are the amounts by which the three springs are extended, and each spring has a spring constant k, then the forces applied to p_1, p_2, and p_3 are, respectively,

$$\begin{aligned}\mathbf{F}_1 &= k\,\Delta_{12}\,\hat{\mathbf{u}}_{12} + k\,\Delta_{31}\,\hat{\mathbf{u}}_{13}\\ \mathbf{F}_2 &= k\,\Delta_{12}\,\hat{\mathbf{u}}_{21} + k\,\Delta_{23}\,\hat{\mathbf{u}}_{23}\\ \mathbf{F}_3 &= k\,\Delta_{23}\,\hat{\mathbf{u}}_{32} + k\,\Delta_{31}\,\hat{\mathbf{u}}_{31}\end{aligned}$$

where unit vectors are as defined by Fig. 6·7.

If we recall from §4·3·5 that the potential energy associated with an extension Δ of a linear spring with spring constant k is given by $V = \frac{1}{2}k\,\Delta^2$, then we might readily record the potential energy of the system of Fig. 6·7 as

$$V = \tfrac{1}{2}k\,\Delta_{12}^2 + \tfrac{1}{2}k\,\Delta_{23}^2 + \tfrac{1}{2}k\,\Delta_{31}^2$$

and indeed we could obtain any of the three forces just recorded by taking a local negative gradient of this expression for V, just as we did in Chap. 4. When only one particle is involved, however, the potential energy changes only with a change in the position of the particle, and we could think of V as the *potential energy of the particle*. Now the potential energy is associated not with the individual particles of the system, but with the *system itself*, since V changes only with changes in the relative distances among the particles.

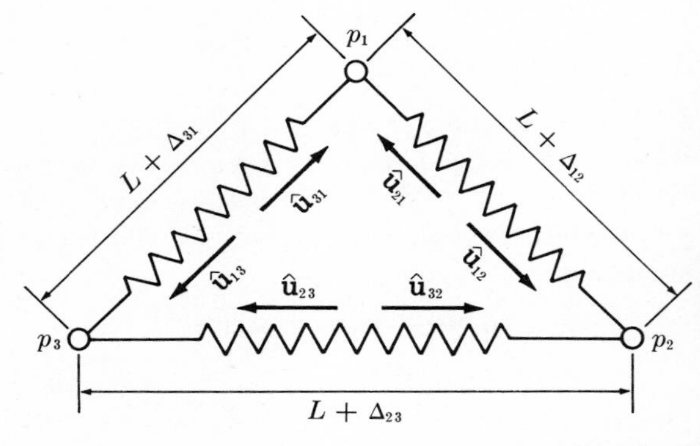

FIGURE 6·7

6·3·3 Equations of constraint

A system of N particles moving about without any constraints in three-dimensional space has $3N$ degrees of freedom; in other words, we must specify $3N$ independent scalar coordinates (we'll call them $x_1, \ldots, x_{3N}$) as functions of time in order to obtain a complete time history of the motions of the particles. As noted when considering a single particle in §4·3·6 and 4·3·7, we frequently wish to simplify the problem by imposing restrictions on the motions of the particles (insisting at the outset, for example, that all N particles will move about on the surface of a sphere). We have called such restrictions *constraints*, and we have called their mathematical expression *constraint equations*, or *equations of constraint*.

Constraint equations can take a variety of forms. The $3N$ scalars $x_1, \ldots, x_{3N}$ might be related by m scalar equations of the form

$$\begin{aligned} \mathfrak{F}_1(x_1, \ldots, x_{3N}, t) &= 0 \\ \mathfrak{F}_2(x_1, \ldots, x_{3N}, t) &= 0 \\ \cdots\cdots\cdots\cdots & \\ \mathfrak{F}_m(x_1, \ldots, x_{3N}, t) &= 0 \end{aligned} \tag{6·24}$$

For example, if x_1, x_2, and x_3 are cartesian coordinates of particle p_1, and x_4, x_5, and x_6 are cartesian coordinates of particle p_2, etc., and if all N particles of the system are constrained to remain on the surface of a sphere of radius R with center at the origin of the coordinate system, then the constraint equations would be

$$\begin{aligned} x_1^2 + x_2^2 + x_3^2 &= R^2 \\ x_4^2 + x_5^2 + x_6^2 &= R^2 \end{aligned}$$

and so on. To write these equations precisely in the form of Eq. (6·24), we simply relocate the R^2 terms, to obtain equations such as

$$\begin{aligned} x_1^2 + x_2^2 + x_3^2 - R^2 &= 0 \\ x_4^2 + x_5^2 + x_6^2 - R^2 &= 0 \end{aligned}$$

If the sphere were a balloon being inflated, we would replace the constant R by some explicit function of time, for example, by $R_0 + At$ if the radius were growing at a constant rate. It is in this fashion that time t might find its way into the constraint equations.

As noted in §4·3·7, when constraint equations have the form of Eq. (6·24), the constraints are referred to as *holonomic;* when the constraint equations *cannot* be written in the form of Eq. (6·24), the constraints are called *nonholonomic.* Two examples of nonholonomic constraints are cited in §4·3·7, where it was noted that the forms of Lagrange's equations provided there for a single particle are inapplicable when nonholonomic constraints are involved. In this chapter too we will establish useful results only for the special case of holonomic constraints. Special classes of nonholonomic constraints are dealt with in more advanced texts.

When the original set of $3N$ coordinates of a system of N particles are related by m *holonomic* constraints, then it is possible to use the m algebraic equations in Eq. (6·24) to solve for m of the original coordinates in terms of the remaining $3N - m$ coordinates. If we let $n \triangleq 3N - m$, then we can call the complete set of independent coordinates $q_1, \ldots, q_n$, and identify them as a set of independent *generalized coordinates*. These might be some set of n of the original coordinates $x_1, \ldots, x_{3N}$, or they might be chosen by inspection of the system. An example will be found in §6·3·5.

6·3·4 Principle of virtual work

We noted in §4·3·6 that it is occasionally instructive when dealing with a single particle to operate on $\mathbf{F} = m\ddot{\mathbf{R}}$ by dot-multiplying both sides by the quantity $\delta\mathbf{R}$, which represents a *virtual displacement*. If we are dealing with a system of N particles $p_1, \ldots, p_N$, with particle p_j identified by inertial position vector $\mathbf{R}_j$, mass m_j, and applied force $\mathbf{F}_j$, we can certainly again imagine a set of N displacements $\delta\mathbf{R}_1, \ldots, \delta\mathbf{R}_N$, and dot-multiply both sides of the corresponding equation of motion, to obtain

$$\mathbf{F}_j \cdot \delta\mathbf{R}_j = m_j\ddot{\mathbf{R}}_j \cdot \delta\mathbf{R}_j \qquad j = 1, \ldots, N$$

Summing these equations produces

$$\sum_{j=1}^{N} \mathbf{F}_j \cdot \delta\mathbf{R}_j = \sum_{j=1}^{N} m_j\ddot{\mathbf{R}}_j \cdot \delta\mathbf{R}_j \tag{6·25}$$

At this point the N vectors $\delta\mathbf{R}_1, \ldots, \delta\mathbf{R}_N$ are completely arbitrary, but we can most easily draw useful conclusions from Eq. (6·25) if we now restrict these virtual displacements in such a way as to make them compatible with any constraints we have imposed on the allowable motions of the particles. The N virtual-displacement vectors are then no longer independent since each of them must be related to the n generalized virtual displacements $\delta q_1, \ldots, \delta q_n$ by

$$\delta\mathbf{R}_j = \sum_{k=1}^{n} \frac{\partial\mathbf{R}_j}{\partial q_k}\,\delta q_k \qquad j = 1, \ldots, N \tag{6·26}$$

If $\delta\mathbf{R}_j$ in Eq. (6·25) is restricted as in Eq. (6·26), then in many problems we will find that some of the forces that combine to produce the resultant forces $\mathbf{F}_j (j = 1, \ldots, N)$ make no contribution to Eq. (6·25). Such forces are called *nonworking constraint forces* because (as we shall see) they are the forces that maintain the constraints, and because the left side of Eq. (6·25) represents the work done by all forces in the course of a virtual displacement. Although in time one comes to recognize nonworking constraint forces by inspection, the supreme test lies in the satisfaction of the requirement

$$\sum_{j=1}^{N} \mathbf{F}_j^c \cdot \frac{\partial \mathbf{R}_j}{\partial q_k} = 0 \qquad k = 1, \ldots, n \tag{6·27}$$

or equivalently [see Eq. (4·77)]

$$\sum_{j=1}^{N} \mathbf{F}_j^c \cdot \frac{\partial \dot{\mathbf{R}}_j}{\partial \dot{q}_k} = 0 \tag{6·28}$$

where $\mathbf{F}_1^c, \ldots, \mathbf{F}_N^c$ represent a system of nonworking constraint forces.

We can now recognize that if the virtual displacements are compatible with constraints, then nonworking constraint forces $\mathbf{F}_j^c (j = 1, \ldots, n)$ make no contribution, and Eq. (6·25) can be written as

$$\sum_{j=1}^{N} (\mathbf{F}_j - \mathbf{F}_j^c) \cdot \delta\mathbf{R}_j = \sum_{j=1}^{N} m_j \ddot{\mathbf{R}}_j \cdot \delta\mathbf{R}_j \tag{6·29}$$

It is often convenient to further distinguish between those forces remaining in Eq. (6·29) that are *internal* to the system of N particles (we'll call them $\bar{\mathbf{f}}_j'$) and those that are *external* (designated $\bar{\mathbf{f}}_j$). Then we can substitute

$$\mathbf{F}_j - \mathbf{F}_j^c = \bar{\mathbf{f}}_j + \bar{\mathbf{f}}_j' \tag{6·30}$$

into Eq. (6·29) to get

$$\sum_{j=1}^{N} \bar{\mathbf{f}}_j \cdot \delta\mathbf{R}_j = \sum_{j=1}^{N} (m_j \ddot{\mathbf{R}}_j - \bar{\mathbf{f}}_j') \cdot \delta\mathbf{R}_j \tag{6·31}$$

The left-hand side of Eq. (6·31) is the work done by external applied forces due to a virtual displacement, and this equation (or more often a restricted version of it) is known as the *principle of virtual work*. In combination with Eq. (6·26), this equation provides, for a system with n degrees of freedom, a complete set of n differential equations that do not involve any nonworking constraint forces. With the help of Eq. (4·77), these equations can be written as

$$\sum_{j=1}^{N} \bar{\mathbf{f}}_j \cdot \frac{\partial \dot{\mathbf{R}}_j}{\partial \dot{q}_\alpha} = \sum_{j=1}^{N} (m_j \ddot{\mathbf{R}}_j - \bar{\mathbf{f}}_j') \cdot \frac{\partial \dot{\mathbf{R}}_j}{\partial \dot{q}_\alpha} \qquad \alpha = 1, \ldots, n \tag{6·32}$$

It should be acknowledged at this point that the traditional concepts of virtual displacement and virtual work are not necessary to the derivation of

Eq. (6·32). It is quite sufficient (and more straightforward) simply to dot-multiply $\mathbf{F}_j = m_j\ddot{\mathbf{R}}_j$ by $\partial\dot{\mathbf{R}}_j/\partial\dot{q}_\alpha$ (for $j = 1, \ldots, N$) and add these N equations together, accomplishing this for each of the n values of α.

6·3·5 Application to a constrained dynamic system

Figure 6·8 illustrates a system consisting of two particles p_1 and p_2 constrained in such a way that p_1 can only slide back and forth along a horizontal line, and the distance between p_1 and p_2 is the constant length a of the pin-ended rod connecting them. Both particles are subjected to a uniform gravitational force in the direction of $\hat{\mathbf{u}}_2$, and a linear spring applies the force $-kx_1\hat{\mathbf{u}}_1$ to p_1, where x_1 is the spring extension. In addition, p_1 receives a force from the frictionless support on which it slides, and both p_1 and p_2 are subjected to forces by the massless rod connecting them. Both particles are constrained to move in the plane of the page.

If we attack this problem in an entirely mechanical way, we might set up a cartesian coordinate system with origin at o and axes parallel to $\hat{\mathbf{u}}_1$, $\hat{\mathbf{u}}_2$, and $\hat{\mathbf{u}}_3$. Then if $\mathbf{R}_1$ and $\mathbf{R}_2$ are position vectors from o to p_1 and p_2, respectively, they may be written in terms of cartesian coordinates as

$$\mathbf{R}_1 = x_1\hat{\mathbf{u}}_1 + x_2\hat{\mathbf{u}}_2 + x_3\hat{\mathbf{u}}_3$$
$$\mathbf{R}_2 = x_4\hat{\mathbf{u}}_1 + x_5\hat{\mathbf{u}}_2 + x_6\hat{\mathbf{u}}_3$$

We see immediately that not all of the six coordinates $x_1, \ldots, x_6$ can have

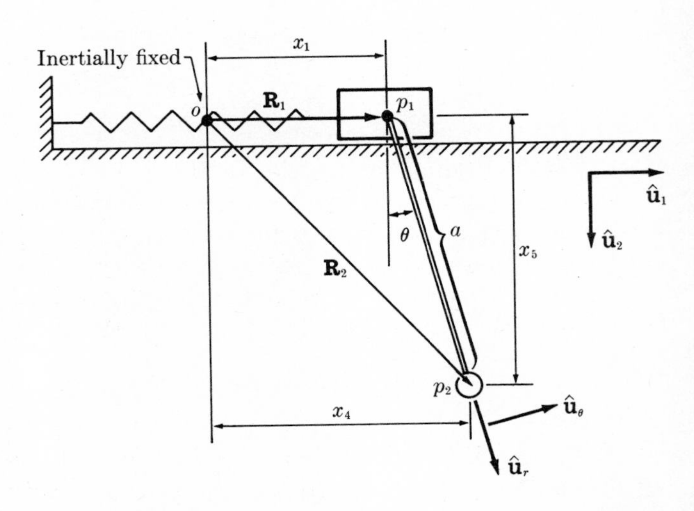

FIGURE 6·8

independent values. Specifically, we have

$$x_2 = 0 \qquad x_3 = 0 \qquad x_6 = 0$$

and, since p_1 and p_2 are a distance a apart,

$$(x_4 - x_1)^2 + (x_5 - 0)^2 - a^2 = 0$$

These four constraint equations are of the form of Eq. (6·24), so the constraints are holonomic. Since the number of degrees of freedom is $n = 6 - 4 = 2$, we should be able to use the constraint equations to reduce the number of independent variables to two. If we choose to retain x_1 and x_4 as our generalized coordinates, then we can replace x_5 by $\pm[a^2 - (x_4 - x_1)^2]$ and set x_2, x_3, and x_6 to zero, leaving only x_1 and x_4. We could then express our equations of motion in terms of these two variables.

Rather than proceed blindly and mechanically to adopt cartesian coordinates at the outset and use the constraint equations to select a set of generalized coordinates, we can with a little thought usually choose better generalized coordinates by inspection of a sketch of the system. In Fig. 6·8, for example, it is fairly clear that with the two independent variables x_1 and θ, we can fully define the system configuration, so these will also suffice as generalized coordinates. Whatever the choice, we will call them q_1 and q_2 for this two-degree-of-freedom problem, and proceed with the business of formulating equations of motion.

In order to apply the principle of virtual work to the system in Fig. 6·8, we must select a set of generalized coordinates and then identify the nonworking constraint forces. If we choose $q_1 \triangleq x_1$ and $q_2 \triangleq \theta$ as our generalized coordinates, then from Fig. 6·8 we can write

$$\mathbf{R}_1 = q_1\hat{\mathbf{u}}_1 \qquad \mathbf{R}_2 = q_1\hat{\mathbf{u}}_1 + a\hat{\mathbf{u}}_r$$

and

$$\dot{\mathbf{R}}_1 = \dot{q}_1\hat{\mathbf{u}}_1 \qquad \dot{\mathbf{R}}_2 = \dot{q}_1\hat{\mathbf{u}}_1 + a\dot{q}_2\hat{\mathbf{u}}_\theta$$

concluding that

$$\frac{\partial\dot{\mathbf{R}}_1}{\partial\dot{q}_1} = \frac{\partial\dot{\mathbf{R}}_2}{\partial\dot{q}_1} = \hat{\mathbf{u}}_1$$

and

$$\frac{\partial\dot{\mathbf{R}}_1}{\partial\dot{q}_2} = 0 \qquad \frac{\partial\dot{\mathbf{R}}_2}{\partial\dot{q}_2} = a\hat{\mathbf{u}}_\theta$$

Equation (6·28) then requires that the set of nonworking constraint forces for this system satisfy

$$\mathbf{F}_1{}^c \cdot \hat{\mathbf{u}}_1 + \mathbf{F}_2{}^c \cdot \hat{\mathbf{u}}_1 = 0$$

and

$$\mathbf{F}_2{}^c \cdot a\hat{\mathbf{u}}_\theta = 0$$

As we examine the forces applied to p_1 and p_2 in Fig. 6·8, we find we can classify them as a spring force, two gravity forces, and a set of constraint forces. Since the spring force on p_1 is in direction $\hat{\mathbf{u}}_1$, it fails the test above; the gravity force on p_2 has a component along $\hat{\mathbf{u}}_\theta$, and also fails. Neither the gravity force on p_1 nor the vertical support force on p_1 has a component along $\hat{\mathbf{u}}_1$, so these forces qualify formally as "nonworking constraint forces" (although the gravity force is not applied by the constraining surface). There remain for consideration the equal and opposite forces transmitted to p_1 and p_2 by the rigid massless rod connecting them. Since these forces are directed along the axis of the rod (direction $\pm\hat{\mathbf{u}}_r$), there is a component along $\hat{\mathbf{u}}_1$ (but not along $\hat{\mathbf{u}}_\theta$). Thus the force applied to p_1 by the rod *does* do work in the course of a virtual displacement δq_1, but that applied to p_2 by the rod does an equal and opposite amount of work, and for the total system there is no work contribution by the forces in the rod. In this problem all of the forces of constraint can therefore be classified as nonworking.

In applying Eq. (6·32) to obtain equations of motion, we can therefore restrict our attention to the spring force on p_1 and the gravity force on p_2, to conclude that

$$\bar{\mathbf{f}}_1 = -kq_1\hat{\mathbf{u}}_1 \qquad \bar{\mathbf{f}}_1' = 0$$
$$\bar{\mathbf{f}}_2 = m_2g\hat{\mathbf{u}}_2 \qquad \bar{\mathbf{f}}_2' = 0$$

Equation (6·32) also requires $\ddot{\mathbf{R}}_1$ and $\ddot{\mathbf{R}}_2$, which are available from $\dot{\mathbf{R}}_1$ and $\dot{\mathbf{R}}_2$ as

$$\ddot{\mathbf{R}}_1 = \ddot{q}_1\hat{\mathbf{u}}_1$$
$$\ddot{\mathbf{R}}_2 = \ddot{q}_1\hat{\mathbf{u}}_1 + a\ddot{q}_2\hat{\mathbf{u}}_\theta - a\dot{q}_2{}^2\hat{\mathbf{u}}_r$$

We can now use the vector partial derivatives obtained previously and write, from Eq. (6·32), the equations of motion

$$(-kq_1\hat{\mathbf{u}}_1)\cdot\hat{\mathbf{u}}_1 + (m_2g\hat{\mathbf{u}}_2)\cdot\hat{\mathbf{u}}_1 = m_1\ddot{q}_1\hat{\mathbf{u}}_1\cdot\hat{\mathbf{u}}_1 + m_2(\ddot{q}_1\hat{\mathbf{u}}_1 + a\ddot{q}_2\hat{\mathbf{u}}_\theta - a\dot{q}_2{}^2\hat{\mathbf{u}}_r)\cdot\hat{\mathbf{u}}_1$$

and

$$(-kq_1\hat{\mathbf{u}}_1)\cdot 0 + (m_2g\hat{\mathbf{u}}_2)\cdot a\hat{\mathbf{u}}_\theta = m_1\ddot{q}_1\hat{\mathbf{u}}_1\cdot 0 + m_2(\ddot{q}_1\hat{\mathbf{u}}_1 + a\ddot{q}_2\hat{\mathbf{u}}_\theta - a\dot{q}_2{}^2\hat{\mathbf{u}}_r)\cdot a\hat{\mathbf{u}}_\theta$$

In reduced scalar form, these two equations are

$$-kq_1 = (m_1 + m_2)\ddot{q}_1 + m_2(a\ddot{q}_2\cos q_2 - a\dot{q}_2{}^2\sin q_2)$$

and

$$-m_2ga\sin q_2 = m_2a(\ddot{q}_1\cos q_2 + a\ddot{q}_2)$$

To obtain these equations of motion by direct application of $\mathbf{F} = m\ddot{\mathbf{R}}$ would have been a much more tedious chore.

6·3·6 Lagrange's equations

In Chap. 4, we found that for a single particle with n degrees of freedom and generalized coordinates $q_1, \ldots, q_n$, equations of motion can be written in the form

$$\frac{d}{dt}\left(\frac{\partial T}{\partial \dot{q}_\alpha}\right) - \frac{\partial T}{\partial q_\alpha} = Q_\alpha \qquad \alpha = 1, \ldots, n \tag{6·33}$$

where T is the kinetic energy of the particle, and the generalized force Q_α is obtained in terms of the force $\mathbf{F}$ on the particle and the inertial velocity $\dot{\mathbf{R}}$ of the particle by

$$Q_\alpha = \mathbf{F} \cdot \frac{\partial \dot{\mathbf{R}}}{\partial \dot{q}_\alpha}$$

The only restriction imposed in Chap. 4 on Eq. (6·33) is the requirement that the coordinates $q_1, \ldots, q_n$ be a complete set of independent generalized coordinates. This restriction can always be satisfied if for any larger set of (redundant) coordinates (say $x_1, \ldots, x_3$ in the case of a single particle), relationships among these coordinates are in the form of *holonomic* constraint equations.

When the system consists of N particles rather than just one, Lagrange's equations in the form of Eq. (6·33) continue to apply, with precisely the same restrictions. Now we might have as many as $3N$ scalar coordinates at the outset (say $x_1, \ldots, x_{3N}$) but again any constraining relationships among them must have the form of holonomic constraint equations, as defined by Eq. (6·24), if Eq. (6·33) is to be applicable. Now the kinetic energy T in Eq. (6·33) is the sum of the kinetic energies of the individual particles, that is,

$$T \triangleq \sum_{j=1}^{N} \frac{1}{2} m_j \dot{\mathbf{R}}_j \cdot \dot{\mathbf{R}}_j \tag{6·34}$$

Similarly the generalized force Q_α becomes the sum

$$Q_\alpha \triangleq \sum_{j=1}^{N} \mathbf{F}_j \cdot \frac{\partial \dot{\mathbf{R}}_j}{\partial \dot{q}_\alpha} \tag{6·35}$$

Proof of the validity of Eq. (6·33) for a system of N particles is entirely parallel to that offered in Chap. 4 for a single particle (following either of the two proofs shown there). If we write $\mathbf{F}_j = m_j \ddot{\mathbf{R}}_j$ for each of the N particles, dot-multiply by $\partial \dot{\mathbf{R}}_j / \partial \dot{q}_\alpha$ for any value of α, and sum over the N particles, we have Q_α on one side of the equation [see Eq. (6·35)], and on the other side we have an expression that reduces to $d/dt\,(\partial T/\partial \dot{q}_\alpha) - \partial T/\partial q_\alpha$ by the same arguments advanced for the single particle in §4·3·7.

Exactly as in Chap. 4, we have the possibility of expressing Eq. (6·33) in the special form

$$\frac{d}{dt}\left(\frac{\partial L}{\partial \dot{q}_\alpha}\right) - \frac{\partial L}{\partial q_\alpha} = 0 \qquad \alpha = 1, \ldots, n \tag{6·36}$$

when all forces (other than nonworking constraint forces) are conservative. Here the lagrangian L is again defined in terms of kinetic energy T and potential energy V by

$$L \triangleq T - V \tag{6·37}$$

Among the primary advantages of Lagrange's equations is their automatic elimination of the unknown and generally unwanted nonworking constraint forces. These forces can always be eliminated from the vector equations obtained from Newton's second law too, but some ingenuity may be required to accomplish the necessary surgery on the equations.

A second important feature of Lagrange's equations is their invariance of form for a wide range of coordinates. The scalars $q_1, \ldots, q_n$ can each be magnitudes of single-particle translations or particle-system rotations, or an individual q_α can describe the magnitude of a displacement in which all particles of the system participate. This feature is illustrated in the second example to follow, but a complete appreciation of the importance of this aspect of Lagrange's equations must await a more advanced treatment of the subject. Once again, it is possible to formulate the equations of motion of N particles by using Newton's second law directly, and *then* introduce a transformation to any coordinates $q_1, \ldots, q_n$ that are of interest. This approach may require some careful thought, however, and each problem would require fresh consideration; Lagrange's equations in contrast are the same for *any* problem with holonomic constraints, whatever choice is made for the generalized coordinates.

To summarize, Lagrange's equations have an advantage in that they provide a procedure for *automatically* formulating equations of motion of systems with holonomic constraints in terms of arbitrary generalized coordinates, without introducing nonworking constraint forces. This automatic feature of this approach is often unimportant to the skillful analyst, since for a given problem he may well accomplish these objectives more efficiently with a more flexible approach, using Newton's second law. The automatic feature of Lagrange's equations is more important to the inexperienced analyst, or (more significantly) to the programmer who is trying to teach the computer to derive the equations. As digital computers become more proficient in symbolic operations (as opposed to numerical operations), we may expect Lagrange's equations to gain an even greater importance in engineering practice.

Lagrange's equations may be extended to systems with nonholonomic constraint equations, but this extension is beyond the proper scope of this book.

The two examples following illustrate some of the features of Lagrange's equations.

6·3·7 Applications of Lagrange's equations

First example The two-particle system shown in Fig. 6·8 was analyzed in §6·3·5; equations of motion were obtained by using the principle of virtual work as represented by Eq. (6·32). To apply Lagrange's equations in the form of Eq. (6·33), we need the kinetic energy

$$T = \tfrac{1}{2}m_1\dot{\mathbf{R}}_1 \cdot \dot{\mathbf{R}}_1 + \tfrac{1}{2}m_2\dot{\mathbf{R}}_2 \cdot \dot{\mathbf{R}}_2$$

With $\dot{\mathbf{R}}_1$ and $\dot{\mathbf{R}}_2$ from §6·3·5, the kinetic energy becomes

$$T = \tfrac{1}{2}m_1\dot{q}_1^2 + \tfrac{1}{2}m_2(\dot{q}_1^2 + a^2\dot{q}_2^2 + 2a\dot{q}_1\dot{q}_2 \cos q_2)$$

We also require Q_1 and Q_2 as provided by Eq. (6·35). For this problem, we have previously shown (§6·3·5) that certain forces applied to p_1 and p_2 are nonworking constraint forces, which make no contribution to Q_1 or Q_2. Excluding these forces from consideration, and using the partial derivatives from §6·3·5, we find

$$Q_1 = (-kq_1\hat{\mathbf{u}}_1) \cdot \hat{\mathbf{u}}_1 + (m_2g\hat{\mathbf{u}}_2) \cdot \hat{\mathbf{u}}_1 = -kq_1$$

and

$$Q_2 = (-kq_1\hat{\mathbf{u}}_1) \cdot 0 + (m_2g\hat{\mathbf{u}}_2) \cdot a\hat{\mathbf{u}}_\theta = -m_2ga \sin q_2$$

Thus Lagrange's equation [Eq. (6·33)] provides

$$\frac{d}{dt}(m_1\dot{q}_1 + m_2\dot{q}_1 + m_2a\dot{q}_2 \cos q_2) = -kq_1$$

and

$$\frac{d}{dt}(m_2a^2\dot{q}_2 + m_2a\dot{q}_1 \cos q_2) + m_2a\dot{q}_1\dot{q}_2 \sin q_2 = -m_2ga \sin q_2$$

In expanded form, these equations are

$$(m_1 + m_2)\ddot{q}_1 + m_2a(\ddot{q}_2 \cos q_2 - \dot{q}_2^2 \sin q_2) = -kq_1$$

and

$$m_2a(a\ddot{q}_2 + \ddot{q}_1 \cos q_2) = -m_2ga \sin q_2$$

These are exactly the same equations as those obtained in §6·3·5 by using Eq. (6·32), which was derived by the method of virtual work (although subsequently noted to be available without recourse to the principle of virtual work).

In comparing the labors of applying these two methods to the problem of Fig. 6·8, we find little to distinguish between them. Equation (6·32) requires the accelerations $\ddot{\mathbf{R}}_1$ and $\ddot{\mathbf{R}}_2$, necessitating more vector differentiation than does the lagrangian approach, but the latter involves more extensive scalar differentiations. In this respect this problem is typical.

Second example Figure 6·9 shows two particles p_1 and p_2, of equal mass m, both constrained to move within the confines of a frictionless tube revolving in a horizontal plane with variable but specified angular velocity $\Omega\hat{\mathbf{u}}_3$. The particles are connected by a spring with spring constant k and undeformed length $2a$, and each particle is also attached to one end of the tube by a linear spring with spring constant $2k$ and undeformed length b. We define $\hat{\mathbf{u}}_1$ and $\hat{\mathbf{u}}_2$ as the unit vectors shown in the figure to be fixed relative to the tube, and let $x\hat{\mathbf{u}}_1$ and $y\hat{\mathbf{u}}_1$ be the displacements of p_1 and p_2, respectively, within the tube from the points at which their springs are undeformed. The position vectors $\mathbf{R}_1$ and $\mathbf{R}_2$ of p_1 and p_2 relative to the stationary midpoint c are given by

$$\mathbf{R}_1 = (a + x)\hat{\mathbf{u}}_1 \qquad \mathbf{R}_2 = (-a + y)\hat{\mathbf{u}}_1$$

The corresponding inertial velocities are

$$\dot{\mathbf{R}}_1 = \dot{x}\hat{\mathbf{u}}_1 + (a + x)\Omega\hat{\mathbf{u}}_3 \times \hat{\mathbf{u}}_1 = \dot{x}\hat{\mathbf{u}}_1 + (a + x)\Omega\hat{\mathbf{u}}_2$$

and

$$\dot{\mathbf{R}}_2 = \dot{y}\hat{\mathbf{u}}_1 + (-a + y)\Omega\hat{\mathbf{u}}_2$$

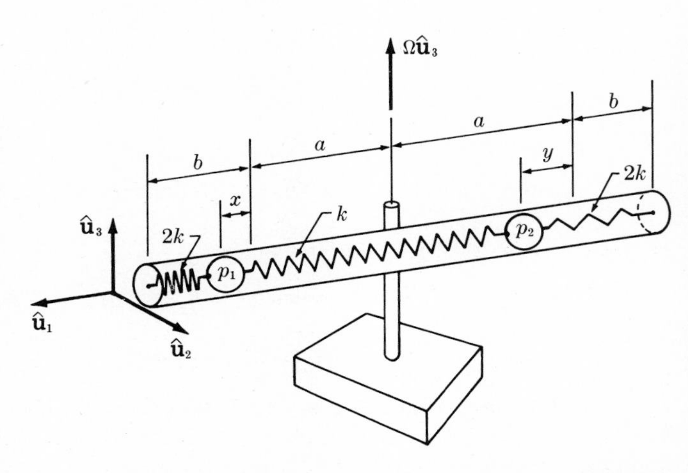

FIGURE 6·9

The system kinetic energy is

$$
\begin{aligned}
T &= \tfrac{1}{2}m\dot{\mathbf{R}}_1 \cdot \dot{\mathbf{R}}_1 + \tfrac{1}{2}m\dot{\mathbf{R}}_2 \cdot \dot{\mathbf{R}}_2 \\
&= \tfrac{1}{2}m[\dot{x}^2 + (a + x)^2\Omega^2 + \dot{y}^2 + (-a + y)^2\Omega^2] \\
&= \tfrac{1}{2}m[\dot{x}^2 + \dot{y}^2 + 2a^2\Omega^2 + 2a(x - y)\Omega + (x^2 + y^2)\Omega^2]
\end{aligned}
$$

Particles p_1 and p_2 are subjected to constraint forces by the tube walls, but we might wonder if these are *nonworking* constraint forces. Certainly they do do work on the particles, and thereby invest them with kinetic energy; but they do *not* do work in the course of virtual displacements compatible with constraints (displacements along the tube), so they do qualify as nonworking constraint forces. An example very much like this one was treated in detail in §4·3·8 (see Fig. 4·11).

Thus, with the exception of nonworking constraint forces, only the conservative spring forces are applied to p_1 and p_2. These we can represent with the potential energy

$$V = \tfrac{1}{2}(2kx^2) + \tfrac{1}{2}(2ky^2) + \tfrac{1}{2}k(x - y)^2 = \tfrac{3}{2}kx^2 + \tfrac{3}{2}ky^2 - kxy$$

There is now no obstacle to the application of Lagrange's equations for conservative holonomic systems [Eq. (6·36)], with $q_1 \triangleq x$ and $q_2 \triangleq y$. In order to illustrate the versatility of the lagrangian approach, however, we will *instead* introduce the new coordinates

$$q_1 \triangleq x - y \qquad \text{and} \qquad q_2 \triangleq x + y$$

In terms of q_1 and q_2, the lagrangian is

$$
\begin{aligned}
L \triangleq T - V &= \tfrac{1}{2}m[\tfrac{1}{2}(\dot{q}_1^2 + \dot{q}_2^2) + 2a^2\Omega^2 + 2a\Omega q_1 + \tfrac{1}{2}\Omega^2(q_1^2 + q_2^2)] \\
&\qquad - \tfrac{1}{2}(2kq_1^2) - \tfrac{1}{2}kq_2^2 \\
&= \tfrac{1}{4}m(\dot{q}_1^2 + \dot{q}_2^2) + \tfrac{1}{4}(m\Omega^2 - 2k)q_2^2 + \tfrac{1}{4}(m\Omega^2 - 4k)q_1^2 \\
&\qquad + ma\Omega q_1 + ma^2\Omega^2
\end{aligned}
$$

Equation (6·36) then provides, for $\alpha = 1$,

$$\tfrac{1}{2}m\ddot{q}_1 - \tfrac{1}{2}(m\Omega^2 - 4k)q_1 = ma\Omega$$

and for $\alpha = 2$,

$$\tfrac{1}{2}m\ddot{q}_2 + \tfrac{1}{2}(m\Omega^2 - 2k)q_2 = 0$$

The special property of these equations (as opposed to those that would have resulted from the choice $q_1 \triangleq x$ and $q_2 \triangleq y$) is in their uncoupled form; here each equation involves one coordinate only. A peculiar feature of the coordinates q_1 and q_2 actually employed in the preceding formulation lies in the fact that *both* particles participate in a displacement of *either* generalized coordinate. This puts q_1 and q_2 in a class sometimes referred to as *distributed coordinates*. We can introduce distributed coordinates in a direct newtonian formulation only after we have obtained scalar equations of motion in terms of some *discrete* coordinates of translation or rotation (such as x and y in this example). With a lagrangian approach, we can introduce distributed coordinates *before* we derive any equations of motion.

6·4 PARTICLE SYSTEMS IN STATIC EQUILIBRIUM

6·4·1 Principle of virtual work in statics

A version of the principle of virtual work much more commonly used in engineering practice than Eq. (6·31) is the special case in which $\ddot{\mathbf{R}}_j = 0$ for $j = 1, \ldots, N$, that is, the special case of the *statics* problem. Then Eq. (6·31) becomes

$$\sum_{j=1}^{N} \bar{\mathbf{f}}_j \cdot \delta\mathbf{R}_j = -\sum_{j=1}^{N} \bar{\mathbf{f}}'_j \cdot \delta\mathbf{R}_j \tag{6·38}$$

which may be verbalized as follows: *For a system of particles at rest in inertial space, the work done during a virtual displacement by the external applied forces is the negative of that done by the internal forces.* This then becomes a necessary and sufficient condition for static equilibrium; it may be simpler to apply than the direct method of finding equilibrating forces (as proposed in Chap. 2) because of the elimination from consideration of nonworking constraint forces. Whereas the methods of Chap. 2 dealt with a group of particles in static equilibrium individually, now we have a tool for dealing collectively with all the particles of a system. The use of this result will be illustrated in subsequent articles, after some implications of Eq. (6·38) in special cases have been noted.

The forces $\bar{\mathbf{f}}_j$ and $\bar{\mathbf{f}}'_j$ in Eqs. (6·32) and (6·38) may or may not be *conservative* forces. If the internal forces $\bar{\mathbf{f}}'_j$ are conservative (as when the particles are interconnected by elastic springs), then each may be expressed as the negative local gradient $\nabla\big|_{p_j}$ of a potential energy function V', that is, the gradient of V' at the location of particle p_j, or

$$\bar{\mathbf{f}}'_j = -\nabla\big|_{p_j} V'$$

Thus if in terms of inertially fixed unit vectors $\hat{\mathbf{i}}_1$, $\hat{\mathbf{i}}_2$, and $\hat{\mathbf{i}}_3$ we have the inertial position vector of p_j given by

$$\mathbf{R}_j = R^j{}_1\hat{\mathbf{i}}_1 + R^j{}_2\hat{\mathbf{i}}_2 + R^j{}_3\hat{\mathbf{i}}_3$$

then the quantity $-\bar{\mathbf{f}}'_j \cdot \delta\mathbf{R}_j$ becomes

$$-\left(-\nabla\Big|_{p_i} V'\right)\cdot\left(\sum_{\alpha=1}^{n}\frac{\partial\mathbf{R}_j}{\partial q_\alpha}\right)\delta q_\alpha = \left(\frac{\partial V'}{\partial R^j{}_1}\hat{\mathbf{i}}_1 + \frac{\partial V'}{\partial R^j{}_2}\hat{\mathbf{i}}_2 + \frac{\partial V'}{\partial R^j{}_3}\hat{\mathbf{i}}_3\right)\Big|_{p_i}$$

$$\cdot \sum_{\alpha=1}^{n}\left(\frac{\partial R^j{}_1}{\partial q_\alpha}\hat{\mathbf{i}}_1 + \frac{\partial R^j{}_2}{\partial q_\alpha}\hat{\mathbf{i}}_2 + \frac{\partial R^j{}_3}{\partial q_\alpha}\hat{\mathbf{i}}_3\right)\delta q_\alpha = \sum_{\alpha=1}^{n}\frac{\partial V'}{\partial q_\alpha}\Big|_{p_i}\delta q_\alpha = \delta V'\Big|_{p_i}$$

and Eq. (6·38) becomes (summing over all particles)

$$\sum_{j=1}^{N}\bar{\mathbf{f}}_j\cdot\delta\mathbf{R}_j = \sum_{j=1}^{N}\delta V'\Big|_{p_i} = \delta V' \tag{6·39}$$

Equation (6·39) says that *for an internally conservative system of particles to be at rest in inertial space, it is necessary and sufficient that the work done by external applied forces in a virtual displacement equals the corresponding change in internal potential energy.* The quantity V' is often referred to as the *strain energy* of an elastic system, and Eq. (6·39) is then succinctly verbalized with the statement that *for static equilibrium, the virtual work of external applied forces equals the virtual change in strain energy.* This is a very commonly used form of the principle of virtual work.

If the external applied forces $\bar{\mathbf{f}}_j$ in Eq. (6·39) are also conservative, then they too can be expressed as the negative local gradient of a potential energy function, that is, there exists some scalar V^e such that

$$\bar{\mathbf{f}}_j = -\nabla\Big|_{p_i} V^e$$

and arguments parallel to those just applied for $\bar{\mathbf{f}}'_j$ produce

$$-\delta V^e = \delta V'$$

or

$$\delta(V^e + V') = 0 \tag{6·40}$$

as a special case of Eq. (6·39). Thus if we have a system for which all forces (internal and external) are conservative, except for any nonworking constraint forces, then the *total* potential energy $V = V^e + V'$ experiences no change in value when the system is subjected to a virtual displacement from its state of rest in inertial space.

In order to apply Eq. (6·40), it is most convenient to select a set of generalized coordinates $q_1, \ldots, q_n$, and then to use the expansion

$$\delta V = \sum_{\alpha=1}^{n}\frac{\partial V}{\partial q_\alpha}\delta q_\alpha$$

together with the independence of the generalized virtual displacements $\delta q_1, \ldots,$ δq_n. This permits Eq. (6·40) to be replaced by the set of n algebraic equations

$$\frac{\partial V}{\partial q_\alpha} = \frac{\partial (V^e + V')}{\partial q_\alpha} = 0 \qquad \alpha = 1, \ldots, n \tag{6·41}$$

to be solved for the n unknowns $q_1, \ldots, q_n$ that establish the state of static equilibrium, or rest in inertial space.

6·4·2 Stability of static equilibrium

It may be helpful in interpreting the character of the static-equilibrium states established by Eq. (6·41) to imagine a geometric representation of V as a surface in the n-dimensional space defined by $q_1, \ldots, q_n$. This is easily visualized only when n is 1 or 2, as in Fig. 6·10*a*, *b*, and *c*. In Fig. 6·10*a*, $n = 1$, and a plot of V versus the single coordinate q is easily constructed as soon as $V(q)$ is obtained. It is apparent that $\partial V/\partial q = 0$ at four locations (called *stationary points of* V) in this example, corresponding to four values of q (labeled $q^{*(1)}, \ldots, q^{*(4)}$) for which static equilibrium is possible. More information can be gleaned from Fig. 6·10*a* if we note that of the four stationary points, $q^{*(2)}$ represents a local *minimum* of $V(q)$; $q^{*(3)}$ is a local *maximum;* $q^{*(1)}$ is a *point of inflection;* and $q^{*(4)}$ is at the transition to a straight horizontal line. If we restrict our attention to systems for which $T + V$ is constant, then we can see that only when V is a mimimum (as for $q^{*(2)}$) can we be sure that if we place the system sufficiently near its equilibrium state (so $q \approx q^{*(2)}$ and $\dot{q} \approx 0$ in this example), it will remain for all time arbitrarily close to this state. The equilibrium motion is then termed *stable.* If V is not a minimum at a given equilibrium state, then even if we start the system very near that state, we face the possibility (depending on our starting values of q and $\dot{q}$) that either q or $\dot{q}$ or both will depart from the given equilibrium state by some amount greater than the arbitrary limits we may impose. The equilibrium motion is then called *unstable.*

The extension to systems with more than one coordinate q is straightforward conceptually, if difficult to visualize geometrically. Figure 6·10*b* and *c* portrays two views of a plot of V versus q_1 and q_2 in the special case $n = 2$; in the second of these views, the q_1q_2 plane is the plane of the paper and solid curves represent lines (*contours*) of constant value of V; the plus sign denotes a maximum value and the minus sign a minimum value of V. Again, stability of equilibrium occurs only when V is minimum for systems with $T + V =$ constant.

A geometrical interpretation of the requirement that V be minimum for stability is conceptually appealing when n is 1 or 2, but awkward for $n > 2$; moreover, it is never easy to use actually to assess stability. As a more useful algebraic equivalent, we can say that, for a system with $T + V$ constant, it is

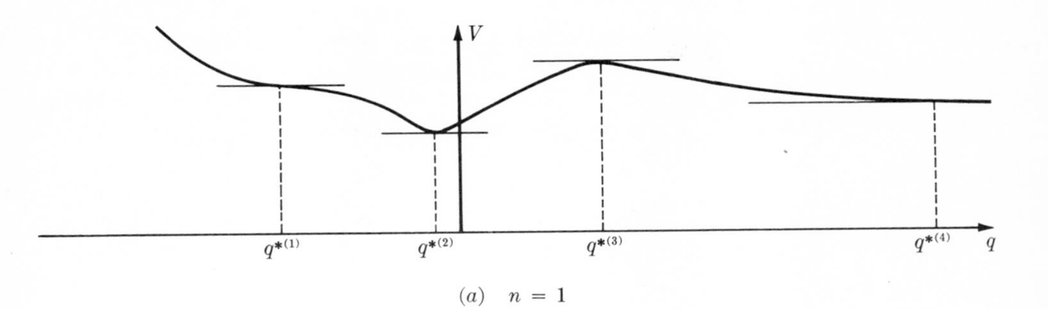

(a) $n = 1$

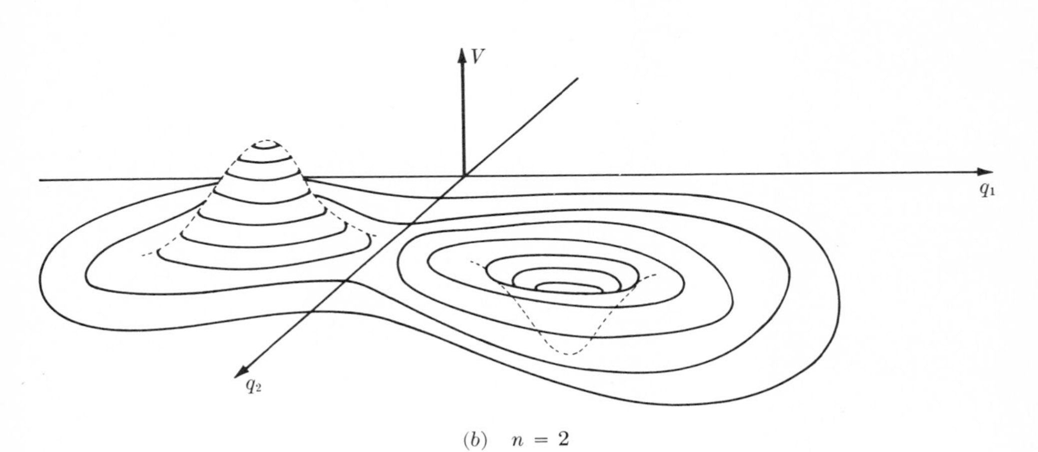

(b) $n = 2$

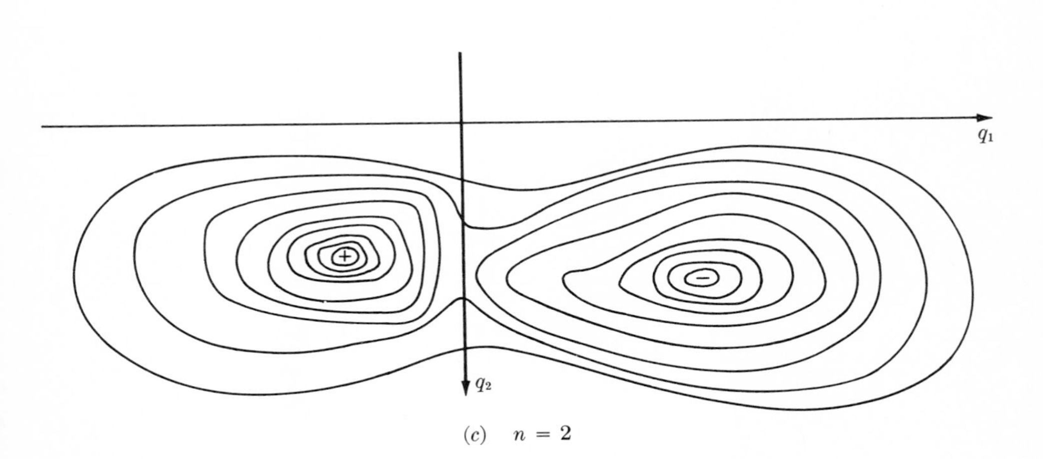

(c) $n = 2$

FIGURE 6·10

necessary and sufficient for stability of a state of static equilibrium that the matrix $[K]$ whose element in row α and column β is $K_{\alpha\beta} = \partial^2 V/\partial q_\alpha \partial q_\beta$ have the property called *positive definiteness.* The matrix $[K]$ is positive definite if and only if all of its principal axis minor determinants are positive, so that

$$K_{11} > 0 \qquad \begin{vmatrix} K_{11} & K_{12} \\ K_{21} & K_{22} \end{vmatrix} > 0, \qquad , \ldots, \qquad |K| > 0 \tag{6·42}$$

6·4·3 Equilibrium and stability of a two-particle system

Figure 6·11 portrays two particles p_1 and p_2 connected by a spring with spring constant k and constrained to remain on the perimeter of a circle of radius a, while subjected to gravity forces. The particle p_1 may be imagined to be a slug sliding inside a frictionless circular tube, while p_2 is a ring sliding without friction on the outside of the circular tube. The connecting spring is unstretched when p_1 and p_2 have coincident centers, so the distance between them is the extension Δ of the spring. As generalized coordinates, we choose the angles q_1 and q_2 shown in the figure for $q_1 > 0$ and $q_2 > 0$. The first objective is to find values of q_1 and q_2 for which static equilibrium is satisfied. Then we'll examine the stability of the various solutions. Both particles have the same mass m.

The nonworking constraint forces are those of the spring and gravity, and both are conservative. Accordingly, we can use the principle of virtual work in the form of Eq. (6·40) or (6·41). The latter equation simply establishes the

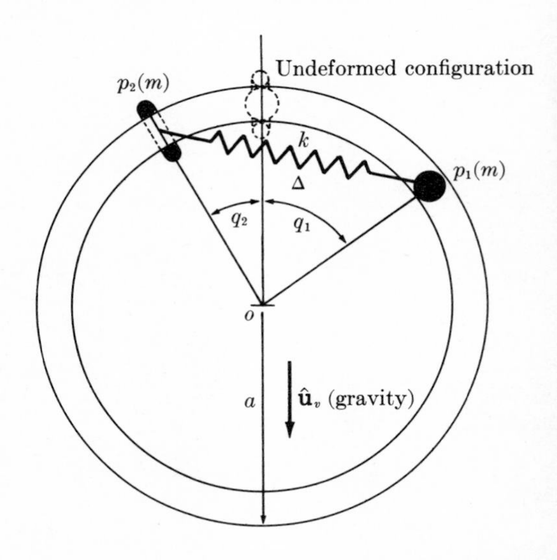

FIGURE 6·11

conditions for stationarity of the total potential energy V. We require only

$$\begin{aligned} V' &= \tfrac{1}{2}k\Delta^2 = \tfrac{1}{2}k[(a\cos q_2 - a\cos q_1)^2 + (a\sin q_2 + a\sin q_1)^2] \\ &= \tfrac{1}{2}ka^2(\cos^2 q_1 + \sin^2 q_1 + \cos^2 q_2 + \sin^2 q_2 - 2\cos q_1 \cos q_2 + 2\sin q_1 \sin q_2) \\ &= ka^2(1 - \cos q_1 \cos q_2 + \sin q_1 \sin q_2) \end{aligned}$$

and

$$V^e = mga\cos q_1 + mga\cos q_2$$

Thus the total potential energy is

$$V = mga(\cos q_1 + \cos q_2) + ka^2(1 - \cos q_1\cos q_2 + \sin q_1 \sin q_2)$$

According to Eq. (6·41), static-equilibrium configurations exist for angles q_1^* and q_2^* such that

$$\left.\frac{\partial V}{\partial q_1}\right|_{q_1^*,\, q_2^*} = -mga\sin q_1^* + ka^2(\sin q_1^* \cos q_2^* + \cos q_1^* \sin q_2^*) = 0$$

and

$$\left.\frac{\partial V}{\partial q_2}\right|_{q_1^*,\, q_2^*} = -mga\sin q_2^* + ka^2(\cos q_1^* \sin q_2^* + \sin q_1^* \cos q_2^*) = 0$$

These equations reveal the equilibrium solutions given by

$$\begin{aligned} q_1^* &= q_2^* = 0 \\ q_1^* &= q_2^* = \pi \end{aligned}$$

and $q_1^* = q_2^* =$ some constant q^* such that

$$-mga + 2ka^2\cos q^* = 0 \qquad \text{or} \qquad \cos q^* = \frac{mg}{2ka}$$

The last of these solutions exists only for $k > mg/2a$. These equilibrium solutions can also be found by assuming $q_1^* = q_2^*$ and equating the magnitudes of components of gravity force and spring force on one particle in the direction tangential to the tube at the particle location.

To evaluate the stability of these solutions, we require the second partial derivatives

$$\frac{\partial^2 V}{\partial q_1{}^2} = -mga\cos q_1 + ka^2(\cos q_1 \cos q_2 - \sin q_1 \sin q_2)$$

$$\frac{\partial^2 V}{\partial q_2{}^2} = -mga\cos q_2 + ka^2(\cos q_1 \cos q_2 - \sin q_1 \sin q_2)$$

$$\frac{\partial^2 V}{\partial q_1 \partial q_2} = \frac{\partial^2 V}{\partial q_2 \partial q_1} = ka^2(\cos q_1 \cos q_2 - \sin q_1 \sin q_2)$$

According to Eq. (6·42), it is necessary and sufficient for stability of equilibrium that

$$K_{11} \triangleq \left.\frac{\partial^2 V}{\partial q_1{}^2}\right|_{q_1^*,\, q_2^*} > 0$$

and

$$|[K]| \triangleq \begin{vmatrix} \left.\dfrac{\partial^2 V}{\partial q_1{}^2}\right|_{q_1^*,\, q_2^*} & \left.\dfrac{\partial^2 V}{\partial q_1 \partial q_2}\right|_{q_1^*,\, q_2^*} \\ \left.\dfrac{\partial^2 V}{\partial q_2 \partial q_1}\right|_{q_1^*,\, q_2^*} & \left.\dfrac{\partial^2 V}{\partial q_2{}^2}\right|_{q_1^*,\, q_2^*} \end{vmatrix} > 0$$

Thus, for the first static-equilibrium solution, we have $q_1^* = q_2^* = 0$, and

$$\begin{aligned} K_{11}{}^{(1)} &= -mga + ka^2 \\ |[K^{(1)}]| &= (-mga + ka^2)^2 - (ka^2)^2 \end{aligned}$$

Since the latter is negative, this solution is *unstable.*

For the second static-equilibrium solution, we have $q_1^* = q_2^* = \pi$, and

$$\begin{aligned} K_{11}{}^{(2)} &= mga + ka^2 \\ |[K^{(2)}]| &= (mga + ka^2)^2 - (ka^2)^2 \end{aligned}$$

Since both are positive, this solution is *stable.*

For the third static-equilibrium solution, we have

$$\cos q_1^* = \cos q_2^* = \cos q^* = \frac{mg}{2ka}$$

so that

$$\begin{aligned} K_{11}{}^{(3)} &= -mga \cos q^* + ka^2(\cos^2 q^* - \sin^2 q^*) \\ &= -mga \cos q^* + ka^2(2 \cos^2 q^* - 1) \\ &= -\frac{m^2 g^2}{2k} + \frac{m^2 g^2}{2k} - ka^2 = -ka^2 \end{aligned}$$

and

$$|[K^{(3)}]| = (-ka^2)^2 - \left(\frac{m^2 g^2}{2k} - ka^2\right)^2$$

Since $K_{11}{}^{(3)}$ is negative, this solution is also *unstable.*

6·4·4 Reaction-force determination using virtual work

Figure 6·12*a* shows a pin-jointed truss whose structural members are to be considered as inextensible rods (so the equilibrium positions of the particles at the joints are known). We require the forces applied to the truss by the wall (the *reaction forces*) at the pinned joint shown at A and the frictionless roller joint shown at B.

In constructing the free-body diagrams of the truss shown in Fig. 6·12*b*, we represent the force applied to the truss by the pin at A by $R_1\hat{\mathbf{u}}_1 + R_2\hat{\mathbf{u}}_2$, while that applied by the roller is simply $R_3\hat{\mathbf{u}}_2$. We are given the applied forces $P\hat{\mathbf{u}}_2$, $T\hat{\mathbf{u}}_2$, and $W_j\hat{\mathbf{u}}_1 (j = 1, \ldots, 6)$. Thus this problem has three scalar unknowns of interest, namely, R_1, R_2, and R_3, as well as eleven unknown and presently undesired member forces.

By the methods of Chap. 2 we would be obliged to consider the particles one by one, progressing from the free end of the truss in toward the supports, determining each of the member forces, until we reached points A and B; only then would we be able to determine the reactions. With the virtual-work approach,[1] we can find R_1, R_2, and R_3 without ever considering the member forces because the latter qualify as *nonworking constraint forces*. It might seem that the forces at A and B are *also* nonworking constraint forces, and you might well wonder how we can find any virtual displacement of this truss that is compatible with constraints. This is a proper concern, which can be dispelled only with a little bit of ingenuity.

Equation (6·38) is the most general expression of the principle of virtual work in statics. In the given problem, the particles are interconnected by rigid rods that constrain each of them to remain a fixed distance from all the others. As illustrated by the example in §6·3·5, the forces transmitted between the particles by the rods do no net work in the course of a virtual displacement compatible with the constraints imposed by the rods, so these are nonworking constraint forces. Since all internal forces are in this category, there is nothing left to serve for the vectors $\bar{\mathbf{f}}'_j$ in Eq. (6·38), and the right-hand side of this equation is zero.

If we are to use Eq. (6·38) to find R_1, R_2, and R_3, we must abandon the notion that Fig. 6·12*a* implies that particle p_7 is fixed to the wall and p_4 can move only vertically; in other words, we must ignore these constraints when we impose virtual displacements on the truss. Since virtual displacements are limited only by the imagination of the analyst, we are free to ignore any constraints we choose, as long as we retain a consistent interpretation of those forces classified as nonworking constraint forces.

Equation (6·38) is a scalar equation, and if we plan to use it to obtain algebraic equations to be solved for the three unknowns R_1, R_2, and R_3, then we must be prepared to use it three times, with three different virtual displacements. In choosing these displacements, it is necessary only that the constraints established by the rigid rods of the truss be respected.

Figure 6·12*c*, *d*, and *e* shows three virtual displacements that will serve

[1] We could do this job even more efficiently by using the methods of force and moment equilibrium, which are developed in Chap. 7.

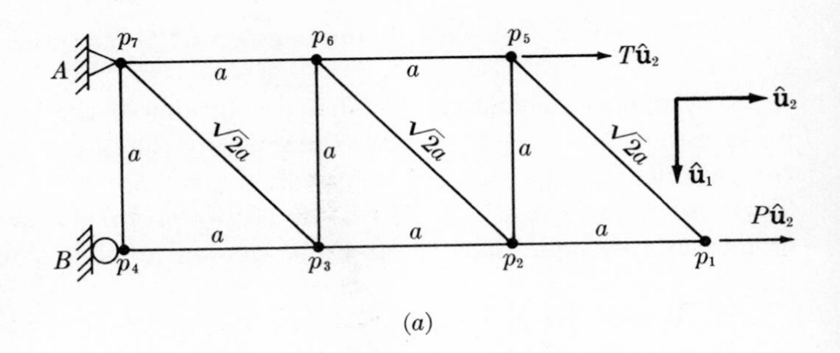

(a)

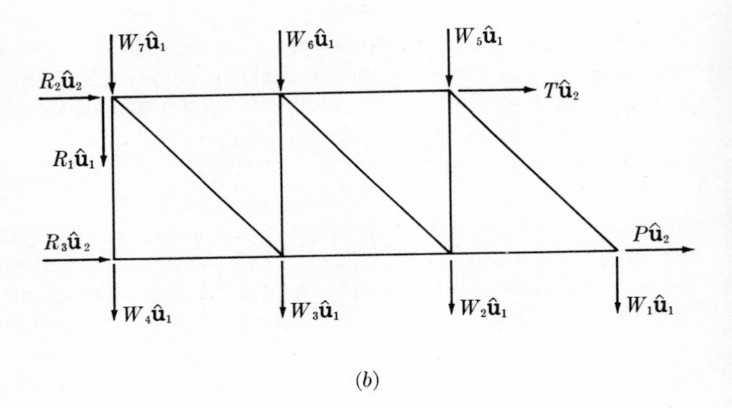

(b)

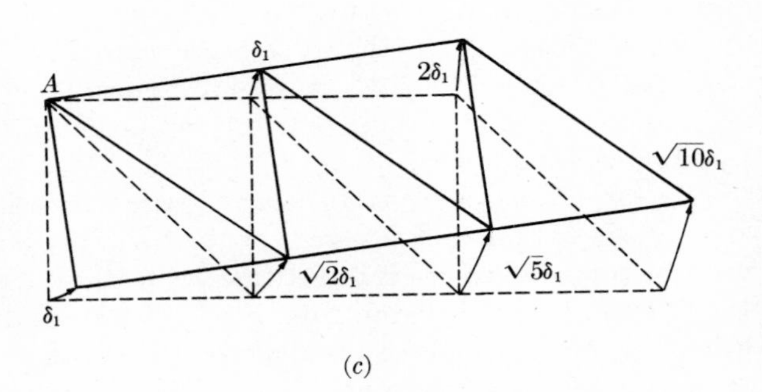

(c)

FIGURE 6·12

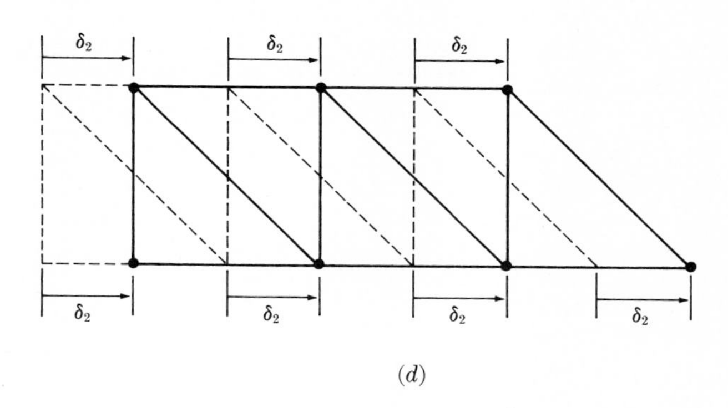

(d)

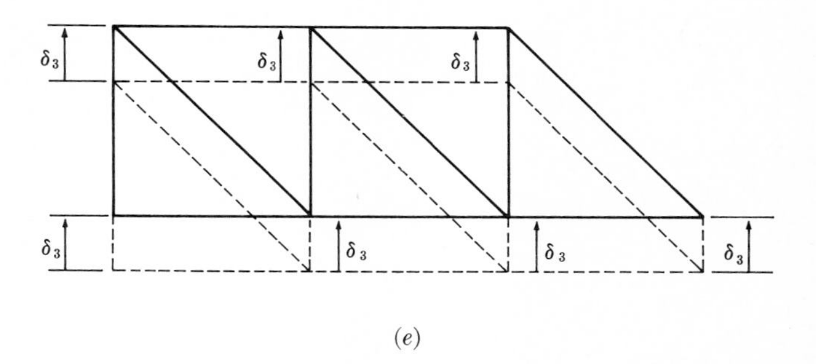

(e)

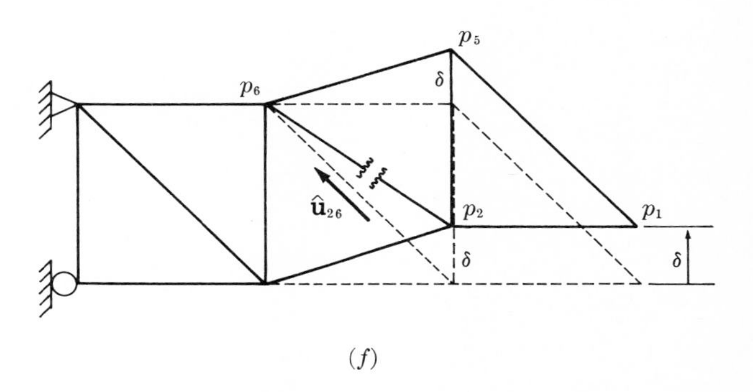

(f)

FIGURE 6·12

our needs. In Fig. 6·12c the truss is rotated about point A, so that p_4 moves a small distance δ_1, and each of the other particles of the system moves through an arc length proportioned to δ_1 as the particle's distance from A is proportioned to the distance a between A and p_4. If the virtual displacement is sufficiently small, the displacement of p_4 is essentially horizontal, and all of the arcs become straight-line segments. Equation (6·38) then becomes

$$(W_1\hat{\mathbf{u}}_1 + P\hat{\mathbf{u}}_2) \cdot \sqrt{10}\,\delta_1\left(-\frac{3}{\sqrt{10}}\hat{\mathbf{u}}_1 + \frac{1}{\sqrt{10}}\hat{\mathbf{u}}_2\right) + W_2\hat{\mathbf{u}}_1 \cdot \sqrt{5}\,\delta_1\left(-\frac{2}{\sqrt{5}}\hat{\mathbf{u}}_1 + \frac{1}{\sqrt{5}}\hat{\mathbf{u}}_2\right) + W_3\hat{\mathbf{u}}_1 \cdot \sqrt{2}\,\delta_1\left(\frac{-1}{\sqrt{2}}\hat{\mathbf{u}}_1 + \frac{1}{\sqrt{2}}\hat{\mathbf{u}}_2\right) + (W_4\hat{\mathbf{u}}_1 + R_3\hat{\mathbf{u}}_2) \cdot \delta_1\,\hat{\mathbf{u}}_2 + (W_5\hat{\mathbf{u}}_1 + T\hat{\mathbf{u}}_2) \cdot (-2\delta_1\,\hat{\mathbf{u}}_1) + W_6\hat{\mathbf{u}}_1 \cdot (-\delta_1\,\hat{\mathbf{u}}_1) + (W_7\hat{\mathbf{u}}_1 + R_1\hat{\mathbf{u}}_1 + R_2\hat{\mathbf{u}}_2) \cdot 0 = 0$$

or

$$-3W_1\delta_1 + P\delta_1 - 2W_2\delta_1 - W_3\delta_1 + R_3\delta_1 - 2W_5\delta_1 - W_6\delta_1 = 0$$

The common term δ_1 can be canceled and the resulting equation solved for R_3, to obtain

$$R_3 = P + 3W_1 + 2W_2 + W_3 + 2W_5 + W_6$$

The second virtual displacement, illustrated in Fig. 6·12d, is a horizontal displacement of the truss through an arbitrary distance δ_2. Equation (6·38) provides

$$P\delta_2 + T\delta_2 + R_3\delta_2 + R_2\delta_2 = 0$$

or

$$R_2 = -P - T - R_3 = -T - 3W_1 - 2W_2 - W_3 - 2W_5 - W_6$$

Finally, by introducing a vertical virtual displacement of magnitude δ_3, as in Fig. 6·12e, and applying Eq. (6·38), we find

$$W_1\delta_3 + W_2\delta_3 + W_3\delta_3 + W_4\delta_3 + W_5\delta_3 + W_6\delta_3 + W_7\delta_3 + R_1\delta_3 = 0$$

or

$$R_1 = -(W_1 + W_2 + W_3 + W_4 + W_5 + W_6 + W_7)$$

Having used the method of virtual work to find R_1, R_2, and R_3, we should try to see our results in a broader perspective than is afforded by the mechanical application of Eq. (6·38). It requires very little thought to recognize that the equations obtained from the virtual displacements illustrated in Fig. 6·12d and e are precisely those that would be obtained by applying $\mathbf{F} = \mathfrak{M}\mathbf{A}^{ci}$ to the total system of seven particles, for the special case in which the acceleration $\mathbf{A}^{ci}$ of the mass center c of the system in inertial space is zero [see Eq. (6·7b)]. It is less obvious that the equation obtained from the virtual displacement illustrated in Fig. 6·12c could alternatively be obtained from the special case of Eq. (6·15) in which the reference point q is point A and both the angular momentum $\mathbf{H}^q$ and the acceleration $\ddot{\mathbf{R}}_q$ are zero. In other words, we could as well have solved this problem by relying on the statements that the resultant moment about point A of all forces applied to the truss is zero, and the resultant force applied to the truss is zero.

6·4·5 Internal-force determination using virtual work

In applying the method of virtual work to find the reaction forces on the truss in Fig. 6·12 we imposed virtual displacements compatible with the constraints imposed by the rigid internal members of the truss, but we freely violated the constraints at the supports. You might prefer to imagine that we replaced the given problem with a new problem in which the supports at A and B were replaced by agencies of force sufficient to accomplish the required equilibrium, and then we introduced virtual displacements compatible with the constraints on the new problem. We can do the same thing if we are required to find the force in one of the internal members of the structure, and we do not wish to proceed as in Chap. 2, particle by particle, from the free end of the truss into the interior, calculating all member forces.

Imagine, for example, that we are required to obtain the force in the member from p_2 to p_6 of the truss shown in Fig. 6·12a. If we relax the constraint requiring that the distance between p_2 and p_6 be fixed, retaining all other constraints, then we can impose the virtual displacement illustrated in Fig. 6·12f, and again use Eq. (6·38).

If the force applied to p_2 by the member between p_2 and p_6 is represented by $\bar{\mathbf{f}}_2' = T\hat{\mathbf{u}}_{26}$, where $\hat{\mathbf{u}}_{26}$ is a unit vector shown in Fig. 6·12, and if δ is the magnitude of the virtual displacement shared by particles p_1, p_2, and p_5 (but by no other particles of the system), then Eq. (6·38) becomes

$$(W_1\hat{\mathbf{u}}_1 + P\hat{\mathbf{u}}_2) \cdot (-\delta\hat{\mathbf{u}}_1) + W_2\hat{\mathbf{u}}_1 \cdot (-\delta\hat{\mathbf{u}}_1) + (W_5\hat{\mathbf{u}}_1 + T\hat{\mathbf{u}}_2) \cdot (-\delta\hat{\mathbf{u}}_1) = -T\hat{\mathbf{u}}_{26} \cdot (-\delta\hat{\mathbf{u}}_1)$$

or

$$-(W_1 + W_2 + W_5)\delta = -T \cos 45° \, \delta$$

Canceling δ and substituting for cos 45° furnishes the tensile force in the chosen member as

$$T = \sqrt{2}\,(W_1 + W_2 + W_5)$$

This result is certainly simpler to obtain than it would have been had we proceeded with the methods of Chap. 2, treating each particle in turn until we worked our way in from p_1 through p_5 to p_2. But with a little thought we can see that even here the method of virtual work provides an answer we could have obtained as easily without it. Had we chosen to focus attention on the system consisting of particles p_1, p_2, and p_5 (cutting the truss with a vertical line intersecting the member from p_2 to p_6), then the application of Newton's second law to *this* system would provide, in the case of static equilibrium, the requirement that the resultant force applied to this system is zero. The resulting vector equation would also involve the unknown forces (call them $F_{23}\hat{\mathbf{u}}_2$ and $F_{56}\hat{\mathbf{u}}_2$) in the members from p_2 to p_3 and p_5 to p_6, respectively, but these unwanted forces could be eliminated by dot-multiplying the resultant force on the p_1, p_2, p_5 system by $\hat{\mathbf{u}}_1$. The resulting equation is

$$(W_1\hat{\mathbf{u}}_1 + P\hat{\mathbf{u}}_2 + W_2\hat{\mathbf{u}}_1 + T\hat{\mathbf{u}}_{26} + F_{23}\hat{\mathbf{u}}_2 + W_5\hat{\mathbf{u}}_1 + F_{56}\hat{\mathbf{u}}_2) \cdot \hat{\mathbf{u}}_1 = 0$$

or

$$W_1 + W_2 - \frac{T}{\sqrt{2}} + W_5 = 0$$

confirming the previous result.

There is a well-established name for the procedure just employed, whereby a portion of a truss is isolated and its equations of motion recorded; this is known as the *method of sections*. In contrast, the procedure of writing equilibrium equations for each of the particles at the truss connections (as in Chap. 2), has been identified as the *method of joints*.

6·4·6 Matrix formulation for small deflections of elastic trusses

As we attempt to apply the methods of Chap. 2 for finding static-equilibrium positions to problems of increasing complexity, we soon realize that unreasonable labors become involved as the size of the structural system grows. We must find a way to deal with the system as a whole, rather than focusing on its particles individually. Consider, for example, the pin-jointed truss of Fig. 6·13, which is only slightly more complex than the example of Fig. 2·14 analyzed in Chap. 2. Again a point of the truss is under load $W\hat{\mathbf{e}}_2$, and again we seek the displacement of that point. Now, however, we must recognize that each of the points labeled ①, ②, and ③ in Fig. 6·13 will displace when the load is applied, so now we have six unknowns instead of two. It becomes convenient to identify these unknown displacement components as x_1, x_2, x_3, x_4, x_5, and x_6, defined so that x_1 and x_2 are the displacement components of point ① in the directions of $\hat{\mathbf{e}}_1$ and $\hat{\mathbf{e}}_2$, respectively, x_3 and x_4 are the corresponding displacement components for point ②, and x_5 and x_6 are the corresponding displacement components for point ③ (see Fig. 6·13*b*).

There is no difficulty if we wish to record symbolically the vector equations for static equilibrium of points ①, ②, and ③. If we accept the characterization of the members of the truss as linear springs, we can easily write force-displacement relationships in the form $F_i = k_i\delta_i$, where F_i is the axial force, k_i is the spring constant, and δ_i is the elongation of the ith member. The great difficulty lies once again in the complexities of geometry. Even if we accept the restrictions of the small-deflection theory developed in Chap. 2, so that the lines of action of the member forces are assumed to parallel the lines of the truss prior to deformation, still we face geometrical difficulties. Somehow we must relate the elongations δ_i of each of the six loaded members of the truss to the six displacements $x_1, \ldots, x_6$. Although we would be satisfied with linear approximations of these relationships, even these are difficult to obtain for this truss. The problem lies in the fact that every displacement x_i may depend on every elongation $\delta_1, \ldots, \delta_6$. (Consider Fig. 6·13, and note, for example, that by elongating the member from ⑤ to ②, we move the remote point at ①.) Imagine the geometrical nightmare we would have if instead of Fig. 6·13 we had a three-dimensional truss with hundreds of members! We'll have to find a better way to handle this class of problems.

We are confronting the general task of expressing the forces applied to the truss points in terms of the unknown displacement components, such as $x_1, \ldots, x_6$. Our job is simplified by the assumption that the structural members are linearly elastic, and by our decision to recognize only linear terms in the displacements. Thus we know in advance that our result must be a system of linear algebraic equations. This is extremely important because it assures the validity of the *principle of superposition:* If we obtain two sets of solutions for the displacements under two separate loads, we know that the solution under

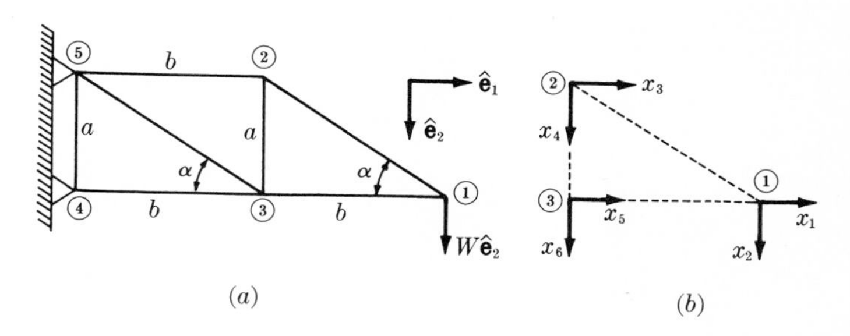

FIGURE 6·13

the two loads combined is just the sum of the solutions under the two loads individually. This makes it possible to obtain general expressions for the (linear) force-displacement relationships by considering the individual displacements one at a time and evaluating the forces applied to the surrounding joints as a result of each displacement. Then expressions for the forces applied to each joint when all displacements occur together can be found by addition, and these expressions can be substituted into the equilibrium equations. Finally, the linear equations for the unknown displacements can be solved.

A systematic method that utilizes the superposition principle is developed for the truss of Fig. 6·13, and then extended to a more general case. Imagine first that all displacement components except x_1 are zero, and that x_1 is unity. We seek the forces that would be applied to each joint of the system in order to accommodate this single displacement. The joints at ② and ③ are not permitted to move since this would make some values of x_i nonzero for $i \neq 1$. In consequence, loads can be induced only in the members actually connected with point ①. Recall that x_1 is a displacement of ① in the $\hat{\mathbf{e}}_1$ direction, so that if x_1 is unity then the member from ③ to ① must stretch by this amount. Since we know its structural properties (its spring constant), we know the load induced in the member from ③ to ① due to the unit displacement of x_1. If we consider Fig. 6·14, we can see that within the restrictions of the small-deflection theory, the elongation of the member from ① to ② due to unit value of x_1 is given by $\cos\alpha$. Thus the load in this member due to this unit displacement is also available.

Because the loads in members ① to ② and ① to ③ due to unit displacement x_1 are now known, we can readily interpret these as forces applied to joints ①, ②, and ③ due to this displacement. It is convenient to work with scalar components of these forces in the directions in which displacements are measured, namely, $\hat{\mathbf{e}}_1$ and $\hat{\mathbf{e}}_2$. Furthermore, we adopt symbols for these forces that involve the numerical indices of both the unit displacement and the direction of the force at the joint in question. For example, the force at joint ② in the direction of $\hat{\mathbf{e}}_1$ due to a unit displacement x_1 is labeled $-k_{31}$, and the force at this joint along $\hat{\mathbf{e}}_2$ induced by unit x_1 is labeled $-k_{41}$. This convention may be explained

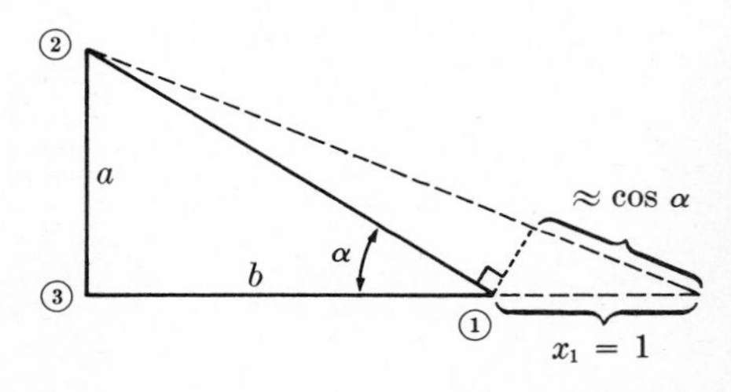

FIGURE 6·14

in terms of Fig. 6·13*b*, which identifies the displacement coordinates of joint ② as x_3 and x_4 in the directions of $\hat{\mathbf{e}}_1$ and $\hat{\mathbf{e}}_2$, respectively. With this labeling convention, we are prepared to say that we have in the foregoing discussion established procedures for finding k_{11}, k_{21}, k_{31}, k_{41}, k_{51}, and k_{61}. Of these values, only k_{61} is zero.

The procedure can now be repeated, taking this time $x_2 = 1$ and all other $x_i = 0$. The same two structural members are loaded, and this time we get k_{12}, k_{22}, k_{32}, k_{42}, k_{52}, and k_{62}.

Next we impose a unit displacement on x_3 and calculate the resulting loads in members ② to ⑤ and ② to ①. (The vertical member ② to ③ is not loaded by a sufficiently small horizontal displacement of ②.) This calculation gives nonzero values for k_{13}, k_{23}, k_{33}, and k_{43}, with zero values for k_{53} and k_{63}. It may now be recognized that certain of these calculations are redundant. For example, the force in member ① to ② is the same whether $x_1 = 1$ or $x_3 = 1$, so it must follow that $k_{13} = k_{31}$, and similarly for all other combinations of indices. (The generality of this conclusion is established in §6·4·7.)

Calculations proceed in the indicated manner until 36 values $k_{ij} (i, j = 1, \ldots, 6)$ are calculated. These scalars are called the *stiffness coefficients* of the structure.

By using the superposition principle we can relate forces at any joint to the displacements $x_1, \ldots, x_6$ quite readily in terms of stiffness coefficients. In addition to the external loads such as W in Fig. 6·13, there are forces due to elastic structural members, which may be expressed as linear combinations of products of stiffness coefficients and displacements. For the truss of Fig. 6·13, this procedure would result in the following set of algebraic equations:

$$\begin{aligned}
-k_{11}x_1 - k_{12}x_2 - k_{13}x_3 - k_{14}x_4 - k_{15}x_5 - k_{16}x_6 &= 0 \\
W - k_{21}x_1 - k_{22}x_2 - k_{23}x_3 - k_{24}x_4 - k_{25}x_5 - k_{26}x_6 &= 0 \\
-k_{31}x_1 - k_{32}x_2 - k_{33}x_3 - k_{34}x_4 - k_{35}x_5 - k_{36}x_6 &= 0 \\
-k_{41}x_1 - k_{42}x_2 - k_{43}x_3 - k_{44}x_4 - k_{45}x_5 - k_{46}x_6 &= 0 \\
-k_{51}x_1 - k_{52}x_2 - k_{53}x_3 - k_{54}x_4 - k_{55}x_5 - k_{56}x_6 &= 0 \\
-k_{61}x_1 - k_{62}x_2 - k_{63}x_3 - k_{64}x_4 - k_{65}x_5 - k_{66}x_6 &= 0
\end{aligned}$$

These scalar equations are represented in more convenient form as the matrix equation

$$[F] - [K][x] = 0$$

where $[x]$ is a column matrix with elements $x_1, \ldots, x_6$ representing the unknown displacements; $[K]$ is a 6×6 matrix of stiffness coefficients $k_{ij} (i, j = 1, \ldots, 6)$; and $[F]$ is a 6×1 matrix of applied forces, say $F_1, \ldots, F_6$. In this case, $F_2 = W$, and all other F_i are zero. As inferred in the calculation of the stiffness coefficients, the matrix $[K]$ is *symmetric*, that is, $k_{ij} = k_{ji}$.

The problem under consideration requires the determination of the displacement matrix $[x]$ under known applied forces in the matrix $[F]$. Solution of the matrix equation for $[x]$ is accomplished by premultiplying the equation by the *inverse matrix* of $[K]$, represented as $[K]^{-1}$. By definition, the inverse of a matrix $[K]$ is that matrix that when premultiplied or postmultiplied by $[K]$ gives the *identity matrix* $[U]$, that is,

$$[K]^{-1}[K] = [K][K]^{-1} = [U]$$

where, for the system of Fig. 6·13, $[U]$ is the 6×6 matrix

$$[U] \triangleq \begin{bmatrix} 1 & 0 & 0 & 0 & 0 & 0 \\ 0 & 1 & 0 & 0 & 0 & 0 \\ 0 & 0 & 1 & 0 & 0 & 0 \\ 0 & 0 & 0 & 1 & 0 & 0 \\ 0 & 0 & 0 & 0 & 1 & 0 \\ 0 & 0 & 0 & 0 & 0 & 1 \end{bmatrix}$$

Thus the solution of the matrix equation can be obtained by premultiplying by $[K]^{-1}$ to find

$$[x] = [K]^{-1}[F]$$

Although procedures for calculating the inverse of a matrix are described in Appendix B, in practice this tedious task is generally relegated to a computer, or else the elements of $[K]^{-1}$ are found by other procedures. Note that these elements may be interpreted as the displacements of individual joints in specified directions due to unit loads at other joints in other specified directions. If a_{ij} is an element of the matrix $[K]^{-1}$, then for the system of Fig. 6·13 we can, for example, interpret a_{35} as the horizontal displacement x_3 of joint ② due to a unit load at joint ③ in the direction of $\hat{\mathbf{e}}_1$ (paralleling x_5). The elements a_{ij} are called *flexibility coefficients*, and their matrix $[A] = [K]^{-1}$ is called the *flexibility matrix*. For the present, we can imagine the elements of $[A]$ to be obtained by inverting $[K]$.

As long as the small-deflection theory is acceptable, the matrix method presented here is (for trusses of typical dimension) preferable to the step-by-step derivation of the equations of equilibrium, force-displacement, and geometry. We should, however, not allow the simplicity of the matrix equation to obscure the fact that a good deal of labor is involved in calculating the stiffness coefficients k_{ij}, and then in inverting the matrix $[K]$. Both of these tasks are well suited for digital-computer operations, but they are not ideal for the manual solution of simple problems. The expression $[x] = [K]^{-1}[F]$ provides us (once $[K]^{-1}$ is calculated) with all of the displacements under any kind of load, and this may be much more information than we really need. In a particular application, we may be interested only in a particular displacement of one joint due to a single known load. In Fig. 6·13*a*, for example, we may require only the vertical deflection of the point to which the load is applied. For this purpose, special procedures can be devised, using extensions of some of the energy methods related to the principle of virtual work.

6·4·7 Maxwell's reciprocity laws

We have noted that when the internal forces in a system are conservative, we can associate with those forces a potential energy V' we call the *strain energy.* Among the very simplest of conservative forces is that transmitted by a linear spring, and we learned in §4·3·5 that if such a spring is extended an amount Δ, then the potential energy stored in the spring is $\frac{1}{2}k\Delta^2$, where k is the spring constant. Now we can conclude (as in the example of §6·3·2) that if we have a system of particles connected by S linear springs, then the strain energy must be

$$V' = \sum_{s=1}^{S} \frac{1}{2} k_s \Delta_s^{\,2} \tag{6·43}$$

This interpretation of V' is consistent with that introduced in §6·3·4, where it was noted that conservative internal forces (excluding any constraint forces) could be expressed in terms of strain energy by

$$\mathbf{f}_j' = -\nabla \Big|_{p_j} V' \tag{6·44}$$

The consistency of these expressions is readily apparent for the four-particle system shown in Fig. 6·15, since for this linear chain of spring-connected particles, the gradient $-\nabla \Big|_{p_j} V'$ becomes simply $-(\partial V'/\partial x_j)\,\hat{\mathbf{u}}$. In terms of the symbols defined in the figure, we have

$$\begin{aligned} V' &= \frac{1}{2} \sum_{s=1}^{3} k_s \Delta_s^{\,2} \\ &= \tfrac{1}{2}k_1(x_2 - x_1)^2 + \tfrac{1}{2}k_2(x_3 - x_2)^2 + \tfrac{1}{2}k_3(x_4 - x_3)^2 \end{aligned}$$

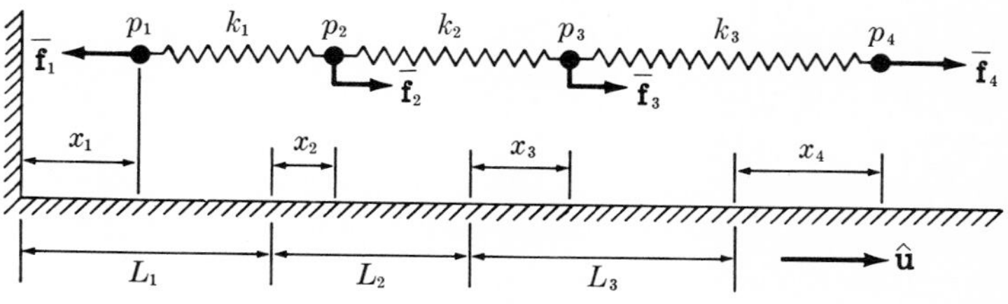

FIGURE 6·15

where it has been assumed that the dimensions shown as L_1, L_2, and L_3 in Fig. 6·15 represent the undeformed lengths of the three springs. Thus the internal forces on the four particles are

$$\begin{aligned}
\bar{\mathbf{f}}_1' &= -\frac{\partial V'}{\partial x_1}\hat{\mathbf{u}} = -k_1(x_2 - x_1)(-1)\hat{\mathbf{u}} = k_1(x_2 - x_1)\hat{\mathbf{u}} \\
\bar{\mathbf{f}}_2' &= -\frac{\partial V}{\partial x_2}\hat{\mathbf{u}} = -k_1(x_2 - x_1)(1)\hat{\mathbf{u}} - k_2(x_3 - x_2)(-1)\hat{\mathbf{u}} \\
&= -[k_1(x_2 - x_1) + k_2(x_2 - x_3)]\hat{\mathbf{u}} \\
\bar{\mathbf{f}}_3' &= -\frac{\partial V}{\partial x_3}\hat{\mathbf{u}} = -k_2(x_3 - x_2)(1)\hat{\mathbf{u}} - k_3(x_4 - x_3)(-1)\hat{\mathbf{u}} \\
&= -[k_2(x_3 - x_2) + k_3(x_3 - x_4)]\hat{\mathbf{u}} \\
\bar{\mathbf{f}}_4' &= -\frac{\partial V}{\partial x_4}\hat{\mathbf{u}} = -k_3(x_4 - x_3)(1)\hat{\mathbf{u}} = -k_3(x_4 - x_3)\hat{\mathbf{u}}
\end{aligned}$$

as we expect to be the case for linear spring forces.

In deriving Eq. (6·39), we noted that the *virtual* change in strain energy due to a *virtual* displacement of the particles of a system is given by

$$\delta V' = -\sum_{j=1}^{N} \bar{\mathbf{f}}_j' \cdot \delta\mathbf{R}_j$$

In contrast, we now see that the *actual* strain energy V' stored in a system whose particles are displaced from their unloaded state by amounts $\Delta\mathbf{R}_j$ is given by

$$V' = -\frac{1}{2}\sum_{j=1}^{N} \bar{\mathbf{f}}_j' \cdot \Delta\mathbf{R}_j \qquad (6·45)$$

where $\bar{\mathbf{f}}_j'$ is the internal force on p_j induced by the loads causing the deformations $\Delta\mathbf{R}_j$. The factor $\frac{1}{2}$ is introduced because the forces $\bar{\mathbf{f}}_j'(j = 1, \ldots, N)$ are proportional to the deformations, so that they build up gradually as the system is loaded, and not all of the final value $\bar{\mathbf{f}}_j'$ experiences the translation $\delta\mathbf{R}_j$. To confirm the presence of the factor $\frac{1}{2}$ in the special case illustrated in Fig. 6·15, we can substitute the internal forces and displacements to obtain

$$\begin{aligned}
V' &= -\tfrac{1}{2}(\bar{\mathbf{f}}_1' \cdot x_1\hat{\mathbf{u}} + \bar{\mathbf{f}}_2' \cdot x_2\hat{\mathbf{u}} + \bar{\mathbf{f}}_3' \cdot x_3\hat{\mathbf{u}} + \bar{\mathbf{f}}_4' \cdot x_4\hat{\mathbf{u}}) \\
&= -\tfrac{1}{2}[k_1(x_2 - x_1)x_1 - k_1(x_2 - x_1)x_2 + k_2(x_3 - x_2)x_2 \\
&\qquad - k_2(x_3 - x_2)x_3 + k_3(x_4 - x_3)x_3 - k_3(x_4 - x_3)x_4] \\
&= \tfrac{1}{2}[k_1(x_2 - x_1)^2 + k_2(x_3 - x_2)^2 + k_3(x_4 - x_3)^2] \\
&= \frac{1}{2}\sum_{s=1}^{3} k_s\Delta_s^{\,2}
\end{aligned}$$

confirming our previous expression for V' [Eq. (6·43)].

Although in our discussions of the system of Fig. 6·15 thus far we have said nothing about the role of the external applied forces $\bar{\mathbf{f}}_1$, $\bar{\mathbf{f}}_2$, $\bar{\mathbf{f}}_3$ and $\bar{\mathbf{f}}_4$ shown in the figure, it is evident both in this special problem and in the general case that $\bar{\mathbf{f}}_j = -\bar{\mathbf{f}}'_j$, so that the equality of strain energy and the work done by external forces is expressed as

$$\frac{1}{2}\sum_{j=1}^{N} \bar{\mathbf{f}}_j \cdot \Delta\mathbf{R}_j = V' \tag{6·46}$$

This expression should be contrasted with Eq. (6·44), which differs in the absence of the factor ½ since there the displacements $\delta\mathbf{R}_j (j = 1, \ldots, N)$ are *virtual.* The concept of strain energy V' of an internally conservative system (such as a collection of spring-connected particles) greatly simplifies the proof of a proposition stated and illustrated in the §6·4·6, namely, that the flexibility matrix $[A]$ and the stiffness matrix $[K]$ of a linearly elastic structure are symmetric. These symmetry properties are known as *Maxwell's reciprocity laws.*

To establish the symmetry of $[A]$, we can consider a system of elastically interconnected particles whose displacements are fully defined by the scalars x_1, $x_2, \ldots, x_n$, and whose external applied forces may be expressed in terms of the corresponding scalar components of force $P_1, P_2, \ldots, P_n$. In terms of these quantities, Eq. (6·46) becomes

$$\frac{1}{2}\sum_{i=1}^{n} P_i x_i = V'$$

If this system, mounted on inertially fixed supports, is subjected to only two forces, as established by the scalars P_α and P_β, then we have

$$\tfrac{1}{2}(P_\alpha x_\alpha + P_\beta x_\beta) = V'$$

This amount of strain energy is the unique result of the application of loads P_α and P_β, whether these loads are applied simultaneously or sequentially.

The deflection x_α due to load P_α is expressed in terms of a *flexibility coefficient* $a_{\alpha\alpha}$ as

$$x_\alpha^{(\alpha)} = a_{\alpha\alpha} P_\alpha$$

while x_α due to P_β is

$$x_\alpha^{(\beta)} = a_{\alpha\beta} P_\beta$$

Similarly

$$x_\beta^{(\alpha)} = a_{\beta\alpha} P_\alpha \qquad \text{and} \qquad x_\beta^{(\beta)} = a_{\beta\beta} P_\beta$$

Substituting above furnishes

$$V' = \tfrac{1}{2}(P_\alpha a_{\alpha\alpha} P_\alpha + P_\alpha a_{\alpha\beta} P_\beta + P_\beta a_{\beta\alpha} P_\alpha + P_\beta a_{\beta\beta} P_\beta)$$

Now let's imagine that the forces are applied sequentially: first P_α and then P_β. The resulting V' must be the same, but the calculation proceeds differently. After load P_α is applied, the strain energy is

$$V_1' = \tfrac{1}{2} P_\alpha a_{\alpha\alpha} P_\alpha$$

Next we apply P_β to the system. The work done by P_β is simply

$$V_2' = \tfrac{1}{2} P_\beta a_{\beta\beta} P_\beta$$

but since the load P_α is still on the structure it does more work as x_α experiences a change due to P_β. We can use Eq. (6·39) to calculate the resulting work since the displacement due to P_β is no different than a virtual displacement. The added term is

$$V_3' = P_\alpha a_{\alpha\beta} P_\beta$$

and the total strain energy for this loading sequence is

$$V' = V_1' + V_2' + V_3' = \tfrac{1}{2}(P_\alpha a_{\alpha\alpha} P_\alpha + 2P_\alpha a_{\alpha\beta} P_\beta + P_\beta a_{\beta\beta} P_\beta)$$

If now we wish to consider the opposite loading sequence, applying first P_β and then P_α, we will find

$$V' = \tfrac{1}{2}(P_\alpha a_{\alpha\alpha} P_\alpha + 2P_\beta a_{\beta\alpha} P_\alpha + P_\beta a_{\beta\beta} P_\beta)$$

We now have three different expressions for the single quantity V'; these can be reconciled only if

$$a_{\alpha\beta} = a_{\beta\alpha} \tag{6·47a}$$

The *flexibility matrix* $[A]$ having the elements $a_{\alpha\beta}$ is thus symmetric, that is, $[A]^T = [A]$.

Recall from §6·4·6 the definition of the stiffness matrix $[K] \triangleq [A]^{-1}$; the element $k_{\alpha\beta}$ of $[K]$ has been interpreted as the negative of the force P_α required to maintain $x_\alpha = 0$ when x_β is given a unit displacement (all $x_\gamma = 0$ for $\gamma \neq \beta$). The symmetry of $[K]$ follows from the commutativity of matrix transposition and matrix inversion (see Appendix B) since

$$[K]^T = ([A]^{-1})^T = ([A]^T)^{-1} = [A]^{-1} = [K]$$

Symmetry of $[K]$ implies the equality of complementary stiffnesses

$$k_{\alpha\beta} = k_{\beta\alpha} \tag{6·47b}$$

6·4·8 Application of Maxwell's reciprocity laws

Let us imagine that for the truss shown in Fig. 6·16 we have completed a deflection analysis for the loading shown in Fig. 6·16*a*, and computed the vertical deflections a_{13}, a_{23}, a_{33}, a_{43}, and a_{53} at points ①, ②, ③, ④, and ⑤ due to a unit vertical load at ③. (Note that these are flexibility coefficients for the truss.) Now we must determine the single deflection Δ_3 at ③ due to a variety of vertical loads located at all of the bottom joints of the truss, as in Fig. 6·16*b*.

By definition, the flexibility coefficient $a_{3\beta}$ is the deflection in coordinate ③ due to a unit load in the direction of coordinate β, so the total deflection Δ_3 is evidently

$$\Delta_3 = W_1 a_{31} + W_2 a_{32} + W_3 a_{33} + W_4 a_{34} + W_5 a_{35}$$

But Maxwell's reciprocity law [Eq. (6·47*a*)] provides $a_{3\beta} = a_{\beta 3} (\beta = 1, \ldots, 5)$, and the latter set we have assumed to be known to us, so the total deflection is available in terms of known quantities as

$$\Delta_3 = W_1 a_{13} + W_2 a_{23} + W_3 a_{33} + W_4 a_{43} + W_5 a_{53}$$

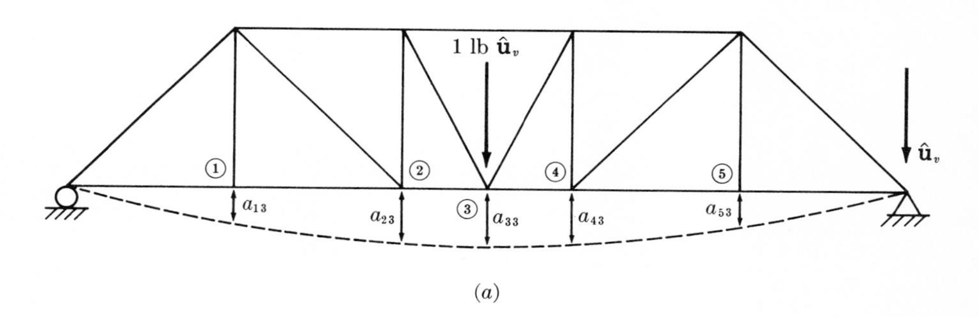

(*a*)

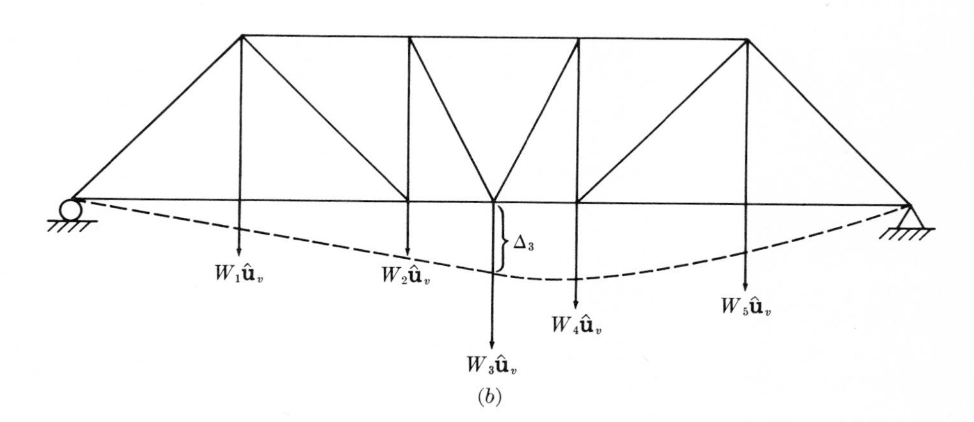

(*b*)

FIGURE 6·16

6·5 APPLICATION TO COLLISIONS

6·5·1 Colliding particles

The concepts of system linear momentum and system kinetic energy prove particularly valuable when used together in application to the problem of particle collisions.

When we imagine two physical bodies in collision, we envisage some deformation of each body, and this we expect to occur over some time interval. It often occurs, however (as in a game of billiards), that the duration of contact is too brief for unassisted human observation, and that the bodies depart from one another after collision with no visible consequence except an abrupt change in velocity.

When we idealize two bodies in collision as particles, we sacrifice the notion of body deformation, and are left with a representation of postcontact behavior that requires careful scrutiny before acceptance in an engineering application. Simply because particle-collision theory is so simple, it is useful as a starting point in any determination of collision response.

A collision between two particles p_1 and p_2 is idealized as providing equal and opposite ideal impulsive loads to the two particles, with these forces applied over a vanishing time interval, and directed along the line joining p_1 and p_2 just prior to contact (see Fig. 6·17, where the impulses parallel the unit vector $\hat{\mathbf{i}}_x$). Since each particle is subjected to an ideal impulsive force, it sustains an instantaneous change in linear momentum. But in order to calculate this change in momentum directly, we would have to know the magnitude of the impulse, and this is generally unavailable.

The linear momentum of the *two-particle system* is, however, unchanged during the vanishingly small time interval of the impact, so that if m_1 and m_2 are the masses of the two particles, $\mathbf{v}_1$ and $\mathbf{v}_2$ are the inertial velocities of the particles prior to impact, and $\mathbf{V}_1$ and $\mathbf{V}_2$ are the inertial velocities subsequent to impact [with the index j identifying the jth particle ($j = 1, 2$)], we have

$$m_1\mathbf{v}_1 + m_2\mathbf{v}_2 = m_1\mathbf{V}_1 + m_2\mathbf{V}_2 \tag{6·48}$$

Our task is to find the postimpact velocities $\mathbf{V}_1$ and $\mathbf{V}_2$, given m_1, m_2, $\mathbf{v}_1$, and $\mathbf{v}_2$. Evidently Eq. (6·48) will not suffice in itself to yield solutions for two unknown vectors, or equivalently, six unknown scalars. Before searching for additional equations, we will rewrite Eq. (6·48) in scalar form.

Let o be a point fixed in inertial space at the point of impact, and let $\hat{\mathbf{i}}_x$, $\hat{\mathbf{i}}_y$, and $\hat{\mathbf{i}}_z$ be a set of three inertially fixed orthogonal unit vectors so defined that $\hat{\mathbf{i}}_x$ parallels the line of action of the impulsive interaction between p_1 and p_2 (see Fig. 6·17). If point o and axes parallel to $\hat{\mathbf{i}}_x$, $\hat{\mathbf{i}}_y$, and $\hat{\mathbf{i}}_z$ establish a cartesian coordinate system, then symbols can be attached to these coordinates such that

$$\begin{aligned}
\mathbf{v}_1 &= \dot{x}_1\hat{\mathbf{i}}_x + \dot{y}_1\hat{\mathbf{i}}_y + \dot{z}_1\hat{\mathbf{i}}_z \\
\mathbf{v}_2 &= \dot{x}_2\hat{\mathbf{i}}_x + \dot{y}_2\hat{\mathbf{i}}_y + \dot{z}_2\hat{\mathbf{i}}_z \\
\mathbf{V}_1 &= \dot{X}_1\hat{\mathbf{i}}_x + \dot{Y}_1\hat{\mathbf{i}}_y + \dot{Z}_1\hat{\mathbf{i}}_z \\
\mathbf{V}_2 &= \dot{X}_2\hat{\mathbf{i}}_x + \dot{Y}_2\hat{\mathbf{i}}_y + \dot{Z}_2\hat{\mathbf{i}}_z
\end{aligned} \tag{6·49}$$

Substituting Eqs. (6·49) into Eq. (6·48) provides three scalar equations in the six unknowns $\dot{X}_1$, $\dot{Y}_1$, $\dot{Z}_1$, $\dot{X}_2$, $\dot{Y}_2$, and $\dot{Z}_2$, specifically,

$$m_1\dot{x}_1 + m_2\dot{x}_2 = m_1\dot{X}_1 + m_2\dot{X}_2 \tag{6·50}$$

$$m_1\dot{y}_1 + m_2\dot{y}_2 = m_1\dot{Y}_1 + m_2\dot{Y}_2 \tag{6·51}$$

$$m_1\dot{z}_1 + m_2\dot{z}_2 = m_1\dot{Z}_1 + m_2\dot{Z}_2 \tag{6·52}$$

The unknowns in Eqs. (6·51) and (6·52) can be determined immediately, due to our judicious choice of a coordinate system. Since the unit vector $\hat{\mathbf{i}}_x$ has been chosen to parallel the impulsive forces applied to p_1 and p_2, and the vector impulse on a particle equals the change in linear momentum of that particle, there can be no change in the components of the linear momenta of particles p_1

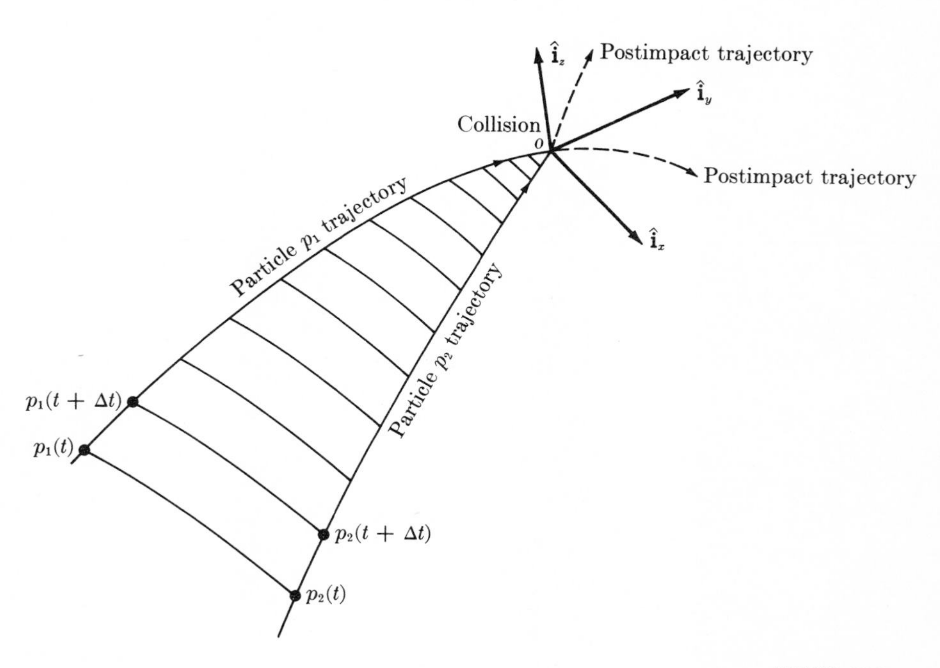

FIGURE 6·17

and p_2 in the directions $\hat{\mathbf{i}}_y$ and $\hat{\mathbf{i}}_z$. Hence

$$\begin{aligned} \dot{Y}_1 &= \dot{y}_1 \\ \dot{Z}_1 &= \dot{z}_1 \\ \dot{Y}_2 &= \dot{y}_2 \\ \dot{Z}_2 &= \dot{z}_2 \end{aligned} \tag{6·53}$$

and only $\dot{X}_1$ and $\dot{X}_2$ remain to be found.

Equation (6·50) must be augmented by one more scalar equation, in order to permit solution. The required equation can be provided in the form of a statement concerning the conservation of mechanical energy.

The form of the work-energy equation developed as Eq. (6·19) might first be considered. There is during the vanishingly small contact duration no impulse provided by any resultant force applied to the two-particle system, so this equation tells us only that the velocity of the mass center of the two-particle system is uninfluenced by the collision between its two constituent parts.

A more general form of the work-energy equation was first written as Eq. (6·20), which for the two-particle system and associated notation takes the form

$$\begin{aligned} \int_{t_1}^{t_2} (\mathbf{F}_1 \cdot \dot{\mathbf{R}}_1 + \mathbf{F}_2 \cdot \dot{\mathbf{R}}_2)\, dt &= \tfrac{1}{2}m_1\mathbf{V}_1 \cdot \mathbf{V}_1 + \tfrac{1}{2}m_2\mathbf{V}_2 \cdot \mathbf{V}_2 \\ &\qquad - \tfrac{1}{2}m_1\mathbf{v}_1 \cdot \mathbf{v}_1 - \tfrac{1}{2}m_2\mathbf{v}_2 \cdot \mathbf{v}_2 \\ &= \tfrac{1}{2}m_1(\dot{X}_1^2 + \dot{Y}_1^2 + \dot{Z}_1^2) \\ &\qquad + \tfrac{1}{2}m_2(\dot{X}_2^2 + \dot{Y}_2^2 + \dot{Z}_2^2) \\ &\qquad - \tfrac{1}{2}m_1(\dot{x}_1^2 + \dot{y}_1^2 + \dot{z}_1^2) \\ &\qquad - \tfrac{1}{2}m_2(\dot{x}_2^2 + \dot{y}_2^2 + \dot{z}_2^2) \\ &= \tfrac{1}{2}[m_1(\dot{X}_1^2 - \dot{x}_1^2) + m_2(\dot{X}_2^2 - \dot{x}_2^2)] \end{aligned} \tag{6·54}$$

Equation (6·54) involves the unknowns $\dot{X}_1$ and $\dot{X}_2$, so that Eqs. (6·50) and (6·54) can be solved in combination for these unknowns, providing only that an expression can be found for the work integral in Eq. (6·54). In general this integral cannot be evaluated, but for the special (idealized) case of a completely *elastic impact*, the work integral, representing the energy dissipated in collision, can be set equal to zero. In this special case, Eq. (6·54) can be written in the form

$$m_1(\dot{X}_1 - \dot{x}_1)(\dot{X}_1 + \dot{x}_1) = -m_2(\dot{X}_2 - \dot{x}_2)(\dot{X}_2 + \dot{x}_2) \tag{6·55}$$

which may seem an obfuscation until compared with Eq. (6·50) in the form

$$m_1(\dot{X}_1 - \dot{x}_1) = -m_2(\dot{X}_2 - \dot{x}_2) \tag{6·56}$$

Dividing the left and right sides of Eq. (6·55) by the left and right sides of Eq.

(6·56) furnishes

$$\dot{X}_1 + \dot{x}_1 = \dot{X}_2 + \dot{x}_2$$

or

$$(\dot{X}_1 - \dot{X}_2) = -(\dot{x}_1 - \dot{x}_2) \qquad (6\cdot57)$$

In verbal terms, Eq. (6·57), applicable only to *elastic* collisions, says that the relative speed of departure of the two particles along the line of the impact force equals the relative speed of approach of the particles along that line.

Of course, if energy is dissipated in impact, Eq. (6·57) is violated; in the limiting case of maximum energy dissipation, we expect the relative speed of departure to be zero. Two colliding balls of clay might illustrate in physical terms this limiting case, referred to as completely *plastic impact.* The full spectrum of possibilities, from elastic impact to plastic impact, can be accommodated by modifying Eq. (6·57) to read

$$(\dot{X}_1 - \dot{X}_2) = -e(\dot{x}_1 - \dot{x}_2) \qquad (6\cdot58)$$

where the constant e, called the *coefficient of restitution,* ranges from *zero* for plastic impact to *one* for elastic impact.

In a given problem involving the collision of two bodies that are to be idealized as particles, we must estimate e [in order to use Eq. (6·58)], on the basis of experience with similar collisions. Although it may be difficult to establish e with much accuracy, this is vastly easier than estimating the energy dissipation represented by the work integral in Eq. (6·54), and it serves the same purpose. Thus it is generally preferable to use Eq. (6·58) [rather than Eq. (6·54)] in conjunction with Eq. (6·50) to find

$$\dot{X}_1 = \frac{(m_1 - m_2 e)\dot{x}_1 + m_2(1 + e)\dot{x}_2}{m_1 + m_2} \qquad (6\cdot59)$$

and

$$\dot{X}_2 = \frac{m_1(1 + e)\dot{x}_1 + (m_2 - m_1 e)\dot{x}_2}{m_1 + m_2} \qquad (6\cdot60)$$

6·5·2 Applications of particle-collision theory

Elastic rods Suppose that an elastic rod of mass m and length L is moving horizontally at speed v along a straight smooth track when it collides with a similar but stationary elastic rod of mass $2m$ and length $2L$ (see Fig. 6·18). The post-impact speeds of the two rods are to be determined.

The first step in this or any other dynamic analysis is the adoption of a mathematical model of the system. The simplest model that we can conceive for this system is one in which each of the rods is replaced by a particle of corresponding mass, and (since the rods are elastic) the coefficient of restitution e is assigned the value unity. Equations (6·59) and (6·60) provide the solutions

$$\dot{X}_1 = \frac{(m - 2m)v}{m + 2m} = -\frac{v}{3}$$

$$\dot{X}_2 = \frac{m(1 + 1)v}{m + 2m} = \frac{2v}{3}$$

Although this is a correct solution for two particles in elastic collision, caution must be exercised in accepting this result for the elastic rods. To provide some basis for assessing the validity of these results, we might resort to experiments, or we might analyze a more complex model that accommodates the deformability of the rods—perhaps portraying each rod as a chain of elastically interconnected particles, or (reaching beyond the scope of this book) idealizing each rod as an elastic continuum. Any of these approaches would provide us with the following results: When the rods separate after collision, the short rod is completely at rest, and the long rod is moving with its mass center traveling at speed $v/2$ and with elastic waves propagating back and forth along the rod.

The inadequacy of the single-particle analysis stems from its failure to account for the fact that, although it may be that virtually no mechanical energy is dissipated as heat when the elastic rods collide, a certain amount of the energy available initially as kinetic energy in the translating short rod is transformed after impact into the kinetic and potential energy of *vibration* of the long rod.

You might in retrospect wonder what coefficient of restitution e might be introduced into Eqs. (6·59) and (6·60) in order to obtain the results $\dot{X}_1 = 0$ and $\dot{X}_2 = v/2$ (as indicated by test or more refined analysis). The answer, in this case, is $e = \frac{1}{2}$, as may be verified by either Eq. (6·59) or (6·60). Thus we see that a two-particle model can produce perfectly satisfactory indications of the postimpact velocities of the mass centers of the colliding rods in Fig. 6·18, if the analyst is clever enough (or sufficiently experienced) to guess the coefficient of restitution e.

Steel spheres Two steel spheres of like mass and diameter moving on orthogonal trajectories collide in space. Figure 6·19 portrays the spheres at time t_0, as they

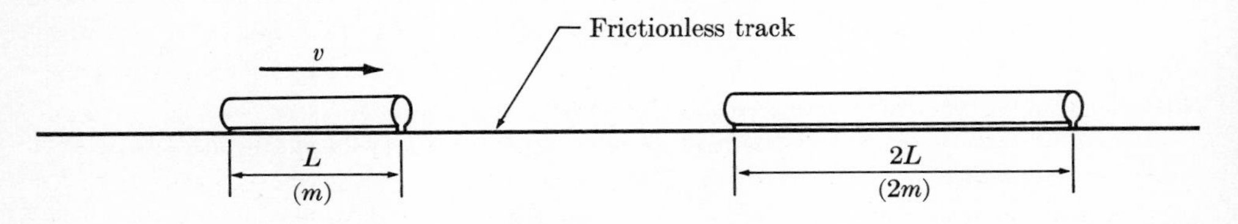

FIGURE 6·18

approach collision at speeds v_1 and v_2 in orthogonal directions, and also at time t_1, when they come into contact. The problem is to determine the postimpact velocities of the mass centers p_1 and p_2 of the spheres.

The simplest approach to this problem is to idealize each sphere as a particle, so that the solutions in Eqs. (6·53), (6·59), and (6·60) can be employed. These solutions are expressed in terms of cartesian coordinates associated with unit vectors $\hat{\mathbf{i}}_x$, $\hat{\mathbf{i}}_y$, and $\hat{\mathbf{i}}_z$, where $\hat{\mathbf{i}}_x$ parallels the line joining the two particles as they approach collision. In the case of the identical colliding spheres, approaching each other at right angles, this line is directed as shown in Fig. 6·19. If $\hat{\mathbf{i}}_y$ is selected so as to lie in the plane established by the trajectories of the spheres, then the conditions prior to impact are given by

$$\dot{x}_1 = v_1 \frac{\sqrt{2}}{2} \qquad \dot{x}_2 = -v_2 \frac{\sqrt{2}}{2}$$

$$\dot{y}_1 = v_1 \frac{\sqrt{2}}{2} \qquad \dot{y}_2 = v_2 \frac{\sqrt{2}}{2}$$

$$\dot{z}_1 = 0 \qquad \dot{z}_2 = 0$$

From Eqs. (6·53), (6·59), and (6·60), we have the postimpact speeds (with $e = 1$)

$$\dot{X}_1 = \dot{x}_2 = -v_2 \frac{\sqrt{2}}{2} \qquad \dot{X}_2 = \dot{x}_1 = v_1 \frac{\sqrt{2}}{2}$$

$$\dot{Y}_1 = \dot{y}_1 = v_1 \frac{\sqrt{2}}{2} \qquad \dot{Y}_2 = \dot{y}_2 = v_2 \frac{\sqrt{2}}{2}$$

$$\dot{Z}_1 = 0 \qquad \dot{Z}_2 = 0$$

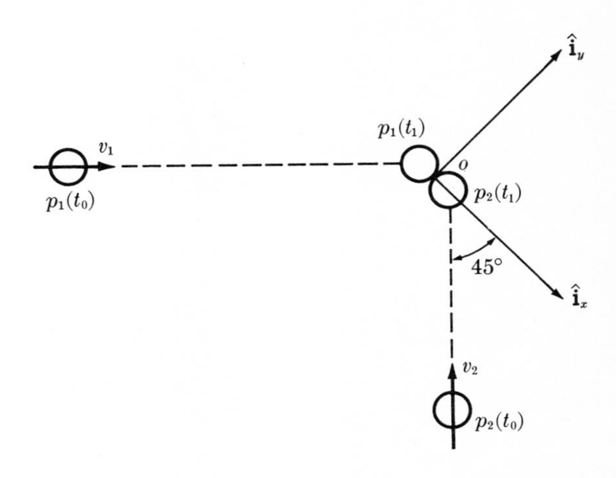

FIGURE 6·19

We might again ask ourselves just what the value of this two-particle idealization and solution is in application to the given problem of two colliding elastic spheres. Would experiments with spheres come closer to matching the particle dynamics results than the experiments with rods noted in the previous example? The answer in this case depends largely on the smoothness of the surfaces of the spheres. A fundamental assumption of the preceding theory of colliding particles limits the line of action of the interaction force to a line defined by $\hat{\mathbf{i}}_x$, leading to the conclusions $\dot{Y}_1 = \dot{y}_1$; $\dot{Z}_1 = \dot{z}_1$; $\dot{Y}_2 = \dot{y}_2$; and $\dot{Z}_2 = \dot{z}_2$. This result would apply to the colliding spheres in Fig. 6·19 only if the contacting surfaces were so smooth as to preclude friction forces, which would act in the direction of $\hat{\mathbf{i}}_y$. Such friction forces would change the velocity components of the spheres in the $\hat{\mathbf{i}}_y$ direction, and they would further diminish the value of the particle solution by investing some of the original kinetic energy in the postimpact rotational kinetic energy of the spheres. The magnitude of the friction forces generated in an actual collision depends on many factors (surface roughness, normal forces, etc.), but among the most critical of these is the angle between $\hat{\mathbf{i}}_x$ and the paths of the approaching spheres. If instead of 45° (as in Fig. 6·19), this angle were zero, so that the spheres would approach one another "head-on," there would be no friction force, and the predictions of the two-particle analysis would agree quite closely with experimental evidence.

Catapult actuation In the illustration in §6·1·4 of the applicability of $\mathbf{F} = m\mathbf{A}$ to a system of particles representing a falling sack of sand, we considered the dynamics of the sand as a catapult actuation device (see Fig. 6·1). Our previous analysis was enough to provide the velocity of the mass center of the sack as it strikes the catapult lever arm, but the dynamic response of the apparatus to the resulting collision has yet to be analyzed. Although our interest is in determining the eventual destination of the "payload" to be delivered, our immediate objective will be to find the "injection velocity" $\mathbf{V}_i$ with which the payload leaves the supporting cup of the catapult when the catapult arm is abruptly halted by the framework. Once this initial velocity is ascertained, we can rely on the methods of §5·2·2 for an elementary determination of the subsequent payload trajectory.

From Eq. (6·8), a sack of sand released from height h at time $t = 0$ reaches the catapult arm at time $t^* = (2h/g)^{1/2}$, arriving with speed $\dot{s}^* = gt^* = \sqrt{2gh}$. For a given sack mass $\mathfrak{M}$, payload mass m, and catapult geometry established by l, L, and α, we are to determine the payload injection velocity $\mathbf{V}_i$.

Before plunging into the solution of this problem, we might pause and take stock of the analytical tools now available to us. We are quite prepared to idealize each grain of sand as a particle, so that $\mathbf{F} = m\mathbf{A}$ is applicable to each particle, but the task of determining interaction forces among these particles has already been judged hopeless, even during the fall preceding the impact. If we attempt, as in §6·1·4, to apply $\mathbf{F} = \mathfrak{M}\mathbf{A}$ to the sack of sand as a *system* of particles, so that $\mathbf{A}$ is the mass-center acceleration, $\mathfrak{M}$ the system mass, and $\mathbf{F}$ the resultant of external forces, we are stymied by our inability to properly assess $\mathbf{F}$ during impact. Our challenge is to find a way to approach the problem so that the forces of collision are not involved. We can accomplish this by defining our material system so as to include the sack of sand and the payload, as well as the catapult arm supporting them both. If we make this choice, however, we find that we have shifted the difficulty to the hinge axis of the catapult arm, where unknown reaction forces are induced by the collision. If, however, we let the intersection of the hinge axis and the catapult arm axis be a reference point o with respect to which we measure the moment $\mathbf{M}^o$ and the angular momentum $\mathbf{H}^o$, then we find that these unknown reaction forces do not become involved when we write $\mathbf{M}^o = \dot{\mathbf{H}}^o$. Thus we can circumvent the problem of our ignorance of the complex pattern of forces induced during collision.

In order to apply the techniques of particle-collision theory (§6·5·1) to this problem, we must assume that the impact causes an instantaneous change in the velocities of all particles of the system. Moreover, we must make an assessment of the energy we expect to be dissipated in collision, and establish something like a coefficient of restitution for the sack of sand. A reasonable and very simple alternative is to imagine that the collision is completely inelastic, so that the sack of sand does not rebound from the catapult receiving cup at all. Thus we have so idealized the problem that just before impact (when $t = t^* - \frac{1}{2}\,\Delta t$ for arbitrarily small Δt), the sack of sand is falling vertically at essentially the speed $\dot{s}^* = \sqrt{2gh}$, and just after impact (when $t = t^* + \frac{1}{2}\,\Delta t$), the catapult arm is rotating at a rate to be called $\dot{\theta}^*$, with both the payload and the sack of sand at rest relative to the catapult arm. Our first task is to determine $\dot{\theta}^*$ from the relationship $\mathbf{M}^o = \dot{\mathbf{H}}^o$.

If the mass center of the system is at the hinge point o (implying $\mathfrak{M}l = mL$ if the catapult arm is for convenience considered massless), then there is no moment applied about the frictionless hinge axis. Then $\mathbf{M}^o = 0$, so that $\mathbf{H}^o$ is constant. In the more general case ($\mathfrak{M}l \neq mL$), the force of gravity at the time of impact contributes $\mathbf{M}^o = \mathfrak{M}gl\hat{\mathbf{e}}_2 - mgL\hat{\mathbf{e}}_2$ (if the arm is massless), but we shall see that $\mathbf{H}^o$ is still unchanged during the arbitrarily small duration of the impulsive collision. To appreciate this fact, integrate $\mathbf{M}^o = \dot{\mathbf{H}}^o$ to find

$$\int_{t^*-\frac{1}{2}\Delta t}^{t^*+\frac{1}{2}\Delta t} \mathbf{M}^o\,dt = \mathbf{H}^o(t^* + \tfrac{1}{2}\,\Delta t) - \mathbf{H}^o(t^* - \tfrac{1}{2}\,\Delta t)$$

If in the interval Δt there is no change in the position of any particles of the system, then

$$\mathbf{H}^o(t^* + \tfrac{1}{2}\,\Delta t) - \mathbf{H}^o(t^* - \tfrac{1}{2}\,\Delta t) = (\mathfrak{M}gl\hat{\mathbf{e}}_2 - mgL\hat{\mathbf{e}}_2)\,\Delta t$$

so that in the limit as $\Delta t \to 0$ (ideal impulse), we find $\mathbf{H}^o$ unchanged by the collision, or

$$\mathbf{H}^o(t^* + \tfrac{1}{2}\,\Delta t) = \mathbf{H}^o(t^* - \tfrac{1}{2}\,\Delta t) \tag{a}$$

Prior to collision, only the sack of sand contributes to $\mathbf{H}^o$, so that

$$\mathbf{H}^o(t^* - \tfrac{1}{2}\,\Delta t) = l\hat{\mathbf{e}}_1 \times (-\mathfrak{M}\dot{s}^*\hat{\mathbf{e}}_3) = \mathfrak{M}l(2gh)^{1/2}\hat{\mathbf{e}}_2 \tag{b}$$

After the inelastic collision, the angular velocity of the catapult arm is $\dot{\theta}^*\hat{\mathbf{e}}_2$, and the angular momentum is

$$\begin{aligned}\mathbf{H}^o(t^* + \tfrac{1}{2}\,\Delta t) &= l\hat{\mathbf{e}}_1 \times (-\mathfrak{M}l\dot{\theta}^*\hat{\mathbf{e}}_3) - L\hat{\mathbf{e}}_1 \times (mL\dot{\theta}^*\hat{\mathbf{e}}_3) \\ &= (\mathfrak{M}l^2 + mL^2)\dot{\theta}^*\hat{\mathbf{e}}_2\end{aligned} \tag{c}$$

Combining Eqs. (a), (b), and (c) provides the scalar equation

$$\mathfrak{M}l(2gh)^{1/2} = (\mathfrak{M}l^2 + mL^2)\dot{\theta}^*$$

or

$$\dot{\theta}^* = \frac{\mathfrak{M}l(2gh)^{1/2}}{\mathfrak{M}l^2 + mL^2} \tag{d}$$

Thus the velocity of the payload immediately after the sack lands is

$$L\dot{\theta}^*\hat{\mathbf{e}}_3 = \left(\frac{\mathfrak{M}lL(2gh)^{1/2}}{\mathfrak{M}l^2 + mL^2}\right)\hat{\mathbf{e}}_3 \tag{e}$$

and the velocity of the sack of sand is

$$-l\dot{\theta}^*\hat{\mathbf{e}}_3 = -\left(\frac{\mathfrak{M}l^2(2gh)^{1/2}}{\mathfrak{M}l^2 + mL^2}\right)\hat{\mathbf{e}}_3 \tag{f}$$

Although we have established the immediate dynamic consequences of the impact of the sack of sand on the catapult arm, we have not yet reached our objective, which is to determine the payload injection velocity $\mathbf{V}_i$. If the angle θ defines the rotation of the catapult arm from the horizontal about the hinge axis, with positive θ corresponding to a dextral rotation in direction $\hat{\mathbf{e}}_2$, then what we want is the value of the payload velocity when $\theta = \alpha$ (see Fig. 6·1).

Once again we face the problem of devising an analytical strategy. We can, of course, simply write $\mathbf{F} = m\mathbf{A}$ for the payload mass center, but we don't yet know enough about the force applied to the payload by the supporting cup. In order to apply $\mathbf{F} = m\mathbf{A}$ successfully to the payload, we would be obliged to determine $\mathbf{F}$ as a function of θ and its time derivatives by finding first expressions for the forces applied to the catapult arm at the hinge and at the cup holding the sand, proceeding thence by equilibrium equations for the catapult arm to an expression for the force applied by the cup to the payload. The job can be done this way, but there is also an easier and more edifying way.

In §6·2·2 we developed in Eqs. (6·19) to (6·22) various forms of work-energy equations, all relating work done by external forces on a system of particles to changes in the kinetic energy of the system. If we are to use any of these equations to advantage now, we must find a way to circumvent the unknown forces of interaction at the catapult hinge and those between the catapult cups and their contents. But Eqs. (6·19) and (6·21) both involve $\mathbf{F}$, the resultant of external forces applied to the system, and Eqs. (6·21) and (6·22) both involve $\mathbf{f}_j'$, the resultant force applied to particle j by other particles internal to the system, and $\mathbf{f}_j$, the external force applied to the jth particle.

If the system is chosen so as to include the payload, the sack of sand, and the catapult arm, then $\mathbf{F}$ includes the unknown reaction force at the hinge, and the presence of this troublesome term in Eqs. (6·19) and (6·21) moves us to reject the use of these equations [$\mathbf{F}$ disappears from these equations when $\dot{\mathbf{R}} = 0$, as is the case when the system mass center is at the hinge point o, but then Eq. (6·19) says nothing at all and Eq. (6·21) reduces to Eq. (6·22)]. The hinge reaction force also appears in Eq. (6·22) in the guise of $\mathbf{f}_j'$ for the particle at the hinge point o, and it appears in Eq. (6·20) as part of $\mathbf{F}_j$ for that particle at o. In Eq. (6·20), however, the offending hinge force is multiplied by the inertial velocity of the point o, which is zero, so in applying this equation we can avoid the unknown hinge force. Thus we choose Eq. (6·20) as the most promising form of work-energy equation, and record it here for further consideration:

$$\sum_{j=1}^{n} \int_{t_1}^{t_2} \mathbf{F}_j \cdot \dot{\mathbf{R}}_j \, dt = \sum_{j=1}^{n} [\tfrac{1}{2} m_j \dot{\mathbf{R}}_j(t_2) \cdot \dot{\mathbf{R}}_j(t_2) - \tfrac{1}{2} m_j \dot{\mathbf{R}}_j(t_1) \cdot \dot{\mathbf{R}}_j(t_1)] \tag{g}$$

where now $t_1 = t^* + \frac{1}{2} \Delta t$, and t_2 is the value of t when $\theta = \alpha$.

The symbol $\mathbf{F}_j$ represents the total force applied to the jth particle, including the forces of interaction with other particles of the system. Again it proves convenient to replace $\mathbf{F}_j$ by $\mathbf{f}_j + \mathbf{f}_j'$, where $\mathbf{f}_j'$ is the resultant force exerted on the jth particle by other particles of the system. By Newton's third law, the forces included in $\mathbf{f}_j'$, $j = 1, \ldots, n$, occur in opposing pairs, which for the system in question may be said to be directed along the line joining the interacting pair of particles. Thus the force $\mathbf{f}_a'$ on particle a will include a term $\mathbf{f}_{ab}' = f'\hat{\mathbf{u}}_{ab}$ due to interaction with particle b, where $\hat{\mathbf{u}}_{ab}$ is a unit vector directed from a toward b, and $\mathbf{f}_b'$ will include $-f'\hat{\mathbf{u}}_{ab}$. Thus the internal forces will contribute to Eq. (g) only as paired terms typified by

$$\int (f'\hat{\mathbf{u}}_{ab} \cdot \dot{\mathbf{R}}_a - f'\hat{\mathbf{u}}_{ab} \cdot \dot{\mathbf{R}}_b) \, dt = \int f'\hat{\mathbf{u}}_{ab} \cdot (\dot{\mathbf{R}}_a - \dot{\mathbf{R}}_b) \, dt \tag{h}$$

But if $\mathbf{R}_j$ is the position vector of particle j from the inertially fixed point o, then by Eq. (3·100)

$$\dot{\mathbf{R}}_j \triangleq \frac{{}^i d}{dt}\mathbf{R}_j = \frac{{}^f d}{dt}\mathbf{R}_j + \boldsymbol{\omega}^{fi} \times \mathbf{R}_j$$

for any reference frame f. In this application it is convenient to choose f as the frame established by the catapult arm, so that if the system does not deform we have

$$\dot{\mathbf{R}}_j = \boldsymbol{\omega}^{fi} \times \mathbf{R}_j = \dot{\theta}\hat{\mathbf{e}}_2 \times \mathbf{R}_j \tag{i}$$

for all particles of the system. In this case Eq. (h) becomes

$$\int f'\hat{\mathbf{u}}_{ab} \cdot (\dot{\mathbf{R}}_a - \dot{\mathbf{R}}_b)\, dt = \int f'\hat{\mathbf{u}}_{ab} \cdot [\dot{\theta}\hat{\mathbf{e}}_2 \times (\mathbf{R}_a - \mathbf{R}_b)]\, dt$$

and if r_{ab} is the distance between a and b, this becomes

$$\int f'\hat{\mathbf{u}}_{ab} \cdot [\dot{\theta}\hat{\mathbf{e}}_2 \times (-r_{ab}\hat{\mathbf{u}}_{ab})]\, dt = 0$$

since the quantity in brackets is orthogonal to $\hat{\mathbf{u}}_{ab}$. Thus the internal interaction forces $\mathbf{f}'_j$ do not contribute to Eq. (g) after all (as long as the system is not deforming), and we may instead write

$$\sum_{j=1}^{n} \int_{t_1}^{t_2} \mathbf{f}_j \cdot \dot{\mathbf{R}}_j\, dt = \sum_{j=1}^{n} \frac{1}{2} m_j [\dot{\mathbf{R}}_j(t_2) \cdot \dot{\mathbf{R}}_j(t_2) - \dot{\mathbf{R}}_j(t_1) \cdot \dot{\mathbf{R}}_j(t_1)] \tag{j}$$

Substituting Eq. (i) into Eq. (j) furnishes

$$\sum_{j=1}^{n} \int_{t_1}^{t_2} \mathbf{f}_j \cdot \dot{\theta}\hat{\mathbf{e}}_2 \times \mathbf{R}_j\, dt = \sum_{j=1}^{n} \frac{1}{2} m_j \{[\dot{\theta}(t_2)\hat{\mathbf{e}}_2 \times \mathbf{R}_j(t_2)] \cdot [\dot{\theta}(t_2)\hat{\mathbf{e}}_2 \times \mathbf{R}_j(t_2)] - [\dot{\theta}(t_1)\hat{\mathbf{e}}_2 \times \mathbf{R}_j(t_1)] \cdot [\dot{\theta}(t_1)\hat{\mathbf{e}}_2 \times \mathbf{R}_j(t_1)]\} \tag{k}$$

The relevant forces $\mathbf{f}_j$ are simply those due to gravity,

$$\mathbf{f}_j = -m_j g \hat{\mathbf{e}}_3 \tag{l}$$

since the only other external force is at the hinge, where $\dot{\mathbf{R}}_j = 0$. Thus Eq. (k) becomes (with all possible extractions from the summation and integration)

$$g\hat{\mathbf{e}}_3 \cdot \left(\hat{\mathbf{e}}_2 \times \sum_{j=1}^{n} m_j \int_{t_1}^{t_2} \mathbf{R}_j\dot{\theta}\, dt\right) = \sum_{j=1}^{n} \frac{1}{2} m_j \{[\dot{\theta}(t_2)\hat{\mathbf{e}}_2 \times \mathbf{R}_j(t_2)] \cdot [\dot{\theta}(t_2)\hat{\mathbf{e}}_2 \times \mathbf{R}_j(t_2)] - [\dot{\theta}(t_1)\hat{\mathbf{e}}_2 \times \mathbf{R}_j(t_1)] \cdot [\dot{\theta}(t_1)\hat{\mathbf{e}}_2 \times \mathbf{R}_j(t_1)]\} \tag{m}$$

At this point we must abandon the cumbersome business of summing over all the particles of the system. Although in Chap. 7 we will develop concepts that will permit us to retain the extended representation of the payload and the sack of sand without continuing to count every particle, for the present we will simply approximate $\mathbf{R}_j$ by $L(\hat{\mathbf{e}}_3 \sin\theta - \hat{\mathbf{e}}_1 \cos\theta)$ for particles in the payload and by $l(\hat{\mathbf{e}}_1 \cos\theta - \hat{\mathbf{e}}_3 \sin\theta)$ for particles of sand. Then the left-hand side of Eq. (m) becomes

$$-g\hat{\mathbf{e}}_3 \cdot \left[\hat{\mathbf{e}}_2 \times (\mathfrak{M}l - mL)\left(\hat{\mathbf{e}}_1 \int_{t_1}^{t_2} \cos\theta\dot{\theta}\,dt - \hat{\mathbf{e}}_3 \int_{t_1}^{t_2} \sin\theta\dot{\theta}\,dt\right)\right]$$
$$= -g(\mathfrak{M}l - mL)\hat{\mathbf{e}}_3 \cdot \left[(-\hat{\mathbf{e}}_3)\int_0^\alpha \cos\theta\,d\theta - \hat{\mathbf{e}}_1 \int_0^\alpha \sin\theta\,d\theta\right]$$
$$= g(\mathfrak{M}l - mL)\sin\alpha \qquad (n)$$

and the right-hand side of Eq. (m) becomes

$$\tfrac{1}{2}\mathfrak{M}[\dot{\theta}(t_2)\hat{\mathbf{e}}_2 \times l(\hat{\mathbf{e}}_1\cos\alpha - \hat{\mathbf{e}}_3\sin\alpha)] \cdot [\dot{\theta}(t_1)\hat{\mathbf{e}}_2 \times l(\hat{\mathbf{e}}_1\cos\alpha - \hat{\mathbf{e}}_3\sin\alpha)]$$
$$+ \tfrac{1}{2}m[\dot{\theta}(t_2)\hat{\mathbf{e}}_2 \times L(\hat{\mathbf{e}}_3\sin\alpha - \hat{\mathbf{e}}_1\cos\alpha)] \cdot [\dot{\theta}(t_2)\hat{\mathbf{e}}_2 \times L(\hat{\mathbf{e}}_3\sin\alpha - \hat{\mathbf{e}}_1\cos\alpha)]$$
$$- \tfrac{1}{2}\mathfrak{M}[\dot{\theta}(t_1)\hat{\mathbf{e}}_2 \times l\hat{\mathbf{e}}_1] \cdot [\dot{\theta}(t_1)\hat{\mathbf{e}}_2 \times l\hat{\mathbf{e}}_1]$$
$$- \tfrac{1}{2}m[\dot{\theta}(t_1)\hat{\mathbf{e}}_2 \times L(-\hat{\mathbf{e}}_1)] \cdot [\dot{\theta}(t_1)\hat{\mathbf{e}}_2 \times L(-\hat{\mathbf{e}}_1)]$$
$$= \tfrac{1}{2}\mathfrak{M}l^2\dot{\theta}^2(t_2) + \tfrac{1}{2}mL^2\dot{\theta}^2(t_2) - \tfrac{1}{2}\mathfrak{M}l^2\dot{\theta}^2(t_1) - \tfrac{1}{2}mL^2\dot{\theta}^2(t_1) \qquad (o)$$

Substituting $\dot{\theta}^*$ for $\dot{\theta}(t_1)$ from Eq. (d) and combining Eqs. (m) to (o) furnishes

$$\tfrac{1}{2}(\mathfrak{M}l^2 + mL^2)\dot{\theta}^2(t_2) = -(mL - \mathfrak{M}l)g\sin\alpha + \tfrac{1}{2}(\mathfrak{M}l^2 + mL^2)\dot{\theta}^{*2}$$

or

$$\dot{\theta}^2(t_2) = \frac{-2g\sin\alpha(mL - \mathfrak{M}l)}{\mathfrak{M}l^2 + mL^2} + \frac{2gh\mathfrak{M}^2l^2}{(\mathfrak{M}l^2 + mL^2)^2} \qquad (p)$$

Since the desired injection velocity $\mathbf{V}_i$ of the payload is given by

$$\mathbf{V}_i = \dot{\theta}(t_2)\hat{\mathbf{e}}_2 \times L(\hat{\mathbf{e}}_3\sin\alpha - \hat{\mathbf{e}}_1\cos\alpha)$$
$$= L\dot{\theta}(t_2)(\hat{\mathbf{e}}_1\sin\alpha + \hat{\mathbf{e}}_3\cos\alpha) \qquad (q)$$

we can attain our final objective by combining Eqs. (p) and (q).

In determining $\dot{\theta}(t_2)$ from the work-energy principle for a system of n particles [Eq. (6·20)], we undertook excursions that were not truly required for the approximate solution we finally obtained as Eq. (p). If we had assumed at the outset that the payload was effectively a single particle of mass m, and the sack of sand a particle of mass M, then the work-energy equation for a system including these two particles and a massless catapult arm would provide the same answer much more quickly, since the change in kinetic energy would more obviously be given by

$$\Delta T = \tfrac{1}{2}\mathfrak{M}[l\dot{\theta}(t_2)]^2 + \tfrac{1}{2}m[L\dot{\theta}(t_2)]^2 - \tfrac{1}{2}\mathfrak{M}(l\dot{\theta}^*)^2 - \tfrac{1}{2}m(L\dot{\theta}^*)^2$$

where the quantities in square brackets are the particle inertial speeds at t_1 and t_2 [see Eqs. (e) and (f)]. As we have chosen to attack the problem, the nature of the approximations involved is more explicit, but the analysis is very inefficient. Analysis is, of course, simplest when all modeling idealizations are made at the outset.

6·6 VARIABLE-MASS SYSTEMS

6·6·1 Definitions and idealizations

The phrase *variable-mass system* has two fundamentally different interpretations: The term may be used in application to a closed material system that by virtue of its motion is changing its mass due to relativistic phenomena, or it may refer to the changing quantity of material within a prescribed geometrical boundary or *control surface*. Since this text is restricted in scope to classical newtonian mechanics, only the second of these two interpretations is admitted here.

According to the definition adopted in Chap. 1, a *particle* is a geometrical point possessing *constant* mass. Thus a material system idealized as consisting of n particles is by definition a system of constant mass. If, however, the system of interest consists only of those particles within a prescribed boundary, and particles are streaming across that boundary, then we are dealing with a variable-mass system.

Although a rocket qualifies as a variable-mass system, the treatment of the dynamics of such systems appropriate at this level is much too highly idealized to permit application to actual rocket engines. The objective here is not to provide a tool for propulsion analysis, but merely to introduce certain rather basic concepts.

In this chapter we are dealing with systems of particles. Although it may be more appropriate to idealize most variable-mass systems as fluid continua, we reserve that treatment for later courses. Most of the basic ideas can be developed with a multiple-particle model.

You need only release a toy balloon with an open air inlet and witness the crazy pattern of its motions to begin to appreciate the complexity of the problem when the system is free to rotate and translate in space. We restrict ourselves here to the translational problem only, imagining our "vehicle" to be running on a straight track fixed in inertial space.

In any physical problem of interest, matter moves *within* a container as well as across its boundaries, and the expulsion or absorption of material causes shifts of the mass center of the material system within the container. We will eliminate both of these problems by restricting attention to the rather artificial device illustrated in Fig. 6·20*a*. It might be helpful to think of this device as a crude model of a jet engine, taking in moving air and expelling exhaust gases. We assume that the solid line establishes the boundary or control surface of our material system, so that particles entering the system are immediately located at the system mass center, where they stick (with no subsequent motion relative to the container, unless they are later expelled).

6·6·2 Derivation from impulse-momentum equation

Perhaps the easiest procedure for deriving the equation of motion of this system makes use of the concept of linear momentum, which for a single particle sustains

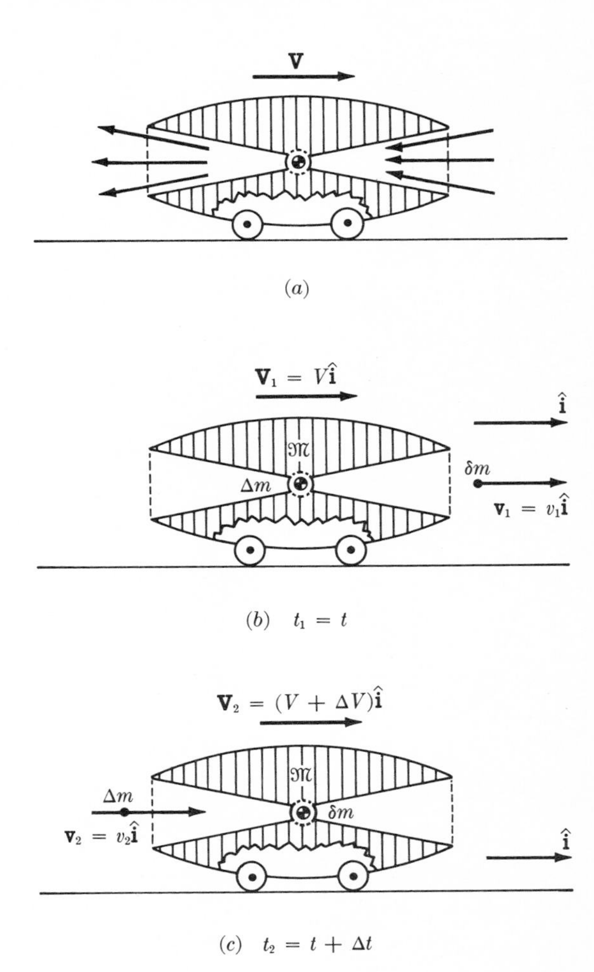

FIGURE 6·20

a change in the time interval $t_2 - t_1$ as given by [see Eqs. (4·23) to (4·25)]

$$\mathbf{I} \triangleq \int_{t_1}^{t_2} \mathbf{F}\, dt \bigg|_i = m\mathbf{V}^{pi}(t_2) - m\mathbf{V}^{pi}(t_1) \tag{6·61}$$

where the inertial time integral of the force $\mathbf{F}$ is called the impulse, and the product of particle mass m and inertial velocity $\mathbf{V}^{pi}$ is called the linear momentum. In §6·2·1 these concepts were extended to systems of particles; this extension involves simply summing Eq. (6·61) over the particles of the system. If over the time interval $t_2 - t_1 = \Delta t$ the external force $\mathbf{F}$ is constant, the external impulse is $\mathbf{I} = \mathbf{F}\,\Delta t$, so that if the system consists of n particles, the impulse-momentum equation provides

$$\mathbf{F}\,\Delta t = \sum_{j=1}^{n} m_j \mathbf{V}^{p_j i}(t_2) - \sum_{j=1}^{n} m_j \mathbf{V}^{p_j i}(t_1) \tag{6·62}$$

In application to the system of Fig. 6·20, we can consider the momentum contribution of the mass δm, which is external to the control surface at time $t_1 = t$ but inside the control surface at time $t_2 = t + \Delta t$; the mass Δm, which is internal at time t_1 but external at time t_2; and the mass $\mathfrak{M}$, which is internal to the control surface during the time interval $t_2 - t_1$. Each of these quantities of material may consist of many particles, but at time t_1 all of the particles of $\mathfrak{M}$ and Δm have inertial velocity $\mathbf{V}_1 = V\hat{\mathbf{i}}$, while all the particles of δm have inertial velocity $\mathbf{v}_1 = v_1\hat{\mathbf{i}}$; and at time t_2 all of the particles of $\mathfrak{M}$ and δm have velocity $\mathbf{V}_2 = (V + \Delta V)\hat{\mathbf{i}}$, while the particles of Δm have velocity $\mathbf{v}_2 = v_2\hat{\mathbf{i}}$. Thus Eq. (6·62) becomes

$$\mathbf{F}\,\Delta t = [(\mathfrak{M} + \delta m)(V + \Delta V) + \Delta m v_2]\hat{\mathbf{i}} - [(\mathfrak{M} + \Delta m)V + \delta m v_1]\hat{\mathbf{i}}$$

The force $\mathbf{F}$ might be an aerodynamic drag or other external force in the direction of the unit vector $\hat{\mathbf{i}}$; let $\mathbf{F}$ be replaced by $F\hat{\mathbf{i}}$ and rewrite the preceding equation in scalar form as

$$F\,\Delta t = (\mathfrak{M} + \Delta m + \delta m)\,\Delta V + \delta m(V - v_1) + \Delta m\,(v_2 - V - \Delta V) \tag{6·63}$$

Define $v_i \triangleq V - v_1$ as the speed of approach of incoming particles relative to the control surface and $v_o \triangleq V + \Delta V - v_2$ as the relative speed of departure of outgoing particles. Dividing Eq. (6·63) by Δt produces

$$F = \mathfrak{M}\frac{\Delta V}{\Delta t} + (\Delta m + \delta m)\frac{\Delta V}{\Delta t} + \frac{\delta m}{\Delta t} v_i - \frac{\Delta m}{\Delta t} v_o \tag{6·64}$$

As the time interval Δt diminishes, the quantities ΔV, Δm, and δm diminish also, so that in the limit as $\Delta t \to 0$ terms such as $(\Delta m + \delta m)\,\Delta V/\Delta t$ go to zero. Because we are dealing with discrete particles rather than a continuous flow of material, certain mathematical problems arise if we take this limit formally, so we may content ourselves with an interpretation of the quantities $\mu_i \triangleq \delta m/\Delta t$ and $\mu_o \triangleq \Delta m/\Delta t$, respectively, as representative in some sense of average rates of flow of particles into and out of the control volume. This is a reasonable step provided that these flow rates are constants. With the additional substitution of dV/dt for $\Delta V/\Delta t$ (a step justified formally only in the limit as $\Delta t \to 0$), Eq. (6·65) takes the final form of

$$F = \mathfrak{M}\frac{dV}{dt} + \mu_i v_i - \mu_o v_o \tag{6·65}$$

6·6·3 Derivation from $\mathbf{F} = m\mathbf{A}$

Perhaps because Newton chose to express his second law in words we symbolize by

$$\mathbf{F} = \frac{{}^i d}{dt}(m\mathbf{V}^{pi}) \tag{6·66}$$

rather than in terms of words symbolized by

$$\mathbf{F} = m\frac{{}^i d}{dt}\mathbf{V}^{pi} \tag{6·67}$$

(see §1·2·1), many students and functioning engineers persist in the notion that Eq. (6·65) for variable-mass systems can be obtained by substituting the variable system mass $\mathfrak{M}$ for the constant particle mass m in Eq. (6·66) and simply differentiating. This is *not correct*, as is easily verified by the substitution

$$\frac{{}^i d}{dt}(\mathfrak{M}V\hat{\mathbf{i}}) = \left(\mathfrak{M}\frac{dV}{dt} + \frac{d\mathfrak{M}}{dt}V\right)\hat{\mathbf{i}}$$

The quantity in parentheses on the right-hand side here matches the right-hand side of Eq. (6·65) only in the extraordinary case when $v_i = v_0 = V$.

It is, of course, still possible to use Newton's second law to derive Eq. (6·65); indeed, the impulse-momentum equation of §6·6·2 is merely an integrated form of Eq. (6·67). Should we wish to apply Eq. (6·67) more directly, then we must apply it separately to each of the constant-mass systems $\mathfrak{M}$, Δm, and δm in Fig. 6·20, and then combine the result to obtain Eq. (6·65). In application to Δm, Eq. (6·67) provides

$$\Delta F\hat{\mathbf{i}} = \Delta m\frac{{}^i d}{dt}(v\hat{\mathbf{i}}) = \Delta m \lim_{\Delta t \to 0}\frac{v_2\hat{\mathbf{i}} - V\hat{\mathbf{i}}}{\Delta t}$$

which may be written in terms of the notation of §6·6·2 as

$$\Delta F = \lim_{\Delta t \to 0} \frac{(v_2 - V)\,\Delta m}{\Delta t} = -\mu_o v_o$$

Similarly in application to δm we find

$$\delta F = \lim_{\Delta t \to 0} \frac{(V + \Delta V - v_1)\,\delta m}{\Delta t} = \mu_i v_i$$

Finally, the material system $\mathfrak{M}$ is subjected to an external force $F\hat{\mathbf{i}}$ in addition to the forces $-\Delta F\hat{\mathbf{i}}$ and $-\delta F\hat{\mathbf{i}}$ due to interactions with Δm and δm, respectively. The result is (in scalar terms)

$$F - \Delta F - \delta F = \mathfrak{M}\frac{dV}{dt}$$

or

$$F = \mathfrak{M}\frac{dV}{dt} + \mu_i v_i - \mu_o v_o$$

confirming Eq. (6·65).

6·7 PERSPECTIVE

In reviewing the material in Chap. 6, you should be conscious of the fact that it is theoretically possible to idealize almost any material system as a collection of particles, so that a complete set of equations of motion is available by writing $\mathbf{F}_j = m_j\mathbf{A}_j$ $(j = 1, \ldots, N)$ for each particle in the system. There are, however, obvious practical obstacles to this approach to mechanics. Although it is perfectly reasonable to idealize a gas as a collection of particles (corresponding perhaps to the molecules), this idealization becomes mathematically awkward in the variable-mass problem when we seek to interpret flow rate as a limit. This little difficulty (which can be circumvented formally only by modifying some basic concepts of calculus to accommodate discrete and instantaneous variations of functions) offers motivation for the alternative idealization of a gas as a continuous distribution of matter.

Similarly, the application in §6·5·2 of the particle-system idealization to the catapult necessitated the manipulation of cumbersome summations over a large and ill-defined number of particles. We shall see that some aspects of this problem can be handled more neatly by introducing the concept of the rigid body.

Nonetheless, the particle-system idealization serves to illustrate most of the basic concepts of mechanics. In later developments we seek primarily to make our tasks easier by introducing more complex material-system idealizations.

PROBLEMS

6·1 Find the mass centers of the systems of particles shown in Fig. P6·1*a*, *b*, and *c* if $m_1 = m_2 = 3m_3 = 2m_4 = 6m$.

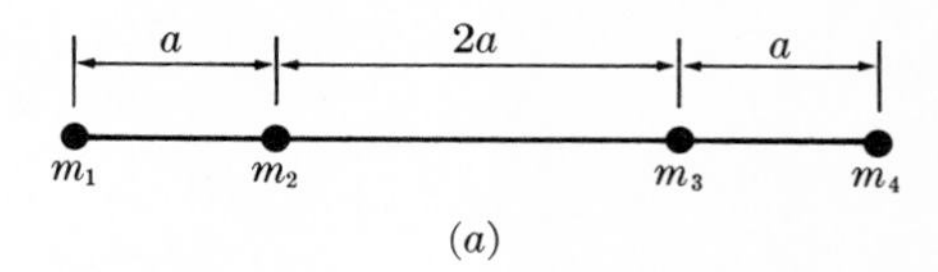

(*a*)

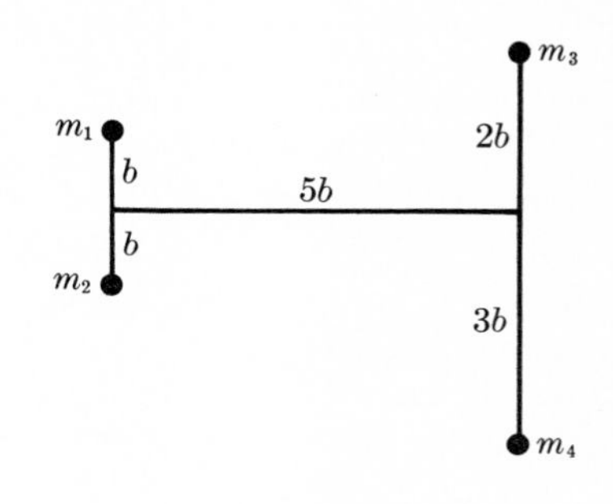

(*b*)

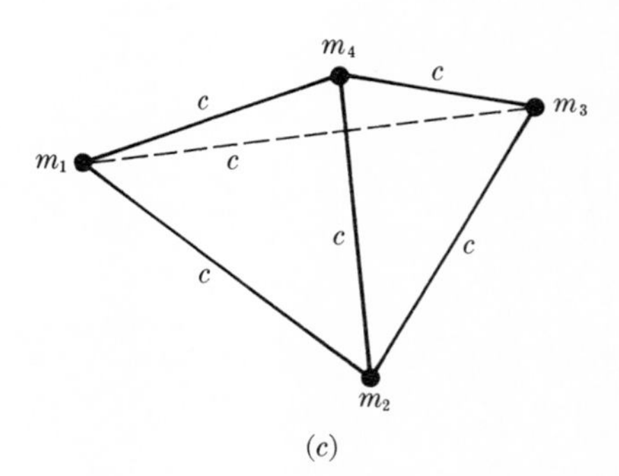

(*c*)

FIGURE P6·1

6·2 A varied group of five balloons are released at ground level simultaneously, and they each rise with a different constant acceleration, as shown in Fig. P6·2. The balloon masses are $m_1 = 0.1$ kg, $m_2 = 0.2$ kg, $m_3 = 0.3$ kg, $m_4 = 0.4$ kg, and $m_5 = 0.5$ kg; and if the altitude above the ground of the balloon with mass m_i is x_i, then the observed acceleration magnitudes are $\ddot{x}_1 = 1$ m/sec², $\ddot{x}_2 = 0.6$ m/sec², $\ddot{x}_3 = 0.5$ m/sec², $\ddot{x}_4 = 0.9$ m/sec², and $\ddot{x}_5 = 1.3$ m/sec². Determine the time history of the altitude $X(t)$ of the mass center of the system of five balloons.

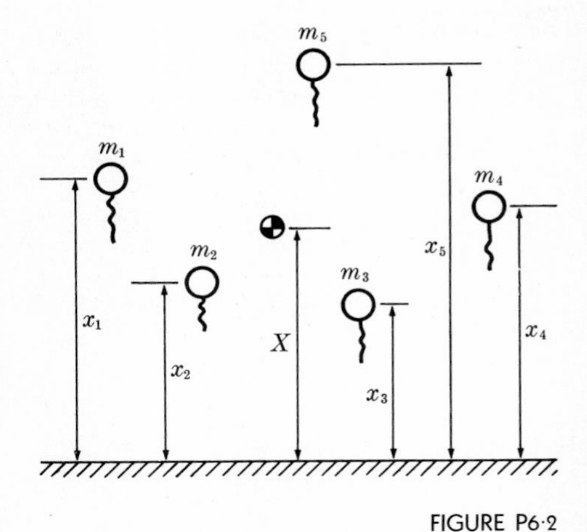

FIGURE P6·2

6·3 A vehicle is traveling through space with the constant inertial velocity **V** when a faulty component causes it to explode violently. Determine the subsequent motion of the mass center of the many bits and pieces of the vehicle, if this is possible, or indicate any additional information required to make this determination.

6·4 Two astronauts connected by a taut tether line of length 100 m are separated from their spacecraft, which is 40 m from the center of the rope (the center of mass of the two-astronaut system), as shown in Fig. P6·4. Both astronauts have depleted their fuel supplies, and have no propulsive capability, so they cannot change the motion of their combined center of mass. Both the astronauts and the spacecraft are at rest in inertial space (they are traveling together at constant velocity relative to the "inertial reference frame" established by the sun and the distant stars). How can they regain the spacecraft?

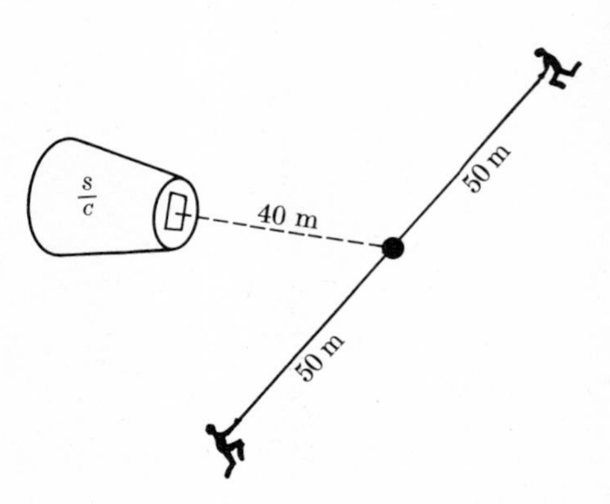

FIGURE P6·4

6·5 The dumbbell shown in Fig. P6·5 is idealized as a pair of particles of mass m connected by a rigid massless rod. A hammer blow applied to one particle in a direction normal to the rod imparts an impulse **I** to the dumbbell. What is the change in the inertial velocity of the *center* of the rod?

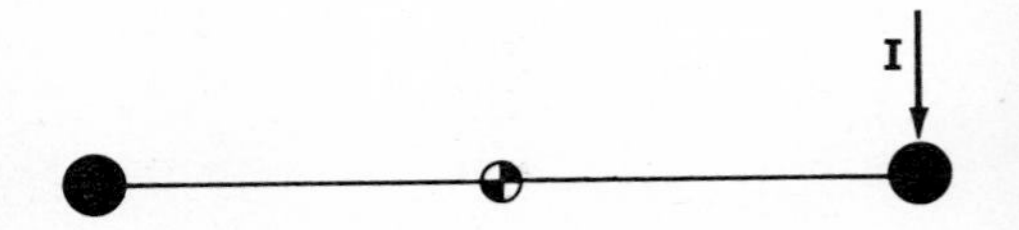

FIGURE P6·5

6·6 The space station shown in Fig. P6·6 is idealized as four identical particles having mass $m = 500$ kg. These are interconnected in a cruciform pattern by rigid massless rods of dimensions $a = 5$ m and $b = 4$ m. A jet on each of the four "particles" applies a thrust of magnitude $T = 50$ newtons normal to the rod and in the plane of the cruciform. If the space station is at rest in inertial space at time $t = 0$ when the jets are turned on, how long does it take before it is spinning at the inertial rate $\Omega = 2$ rad/sec?

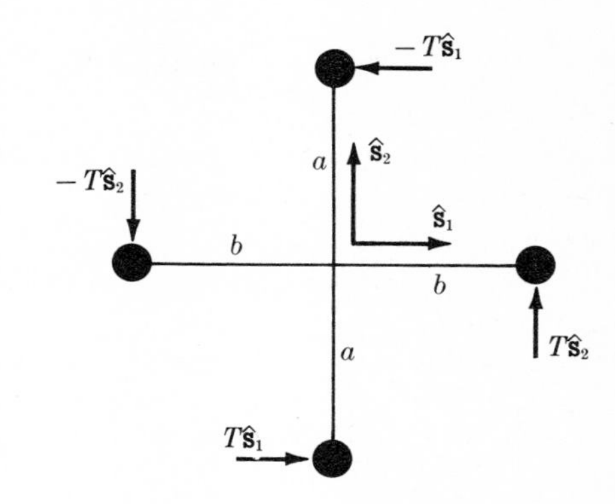

FIGURE P6·6

6·7 A dumbbell, consisting of two particles, each of mass m, separated by a massless

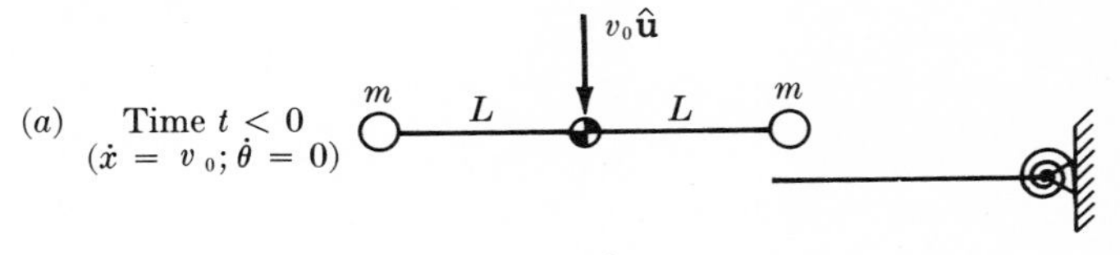

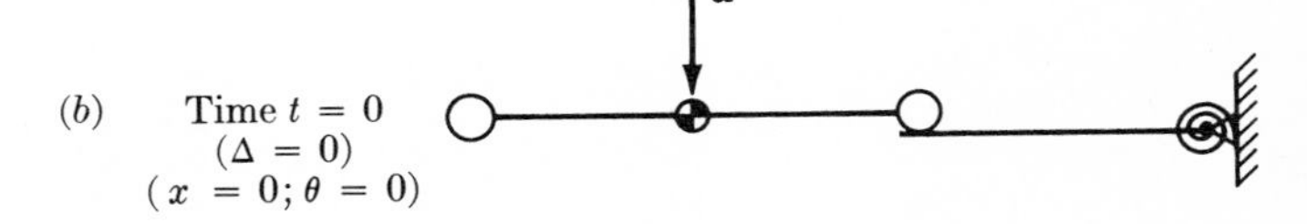

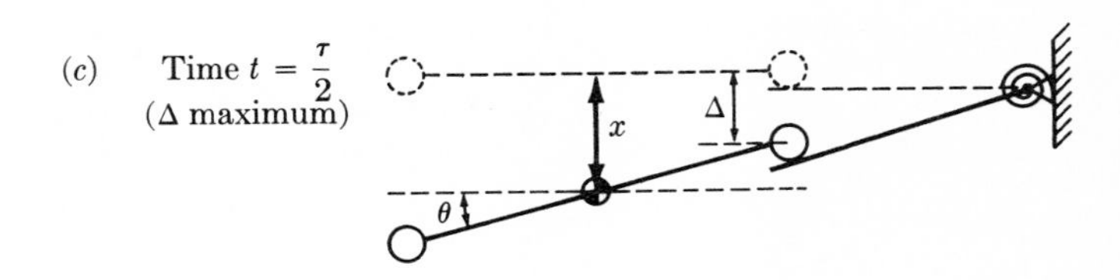

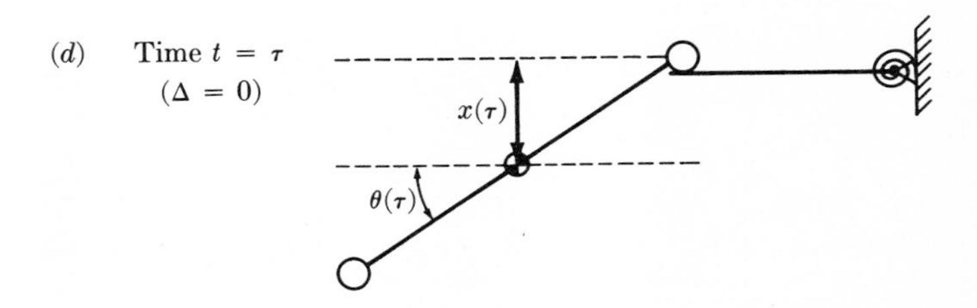

FIGURE P6·7

rigid rod of length $2L$, is translating freely through space (no gravity) with velocity $v_0\hat{\mathbf{u}}$ when it encounters at one end a linear spring that applies a force $-k\Delta\hat{\mathbf{u}}$, where Δ is the displacement in direction $\hat{\mathbf{u}}$ of the particle touching the spring. Let x be a coordinate describing the displacement of the dumbbell center in direction $\hat{\mathbf{u}}$ after contact at $t = 0$, and let θ be a coordinate describing the rotation of the dumbbell from its original orientation. Figure P6·7 shows the system at four points in time, with $t = \tau$ being the time when the spring returns to its undeflected position and contact ceases.

Assuming that θ remains small, so $\sin \theta \approx \theta$ and $\cos \theta \approx 1$, find $\dot{x}$ and $\dot{\theta}$ for time $t = \tau$. (Since contact ceases at this time, these values will persist until other forces are applied.)

6·8 Trouble was experienced in early 1966 during the first attempt by United States astronauts to rendezvous their spacecraft (*Gemini 8*) with another vehicle and dock. Before an explanation of the malfunction was publicly announced, various hypotheses were advanced in the public press, among which was the following advice in a letter to the editor of an aerospace industry weekly newsmagazine: "I believe the simple law of motion is being overlooked in the orbital linkup of two vehicles. It is my estimation that when the *Gemini 8* spacecraft made a rendezvous with the *Agena* target vehicle, in preparation for docking, their orbits were not a perfect match. After linkup, each vehicle, naturally, tried to continue in its respective orbit with the resultant crisis. In space, as on the earth, it is a good idea to stay within the law."

A second letter in the same magazine advanced the same theory, only with much more elaboration, concluding that "a major and possibly unsolved problem exists in solid docking of two objects from separate orbits because Newton's basic laws of motion are violated."

Your assignment is to write a letter of thanks to these readers, explaining in detail your agreement or disagreement with their hypothesis.

6·9 Apply an appropriate work-energy equation to obtain an expression for the speed of the mass center of a 60-lb sack of sand which at time $t = 0$ is released with its mass center a distance 100 ft above the earth, and with the mass center traveling horizontally at speed 50 ft/sec and the sack rotating without distortion with angular speed 2 rad/sec about a vertical axis. Assume a uniform (constant) gravitational attraction.

6·10 Particles p_1 and p_2, of masses m and $2m$, respectively, are released at time $t = 0$ from a position of rest in inertial space, as shown in Fig. P6·10, where they are confined to a massless tube and connected by a linear spring with spring constant k and undeformed length L. Prior to release at $t = 0$ the particles are held in the tube a distance $L - \Delta$ apart. Determine the maximum inertial speed subsequently experienced by either particle.

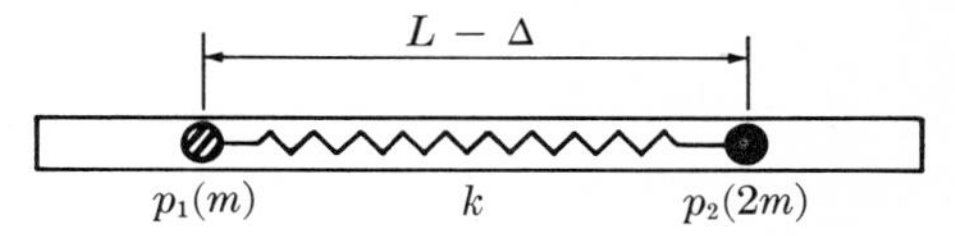

FIGURE P6·10

6·11 A train of length 50 π m is shown in Fig. P6·11 parked at the top of a hill. All brakes and other friction restraints

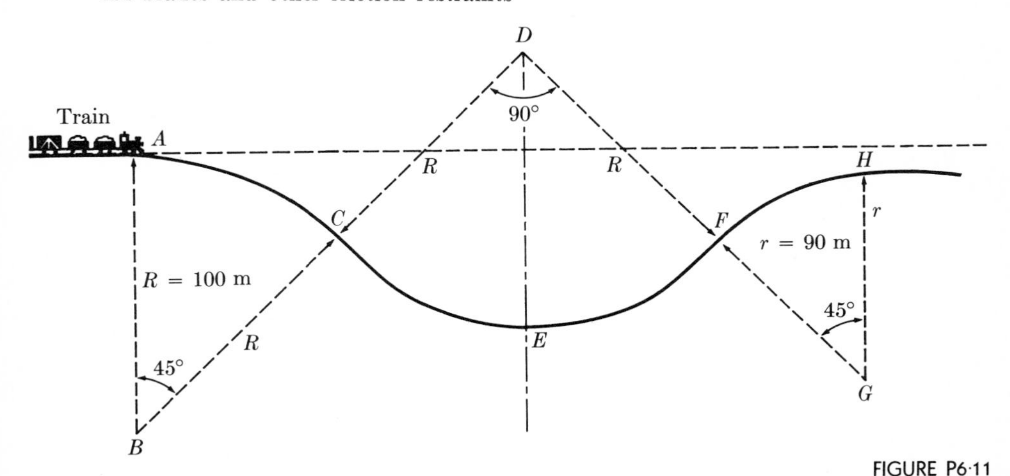

FIGURE P6·11

are removed from the wheels and couplings of the train, so with a small push it rolls down the hill, dissipating no energy. As shown in Fig. P6·11, the track consists of circular arcs. Where will the front of the engine be when the train reaches maximum speed? How fast will it go? (Hint: Treat the train as a chain of particles.)

6·12 A rigid hoop idealized as a ring of particles starts from rest at the top of a 30° slope in a uniform gravity field, as shown in Fig. P6·12; the hoop rolls straight down the hill without slipping, skidding, or deviating from planar motion. Determine the speed of the center of the hoop after the hoop has rolled a distance L down the slope.

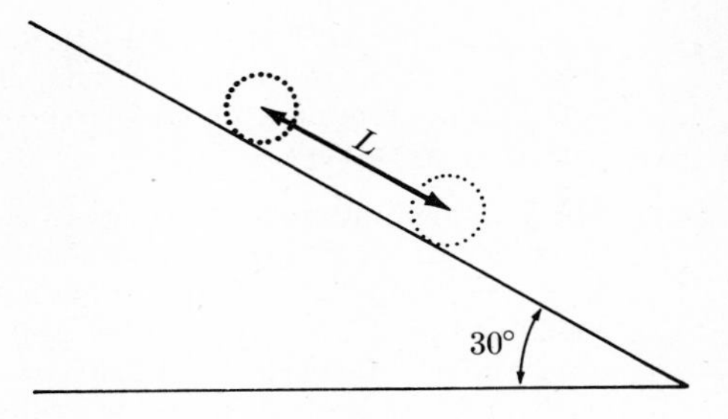

FIGURE P6·12

6·13 Classify the following equations of constraint as holonomic or nonholonomic (the unknown coordinates are denoted by x_1, etc.).

a. $x_1^2 + x_1x_2 + Rx_3 = 5R^2$
b. $\dot{x}_1 + \dot{x}_2 + \dot{x}_3 = C$
c. $x_1^2 > x_2^2$
d. $x_1\dot{x}_1 + x_2\dot{x}_2 + x_3\dot{x}_3 = 0$
e. $x_1\ddot{x}_2 + \dot{x}_2^2 = 0$
f. $t^2x_1 + x_2 = Ct^2$

6·14 Classify as holonomic or nonholonomic the constraint equations implied for the N-particle systems pictured in Fig. P6·14, in which a solid line is an undeformable line or surface.

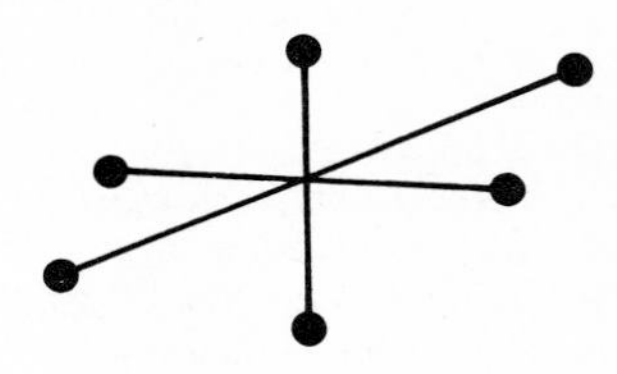

(a)

Interior particles

(b)

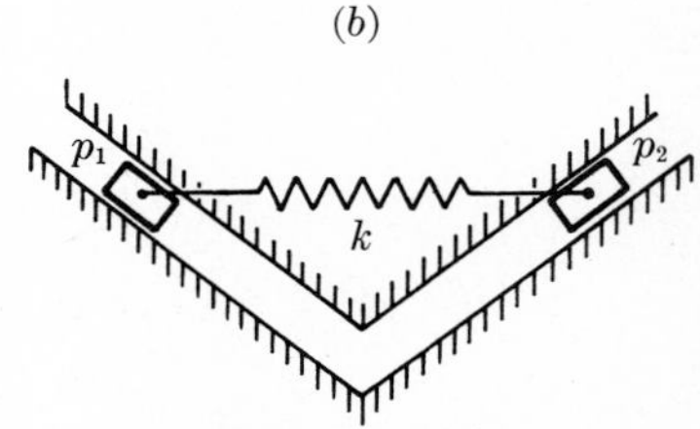

Frictionless surfaces

(c)

FIGURE P6·14

6·15 Two particles p_1 and p_2, each having mass m, are attached by a massless pin-jointed rigid rod of length r and constrained to remain in the plane of the paper, as shown in Fig. P6·15. Particle p_1 is further constrained to move on a frictionless surface inclined 30° from the horizontal; it is resisted in its motion on this surface by a linear spring having undeformed length L and spring constant k. Gravity force $mg\hat{\mathbf{u}}_y$ is exerted on each particle. Let w be the extension of the linear spring, and let θ be the inclination of the rigid rod from the vertical, with $\theta > 0$ in the figure. Apply $\mathbf{F} = m\mathbf{A}$ to each of the particles to obtain equations of motion in the form of two second-order scalar differential equations in w and θ.

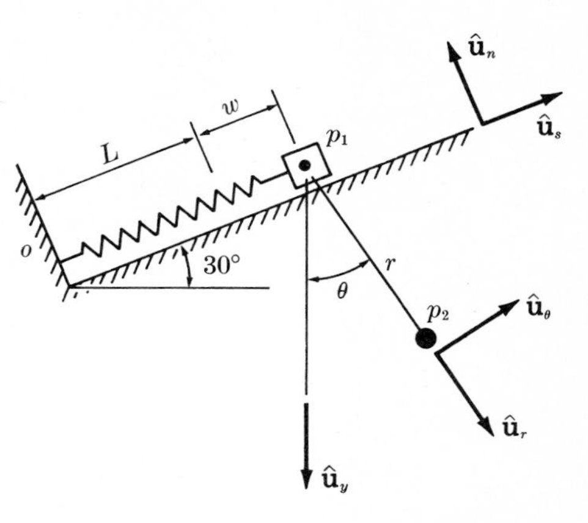

FIGURE P6·15

6·16 Repeat Prob. 6·15, this time using Eq. (6·32) rather than $\mathbf{F} = m\mathbf{A}$ to obtain the required equations of motion.

6·17 Repeat Prob. 6·15, this time using Lagrange's equations in the form of Eq. (6·33) to obtain the required equations of motion.

6·18 Repeat Prob. 6·15, this time using Lagrange's equations in the form of Eq. (6·36) to obtain the required equations of motion.

6·19 Repeat Prob. 6·7, using an appropriate form of Lagrange's equations.

6·20 Determine the reactions at A and B for static equilibrium of the simple pin-jointed truss shown in Fig. P6·20, using the method of virtual work. Check your answer with some other method. Assume $W = 500$ lb, $T = 100$ lb, and $a = 10$ ft.

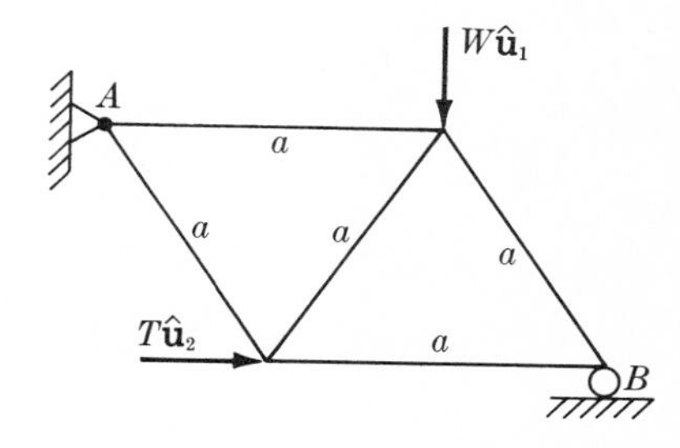

FIGURE P6·20

6·21 Determine the forces in the members aB and iH of the truss shown in Fig. P6·21 using the method of virtual work. Check your answer with some other method. Assume that $W = 1000$ lb, and each span has width L and height $\sqrt{3}L$.

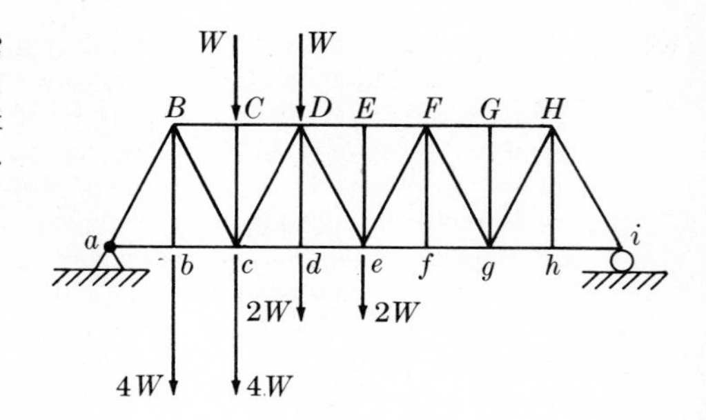

FIGURE P6·21

6·22 Three particles of mass m are supported by four massless rods connected by pin joints, as shown in Fig. P6·22. Solve for the values of θ_1 and θ_2 for which the system is in static equilibrium under the uniform attraction of gravity. Total span is $2L$.

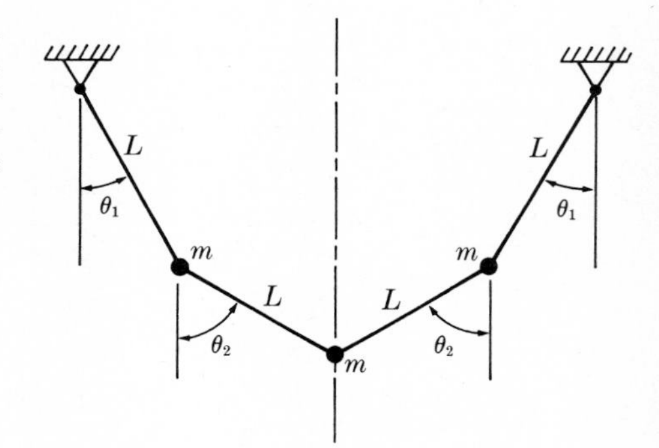

FIGURE P6·22

6·23 Use the fact that potential energy V has a stationary value for a condition of static equilibrium to find an algebraic equation that when solved provides a static-equilibrium value of the vertical displacement δ for the system shown in Fig. P6·23. Compare your result to Eq. (2·26). Each spring has spring constant k and undeformed length a.

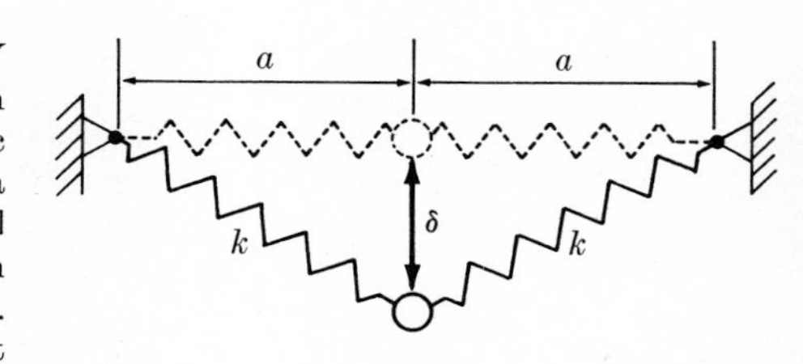

FIGURE P6·23

6·24 A particle p of weight W is constrained by a rigid rod of length L to remain on the surface of a circular ring supported in a vertical plane, as shown in Fig. P6·24. The particle moves without frictional restraint on the ring, but the rod is attached to a rotary spring with spring constant K at the center of the ring, so that when p is displaced through the angle θ rad, the rod applies to p a restoring force of magnitude $K\theta/L$. Assuming $K = WL/1.2$ leads to three equilibrium values of θ (call them $\theta_1{}^*$, $\theta_2{}^*$, and $\theta_3{}^*$):

a. Derive algebraic equations that can be solved for $\theta_\alpha{}^*$ ($\alpha = 1, 2, 3$)
b. Assess the stability of the solutions $\theta_1{}^*$, $\theta_2{}^*$, and $\theta_3{}^*$

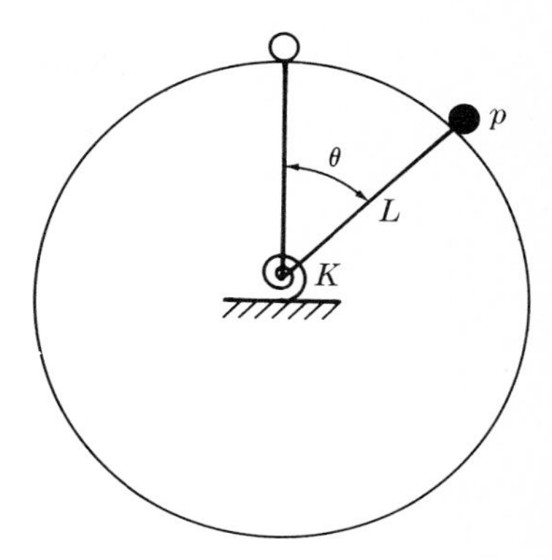

FIGURE P6·24

6·25 Six rigid rods are radially mounted like the spokes of a wheel, meeting together at the center C of a circle of radius R, as shown in Fig. P6·25. Each rod is pinned to the rigid structure serving as the rim of the wheel, and rotation at the pin is resisted by a rotary spring with spring constant K, such that a single vertical force of 1 lb at C causes each of the six rods to rotate through an angle of $R/6K$ rad. (Adopt the numerical value 0.006 for the ratio R/K.) At the center of each rod there is a particle, and the individual weights are 6 lb, 30 lb, 12 lb, 60 lb, 36 lb, and 48 lb. What is the vertical deflection at point C due to this loading?

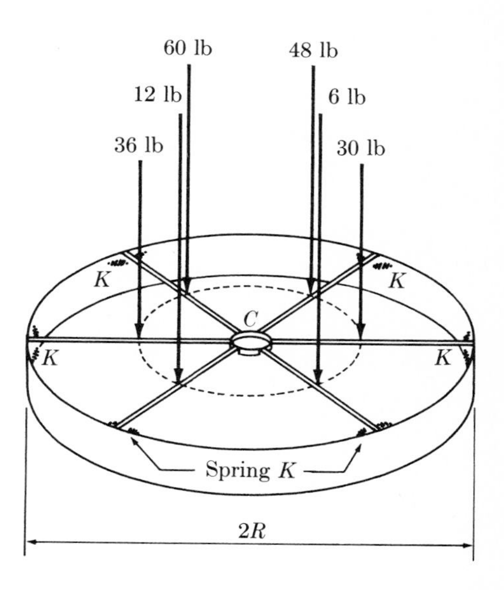

FIGURE P6·25

6·26 Calculate the 36 elements k_{ij} $(i, j = 1, \ldots, 6)$ of the stiffness matrix $[K]$ for the pin-jointed truss of Fig. P6·26 (compare to Fig. 6·13a).

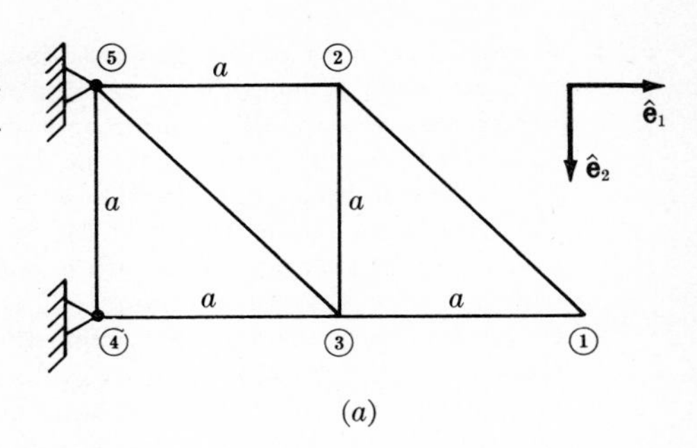

(a)

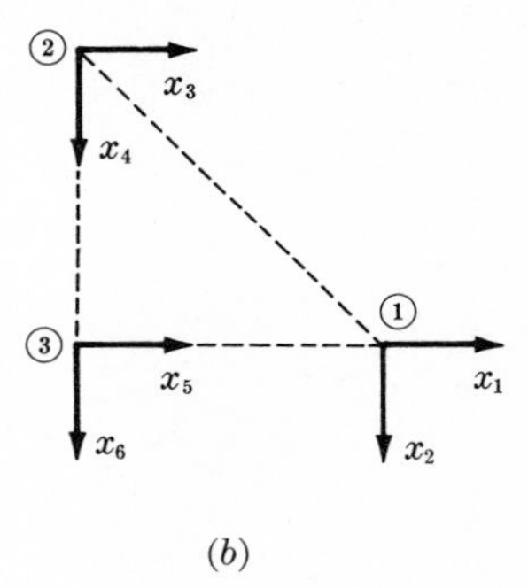

(b)

FIGURE P6·26

6·27 For the pin-jointed truss of Fig. P6·27, calculate the 2×2 stiffness matrix $[K]$, find $[K]^{-1}$ (see Appendix B), and use the result to determine the deflections x and y of point P, given the numerical data: $k_A = 1000$ lb/in., $k_B = 1500$ lb/in., $|F| = 2000$ lb, and $b = 120$ in.

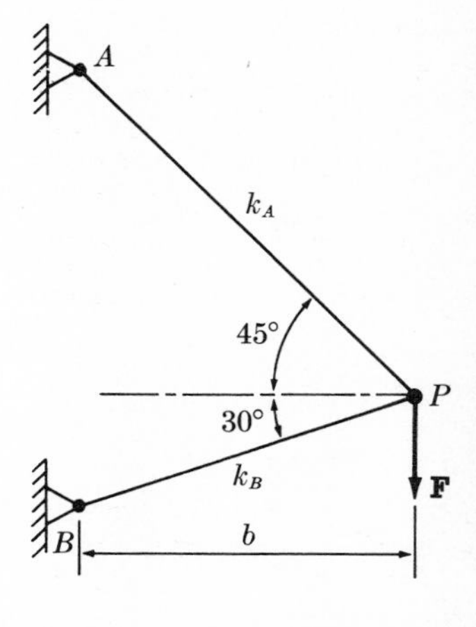

FIGURE P6·27

6·28 Our Hero is chained to the wall of a bulletproof, enclosed boxcar waiting on the tracks ahead of the oncoming royal train (see Fig. P6·28). The Villain has placed explosives on the track in a plot to assassinate the King. Our Hero has a plan to save the King, at the cost of his own life. He takes out his concealed revolver and fires into the opposite wall of the boxcar, reasoning that the kinetic energy of the bullets must be absorbed by the wall, thereby imparting to the boxcar a motion that would cause it to cross the explosive charge first.

If the wheel bearings of the box car are frictionless, will our Hero's plan succeed? Explain carefully.

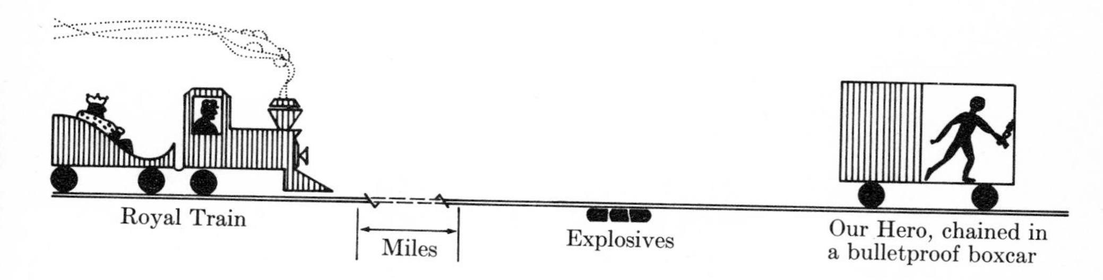

FIGURE P6·28

6·29 An airplane of mass $3m$ flying horizontally at speed v (known) picks up a stationary glider of mass m by snagging a tow rope. The tow line, which is assumed massless, transmits an impulse corresponding to an impact with coefficient of restitution $e = 0.5$. As shown in Fig. P6·29, the tow line becomes taut at an angle of 30 degrees from the horizontal. What are the velocities of airplane and glider immediately after the pickup?

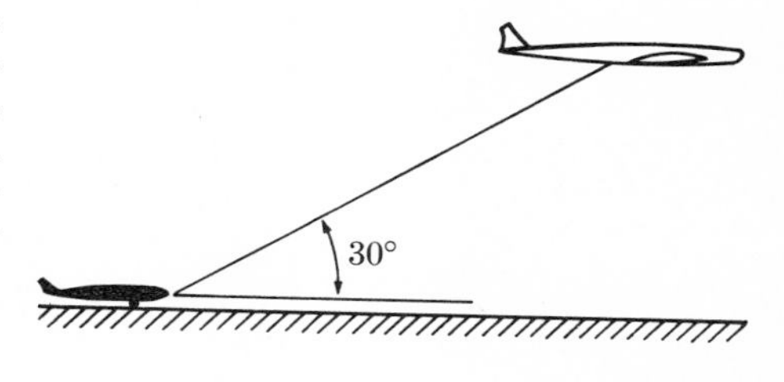

FIGURE P6·29

6·30 Two particles, each having mass m, approach each other along a straight line, each having inertial speed v, and collide with a coefficient of restitution e. What is the percentage of reduction in the system kinetic energy?

6·31 A ball dropped from rest a distance h from the floor in a uniform gravity field rebounds to height $0.75h$. What is the coefficient of restitution?

6·32 A cart of mass $3m$ is struck by the bob of a pendulum of mass m and length L, as shown in Fig. P6·32. Determine the angle φ_0 to which the pendulum must be elevated if it is to be released from rest so that it strikes the cart with sufficient impact to push the cart over a rise of height $h = L/8$. Assume that no energy is dissipated anywhere.

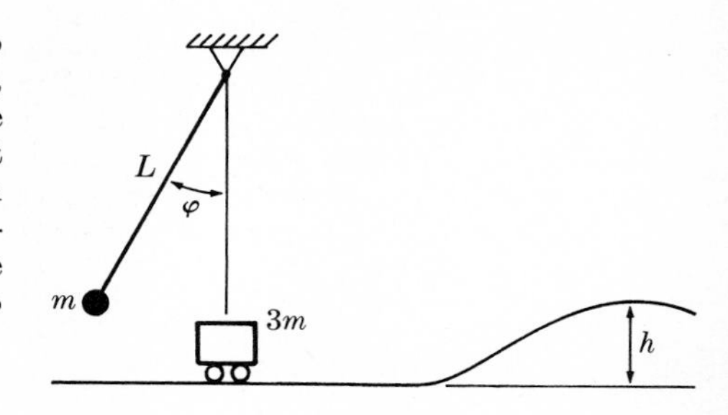

FIGURE P6·32

6·33 A birdhouse weighing 6.44 lb is mounted on a slender (massless) elastic pole, as shown in Fig. P6·33. When a horizontal force of 22.5 lb is applied statically to the birdhouse, it deflects horizontally 1 in., providing a measure of its elastic spring constant.

A fat, careless pigeon (weighing 3.22 lb) flying horizontally at 45.0 ft/sec crashes into the birdhouse, and falls unconscious on the landing porch.

a. What is the frequency of the ensuing oscillation?

b. Estimate the maximum horizontal displacement of the birdhouse after impact, making convenient assumptions.

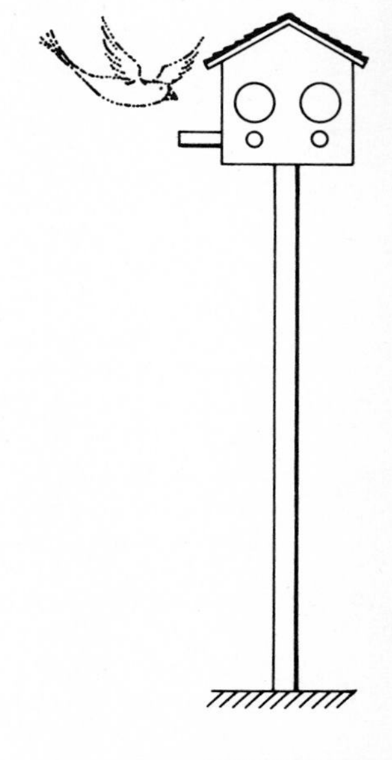

FIGURE P6·33

6·34 A sack of sand of mass m is dropped from rest at a height h directly above one end of a rigid, massless board of length $2b$; at the opposite end, the centrally supported board carries a second sack of mass m (see Fig. P6·34). Assume that the sack makes a plastic collision with the board (no elastic rebound), and as soon as the board rotates from its initial small inclination angle α to a horizontal orientation, the board collides inelastically with the ground. How high will the sack initially on the ground go after the collisions (assuming that it leaves the board with vertical velocity)?

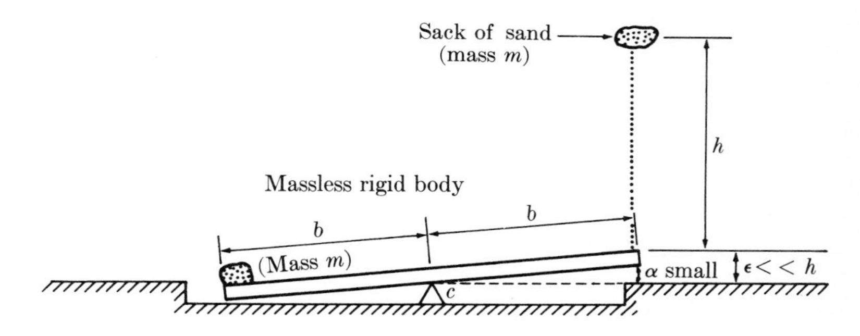

FIGURE P6·34

6·35 As he approaches the edge of a 20-ft precipice, Tarzan sees that Jane is in danger of being attacked by an alligator (see Fig. P6·35). Assessing the situation in an instant, he sees that he must make use of a nearby vine to swoop down, grasp his mate, and swing with her to safety on the opposite shore of the lagoon. When performing such feats, Tarzan is accustomed to letting go of the vine when the angle θ between the vine and a vertical line is 30 degrees. He then flies through space to the desired landing spot. Tarzan quickly compiles the following data relevant to this critical situation: length of vine, 30 ft; width of lagoon, 60 ft; height of precipice, 20 ft;

Jane's weight, 120 lb; and Tarzan's weight, 200 lb. Noting further that the vine's support point is 30 ft directly above Jane, he realizes that one more piece of information is required before he can act. You must help him obtain this information.

How many stones must Tarzan carry in his loincloth so that he and Jane will land safely on the shore of the lagoon? Each stone weighs 20 lb.

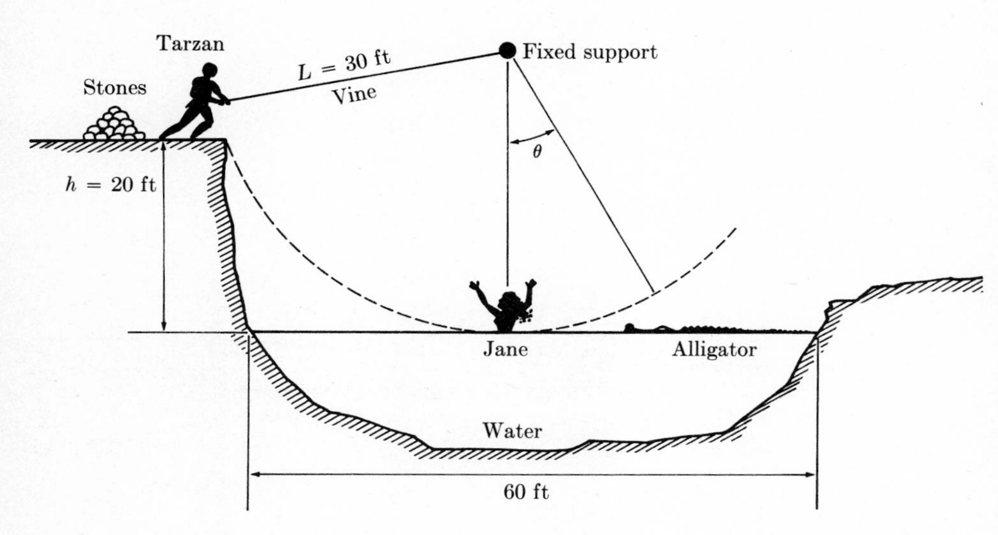

FIGURE P6·35

PART TWO
MECHANICS OF RIGID BODIES

CHAPTER SEVEN
STATICS: FORCE AND MOMENT EQUILIBRIUM

7·1 EQUIVALENCE OF SYSTEMS OF FORCES

7·1·1 Force resultant and moment resultant

In anticipation of the transition from *systems of particles* to *rigid bodies* for the mathematical models of mechanical systems, a brief digression is required for the more efficient treatment of *systems of forces.*

In Part One each force is always associated with a particle, and the *line of action* of the force as a *bound vector* always passes through the corresponding particle. Indeed, the very *definition* of a force as the product $m\mathbf{A}$ (see Chap. 1) involves a particle of mass m and inertial acceleration $\mathbf{A}$. The extension of Newton's laws from *particles* to *rigid bodies* introduces certain conceptual difficulties, and among them is the interpretation of the word *force;* but for the moment such problems can be avoided if we conceive of a rigid body as a system of particles (finite in number) constrained so that each particle maintains a fixed distance from every other particle. This concept of the rigid body is broadened in the following section.

We can imagine then a *system of forces* (associated if you wish with a system of particles), and we can search for properties of such force systems that will be useful in rigid-body mechanics. We have already encountered in the restricted framework of Part One two invaluable concepts in formulating the laws of rigid-body mechanics; these are the concepts of the *moment* of a bound vector and the *resultant* of a system or set of vectors. Instead of the restricted applications encountered in Part One, we now adopt the following general definitions:

Definitions

1 The moment of a bound vector $\mathbf{V}$ with respect to a reference point q is given by $\mathbf{Q} \times \mathbf{V}$, where $\mathbf{Q}$ is a position vector from point q to any point on the line of action of $\mathbf{V}$.

2 The resultant of a set of vectors $\mathbf{V}_1, \mathbf{V}_2, \ldots, \mathbf{V}_N$ is their vector sum $\mathbf{V}_1 + \mathbf{V}_2 + \cdots + \mathbf{V}_N$.

Neither the moment of a bound vector nor the resultant of a system of bound vectors is itself bound to any particular point or line of action.

In §4·2·5, we encountered the moment of the force $\mathbf{F}$ on a particle p with respect to an inertially fixed reference point o, and in that context we replaced $\mathbf{Q}$—in the present definition of moment—by $\mathbf{R}$, the vector from o to p. It is always possible when calculating the moment of a force to select for $\mathbf{Q}$ the vector from the reference point to the specific point of application of the force (in this case to the particle subjected to the force).

The concept of the moment of a vector also underlies the definition in §4·2·5 of the *moment of momentum* of a particle p with respect to an arbitrary point q, as given by Eq. (4·38): $\mathbf{h}^q \triangleq \mathbf{r} \times m\dot{\mathbf{R}}$. Here $\mathbf{r}$ is the vector from q to p, m is the mass of p, and $\dot{\mathbf{R}}$ is the velocity of p with respect to an inertial reference frame. From the definition of (inertial) linear momentum in Eq. (4·24), we are reminded that $\mathbf{h}^q$ is the *moment* with respect to point q of the (inertial) linear momentum. [The *angular momentum* $\mathbf{H}^q \triangleq m\mathbf{r} \times \dot{\mathbf{r}}$ defined by Eq. (4·39) is, in contrast, the moment with respect to point q of a quantity $m\dot{\mathbf{r}}$ to which we have not assigned a name in this text.] Again we have in defining the moment of momentum chosen to cross multiply $m\dot{\mathbf{R}}$ by the specific vector $\mathbf{r}$ rather than the more general vector $\mathbf{Q}$ used in the definition of *moment;* but the particularization of the generic vector $\mathbf{Q}$ is simply a convenience, and not a significant deviation from the general definition of moment.

The concept of a *vector resultant* has also been used previously in this text in a restricted manner. The term *force resultant* was used for the vector sum of a set of concurrent forces as early as §1·3·5, where it was observed that a system of forces applied to a single particle could for dynamic analysis be replaced by a single force equal to the force resultant and having a line of action also passing through p. This interpretation is in the single-particle context justified by the superposition axiom (axiom 4 in §1·3·1), but it was explicitly noted in Chap. 1 that the term force resultant does not imply any particular line of action for the vector sum.

Since the resultant of a set of bound vectors is itself not bound to any particular line of action (unless we *assign* the resultant to some arbitrarily selected line), the resultant of a set of forces is not itself a force; by definition a force must be a bound vector. Generally we do find it convenient to *assign* the force resultant some line of action (as we did in Chap. 1 for the system of concurrent forces), and only then can we call it a force.

Definitions 1 and 2 combine as follows: *The moment resultant of a system of forces with respect to a given point is the vector sum or resultant of the moments of the individual forces.* The moment resultant is a free vector, not bound to any particular line of action.

If a given system of bound vectors $\mathbf{V}_1, \ldots, \mathbf{V}_N$ has a resultant $\mathbf{V} = \mathbf{V}_1 + \cdots + \mathbf{V}_N$ and a moment resultant $\mathbf{M}^q$ with respect to point q, then the moment of the system with respect to a new point q' is given by

$$\mathbf{M}^{q'} = \mathbf{M}^q + \mathbf{S} \times \mathbf{V} \tag{7·1}$$

where $\mathbf{S}$ is the position vector of q with respect to q'. The result is easily established by letting $\mathbf{Q}_1, \ldots, \mathbf{Q}_N$ be position vectors from q to selected points on the lines of action of $\mathbf{V}_1, \ldots, \mathbf{V}_N$, respectively (as in Fig. 7·1, where $N = 4$), similarly letting $\mathbf{Q}'_1, \ldots, \mathbf{Q}'_N$ be position vectors to the selected points from q', and substituting $\mathbf{Q}'_j = \mathbf{S} + \mathbf{Q}_j$ into the moment definitions for $j = 1, \ldots, N$. The result is

$$\begin{aligned}\mathbf{M}^{q'} &= \sum_{j=1}^{N} \mathbf{Q}'_j \times \mathbf{V}_j = \sum_{j=1}^{N} (\mathbf{S} + \mathbf{Q}_j) \times \mathbf{V}_j = \mathbf{S} \times \sum_{j=1}^{N} \mathbf{V}_j + \sum_{j=1}^{N} \mathbf{Q}_j \times \mathbf{V}_j \\ &= \mathbf{S} \times \mathbf{V} + \mathbf{M}^q\end{aligned}$$

confirming Eq. (7·1).

The concepts of force resultant, moment, and moment resultant can be illustrated for the system portrayed in Fig. 7·2 as follows. The force resultant $\mathbf{F}$

for the four forces shown is

$$\mathbf{F} = \mathbf{F}_1 + \mathbf{F}_2 + \mathbf{F}_3 + \mathbf{F}_4$$
$$= 2F\hat{\mathbf{u}}_3 - F\hat{\mathbf{u}}_3 - 2F\hat{\mathbf{u}}_3 + F\hat{\mathbf{u}}_2 = -F\hat{\mathbf{u}}_3 + F\hat{\mathbf{u}}_2$$

The moment $\mathbf{M}_j{}^q$ for the central point q of force $\mathbf{F}_j$ is given for $j = 1, 2, 3, 4$ by

$$\mathbf{M}_1{}^q = (a\hat{\mathbf{u}}_1 + b\hat{\mathbf{u}}_2) \times 2F\hat{\mathbf{u}}_3 = 2bF\hat{\mathbf{u}}_1 - 2aF\hat{\mathbf{u}}_2$$
$$\mathbf{M}_2{}^q = (-a\hat{\mathbf{u}}_1 + b\hat{\mathbf{u}}_2) \times (-F\hat{\mathbf{u}}_3) = -bF\hat{\mathbf{u}}_1 - aF\hat{\mathbf{u}}_2$$
$$\mathbf{M}_3{}^q = (-a\hat{\mathbf{u}}_1 - b\hat{\mathbf{u}}_2) \times (-2F\hat{\mathbf{u}}_3) = 2bF\hat{\mathbf{u}}_1 - 2aF\hat{\mathbf{u}}_2$$
$$\mathbf{M}_4{}^q = (a\hat{\mathbf{u}}_1 - c\hat{\mathbf{u}}_3) \times F\hat{\mathbf{u}}_2 = cF\hat{\mathbf{u}}_1 + aF\hat{\mathbf{u}}_3$$

Note that in each of the preceding calculations an arbitrary choice is made for a particular vector from q to the line of action of the force. For example, $\mathbf{M}_1{}^q$ could

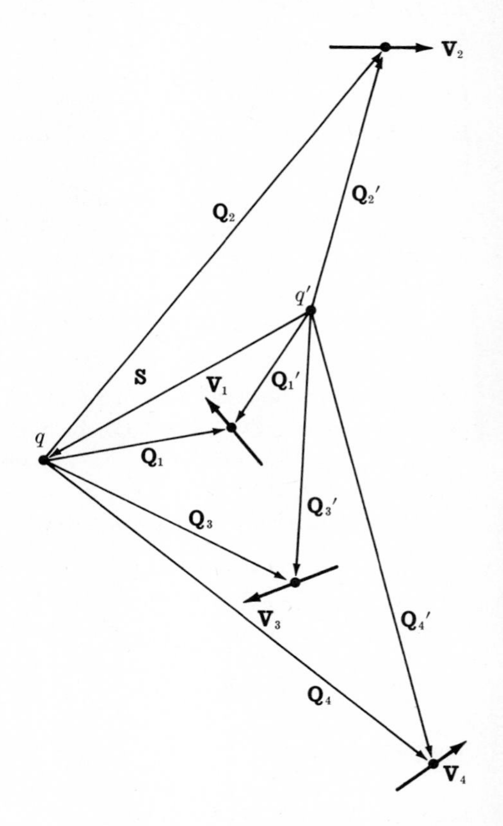

FIGURE 7·1

as well have been calculated from

$$\mathbf{M}_1{}^q = (a\hat{\mathbf{u}}_1 + b\hat{\mathbf{u}}_2 - c\hat{\mathbf{u}}_3) \times 2F\hat{\mathbf{u}}_3$$

and the result would have been the same. The choice is governed by the analyst's convenience.

The moment resultant $\mathbf{M}^q$ of the four moments in question is given by

$$\begin{aligned}\mathbf{M}^q &= \mathbf{M}_1{}^q + \mathbf{M}_2{}^q + \mathbf{M}_3{}^q + \mathbf{M}_4{}^q \\ &= (3b + c)F\hat{\mathbf{u}}_1 - 5aF\hat{\mathbf{u}}_2 + aF\hat{\mathbf{u}}_3\end{aligned}$$

To find the moment resultant $\mathbf{M}^{q'}$ of forces $\mathbf{F}_1, \ldots, \mathbf{F}_4$ with respect to point q', we can either calculate and sum $\mathbf{M}_1{}^{q'}, \ldots, \mathbf{M}_4{}^{q'}$ or use Eq. (7·1). The latter choice leads to

$$\begin{aligned}\mathbf{M}^{q'} &= \mathbf{M}^q + (a\hat{\mathbf{u}}_1 + b\hat{\mathbf{u}}_2 + c\hat{\mathbf{u}}_3) \times \mathbf{F} \\ &= (3b + c)F\hat{\mathbf{u}}_1 - 5aF\hat{\mathbf{u}}_2 + aF\hat{\mathbf{u}}_3 \\ &\qquad + (a\hat{\mathbf{u}}_1 + b\hat{\mathbf{u}}_2 + c\hat{\mathbf{u}}_3) \times (-F\hat{\mathbf{u}}_3 + F\hat{\mathbf{u}}_2) \\ &= 2bF\hat{\mathbf{u}}_1 - 4aF\hat{\mathbf{u}}_2 + 2aF\hat{\mathbf{u}}_3\end{aligned}$$

In this particular case, it is probably easier to find the same result by calculating

$$\begin{aligned}\mathbf{M}^{q'} &= \mathbf{M}_1{}^{q'} + \mathbf{M}_2{}^{q'} + \mathbf{M}_3{}^{q'} + \mathbf{M}_4{}^{q'} \\ &= (2a\hat{\mathbf{u}}_1 + 2b\hat{\mathbf{u}}_2) \times 2F\hat{\mathbf{u}}_3 + 2b\hat{\mathbf{u}}_2 \times (-F\hat{\mathbf{u}}_3) + 0 + 2a\hat{\mathbf{u}}_1 \times F\hat{\mathbf{u}}_2 \\ &= -4aF\hat{\mathbf{u}}_2 + 4bF\hat{\mathbf{u}}_1 - 2bF\hat{\mathbf{u}}_1 + 2aF\hat{\mathbf{u}}_3 \\ &= 2bF\hat{\mathbf{u}}_1 - 4aF\hat{\mathbf{u}}_2 + 2aF\hat{\mathbf{u}}_3\end{aligned}$$

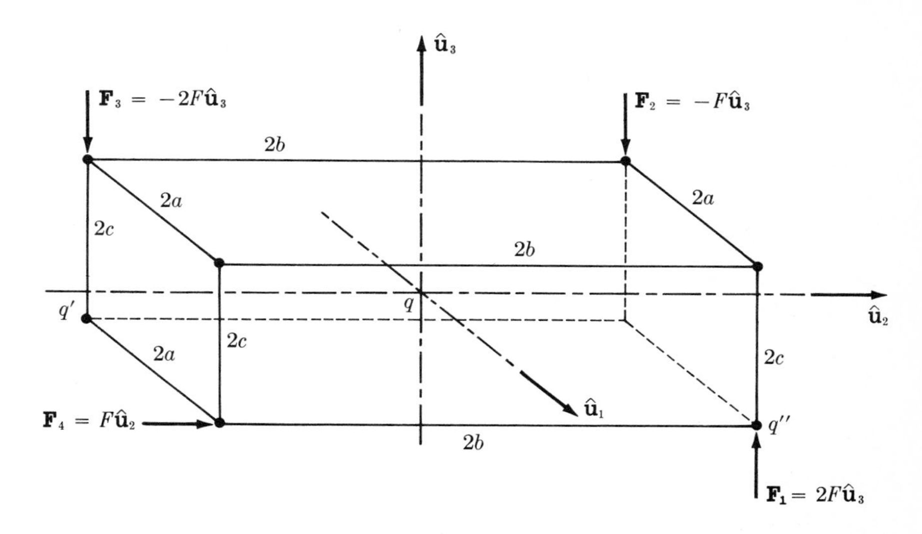

FIGURE 7·2

7·1·2 Couples and torques

The word couple is applied to a pair of equal and opposite forces[1] *having different lines of action;* thus forces $\mathbf{F}_1$ and $\mathbf{F}_3$ in Fig. 7·2 constitute a couple.

The force resultant of a couple is of course zero (as in Fig. 7·2, where $\mathbf{F}_1 + \mathbf{F}_3 = 0$), but the moment resultant is not zero (as the resultant

$$\mathbf{M}_1{}^q + \mathbf{M}_3{}^q = 4bF\hat{\mathbf{u}}_1 - 4aF\hat{\mathbf{u}}_2 \neq 0$$

for that system).

It is apparent from Eq. (7·1) that, because a couple has zero resultant force, *the resultant moment of a couple is the same for all reference points.* It is this important property of a couple that explains its usefulness. The moment of a couple (for any and every point) is a free vector called the *torque* of the couple.

To calculate the torque of a couple, we can, of course, choose any convenient reference point and calculate the moment of the couple; it is most often convenient to pick a point on the line of action of one of the forces in the couple. For example, to calculate the torque $\mathbf{T}_{1,3}$ of the couple $\mathbf{F}_1$ and $\mathbf{F}_3$ in Fig. 7·2, it is most convenient to choose a reference point such as q', to find

$$\begin{aligned}\mathbf{T}_{1,3} &= \mathbf{M}_1{}^{q'} + \mathbf{M}_3{}^{q'} = (2a\hat{\mathbf{u}}_1 + 2b\hat{\mathbf{u}}_2) \times 2F\hat{\mathbf{u}}_3 + 0 \\ &= 4bF\hat{\mathbf{u}}_1 - 4aF\hat{\mathbf{u}}_2\end{aligned}$$

This result is (and must be) the same as $\mathbf{M}_1{}^q + \mathbf{M}_3{}^q$ recorded previously.

7·1·3 Equivalence

In dealing with the mechanics of a single particle, we found that for the purposes of dynamic analysis we could replace a system of forces $\mathbf{F}_1, \ldots, \mathbf{F}_N$ applied to a given particle p by a single force $\mathbf{F}$ applied to p and equal to the system force resultant. This is an example of the *equivalence* of the system of forces $\mathbf{F}_1, \ldots, \mathbf{F}_N$ and the single force $\mathbf{F}$. A general definition of equivalence is: *Two systems of forces are said to be equivalent if they have the same force resultant and the same moment with respect to at least one reference point.*

Equation (7·1) provides assurance that *if two systems of forces are equivalent then their resultant moments with respect to any reference point are the same.*

In the chapters that follow, you will find that the dynamic response of a physical system idealized as a single rigid body depends only on the force resultant applied to the body and the moment resultant of the force system for various

[1] The definition of the word *couple* generally encompasses all such pairs of bound vectors, and not merely pairs of equal and opposite forces; but only force couples will be used in this text.

reference points. It is for this reason that the concept of equivalence is important; rather than struggle with the complex pattern of forces actually applied to a rigid body, we will first replace the actual force system by a simpler equivalent system, and then complete the analysis of the dynamic response to the simpler system of forces. We are therefore interested in finding general observations that permit in any given problem the construction of the simplest possible equivalent system of forces.

Any system of forces with force resultant $\mathbf{F}$ *is equivalent to a system consisting of a couple and a single force equal to* $\mathbf{F}$ *and having a line of action passing through an arbitrarily selected point* q; *the torque of the couple is the resultant moment of the system of forces for the point* q. The validity of this proposition is apparent since the two force systems in question have the same force resultant and the same moment resultant for point q.

To illustrate this pattern of constructing simple equivalent force systems consisting of a couple and a force, consider again the loaded block in Fig. 7·2. The system of four forces shown can be replaced by an infinite variety of equivalent systems of forces. Recalling that the force resultant is $\mathbf{F} = -F\hat{\mathbf{u}}_3 + F\hat{\mathbf{u}}_2$, and the moment resultant for point q is $\mathbf{M}^q = (3b + c)F\hat{\mathbf{u}}_1 - 5aF\hat{\mathbf{u}}_2 + aF\hat{\mathbf{u}}_3$, we can, for example, devise an equivalent system consisting of a force $\mathbf{F}$ having a line of action through q together with a couple whose torque is $\mathbf{T}^q = \mathbf{M}^q$. (The superscript q in $\mathbf{T}^q$ is a reminder that this is the torque of the couple associated with a force $\mathbf{F}$ passing through q; it should not be imagined that $\mathbf{T}^q$ has in any sense a line of action passing through q, since $\mathbf{T}^q$ is not a bound vector.)

Alternatively, we could replace the system of four forces in Fig. 7·2 by a single force $\mathbf{F}$ having a line of action through q' together with a couple whose torque is $\mathbf{T}^{q'} = \mathbf{M}^{q'} = 2bF\hat{\mathbf{u}}_1 - 4aF\hat{\mathbf{u}}_2 + 2aF\hat{\mathbf{u}}_3$.

If we pursue the practice of replacing all force systems we encounter with a single force through an arbitrary point plus a couple, we'll find that sometimes the force is zero and sometimes the couple is zero. For example, if the forces $\mathbf{F}_2$ and $\mathbf{F}_4$ were omitted from Fig. 7·2, the resulting force system would have resultant $\mathbf{F}_1 + \mathbf{F}_3 = 0$. The simplest equivalent system then consists simply of a couple with torque $\mathbf{T}_{1,3}$ as established in §7·1·2.

As another example, consider the force system that would remain in Fig. 7·2 if $\mathbf{F}_2$ and $\mathbf{F}_3$ were removed. The remaining system would have force resultant $\mathbf{F}_{1,4} = \mathbf{F}_1 + \mathbf{F}_4 = 2F\hat{\mathbf{u}}_3 + F\hat{\mathbf{u}}_2$. We could replace $\mathbf{F}_1$ and $\mathbf{F}_4$ with the single force $\mathbf{F}_{1,4}$ passing through point q plus a couple having the torque

$$\begin{aligned}\mathbf{T}^q_{1.4} &= (a\hat{\mathbf{u}}_1 + b\hat{\mathbf{u}}_2) \times 2F\hat{\mathbf{u}}_3 + (a\hat{\mathbf{u}}_1 - c\hat{\mathbf{u}}_3) \times F\hat{\mathbf{u}}_2 \\ &= (2b + c)F\hat{\mathbf{u}}_1 - 2aF\hat{\mathbf{u}}_2 + aF\hat{\mathbf{u}}_3\end{aligned}$$

Alternatively, we might select point q'' in Fig. 7·2 as the point through which to pass the line of action of the force resultant $\mathbf{F}_{1,4}$, and then (since $\mathbf{F}_1$ and $\mathbf{F}_4$ both pass through q'') the associated couple would be zero.

This last-mentioned example raises an interesting general question: Under what circumstances can we replace a system of forces by a *single force*, with no associated couple? Such an equivalence obviously exists for a system of *concurrent forces*, and in the two propositions following, other force systems are included in this category.

Propositions

1 A system of *coplanar* forces with resultant $\mathbf{F} \neq 0$ can always be replaced by an equivalent system consisting of the single force $\mathbf{F}$ having a specific line of action called the *central axis of the force system.*

2 A system of *parallel* forces with resultant $\mathbf{F} \neq 0$ can always be replaced by an equivalent system consisting of the single force $\mathbf{F}$ having a specific line of action called the central axis of the force system.

Before proving these closely related propositions, we might assess their implications for the specific problem in Fig. 7·2. We have already seen that the system of coplanar forces $\mathbf{F}_1$ and $\mathbf{F}_4$ is equivalent to the single force $\mathbf{F}_{1,4} = \mathbf{F}_1 + \mathbf{F}_4$ passing through q''. The generalization to any system of concurrent forces is straightforward. But proposition 1 says that it should also be possible to replace the system of nonconcurrent but coplanar forces $\mathbf{F}_2$ and $\mathbf{F}_3$ by a single force $\mathbf{F}_2 + \mathbf{F}_3$ for a specific line of action. Only a little thought is required to confirm this claim: The indicated line of action must intersect the line jointing the points of application of $\mathbf{F}_2$ and $\mathbf{F}_3$ at a point twice as far from $\mathbf{F}_3$ as from $\mathbf{F}_2$.

Proposition 2 indicates that forces $\mathbf{F}_1$, $\mathbf{F}_2$, and $\mathbf{F}_3$ in Fig. 7·2 comprise a system of parallel forces that can be replaced by a single force equal to $\mathbf{F}_1 + \mathbf{F}_2 + \mathbf{F}_3$ and having a specific line of action. Finding the right line of action is not so simple this time, even though this special case is simplified by the fact that $\mathbf{F}_1$ and $\mathbf{F}_3$ together comprise a couple, whose torque has been shown (in §7·1·2) to be $\mathbf{T}_{1,3} = 4bF\hat{\mathbf{u}}_1 - 4aF\hat{\mathbf{u}}_2$. Thus we know that if we let $\mathbf{P}$ be the vector from the point of application of $\mathbf{F}_2$ to a point O on the required line of action, we have

$$\mathbf{M}^O = \mathbf{T}_{1,3} - \mathbf{P} \times \mathbf{F}_2 = 0$$

or

$$\mathbf{P} \times \mathbf{F}_2 = \mathbf{T}_{1,3}$$

or

$$\mathbf{P} \times \hat{\mathbf{u}}_3 = -4b\hat{\mathbf{u}}_1 + 4a\hat{\mathbf{u}}_2$$

Thus $\mathbf{P}$ is given by

$$\mathbf{P} = -4a\hat{\mathbf{u}}_1 - 4b\hat{\mathbf{u}}_2$$

We cannot expect continued success in finding the appropriate line of action of the single force equivalent called for in propositions 1 and 2 without establishing a general formula; we shall seek this result in the proofs that follow.

In order to establish the required proofs, we can rewrite Eq. (7·1) with $\mathbf{V}$, $\mathbf{S}$, and $\mathbf{M}^q$ replaced, respectively, by $\mathbf{F}$, $\mathbf{S}^*$, and $\mathbf{M}^{q^*}$ for some point q^*. Then

$$\mathbf{M}^{q'} = \mathbf{M}^{q^*} + \mathbf{S}^* \times \mathbf{F} \tag{7·2}$$

so that $\mathbf{M}^{q^*}$ must have the same component in the direction of $\mathbf{F}$ as has $\mathbf{M}^{q'}$ for an arbitrarily chosen point q', this because $\mathbf{S}^* \times \mathbf{F}$ has no component in the direction of $\mathbf{F}$. Moreover, if q^* is that point for which the moment is of minimum magnitude, then $\mathbf{M}^{q^*}$ cannot have any component normal to $\mathbf{F}$, since if it had there would exist some vector $\mathbf{S}^{**}$ locating some point q^{**} from point q' such that $\mathbf{S}^{**} \times \mathbf{F}$ would cancel the hypothesized component of $\mathbf{M}^{q^*}$ normal to $\mathbf{F}$, implying the contradictory conclusion that $|\mathbf{M}^{q^{**}}| < |\mathbf{M}^{q^*}|$. Hence we can easily find the minimum magnitude of moment $\mathbf{M}^{q^*}$ by calculating the moment $\mathbf{M}^{q'}$ for a wholly arbitrary point q' and taking its vector component in the direction of $\mathbf{F}$. Formally, this produces an expression for the minimum moment given by

$$\mathbf{M}^{q^*} = \frac{\mathbf{M}^{q'} \cdot \mathbf{F}\mathbf{F}}{\mathbf{F} \cdot \mathbf{F}} \tag{7·3}$$

Now we can observe that if we substitute for a system of coplanar forces an equivalent system consisting of a couple plus a force resultant $\mathbf{F}$ bound to a line of action through some point q' of the plane, then the torque $\mathbf{M}^{q'}$ of the couple is normal to the plane. Then $\mathbf{M}^{q'} \cdot \mathbf{F} = 0$, and from Eq. (7·3) the minimum moment $\mathbf{M}^{q^*} = 0$.

Similarly, if we substitute for a system of parallel forces an equivalent system consisting of a couple plus a force resultant $\mathbf{F}$ bound to a line of action through any point q', then the torque $\mathbf{M}^{q'}$ of the couple is normal to $\mathbf{F}$, and again $\mathbf{M}^{q^*} = 0$ in Eq. (7·3).

Thus we have confirmed the claims in propositions 1 and 2 that there exist in these cases equivalent systems consisting of single forces, with zero associated couples.

In order to find those single-force equivalent systems noted to exist for systems of coplanar forces and systems of parallel forces, we must find an expression locating the indicated line of action, which is called the central axis of the force system. This can be accomplished by using Eq. (7·2) together with (7·3) to conclude that the vector $\mathbf{S}^*$ from an arbitrary point q' to that point q^* for which

$|\mathbf{M}|$ is minimum must satisfy

$$\mathbf{S}^* \times \mathbf{F} = \mathbf{M}^{q'} - \frac{\mathbf{M}^{q'} \cdot \mathbf{F}\mathbf{F}}{\mathbf{F} \cdot \mathbf{F}}$$
$$= \frac{\mathbf{F} \cdot \mathbf{F}\mathbf{M}^{q'} - \mathbf{F} \cdot \mathbf{M}^{q'}\mathbf{F}}{\mathbf{F} \cdot \mathbf{F}}$$
$$= \frac{\mathbf{F} \times (\mathbf{M}^{q'} \times \mathbf{F})}{\mathbf{F} \cdot \mathbf{F}} \tag{7·4}$$

implying that

$$\mathbf{S}^* = \frac{\mathbf{F} \times \mathbf{M}^{q'}}{\mathbf{F} \cdot \mathbf{F}} \tag{7·5}$$

A force system with zero force resultant and zero moment resultant for *one* reference point must, by Eq. (7·2), also have zero moment resultant for *every* reference point; such a force system is equivalent to a system consisting of no forces at all, and is called a *zero force system.* In the balance of this chapter on static equilibrium we will be working exclusively with zero force systems.

The word *equivalent* as defined here has a more restricted meaning than you will find in the dictionary. For example, the three equivalent force systems applied to the man in Fig. 7·3 are all *zero force systems,* with zero force resultant and zero moment resultant for any reference point chosen, but they do not appear to the man in the figure to be "equivalent" in every sense of that word. Equivalent systems of forces do not produce an equal distribution of loads within a body, but we'll see that they have the same influence on the motion of a rigid body.

The equivalent force system most commonly used in mechanics involves the substitution of a single concentrated force along the central axis for a system of gravity forces distributed over a body. When the gravitational forces on a body are all considered to be directed toward a common point at the attracting center, the system of forces is a *concurrent-force system,* and the central axis passes through the attracting center. The point on the central axis at which the entire body mass could be concentrated as a particle whose gravitational attraction would match

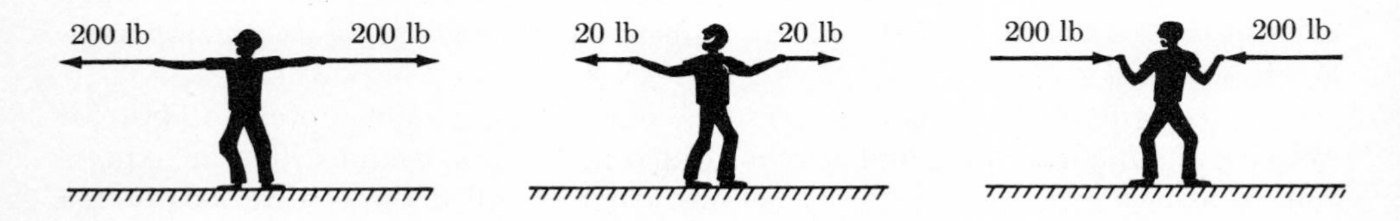

FIGURE 7·3

the resultant of the gravity forces on the body is called the *center of gravity*. For a body of arbitrary shape, the center of gravity will not be a point fixed in the body, but will depend for its location on the orientation of the body relative to the attracting center.

When the forces of gravity are modeled as *parallel forces* rather than concurrent forces, the central axis passes through the body-fixed point called the *center of mass*. For this reason the terms center of mass and center of gravity are often erroneously used interchangeably.

7·2 STATIC EQUILIBRIUM OF A RIGID BODY

7·2·1 The rigid-body idealization

In Sec. 7·1 we adopted for our rigid-body idealization the special case of the particle-system idealization for which the distance between any two particles of the system remains constant. Although this is a perfectly satisfactory concept for a rigid body, it would for realistic modeling of deformable physical systems often require a great many densely packed particles, and this would present a serious obstacle to efficient analysis.

The rigid-body idealization can alternatively be interpreted as a special case of a mathematical model quite different from the particle-system model; the rigid body can be considered as a *rigid continuum*. This model can easily be generalized to allow deformation. Just as we might reasonably idealize a body of water as a continuous distribution of matter, ignoring for the purposes of engineering the subdivision of the water into molecules and smaller discrete objects, so also we might treat a block of ice as a rigid and undeformable continuous distribution of matter. If we take this option (and we will), we must give some serious thought to the formulation of laws of mechanics for our new continuum models, since Newton's laws have been stated quite explicitly for particles.

Whereas in dealing with a particle-system model we built up our laws of mechanics from the laws applicable to a single particle, now we must establish rules for the mechanical behavior of continuous models by first formulating laws of mechanics for the constituent parts of the system and then combining our results for the total system. In the case of the continuum, the building block is the *differential element of mass* rather than the particle, and the combining process is *integration*, rather than summation. As illustrated in Fig. 7·4, the differential element of mass is, for a system whose properties are defined in terms of cartesian coordinates x_1, x_2, and x_3, given in terms of a mass-density function $\mu(x_1, x_2, x_3)$ and the differential element of volume $dv = dx_1\,dx_2\,dx_3$ by $dm = \mu\,dv$. Although more general examples might be imagined (noncartesian coordinates might be used, and μ might depend also on time t), the portrayal in Fig. 7·4 will suffice for present purposes; we wish only to establish a starting point in the formulation of equations of motion for continuous systems.

A particle, you will recall, is a point having mass; as such it has no dimensions, even though its mass might be very large. In contrast, a differential element of mass has dimensions, such as dx_1, dx_2, and dx_3 in Fig. 7·4, and since these differential lengths vanish in the limit so also does the mass dm of the element. Thus a differential element of mass is quite different from a particle.

Nonetheless, motivated by the feeling that equations of motion of a mathematical idealization consisting of millions of particles cannot predict very different behavior from those for a model involving an infinite number of differential mass elements, we adopt the following proposition: *Newton's laws apply to differential elements of mass, just as they do to particles.* Newton's second law then relates the inertial acceleration $\mathbf{A}$ of a differential element[1] of mass dm to the force $d\mathbf{F}$ applied to the differential element by

$$d\mathbf{F} = A\ dm \tag{7·6}$$

We accept this proposition as an axiom, just as we accepted Newton's laws for particles in Chap. 1. The correctness of this decision is established only by the success with which it permits the prediction of the motions of mechanical systems.

Thus we have two interpretations of the term *rigid body:* We can imagine

[1] Since the dimensions of a differential element of volume approach zero as a limit, all points in the element have the same inertial acceleration $\mathbf{A}$.

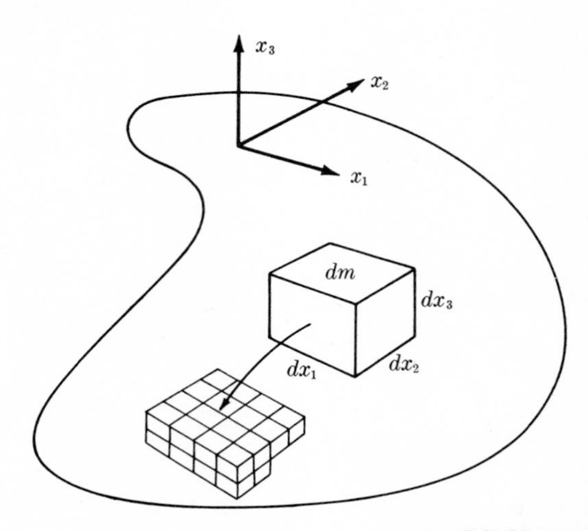

FIGURE 7·4

a rigid body either as a severely constrained set of particles or as an undeformable continuum. The latter option is almost invariably the most convenient.

7·2·2 Moments and centroids of geometrical figures

When a physical object is mathematically modeled as a continuous distribution of matter, the geometrical figure occupied by that matter at any time assumes an importance in analysis. When the object is modeled as a rigid continuum, the properties of that figure establish permanent characteristics of the mathematical model of the physical object. We are therefore interested in certain properties of geometrical figures that prove useful in the study of mechanics, and we digress briefly at this point to establish some results that will be used as our study of rigid-body dynamics unfolds.

In what follows we imagine that a cartesian coordinate system, with coordinate axes x_1, x_2, and x_3 parallel to dextral orthogonal unit vectors $\hat{\mathbf{u}}_1$, $\hat{\mathbf{u}}_2$, and $\hat{\mathbf{u}}_3$, has been established in a fixed relationship to the geometrical figure, as shown in Fig. 7·5, and we assume that the geometry of the figure is known in terms of those coordinates. Moreover, we imagine that the geometry and other functions to be introduced here do not depend on time. This restriction and the specific coordinate system adopted here are intended simply to focus attention on the basic concepts being introduced; in many cases other coordinate systems might be preferable, and in general all quantities depend on time.

Volume The volume v of a geometrical figure is the integral

$$v \triangleq \int dv = \int_{\bar{x}_1}^{\bar{\bar{x}}_1} \int_{\bar{x}_2(x_1)}^{\bar{\bar{x}}_2(x_1)} \int_{\bar{x}_3(x_1,x_2)}^{\bar{\bar{x}}_3(x_1,x_2)} dx_3\, dx_2\, dx_1 \tag{7·7}$$

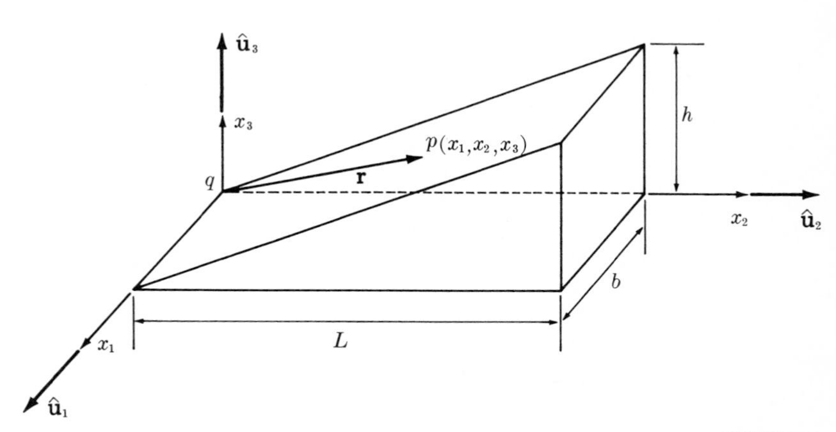

FIGURE 7·5

where overbars denote minimum value and double overbars denote maximum value of a function. This expression is illustrated for Fig. 7·5 as follows:

$$\begin{aligned} v &= \int_0^b \int_0^L \int_0^{x_2h/L} dx_3\, dx_2\, dx_1 \\ &= \int_0^b \int_0^L x_2 \frac{h}{L}\, dx_2\, dx_1 = \left(\frac{h}{L}\right) \int_0^b \frac{x_2{}^2}{2}\Big|_0^L dx_1 \\ &= \frac{hL}{2} \int_0^b dx_1 = \frac{hLb}{2} \end{aligned}$$

Often of greater interest in mechanics than the volume itself is the *weighted volume* $\bar{v}$, which differs from v only in the incorporation of a *weighting function* w within the integral, that is,

$$\bar{v} \triangleq \int w\, dv = \int_{\bar{x}_1}^{\bar{\bar{x}}_1} \int_{\bar{x}_2(x_1)}^{\bar{\bar{x}}_2(x_1)} \int_{\bar{x}_3(x_1,x_2)}^{\bar{\bar{x}}_3(x_1,x_2)} w(x_1, x_2, x_3)\, dx_3\, dx_2\, dx_1 \tag{7·8}$$

For example, we might replace the weighting function w by the *mass density* or mass per unit volume μ, to obtain the mass $\mathfrak{M}$ of the body occupying the figure as

$$\mathfrak{M} = \int \mu\, dv \tag{7·9}$$

so that if the constant μ is the uniform mass density of the block in Fig. 7·5, the mass of the block is $\mathfrak{M} = \mu\, hLb/2$.

First moment The word *moment* as defined in §7·1·1 refers only to the moment of a bound vector with respect to a reference point. We now offer a related definition of the first moment of a geometrical figure with respect to a point q as

$$\mathbf{m}^q \triangleq \int \mathbf{r}\, dv \tag{7·10}$$

where $\mathbf{r}$ is the position vector of a generic point p with respect to q.

For example, if q is selected as the origin of the coordinate system in Fig. 7·5, the block has the property

$$\begin{aligned} \mathbf{m}^q &= \int_0^b \int_0^L \int_0^{x_2h/L} (x_1\hat{\mathbf{u}}_1 + x_2\hat{\mathbf{u}}_2 + x_3\hat{\mathbf{u}}_3)\, dx_3\, dx_2\, dx_1 \\ &= \hat{\mathbf{u}}_1 \int_0^b \int_0^L \int_0^{x_2h/L} x_1\, dx_3\, dx_2\, dx_1 + \hat{\mathbf{u}}_2 \int_0^b \int_0^L \int_0^{x_2h/L} x_2\, dx_3\, dx_2\, dx_1 \\ &\qquad + \hat{\mathbf{u}}_3 \int_0^b \int_0^L \int_0^{x_2h/L} x_3\, dx_3\, dx_2\, dx_1 \\ &= \hat{\mathbf{u}}_1 \frac{hL}{2} \int_0^b x_1\, dx_1 + \hat{\mathbf{u}}_2 \int_0^b \int_0^L x_2{}^2 \frac{h}{L}\, dx_2\, dx_1 \\ &\qquad + \hat{\mathbf{u}}_3 \int_0^b \int_0^L \frac{1}{2}\left(x_2 \frac{h}{L}\right)^2 dx_2\, dx_1 \end{aligned}$$

$$= \frac{\hat{\mathbf{u}}_1 hLb^2}{4} + \hat{\mathbf{u}}_2 \frac{h}{L} \int_0^b \frac{L^3}{3} dx_1 + \hat{\mathbf{u}}_3 \frac{h^2}{2L^2} \int_0^b \frac{L^3}{3} dx_1$$

$$= \frac{\hat{\mathbf{u}}_1 hLb^2}{4} + \frac{\hat{\mathbf{u}}_2 hL^2 b}{3} + \frac{\hat{\mathbf{u}}_3 h^2 Lb}{6}$$

Again, the *weighted first moment* $\bar{\mathbf{m}}^q$ defined in terms of the weighting function w as

$$\bar{\mathbf{m}}^q \triangleq \int w\mathbf{r}\, dv \tag{7·11}$$

is more often useful in mechanics than the simple first moment of the figure. If we choose the mass density μ for the weighting function, the weighted first moment is called the *first mass moment,* and is given by

$$\boldsymbol{\mathfrak{u}}^q \triangleq \int \mu\mathbf{r}\, dv \tag{7·12}$$

Note that the first moment and the weighted first moment of a figure are both *vectors.*

Centroid The centroid c^* of a geometrical figure is that point for which the first moment of the figure is zero. Thus if $\boldsymbol{\rho}^*$ is a generic position vector from c^* to an arbitrary point of a figure, we have

$$\mathbf{m}^{c^*} = \int \boldsymbol{\rho}^*\, dv = 0$$

Replacing $\mathbf{r}$ in Eq. (7·10) by the vector $\mathbf{r}_{c^*}$ from q to c^* plus the generic vector $\boldsymbol{\rho}^*$ provides

$$\mathbf{m}^q = \int (\mathbf{r}_{c^*} + \boldsymbol{\rho}^*)\, dv = \mathbf{r}_{c^*} \int dv = \mathbf{r}_{c^*} v$$

so that the position vector $\mathbf{r}_{c^*}$ of the centroid of a figure with respect to an arbitrarily chosen point q is given by

$$\mathbf{r}_{c^*} = \frac{\mathbf{m}^q}{v} \tag{7·13}$$

Thus the centroid of the block in Fig. 7·5 is located from q by

$$\mathbf{r}_{c^*} = \frac{b}{2}\hat{\mathbf{u}}_1 + \frac{2}{3}L\hat{\mathbf{u}}_2 + \frac{1}{3}h\hat{\mathbf{u}}_3$$

It must be expected in general that the *weighted* first moment $\bar{\mathbf{m}}^{c^*}$ with respect to the centroid c^* will be nonzero, but that there will exist some other point for which any given weighted first moment will be zero. In particular, when the weighting function is the mass density μ, the reference point for which the first mass moment is zero is called the center of mass or the *mass center,* here designated c. If the generic vector from c to a point in the figure is denoted by $\boldsymbol{\rho}$, Eq. (7·12) provides

$$\boldsymbol{\mu}^c = \int \mu \boldsymbol{\rho}\, dv = 0 \tag{7·14}$$

Equations (7·9), (7·12), and (7·14) then permit the position vector $\mathbf{r}_c$ of c with respect to an arbitrarily selected point q to be expressed in parallel with Eq. (7·13) as

$$\mathbf{r}_c = \frac{\boldsymbol{\mu}^q}{\mathfrak{M}} \tag{7·15}$$

Evidently when μ is constant, the mass center c and the centroid c^* are coincident, as in the case of the uniform block in Fig. 7·5.

You should recall that the term mass center was introduced in §6·1·2 as a property of a system of particles; these two usages are consistent, representing parallel concepts for the particle and continuum idealizations of matter. Equations (7·15) and (6·5) are analogous representations for the position vectors that locate the mass centers of idealized systems with respect to an arbitrary point q.

(Centroid locations for some frequently encountered geometrical figures are recorded in Appendix C of this text.)

Second moment Useful extensions of the first-moment concepts defined by Eqs. (7·10) to (7·12) are the second moment with respect to point q,

$$\mathbf{m}^q \triangleq \int \mathbf{r}\mathbf{r}\, dv \tag{7·16a}$$

the weighted second moment with respect to q,

$$\bar{\mathbf{m}}^q = \int w\mathbf{r}\mathbf{r}\, dv \tag{7·16b}$$

and the second mass moment with respect to q,

$$\boldsymbol{\mu}^q \triangleq \int \mu \mathbf{r}\mathbf{r}\, dv \tag{7·16c}$$

We will not dwell on the properties of these objects until they arise naturally in the course of our proposed formulation of the laws of rigid-body dynamics. We might at this point note, however, that the combination of two vectors side by side (without intervening dot or cross) is called a *dyad*, and linear combinations of dyads, such as the integrals in Eqs. (7·16), are called *dyadics* (see Appendix B).

7·2·3 The zero force system

If a continuum is in a condition of static equilibrium (at rest in inertial space), then, for each differential element of mass, the acceleration $\mathbf{A}$ in Eq. (7·6) is zero, so that $d\mathbf{F}$ is also zero for every element. Thus the system of forces applied to an individual differential element of a continuous body in static equilibrium is a zero force system.

Considered as a whole, a continuous body is subjected to a system of forces that can be regarded as the sum (or integral over the body) of the individual force systems associated with the infinite number of differential elements of the body. Since each of the individual element force systems is a zero force system, so also is the total system of forces on the continuum. Newton's third law (assumed applicable to differential mass elements as well as to particles) indicates that forces between differential elements within a continuum occur in equal and opposite pairs, so that the system of internal forces in a continuum is also a zero force system. These observations combine to permit the following important conclusion: *The system of external forces applied to a continuum in static equilibrium is a zero force system.*

Thus if we are told that a continuous body is in static equilibrium, we know that the force resultant for the system of external forces on the body is zero, and so is the moment resultant for any reference point we may choose. If the body is rigid, then from the discussion of rigid-body kinematics in §3·2·1 we know that it can have no more than six degrees of freedom, three in translation and three in rotation. This means that there can be no more than six independent scalar equations of motion, or in this case six independent scalar equations establishing static equilibrium. Thus it is apparent that, for a rigid body in static equilibrium under the influence of external forces $\mathbf{F}_1, \ldots, \mathbf{F}_N$, a complete set of equations of mechanics can be obtained by recording the vector equation

$$\sum_{j=1}^{N} \mathbf{F}_j = 0 \tag{7·17}$$

and, for an arbitrarily chosen point q, the vector equation

$$\mathbf{M}^q = \sum_{j=1}^{N} \mathbf{Q}_j \times \mathbf{F}_j = 0 \tag{7·18}$$

where $\mathbf{Q}_j$ is the position vector from q to the point of application of $\mathbf{F}_j$ (or equivalently to any point on the line of action of $\mathbf{F}_j$).

Equations (7·17) and (7·18) provide the required six independent scalar equations. In some cases it may be convenient after using the three equations implied by Eq. (7·17) to select three more scalar equations from a variety of vector equations typified by Eq. (7·18) but having different choices for the point q.

Procedures for the practical application of Eqs. (7·17) and (7·18) are described in the following articles.

7·2·4 Trusses

In §2·2·3 and also in Sec. 6·4 a *truss* is defined as a collection of strings, springs, or rods with their end points interconnected by frictionless pinned joints where all loads are concentrated. A truss is defined as an idealized device that supports the system of particles at its joints by transmitting forces to the joints *along the structural members of the truss.* The idea that the line of action of the force transmitted by a given member has no component perpendicular to that member was embodied in the *definition* of a truss. This restriction is also inherent in the idealizations called *strings* and *springs* since the definitions of these terms in §2·1·2 incorporate the restriction to force transmission along the length of the member. The term *rod* is, however, not so restricted, and since rods may also be found among the members of a truss it was necessary in Chap. 2 to state explicitly that the word truss implies that all of its structural members transmit only axial forces.

Now that we have developed the basic concepts of rigid-body statics in Eqs. (7·17) and (7·18), we can justify the declaration that only axial forces are transmitted by the members of a truss by relying on the characterization of all connections as frictionless pinned joints at which external forces are concentrated. By this phrase we mean that the system of forces applied to a structural member of a truss is equivalent to a single force passing through each end joint, with no associated couple. These end loads must be directed along the axis of the member if moment equilibrium about an end point is to be satisfied [see Eq. (7·18)]. Thus the restriction to axial forces in truss members is justified by the restriction to frictionless pinned points at which all external forces are concentrated.

7·2·5 Distributed forces

By extending the idealized mathematical models used in mechanics from systems of particles to extended and continuous bodies, such as rigid bodies, we have also broadened the concept of force, so that a force can now be represented not only by a discrete load concentrated at a point but also by a continuous function in space.

The symbol $d\mathbf{F}$ in Eq. (7·6) is a differential element of force, requiring integration over a volume to represent a load applied to a physical object. Such volume integrals provide an appropriate representation of certain physical phenomena, such as gravitational attraction; the corresponding forces are called *body forces.*

We are also interested in loads such as those due to wind pressure on the surface of a building, and forces due to sliding contact between bodies. To accommodate such phenomena we invent a new representation of a force in terms of a *surface* integral, distributing what we call a *traction* $\boldsymbol{\tau}$ over an area A to obtain a force

$$\mathbf{F} = \int \boldsymbol{\tau}\, dA \tag{7·19}$$

This integral should be considered to imply the definition of the traction $\boldsymbol{\tau}$ at a point on the surface of a body: Traction $\boldsymbol{\tau}$ is the limit $\mathbf{F}/A$ as A goes to zero. Thus force is still the fundamental quantity, and our previous definitions of force still stand. Those forces represented as integrals of tractions over areas are called *surface forces*.

It is often convenient to adopt a geometrical representation of scalar components of surface-traction functions, representing the distribution of forces in a given direction over a specified surface in terms of a three-dimensional geometrical figure overlaid on the surface and having a local height normal to the surface and equal to the local value of the traction. Such geometrical figures are called *load diagrams*. Figure 7·6 illustrates the load-diagram concept for two examples; Fig. 7·6*a* shows the distribution of normal forces on a flat surface and Fig. 7·6*b* shows two superimposed plots of the distribution of two scalar components σ and p of the traction $\boldsymbol{\tau} = -\sigma\hat{\mathbf{u}}_3 + p\hat{\mathbf{u}}_2$ on a flat surface.

Equation (7·19) implies that the volume of each of the load diagrams typified by Fig. 7·6 equals the magnitude of the corresponding force resultant scalar component in the given direction; Fig. 7·6*a*, for example, indicates that the force resultant scalar component F_3 in the direction of $\hat{\mathbf{u}}_3$ is given by minus the volume under the curve, namely,

$$F_3 = -\int_0^L \int_0^b p\, dx_1\, dx_2 = -\frac{P}{L}\int_0^L \int_0^b x_2\, dx_1\, dx_2 = -\frac{Pb}{L}\int_0^L x_2\, dx_2$$

$$= -\frac{Pb}{L}\frac{L^2}{2} = -\frac{PbL}{2}$$

Proposition 2 in §7·1·3 indicates that a system of parallel forces can be replaced by a single force equal to the force resultant and having a line of action through the *central axis* of the force system. This proposition is applicable to any of the distributed force components represented by load diagrams such as those illustrated in Fig. 7·6, and the central axis is a property of every such geometrical figure. For any point q^* on the central axis, the moment resultant $\mathbf{M}^{q^*}$ for the bound vectors comprising the system of forces is zero. When this observation is considered in the light of the definition of the centroid of a geometrical figure (see §7·2·2), it becomes apparent that the central axis of a system of distributed parallel forces passes through the centroid of the corresponding load diagram.

7·2·6 Classification of problems

Having established in Eqs. (7·17) and (7·18) the equations of static equilibrium of a rigid body, just as in Chap. 2 the counterpart to Eq. (7·17) alone provided the static-equilibrium equation, we can recognize a precise parallel between problems of statics of rigid bodies and the problems of particle statics treated extensively in Chap. 2. Once again we can distinguish two categories of statics problems, depending on what is given and what is to be determined.

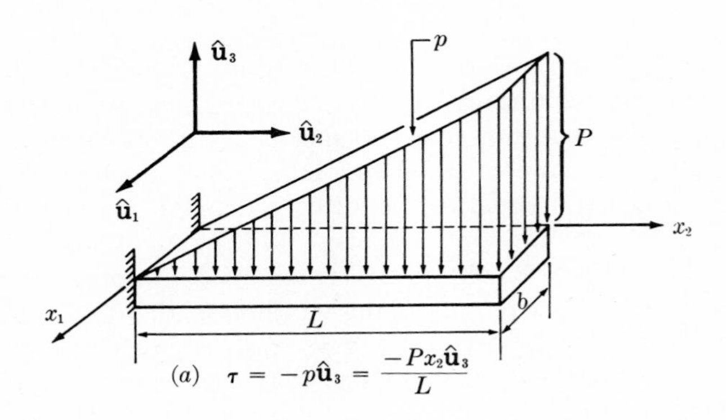

(a) $\tau = -p\hat{\mathbf{u}}_3 = \dfrac{-Px_2\hat{\mathbf{u}}_3}{L}$

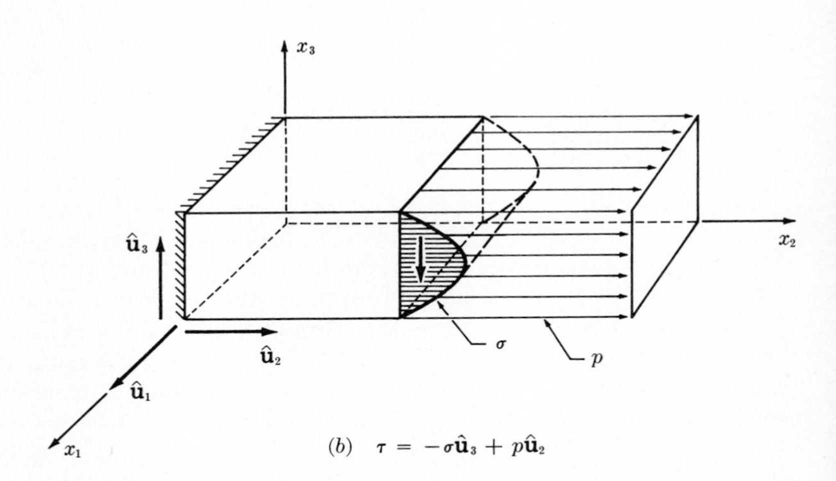

(b) $\tau = -\sigma\hat{\mathbf{u}}_3 + p\hat{\mathbf{u}}_2$

FIGURE 7·6

1 Given static-equilibrium positions of rigid bodies subjected to certain prescribed forces, determine the additional forces required to sustain static equilibrium.

2 Given all applied forces as functions of position and/or orientation, determine the state (or states) of static equilibrium.

In addition, it may be required in either class of problem that the *stability* of the equilibrium state be established, as that term is employed in §6·4·2.

As noted previously (see Chap. 2), there is very little difficulty to be found with the *mechanics* aspect of any statics problem; Eqs. (7·17) and (7·18) say all there is to say. The problems of geometry and algebra can be quite substantial, but increasingly these difficulties are relegated to the digital computer. The following sections are designed to reinforce by illustration the few basic ideas involved in a problem of rigid-body statics; complexities of geometry and algebra are avoided wherever possible in these examples.

7·3 DETERMINATION OF EQUILIBRIUM FORCES

7·3·1 Equilibrating force and torque

If we are given the fact that a rigid body is in static equilibrium, and we are given a system of applied forces with force resultant $\mathbf{F}$ and moment resultant $\mathbf{M}^q$ for some arbitrary point q, then we can assure the satisfaction of Eqs. (7·17) and (7·18) by exerting on the body *any* force system with force resultant $-\mathbf{F}$ and moment resultant $-\mathbf{M}^q$ for point q; the simplest such system consists of a force $-\mathbf{F}$ passing through q and a couple having torque $-\mathbf{M}^q$. Only by examining the specific force-transmission capabilities of the body-support system can actual reaction forces be determined. Several examples of this kind are considered in the next article.

7·3·2 Solving for specific reaction forces

First example Figure 7·7*a* portrays a rigid body subjected to the applied forces, in pounds, $\mathbf{F}_1 = (50\sqrt{2}\,\hat{\mathbf{u}}_1 - 50\sqrt{2}\,\hat{\mathbf{u}}_2)$, $\mathbf{F}_2 = -100\hat{\mathbf{u}}_2$, $\mathbf{F}_3 = (50\sqrt{3}\,\hat{\mathbf{u}}_1 + 50\hat{\mathbf{u}}_2)$, and $\mathbf{F}_4 = 100\hat{\mathbf{u}}_1$. These forces are equilibrated by the unknown reactions at A and B, which are to be found. Dimensions a and b are 2 in. and 4 in., respectively.

As in Chap. 2, the first step is the construction of a *free-body diagram,* as shown in Fig. 7·7*b*. The schematic representations of the pinned-joint and frictionless roller supports in Fig. 7·7*a* communicate the capacity of the supports at A and B for forces of the sort shown in Fig. 7·7*b*; only the scalars H_A, V_A, and V_B need be found.

Equation (7·17) provides

$$H_A\hat{\mathbf{u}}_1 + V_A\hat{\mathbf{u}}_2 + V_B\hat{\mathbf{u}}_2 + \mathbf{F}_1 + \mathbf{F}_2 + \mathbf{F}_3 + \mathbf{F}_4 = 0$$

or

$$H_A\hat{\mathbf{u}}_1 + (V_A + V_B)\hat{\mathbf{u}}_2 = -50(2 + \sqrt{2} + \sqrt{3})\hat{\mathbf{u}}_1 \text{ lb} + 50(1 + \sqrt{2})\hat{\mathbf{u}}_2 \text{ lb}$$

with the implications

$$H_A = -50(2 + \sqrt{2} + \sqrt{3}) \text{ lb}$$

and

$$V_A + V_B = 50(1 + \sqrt{2}) \text{ lb}$$

Equation (7·18) must now be used to obtain the additional scalar equa-

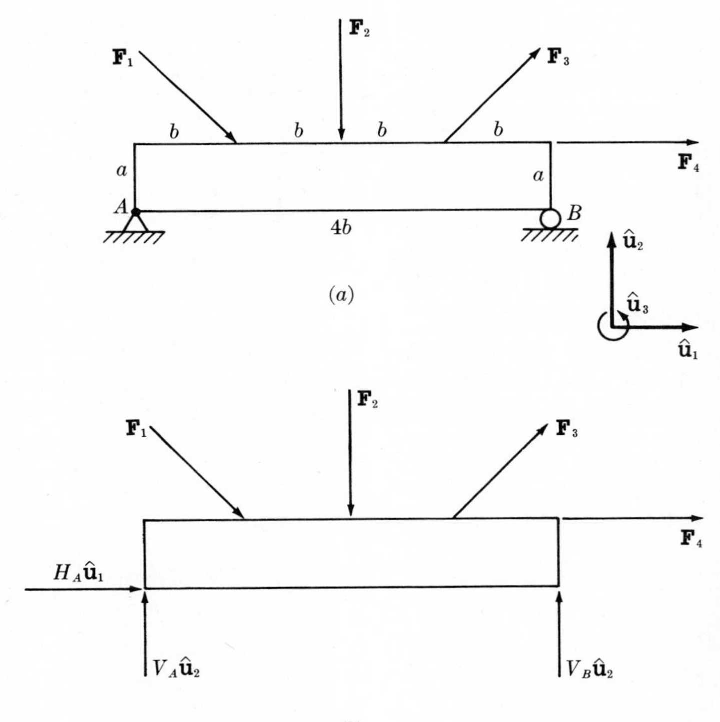

FIGURE 7·7

tion needed to solve for V_A and V_B individually. Although any reference point can be chosen, selecting point A yields

$$0 + 4b\hat{\mathbf{u}}_1 \times V_B\hat{\mathbf{u}}_2 + (b\hat{\mathbf{u}}_1 + a\hat{\mathbf{u}}_2) \times \mathbf{F}_1 + 2b\hat{\mathbf{u}}_1 \times \mathbf{F}_2 + (3b\hat{\mathbf{u}}_1 + a\hat{\mathbf{u}}_2) \times \mathbf{F}_3 + a\hat{\mathbf{u}}_2 \times \mathbf{F}_4 = 0$$

or

$$4bV_B\hat{\mathbf{u}}_3 = -(-50b\sqrt{2} - 50a\sqrt{2})\hat{\mathbf{u}}_3 + 200b\hat{\mathbf{u}}_3 - (150b - 50a\sqrt{3})\hat{\mathbf{u}}_3 + 100a\hat{\mathbf{u}}_3$$

Dropping $\hat{\mathbf{u}}_3$ and substituting $b = 2a = 4$ in. produces

$$V_B = \tfrac{1}{8}(200 + 150\sqrt{2} + 50\sqrt{3}) \text{ lb}$$

so that

$$V_A = \tfrac{1}{8}(200 - 250\sqrt{2} + 350\sqrt{3}) \text{ lb}$$

Second example Figure 7·8 shows a horizontal right triangular plate under a uniform vertical load of $p = 10$ lb/in.^{2}; the right angle vertex C of the triangle is supported by a pinned joint, and the acute angle vertices A and B are on frictionless rollers. We can ascertain by inspection that no horizontal force will be applied by the pin connection at C, so only the vertical reactions $V_A\hat{\mathbf{u}}_3$, $V_B\hat{\mathbf{u}}_3$, and $V_C\hat{\mathbf{u}}_3$ are required.

Equation (7·17) provides

$$V_A\hat{\mathbf{u}}_3 + V_B\hat{\mathbf{u}}_3 + V_C\hat{\mathbf{u}}_3 - \tfrac{1}{2}abp\hat{\mathbf{u}}_3 = 0$$

or

$$V_A + V_B + V_C = \tfrac{1}{2}pab = 6000 \text{ lb} \qquad (a)$$

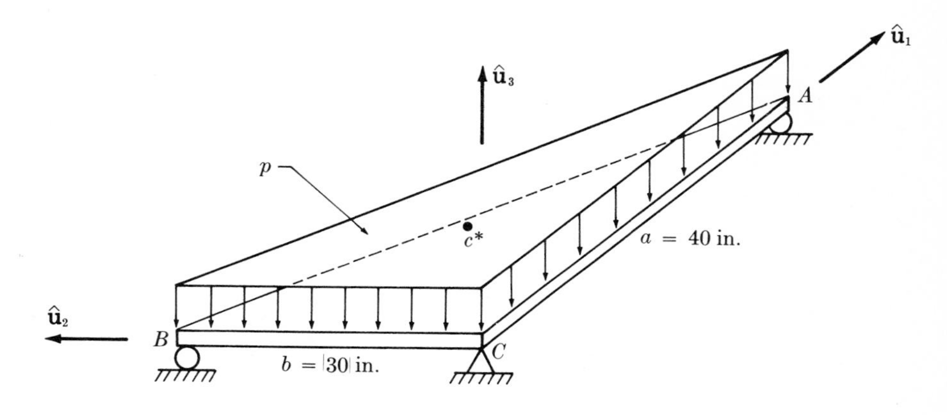

FIGURE 7·8

Equation (7·18) must be used to provide the two additional scalar equations required. Although any reference point may be chosen for q, the choice $q = C$ is convenient. The result then becomes

$$a\hat{\mathbf{u}}_1 \times V_A\hat{\mathbf{u}}_3 + b\hat{\mathbf{u}}_2 \times V_B\hat{\mathbf{u}}_3 = \int_0^a \int_0^{b - x_1(b/a)} (x_1\hat{\mathbf{u}}_1 + x_2\hat{\mathbf{u}}_2) \times p\hat{\mathbf{u}}_3 \, dx_2 \, dx_1$$

or

$$-aV_A\hat{\mathbf{u}}_2 + bV_B\hat{\mathbf{u}}_1 = \int_0^a \int_0^{(a - x_1)b/a} (-px_1\hat{\mathbf{u}}_2 + px_2\hat{\mathbf{u}}_1) \, dx_2 \, dx_1$$

or equivalently

$$\begin{aligned} V_A &= \frac{p}{a}\int_0^a \int_0^{(a-x_1)b/a} x_1 \, dx_2 \, dx_1 = \frac{p}{a}\int_0^a x_1 x_2 \Big|_0^{(a-x_1)b/a} dx_1 \\ &= \frac{pb}{a^2}\int_0^a x_1(a - x_1)\, dx_1 = \frac{pb}{a^2}\left(a\,\frac{x_1^2}{2}\Big|_0^a - \frac{x_1^3}{3}\Big|_0^a\right) \\ &= \frac{pb}{a^2}\left(\frac{a^3}{2} - \frac{a^3}{3}\right) = \tfrac{1}{6}pab = 2000 \text{ lb} \end{aligned} \tag{b}$$

and

$$\begin{aligned} V_B &= \frac{p}{b}\int_0^a \int_0^{(a-x_1)b/a} x_2 \, dx_2 \, dx_1 = \frac{p}{b}\int_0^a \frac{x_2^2}{2}\Big|_0^{(a-x_1)b/a} dx_1 \\ &= \frac{pb^2}{2ba^2}\int_0^a (a - x_1)^2 \, dx_1 = \frac{pb}{2a^2}\left(a^2x_1 - 2a\,\frac{x_1^2}{2} + \frac{x_1^3}{3}\right)\Big|_0^a \\ &= \frac{pb}{2a^2}\left(a^3 - a^3 + \frac{a^3}{3}\right) = \frac{pab}{6} = 2000 \text{ lb} \end{aligned} \tag{c}$$

Combining Eqs. (a), (b), and (c) produces

$$V_C = (6000 - 2000 - 2000) \text{ lb} = 2000 \text{ lb} \tag{d}$$

Rather than work directly with the distributed forces given for the loaded plate in Fig. 7·8, we might have avoided some of the ensuing integration by first replacing the given system of forces with a simpler equivalent system. By virtue of proposition 2 of §7·1·3, we can expect to replace the triangular distribution of parallel forces by a single force equal to the force resultant **F** of the given system of forces and having a line of action along the central axis of the force system. For the force system in Fig. 7·8, the force resultant is

$$\mathbf{F} = \tfrac{1}{2}abp\hat{\mathbf{u}}_3 = 6000\hat{\mathbf{u}}_3 \text{ lb}$$

By the argument in §7·2·5, the central axis of the distributed load in Fig. 7·8 is parallel to the forces on the plate and passes through the centroid of the loading diagram. Appendix C shows the centroid c^* of the right triangular laminar volume of the loading diagram in Fig. 7·8 to be on a vertical axis located with respect to point C by

$$\mathbf{S}^* = \tfrac{1}{3}a\hat{\mathbf{u}}_1 + \tfrac{1}{3}b\hat{\mathbf{u}}_2 \tag{e}$$

Since the moment about any point on the central axis is zero, the applied moment $\bar{\mathbf{M}}^C$ about C can be obtained from Eq. (7·2) as

$$\begin{aligned}\bar{\mathbf{M}}^C &= 0 + \mathbf{S}^* \times \mathbf{F} \\ &= (\tfrac{1}{3}a\hat{\mathbf{u}}_1 + \tfrac{1}{3}b\hat{\mathbf{u}}_2) \times \tfrac{1}{2}pab\hat{\mathbf{u}}_3 \\ &= \tfrac{1}{6}pab^2\hat{\mathbf{u}}_1 - \tfrac{1}{6}pa^2b\hat{\mathbf{u}}_2\end{aligned} \tag{f}$$

Now we can return to Eq. (7·8) and write

$$\mathbf{M}^C = a\hat{\mathbf{u}}_1 \times V_A\hat{\mathbf{u}}_3 + b\hat{\mathbf{u}}_2 \times V_B\hat{\mathbf{u}}_3 + \bar{\mathbf{M}}^C$$

to obtain, in agreement with Eqs. (b) and (c),

$$V_A = \tfrac{1}{6}pab \qquad V_B = \tfrac{1}{6}pab$$

Third example Figure 7·9 portrays a rigid body of uniform thickness and T-shaped plan view, supported with the stem of the T fixed at B in a rigid wall,[1] and uniformly loaded by a traction $\boldsymbol{\tau} = -p\hat{\mathbf{u}}_3$ on the top segment of the T. The body is in static equilibrium, and the reaction forces applied to the body at the support at B are to be sought.

[1] A body is said to have a *cantilever support* when, as in this example, one face of the body is fixed with respect to an adjoining rigid body.

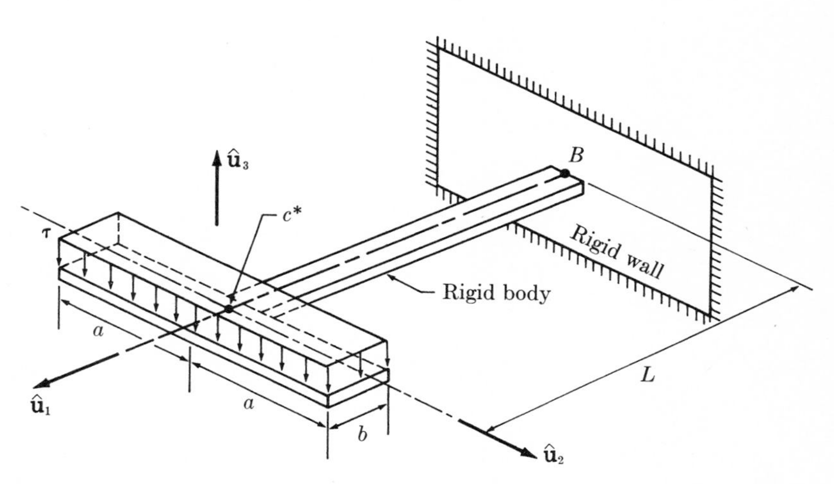

FIGURE 7·9

As a first step, we replace the distributed force system with the simplest possible equivalent system, namely, a concentrated force $\mathbf{F} = -2abp\hat{\mathbf{u}}_3$ having a line of action through the centroid c^* of the load diagram.

The second step is the determination of an equivalent system for the support force from Eqs. (7·17) and (7·18), with B replacing the arbitrary reference point q in the latter. The simplest system of forces equivalent to those applied to the body by the wall at B is a force $\mathbf{R}$ passing through B and a couple having torque $\mathbf{T}^B$. From Eq. (7·17), $\mathbf{R} = -\mathbf{F} = 2abp\hat{\mathbf{u}}_3$, and from Eq. (7·18),

$$\mathbf{T}^B = -L\hat{\mathbf{u}}_1 \times \mathbf{F} = 2abpL\hat{\mathbf{u}}_2$$

Having found a simple system of forces equivalent to the presumably more complex system of forces actually applied to the rigid body by the wall, we might next want to know what those actual forces are. This information can never be obtained, however, as long as we persist in idealizing both the T-shaped structure and the wall as rigid bodies constrained against relative motion. We have adopted the rigid-body idealization because it simplifies the problem, limiting the number of degrees of freedom of the body to six. For this simplification of the mathematical model of the physical body, we must pay a price: We can expect to find no more than six scalar measures of the reaction forces applied to the body. The three components of $\mathbf{R}$ and $\mathbf{T}^B$ are all that we can find, unless we adopt a more complicated model. In conformity with the nomenclature introduced in Chap. 2, we say that the distribution of forces over the body at the support is statically indeterminate.

7·3·3 Truss analysis using force- and moment-equilibrium equations

Section 6·4, containing applications of virtual work in statics, includes examples of the use of a special case of the virtual-work equation to find reaction forces and member forces for trusses. At the conclusion of the examples in §6·4·4, we noted that the virtual displacements considered had the effect of extracting three scalar equations of static equilibrium to be solved for the three scalar unknowns characterizing the reactions. Similar observations were made in §6·4·5. Now we can see even more clearly that the virtual-work calculations were producing results available directly from Eqs. (7·17) and (7·18). This should not be surprising since the trusses considered in Chap. 6 consisted of fully constrained sets of particles, and as such they qualify as rigid bodies in the first sense considered in Part Two.

These observations point to the fact that the virtual-work methods introduced in Chap. 6 are directly applicable to the rigid continuous bodies introduced in Chap. 7, just as the force- and moment-equilibrium equations [Eqs. (7·17) and

(7·18)] apply equally well to both the continuous rigid bodies in Chap. 7 and the discretized rigid bodies described as constrained sets of particles in Chap. 6. Although selection between virtual-work methods and the direct equilibrium methods employing Eqs. (7·17) and (7·18) is a subjective process, the latter procedure seems more straightforward. In applying these equations, you should take advantage of the fact that the *method of sections* introduced in §6·4·5 permits the application of Eqs. (7·17) and (7·18) to any *portion* of the truss that can be isolated by the analyst.

7·4 DETERMINATION OF EQUILIBRIUM POSITION

7·4·1 Functional representation of forces

In Sec. 7·3 we considered problems in which we were given a rigid body in a known position of static equilibrium, subject to specified applied forces, and we sought to evaluate the reaction forces applied to the body by its supports. Now we consider the second major class of rigid-body statics problems: We are given all forces on the body as functions of the kinematical variables of the body, and we seek those values of the kinematical variables for which static equilibrium is possible.

The useful equations of mechanics are the same in all cases; Eqs. (7·17) and (7·18) contain all the information that can be extracted from Newton's laws for application to a rigid body. The difference in the two classes of problems lies in the choice of unknowns, which are scalar components of forces in one case and scalar kinematical variables in the other.

As long as the idealized mechanical system is a single rigid body, these two classes of problems are of the same order of difficulty; in either case we must solve a set of up to six algebraic equations in a corresponding number of unknowns.[1] However, when the idealized system is a collection of N interconnected rigid bodies, the determination of equilibrium states involves the simultaneous solution of as many as $6N$ algebraic equations, and this is invariably a task for the digital computer. The problem of reaction-force determination is, on the other hand, vastly simpler when many bodies are involved, since (for a statically determinate problem) the reaction forces on each body can be determined individually.

In §7·4·2 two very simple examples are developed, in order to illustrate the use of Eqs. (7·17) and (7·18) with minimal geometric and algebraic complexity.

[1] If in the equilibrium-position determination problem the forces are not given as *explicit* functions of the six independent kinematical variables of the body (as is the case in the first example in §7·4·2), then an additional set of kinematical equations is required.

7·4·2 Solving the zero equivalent system equations

First example Figure 7·10 shows a flat, rigid, rectangular plate supported by four linear springs, each having the spring constant $k = 1000$ lb/in. The problem is to determine the distances Δ_1, Δ_2, Δ_3, and Δ_4 by which the four springs are compressed when the plate is loaded as shown in Fig. 7·10*a*, with $p = 1$ lb/in.^{2}, $P = 10$ lb/in.^{2}, and $F = 1000$ lb. The plate dimensions are $a = 20$ in. and $b = 10$ in.

The first step is the replacement of the given load by a simpler equivalent force system, and the construction of a free-body diagram, as shown in Fig. 7·10*b*. Here the rectangular distributed load diagram has been replaced by a concentrated force through its centroid c_1^*, having the direction of the load $(-\hat{\mathbf{u}}_3)$ and the magnitude $4pab$ of its force resultant, which is the volume of the load diagram. The triangular load diagram has been similarly replaced by a concentrated force of magnitude Pab and direction $-\hat{\mathbf{u}}_3$ through the centroid c_2^*. All support springs are assumed to be in compression for the construction of the free-body diagram.

The second step is the recording of all relationships between forces and displacements. In this case the four spring forces are directly proportional to the corresponding displacements; that is, the force at support A is given by $k\Delta_1 \hat{\mathbf{u}}_3$, etc.

The next objective is the application of Eqs. (7·17) and (7·18) to obtain the necessary scalar equilibrium equations. The kinematical variables Δ_1, Δ_2, Δ_3, and Δ_4 are not independent, so we don't need four scalar equations. Before writing the equilibrium equations, we select the three independent scalars w, θ_1, and θ_2, representing, respectively, the displacement of c_1^* in the $\hat{\mathbf{u}}_3$ direction and the assumedly small rotations of the plate about axes parallel to $\hat{\mathbf{u}}_1$ and $\hat{\mathbf{u}}_2$. Then we record the kinematical equations (based on small-angle approximations)

$$\Delta_1 = -w + b\theta_1 - a\theta_2 \qquad (a)$$
$$\Delta_2 = -w + b\theta_1 + a\theta_2 \qquad (b)$$
$$\Delta_3 = -w - b\theta_1 + a\theta_2 \qquad (c)$$
$$\Delta_4 = -w - b\theta_1 - a\theta_2 \qquad (d)$$

Now Eq. (7·17) provides (with appropriate representations of forces applied to the plate by the four supporting springs)

$$(-F - 4pab - Pab)\hat{\mathbf{u}}_3 + k\,(\Delta_1 + \Delta_2 + \Delta_3 + \Delta_4)\,\hat{\mathbf{u}}_3 = 0$$

or

$$\Delta_1 + \Delta_2 + \Delta_3 + \Delta_4 = \frac{1}{k}\,(F + 4pab + Pab) \qquad (e)$$

Equation (7·18) can be applied for any reference point q. If we select

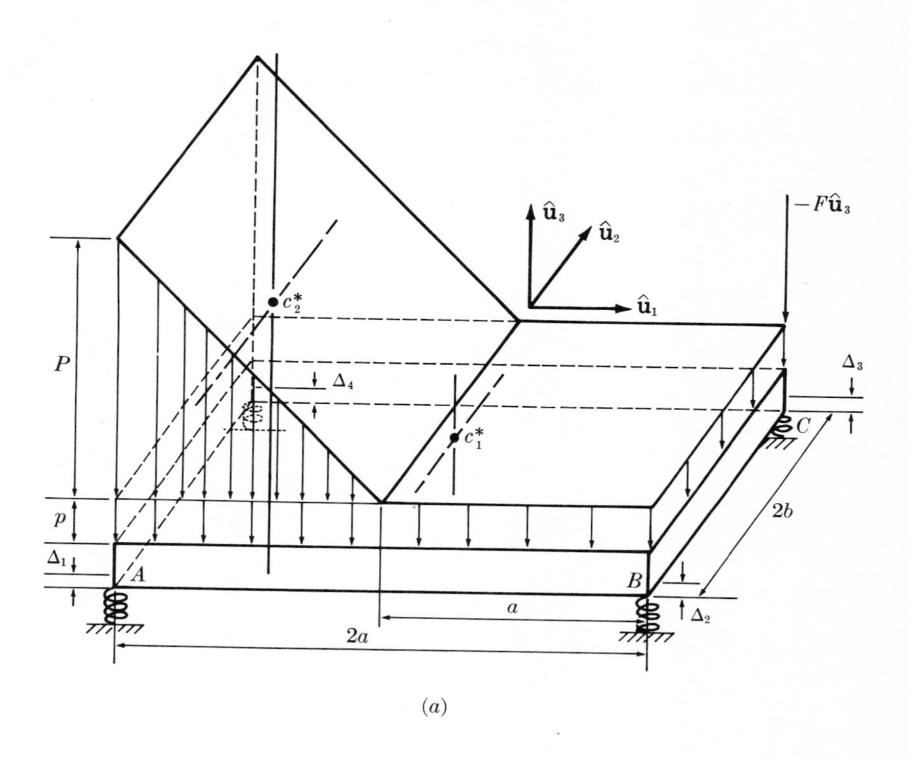

(a)

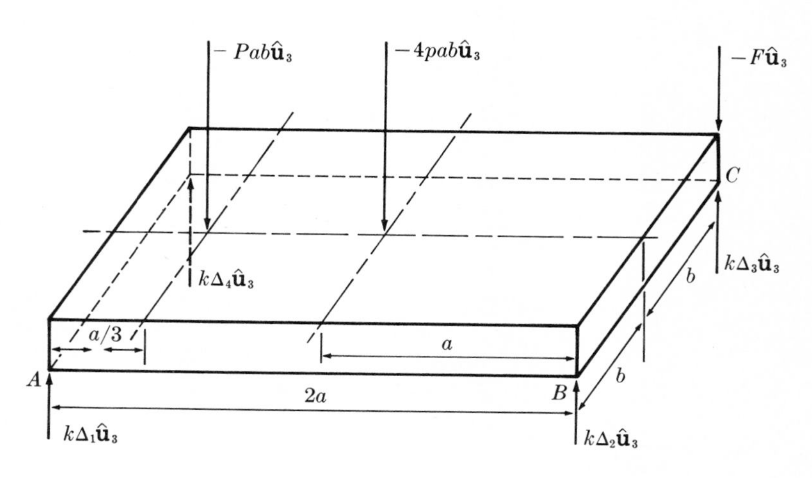

(b)

FIGURE 7·10

corner C for q, we find

$$(-a\hat{\mathbf{u}}_1 - b\hat{\mathbf{u}}_2) \times (-4pab\hat{\mathbf{u}}_3) + \left(-\frac{5a}{3}\hat{\mathbf{u}}_1 - b\hat{\mathbf{u}}_2\right) \times (-Pab\hat{\mathbf{u}}_3)$$
$$+ (-2a\hat{\mathbf{u}}_1 - 2b\hat{\mathbf{u}}_2) \times (k\,\Delta_1\,\hat{\mathbf{u}}_3) - 2b\hat{\mathbf{u}}_2 \times k\,\Delta_2\,\hat{\mathbf{u}}_3$$
$$+ (-2a\hat{\mathbf{u}}_1) \times k\,\Delta_4\,\hat{\mathbf{u}}_3 = 0$$

or equivalently

$$4pab^2 + Pab^2 - 2bk\,\Delta_1 - 2bk\,\Delta_2 = 0 \tag{f}$$

and

$$-4pa^2b - \tfrac{5}{3}Pa^2b + 2ak\,\Delta_1 + 2ak\,\Delta_4 = 0 \tag{g}$$

Equations (a) to (g) comprise seven independent algebraic equations, which can be solved for the seven unknowns Δ_1, Δ_2, Δ_3, Δ_4, w, θ_1, and θ_2. In this problem it is not difficult to perceive that by comparing Eq. (e) to the sum of Eqs. (a) to (d),

$$\begin{aligned} w &= -\frac{1}{4k}(F + 4pab + Pab) \\ &= -\tfrac{1}{4000}(1000 + 800 + 2000) = -0.95 \text{ in.} \end{aligned} \tag{h}$$

Combining Eq. (f) with the sum of Eqs. (a) and (b) furnishes

$$4pab^2 + Pab^2 = 2bk\,(\Delta_1 + \Delta_2) = 2bk(-2w + 2b\theta_1) \tag{i}$$

Similarly, the combination of Eq. (g) and the sum of Eqs. (a) and (d) provides

$$4pa^2b + \tfrac{5}{3}Pa^2b = 2ak(\Delta_1 + \Delta_4) = 2ak(-2w - 2a\theta_2) \tag{j}$$

Equations (i) and (j) permit explicit representations of θ_1 and θ_2, which with Eq. (h) become, respectively,

$$\begin{aligned} \theta_1 &= \frac{1}{4b^2k}(4bkw + 4pab^2 + Pab^2) \\ &= \frac{1}{4b^2k}(-bF - 4pab^2 - Pab^2 + 4pab^2 + Pab^2) \\ &= \frac{-F}{4bk} = -\tfrac{1}{40} = -0.025 \text{ rad} \end{aligned} \tag{k}$$

and

$$\begin{aligned} \theta_2 &= -\frac{1}{4a^2k}(4akw + 4pa^2b + \tfrac{5}{3}Pa^2b) \\ &= \frac{1}{4a^2k}(aF + 4pa^2b + Pa^2b - 4pa^2b - \tfrac{5}{3}Pa^2b) \\ &= \frac{1}{4ak}(F - \tfrac{2}{3}Pab) = -\tfrac{1}{240} = -0.00417 \text{ rad} \end{aligned} \tag{l}$$

Returning to Eqs. (a) to (d) with the results established as Eqs. (h), (k), and (l) yields the answers originally sought:

$$\Delta_1 = \frac{1}{4k}(F + 4pab + Pab) - \frac{F}{4k} - \frac{1}{4k}(F - \tfrac{2}{3}Pab)$$

$$= \frac{1}{4k}(-F + 4pab + \tfrac{5}{3}Pab) \cong 0.7833 \text{ in.}$$

$$\Delta_2 = \frac{1}{4k}(F + 4pab + Pab) - \frac{F}{4k} + \frac{1}{4k}(F - \tfrac{2}{3}Pab)$$

$$= \frac{1}{4k}(F + 4pab + \tfrac{1}{3}Pab) = 0.6167 \text{ in.}$$

$$\Delta_3 = \frac{1}{4k}(F + 4pab + Pab) + \frac{F}{4k} + \frac{1}{4k}(F - \tfrac{2}{3}Pab)$$

$$= \frac{1}{4k}(3F + 4pab + \tfrac{1}{3}Pab) = 1.1167 \text{ in.}$$

$$\Delta_4 = \frac{1}{4k}(F + 4pab + Pab) + \frac{F}{4k} - \frac{1}{4k}(F - \tfrac{2}{3}Pab)$$

$$= \frac{1}{4k}(F + 4pab + \tfrac{5}{3}Pab) = 1.2833 \text{ in.}$$

As a check on the algebra, note that $\Delta_1 = \Delta_2 = \Delta_3 = \Delta_4$ if F and P are zero; if only F is zero, then $\Delta_2 = \Delta_3$ and $\Delta_1 = \Delta_4$. These results can be expected, due to the patterns of symmetry in the distributed loads. Note also that if p and P are zero, we have $\Delta_1 + \Delta_2 + \Delta_3 + \Delta_4 = (1/4k)(-F + F + 3F + F) = F/k$, as required for equilibrium.

As you reflect on this example, you should discern that the following systematic procedure has been pursued:

1. Replace the given system of applied forces by a simpler equivalent system of forces if possible, and construct a free-body diagram.
2. Establish the force-displacement relationships.
3. Record any kinematical equations that establish constraints among the kinematical variables used in the force-displacement relationships.
4. Record the equilibrium equations [Eqs. (7·17) and (7·18)].
5. Solve the system of equations developed in steps 2 to 4.

Second example Figure 7·11a shows a rigid square plate that has six degrees of freedom; it is free to translate in any direction and to rotate about any axis, and

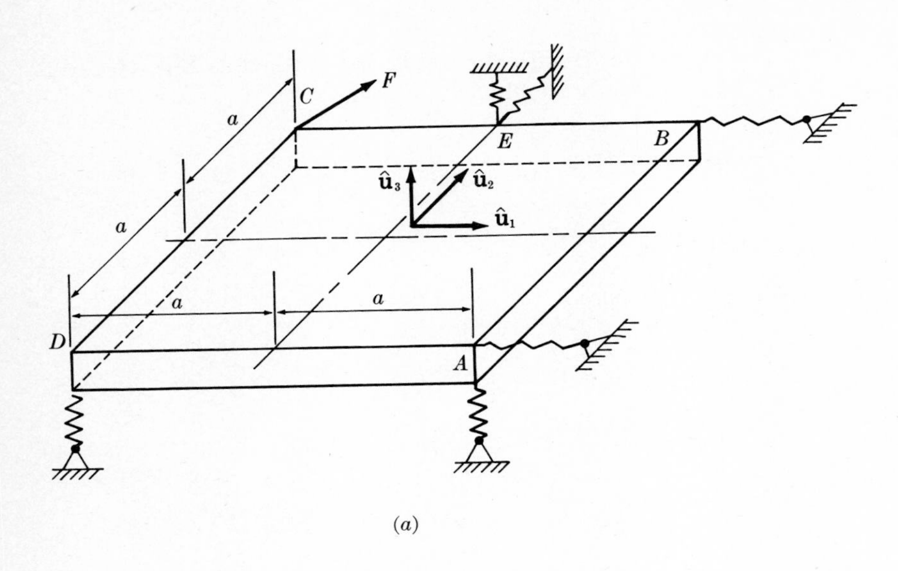

(a)

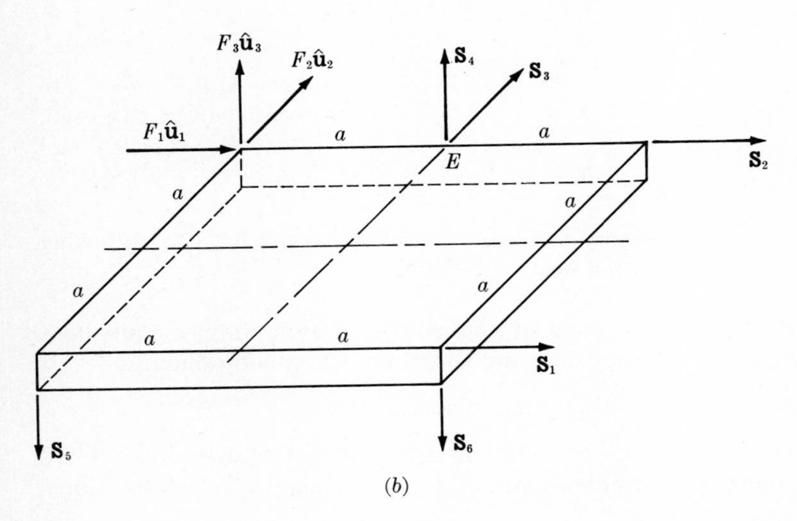

(b)

FIGURE 7·11

under the single concentrated force applied it will be displaced in a rather complicated way. Each side of the square has length $2a = 20$ in.

In terms of the unit vectors shown in Fig. 7·11, the applied force is given by

$$\mathbf{F} = F_1\hat{\mathbf{u}}_1 + F_2\hat{\mathbf{u}}_2 + F_3\hat{\mathbf{u}}_3 = 100\hat{\mathbf{u}}_1 + 80\hat{\mathbf{u}}_2 + 40\hat{\mathbf{u}}_3 \text{ lb}$$

Step 1 in the systematic procedure involves only the construction of the free-body diagram shown in Fig. 7·11*b*, since the system of applied forces cannot easily be replaced by a simpler equivalent system. For convenience in constructing the free-body diagram, all springs are assumed to be in tension. Any error in this assumption will be revealed by a minus sign.

Step 2 is the establishment of force-displacement relationships. Again in this problem the supports are all linear springs having spring constant $k =$ 1000 lb, and the spring forces are proportional to their deformations. If we (quite arbitrarily) let $\Delta_1, \ldots, \Delta_6$ be the *elongations* of the six springs, and assume that displacements are so small that the line of action of each spring force is uninfluenced by the displacements under load, we can then record the spring forces shown in Fig. 7·11*b* as:

$$\mathbf{S}_1 = k\,\Delta_1\,\hat{\mathbf{u}}_1 \qquad \mathbf{S}_2 = k\,\Delta_2\,\hat{\mathbf{u}}_1 \qquad \mathbf{S}_3 = k\,\Delta_3\,\hat{\mathbf{u}}_2$$
$$\mathbf{S}_4 = k\,\Delta_4\,\hat{\mathbf{u}}_3 \qquad \mathbf{S}_5 = -k\,\Delta_5\,\hat{\mathbf{u}}_3 \qquad \mathbf{S}_6 = -k\,\Delta_6\,\hat{\mathbf{u}}_3$$

If equilibrium requires that any of the six springs be in compression, that result will manifest itself as a negative value of the corresponding Δ_j ($j = 1, \ldots, 6$).

Step 3 can be foregone here. The plate has six degrees of freedom and there are six kinematical variables $\Delta_1, \ldots, \Delta_6$ appearing in the force-displacement equations; these spring elongations are therefore independent, and need not be related by kinematical constraint equations.

Step 4 is the statement of the equilibrium equations. Equation (7·17) for force equilibrium furnishes

$$\mathbf{F} + \mathbf{S}_1 + \mathbf{S}_2 + \mathbf{S}_3 + \mathbf{S}_4 + \mathbf{S}_5 + \mathbf{S}_6 = 0$$

or in scalar terms

$$F_1 + k\,\Delta_1 + k\,\Delta_2 = 0 \qquad (a)$$
$$F_2 + k\,\Delta_3 = 0 \qquad (b)$$
$$F_3 + k\,\Delta_4 - k\,\Delta_5 - k\,\Delta_6 = 0 \qquad (c)$$

Equation (7·18) for moment equilibrium can be written for any reference point q. If we choose $q = E$, we remove forces $\mathbf{S}_2$, $\mathbf{S}_3$, $\mathbf{S}_4$, and $F_1\,\hat{\mathbf{u}}_1$ from the moment-equilibrium equation, which becomes

$$\mathbf{M}^E = \hat{\mathbf{u}}_1(2ak\,\Delta_6 + 2ak\,\Delta_5) + \hat{\mathbf{u}}_2(aF_3 - ak\,\Delta_5 + ak\,\Delta_6) + \hat{\mathbf{u}}_3(2ak\,\Delta_1 - aF_2) = 0$$

or equivalently

$$\Delta_5 + \Delta_6 = 0 \tag{d}$$

$$F_3 - k\,\Delta_5 + k\,\Delta_6 = 0 \tag{e}$$

$$2k\,\Delta_1 - F_2 = 0 \tag{f}$$

Step 5 is the solution of a system of simultaneous algebraic equations—in this case Eqs. (*a*) to (*f*). Equations (*d*) to (*f*) alone yield the solutions

$$\Delta_1 = \frac{F_2}{2k} = 0.04 \text{ in.} \tag{g}$$

$$\Delta_5 = \frac{F_3}{2k} = 0.02 \text{ in.} \tag{h}$$

$$\Delta_6 = \frac{-F_3}{2k} = -0.02 \text{ in.} \tag{i}$$

and Eq. (*b*) provides

$$\Delta_3 = \frac{-F_2}{k} = -0.08 \text{ in.} \tag{j}$$

With these results, Eqs. (*a*) and (*c*) yield

$$\Delta_2 = \frac{-F_1 - \frac{1}{2}F_2}{k} = -0.14 \text{ in.} \tag{k}$$

and

$$\Delta_4 = \frac{-F_3}{k} = -0.04 \text{ in.} \tag{l}$$

7·4·3 The potential function approach

The procedure outlined and illustrated in §7·4·2 can be used to find the static-equilibrium state of any rigid body or system of rigid bodies. This approach may, however, require a good deal of tedious algebra, and while for some problems this is unavoidable, for other problems more efficient procedures can be found. If the system has many degrees of freedom, then there is no escape from the task of writing and solving a correspondingly large set of algebraic equations. But when the rigid bodies in a complex system are constrained so as to have relatively few degrees of freedom, then methods developed in Chap. 6 may prove more efficient than the direct application of the vector equilibrium equations [Eqs. (7·17) and (7·18)].

As an example of the kind of problem for which the method of §7·4·2 proves cumbersome, consider the system of two rigid bodies portrayed in Fig.

7·12. Both bodies are uniform rigid rods of length L and weight W. One end of body b_1 is attached to the center of body b_2 by a frictionless pinned joint, and the other end of b_1 is pinned to a fixed support at the center C of a smooth-walled hemispherical cup; a rotary spring at C applies to b_1 a torque $\mathbf{T} = k\theta\hat{\mathbf{u}}_3$, where k is a spring constant and θ is the angle between the centerline of b_1 and a horizontal line. The centerlines of b_1 and b_2 define a plane in which $\hat{\mathbf{u}}_1$ and $\hat{\mathbf{u}}_2$ are fixed. We seek an expression for θ^*, the equilibrium value of θ, in terms of k, L, and W.

Although this system has but one degree of freedom in the $\hat{\mathbf{u}}_1\hat{\mathbf{u}}_2$ plane, as defined by θ, the method of §7·4·2 would appear to require the construction of four vector equations, Eqs. (7·17) and (7·18) once for each body, implying twelve scalar equations. If we were to attempt this approach, we would soon find that seven of these twelve equations were useless, and could be discarded, but we would be stuck with five algebraic equations involving five unknowns including θ and two scalar components for each of the constraint forces applied to the ends of b_1. We could solve these equations for θ, but it seems that there must be a better way.

You may recall that in §6·3·4 on virtual work we established several variations of static-equilibrium equations [Eqs. (6·38) to (6·41)] that were equal in number to the degrees of freedom in the system. In §6·4·3 an example was presented to illustrate the use of Eq. (6·41) to find equilibrium states. Although the derivation of Eq. (6·41) in Chap. 6 is formally limited to mechanical systems represented by a finite number of particles, an entirely parallel proof could be devised for a continuous distribution of matter (replacing particles by differential

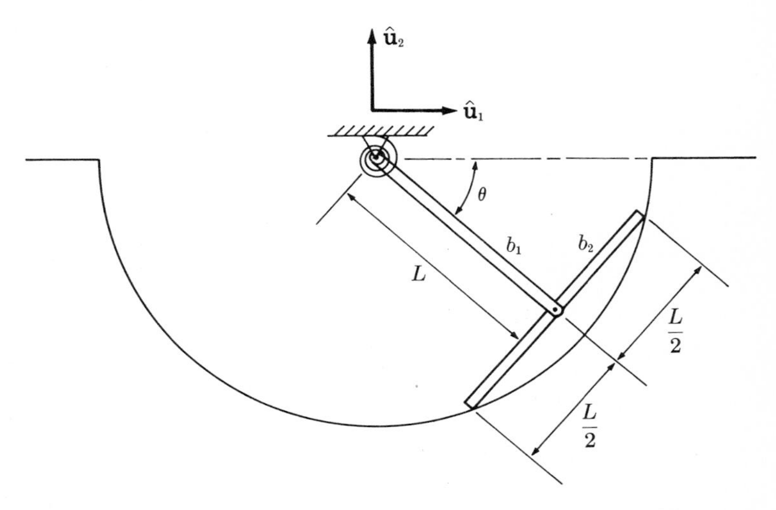

FIGURE 7·12

mass elements, forces by force differentials, and summations by integrations). On the strength of this observation, we shall apply Eq. (6·41) to the rigid-body statics problem posed by Fig. 7·12.

Since there is no internal strain energy in this system, Eq. (6·41) becomes (for $q_1 = \theta$ and $n = 1$)

$$\frac{\partial V}{\partial \theta} = 0 \tag{a}$$

The potential energy V consists of the spring contribution $\frac{1}{2}k\theta^2$ plus the contributions of the gravity force on each body. In a uniform gravity field as postulated here, the gravitational potential energy of a body is simply the product of its weight and the height of its center of gravity above some datum. If we select as a datum the horizontal line through C, and note that for a uniform body in a uniform gravity field the center of gravity, mass center, and centroid are all coincident, we can record

$$\begin{aligned} V &= \tfrac{1}{2}k\theta^2 - \tfrac{1}{2}WL \sin\theta - WL \sin\theta \\ &= \tfrac{1}{2}k\theta^2 - \tfrac{3}{2}WL \sin\theta \end{aligned} \tag{b}$$

Combining Eqs. (*b*) and (*a*) yields

$$\left.\frac{\partial V}{\partial \theta}\right|_{\theta^*} = k\theta^* - \tfrac{3}{2}WL \cos\theta^* = 0 \tag{c}$$

and this transcendental equation can be solved (graphically or numerically) for the equilibrium value θ^*.

A special feature of the potential function approach illustrated here is that, with minor extension, it permits the assessment of the *stability* of the equilibrium solution, as shown in §6·4·2. Applying to this example the stability criterion previously derived as Eq. (6·42), we find that

$$\left.\frac{\partial^2 V}{\partial \theta^2}\right|_{\theta^*} = k + \tfrac{3}{2}WL \sin\theta^* > 0 \tag{d}$$

is required for stability. As long as the rods in Fig. 7·12 remain in the hemispherical cup, we are assured that Eq. (*d*) is satisfied and any equilibrium solution for θ^* produced by Eq. (*c*) is stable.

7·5 PERSPECTIVE

Chapter 7 serves a dual purpose: It introduces the concept of the continuous body, and it illustrates procedures for solving problems involving rigid bodies in a state of static equilibrium, or rest in inertial space.

In preparation for the analysis of continuous rigid bodies, Sec. 7·1 provides several new concepts that permit systems of forces to be treated with an efficiency that greatly facilitates the solution of engineering problems. To keep this material in proper perspective, however, you should note that none of it is *necessary* for the proper formulation of problems in mechanics.

In fact, the concept of the rigid body as a continuous distribution of matter is not essential to the theoretical predictions of the motions of physical bodies—we could always fall back on the notion of the particle-system model, constraining the particles to obtain a "rigid body" if we chose. We shall see, however, in the last two chapters that by modeling a rigid body as a continuum we greatly simplify our analytical tasks, and in Appendix A you will find an intimation of the importance of the continuum model in analyzing deformable solids and fluids.

PROBLEMS

7·1 For each of the systems of forces shown in Fig. P7·1, determine the force resultant, the moment resultant for point q, and the moment resultant for point q' [use Eq. (7·1)].

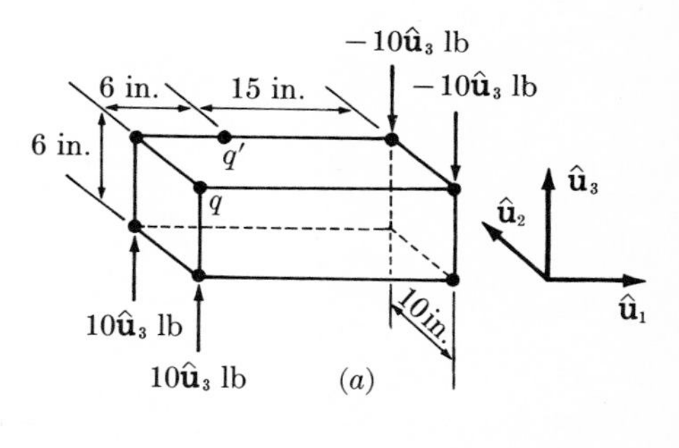

(*a*)

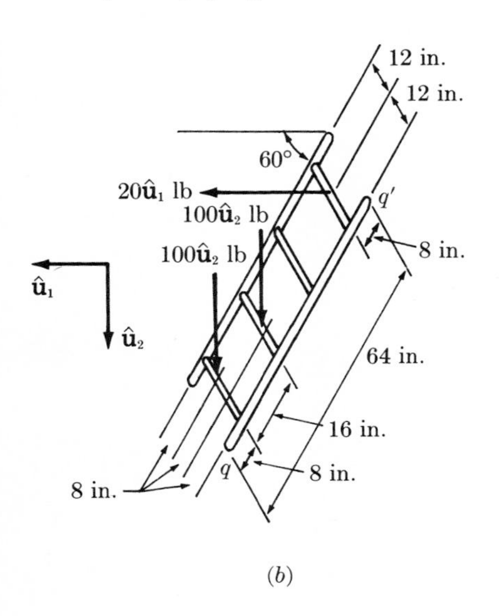

(*b*)

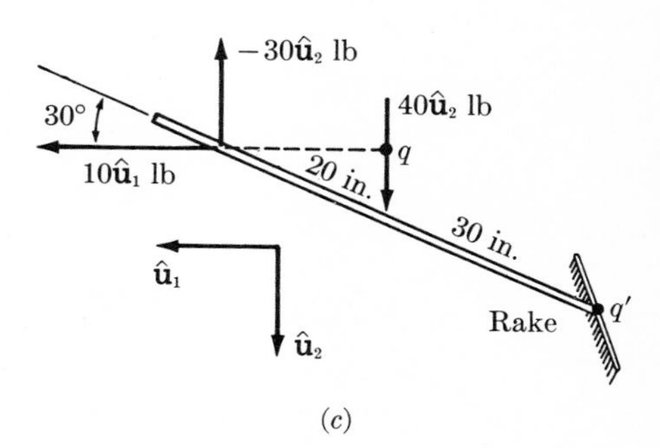

(*c*)

FIGURE P7·1

7·2 Identify any couples among the applied loads shown on the three systems in Fig. P7·1.

7·3 For each of the three systems in Fig. P7·1, construct three equivalent systems of forces.

7·4 Identify any of the three systems in Fig. P7·1 for which there exists an equivalent force system consisting of a single couple.

7·5 Identify any of the three systems in Fig. P7·1 for which there exists an equivalent force system consisting of a single force, and identify the line of action of that force.

7·6 Find the force-couple system of forces equivalent to the applied forces shown in Fig. P7·6, and having the smallest possible couple torque.

(a)

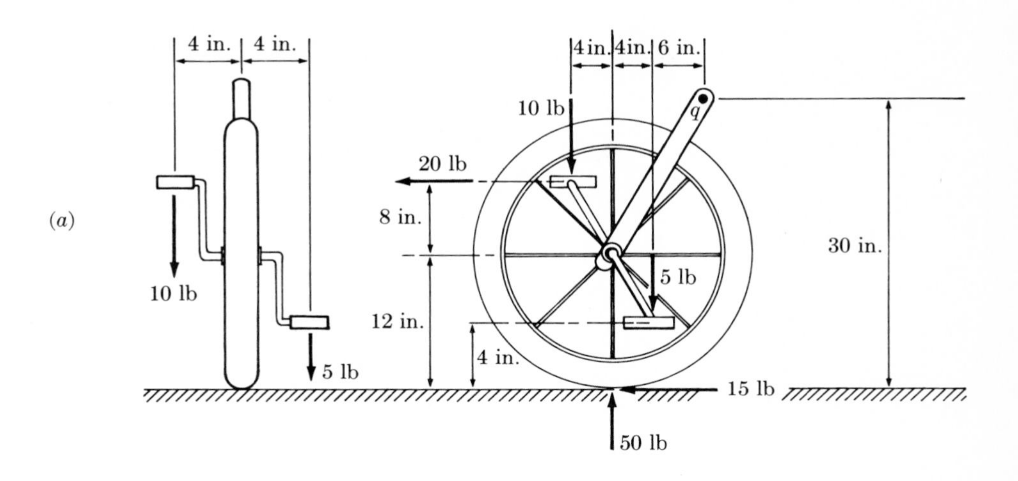

(b)

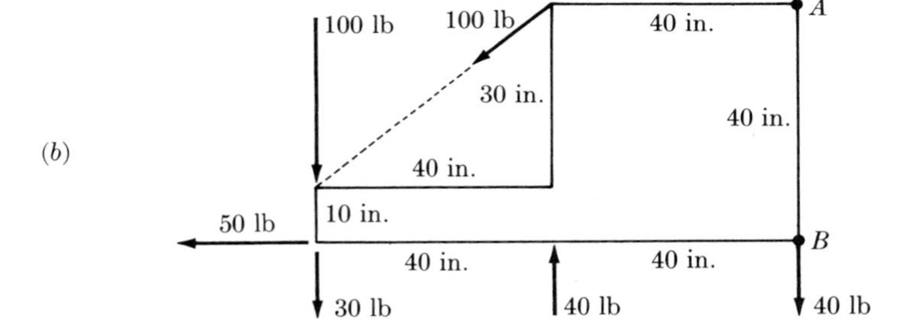

(c)

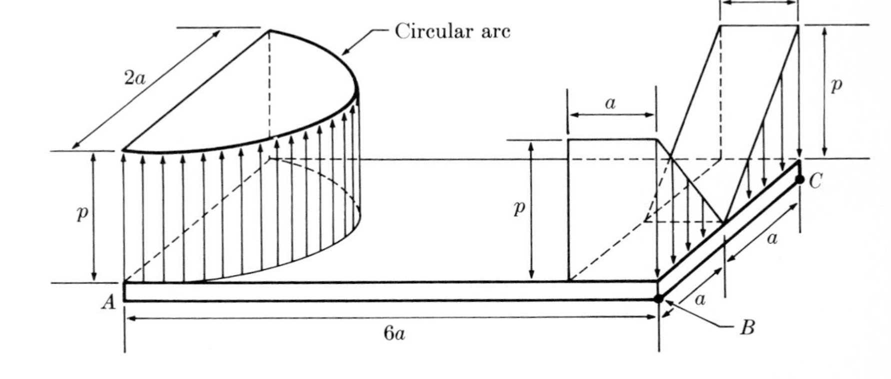

FIGURE P7·6

7·7 For the figures shown in Fig. P7·7, find the volume v and the centroid c^*.

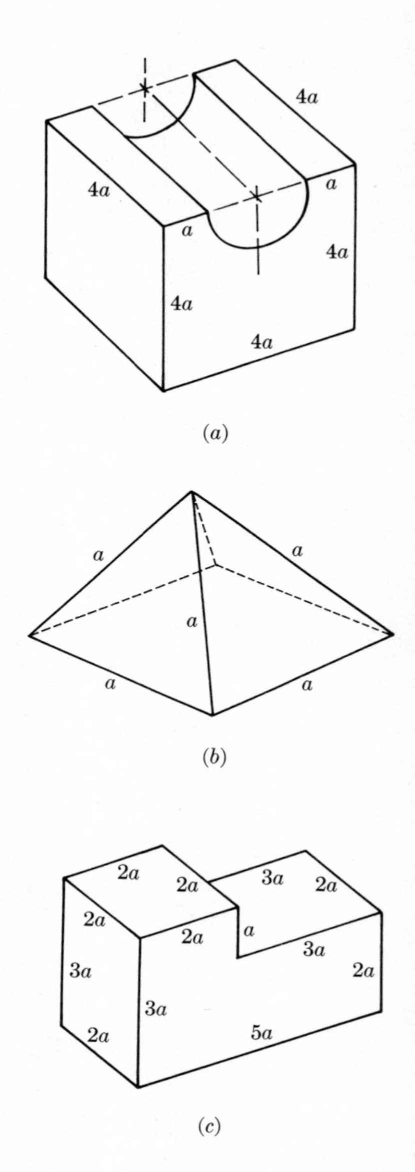

FIGURE P7·7

7·8 For the figures shown in Fig. P7·8, given mass density $\mu = \mu_o (3a - x_3)/3a$, find the mass $\mathfrak{M}$ and the first mass moment $\mathbf{u}^q$ with respect to point q; and finally, find the mass center c.

7·9 Consider the loaded ladder in Fig. P7·1*b*, and determine the reaction forces at the ladder supports required for static equilibrium, under the following conditions:

a. The two feet of the ladder are in contact with a rough horizontal floor, so they cannot slide; the end at q' is supported on a smooth wall, which can apply only a horizontal force to the ladder; and the fourth end is free.

b. The foot at q is securely welded to a metal floor, and all other ends of the ladder are free.

c. All four ends of the ladder rest against smooth surfaces, capable only of applying normal loads.

d. Both feet rest against a rough floor, and both upper ends rest against a smooth wall.

Note: Identify any of the above for which equilibrium is impossible, or for which your answer is not unique (that is, statically indeterminate).

7·10 Consider the rake in Fig. P7·1*c* and determine a force system that would establish static equilibrium of the rake if applied to the tines of the rake by the ground.

If the rake is on a hard surface, the force system applied to the rake by the ground must be equivalent to a single force passing through the line established by the cutting edge of the tines; static equilibrium may then be impossible. Assume that the tines are 3 in. long and in a plane normal to the handle, and seek the single force applied by the ground to the tines such that moment equilibrium is satisfied and force equilibrium is satisfied. Is this equilibrium possible?

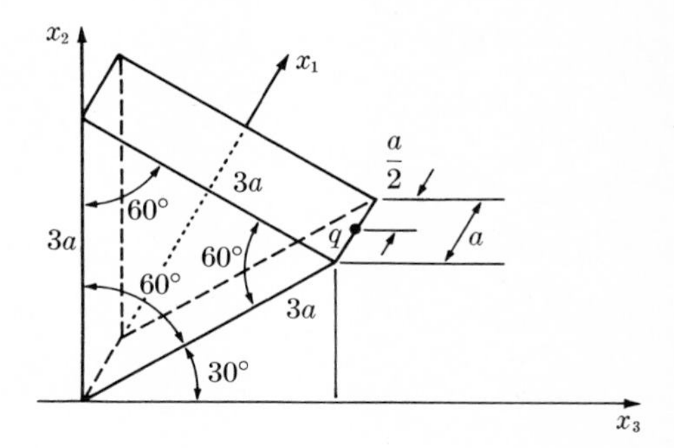

(*a*)

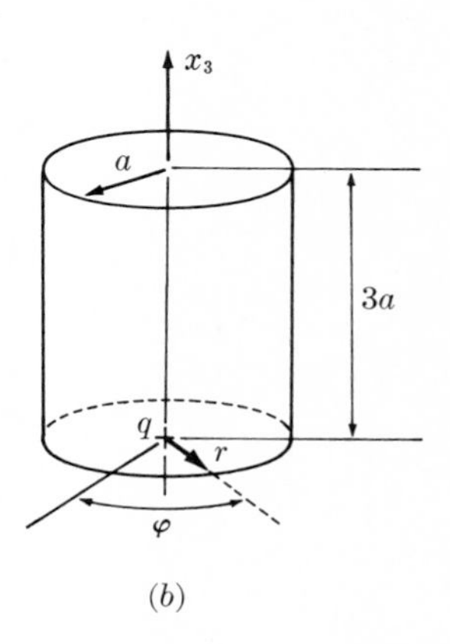

(*b*)

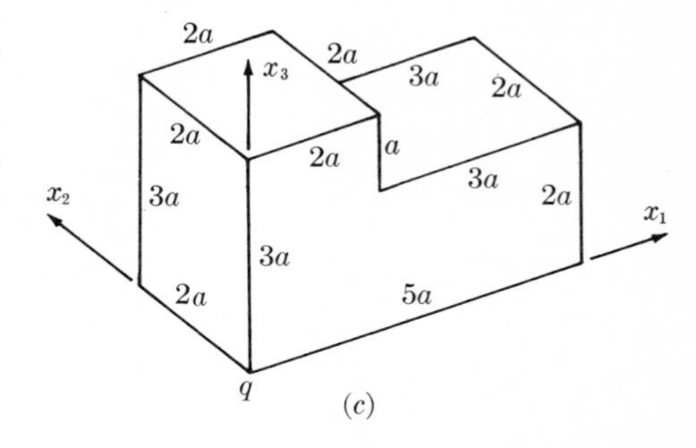

(*c*)

FIGURE P7·8

7·11 Consider the tricycle front-wheel assembly shown in Fig. P7·6*a*. Assume that the fork extends from a horizontal frictionless line hinge connection at the wheel center *o* to a welded connection at point *q* with the tricycle frame. Determine a force-couple system for point *q* that will maintain both the tricycle wheel and its fork in a state of static equilibrium. Also find a force-couple system for point *o* equivalent to the system of forces applied to the wheel by the fork.

7·12 Consider Fig. P7·6*b*, which shows a rigid plate subjected to loads in its plane. Imagine that point *A* of the plate is pinned to a wall, and point *B* bears against a frictionless roller support on that wall. Determine the forces applied to the plate at *A* and *B* for static equilibrium.

7·13 Consider Fig. P7·6*c*, which shows a rigid plate subjected to distributed loads normal to its plane. Imagine that the plate is supported at point *A* by a pinned joint and at points *B* and *C* by frictionless rollers. Find the forces applied at *A*, *B*, and *C* for static equilibrium.

7·14 Figure P7·14 shows a massless rigid block of length $2\sqrt{3}a$ and of width and depth a, loaded with the triangular wedge of material portrayed in Fig. P7·8*a*. [Note that, from Prob. 7·8, the mass density μ is given by $\mu = \mu_0 (3a - x_3)/2\sqrt{3}a$]. Find a force system equivalent to the forces applied to the block by the wall.

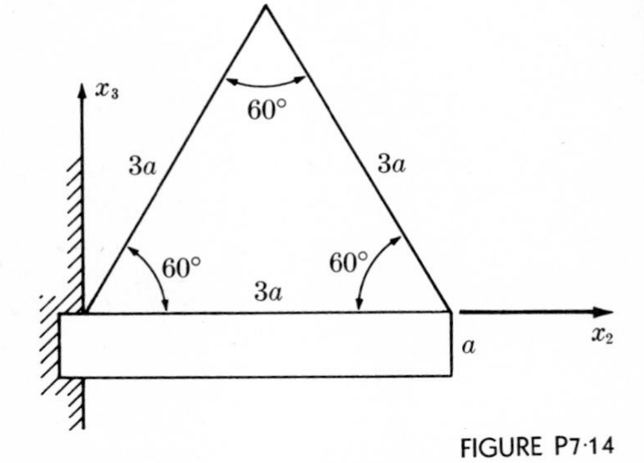

FIGURE P7·14

7·15 Figure P7·15*a* shows a massless inextensible cable connected at either end to a spring with spring constant k; the cable passes over four frictionless pulleys, and supports a block of weight W in the cable span between each pair of pulleys. The end segments of the cable are vertical, and the spans between the pulleys have a slope of ± 30 degrees. Figure P7·15*b* shows a similar arrangement, but with one weight and two pulleys. You are to find, for both systems shown, the spring extensions required for static equilibrium in terms of k and W.

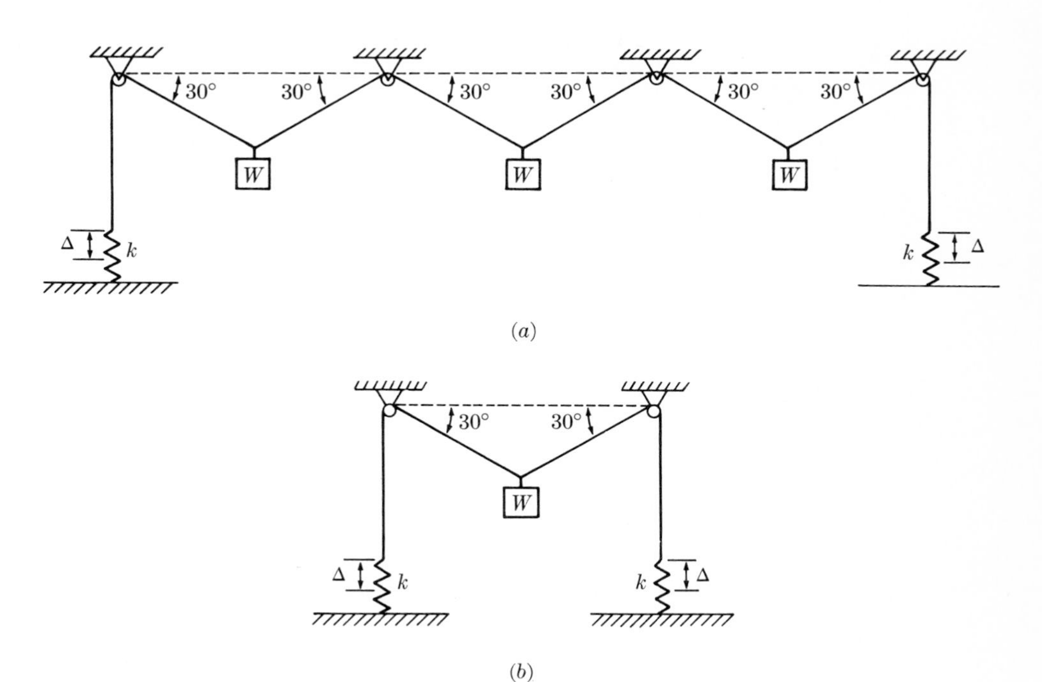

FIGURE P7·15

7·16 Repeat Prob. 7·15, using the method of virtual work.

7·17 The uniform rigid body shown in Fig. P7·17 is supported at point A by a frictionless pinned joint and at B by a linear spring of spring constant k; when the spring is undeformed, the body's top surface is level in a uniform gravity field. If the body weight is $W = 50$ lb, its length L is 10 ft, and the spring constant k is 200 lb/in., what is the value of the spring deformation Δ for static equilibrium, assuming small Δ/L?

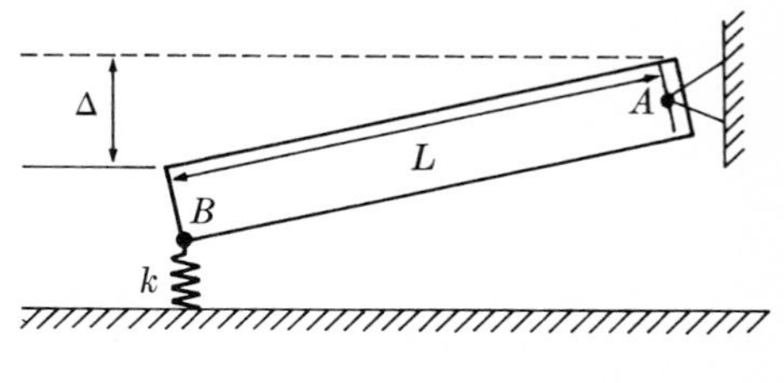

FIGURE P7·17

7·18 Determine the spring extensions for the loaded systems in Fig. P7·18a, b, and

c, assuming in every case that $k = 1000$ lb/in., and that the indicated loads produce small deformations of the spring.

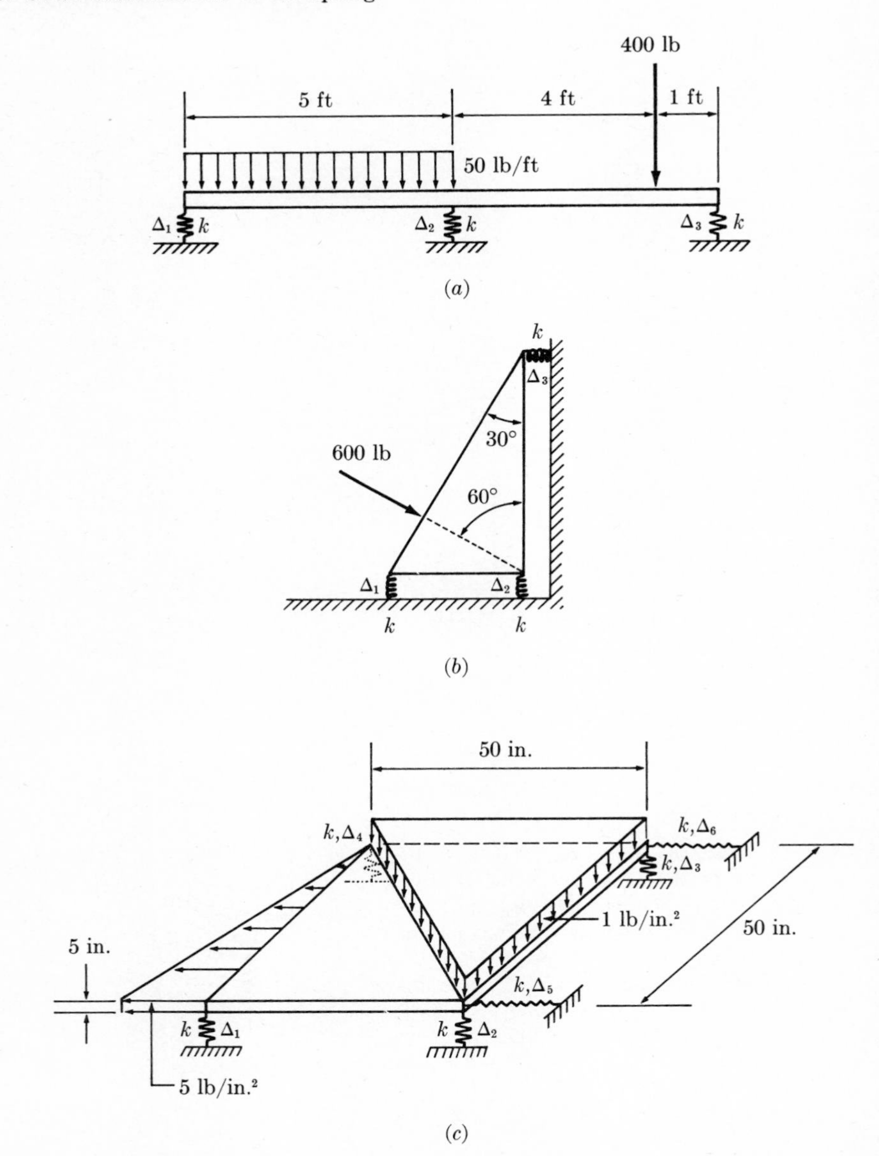

FIGURE P7·18

7·19 Figure P7·19a and b shows four balls, each of weight $W = 20$ lb, in a uniform gravity field, constrained by surfaces against which they slide without friction, and by springs. In each case the spring constant k is 10 lb/in. and the distance L is 10 in.; in Fig. P7·19a the rotary spring constant is $K = 200$ in. lb/rad, and in Fig. P7·19b the radius $R = 50$ in. The dashed lines show the configurations as they appear with unstretched springs, with (a) length $L/2$.

In each case, find values of x and θ for static equilibrium, and assess the stability of the equilibrium, using the potential function approach.

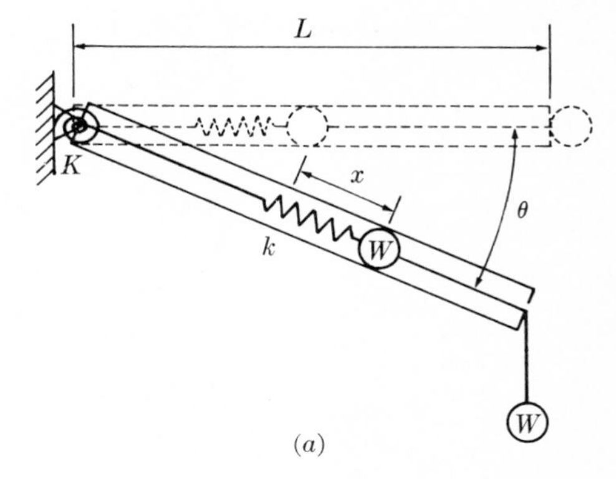

(a)

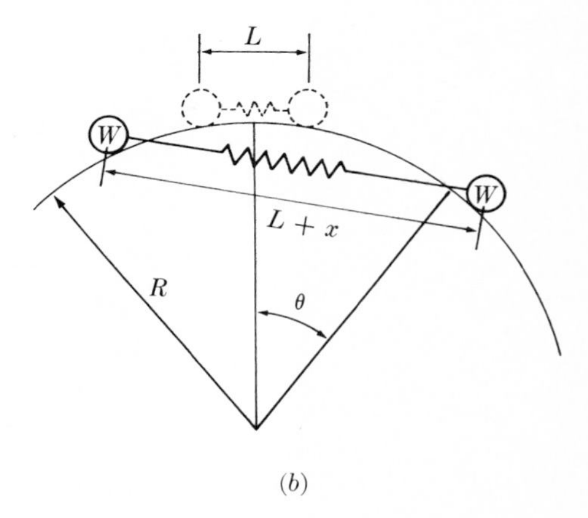

(b)

FIGURE P7·19

CHAPTER EIGHT
RIGID-BODY DYNAMICS: FORMULATIONS

8·1 DYNAMICS OF MASS–CENTER MOTION

8·1·1 Vector equations of motion

In Chap. 7 we adopted as an axiom the proposition that Newton's laws, as presented in Chap. 1, apply to differential elements of mass just as they do to particles. Newton's second law then takes the form

$$d\mathbf{F} = \mathbf{A}\,dm \triangleq \ddot{\mathbf{R}}\,dm \tag{8·1}$$

where $d\mathbf{F}$ is the force applied to a differential quantity of matter having mass dm and mass-center acceleration $\mathbf{A} \triangleq \ddot{\mathbf{R}}$ in an inertial reference frame. (Here $\mathbf{R}$ is the element position vector from some point o fixed in inertial space, and dots denote time differentiation in inertial space.)

A differential version of Newton's laws is necessary if we are to make the transition from mathematical models comprised exclusively of particles to continuous distributions of matter. Once this transition has been made, it is a straightforward task to repeat much of the development previously accomplished for systems of particles (see Chap. 6); the new equations would involve integrals over differential forces and differential masses rather than finite sums over discrete forces and discrete masses.

In anticipation of the integration of Eq. (8·1) over a continuous body, we separate $d\mathbf{F}$ into the resultant $d\mathbf{f}'$ of those forces applied to the element dm by other mass elements of the body, and the resultant $d\mathbf{f}$ of the forces applied externally. Substituting $d\mathbf{f}' + d\mathbf{f}$ for $d\mathbf{F}$ and integrating Eq. (8·1) furnishes

$$\int(d\mathbf{f}' + d\mathbf{f}) = \int\ddot{\mathbf{R}}\, dm = \int\ddot{\mathbf{R}}\, \mu\, dv \tag{8·2}$$

where μ is the mass-density function and dv is a differential element of volume.

By Newton's third law (the action-reaction law), the internal forces exist in equal and opposite pairs, so that

$$\int d\mathbf{f}' = 0 \tag{8·3}$$

If now we let the symbol $\mathbf{F}$ represent $\int d\mathbf{f}$ (and call $\mathbf{F}$ the *force resultant,* generalizing the definition in §7·1·1 to admit integrals as well as sums), we can write Eq. (8·2) as

$$\mathbf{F} = \int\ddot{\mathbf{R}}\, \mu\, dv \tag{8·4}$$

Parenthetically, note that the right side of Eq. (8·4) can be expressed in terms of the first mass moment defined by Eq. (7·12), to obtain

$$\mathbf{F} = \ddot{\boldsymbol{\mathfrak{u}}}^{o} \tag{8·5}$$

A concept of Chap. 7 that is, however, more useful at this point than the first mass moment is the *mass center* c, as defined by Eq. (7·15). If $\mathbf{R}_c$ is the position vector of c relative to the inertially fixed point o, and $\boldsymbol{\rho}$ is the generic position vector to a differential mass element with respect to c, then we can replace $\mathbf{R}$ in Eq. (8·4) by $\mathbf{R}_c + \boldsymbol{\rho}$. Moreover, since the integration limits in Eq. (8·4) are presumed time-independent (no mass is flowing through the boundaries defined by the integration limits), we can reverse the integration-differentiation sequence. The result is

$$\begin{aligned}\mathbf{F} &= \frac{{}^{i}d^2}{dt^2}\int\mathbf{R}\, \mu\, dv = \frac{{}^{i}d^2}{dt^2}\left[\int(\mathbf{R}_c + \boldsymbol{\rho})\mu\, dv\right] \\ &= \frac{{}^{i}d^2}{dt^2}\left(\mathbf{R}_c\int\mu\, dv\right) + \frac{{}^{i}d^2}{dt^2}\left(\int\boldsymbol{\rho}\mu\, dv\right)\end{aligned}$$

The second term is zero since $\boldsymbol{\rho}$ is measured from the mass center [see Eq. (7·14)], and the integral in the first term is the total mass $\mathfrak{M}$ [see Eq. (7·9)]; hence the vector equation of motion of the continuum becomes

$$\mathbf{F} = \mathfrak{M}\ddot{\mathbf{R}}_c \tag{8·6a}$$

or, in the more explicit notation introduced in Chap. 3,

$$\mathbf{F} = \mathfrak{M}\mathbf{A}^{ci} \tag{8·6b}$$

Since exactly the same equations emerged as Eqs. (6·7*a*) and (6·7*b*) for physical objects modeled as systems of particles, this result offers no surprises.

8·2 ATTITUDE DYNAMICS

8·2·1 Background

Once the method Sec. 8·1 has been used to formulate equations of motion of the mass center of a continuum, there remain the much more complex problems of establishing equations that determine the motions of each point in the continuum relative to the mass center. In describing these motions, it is generally advantageous to imagine some reference frame to be associated with the continuum, and then to describe the motions of points in the continuum firstly as motions relative to the newly introduced reference frame and secondly as motions attributed to motions of the reference frame in inertial space. If the continuum of interest is a fluid in a rigid container, then the new reference frame could well be that established by the container. If the continuum is a collection of rigid and deformable bodies, then it might be desirable to associate the indicated reference frame with one of the rigid bodies, or it might be better to generate some rather abstract definition of a reference frame whose motion represents in some sense the "mean motion" of the system. As you can well imagine, this problem of selecting an appropriate reference frame and then formulating equations describing its motion can be a complicated business in itself, but it is no more than a first step in the general problem of deriving equations of motion of a continuum.

In the special case of the *rigid* continuum, the selection of the above-mentioned reference frame is obvious—we simply choose the reference frame established by the body itself, which we call the *body frame*. Moreover, once equations of motion for this reference frame are established, the job is done, since there are by definition no motions of points of the body relative to this reference frame. Since the mass center is fixed relative to the body frame, we must augment the equations of motion for the mass center only by a set of equations sufficient in number to establish the *orientation* or *attitude* of the body frame relative to an inertial reference frame. As we discovered in Chap. 3, the specification of attitude can be accomplished with as few as three scalars (such as the three Euler angles), so the equations of attitude dynamics might consist of three second-order scalar differential equations. As we shall see, this is not always the most convenient alternative, but we can accept this option as an indication of the modest magnitude of the task at hand; if we idealize a material system as a *rigid continuum*, the equations of attitude dynamics might be no different in number or order than the three second-order scalar equations of mass-center motion that can be obtained from Eq. (8·6). Unfortunately, however, they are generally much more complicated in structure, and much more difficult to solve.

8·2·2 Angular momentum/moment of momentum

The concepts of *angular momentum* and *moment of momentum* of a single particle with respect to a point were introduced in §4·2·4, and in §6·2·2 these concepts were applied to systems of particles. Now we extend our definitions to apply to continua.

The *angular momentum* $\mathbf{H}^q$ of a material continuum of constant mass $\mathfrak{M}$ with respect to an arbitrary point q is [in parallel with Eq. (6·12)] defined by

$$\mathbf{H}^q \triangleq \int \mathbf{r} \times \dot{\mathbf{r}}\, dm \tag{8·7}$$

where $\mathbf{r}$ is a generic position vector from point q to a differential element of mass dm, and where integration extends over all differential elements of mass in the continuum, so that

$$\int dm = \mathfrak{M} \tag{8·8}$$

The *moment of momentum* $\mathbf{h}^q$ of a material continuum of constant mass $\mathfrak{M}$ with respect to an arbitrary point q is [in parallel with Eq. (6·13)] defined by

$$\mathbf{h}^q \triangleq \int \mathbf{r} \times \dot{\mathbf{R}}\, dm \tag{8·9}$$

where $\mathbf{R}$ is a generic position vector from any inertially fixed point (say point o) to a differential element of mass dm, and other symbols are as previously defined (see Fig. 8·1).

On the basis of experience gained with systems of particles, we might expect that there will be useful relationships between the moment of external forces about point q and the inertial time derivatives $\dot{\mathbf{H}}^q$ and $\dot{\mathbf{h}}^q$.

8·2·3 Vector equations for rotations

By forming the time derivative of the angular momentum $\mathbf{H}^q$ in an inertial frame of reference, we find

$$\dot{\mathbf{H}}^q = \frac{{}^i d}{dt} \int \mathbf{r} \times \dot{\mathbf{r}}\, dm = \int \mathbf{r} \times \ddot{\mathbf{r}}\, dm$$

From Fig. 8·1 we note the relationship $\mathbf{r} = \mathbf{R} - \mathbf{R}_q$, and substitute to find

$$\dot{\mathbf{H}}^q = \int \mathbf{r} \times \ddot{\mathbf{R}}\, dm - \int \mathbf{r} \times \ddot{\mathbf{R}}_q\, dm$$

By substituting $\mathbf{dF}$ for $\ddot{\mathbf{R}}\, dm$ in the first term [see Eq. (8·1)], and removing $\ddot{\mathbf{R}}_q$ from the second integral, we find

$$\dot{\mathbf{H}}^q = \int \mathbf{r} \times \mathbf{dF} + \ddot{\mathbf{R}}_q \times \int \mathbf{r}\, dm \tag{8·10}$$

The first integral in Eq. (8·10) is the moment resultant with respect to point q of all the differential forces in the continuum, designated

$$\mathbf{M}^q = \int \mathbf{r} \times \mathbf{dF} \tag{8·11}$$

The second integral appearing in Eq. (8·10) we might recognize as the first mass moment $\boldsymbol{\mu}^q$, as defined in Eq. (7·12), or alternatively [see Eq. (7·15)] as the product of the total mass $\mathfrak{M}$ and the vector $\mathbf{r}_c$ from q to the center of mass c, as defined in Eq. (7·14).

Thus Eq. (8·10) becomes

$$\dot{\mathbf{H}}^q = \mathbf{M}^q + \mathfrak{M}\ddot{\mathbf{R}}_q \times \mathbf{r}_c \tag{8·12}$$

This equation is identical to Eq. (6·15), which was derived for a system of particles. Equation (6·16) provides a parallel expression for the inertial frame time

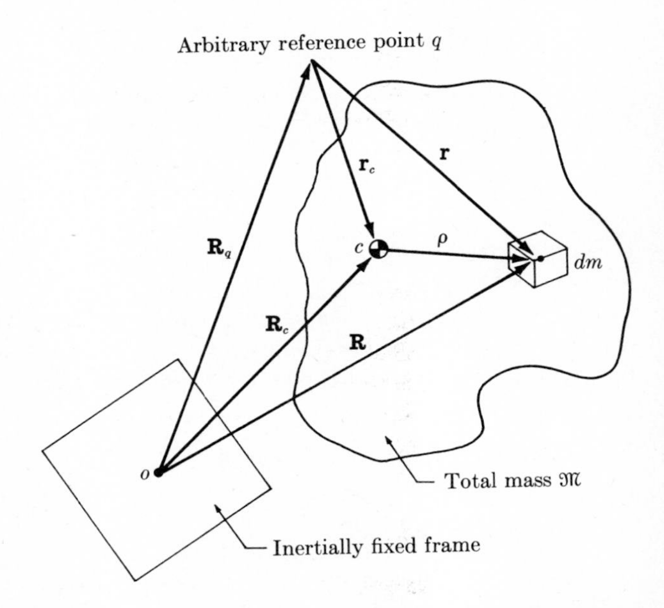

FIGURE 8·1

derivative of the moment of momentum $\mathbf{h}^q$ for a system of particles. This result too we could obtain for a continuum, simply replacing summations by integrations as before, to find

$$\dot{\mathbf{h}}^q = \mathbf{M}^q + \dot{\mathbf{r}}_c \times \mathfrak{M}\dot{\mathbf{R}}_q \tag{8·13}$$

As in Chap. 6, our expression for $\mathbf{M}^q$ [see Eq. (8·11)] can almost always be greatly simplified. As shown in the argument producing Eq. (6·18) from (6·17), we can construct $\mathbf{M}^q$ from external forces only as

$$\mathbf{M}^q = \int \mathbf{r} \times d\mathbf{f} \tag{8·14}$$

whenever all internal forces $d\mathbf{f}'$ exist in pairs that are equal in magnitude and opposite in direction, and have a common line of action.

8·2·4 Special cases of practical interest

When point q is fixed in inertial space, both $\dot{\mathbf{R}}_q$ and $\ddot{\mathbf{R}}_q$ are zero, and both alternative vector equations of rotation [Eqs. (8·12) and (8·13)] simplify. We can then designate the inertially fixed reference point q as o, so that $\mathbf{H}^q$ and $\mathbf{h}^q$ are identically $\mathbf{H}^o$, and we can write

$$\dot{\mathbf{H}}^o = \mathbf{M}^o \tag{8·15}$$

An even more important special case of the vector rotational equations arises when point q is identical to the center of mass c. Then the vector $\mathbf{r}_c$ is identically zero (see Fig. 8·1), and we can replace both $\mathbf{h}^q$ and $\mathbf{H}^q$ by $\mathbf{H}^c$ and write

$$\dot{\mathbf{H}}^c = \mathbf{M}^c \tag{8·16}$$

8·2·5 The compound pendulum

Consider Fig. 8·2, which shows a uniform rigid rod supported on a frictionless hinge at point o, with hinge axis normal to the page and fixed in inertial space. A uniform gravitational force is applied in the direction of $\hat{\mathbf{u}}_v$. This device is called a *compound pendulum*, as distinguished from the *simple pendulum* shown in Fig. 5·1; the term compound pendulum would be applied to the system in Fig. 8·2 even if the uniform rod were replaced by an arbitrary rigid body.

To apply Eq. (8·15), we use Eq. (8·14) to find

$$\mathbf{M}^o = \int \mathbf{R} \times d\mathbf{f} = \int_0^L (y\hat{\mathbf{b}}_1) \times (\hat{\mathbf{u}}_v a\mu g\, dy)$$

where a is the cross-sectional area of the rod, and μ is the mass per unit volume, so that $a\mu\, dy$ is dm, the differential element of mass. Unit vectors $\hat{\mathbf{b}}_1$ and $\hat{\mathbf{u}}_v$ are as shown in Fig. 8·2, and g is the gravity constant (≈ 32 ft/sec^2). Integration yields

$$\mathbf{M}^o = -ga\mu \sin\theta\, \hat{\mathbf{b}}_3 \int_0^L y\, dy = -ga\mu \sin\theta \left(\frac{L^2}{2}\right) \hat{\mathbf{b}}_3$$

where $\hat{\mathbf{b}}_3$ is directed out of the page and θ is as shown in Fig. 8·2. Substituting the total mass $\mathfrak{M}$ for $a\mu L$ provides

$$\mathbf{M}^o = -\tfrac{1}{2}\mathfrak{M}gL \sin\theta\, \hat{\mathbf{b}}_3$$

Equation (8·15) also requires

$$\mathbf{H}^o = \int \mathbf{R} \times \dot{\mathbf{R}}\, dm = \int \mathbf{R} \times (\boldsymbol{\omega} \times \mathbf{R})\, dm$$

Here we have once again used Eq. (3·100) to replace a vector time derivative by a vector cross product. In this particular case, we can substitute $y\hat{\mathbf{b}}_1$ for $\mathbf{R}$; $\dot{\theta}\hat{\mathbf{b}}_3$ for $\boldsymbol{\omega}$; and $a\mu\, dy$ for dm, to obtain

$$\begin{aligned}\mathbf{H}^o &= \int_0^L y\hat{\mathbf{b}}_1 \times (\dot{\theta}\hat{\mathbf{b}}_3 \times y\hat{\mathbf{b}}_1) a\mu\, dy = a\mu\dot{\theta} \int_0^L y^2\, dy\, \hat{\mathbf{b}}_3 \\ &= \tfrac{1}{3} a\mu L^3 \dot{\theta}\hat{\mathbf{b}}_3 = \tfrac{1}{3}\mathfrak{M}L^2\dot{\theta}\hat{\mathbf{b}}_3\end{aligned}$$

Since $\hat{\mathbf{b}}_3$ is inertially fixed, $\dot{\mathbf{H}}^o$ is easily found, and Eq. (8·15) provides for the compound pendulum

$$-\tfrac{1}{2}\mathfrak{M}gL \sin\theta\, \hat{\mathbf{b}}_3 = \tfrac{1}{3}\mathfrak{M}L^2\ddot{\theta}\hat{\mathbf{b}}_3$$

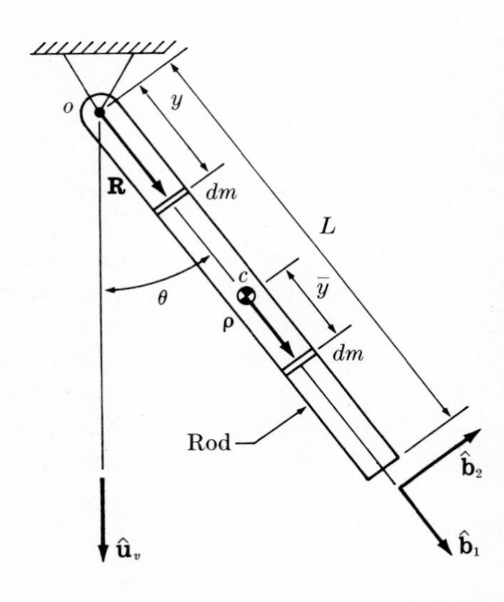

FIGURE 8·2

with the consequence

$$\ddot{\theta} + \frac{2g}{3L} \sin \theta = 0$$

(An equation of the same form is found for the simple pendulum in §5·1·3, where a partial solution may be found.)

Equation (8·16) can also be applied to the compound pendulum in Fig. 8·2, although not quite so easily. This time the moment is

$$\mathbf{M}^c = \int \boldsymbol{\rho} \times \mathbf{df} - \frac{L}{2} \hat{\mathbf{b}}_1 \times \mathbf{F}^o$$

where $\boldsymbol{\rho}$ is the generic position vector to dm from c, and $\mathbf{F}^o$ is the force applied to the bar at point o. As in the previous example, the differential external force is the gravitational attraction, but this time the integral of such forces is zero since the reference point c is the center of gravity of the bar. Only the reaction force $\mathbf{F}^o$ contributes to $\mathbf{M}^c$, and this force is unknown. To find an expression for $\mathbf{F}^o$, we can, from Eq. (8·6a), write

$$\begin{aligned} \mathbf{F}^o + \mathfrak{M} g \hat{\mathbf{u}}_v &= \mathfrak{M} \ddot{\mathbf{R}}_c = \mathfrak{M} \frac{{}^i d^2}{dt^2} (\tfrac{1}{2} L \hat{\mathbf{b}}_1) \\ &= \mathfrak{M}(\tfrac{1}{2} L \ddot{\theta} \hat{\mathbf{b}}_2 - \tfrac{1}{2} L \dot{\theta}^2 \hat{\mathbf{b}}_1) \end{aligned}$$

or

$$\mathbf{F}^o = \mathfrak{M}[(\tfrac{1}{2} L \ddot{\theta} - g \sin \theta) \hat{\mathbf{b}}_2 + (g \cos \theta - \tfrac{1}{2} L \dot{\theta}^2) \hat{\mathbf{b}}_1]$$

From this expression we have

$$\begin{aligned} \mathbf{M}^c &= -\tfrac{1}{2} \mathfrak{M} L \hat{\mathbf{b}}_1 \times (\tfrac{1}{2} L \ddot{\theta} - g \sin \theta) \hat{\mathbf{b}}_2 \\ &= -(\tfrac{1}{4} \mathfrak{M} L^2 \ddot{\theta} - \tfrac{1}{2} \mathfrak{M} g L \sin \theta) \hat{\mathbf{b}}_3 \end{aligned}$$

The angular momentum for the mass center is given by

$$\begin{aligned} \mathbf{H}^c &= \int \boldsymbol{\rho} \times \dot{\boldsymbol{\rho}}\, dm = \int \boldsymbol{\rho} \times (\boldsymbol{\omega} \times \boldsymbol{\rho})\, dm \\ &= \int_{-L/2}^{L/2} \bar{y} \hat{\mathbf{b}}_1 \times (\dot{\theta} \hat{\mathbf{b}}_3 \times \bar{y} \hat{\mathbf{b}}_1) a\mu \, d\bar{y} \end{aligned}$$

where $\bar{y}$ is the coordinate from c along the length of the rod. Multiplication and integration provide

$$\begin{aligned} \mathbf{H}^c &= a\mu\dot{\theta} \int_{-L/2}^{L/2} \bar{y}^2 \, d\bar{y} \, \hat{\mathbf{b}}_3 = \tfrac{1}{3} a\mu\dot{\theta}\bar{y}^3 \Big|_{-L/2}^{L/2} \hat{\mathbf{b}}_3 \\ &= \tfrac{1}{12} a\mu L^3 \dot{\theta} \hat{\mathbf{b}}_3 = \tfrac{1}{12} \mathfrak{M} L^2 \dot{\theta} \hat{\mathbf{b}}_3 \end{aligned}$$

Equating $\mathbf{M}^c$ with $\dot{\mathbf{H}}^c$ then produces

$$\tfrac{1}{12}\mathfrak{M}L^2\ddot{\theta}\hat{\mathbf{b}}_3 = (\tfrac{1}{2}\mathfrak{M}gL \sin\theta - \tfrac{1}{4}\mathfrak{M}L^2\ddot{\theta})\hat{\mathbf{b}}_3$$

or, once again,

$$\ddot{\theta} + \frac{3g}{2L}\sin\theta = 0$$

By the example of the compound pendulum, we have seen that there is more than one acceptable way to use the results of §8·2·4 to obtain correct equations of rotational motion. Of course, we could have reached the same equation of motion for the compound pendulum if we had selected some reference point q other than the support point o or the mass center c; we could, for example, have established q at the free end of the rod. Had we made this peculiar choice, we could either have employed the angular-momentum concept and Eq. (8·12), or the moment-of-momentum concept and Eq. (8·13); either starting point would have brought us to the correct answer.

With a diversity of opportunities comes the burden of selection. You may well be asking, "How do I choose a reference point q?" or "Should I use angular momentum or moment of momentum?" The answers, of course, depend on the difficulties of constructing $\mathbf{M}^q$ and the corresponding momenta. For reasons that become clear only in the next section, it is often easier to construct angular momentum $\mathbf{H}^q$ than moment of momentum $\mathbf{h}^q$, particularly for rigid bodies. This choice will therefore be made in this text. As for the reference point, whenever there exists a point fixed in a rigid body and also fixed in inertial space, it is probably most efficient to call that point o and use Eq. (8·15). In other problems of rigid-body dynamics, point q is generally selected as the mass center c so that Eq. (8·16) can be used, but in special cases (such as in Probs. 8·2 to 8·4) it might be simpler to select some other point q for which $\mathbf{M}^q$ is more easily calculated.

8·2·6 Angular-impulse-momentum equation

Equations (8·15) and (8·16) are the most useful forms of the vector rotational equations, and they are, at least conceptually, easily integrated over time. Integration of Eq. (8·16) over time in an inertial reference frame provides

$$\int_{t_1}^{t_2} \mathbf{M}^c\,dt\,\Big|_i = \int_{t_1}^{t_2} \frac{{}^i d\mathbf{H}^c}{dt}\,dt\,\Big|_i = \mathbf{H}^c(t_2) - \mathbf{H}^c(t_1) \tag{8·17}$$

The left side of this equation is known as the (inertial) *angular impulse*, and Eq. (8·17) is an angular-impulse-momentum equation [in parallel with the linear impulse-momentum equation (4·25)]. This equation is most useful when the angular impulse is zero (so that angular momentum is conserved), or when the time interval $t_2 - t_1$ is considered to vanish in the limiting case characterized as an *ideal angular impulse.*

We could, of course, have begun with Eq. (8·15) rather than (8·16), so that (8·17) applies also when the reference point is o rather than c.

8·2·7 Application to spinning spacecraft attitude control

Many spin-stabilized space vehicles include pairs of gas jets mounted as shown in Fig. 8·3, and capable of applying moments about the spacecraft mass center for the purpose of reorienting the spin axis. The jets are designed to fire bursts of gas simultaneously; the two valves are open and apply a force of nearly constant magnitude F for perhaps 0.1 sec, and it may take a full second for the spacecraft to revolve. During the brief interval Δt of jet actuation following time t_1, the actual (inertial) angular impulse imparted to the spacecraft is given by

$$-\int_{t_1}^{t_1+\Delta t} 2(b\hat{\mathbf{s}}_1 \times F\hat{\mathbf{s}}_3)\, dt \bigg|_i = 2bF \int_{t_1}^{t_1+\Delta t} \hat{\mathbf{s}}_2\, dt \bigg|_i$$

where b is the "lever arm" of the jets and $\hat{\mathbf{s}}_1$, $\hat{\mathbf{s}}_2$, and $\hat{\mathbf{s}}_3$ are fixed in the spacecraft. This integral is very difficult to evaluate exactly because $\hat{\mathbf{s}}_2$ is changing its

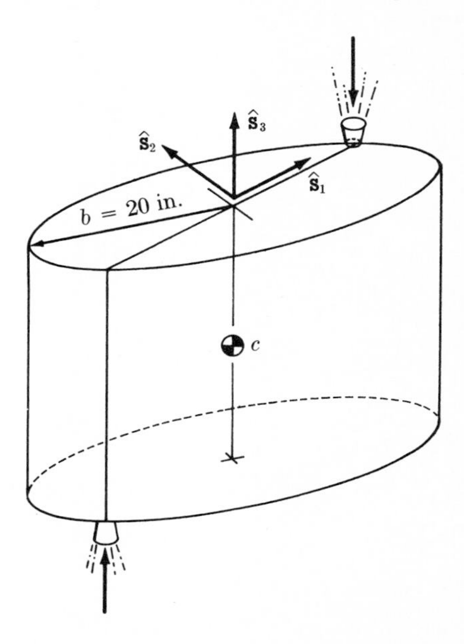

FIGURE 8·3

inertial orientation during the time interval Δt, and we don't know exactly how it changes. By treating the jet actuation as an *instantaneous* event, we circumvent this problem, and apply to the spacecraft an ideal impulse of magnitude $2bF\,\Delta t$ and of the direction established by $\hat{\mathbf{s}}_2$ at the time the valve is opened. With this knowledge of the left side of Eq. (8·17), we can immediately assess the resulting change in angular momentum. In order to proceed from knowledge of angular-momentum change, however, to an understanding of the resulting vehicle motion, we would need some of the results of Chap. 9. With this information, we would be equipped to make a preliminary estimate of the response of a spinning spacecraft to the actuation of a pair of attitude-control jets. We might use such rough analysis to design our vehicle, and then confirm our predictions by digital-computer numerical integration of more exact equations of motion that account for the duration Δt of the jet firings.

8·3 ANGULAR-MOMENTUM EXPANSIONS

8·3·1 Angular momentum in terms of scalars and unit vectors

The derivative $\dot{\mathbf{r}} \triangleq {}^i d\mathbf{r}/dt$ appearing in the angular-momentum definition found in Eq. (8·7) can, of course, be expanded in the fashion of Eq. (3·100), writing for any reference frame f

$$\frac{{}^i d}{dt}\mathbf{r} = \frac{{}^f d}{dt}\mathbf{r} + \boldsymbol{\omega}^{fi} \times \mathbf{r} \tag{8·18a}$$

In what follows, this equation will be written in more abbreviated notation as

$$\dot{\mathbf{r}} = \mathring{\mathbf{r}} + \boldsymbol{\omega} \times \mathbf{r} \tag{8·18b}$$

Be careful not to lose sight of the significance of the symbols $\boldsymbol{\omega}$ and $\mathring{\mathbf{r}}$; these both depend on the choice of frame f, which has yet to be specified. Substitution of Eq. (8·18b) into (8·7) produces

$$\mathbf{H}^q = \int \mathbf{r} \times \mathring{\mathbf{r}}\, dm + \int \mathbf{r} \times (\boldsymbol{\omega} \times \mathbf{r})\, dm \tag{8·19}$$

Whenever possible, the frame f should be chosen so that $\mathring{\mathbf{r}} \equiv 0$. In the applications to the compound pendulum in §8·2·5, the rigid bar itself established the frame f, and since both o and c were fixed in f, the angular momenta $\mathbf{H}^o$ and $\mathbf{H}^c$ both reduced to the equivalent of the second integral in Eq. (8·19). This will always be the case when the material system is a rigid body and the reference point is fixed in the reference frame established by that body. Because of the importance of this integral, we'll give it a label, defining

$$\mathbf{H}^* \triangleq \int \mathbf{r} \times (\boldsymbol{\omega} \times \mathbf{r})\, dm \tag{8·20}$$

Now we can say that *the angular momentum of a rigid body with respect to a point fixed relative to that body is simply* $\mathbf{H}^*$, with $\boldsymbol{\omega}$ the inertial angular velocity of the body, and that even for an entirely general combination of material system and reference point, $\mathbf{H}^*$ is an important part of $\mathbf{H}^q$. (Such a sweeping statement cannot be made for the moment of momentum $\mathbf{h}^q$, and this makes it a less useful concept in application.)

Although it is possible, as in the two applications to the compound pendulum in §8·2·5, to formulate specific values of $\mathbf{H}^*$ by evaluating the integral in Eq. (8·20) explicitly, it is also possible to expand this integral in generic terms; once this is accomplished, we can solve specific problems without worrying about evaluating integrals.

The most direct way to expand $\mathbf{H}^*$ is by introducing a set of orthogonal unit vectors $\hat{\mathbf{u}}_1$, $\hat{\mathbf{u}}_2$, and $\hat{\mathbf{u}}_3$ and expanding $\mathbf{H}^*$, $\boldsymbol{\omega}$, and $\mathbf{r}$ as follows:

$$\mathbf{H}^* = H_1^*\hat{\mathbf{u}}_1 + H_2^*\hat{\mathbf{u}}_2 + H_3^*\hat{\mathbf{u}}_3 \tag{8·21a}$$
$$\boldsymbol{\omega} = \omega_1\hat{\mathbf{u}}_1 + \omega_2\hat{\mathbf{u}}_2 + \omega_3\hat{\mathbf{u}}_3 \tag{8·21b}$$
$$\mathbf{r} = r_1\hat{\mathbf{u}}_1 + r_2\hat{\mathbf{u}}_2 + r_3\hat{\mathbf{u}}_3 \tag{8·21c}$$

Equation (8·20) then becomes

$$\begin{aligned} H_1^*\hat{\mathbf{u}}_1 + H_2^*\hat{\mathbf{u}}_2 + H_3^*\hat{\mathbf{u}}_3 &= \int \mathbf{r} \times (\boldsymbol{\omega} \times \mathbf{r})\, dm \\ &= \int (\mathbf{r}\cdot\mathbf{r}\boldsymbol{\omega} - \mathbf{r}\cdot\boldsymbol{\omega}\mathbf{r})\, dm \\ &= \int [(r_1^2 + r_2^2 + r_3^2)(\omega_1\hat{\mathbf{u}}_1 + \omega_2\hat{\mathbf{u}}_2 + \omega_3\hat{\mathbf{u}}_3) \\ &\qquad - (r_1\omega_1 + r_2\omega_2 + r_3\omega_3)(r_1\hat{\mathbf{u}}_1 + r_2\hat{\mathbf{u}}_2 + r_3\hat{\mathbf{u}}_3)]\, dm \\ &= \hat{\mathbf{u}}_1[\int (r_2^2 + r_3^2)\, dm\, \omega_1 - \int r_1r_2\, dm\, \omega_2 - \int r_1r_3\, dm\, \omega_3] \\ &\quad + \hat{\mathbf{u}}_2[-\int r_2r_1\, dm\, \omega_1 + \int (r_3^2 + r_1^2)\, dm\, \omega_2 - \int r_2r_3\, dm\, \omega_3] \\ &\quad + \hat{\mathbf{u}}_3[-\int r_3r_1\, dm\, \omega_1 - \int r_3r_2\, dm\, \omega_2 + \int (r_1^2 + r_2^2)\, dm\, \omega_3] \end{aligned} \tag{8·21d}$$

It is customary to introduce the symbols

$$I_{11} \triangleq \int (r_2^2 + r_3^2)\, dm \tag{8·22a}$$
$$I_{12} \triangleq -\int r_1r_2\, dm \tag{8·22b}$$
$$I_{13} \triangleq -\int r_1r_3\, dm \tag{8·22c}$$
$$I_{21} \triangleq -\int r_2r_1\, dm \tag{8·22d}$$
$$I_{22} \triangleq \int (r_3^2 + r_1^2)\, dm \tag{8·22e}$$
$$I_{23} \triangleq -\int r_2r_3\, dm \tag{8·22f}$$
$$I_{31} \triangleq -\int r_3r_1\, dm \tag{8·22g}$$
$$I_{32} \triangleq -\int r_3r_2\, dm \tag{8·22h}$$
$$I_{33} \triangleq \int (r_1^2 + r_2^2)\, dm \tag{8·22i}$$

so that $\mathbf{H}^*$ may be written

$$\begin{aligned}\mathbf{H}^* &= H_1^*\hat{\mathbf{u}}_1 + H_2^*\hat{\mathbf{u}}_2 + H_3^*\hat{\mathbf{u}}_3 \\ &= \hat{\mathbf{u}}_1(I_{11}\omega_1 + I_{12}\omega_2 + I_{13}\omega_3) + \hat{\mathbf{u}}_2(I_{21}\omega_1 + I_{22}\omega_2 + I_{23}\omega_3) \\ &\qquad + \hat{\mathbf{u}}_3(I_{31}\omega_1 + I_{32}\omega_2 + I_{33}\omega_3)\end{aligned} \tag{8·23}$$

Note that the definitions of I_{12} and I_{21} are identical, and, in fact, $I_{ij} \triangleq I_{ji}$ for any values of i and j; separate symbols are used simply to preserve a certain symmetry in Eq. (8·23). It should also be carefully noted that the definitions of I_{ij} $(i, j = 1, 2, 3)$ imply a particular choice of reference point q and a particular selection of unit vectors $\hat{\mathbf{u}}_1$, $\hat{\mathbf{u}}_2$, and $\hat{\mathbf{u}}_3$. If we must distinguish between these quantities for two different reference points (say c and o) we will use postsuperscripts (such as I_{ij}^c and I_{ij}^o); and to distinguish between these quantities for two different vector bases (say $\hat{\mathbf{u}}_1$, $\hat{\mathbf{u}}_2$, and $\hat{\mathbf{u}}_3$ and $\hat{\mathbf{a}}_1$, $\hat{\mathbf{a}}_2$, and $\hat{\mathbf{a}}_3$), we'll use presuperscripts (such as ${}^uI_{ij}$ and ${}^aI_{ij}$). Only rarely will this complexity of notation be necessary.

The scalars I_{11}, I_{22} and I_{33} are known as *moments of inertia* and the quantities $-I_{ij} = \int r_i r_j \, dm$ for $i \neq j$ are often called *products of inertia* (note the minus sign in the latter). We may also speak of I_{11}, for example, as the *moment of inertia about a line* passing through the given reference point and parallel to the unit vector with subscript 1.

In order to establish a set of moments and products of inertia for any material system, we must choose some reference point q and then make an arbitrary choice of vector basis $\hat{\mathbf{u}}_1$, $\hat{\mathbf{u}}_2$, and $\hat{\mathbf{u}}_3$. These concepts are most useful when the scalars I_{ij} $(i, j = 1, 2, 3)$ are constant, since $\mathbf{H}^*$ in Eq. (8·23) must be differentiated in time to obtain equations of rotation. The moments and products of inertia remain constant if the material system is a rigid body, provided that the reference point q and the unit vectors $\hat{\mathbf{u}}_1$, $\hat{\mathbf{u}}_2$, and $\hat{\mathbf{u}}_3$ are chosen so as to be fixed in that body; this special case receives primary attention in what follows.

8·3·2 Angular momentum in terms of matrices and vector arrays

Equation (8·23) can be written in terms of arrays of vectors and arrays of scalars as follows:

$$\mathbf{H}^* = [\hat{\mathbf{u}}_1\hat{\mathbf{u}}_2\hat{\mathbf{u}}_3]\begin{bmatrix} H_1^* \\ H_2^* \\ H_3^* \end{bmatrix} = [\hat{\mathbf{u}}_1\hat{\mathbf{u}}_2\hat{\mathbf{u}}_3]\begin{bmatrix} I_{11} & I_{12} & I_{13} \\ I_{21} & I_{22} & I_{23} \\ I_{31} & I_{32} & I_{33} \end{bmatrix}\begin{bmatrix} \omega_1 \\ \omega_2 \\ \omega_3 \end{bmatrix} \tag{8·24a}$$

or, more compactly, as

$$\mathbf{H}^* = \{\hat{\mathbf{u}}\}^T[H^*] = \{\hat{\mathbf{u}}\}^T[I][\omega] \tag{8·24b}$$

Here we are applying the rules of matrix multiplication to the vector array $\{\hat{\mathbf{u}}\}$ as well as to the matrices $[I]$ and $[\omega]$. (This practice was useful in Chap. 3 also.)

The matrix $[I]$ is called the *inertia matrix;* again it should be emphasized that the elements of the inertia matrix depend on the choice of a reference point and a vector basis, and when it is necessary to identify the reference point (say q) and the vector basis (say $\{\hat{\mathbf{u}}\}$), we will adopt an explicit notation, for example, $[^uI^q]$.

8·3·3 Angular momentum in terms of vectors and dyadics

Whether we expand angular momentum in the form of scalars and vectors [as in Eq. (8·23)] or matrices and vector arrays [as in Eqs. (8·24)], we are in either case sacrificing one of the most valued features of a vector—its freedom from commitment to expression in any particular vector basis. The symbol $\mathbf{H}^*$ has the same meaning as the ordered array of symbols H_1^*, H_2^*, and H_3^*, but in the latter case it must be understood that unit vectors $\hat{\mathbf{u}}_1$, $\hat{\mathbf{u}}_2$, and $\hat{\mathbf{u}}_3$ are implied [see Eq. (8·24*a*)]. We would like to avoid the introduction of any particular unit vectors, and still expand the integral definition of $\mathbf{H}^*$ in Eq. (8·20) in a way that separates the mass-distribution properties of the material system from the angular-velocity vector $\boldsymbol{\omega}$. We can accomplish this with the help of the *dyadic* (see Appendix B).

Once the integral for $\mathbf{H}^*$ is written in the form

$$\mathbf{H}^* \triangleq \int \mathbf{r} \times (\boldsymbol{\omega} \times \mathbf{r})\, dm = \int (\mathbf{r} \cdot \mathbf{r}\boldsymbol{\omega} - \mathbf{r} \cdot \boldsymbol{\omega}\mathbf{r})\, dm$$

or

$$\mathbf{H}^* = \int (\mathbf{r} \cdot \mathbf{r}\boldsymbol{\omega} - \mathbf{r}\mathbf{r} \cdot \boldsymbol{\omega})\, dm$$

it becomes tempting to try to remove the vector $\boldsymbol{\omega}$ from the integral, factoring it out to the right. The mathematical obstacle to this factorization lies in the fact that $\boldsymbol{\omega}$ is preceded by a dot in one location within the integral, and not in the other. In order to circumvent this obstacle, we introduce a new concept, called the *unit dyadic* $\mathbb{U}$, so defined that its dot product (in either sequence) with any vector gives that vector back again. Then we replace $\boldsymbol{\omega}$ in $\mathbf{H}^*$ by $\mathbb{U} \cdot \boldsymbol{\omega}$, to obtain

$$\begin{aligned} \mathbf{H}^* &= \int (\mathbf{r} \cdot \mathbf{r}\mathbb{U} \cdot \boldsymbol{\omega} - \mathbf{r}\mathbf{r} \cdot \boldsymbol{\omega})\, dm \\ &= \int (\mathbf{r} \cdot \mathbf{r}\mathbb{U} - \mathbf{r}\mathbf{r})\, dm \cdot \boldsymbol{\omega} \end{aligned} \tag{8·25a}$$

The integral in Eq. (8·25*a*) has at least one of the features of the new unit dyadic: when dot-multiplied by a vector, it produces a vector. (In contrast, one vector dot-multiplied by another produces a scalar.) This integral is further characterized by the fact that it includes two vectors side by side without an intervening dot or cross; this combination is called a *dyad,* and any linear combination of dyads is called a *dyadic.* Specifically, the integral in Eq. (8·25*a*) is called the *inertia dyadic* of the system for the given reference point q, written $\mathbb{I}$ or (if the reference point must be explicit) $\mathbb{I}^q$. Thus we can write the vector $\mathbf{H}^*$ in terms of the vector $\boldsymbol{\omega}$ and the dyadic $\mathbb{I}$ as

$$\mathbf{H}^* = \mathbb{I} \cdot \boldsymbol{\omega} \tag{8·25b}$$

It should be noted, if only in passing, that the inertia dyadic $\mathbb{I}^q$ defined in Eq. (8·25a) can be written in terms of the second moment $\boldsymbol{\mu}^q$ defined in Eq. (7·16c) and the scalar $\int \mathbf{r} \cdot \mathbf{r}\, dm = \int (r_1^2 + r_2^2 + r_3^2)\, dm = \frac{1}{2} \operatorname{tr} (I^q)$, where tr (I^q), or the trace of I^q, means the sum of the diagonal elements in I^q. The result is

$$\mathbb{I}^q \triangleq \int (\mathbf{r} \cdot \mathbf{r}\mathbb{U} - \mathbf{rr}) dm = \tfrac{1}{2} \operatorname{tr} (I^q)\, \mathbb{U} - \boldsymbol{\mu}^q \tag{8·26a}$$

Just as any vector, such as $\boldsymbol{\omega}$, can be written in terms of any particular set of orthogonal unit vectors, as illustrated by

$$\boldsymbol{\omega} = \omega_1 \hat{\mathbf{u}}_1 + \omega_2 \hat{\mathbf{u}}_2 + \omega_3 \hat{\mathbf{u}}_3 = \sum_{\alpha=1}^{3} \omega_\alpha \hat{\mathbf{u}}_\alpha = [\hat{\mathbf{u}}_1 \hat{\mathbf{u}}_2 \hat{\mathbf{u}}_3] \begin{bmatrix} \omega_1 \\ \omega_2 \\ \omega_3 \end{bmatrix} = \{\hat{\mathbf{u}}\}^T [\omega]$$

so also any dyadic, such as $\mathbb{I}$, can be written in terms of any particular set of orthogonal unit vectors, as illustrated by

$$\begin{aligned} \mathbb{I} &= I_{11} \hat{\mathbf{u}}_1 \hat{\mathbf{u}}_1 + I_{12} \hat{\mathbf{u}}_1 \hat{\mathbf{u}}_2 + I_{13} \hat{\mathbf{u}}_1 \hat{\mathbf{u}}_3 + I_{21} \hat{\mathbf{u}}_2 \hat{\mathbf{u}}_1 + I_{22} \hat{\mathbf{u}}_2 \hat{\mathbf{u}}_2 + I_{23} \hat{\mathbf{u}}_2 \hat{\mathbf{u}}_3 \\ &\qquad + I_{31} \hat{\mathbf{u}}_3 \hat{\mathbf{u}}_1 + I_{32} \hat{\mathbf{u}}_3 \hat{\mathbf{u}}_2 + I_{33} \hat{\mathbf{u}}_3 \hat{\mathbf{u}}_3 = \sum_{\alpha=1}^{3} \sum_{\beta=1}^{3} I_{\alpha\beta} \hat{\mathbf{u}}_\alpha \hat{\mathbf{u}}_\beta \\ &= [\hat{\mathbf{u}}_1 \hat{\mathbf{u}}_2 \hat{\mathbf{u}}_3] \begin{bmatrix} I_{11} & I_{12} & I_{13} \\ I_{21} & I_{22} & I_{23} \\ I_{31} & I_{32} & I_{33} \end{bmatrix} \begin{bmatrix} \hat{\mathbf{u}}_1 \\ \hat{\mathbf{u}}_2 \\ \hat{\mathbf{u}}_3 \end{bmatrix} = \{\hat{\mathbf{u}}\}^T [I] \{\hat{\mathbf{u}}\} \end{aligned} \tag{8·26b}$$

This choice of symbols I_{ij} $(i, j = 1, 2, 3)$ is reconciled with the previous use of these symbols by the following rule for the multiplication of dyads: When a dyad is (dot- or cross-) multiplied by a vector or another dyad, simply perform the usual vector-multiplication operations on the vectors immediately separated by the multiplication symbols (dot or cross). This process is illustrated for the dyad **ab**, the vector **c**, and the dyad **de** by

$\mathbf{ab} \cdot \mathbf{c} = \mathbf{a}(\mathbf{b} \cdot \mathbf{c}) = (\mathbf{b} \cdot \mathbf{c})\mathbf{a}$ vector
$\mathbf{c} \cdot \mathbf{ab} = (\mathbf{c} \cdot \mathbf{a})\mathbf{b}$ vector
$\mathbf{ab} \times \mathbf{c} = \mathbf{a}(\mathbf{b} \times \mathbf{c})$ dyad
$\mathbf{c} \times \mathbf{ab} = (\mathbf{c} \times \mathbf{a})\mathbf{b}$ dyad
$\mathbf{ab} \cdot \mathbf{de} = \mathbf{a}(\mathbf{b} \cdot \mathbf{d})\mathbf{e} = (\mathbf{b} \cdot \mathbf{d})\mathbf{ae}$ dyad

The quantity $\mathbf{ab} \times \mathbf{de} = \mathbf{a}(\mathbf{b} \times \mathbf{d})\mathbf{e}$ is, not surprisingly, called a *triad*—but this concept has no immediate utility in mechanics.

By simply applying the rules for dot multiplication to the dyadic $\mathbb{I}$ as expanded in Eq. (8·26b), we find that the scalars I_{ij} $(i, j = 1, 2, 3)$ can be obtained from the dyadic $\mathbb{I}$ as follows:

$$I_{ij} = \hat{\mathbf{u}}_i \cdot \mathbb{I} \cdot \hat{\mathbf{u}}_j \tag{8·26c}$$

Application of the multiplication rules to Eq. (8·26*b*) produces reconciliation with Eqs. (8·24), since

$$\mathbf{H}^* = \mathbb{I} \cdot \boldsymbol{\omega} = \{\hat{\mathbf{u}}\}^T[I]\{\hat{\mathbf{u}}\} \cdot \{\hat{\mathbf{u}}\}^T[\omega] = \{\hat{\mathbf{u}}\}^T[I][\omega] \tag{8·25c}$$

where the product

$$\{\hat{\mathbf{u}}\} \cdot \{\hat{\mathbf{u}}\}^T = \begin{bmatrix} \hat{\mathbf{u}}_1 \\ \hat{\mathbf{u}}_2 \\ \hat{\mathbf{u}}_3 \end{bmatrix} \cdot [\hat{\mathbf{u}}_1\hat{\mathbf{u}}_2\hat{\mathbf{u}}_3]$$

$$= \begin{bmatrix} \hat{\mathbf{u}}_1 \cdot \hat{\mathbf{u}}_1 & \hat{\mathbf{u}}_1 \cdot \hat{\mathbf{u}}_2 & \hat{\mathbf{u}}_1 \cdot \hat{\mathbf{u}}_3 \\ \hat{\mathbf{u}}_2 \cdot \hat{\mathbf{u}}_1 & \hat{\mathbf{u}}_2 \cdot \hat{\mathbf{u}}_2 & \hat{\mathbf{u}}_2 \cdot \hat{\mathbf{u}}_3 \\ \hat{\mathbf{u}}_3 \cdot \hat{\mathbf{u}}_1 & \hat{\mathbf{u}}_3 \cdot \hat{\mathbf{u}}_2 & \hat{\mathbf{u}}_3 \cdot \hat{\mathbf{u}}_3 \end{bmatrix} = \begin{bmatrix} 1 & 0 & 0 \\ 0 & 1 & 0 \\ 0 & 0 & 1 \end{bmatrix}$$

can be omitted from the expression entirely without changing anything.

The multiplication rules also tell us that in terms of an arbitrarily selected set of orthogonal unit vectors $\hat{\mathbf{u}}_1$, $\hat{\mathbf{u}}_2$, and $\hat{\mathbf{u}}_3$, the unit dyadic $\mathbb{U}$ may be expanded in terms of the Kronecker delta $\delta_{\alpha\beta}$ or the unit matrix $[U]$ as

$$\mathbb{U} = \sum_{\alpha=1}^{3} \sum_{\beta=1}^{3} \delta_{\alpha\beta}\, \hat{\mathbf{u}}_\alpha \hat{\mathbf{u}}_\beta = \hat{\mathbf{u}}_1\hat{\mathbf{u}}_1 + \hat{\mathbf{u}}_2\hat{\mathbf{u}}_2 + \hat{\mathbf{u}}_3\hat{\mathbf{u}}_3$$

$$= [\hat{\mathbf{u}}_1\hat{\mathbf{u}}_2\hat{\mathbf{u}}_3] \begin{bmatrix} 1 & 0 & 0 \\ 0 & 1 & 0 \\ 0 & 0 & 1 \end{bmatrix} \begin{bmatrix} \hat{\mathbf{u}}_1 \\ \hat{\mathbf{u}}_2 \\ \hat{\mathbf{u}}_3 \end{bmatrix} = \{\hat{\mathbf{u}}\}^T[U]\{\hat{\mathbf{u}}\} \tag{8·27}$$

8·4 INERTIA PROPERTIES

8·4·1 Calculation of an inertia matrix

Consider the uniform rectangular parallelepiped of mass density μ and side dimensions $2a$, $2b$, and $2d$, as shown in Fig. 8·4. In order to calculate the moments and products of inertia for the mass center c and unit vectors $\hat{\mathbf{u}}_1$, $\hat{\mathbf{u}}_2$, and $\hat{\mathbf{u}}_3$ parallel to the body axes, we can apply Eqs. (8·22) (changing r's to ρ's) to find

$$I_{11} = \int (\rho_2{}^2 + \rho_3{}^2)\, dm = \int_{-d}^{d} \int_{-b}^{b} \int_{-a}^{a} (\rho_2{}^2 + \rho_3{}^2)\mu\, d\rho_1\, d\rho_2\, d\rho_3$$

$$= \tfrac{8}{3}\mu d^3\, ba + \tfrac{8}{3}\mu d\, b^3 a = \tfrac{1}{3}\mathfrak{M}(d^2 + b^2)$$

and similarly

$$I_{22} = \tfrac{1}{3}\mathfrak{M}(a^2 + d^2) \qquad \text{and} \qquad I_{33} = \tfrac{1}{3}\mathfrak{M}(a^2 + b^2)$$

while

$$I_{12} = I_{21} = I_{13} = I_{31} = I_{23} = I_{32} = 0$$

8·4·2 Transfer of reference point for an inertia matrix

The inertia matrix $[I^q]$ of a given material system of mass $\mathfrak{M}$ for a given reference point q and a given vector basis $\hat{\mathbf{u}}_1$, $\hat{\mathbf{u}}_2$, and $\hat{\mathbf{u}}_3$ can be related to the inertia matrix $[I^c]$ for the mass center c and *the same vector basis* by

$$[I^q] = [I^c] - \mathfrak{M}[\tilde{r}_c][\tilde{r}_c] \tag{8·28}$$

where

$$[\tilde{r}_c] \triangleq \begin{bmatrix} 0 & -r_{c3} & r_{c2} \\ r_{c3} & 0 & -r_{c1} \\ -r_{c2} & r_{c1} & 0 \end{bmatrix}$$

and r_{c1}, r_{c2}, and r_{c3} are defined in terms of the vector $\mathbf{r}_c$ from q to c by

$$r_{c1} \triangleq \mathbf{r}_c \cdot \hat{\mathbf{u}}_1 \qquad r_{c2} \triangleq \mathbf{r}_c \cdot \hat{\mathbf{u}}_2 \qquad r_{c3} \triangleq \mathbf{r}_c \cdot \hat{\mathbf{u}}_3$$

This result is called the *inertia-matrix reference-point transfer theorem.*

Before proving Eq. (8·28), we note the scalar implications

$$I_{11}^q = I_{11}^c + \mathfrak{M}(r_{c2}^2 + r_{c3}^2) \tag{8·29a}$$
$$I_{22}^q = I_{22}^c + \mathfrak{M}(r_{c3}^2 + r_{c1}^2) \tag{8·29b}$$
$$I_{33}^q = I_{33}^c + \mathfrak{M}(r_{c1}^2 + r_{c2}^2) \tag{8·29c}$$
$$I_{12}^q = I_{21}^q = I_{12}^c - \mathfrak{M}r_{c1}r_{c2} = I_{21}^c - \mathfrak{M}r_{c1}r_{c2} \tag{8·29d}$$
$$I_{13}^q = I_{31}^q = I_{13}^c - \mathfrak{M}r_{c1}r_{c3} = I_{31}^c - \mathfrak{M}r_{c1}r_{c3} \tag{8·29e}$$
$$I_{23}^q = I_{32}^q = I_{23}^c - \mathfrak{M}r_{c2}r_{c3} = I_{32}^c - \mathfrak{M}r_{c2}r_{c3} \tag{8·29f}$$

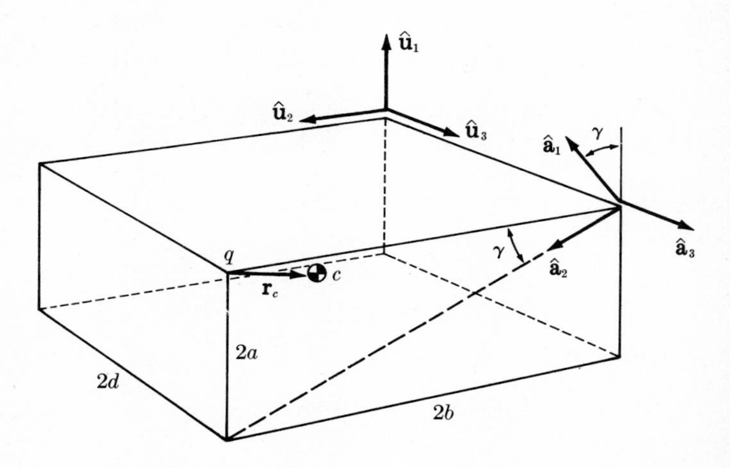

FIGURE 8·4

The most straightforward proof of Eqs. (8·28) and (8·29) stems from direct expansions, as typified by

$$\begin{aligned} I_{11}^q &\triangleq \int (r_2^2 + r_3^2)\,dm = \int[(r_{c2} + \rho_2)^2 + (r_{c3} + \rho_3)^2]\,dm \\ &= \int (\rho_2^2 + \rho_3^2)\,dm + 2r_{c2}\int \rho_2\,dm + 2r_{c3}\int \rho_3\,dm + \int (r_{c2}^2 + r_{c3}^2)\,dm \\ &= I_{11}^c + (r_{c2}^2 + r_{c3}^2)\int dm = I_{11}^c + \mathfrak{M}(r_{c2}^2 + r_{c3}^2) \end{aligned}$$

and

$$\begin{aligned} I_{12}^q &= -\int r_1 r_2\,dm = -\int (r_{c1} + \rho_1)(r_{c2} + \rho_2)\,dm \\ &= -\int \rho_1\rho_2\,dm - r_{c1}r_{c2}\int dm - r_{c1}\int \rho_2\,dm - r_{c2}\int \rho_1\,dm \\ &= I_{12}^c - \mathfrak{M}r_{c1}r_{c2} \end{aligned}$$

Here we have made repeated use of the fact that $\int \boldsymbol{\rho}\,dm = 0$, where $\boldsymbol{\rho}$ is the generic position vector to a differential mass element from the mass center c.

8·4·3 Application of reference-point transfer theorem

If in reference to the rectangular parallelepiped shown in Fig. 8·4, the inertia matrix is required for corner point q and unit vectors $\hat{\mathbf{u}}_1$, $\hat{\mathbf{u}}_2$, and $\hat{\mathbf{u}}_3$, this is most easily obtained by applying §8·4·1 to obtain

$$[I^c] = \tfrac{1}{3}\mathfrak{M}\begin{bmatrix} b^2 + d^2 & 0 & 0 \\ 0 & a^2 + d^2 & 0 \\ 0 & 0 & a^2 + b^2 \end{bmatrix}$$

and then applying Eq. (8·28), with

$$\mathbf{r}_c = -a\hat{\mathbf{u}}_1 - b\hat{\mathbf{u}}_2 - d\hat{\mathbf{u}}_3$$

The result is

$$\begin{aligned} [I^q] &= \tfrac{1}{3}\mathfrak{M}\begin{bmatrix} b^2 + d^2 & 0 & 0 \\ 0 & a^2 + d^2 & 0 \\ 0 & 0 & a^2 + b^2 \end{bmatrix} - \mathfrak{M}\begin{bmatrix} 0 & d & -b \\ -d & 0 & a \\ b & -a & 0 \end{bmatrix}\begin{bmatrix} 0 & d & -b \\ -d & 0 & a \\ b & -a & 0 \end{bmatrix} \\ &= \mathfrak{M}\begin{bmatrix} \tfrac{4}{3}(b^2 + d^2) & -ab & -ad \\ -ab & \tfrac{4}{3}(a^2 + d^2) & -bd \\ -ad & -bd & \tfrac{4}{3}(a^2 + b^2) \end{bmatrix} \end{aligned}$$

8·4·4 Inertia matrix calculated with transfer theorem

It is often convenient to use the transfer theorem for shifting inertia-matrix reference points when assembling the inertia matrix of an object as the sum (or integral) of the inertia matrices of its composite parts. This practice is illustrated here for the uniform, solid, right circular cylinder of mass density μ, radius a, and height h, as shown in Fig. 8·5. We require the inertia matrix for the central reference point C and the vector basis $\hat{\mathbf{u}}_1$, $\hat{\mathbf{u}}_2$, and $\hat{\mathbf{u}}_3$.

Rather than jump immediately into Eqs. (8·22) to obtain integral definitions for moments and products of inertia for the cylinder, we can much more easily use these definitions in application to a typical laminar element of the cylinder, and then apply the transfer theorem to find the contribution of the lamina to the total inertia matrix for point C. The inertia matrix for the cylinder will then be obtained by integrating over all of the lamina.

A typical lamina is a uniform circular plate of radius a and infinitesimal thickness $d\rho_3$. If the center of the lamina is called c, the differential moment of

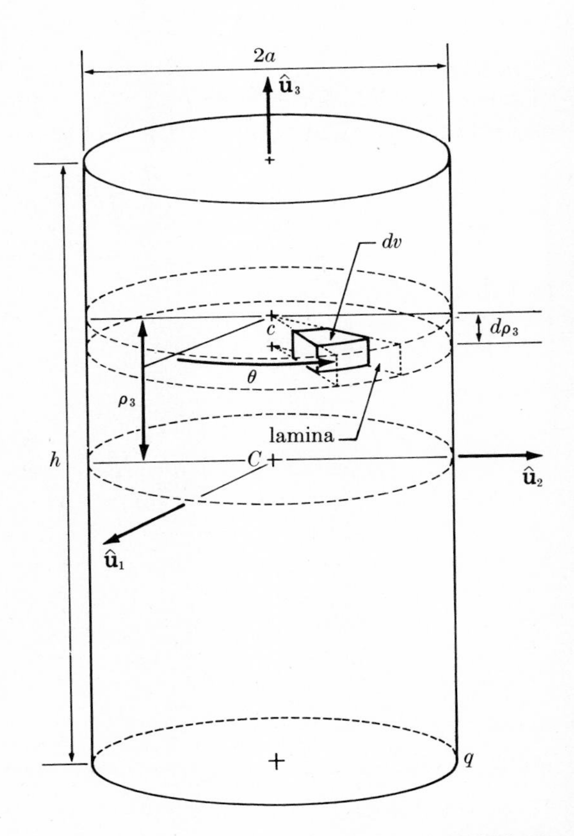

FIGURE 8·5

inertia for c about the axis of symmetry is

$$dI_{33} = d[\int (\rho_1{}^2 + \rho_2{}^2)\, dm] = d(\int r^2\, dm) = d(\int r^2 \mu\, dv)$$

where r is the radial distance to the differential element of volume dv. If θ is introduced (see Fig. 8·5), then we have

$$dI_{33} = \int_0^a \int_0^{2\pi} r^2 \mu r d\theta\, dr\, d\rho_3 = 2\pi\mu \int_0^a r^3\, dr\, d\rho_3$$

or

$$dI_{33} = \frac{\pi\mu a^4}{2}\, d\rho_3$$

For a lamina whose plane is normal to $\hat{\mathbf{u}}_3$, and contains reference point c, the definitions in Eqs. (8·22) indicate that

$$dI_{33}^c = dI_{11}^c + dI_{22}^c$$

In this case symmetry establishes $dI_{11}^c = dI_{22}^c$, so we have

$$dI_{11}^c = dI_{22}^c = \tfrac{1}{2} dI_{33}^c = \tfrac{1}{4}\pi\mu a^4\, d\rho_3$$

The differential products of inertia of the lamina in vector basis $\hat{\mathbf{u}}_1$, $\hat{\mathbf{u}}_2$, and $\hat{\mathbf{u}}_3$ are zero.

To obtain the inertia matrix of the total cylinder for point C, the transfer theorem can be applied to each lamina; this is most easily accomplished by inspection of Eqs. (8·29). Evidently none of the lamina makes any contribution to any product of inertia [see Eqs. (8·29d, e, and f)]. The term I_{33}^C is simply

$$I_{33}^C = \int dI_{33}^c = \int_{-\frac{1}{2}h}^{\frac{1}{2}h} \tfrac{1}{2}\pi\mu a^4\, d\rho_3 = \tfrac{1}{2}\pi\mu a^4 h = \tfrac{1}{2}\mathfrak{M} a^2$$

where $\mathfrak{M}$ is the total cylinder mass. In calculating I_{11}^C and I_{22}^C from Eqs. (8·29a and b), we must not only integrate over the differential contributions dI_{11}^c and dI_{22}^c, but also add the transfer terms, using $d\mathfrak{M} = \pi a^2 \mu\, d\rho_3$ for the lamina mass, and ρ_3 for r_{c3}, with r_{c1} and r_{c2} both zero. The result is

$$\begin{aligned} I_{22}^C = I_{11}^C &= \int_{-\frac{1}{2}h}^{\frac{1}{2}h} \tfrac{1}{4}\pi\mu a^4\, d\rho_3 + \int_{-h/2}^{h/2} \rho_3{}^2 \pi a^2 \mu\, d\rho_3 \\ &= \tfrac{1}{4}\pi\mu a^4 h + \tfrac{1}{12}\pi\mu a^2 h^3 = \tfrac{1}{4}\mathfrak{M} a^2 + \tfrac{1}{12}\mathfrak{M} h^2 \end{aligned}$$

Thus the required inertia matrix is

$$[I^C] = \tfrac{1}{12}\mathfrak{M} \begin{bmatrix} 3a^2 + h^2 & 0 & 0 \\ 0 & 3a^2 + h^2 & 0 \\ 0 & 0 & 6a^2 \end{bmatrix}$$

8·4·5 Change of vector basis for an inertia matrix

Suppose that for some material system we are given a set of moments and products of inertia for some specified reference point q and specified axes parallel to $\hat{\mathbf{u}}_1$, $\hat{\mathbf{u}}_2$, and $\hat{\mathbf{u}}_3$. It may be of interest to find new moments and products of inertia for the *same* reference point q and a different set of unit vectors $\hat{\mathbf{a}}_1$, $\hat{\mathbf{a}}_2$, and $\hat{\mathbf{a}}_3$. To this end, we first assemble the given inertia quantities [as defined by Eqs. (8·22)] as the elements of an inertia matrix $[{}^uI]$ [as defined by Eqs. (8·24), but with presuperscript u added to make the vector basis explicit] and assemble the unknown moments and products of inertia for vector basis $\{\hat{\mathbf{a}}\}$ into an unknown inertia matrix $[{}^aI]$. We can rewrite Eq. (8·24b) as the matrix equation

$$[{}^uH^*] = [{}^uI][{}^u\omega] \tag{8·30a}$$

where the matrices $[{}^uH^*]$ and $[{}^u\omega]$ are defined by

$$\mathbf{H}^* = \{\hat{\mathbf{u}}\}^T[{}^uH^*] \qquad \text{and} \qquad \boldsymbol{\omega} = \{\hat{\mathbf{u}}\}^T[{}^u\omega]$$

If we replace the unit vectors in $\{\hat{\mathbf{u}}\}$ by the set $\hat{\mathbf{a}}_1$, $\hat{\mathbf{a}}_2$, and $\hat{\mathbf{a}}_3$ arranged in the vector array $\{\hat{\mathbf{a}}\}$, and write

$$\mathbf{H}^* = \{\hat{\mathbf{a}}\}^T[{}^aH^*] \qquad \text{and} \qquad \boldsymbol{\omega} = \{\hat{\mathbf{a}}\}^T[{}^a\omega]$$

then in parallel with Eq. (8·30a) we can record

$$[{}^aH^*] = [{}^aI][{}^a\omega] \tag{8·30b}$$

The transformation of vector basis for a column matrix representing a gibbsian vector is a familiar process; in Eqs. (3·64) we noted that if two sets of unit vectors $\{\hat{\mathbf{u}}\}$ and $\{\hat{\mathbf{a}}\}$ are related by [see Eqs. (3·59)]

$$\{\hat{\mathbf{a}}\} = [C^{au}]\{\hat{\mathbf{u}}\} \tag{8·31a}$$

and

$$\{\hat{\mathbf{u}}\} = [C^{au}]^{-1}\{\hat{\mathbf{a}}\} = [C^{au}]^T\{\hat{\mathbf{a}}\} = [C^{ua}]\{\hat{\mathbf{a}}\} \tag{8·31b}$$

then the matrix representations of any vector, such as $\boldsymbol{\omega}$, are related as in

$$[{}^a\omega] = [C^{au}][{}^u\omega] \tag{8·32a}$$

or

$$[{}^u\omega] = [C^{au}]^{-1}[{}^a\omega] = [C^{au}]^T[{}^a\omega] \tag{8·32b}$$

Similarly

$$[{}^aH^*] = [C^{au}][{}^uH^*] \tag{8·33a}$$

and

$$[{}^uH^*] = [C^{au}]^{-1}[{}^aH^*] = [C^{au}]^T[{}^aH^*] \tag{8·33b}$$

Substituting Eqs. (8·33*b*) and (8·32*b*) into (8·30*a*) yields

$$[C^{au}]^{-1}[{}^aH^*] = [{}^uI][C^{au}]^T[{}^a\omega]$$

or, after premultiplication by $[C^{au}]$,

$$[{}^aH^*] = [C^{au}][{}^uI][C^{au}]^T[{}^a\omega] \tag{8·34}$$

Comparison of Eqs. (8·34) and (8·30*b*) produces the transformation sought, that is,

$$[{}^aI] = [C^{au}][{}^uI][C^{au}]^T \tag{8·35a}$$

or

$$[{}^aI] = [C^{au}][{}^uI][C^{ua}] \tag{8·35b}$$

This result can be obtained even more simply if we take advantage of the *dyadic* concept introduced in §8·3·3. We can then combine Eqs. (8·26*b*) and (8·31*b*) to obtain

$$\begin{aligned}\mathbb{I} &= \{\hat{\mathbf{u}}\}^T[{}^uI]\{\hat{\mathbf{u}}\} = ([C^{au}]^T\{\hat{\mathbf{a}}\})^T[{}^uI][C^{au}]^T\{\hat{\mathbf{a}}\} \\ &= \{\hat{\mathbf{a}}\}^T[C^{au}][{}^uI][C^{au}]^T\{\hat{\mathbf{a}}\}\end{aligned} \tag{8·36}$$

Comparison with a representation of $\mathbb{I}$ as in Eq. (8·26*b*), but in vector basis $\{\hat{\mathbf{a}}\}$, namely,

$$\mathbb{I} = \{\hat{\mathbf{a}}\}^T[{}^aI]\{\hat{\mathbf{a}}\} \tag{8·37}$$

confirms the result found previously as Eq. (8·35).

8·4·6 Inertia-matrix vector-basis transformation

Consider once again the uniform rectangular parallelepiped shown in Fig. 8·4, and recall from §8·4·1 that the inertia matrix for point c and vector basis $\{\hat{\mathbf{u}}\}$ is given by

$$[{}^uI^c] = \tfrac{1}{3}\mathfrak{M}\begin{bmatrix} b^2+d^2 & 0 & 0 \\ 0 & a^2+d^2 & 0 \\ 0 & 0 & a^2+b^2 \end{bmatrix}$$

Imagine now that we require $[{}^aI^c]$, the inertia matrix of the block for point c and vector basis $\{\hat{\mathbf{a}}\}$. From Fig. 8·4, we have

$$\begin{Bmatrix}\hat{\mathbf{a}}_1 \\ \hat{\mathbf{a}}_2 \\ \hat{\mathbf{a}}_3\end{Bmatrix} = \begin{bmatrix} \cos\gamma & \sin\gamma & 0 \\ -\sin\gamma & \cos\gamma & 0 \\ 0 & 0 & 1 \end{bmatrix}\begin{Bmatrix}\hat{\mathbf{u}}_1 \\ \hat{\mathbf{u}}_2 \\ \hat{\mathbf{u}}_3\end{Bmatrix}$$

which we write in the form $\{\hat{\mathbf{a}}\} = [C^{au}]\{\hat{\mathbf{u}}\}$ to define $[C^{au}]$. From the transformation equations (8·35), we then have

$$[{}^aI^c] = \frac{\mathfrak{M}}{3}\begin{bmatrix} c\gamma & s\gamma & 0 \\ -s\gamma & c\gamma & 0 \\ 0 & 0 & 1 \end{bmatrix}\begin{bmatrix} b^2+d^2 & 0 & 0 \\ 0 & a^2+d^2 & 0 \\ 0 & 0 & a^2+b^2 \end{bmatrix}\begin{bmatrix} c\gamma & -s\gamma & 0 \\ s\gamma & c\gamma & 0 \\ 0 & 0 & 1 \end{bmatrix}$$
$$= \frac{\mathfrak{M}}{3}\begin{bmatrix} (b^2+d^2)c^2\gamma + (a^2+d^2)s^2\gamma & (a^2-b^2)s\gamma c\gamma & 0 \\ (a^2-b^2)s\gamma c\gamma & (a^2+d^2)c^2\gamma + (b^2+d^2)s^2\gamma & 0 \\ 0 & 0 & a^2+b^2 \end{bmatrix}$$

where $s\gamma$ denotes $\sin\gamma$ and $c\gamma$ denotes $\cos\gamma$. It should be noted that unless $a = b$, the inertia matrix $[{}^aI^c]$ has more nonzero elements than does $[{}^uI^c]$. Since we plan to use these inertia matrices to calculate angular momenta (see Sec. 8·3) and eventually to obtain equations of motion (as in Sec. 8·2), it is to our advantage to select the vector basis of each inertia matrix so that it has a minimum number of nonzero elements.

8·4·7 Principal axes of inertia by symmetries

It is evident from examples in the preceding articles that some or all of the products of inertia may be zero for a given combination of material system, reference point, and vector basis. When for an axis of a material system defined by a reference point q and a unit vector $\hat{\mathbf{u}}_i$ there are no corresponding nonzero products of inertia (so $I_{ij} = I_{ji} = 0$ for $j = 1, 2, 3$, but $j \neq i$), that axis is called a *principal axis of inertia* of the material system for the given reference point. Thus we can conclude on the basis of §8·4·1 that, for Fig. 8·4, axes through c and parallel to $\hat{\mathbf{u}}_1$, $\hat{\mathbf{u}}_2$, and $\hat{\mathbf{u}}_3$ are principal axes. From the example in §8·4·3 we know that, for the same Fig. 8·4, axes through q and parallel to $\hat{\mathbf{u}}_1$, $\hat{\mathbf{u}}_2$, and $\hat{\mathbf{u}}_3$ are *not* principal axes, while, for Fig. 8·5, the axes through C and parallel to $\hat{\mathbf{u}}_1$, $\hat{\mathbf{u}}_2$, and $\hat{\mathbf{u}}_3$ *are* principal axes. The example in §8·4·6 indicates that for the rectangular parallelepiped in Fig. 8·4, an axis through c and parallel to $\hat{\mathbf{a}}_3$ *is* a principal axis, while axes through c and parallel to $\hat{\mathbf{a}}_1$ and $\hat{\mathbf{a}}_2$ are *not.*

Consideration of the definitions of the scalars $I_{ij}(i, j = 1, 2, 3)$ in Eqs. (8·22) indicates in general terms that the products of inertia can be zero (whereas the moments of inertia are always positive for any real, three-dimensional material system). These definitions also reveal certain special cases in which principal axes can be recognized by consideration of symmetries of the system. For example, the quantity $I_{12} \triangleq -\int r_1 r_2\, dm$ will be zero if for each point of the material system with a given value of r_1, r_2, and r_3 (say r_1^*, r_2^*, r_3^*) there is either another point with the same r_2 and r_3 and an equal and opposite r_1 (that is, $-r_1^*$, r_2^*, r_3^*), or another point with the same r_1 and r_3 and an equal and opposite r_2

(that is, r_1^*, $-r_2^*$, r_3^*). In the former case, not only I_{12} but also I_{13} is zero (so the 1 axis is a principal axis), while in the second case, both I_{12} and I_{23} are zero (so the 2 axis is a principal axis). These and similar observations permit the formulation of the following general rule: *If a material system has a plane of symmetry passing through a reference point* q (so the material on one side of the plane is a mirror image of that on the other side), *then an axis through* q *and normal to that plane is a principal axis of inertia for point* q.

The examples of the preceding articles should be considered again in the light of this general rule. You will find in every case considered thus far that the axes identified as principal axes are normal to planes of symmetry through the corresponding reference points.

8·4·8 Principal axes of inertia by eigenvector solution

It will not always be possible to find principal axes by looking for symmetries. Yet the question remains, "Do there exist principal axes even in the absence of any symmetries?" The answer is important because the labors of calculating $\dot{\mathbf{H}}^q$ for the equations of motion in §8·2·3 will be greatly reduced if we can manage to work in terms of a vector basis corresponding to principal axes of inertia for our chosen reference point q.

To pursue this question, let's *assume* that for any reference point q there exists a set of orthogonal unit vectors $\hat{\mathbf{e}}_1$, $\hat{\mathbf{e}}_2$, and $\hat{\mathbf{e}}_3$ that parallel principal axes through q. The products of inertia for these axes would be zero, and the moments of inertia we'll designate simply as I_1, I_2, and I_3 (using a single subscript to denote a *moment of inertia for a principal axis*). Thus the inertia dyadic would appear in basis $\hat{\mathbf{e}}_1$, $\hat{\mathbf{e}}_2$, and $\hat{\mathbf{e}}_3$ as [see Eq. (8·26*b*)]

$$\begin{aligned}\mathbb{I} &= I_1\hat{\mathbf{e}}_1\hat{\mathbf{e}}_1 + I_2\hat{\mathbf{e}}_2\hat{\mathbf{e}}_2 + I_3\hat{\mathbf{e}}_3\hat{\mathbf{e}}_3 \\ &= [\hat{\mathbf{e}}_1\hat{\mathbf{e}}_2\hat{\mathbf{e}}_3]\begin{bmatrix} I_1 & 0 & 0 \\ 0 & I_2 & 0 \\ 0 & 0 & I_3 \end{bmatrix}\begin{bmatrix}\hat{\mathbf{e}}_1 \\ \hat{\mathbf{e}}_2 \\ \hat{\mathbf{e}}_3\end{bmatrix}\end{aligned} \tag{8·38a}$$

and the corresponding inertia matrix would be

$$[{}^eI] = \begin{bmatrix} I_1 & 0 & 0 \\ 0 & I_2 & 0 \\ 0 & 0 & I_3 \end{bmatrix} \tag{8·38b}$$

In some nonprincipal vector basis $\hat{\mathbf{u}}_1$, $\hat{\mathbf{u}}_2$, and $\hat{\mathbf{u}}_3$ for which the moments and products of inertia are known, the inertia dyadic would be

$$\mathbb{I} = [\hat{\mathbf{u}}_1\hat{\mathbf{u}}_2\hat{\mathbf{u}}_3]\begin{bmatrix} I_{11} & I_{12} & I_{13} \\ I_{21} & I_{22} & I_{23} \\ I_{31} & I_{32} & I_{33} \end{bmatrix}\begin{bmatrix}\hat{\mathbf{u}}_1 \\ \hat{\mathbf{u}}_2 \\ \hat{\mathbf{u}}_3\end{bmatrix} \tag{8·39a}$$

and the corresponding inertia matrix would be

$$[{}^uI] = \begin{bmatrix} I_{11} & I_{12} & I_{13} \\ I_{21} & I_{22} & I_{23} \\ I_{31} & I_{32} & I_{33} \end{bmatrix} \tag{8·39b}$$

Dot-multiplying $\mathbb{I}$ by $\hat{\mathbf{e}}_j$ would produce [from Eq. (8·38*a*)] (for $j = 1, 2,$ or 3)

$$\mathbb{I} \cdot \hat{\mathbf{e}}_j = (I_1\hat{\mathbf{e}}_1\hat{\mathbf{e}}_1 \cdot \hat{\mathbf{e}}_j + I_2\hat{\mathbf{e}}_2\hat{\mathbf{e}}_2 \cdot \hat{\mathbf{e}}_j + I_3\hat{\mathbf{e}}_3\hat{\mathbf{e}}_3 \cdot \hat{\mathbf{e}}_j) = I_j\hat{\mathbf{e}}_j \tag{8·40a}$$

or

$$\mathbb{I} \cdot \hat{\mathbf{e}}_j = [\hat{\mathbf{e}}_1\hat{\mathbf{e}}_2\hat{\mathbf{e}}_3] \begin{bmatrix} I_1 & 0 & 0 \\ 0 & I_2 & 0 \\ 0 & 0 & I_3 \end{bmatrix} \begin{bmatrix} \hat{\mathbf{e}}_1 \cdot \hat{\mathbf{e}}_j \\ \hat{\mathbf{e}}_2 \cdot \hat{\mathbf{e}}_j \\ \hat{\mathbf{e}}_3 \cdot \hat{\mathbf{e}}_j \end{bmatrix} = I_j\hat{\mathbf{e}}_j \tag{8·40b}$$

This result would have to be the same if we expanded $\mathbb{I}$ in vector basis $\{\hat{\mathbf{u}}\}$; we would then find

$$\mathbb{I} \cdot \hat{\mathbf{e}}_j = [\hat{\mathbf{u}}_1\hat{\mathbf{u}}_2\hat{\mathbf{u}}_3] \begin{bmatrix} I_{11} & I_{12} & I_{13} \\ I_{21} & I_{22} & I_{23} \\ I_{31} & I_{32} & I_{33} \end{bmatrix} \begin{bmatrix} \hat{\mathbf{u}}_1 \cdot \hat{\mathbf{e}}_j \\ \hat{\mathbf{u}}_2 \cdot \hat{\mathbf{e}}_j \\ \hat{\mathbf{u}}_3 \cdot \hat{\mathbf{e}}_j \end{bmatrix} = I_j\hat{\mathbf{e}}_j \tag{8·40c}$$

If we define $e_1^j \triangleq \hat{\mathbf{u}}_1 \cdot \hat{\mathbf{e}}_j$, $e_2^j \triangleq \hat{\mathbf{u}}_2 \cdot \hat{\mathbf{e}}_j$, and $e_3^j \triangleq \hat{\mathbf{u}}_3 \cdot \hat{\mathbf{e}}_j$ as the scalar components of $\hat{\mathbf{e}}_j$ in the vector basis $\{\hat{\mathbf{u}}\}$, then we have

$$\hat{\mathbf{e}}_j = e_1^j\hat{\mathbf{u}}_1 + e_2^j\hat{\mathbf{u}}_2 + e_3^j\hat{\mathbf{u}}_3 = [\hat{\mathbf{u}}_1\hat{\mathbf{u}}_2\hat{\mathbf{u}}_3] \begin{bmatrix} e_1^j \\ e_2^j \\ e_3^j \end{bmatrix} \tag{8·41}$$

and Eq. (8·40*c*) becomes

$$[\hat{\mathbf{u}}_1\hat{\mathbf{u}}_2\hat{\mathbf{u}}_3] \begin{bmatrix} I_{11} & I_{12} & I_{13} \\ I_{21} & I_{22} & I_{23} \\ I_{31} & I_{32} & I_{33} \end{bmatrix} \begin{bmatrix} e_1^j \\ e_2^j \\ e_3^j \end{bmatrix} = [\hat{\mathbf{u}}_1\hat{\mathbf{u}}_2\hat{\mathbf{u}}_3]I^j \begin{bmatrix} e_1^j \\ e_2^j \\ e_3^j \end{bmatrix} \tag{8·42a}$$

Satisfaction of this vector equation requires that

$$\begin{bmatrix} I_{11} & I_{12} & I_{13} \\ I_{21} & I_{22} & I_{23} \\ I_{31} & I_{32} & I_{33} \end{bmatrix} \begin{bmatrix} e_1^j \\ e_2^j \\ e_3^j \end{bmatrix} = I_j \begin{bmatrix} e_1^j \\ e_2^j \\ e_3^j \end{bmatrix} \qquad j = 1, 2, 3 \tag{8·42b}$$

or

$$\begin{bmatrix} I_{11} - I_j & I_{12} & I_{13} \\ I_{21} & I_{22} - I_j & I_{23} \\ I_{31} & I_{32} & I_{33} - I_j \end{bmatrix} \begin{bmatrix} e_1^j \\ e_2^j \\ e_3^j \end{bmatrix} = \begin{bmatrix} 0 \\ 0 \\ 0 \end{bmatrix} \qquad j = 1, 2, 3 \tag{8·42c}$$

Equations (8·42) have been written on the basis of the *assumption* that for any point q a set of orthogonal principal axis unit vectors $\hat{\mathbf{e}}_1$, $\hat{\mathbf{e}}_2$, and $\hat{\mathbf{e}}_3$ exists;

at this point we can check the validity of our assumption by attempting to solve for e_1^j, e_2^j, and e_3^j ($j = 1, 2, 3$). We have in effect *assumed* that for $j = 1, 2,$ or 3, there will exist real solutions for e_1^j, e_2^j, and e_3^j. Now we must see under what conditions this assumption is valid.

Equation (8·42*c*) represents three scalar homogeneous algebraic equations in the four unknowns I_j, e_1^j, e_2^j, and e_3^j. These equations are evidently satisfied for any I_j by the trivial solution $e_1^j = e_2^j = e_3^j = 0$, but they will admit a nontrivial solution only if I_j has some value such that

$$\begin{vmatrix} I_{11} - I_j & I_{12} & I_{13} \\ I_{21} & I_{22} - I_j & I_{23} \\ I_{31} & I_{32} & I_{33} - I_j \end{vmatrix} = 0 \tag{8·43}$$

(This is evident from Cramer's rule, or any other procedure for solving algebraic equations.)

Equation (8·43) is a cubic equation in I_j, known as the *characteristic equation*. It has three solutions, all of which can be shown to be real by assuming a complex solution and demonstrating a contradiction. Therefore these solutions can be identified as I_1, I_2, and I_3, the principal axis moments of inertia. Since I_j in Eq. (8·42*c*) is real, so also are e_1^j, e_2^j, and e_3^j, satisfying our requirements and confirming the validity of our assumption that orthogonal principal axes parallel to $\hat{\mathbf{e}}_1$, $\hat{\mathbf{e}}_2$, and $\hat{\mathbf{e}}_3$ exist for any reference point q.

In order to actually find the principal axis unit vectors $\hat{\mathbf{e}}_1$, $\hat{\mathbf{e}}_2$, and $\hat{\mathbf{e}}_3$, we must solve Eqs. (8·42) for e_1^j, e_2^j, and e_3^j, for $j = 1, 2, 3$, and then use Eq. (8·41). To accomplish this, we first solve Eq. (8·43) for I_1, I_2, and I_3, and then substitute each one of these in turn into Eq. (8·42*c*) and attempt to solve for the corresponding e_1^j, e_2^j, and e_3^j. We would soon find, however, that of the three scalar equations in Eq. (8·42*c*), no more than two are independent [this is the significance of Eq. (8·43)], so we cannot find unique values of e_1^j, e_2^j, and e_3^j from Eq. (8·42*c*) alone. If the roots I_1, I_2, and I_3 are distinct from each other, we can circumvent this problem by remembering that $\hat{\mathbf{e}}_j = e_1^j\hat{\mathbf{u}}_1 + e_2^j\hat{\mathbf{u}}_2 + e_3^j\hat{\mathbf{u}}_3$ is a *unit* vector, so that its scalar components are related by

$$(e_1^j)^2 + (e_2^j)^2 + (e_3^j)^2 = 1 \tag{8·44}$$

Then by considering Eq. (8·44) in conjunction with any two of the three scalar equations in (8·42*c*), we can obtain three equations to be solved for the three unknowns e_1^j, e_2^j, and e_3^j. If, however, two of the roots I_1, I_2, and I_3 have the same value, say $I_1 = I_2 \neq I_3$, then we can find only the third principal axis ($\hat{\mathbf{e}}_3$) in the indicated fashion, since then for $I_j = I_1 = I_2$ in Eq. (8·42*c*) only *one* independent equation emerges. This means that *any* pair of unit vectors orthogonal to the one principal axis found (here $\hat{\mathbf{e}}_3$) will qualify as principal axis unit vectors. Finally, if all of the scalars I_1, I_2, and I_3 emerging from Eq. (8·43) are the same, then *any* unit vector defines a principal axis for the given reference point.

It should be noted that Eqs. (8·42*b*) and (8·42*c*) have a classical form known as the *eigenvalue equation;* the solutions I_1, I_2, and I_3 of the characteristic equation [Eq. (8·43)] are called the *eigenvalues* of the matrix $[I]$, and the 3×1 column matrices $[e_1^j e_2^j e_3^j]^T$ are known as the corresponding *eigenvectors.* In the general problem, eigenvectors need not be unit vectors, but may have any magnitude; moreover, they need not be 3×1 matrices representing gibbsian vectors at all, but can be of dimension $n \times 1$ for any n (see Appendix B). The eigenvalue equation arises in many contexts, but here it serves to provide a general procedure for finding principal axes and principal axis moments of inertia.

When we consider the results of the present section in the light of §8·4·5 on vector-basis transformations, we see that we could as well have used Eq. (8·35*a*) as the basis for the question "Under what circumstances does there exist a direction-cosine matrix $[C^{eu}]$ relating a given vector basis $\{\hat{\mathbf{u}}\}$ to a new vector basis $\{\hat{\mathbf{e}}\}$ such that $\{\hat{\mathbf{e}}\} = [C^{eu}]\{\hat{\mathbf{u}}\}$,

$$[{}^eI] = [C^{eu}][{}^uI][C^{eu}]^T \tag{8·45}$$

and $[{}^eI]$ and $[{}^uI]$ appear as in Eqs. (8·39*a*) and (8·39*b*), with $[{}^eI]$ diagonal?" By premultiplying Eq. (8·42*b*) by $[e_1^i e_2^i e_3^i]$ and noting that

$$[e_1^i e_2^i e_1^i] \begin{bmatrix} e_1^j \\ e_2^j \\ e_3^j \end{bmatrix} = \delta_{ij} \triangleq \begin{cases} 1 & \text{for } i = j \\ 0 & \text{for } i \neq j \end{cases}$$

we can deduce that $[C^{eu}]$ in Eq. (8·45) is given by

$$[C^{eu}] = \begin{bmatrix} e_1^1 & e_2^1 & e_3^1 \\ e_1^2 & e_2^2 & e_3^2 \\ e_1^3 & e_2^3 & e_3^3 \end{bmatrix} \tag{8·46}$$

8·4·9 Inertia-matrix diagonalization

For an example of the calculation of a diagonal inertial matrix, let's return to the right circular cylinder in Fig. 8·5, and find the principal axes and principal axis inertias for point q on the bottom edge. To simplify the problem, we'll let $h = a\sqrt{3}/2$. We know from the example in §8·4·4 that for the center C of the cylinder and vector basis $\hat{\mathbf{u}}_1$, $\hat{\mathbf{u}}_2$, and $\hat{\mathbf{u}}_3$ the inertia matrix is

$$[I^C] = \tfrac{1}{12}\mathfrak{M} \begin{bmatrix} 3a^2 + h^2 & 0 & 0 \\ 0 & 3a^2 + h^2 & 0 \\ 0 & 0 & 6a^2 \end{bmatrix}$$

The inertia matrix for point q and axes parallel to $\hat{\mathbf{u}}_1$, $\hat{\mathbf{u}}_2$, and $\hat{\mathbf{u}}_3$ can be obtained from Eq. (8·28), after noting that $\mathbf{r}_c = -a\hat{\mathbf{u}}_2 + \frac{1}{2}h\hat{\mathbf{u}}_3$, so that

$$[I^q] = \tfrac{1}{12}\mathfrak{M}\begin{bmatrix} 3a^2 + h^2 & 0 & 0 \\ 0 & 3a^2 + h^2 & 0 \\ 0 & 0 & 6a^2 \end{bmatrix} - \mathfrak{M}\begin{bmatrix} 0 & -\frac{1}{2}h & -a \\ \frac{1}{2}h & 0 & 0 \\ a & 0 & 0 \end{bmatrix}\begin{bmatrix} 0 & -\frac{1}{2}h & -a \\ \frac{1}{2}h & 0 & 0 \\ a & 0 & 0 \end{bmatrix}$$

$$= \tfrac{1}{12}\mathfrak{M}\begin{bmatrix} 15a^2 + 4h^2 & 0 & 0 \\ 0 & 3a^2 + 4h^2 & 6ha \\ 0 & 6ha & 18a^2 \end{bmatrix}$$

In the special case for which $h = \frac{1}{2}\sqrt{3a}$, we have

$$[I^q] = \begin{bmatrix} \frac{3}{2}\mathfrak{M}a^2 & 0 & 0 \\ 0 & \frac{1}{2}\mathfrak{M}a^2 & \frac{\sqrt{3}}{4}\mathfrak{M}a^2 \\ 0 & \frac{\sqrt{3}}{4}\mathfrak{M}a^2 & \frac{3}{2}\mathfrak{M}a^2 \end{bmatrix}$$

Now we need to find a set of orthogonal unit vectors $\hat{\mathbf{e}}_1$, $\hat{\mathbf{e}}_2$, and $\hat{\mathbf{e}}_3$ that parallel the principal axes through q, and then find the corresponding principal axis moments of inertia I_1, I_2, and I_3.

If we proceed formally by the method of §8·4·8, we can use Eq. (8·43) to obtain I_1, I_2, and I_3 from

$$\begin{vmatrix} \frac{3}{2}\mathfrak{M}a^2 - I_j & 0 & 0 \\ 0 & \frac{1}{2}\mathfrak{M}a^2 - I_j & \frac{\sqrt{3}}{4}\mathfrak{M}a^2 \\ 0 & \frac{\sqrt{3}}{4}\mathfrak{M}a^2 & \frac{3}{2}\mathfrak{M}a^2 - I_j \end{vmatrix} = 0$$

or

$$(\tfrac{3}{2}\mathfrak{M}a^2 - I_j)[(\tfrac{1}{2}\mathfrak{M}a^2 - I_j)(\tfrac{3}{2}\mathfrak{M}a^2 - I_j) - \tfrac{3}{16}\mathfrak{M}^2a^4] = 0$$

This equation has a solution $\frac{3}{2}\mathfrak{M}a^2$, which we'll arbitrarily call I_1, and then we can get I_2 and I_3 from the bracketed expression, or

$$I_j^2 - 2\mathfrak{M}a^2 I_j + \tfrac{9}{16}\mathfrak{M}^2a^4 = 0$$

This quadratic equation has the roots (arbitrarily labeled)

$$I_2 = (1 - \tfrac{1}{4}\sqrt{7})\mathfrak{M}a^2$$
$$I_3 = (1 + \tfrac{1}{4}\sqrt{7})\mathfrak{M}a^2$$

In order to locate the principal axes, we must solve for e_1^j, e_2^j, and e_3^j in

$$\begin{bmatrix} \frac{3}{2}\mathfrak{M}a^2 - I_j & 0 & 0 \\ 0 & \frac{1}{2}\mathfrak{M}a^2 - I_j & \frac{\sqrt{3}}{4}\mathfrak{M}a^2 \\ 0 & \frac{\sqrt{3}}{4}\mathfrak{M}a^2 & \frac{3}{2}\mathfrak{M}a^2 - I_j \end{bmatrix} \begin{bmatrix} e_1^j \\ e_2^j \\ e_3^j \end{bmatrix} = \begin{bmatrix} 0 \\ 0 \\ 0 \end{bmatrix}$$

for $j = 1, 2, 3$. Substituting $I_1 = \tfrac{3}{2}\mathfrak{M}a^2$ for I_j satisfies the first of the three scalar equations implied here for any values of e_1^1, e_2^1, and e_3^1, and the next two equations require $e_2^j = e_3^j = 0$. Equation (8·44) then indicates that $e_1^1 = 1$, so that $\hat{\mathbf{e}}_1 = \hat{\mathbf{u}}_1$ [from Eq. (8·41)].

To find $\hat{\mathbf{e}}_2$, substitute $j = 2$ in the preceding matrix equation, to obtain the three scalar equations

$$(\tfrac{1}{2} + \tfrac{1}{4}\sqrt{7})\mathfrak{M}a^2 e_1^2 = 0$$
$$(-\tfrac{1}{2} + \tfrac{1}{4}\sqrt{7})\mathfrak{M}a^2 e_2^2 + \frac{\sqrt{3}}{4}\mathfrak{M}a^2 e_3^2 = 0$$
$$\frac{\sqrt{3}}{4}\mathfrak{M}a^2 e_2^2 + (\tfrac{1}{2} + \tfrac{1}{4}\sqrt{7})\mathfrak{M}a^2 e_3^2 = 0$$

The first of these requires $e_1^2 = 0$, while *both* the second and the third indicate

$$e_3^2 = \tfrac{1}{3}\sqrt{3}\,(2 - \sqrt{7})e_2^2$$

Since only two of these three equations are independent, they can be solved only by incorporating Eq. (8·44), namely,

$$(e_1^2)^2 + (e_2^2)^2 + (e_3^2)^2 = 1$$

In combination these equations provide

$$e_1^2 = 0 \qquad e_2^2 = \left(\frac{3}{14 - 4\sqrt{7}}\right)^{1/2} \qquad e_3^2 = \frac{2 - \sqrt{7}}{(14 - 4\sqrt{7})^{1/2}}$$

so that

$$\hat{\mathbf{e}}_2 = \left(\frac{3}{14 - 4\sqrt{7}}\right)^{1/2} \hat{\mathbf{u}}_2 + \frac{2 - \sqrt{7}}{(14 - 4\sqrt{7})^{1/2}} \hat{\mathbf{u}}_3$$

Similar calculations provide

$$\hat{\mathbf{e}}_3 = \left(\frac{3}{14 + 4\sqrt{7}}\right)^{1/2} \hat{\mathbf{u}}_2 + \frac{2 + \sqrt{7}}{(14 + 4\sqrt{7})^{1/2}} \hat{\mathbf{u}}_3$$

Having proceeded formally to obtain principal axis base vectors $\hat{\mathbf{e}}_1$, $\hat{\mathbf{e}}_2$, and $\hat{\mathbf{e}}_3$ and principal axis inertias I_1, I_2, and I_3 by the methods of §8·4·8, we can now look again at Fig. 8·5, and observe that a plane through q and normal to $\hat{\mathbf{u}}_1$ is a plane of symmetry. Hence we could have identified $\hat{\mathbf{u}}_1 = \hat{\mathbf{e}}_1$ as a principal axis unit vector by the criterion noted in §8·4·7.

8·4·10 Special properties of the inertia matrix

In checking calculations and developing an understanding of physical systems, it is helpful to maintain an awareness of the following relationships among the elements of the inertia matrix: *The sum of the moments of inertia about any two orthogonal axes is greater than or equal to the moment of inertia about the third axis of the orthogonal triad*, so that for any point q and vector basis $\hat{\mathbf{u}}_1$, $\hat{\mathbf{u}}_2$, and $\hat{\mathbf{u}}_3$ we must have

$$I_{11} + I_{22} \geq I_{33} \tag{8·47}$$

Proof follows directly from the definitions in Eqs. (8·22), which indicate that

$$I_{11} + I_{22} = \int (r_2^2 + r_3^2 + r_3^2 + r_1^2)\, dm$$

and

$$I_{33} = \int (r_1^2 + r_2^2)\, dm$$

Thus we have

$$I_{11} + I_{22} - I_{33} = 2\int r_3^2\, dm \geq 0$$

with the equality holding only for the nonphysical idealization of the material lamina or layer of zero thickness in the $\hat{\mathbf{u}}_1\hat{\mathbf{u}}_2$ plane.

The trace (or sum of main diagonal elements) *of an inertia matrix is invariant to a change of vector basis*, so that for any reference point q and vector bases $\{\hat{\mathbf{u}}\}$ and $\{\hat{\mathbf{a}}\}$ we have

$$\text{tr}\,[{}^uI] \triangleq {}^uI_{11} + {}^uI_{22} + {}^uI_{33} = {}^aI_{11} + {}^aI_{22} + {}^aI_{33} \triangleq \text{tr}\,[{}^aI] \tag{8·48}$$

Proof requires only the expansion [from Eqs. (8·22)]

$$\begin{aligned} {}^uI_{11} + {}^uI_{22} + {}^uI_{33} &= \int (r_2^2 + r_3^2 + r_3^2 + r_1^2 + r_1^2 + r_2^2)\, dm \\ &= 2\int (r_1^2 + r_2^2 + r_3^2)\, dm \\ &= 2\int \mathbf{r} \cdot \mathbf{r}\, dm = {}^aI_{11} + {}^aI_{22} + {}^aI_{33} \end{aligned}$$

Valuable insight into the physical and geometrical significance of the inertia dyadic and matrix can be obtained by construction of a geometrical figure characterizing the inertia properties of a material system for a given reference point q. If we define I_s as the moment of inertia of a body for q in the direction of unit vector $\hat{\mathbf{s}}$, then from Eq. (8·26c) we have

$$I_s = \hat{\mathbf{s}} \cdot \mathbb{I} \cdot \hat{\mathbf{s}}$$

If now we introduce a new vector

$$\mathbf{s} = \frac{\hat{\mathbf{s}}}{\sqrt{I_s}}$$

and expand it in terms of unit vectors $\hat{\mathbf{u}}_1$, $\hat{\mathbf{u}}_2$, and $\hat{\mathbf{u}}_3$ fixed in the body as

$$\mathbf{s} = s_1\hat{\mathbf{u}}_1 + s_2\hat{\mathbf{u}}_2 + s_3\hat{\mathbf{u}}_3,$$

then we can expand $\mathbb{I}$ in the same basis and write

$$\begin{aligned} \mathbf{s} \cdot \mathbb{I} \cdot \mathbf{s} &= [s_1 s_2 s_3] \begin{bmatrix} \hat{\mathbf{u}}_1 \\ \hat{\mathbf{u}}_2 \\ \hat{\mathbf{u}}_3 \end{bmatrix} \cdot [\hat{\mathbf{u}}_1\hat{\mathbf{u}}_2\hat{\mathbf{u}}_3] \begin{bmatrix} I_{11} & I_{12} & I_{13} \\ I_{21} & I_{22} & I_{23} \\ I_{31} & I_{32} & I_{33} \end{bmatrix} \begin{bmatrix} \hat{\mathbf{u}}_1 \\ \hat{\mathbf{u}}_2 \\ \hat{\mathbf{u}}_3 \end{bmatrix} \cdot [\hat{\mathbf{u}}_1\hat{\mathbf{u}}_2\hat{\mathbf{u}}_3] \begin{bmatrix} s_1 \\ s_2 \\ s_3 \end{bmatrix} \\ &= [s_1 s_2 s_3] \begin{bmatrix} 1 & 0 & 0 \\ 0 & 1 & 0 \\ 0 & 0 & 1 \end{bmatrix} \begin{bmatrix} I_{11} & I_{12} & I_{13} \\ I_{21} & I_{22} & I_{13} \\ I_{31} & I_{32} & I_{33} \end{bmatrix} \begin{bmatrix} 1 & 0 & 0 \\ 0 & 1 & 0 \\ 0 & 0 & 1 \end{bmatrix} \begin{bmatrix} s_1 \\ s_2 \\ s_3 \end{bmatrix} \\ &= I_{11}s_1^2 + I_{22}s_2^2 + I_{33}s_3^2 + 2I_{12}s_1s_2 + 2I_{13}s_1s_3 + 2I_{23}s_2s_3 \end{aligned}$$

But

$$\mathbf{s} \cdot \mathbb{I} \cdot \mathbf{s} = \frac{1}{\sqrt{I_s}}\, \hat{\mathbf{s}} \cdot \mathbb{I} \cdot \hat{\mathbf{s}}\, \frac{1}{\sqrt{I_s}} = 1$$

so we have

$$I_{11}s_1^2 + I_{22}s_2^2 + I_{33}s_3^2 + 2I_{12}s_1s_2 + 2I_{13}s_1s_3 + 2I_{23}s_2s_3 = 1 \tag{8·49}$$

This equation can be interpreted geometrically as an ellipsoid, as shown in Fig. 8·6. Here $\mathbf{s} = s_1\hat{\mathbf{u}}_1 + s_2\hat{\mathbf{u}}_2 + s_3\hat{\mathbf{u}}_3$ is the vector from q to a point on the surface of the ellipsoid, and the length of $\mathbf{s}$ is $1/\sqrt{I_s}$, a measure of the moment of inertia about the axis through q and parallel to $\mathbf{s}$.

If instead of $\hat{\mathbf{u}}_1$, $\hat{\mathbf{u}}_2$, and $\hat{\mathbf{u}}_3$, we had chosen the principal axis unit vectors $\hat{\mathbf{e}}_1$, $\hat{\mathbf{e}}_2$, and $\hat{\mathbf{e}}_3$ to represent $\mathbf{s}$, we would have the special case of Eq. (8·49) corresponding to the principal axes of the ellipsoid (see Fig. 8·6). From the figure and the familiar properties of the ellipsoid, it becomes apparent that for a given point q, the axes of maximum moment of inertia and minimum moment of inertia are both principal axes (in Fig. 8·6, $\hat{\mathbf{e}}_1$ parallels the axis of minimum inertia, for example).

The inertia ellipsoid helps us to visualize the inertial and dynamical equivalence of pairs of bodies that are geometrically dissimilar. It is clear from the ellipsoid, for example, that if two principal axis inertias are identical (say $I_2 = I_3$ in Fig. 8·6), then the ellipsoid is a figure of revolution, and the moments of inertia will be the same for any axis through q in the plane of the two axes (the $\hat{\mathbf{e}}_2\hat{\mathbf{e}}_3$ plane in Fig. 8·6, for example). This will be the case not only for the mass-center inertia ellipsoid of a body such as the right circular cylinder in Fig. 8·5, but also for the mass-center inertia ellipsoid of a rectangular parallelepiped with two equal edges, such as Fig. 8·4 with $d = a$.

The inertia ellipsoid for the mass center of a uniform sphere is a sphere of radius $1/(\frac{2}{5}\mathfrak{M}a^2)^{1/2}$, where $\mathfrak{M}$ is the mass and a the radius of the physical sphere (see Appendix C for the inertia of a sphere). More surprising, perhaps, is the realization that the inertia matrix for the mass center of a uniform cube of mass $\mathfrak{M}$ and side dimension a is *also* a sphere, this one (from Appendix C) having radius $1/(\frac{1}{6}\mathfrak{M}a^2)^{1/2}$.

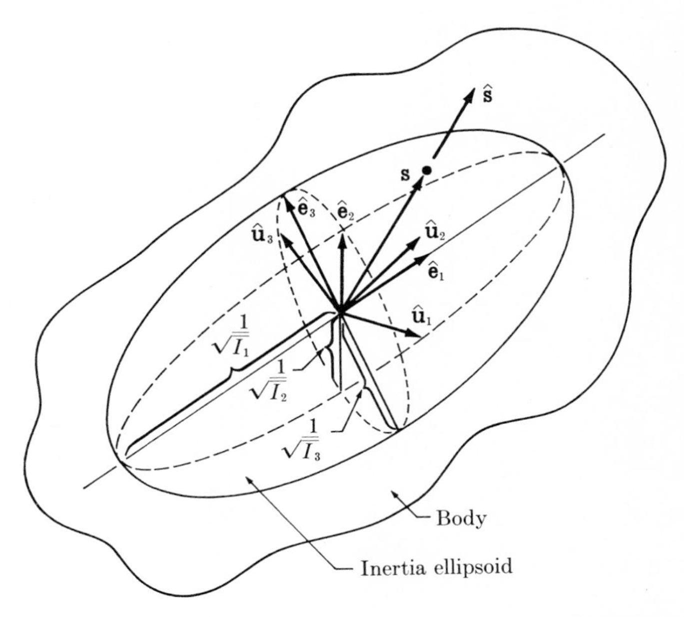

FIGURE 8·6

8·5 EQUATIONS OF ROTATIONAL MOTION OF A RIGID BODY

8·5·1 Euler's scalar equations for principal axes

Vector equations for rotational motion were derived in §8·2·3 in the form [see Eq. (8·12)]

$$\dot{\mathbf{H}}^q = \mathbf{M}^q + \mathfrak{M}\ddot{\mathbf{R}}_q \times \mathbf{r}_c \tag{8·50}$$

This equation applies to any material system having constant mass $\mathfrak{M}$ and angular momentum $\mathbf{H}^q$ with respect to an arbitrary reference point q, which has inertial acceleration $\ddot{\mathbf{R}}_q$. The vector $\mathbf{r}_c$ locates the center of mass of the system with respect to point q, and $\mathbf{M}^q$ is the moment with respect to q of all forces applied to the system.

Only very rarely is the full generality of Eq. (8·50) utilized; as noted in §8·2·4, point q is usually selected as the inertially fixed point o or the mass center c, resulting in the simplifications

$$\dot{\mathbf{H}}^o = \mathbf{M}^o \qquad \text{and} \qquad \dot{\mathbf{H}}^c = \mathbf{M}^c \tag{8·51}$$

A much more substantial reduction of difficulty occurs when application is restricted to rigid bodies; in this case we normally select for q some point fixed in the body, so that $\mathbf{H}^q$ becomes $\mathbf{H}^*$ as defined in Eq. (8·20). The vector $\mathbf{H}^*$ is expanded in terms of scalars and unit vectors in Eq. (8·23), in terms of matrices and vector arrays in Eqs. (8·24), and in terms of the inertia dyadic and the angular velocity vector in Eq. (8·25*b*).

In all that follows, attention is restricted to the special case of a rigid body whose moment and inertia properties are referred to a reference point fixed in the reference frame b established by the body, being also either coincident with the mass center or fixed in inertial space. We shall at this point divest ourselves of the superscripts that we've been employing as reminders of the restrictions of various equations, and write simply

$$\mathbf{M} = \dot{\mathbf{H}} \tag{8·52}$$

instead of one of the forms in Eq. (8·51), and

$$\mathbf{H} = \mathbb{I} \cdot \boldsymbol{\omega} \tag{8·53a}$$

$$\mathbf{H} = [\hat{\mathbf{e}}_1\hat{\mathbf{e}}_2\hat{\mathbf{e}}_3] \begin{bmatrix} I_1 & 0 & 0 \\ 0 & I_2 & 0 \\ 0 & 0 & I_3 \end{bmatrix} \begin{bmatrix} \omega_1 \\ \omega_2 \\ \omega_3 \end{bmatrix} \tag{8·53b}$$

$$\mathbf{H} = I_1\omega_1\hat{\mathbf{e}}_1 + I_2\omega_2\hat{\mathbf{e}}_2 + I_3\omega_3\hat{\mathbf{e}}_3 \tag{8·53c}$$

in the place of Eqs. (8·23) to (8·25). Note that in Eqs. (8·53*b* and *c*) we have made a commitment to a principal axis vector basis $\hat{\mathbf{e}}_1$, $\hat{\mathbf{e}}_2$, and $\hat{\mathbf{e}}_3$, and that these unit vectors are taken as fixed relative to the rigid body that constitutes our system. The inertial angular velocity of that body is then

$$\boldsymbol{\omega} = \omega_1 \hat{\mathbf{e}}_1 + \omega_2 \hat{\mathbf{e}}_2 + \omega_3 \hat{\mathbf{e}}_3 \tag{8·54}$$

If now we expand $\mathbf{M}$ in the same vector basis, we have

$$\mathbf{M} = M_1 \hat{\mathbf{e}}_1 + M_2 \hat{\mathbf{e}}_2 + M_3 \hat{\mathbf{e}}_3 \tag{8·55}$$

and Eq. (8·52) can be written as

$$\begin{aligned} M_1 \hat{\mathbf{e}}_1 + M_2 \hat{\mathbf{e}}_2 + M_3 \hat{\mathbf{e}}_3 &= \frac{{}^i d}{dt} \mathbf{H} \\ &= \frac{{}^b d}{dt} \mathbf{H} + \omega \times \mathbf{H} \\ &= I_1 \dot{\omega}_1 \hat{\mathbf{e}}_1 + I_2 \dot{\omega}_2 \hat{\mathbf{e}}_2 + I_3 \dot{\omega}_3 \hat{\mathbf{e}}_3 \\ &\qquad + (\omega_1 \hat{\mathbf{e}}_1 + \omega_2 \hat{\mathbf{e}}_2 + \omega_3 \hat{\mathbf{e}}_3) \\ &\qquad \times (I_1 \omega_1 \hat{\mathbf{e}}_1 + I_2 \omega_2 \hat{\mathbf{e}}_2 + I_3 \omega_3 \hat{\mathbf{e}}_3) \\ &= [I_1 \dot{\omega}_1 - \omega_2 \omega_3 (I_2 - I_3)] \hat{\mathbf{e}}_1 \\ &\qquad + [I_2 \dot{\omega}_2 - \omega_3 \omega_1 (I_3 - I_1)] \hat{\mathbf{e}}_2 \\ &\qquad + [I_3 \dot{\omega}_3 - \omega_1 \omega_2 (I_1 - I_2)] \hat{\mathbf{e}}_3 \end{aligned} \tag{8·56}$$

From this vector equation we can write the three scalar equations known as *Euler's equations:*

$$M_1 = I_1 \dot{\omega}_1 - \omega_2 \omega_3 (I_2 - I_3) \tag{8·57a}$$
$$M_2 = I_2 \dot{\omega}_2 - \omega_3 \omega_1 (I_3 - I_1) \tag{8·57b}$$
$$M_3 = I_3 \dot{\omega}_3 - \omega_1 \omega_2 (I_1 - I_2) \tag{8·57c}$$

These equations are the starting point for much (perhaps most) of the dynamic analysis of rotating rigid bodies. As previously noted, *they are restricted in application to a rigid body in whose reference frame b is fixed the reference point q to which the external moment and the inertia properties are referred. Moreover, q must be either the mass center c or a point fixed in inertial space, and the body-fixed unit vectors $\hat{\mathbf{e}}_1$, $\hat{\mathbf{e}}_2$, and $\hat{\mathbf{e}}_3$ which establish the numerical subscripts must parallel axes of the body for point q.*

8·5·2 Vector-dyadic equations

A less restricted version of the rotational equations of a rigid body can be obtained by avoiding a commitment to a principal axis vector basis. We can retain the restrictions of Eqs. (8·52) and (8·53) but avoid this vector-basis commitment by sticking with vectors and dyadics, writing

$$\mathbf{M} = \dot{\mathbf{H}} \triangleq \frac{{}^i d}{dt} (\mathbb{I} \cdot \boldsymbol{\omega}) = \frac{{}^b d}{dt} (\mathbb{I} \cdot \boldsymbol{\omega}) + \boldsymbol{\omega} \times \mathbb{I} \cdot \boldsymbol{\omega} \tag{8·58}$$

Since the inertia dyadic $\mathbb{I}$ is unchanging with respect to the reference frame b established by the body, this equation simplifies to

$$\mathbf{M} = \mathbb{I} \cdot \frac{{}^b d}{dt} \boldsymbol{\omega} + \boldsymbol{\omega} \times \mathbb{I} \cdot \boldsymbol{\omega} \tag{8·59}$$

and since

$$\frac{{}^b d}{dt} \boldsymbol{\omega} = \frac{{}^i d}{dt} \boldsymbol{\omega} + \boldsymbol{\omega} \times \boldsymbol{\omega} = \frac{{}^i d}{dt} \boldsymbol{\omega} \triangleq \dot{\boldsymbol{\omega}} \tag{8·60}$$

we can as well write Eq. (8·59) as

$$\mathbf{M} = \mathbb{I} \cdot \dot{\boldsymbol{\omega}} + \boldsymbol{\omega} \times \mathbb{I} \cdot \boldsymbol{\omega} \tag{8·61}$$

Equation (8·61) *is the vector-dyadic equation of rotational motion of a rigid body, so restricted that* $\mathbf{M}$ *and* $\mathbb{I}$ *are taken with respect to a point* q *which is fixed in the reference frame* b *of the rigid body and also either coincident with the mass center* c *of the rigid body or fixed in inertial space.*

8·5·3 Matrix equations

The vector-dyadic equation (8·61) implies no commitment to any vector basis, but it is not expressed in terms of the scalars that people (and computers) must ultimately deal with in the solution of problems. We can take a step toward greater utility by representing all the vectors and dyadics in Eq. (8·61) in terms of matrices and vector arrays; in particular, we shall restrict attention to a vector basis $\{\hat{\mathbf{b}}\} \triangleq \{\hat{\mathbf{b}}_1\hat{\mathbf{b}}_2\hat{\mathbf{b}}_3\}^T$ that by definition is in some way fixed in the reference frame b of the rigid body. This restriction simplifies the differentiation

$$\dot{\boldsymbol{\omega}} \triangleq \frac{{}^i d}{dt} \boldsymbol{\omega} = \frac{{}^b d}{dt} \boldsymbol{\omega} + \boldsymbol{\omega} \times \boldsymbol{\omega} = \frac{{}^b d}{dt} \boldsymbol{\omega} \tag{8·62}$$

Substituting the expansions

$$\mathbf{M} = \{\hat{\mathbf{b}}_1\hat{\mathbf{b}}_2\hat{\mathbf{b}}_3\} \begin{bmatrix} {}^bM_1 \\ {}^bM_2 \\ {}^bM_3 \end{bmatrix} \triangleq \{\hat{\mathbf{b}}\}^T[{}^bM] \tag{8·63}$$

$$\boldsymbol{\omega} = \{\hat{\mathbf{b}}_1\hat{\mathbf{b}}_2\hat{\mathbf{b}}_3\} \begin{bmatrix} {}^b\omega_1 \\ {}^b\omega_2 \\ {}^b\omega_3 \end{bmatrix} \triangleq \{\hat{\mathbf{b}}\}^T[{}^b\omega] \tag{8·64}$$

$$\mathbb{I} = \{\hat{\mathbf{b}}_1\hat{\mathbf{b}}_2\hat{\mathbf{b}}_3\} \begin{bmatrix} I_{11} & I_{12} & I_{13} \\ I_{21} & I_{22} & I_{23} \\ I_{31} & I_{32} & I_{33} \end{bmatrix} \begin{Bmatrix} \hat{\mathbf{b}}_1 \\ \hat{\mathbf{b}}_2 \\ \hat{\mathbf{b}}_3 \end{Bmatrix} \triangleq \{\hat{\mathbf{b}}\}^T[{}^bI]\{\hat{\mathbf{b}}\} \tag{8·65}$$

into Eq. (8·61) produces

$$\{\hat{\mathbf{b}}\}^T[{}^bM] = \{\hat{\mathbf{b}}\}^T[{}^bI]\{\hat{\mathbf{b}}\} \cdot \{\hat{\mathbf{b}}\}^T[{}^b\dot{\omega}] + \{\hat{\mathbf{b}}\}^T[{}^b\omega] \times \{\hat{\mathbf{b}}\}^T[{}^bI]\{\hat{\mathbf{b}}\} \cdot \{\hat{\mathbf{b}}\}^T[{}^b\omega] \tag{8·66}$$

The dot products

$$\{\hat{\mathbf{b}}\} \cdot \{\hat{\mathbf{b}}\}^T \triangleq \begin{Bmatrix} \hat{\mathbf{b}}_1 \\ \hat{\mathbf{b}}_2 \\ \hat{\mathbf{b}}_3 \end{Bmatrix} \cdot \{\hat{\mathbf{b}}_1\hat{\mathbf{b}}_2\hat{\mathbf{b}}_3\} = \begin{bmatrix} \hat{\mathbf{b}}_1\cdot\hat{\mathbf{b}}_1 & \hat{\mathbf{b}}_1\cdot\hat{\mathbf{b}}_2 & \hat{\mathbf{b}}_1\cdot\hat{\mathbf{b}}_3 \\ \hat{\mathbf{b}}_2\cdot\hat{\mathbf{b}}_1 & \hat{\mathbf{b}}_2\cdot\hat{\mathbf{b}}_2 & \hat{\mathbf{b}}_2\cdot\hat{\mathbf{b}}_3 \\ \hat{\mathbf{b}}_3\cdot\hat{\mathbf{b}}_1 & \hat{\mathbf{b}}_3\cdot\hat{\mathbf{b}}_2 & \hat{\mathbf{b}}_3\cdot\hat{\mathbf{b}}_3 \end{bmatrix} = \begin{bmatrix} 1 & 0 & 0 \\ 0 & 1 & 0 \\ 0 & 0 & 1 \end{bmatrix} = [U] \tag{8·67}$$

can be removed from Eq. (8·66) without effect. The cross product

$$\{\hat{\mathbf{b}}\}^T[{}^b\omega] \times \{\hat{\mathbf{b}}\}^T$$

can be expanded as

$$\begin{aligned} \{\hat{\mathbf{b}}\}^T[{}^b\omega] \times \{\hat{\mathbf{b}}\}^T &= \{\hat{\mathbf{b}}_1\hat{\mathbf{b}}_2\hat{\mathbf{b}}_3\} \begin{bmatrix} {}^b\omega_1 \\ {}^b\omega_2 \\ {}^b\omega_3 \end{bmatrix} \times \{\hat{\mathbf{b}}_1\hat{\mathbf{b}}_2\hat{\mathbf{b}}_3\} \\ &= \{{}^b\omega_1\hat{\mathbf{b}}_1 + {}^b\omega_2\hat{\mathbf{b}}_2 + {}^b\omega_3\hat{\mathbf{b}}_3\} \times \{\hat{\mathbf{b}}_1\hat{\mathbf{b}}_2\hat{\mathbf{b}}_3\} \\ &= \{({}^b\omega_3\hat{\mathbf{b}}_2 - {}^b\omega_2\hat{\mathbf{b}}_3) \quad ({}^b\omega_1\hat{\mathbf{b}}_3 - {}^b\omega_3\hat{\mathbf{b}}_1) \quad ({}^b\omega_2\hat{\mathbf{b}}_1 - {}^b\omega_1\hat{\mathbf{b}}_2)\} \\ &= \{\hat{\mathbf{b}}_1\hat{\mathbf{b}}_2\hat{\mathbf{b}}_3\} \begin{bmatrix} 0 & -{}^b\omega_3 & {}^b\omega_2 \\ {}^b\omega_3 & 0 & -{}^b\omega_1 \\ -{}^b\omega_2 & {}^b\omega_1 & 0 \end{bmatrix} \triangleq \{\hat{\mathbf{b}}\}^T[{}^b\tilde{\omega}] \end{aligned} \tag{8·68}$$

where the symbol $[{}^b\tilde{\omega}]$ has been adopted to represent the skew-symmetric matrix with elements $\pm{}^b\omega_1$, $\pm{}^b\omega_2$, and $\pm{}^b\omega_3$ distributed as in Eq. (8·68).

Combining Eqs. (8·66) to (8·68) produces

$$\{\hat{\mathbf{b}}\}^T[{}^bM] = \{\hat{\mathbf{b}}\}^T[{}^bI][{}^b\dot{\omega}] + \{\hat{\mathbf{b}}\}^T[{}^b\tilde{\omega}][{}^bI][{}^b\omega] \tag{8·69}$$

This equation implies the matrix equation

$$[{}^bM] = [{}^bI][{}^b\dot{\omega}] + [{}^b\tilde{\omega}][{}^bI][{}^b\omega] \tag{8·70}$$

If as a special case of the body-fixed vector basis $\{\hat{\mathbf{b}}\}$ we adopt the principal axis vector basis $\{\hat{\mathbf{e}}\}$, then in the notation of §8·5·1 we have

$$\begin{bmatrix} M_1 \\ M_2 \\ M_3 \end{bmatrix} = \begin{bmatrix} I_1 & 0 & 0 \\ 0 & I_2 & 0 \\ 0 & 0 & I_3 \end{bmatrix} \begin{bmatrix} \dot{\omega}_1 \\ \dot{\omega}_2 \\ \dot{\omega}_3 \end{bmatrix} + \begin{bmatrix} 0 & -\omega_3 & \omega_2 \\ \omega_3 & 0 & -\omega_1 \\ -\omega_2 & \omega_1 & 0 \end{bmatrix} \begin{bmatrix} I_1 & 0 & 0 \\ 0 & I_2 & 0 \\ 0 & 0 & I_3 \end{bmatrix} \begin{bmatrix} \omega_1 \\ \omega_2 \\ \omega_3 \end{bmatrix} \tag{8·71}$$

By multiplying out these matrices, we could obtain once again the three scalar equations written previously as Eqs. (8·57), and identified as Euler's equations.

8·5·4 Rotations about a fixed axis

When a line is fixed in inertial space and also fixed in a rigid body, then each point of the body can move only in a plane normal to that line, which is called the *axis of rotation*. Figure 8·7 illustrates a typical application, in which a rigid body is supported on a rigid shaft supported on bearings at inertially fixed points A and B; a motor at B can apply to the shaft a couple having torque $\tau\hat{\mathbf{b}}_3$.

We can approach this problem in several different ways. We might wish to apply Euler's equations [Eqs. (8·57)], choosing as our reference point either the mass center c or some point fixed on the axis of rotation, but we must expect in general that the axis of rotation will not parallel a principal axis for any reference point we might choose. As a simpler alternative, we could employ the

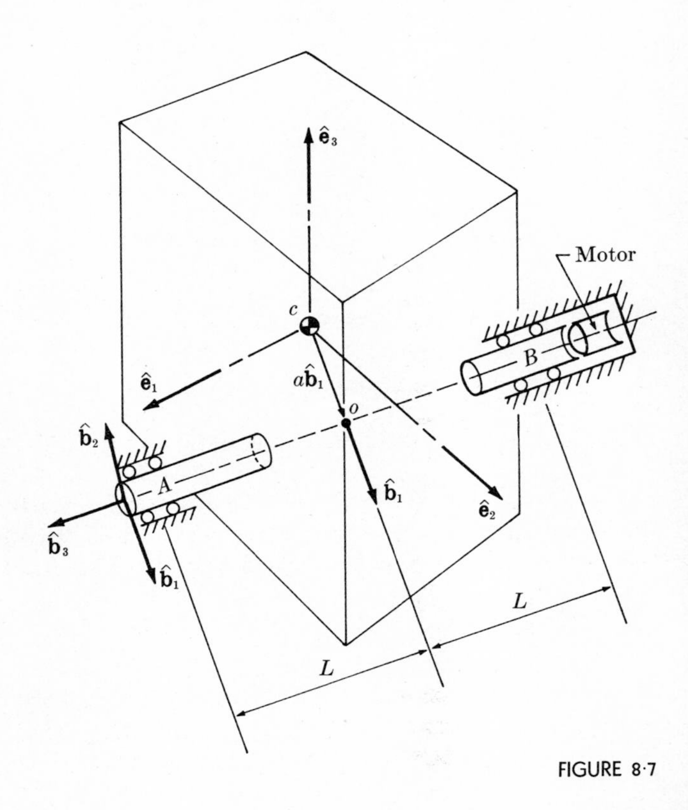

FIGURE 8·7

matrix equation recorded as Eq. (8·70), choosing our body-fixed vector-basis unit vectors $\hat{\mathbf{b}}_1$, $\hat{\mathbf{b}}_2$, and $\hat{\mathbf{b}}_3$ in such a way that one of them (say $\hat{\mathbf{b}}_3$) parallels the axis of rotation. Of course, we could also go back to the beginning and apply $\dot{\mathbf{H}}^o = \mathbf{M}^o$ or $\dot{\mathbf{H}}^c = \mathbf{M}^c$ to the body, choosing o as some point on the axis of rotation. (This last option was examined in §8·2·5).

If in applying the matrix equation (8·70), we choose $\hat{\mathbf{b}}_1$, $\hat{\mathbf{b}}_2$, and $\hat{\mathbf{b}}_3$ such that

$$\boldsymbol{\omega} = \Omega\hat{\mathbf{b}}_3 \qquad (8{\cdot}72a)$$

then

$$[{}^b\omega] \triangleq \begin{bmatrix} {}^b\omega_1 \\ {}^b\omega_2 \\ {}^b\omega_3 \end{bmatrix} = \begin{bmatrix} 0 \\ 0 \\ \Omega \end{bmatrix} \qquad (8{\cdot}72b)$$

and Eq. (8·70) becomes

$$\begin{bmatrix} {}^bM_1 \\ {}^bM_2 \\ {}^bM_3 \end{bmatrix} = \begin{bmatrix} I_{11} & I_{12} & I_{13} \\ I_{21} & I_{22} & I_{23} \\ I_{31} & I_{32} & I_{33} \end{bmatrix} \begin{bmatrix} 0 \\ 0 \\ \dot{\Omega} \end{bmatrix} + \begin{bmatrix} 0 & -\Omega & 0 \\ \Omega & 0 & 0 \\ 0 & 0 & 0 \end{bmatrix} \begin{bmatrix} I_{11} & I_{12} & I_{11} \\ I_{21} & I_{22} & I_{23} \\ I_{31} & I_{32} & I_{33} \end{bmatrix} \begin{bmatrix} 0 \\ 0 \\ \Omega \end{bmatrix} \qquad (8{\cdot}73)$$

The three scalar equations implied here are

$$\begin{aligned} {}^bM_1 &= I_{13}\dot{\Omega} - I_{23}\Omega^2 && (8{\cdot}74a) \\ {}^bM_2 &= I_{23}\dot{\Omega} + I_{13}\Omega^2 && (8{\cdot}74b) \\ {}^bM_3 &= I_{33}\dot{\Omega} && (8{\cdot}74c) \end{aligned}$$

Equations (8·74) are valid for a reference point either at c or fixed in the body and in inertial space; we elect in what follows to choose a point o on the axis of rotation and as close as possible to the mass center c. If gravity and other external influences are absent, the moment $\mathbf{M}^o = {}^bM_1\hat{\mathbf{b}}_1 + {}^bM_2\hat{\mathbf{b}}_2 + {}^bM_3\hat{\mathbf{b}}_3$ is due solely to the motor torque $\tau\hat{\mathbf{b}}_3$ and the moment due to the bearing reaction forces at A and B. Since the forces applied at the bearings cannot contribute to ${}^bM_3$, Eq. (8·74c) is all that is required to determine the dynamic response to torque $\tau\hat{\mathbf{b}}_3$; only the scalar equation

$$\tau = I_{33}\dot{\Omega} \qquad (8{\cdot}75)$$

is needed to find that

$$\Omega = \frac{1}{I_{33}} \int_{t_0}^{t} \tau \, dt \qquad (8{\cdot}76)$$

Once Ω is known as a function of time, there remains the practical task of finding the reaction forces applied to the shaft at the bearings. If point o is an equal distance L from points A and B, then the reaction forces $\mathbf{F}_A$ and $\mathbf{F}_B$ must provide the moment

$$L\hat{\mathbf{b}}_3 \times \mathbf{F}_A - L\hat{\mathbf{b}}_3 \times \mathbf{F}_B = {}^bM_1\hat{\mathbf{b}}_1 + {}^bM_2\hat{\mathbf{b}}_2 \tag{8·77}$$

In addition, the reaction forces $\mathbf{F}_A$ and $\mathbf{F}_B$ must provide the mass-center acceleration in inertial space. If the distance from o to c is called a, and the unit vector $\hat{\mathbf{b}}_1$ is selected so that $a\hat{\mathbf{b}}_1$ is the vector from c to o, then the inertial acceleration of the mass center is

$$\mathbf{A}^{ci} = \frac{^id^2}{dt^2}(-a\hat{\mathbf{b}}_1) = -a\frac{^id}{dt}(\Omega\hat{\mathbf{b}}_3 \times \hat{\mathbf{b}}_1) = -a\dot{\Omega}\hat{\mathbf{b}}_2 + a\Omega^2\hat{\mathbf{b}}_1$$

and the equation of mass-center translation [see Eqs. (8·6)] is

$$\mathbf{F}_A + \mathbf{F}_B = \mathfrak{M}a(\Omega^2\hat{\mathbf{b}}_1 - \dot{\Omega}\hat{\mathbf{b}}_2) \tag{8·78}$$

Equations (8·78) and (8·77) are both satisfied by the reaction forces

$$\mathbf{F}_A = \tfrac{1}{2}\mathfrak{M}a(\Omega^2\hat{\mathbf{b}}_1 - \dot{\Omega}\hat{\mathbf{b}}_2) + \frac{1}{L}({}^bM_2\hat{\mathbf{b}}_1 - {}^bM_1\hat{\mathbf{b}}_2) \tag{8·79a}$$

$$\mathbf{F}_B = \tfrac{1}{2}\mathfrak{M}a(\Omega^2\hat{\mathbf{b}}_1 - \dot{\Omega}\hat{\mathbf{b}}_2) - \frac{1}{L}({}^bM_2\hat{\mathbf{b}}_1 - {}^bM_1\hat{\mathbf{b}}_2) \tag{8·79b}$$

Combining Eqs. (8·79) and (8·74) produces

$$\mathbf{F}_A = \tfrac{1}{2}\mathfrak{M}a(\Omega^2\hat{\mathbf{b}}_1 - \dot{\Omega}\hat{\mathbf{b}}_2) + \frac{1}{L}[(I_{23}\dot{\Omega} + I_{13}\Omega^2)\hat{\mathbf{b}}_1 - (I_{13}\dot{\Omega} - I_{23}\Omega^2)\hat{\mathbf{b}}_2] \tag{8·80a}$$

$$\mathbf{F}_B = \tfrac{1}{2}\mathfrak{M}a(\Omega^2\hat{\mathbf{b}}_1 - \dot{\Omega}\hat{\mathbf{b}}_2) - \frac{1}{L}[(I_{23}\dot{\Omega} + I_{13}\Omega^2)\hat{\mathbf{b}}_1 - (I_{13}\dot{\Omega} - I_{23}\Omega^2)\hat{\mathbf{b}}_2] \tag{8·80b}$$

We might note in passing that we could now analyze the compound pendulum examined in §8·2·5 (see Fig. 8·2) more easily by using Eq. (8·74*c*). If we adopt the support point o as our reference point, then ${}^bM_3$ in Eq. (8·74*c*) is given by

$${}^bM_3 = \hat{\mathbf{b}}_3 \cdot \mathbf{M}^o = \hat{\mathbf{b}}_3 \cdot \tfrac{1}{2}L\hat{\mathbf{b}}_1 \times \mathfrak{M}g\hat{\mathbf{u}}_v = -\tfrac{1}{2}\mathfrak{M}gL \sin\theta$$

and the angular speed Ω in that equation is $\dot{\theta}$. Therefore Eq. (8·74*c*) provides immediately

$$-\tfrac{1}{2}\mathfrak{M}gL \sin\theta = I_{33}\,\ddot{\theta}$$

The moment of inertia I_{33} for the rod of mass $\mathfrak{M}$ and length L about its end point is easily shown (see definition or Appendix C) to be $I_{33} = \frac{1}{3}\mathfrak{M}L^2$, so that the equation of motion becomes

$$\ddot{\theta} + \frac{3g}{2L}\sin\theta = 0$$

confirming the lengthier derivation found in §8·2·5.

This same equation could have been obtained by applying Eq. (8·74c) with the mass center c chosen for the reference point. This application would, however, necessitate the calculation of the reaction force at o in order to obtain the moment $\mathbf{M}^c$, so this would be a less efficient procedure.

8·5·5 Physical significance of moments and products of inertia

If we are asked to characterize the "physical significance" of the mass of a particle, we might consider the equation $\mathbf{F} = m\mathbf{A}^{pi}$ and describe mass as a measure of resistance to acceleration in inertial space. In a similar way, we might consider the scalar equation $\tau = I_{33}\dot{\Omega}$ [Eq. (8·75)] and describe the moment of inertia of a rigid body about a given inertially fixed axis of the body as a measure of the resistance of that body to angular acceleration about that axis.

The products of inertia are a measure of the resistance of a rigid body to constant angular-speed rotation about an inertially fixed axis, as reflected in the bearing reactions shown in Eqs. (8·80). Even if the angular-acceleration magnitude $\dot{\Omega}$ is zero, these equations indicate the need for forces applied at the bearings in order to permit the simple rotation to persist, as indicated specifically by

$$\mathbf{F}_A = \tfrac{1}{2}\mathfrak{M}a\Omega^2\hat{\mathbf{b}}_1 + \frac{I_{13}}{L}\Omega^2\hat{\mathbf{b}}_1 + \frac{I_{23}}{L}\Omega^2\hat{\mathbf{b}}_2 \qquad (8\cdot81a)$$

$$\mathbf{F}_B = \tfrac{1}{2}\mathfrak{M}a\Omega^2\hat{\mathbf{b}}_1 - \frac{I_{13}}{L}\Omega^2\hat{\mathbf{b}}_1 - \frac{I_{23}}{L}\Omega^2\hat{\mathbf{b}}_2 \qquad (8\cdot81b)$$

The first term appearing in each of these expressions is a consequence of the displacement of the mass center c from the axis of rotation; this is often called the *static unbalance* term because it can be detected by observing the stable rest position of the body on its (frictionless) bearings in the presence of gravity. The second and third terms are called the *dynamic unbalance* terms, and are proportional to the products of inertia with indices including that of the spin axis. By conducting a number of steady-state spin tests for the body about axes passing through c and parallel to the unit vectors $\hat{\mathbf{b}}_1$, $\hat{\mathbf{b}}_2$, and $\hat{\mathbf{b}}_3$, we could determine all the products of inertia from measurements of the bearing forces.

8·5·6 Kinetic energy of a rigid body

The kinetic energy of a differential element of mass dm is defined, in parallel with that of a particle, as one-half dm times the inertial velocity squared, so that the kinetic energy of a continuum is

$$T = \tfrac{1}{2}\int \dot{\mathbf{R}} \cdot \dot{\mathbf{R}}\, dm \tag{8·82}$$

In terms of the notation illustrated in Fig. 8·1, we can replace $\mathbf{R}$ by $\mathbf{R}_c + \boldsymbol{\rho}$, to obtain

$$\begin{aligned} T &= \tfrac{1}{2}\int (\dot{\mathbf{R}}_c + \dot{\boldsymbol{\rho}}) \cdot (\dot{\mathbf{R}}_c + \dot{\boldsymbol{\rho}})\, dm \\ &= \tfrac{1}{2}\dot{\mathbf{R}}_c \cdot \dot{\mathbf{R}}_c \int dm + \dot{\mathbf{R}}_c \int \dot{\boldsymbol{\rho}}\, dm + \tfrac{1}{2}\int \dot{\boldsymbol{\rho}} \cdot \dot{\boldsymbol{\rho}}\, dm \end{aligned}$$

But $\int dm \triangleq \mathfrak{M}$ and $\int \boldsymbol{\rho}\, dm = 0$, so T becomes

$$T = \tfrac{1}{2}\mathfrak{M}\dot{\mathbf{R}}_c \cdot \dot{\mathbf{R}}_c + \tfrac{1}{2}\int \dot{\boldsymbol{\rho}} \cdot \dot{\boldsymbol{\rho}}\, dm \tag{8·83}$$

If the continuum is a rigid body, then [by Eq. (3·100)] $\dot{\boldsymbol{\rho}} = \boldsymbol{\omega} \times \boldsymbol{\rho}$, where $\boldsymbol{\omega}$ is the inertial angular velocity of the body. Thus a rigid body has kinetic energy

$$\begin{aligned} T &= \tfrac{1}{2}\mathfrak{M}\dot{\mathbf{R}}_c \cdot \dot{\mathbf{R}}_c + \tfrac{1}{2}\int (\boldsymbol{\omega} \times \boldsymbol{\rho}) \cdot (\boldsymbol{\omega} \times \boldsymbol{\rho})\, dm \\ &= \tfrac{1}{2}\mathfrak{M}\dot{\mathbf{R}}_c \cdot \dot{\mathbf{R}}_c + \tfrac{1}{2}\int \boldsymbol{\omega} \cdot \boldsymbol{\rho} \times (\boldsymbol{\omega} \times \boldsymbol{\rho})\, dm \\ &= \tfrac{1}{2}\mathfrak{M}\dot{\mathbf{R}}_c \cdot \dot{\mathbf{R}}_c + \tfrac{1}{2}\boldsymbol{\omega} \cdot \int \boldsymbol{\rho} \times (\boldsymbol{\omega} \times \boldsymbol{\rho})\, dm \end{aligned}$$

or in terms of $\mathbf{H}^c$, the angular momentum with respect to c,

$$T = \tfrac{1}{2}\mathfrak{M}\dot{\mathbf{R}}_c \cdot \dot{\mathbf{R}}_c + \tfrac{1}{2}\boldsymbol{\omega} \cdot \mathbf{H}^c \tag{8·84}$$

We can express T in terms of the inertia dyadic $\mathbb{I}^c$ [using Eq. (8.53)] as

$$T = \tfrac{1}{2}\mathfrak{M}\dot{\mathbf{R}}_c \cdot \dot{\mathbf{R}}_c + \tfrac{1}{2}\boldsymbol{\omega} \cdot \mathbb{I}^c \cdot \boldsymbol{\omega} \tag{8·85}$$

The kinetic energy is a scalar, and we can express it in terms of other scalars by letting $v_c \triangleq |\dot{\mathbf{R}}_c|$ represent the inertial *speed* of c, and expanding $\boldsymbol{\omega}$ and $\mathbb{I}^c$ in terms of a convenient vector basis. If we choose the principal axis vector basis $\hat{\mathbf{e}}_1$, $\hat{\mathbf{e}}_2$, and $\hat{\mathbf{e}}_3$, then we have

$$\boldsymbol{\omega} = \omega_1\hat{\mathbf{e}}_1 + \omega_2\hat{\mathbf{e}}_2 + \omega_3\hat{\mathbf{e}}_3 = [\omega_1\omega_2\omega_3] \begin{Bmatrix} \hat{\mathbf{e}}_1 \\ \hat{\mathbf{e}}_2 \\ \hat{\mathbf{e}}_3 \end{Bmatrix} = [\omega]^T\{\hat{\mathbf{e}}\}$$

or (if we prefer)

$$\omega = \{\hat{\mathbf{e}}_1\hat{\mathbf{e}}_2\hat{\mathbf{e}}_3\} \begin{bmatrix} \omega_1 \\ \omega_2 \\ \omega_3 \end{bmatrix} = \{\hat{\mathbf{e}}\}^T[\omega]$$

and

$$\mathbb{I} = \{\hat{\mathbf{e}}_1\hat{\mathbf{e}}_2\hat{\mathbf{e}}_3\} \begin{bmatrix} I_1 & 0 & 0 \\ 0 & I_2 & 0 \\ 0 & 0 & I_3 \end{bmatrix} \begin{Bmatrix} \hat{\mathbf{e}}_1 \\ \hat{\mathbf{e}}_2 \\ \hat{\mathbf{e}}_3 \end{Bmatrix} \triangleq \{\hat{\mathbf{e}}\}^T[I]\{\hat{\mathbf{e}}\}$$

Substitution into Eq. (8·85) and recognition of Eq. (8·67) provides

$$\begin{aligned} T &= \tfrac{1}{2}v_c^2 + \tfrac{1}{2}[\omega]^T\{\hat{\mathbf{e}}\} \cdot \{\hat{\mathbf{e}}\}^T[I]\{\hat{\mathbf{e}}\} \cdot \{\hat{\mathbf{e}}\}^T[\omega] \\ &= \tfrac{1}{2}v_c^2 + \tfrac{1}{2}[\omega]^T[I][\omega] \end{aligned} \tag{8·86}$$

Matrix multiplication then leads to

$$T = \tfrac{1}{2}\mathfrak{M}v_c^2 + \tfrac{1}{2}I_1\omega_1^2 + \tfrac{1}{2}I_2\omega_2^2 + \tfrac{1}{2}I_3\omega_3^2 \tag{8·87}$$

If an expression should be required in terms of some vector basis $\hat{\mathbf{u}}_1$, $\hat{\mathbf{u}}_2$, and $\hat{\mathbf{u}}_3$ composed of unit vectors that do not parallel principal axes for c, a development analogous to that leading to Eq. (8·86) yields

$$T = \tfrac{1}{2}\mathfrak{M}v_c^2 + \tfrac{1}{2}[{}^u\omega]^T[{}^uI^c][{}^u\omega] \tag{8·88}$$

In the special case in which there is a point o fixed in a rigid body and fixed also in inertial space, then the vector $\mathbf{R}$ in Eq. (8·82) can be measured from o and still be fixed in the body. The basic equation (8·82) can then be expanded as

$$\begin{aligned} T &= \tfrac{1}{2}\int (\boldsymbol{\omega} \times \mathbf{R}) \cdot (\boldsymbol{\omega} \times \mathbf{R})\, dm \\ &= \tfrac{1}{2}\boldsymbol{\omega} \cdot \int \mathbf{R} \times (\boldsymbol{\omega} \times \mathbf{R})\, dm \\ &= \tfrac{1}{2}\boldsymbol{\omega} \cdot \mathbb{I}^o \cdot \boldsymbol{\omega} \end{aligned} \tag{8·89}$$

This result should be compared to Eq. (8·85); it can be rewritten in terms of scalars just as Eq. (8·85) was rewritten in Eqs. (8·86) to (8·88).

8·5·7 The rotational work-energy equation

The most useful form of the vector equation for rotational motion is Eq. (8·52), which says, with stated restrictions on the reference point

$$\mathbf{M} = \dot{\mathbf{H}} \tag{8·90}$$

In §8·2·6 we discovered that by integrating over time we could obtain from this equation an angular-impulse-momentum equation, precisely in parallel with the linear-impulse-momentum equations obtained earlier from $\mathbf{F} = m\mathbf{A}$.

Now we seek from $\mathbf{M} = \dot{\mathbf{H}}$ a form of *work-energy equation*, just as we found work-energy equations from $\mathbf{F} = m\mathbf{A}$ in Chaps. 4 and 6. Just as previously we dot-multiplied $\mathbf{F}$ by the inertial linear velocity $\mathbf{V}$ and integrated over time, now we write from Eq. (8·90)

$$\int_{t_1}^{t_2} \mathbf{M} \cdot \boldsymbol{\omega}\, dt = \int_{t_1}^{t_2} \dot{\mathbf{H}} \cdot \boldsymbol{\omega}\, dt = \int_{t_1}^{t_2} \boldsymbol{\omega} \cdot \dot{\mathbf{H}}\, dt \tag{8·91}$$

We can expand the integrand with the relationship

$$\frac{d}{dt}(\tfrac{1}{2}\boldsymbol{\omega}\cdot\mathbf{H}) = \frac{{}^{i}d}{dt}(\tfrac{1}{2}\boldsymbol{\omega}\cdot\mathbf{H}) = \tfrac{1}{2}\dot{\boldsymbol{\omega}}\cdot\mathbf{H} + \tfrac{1}{2}\boldsymbol{\omega}\cdot\dot{\mathbf{H}}$$
$$= \tfrac{1}{2}\dot{\boldsymbol{\omega}}\cdot\mathbf{H} - \tfrac{1}{2}\boldsymbol{\omega}\cdot\dot{\mathbf{H}} + \boldsymbol{\omega}\cdot\dot{\mathbf{H}}$$

Judicious substitution of $\mathbb{I}\cdot\boldsymbol{\omega}$ for $\mathbf{H}$ [as in Eq. (8·53a)] then yields for the integrand of interest (noting as in Prob. 8·10 that the symmetry of $\mathbb{I}$ makes $\mathbb{I}\cdot\boldsymbol{\omega} = \boldsymbol{\omega}\cdot\mathbb{I}$)

$$\boldsymbol{\omega}\cdot\dot{\mathbf{H}} = \frac{d}{dt}(\tfrac{1}{2}\boldsymbol{\omega}\cdot\mathbb{I}\cdot\boldsymbol{\omega}) - \tfrac{1}{2}\dot{\boldsymbol{\omega}}\cdot\mathbb{I}\cdot\boldsymbol{\omega} + \tfrac{1}{2}(\boldsymbol{\omega}\cdot\dot{\mathbb{I}}\cdot\boldsymbol{\omega} + \boldsymbol{\omega}\cdot\mathbb{I}\cdot\dot{\boldsymbol{\omega}})$$
$$= \frac{d}{dt}(\tfrac{1}{2}\boldsymbol{\omega}\cdot\mathbb{I}\cdot\boldsymbol{\omega}) - \tfrac{1}{2}\boldsymbol{\omega}\cdot\dot{\mathbb{I}}\cdot\boldsymbol{\omega} \tag{8·92}$$

The beauty of this expression lies in the fact that the final term is zero. To prove this remarkable identity, we must know how to expand the inertial time derivative $\dot{\mathbb{I}}$. Since a dyadic is just an ordered collection of vectors with certain operational rules, it should not be surprising that we can replace a dyadic time derivative in one frame by the time derivative in another frame plus some combination of cross products with angular-velocity vectors, just as we did in Eq. (3·100) for vector derivatives. In fact, since any dyadic is a linear combination of dyads, and any dyad can be written in the form $\mathbb{D} = \mathbf{uv}$ for some vectors $\mathbf{u}$ and $\mathbf{v}$, we can use Eq. (3·100) directly to obtain for any reference frames a and b

$$\frac{{}^{a}d}{dt}\mathbb{D} = \frac{{}^{a}d}{dt}(\mathbf{uv}) = \left(\frac{{}^{a}d}{dt}\mathbf{u}\right)\mathbf{v} + \mathbf{u}\frac{{}^{a}d}{dt}\mathbf{v}$$
$$= \left(\frac{{}^{b}d}{dt}\mathbf{u} + \boldsymbol{\omega}^{ba}\times\mathbf{u}\right)\mathbf{v} + \mathbf{u}\left(\frac{{}^{b}d}{dt}\mathbf{v} + \boldsymbol{\omega}^{ba}\times\mathbf{v}\right)$$
$$= \frac{{}^{b}d}{dt}\mathbf{uv} + \mathbf{u}\frac{{}^{b}d}{dt}\mathbf{v} + \boldsymbol{\omega}^{ba}\times\mathbf{uv} - \mathbf{uv}\times\boldsymbol{\omega}^{ba}$$
$$= \frac{{}^{b}d}{dt}\mathbb{D} + \boldsymbol{\omega}^{ba}\times\mathbb{D} - \mathbb{D}\times\boldsymbol{\omega}^{ba} \tag{8·93}$$

The term $\dot{\mathbb{I}}$ in Eq. (8·92) can be expanded as in (8·93) to find for reference frame b fixed in the rigid body,

$$\dot{\mathbb{I}} \triangleq \frac{{}^{i}d}{dt}\mathbb{I} = \frac{{}^{b}d}{dt}\mathbb{I} + \boldsymbol{\omega}^{bi}\times\mathbb{I} - \mathbb{I}\times\boldsymbol{\omega}^{bi}$$
$$= 0 + \boldsymbol{\omega}\times\mathbb{I} - \mathbb{I}\times\boldsymbol{\omega} \tag{8·94}$$

Thus the final term in Eq. (8·92) becomes

$$-\tfrac{1}{2}\boldsymbol{\omega} \cdot \dot{\mathbb{I}} \cdot \boldsymbol{\omega} = -\tfrac{1}{2}\boldsymbol{\omega} \cdot (\boldsymbol{\omega} \times \mathbb{I} - \mathbb{I} \times \boldsymbol{\omega}) \cdot \boldsymbol{\omega}$$

But $\boldsymbol{\omega} \cdot \boldsymbol{\omega} \times \mathbb{I} = 0$ because $\boldsymbol{\omega} \times \mathbb{I}$ is perpendicular to $\boldsymbol{\omega}$, and $\mathbb{I} \times \boldsymbol{\omega} \cdot \boldsymbol{\omega} = 0$ because $\mathbb{I} \times \boldsymbol{\omega}$ is perpendicular to $\boldsymbol{\omega}$, so that the final term in Eq. (8·92) is zero (as promised) and Eq. (8·91) becomes

$$\int_{t_1}^{t_2} \mathbf{M} \cdot \boldsymbol{\omega}\, dt = \int_{t_1}^{t_2} \frac{d}{dt} (\tfrac{1}{2}\boldsymbol{\omega} \cdot \mathbb{I} \cdot \boldsymbol{\omega})\, dt$$

$$= \tfrac{1}{2}\boldsymbol{\omega} \cdot \mathbb{I} \cdot \boldsymbol{\omega} \Big|_{t_1}^{t_2} \qquad (8{\cdot}95)$$

Because the term $\tfrac{1}{2}\boldsymbol{\omega} \cdot \mathbb{I} \cdot \boldsymbol{\omega}$ is either all or part of the kinetic energy, depending on the reference point for $\mathbb{I}$ [see Eqs. (8·89) and (8·85)], and in either case this term is the part of the kinetic energy associated with rotation, the result in Eq. (8·95) is sometimes called the *rotational work-energy equation*. Its utility is demonstrated in the following example.

8·5·8 Work calculation for grinding-wheel spin-up

Figure 8·8*a* shows a grinding wheel driven ultimately by a motor with torque capacity $\tau\hat{\mathbf{u}}_n$. The wheel has all of its mass $\mathfrak{M}$ on its rim, which has radius r; the wheel rolls without slip on a circular track of radius R, being driven by a massless horizontal rigid shaft of length R caused to rotate by the torque motor shaft.

We can use Eq. (8·95) to calculate the work required of the torque motor in order to bring the angular speed of the horizontal shaft from zero to some speed Ω. This calculation requires the recognition that the friction force applied at the contact point of the wheel rim is a "nonworking constraint force" (see §4·3·7); this is evident from the fact that the inertial velocity of the contact point of the wheel is zero. Thus the only work done on the wheel is that due to the motor torque, and this work is [by Eqs. (8·95) and (8·89)]

$$W = \tfrac{1}{2}\boldsymbol{\omega} \cdot \mathbb{I}^o \cdot \boldsymbol{\omega} = T$$

where $\mathbb{I}^o$ is the inertia dyadic relative to the inertially fixed point o. By the chain rule for angular velocities [see Eq. (3·102)]

$$\boldsymbol{\omega} = \dot{\theta}\hat{\mathbf{u}}_r + \Omega\hat{\mathbf{u}}_n$$

The requirement that the wheel rolls without slip implies the constraint

$$-r\dot{\theta} = R\Omega$$

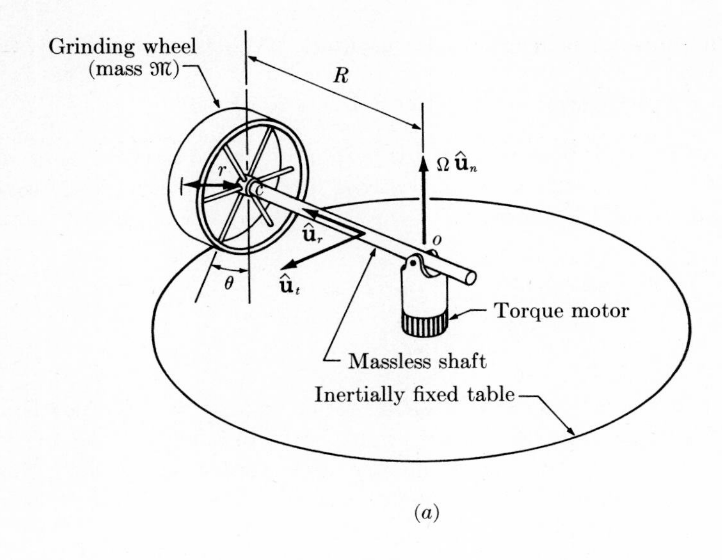
Grinding wheel
(mass $\mathfrak{M}$)
R
$\Omega\,\hat{\mathbf{u}}_n$
r
c
$\hat{\mathbf{u}}_r$
o
θ
$\hat{\mathbf{u}}_t$
Torque motor
Massless shaft
Inertially fixed table

(a)

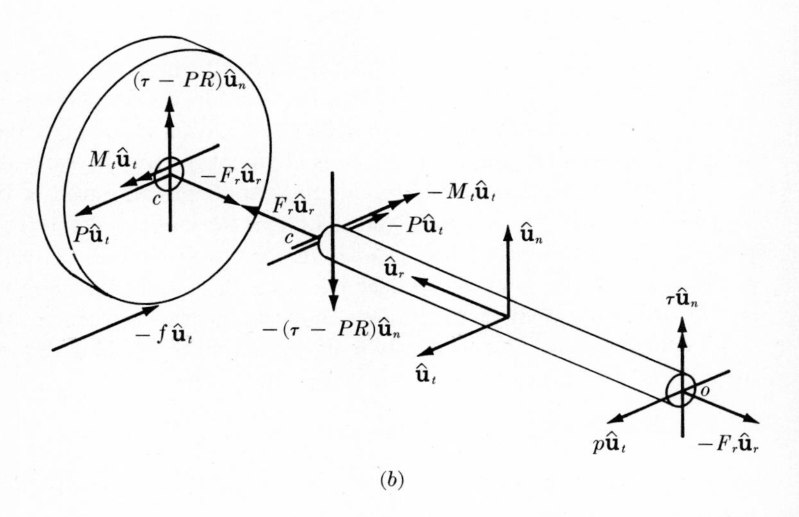
$(\tau - PR)\hat{\mathbf{u}}_n$
$M_t\hat{\mathbf{u}}_t$
$-F_r\hat{\mathbf{u}}_r$
c
$P\hat{\mathbf{u}}_t$
$F_r\hat{\mathbf{u}}_r$
$-M_t\hat{\mathbf{u}}_t$
$-P\hat{\mathbf{u}}_t$
c
$\hat{\mathbf{u}}_n$
$\hat{\mathbf{u}}_r$
$\tau\hat{\mathbf{u}}_n$
$-f\,\hat{\mathbf{u}}_t$
$-(\tau - PR)\hat{\mathbf{u}}_n$
$\hat{\mathbf{u}}_t$
o
$p\hat{\mathbf{u}}_t$
$-F_r\hat{\mathbf{u}}_r$

(b)

FIGURE 8·8

Since, relative to the inertially fixed and body-fixed support point o, $\mathbb{I}^o$ is given by

$$\mathbb{I}^o = \mathfrak{M}r^2\hat{\mathbf{u}}_r\hat{\mathbf{u}}_r + (\tfrac{1}{2}\mathfrak{M}r^2 + \mathfrak{M}R^2)\hat{\mathbf{u}}_n\hat{\mathbf{u}}_n + (\tfrac{1}{2}\mathfrak{M}r^2 + \mathfrak{M}R^2)\hat{\mathbf{u}}_t\hat{\mathbf{u}}_t$$

we have

$$W = \tfrac{1}{2}\left(\dot{\theta}\hat{\mathbf{u}}_r - \frac{r\dot{\theta}}{R}\hat{\mathbf{u}}_n\right) \cdot [\mathfrak{M}r^2\hat{\mathbf{u}}_r\hat{\mathbf{u}}_r + (\tfrac{1}{2}\mathfrak{M}r^2 + \mathfrak{M}R^2)(\hat{\mathbf{u}}_n\hat{\mathbf{u}}_n + \hat{\mathbf{u}}_t\hat{\mathbf{u}}_t)] \cdot \left(\dot{\theta}\hat{\mathbf{u}}_r - \frac{r\dot{\theta}}{R}\hat{\mathbf{u}}_n\right)$$

$$= \tfrac{1}{2}\mathfrak{M}r^2\dot{\theta}^2 + \tfrac{1}{2}\mathfrak{M}\dot{\theta}^2\frac{r^2}{R^2}(\tfrac{1}{2}r^2 + R^2) = \mathfrak{M}r^2\dot{\theta}^2\left(1 + \frac{r^2}{4R^2}\right)$$

$$= \mathfrak{M}R^2\Omega^2\left(1 + \frac{r^2}{4R^2}\right)$$

With this information, we can begin to make a selection of motor characteristics required to accomplish the spin-up task.

Note that this problem can be analyzed without recourse to a work-energy equation; the direct application of $\mathbf{F} = \mathfrak{M}\mathbf{A}$ and $\mathbf{M}^c = \dot{\mathbf{H}}^c$ would also suffice. These vectorial methods would, however, involve the calculation of the constraint force at the contact point of the wheel, and this labor is avoided with the use of Eq. (8·95).

8·5·9 Lagrange's equations

The scalar equations of motion known as Lagrange's equations were derived in Chap. 4 for a single particle [see Eq. (4·80)], and noted in Chap. 6 to be applicable also to a system of particles. Now we shall note the extension of scope to a single rigid body, and finally to a system of particles and rigid bodies.

For a single rigid body, a complete set of equations of motion is available from the combination

$$\mathbf{F} = \mathfrak{M}\mathbf{A} \qquad \text{and} \qquad \mathbf{M}^c = \dot{\mathbf{H}}^c$$

where the reference point for $\mathbf{M}^c$ and $\dot{\mathbf{H}}^c$ is the mass center c, and $\mathbf{A}$ is the inertial acceleration of c.

The concept of the generalized coordinate remains applicable to rigid bodies: The coordinates $q_1, \ldots, q_n$ qualify as generalized coordinates if they constitute a set of variables completely characterizing the configuration at any given time. A single rigid body can therefore have as many as six independent generalized coordinates (three in translation and three in rotation).

If $\mathbf{R}_c$ is the position vector to the body mass center (so that $\mathbf{A} \triangleq \ddot{\mathbf{R}}_c$), and $\boldsymbol{\omega}$ is the inertial angular velocity of the body, we can certainly write n independent scalar equations of motion by combining $\mathbf{F} = \mathfrak{M}\mathbf{A}$ and $\mathbf{M}^c = \dot{\mathbf{H}}^c$ in the form

$$\mathbf{F} \cdot \frac{\partial \dot{\mathbf{R}}_c}{\partial \dot{q}_\alpha} + \mathbf{M}^c \cdot \frac{\partial \boldsymbol{\omega}}{\partial \dot{q}_\alpha} = \mathfrak{M}\ddot{\mathbf{R}}_c \cdot \frac{\partial \dot{\mathbf{R}}_c}{\partial \dot{q}_\alpha} + \dot{\mathbf{H}}^c \cdot \frac{\partial \boldsymbol{\omega}}{\partial \dot{q}_\alpha} \qquad \alpha = 1, \ldots, n \tag{8·96}$$

We could stop at this point, satisfied that we have obtained the necessary number of scalar equations and in the process have avoided the introduction of any extraneous unknowns by eliminating nonworking constraint forces [see the definition following Eq. (4·85)]. We shall see, however, that Eq. (8·96) can alternatively be written in the more convenient form

$$Q_\alpha = \frac{d}{dt}\left(\frac{\partial T}{\partial \dot{q}_\alpha}\right) - \frac{\partial T}{\partial q_\alpha} \qquad \alpha = 1, \ldots, n \tag{8·97}$$

where T is the kinetic energy of the rigid body [see Eqs. (8·84) to (8·89)], and Q_α is the generalized force defined in Eq. (6·45) as

$$Q_\alpha \triangleq \sum_{j=1}^{N} \mathbf{F}_j \cdot \frac{\partial \dot{\mathbf{R}}_j}{\partial \dot{q}_\alpha} \qquad \alpha = 1, \ldots, n \tag{8·98}$$

Here $\mathbf{F}_j$ is one of N concentrated forces applied to the rigid body, and $\mathbf{R}_j$ is the inertially referenced position vector of the point of the body to which $\mathbf{F}_j$ is applied. If the body is subjected to distributed forces rather than concentrated forces, application of Eq. (8·98) must be preceded by the introduction of an equivalent system of concentrated forces (see Sec. 7·1).

Equation (8·97) was written previously as Eq. (6·43), and identified as Lagrange's equation; in that context it was shown to apply to a collection of N particles, for any finite number N. To establish its validity in application to a rigid body, it is necessary only to replace the previous sums over N particles by integrals over the body; all arguments developed in Chap. 4 for a single particle and in Chap. 6 for a set of particles remain intact, and will not be repeated here. Extension to a system of rigid bodies, or a system of particles and rigid bodies, is also straightforward.

To establish the equivalence of the Eqs. (8·96) and (8·97), it is sufficient to establish that their left sides are equal, that is, that

$$Q_\alpha = \mathbf{F} \cdot \frac{\partial \dot{\mathbf{R}}_c}{\partial \dot{q}_\alpha} + \mathbf{M}^c \cdot \frac{\partial \boldsymbol{\omega}}{\partial \dot{q}_\alpha} \qquad \alpha = 1, \ldots, n \tag{8·99}$$

Proof of Eq. (8·99) requires only the expansion of Q_α as defined in Eq. (8·98), using the symbols of Fig. 8·9. Thus

$$Q_\alpha \triangleq \sum_{j=1}^{N} \mathbf{F}_j \cdot \frac{\partial \dot{\mathbf{R}}_j}{\partial \dot{q}_\alpha} = \sum_{j=1}^{N} \mathbf{F}_j \cdot \frac{\partial}{\partial \dot{q}_\alpha} (\dot{\mathbf{R}}_c + \dot{\boldsymbol{\rho}}_j)$$

$$= \sum_{j=1}^{N} \mathbf{F}_j \cdot \frac{\partial \dot{\mathbf{R}}_c}{\partial \dot{q}_\alpha} + \sum_{j=1}^{N} \mathbf{F}_j \cdot \frac{\partial}{\partial \dot{q}_\alpha} (\boldsymbol{\omega} \times \boldsymbol{\rho}_j)$$

But

$$\mathbf{F}_j \cdot \frac{\partial}{\partial \dot{q}_\alpha} (\boldsymbol{\omega} \times \boldsymbol{\rho}_j) = \mathbf{F}_j \cdot \left(\frac{\partial \boldsymbol{\omega}}{\partial \dot{q}_\alpha} \times \boldsymbol{\rho}_j \right)$$

$$= -\mathbf{F}_j \cdot \left(\boldsymbol{\rho}_j \times \frac{\partial \boldsymbol{\omega}}{\partial \dot{q}_\alpha} \right) = -(\mathbf{F}_j \times \boldsymbol{\rho}_j) \cdot \frac{\partial \boldsymbol{\omega}}{\partial \dot{q}_\alpha}$$

$$= (\boldsymbol{\rho}_j \times \mathbf{F}_j) \cdot \frac{\partial \boldsymbol{\omega}}{\partial \dot{q}_\alpha}$$

so that

$$Q_\alpha = \sum_{j=1}^{N} \mathbf{F}_j \cdot \frac{\partial \dot{\mathbf{R}}_c}{\partial \dot{q}_\alpha} + \sum_{j=1}^{N} (\boldsymbol{\rho}_j \times \mathbf{F}_j) \cdot \frac{\partial \boldsymbol{\omega}}{\partial \dot{q}_\alpha}$$

This expression combines with the definitions

$$\mathbf{F} \triangleq \sum_{j=1}^{N} \mathbf{F}_j \qquad \mathbf{M}^c \triangleq \sum_{j=1}^{N} \boldsymbol{\rho}_j \times \mathbf{F}_j \tag{8·100}$$

to establish the validity of Eq. (8·99), and hence the equivalence of Eqs. (8·96) and (8·97).

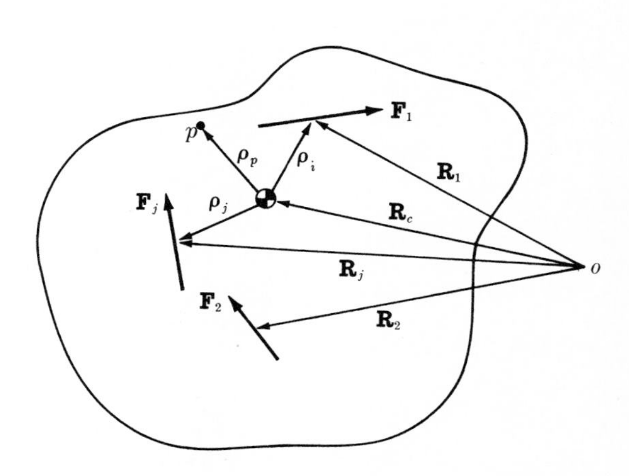

FIGURE 8·9

Moreover, the restriction to reference point c in Eq. (8·99) can be seen to be unnecessary. If we choose *any* point p fixed in the rigid body and note the relationships

$$\mathbf{R}_p = \mathbf{R}_c + \boldsymbol{\rho}_p$$

and

$$\mathbf{M}^p = \sum_{j=1}^{N} (-\boldsymbol{\rho}_p + \boldsymbol{\rho}_j) \times \mathbf{F}_j = -\boldsymbol{\rho}_p \times \mathbf{F} + \mathbf{M}^c$$

we deduce from Eq. (8·99) that

$$Q_\alpha = \mathbf{F} \cdot \frac{\partial \dot{\mathbf{R}}_p}{\partial \dot{q}_\alpha} + \mathbf{M}^p \cdot \frac{\partial \boldsymbol{\omega}}{\partial \dot{q}_\alpha} \tag{8·101}$$

If we are dealing with a collection of ν rigid bodies, then we can sum terms such as those in Eq. (8·101) directly, to obtain

$$Q_\alpha = \sum_{k=1}^{\nu} \mathbf{F}^k \cdot \frac{\partial \dot{\mathbf{R}}_{p_k}}{\partial \dot{q}_\alpha} + \mathbf{M}^{p_k} \cdot \frac{\partial \boldsymbol{\omega}^k}{\partial \dot{q}_\alpha} \tag{8·102}$$

where $\mathbf{F}^k$ = resultant of forces applied to kth body
p_k = point fixed in kth body
$\boldsymbol{\omega}^k$ = inertial angular velocity of kth body
$\mathbf{M}^{p_k}$ = resultant of moments about point p_k of forces applied to kth body

As in Chap. 6, when the generalized force Q_α found in Eq. (8·99) is related to some potential energy function V by $Q_\alpha = -\partial V/\partial q_\alpha$, Lagrange's equations can be written in the form of Eq. (6·46).

8·5·10 Application of Lagrange's equations to the grinding wheel

The grinding wheel examined in §8·5·8 and illustrated in Fig. 8·8 provides an example of a rigid body forced by a torque motor and mechanical constraints to move in a manner described by the single generalized coordinate θ.

In §8·5·9 the angular velocity $\boldsymbol{\omega}$ and kinetic energy $T = \frac{1}{2}\boldsymbol{\omega} \cdot \mathbf{I}^o \cdot \boldsymbol{\omega}$ are recorded; Eq. (8·97) thus becomes

$$\begin{aligned} Q_\alpha &= \frac{d}{dt}\frac{\partial}{\partial \dot{\theta}}\left[\mathfrak{M}r^2\dot{\theta}^2\left(1 + \frac{r^2}{4R^2}\right)\right] - \frac{\partial}{\partial \theta}\left[\mathfrak{M}r^2\dot{\theta}^2\left(1 + \frac{r^2}{4R^2}\right)\right] \\ &= 2\mathfrak{M}r^2\ddot{\theta}\left(1 + \frac{r^2}{4R^2}\right) \end{aligned} \tag{a}$$

The generalized force Q_α must be calculated from Eq. (8·99), which requires $\mathbf{F}$ and $\mathbf{M}^c$. Figure 8·8b shows free-body diagrams of both the massless shaft and the massive wheel; if the resultant force component applied to the wheel by the rod in the direction of $\hat{\mathbf{u}}_t$ is $P\hat{\mathbf{u}}_t$, then equilibrium of the massless shaft requires the application at end c of a force system equivalent to a force $-P\hat{\mathbf{u}}_t$, and a couple having torque $-(\tau - PR)\hat{\mathbf{u}}_n$, as well as some unspecified force $F_r\hat{\mathbf{u}}_r$, and unspecified couple having torque $-M_t\hat{\mathbf{u}}_t$. Newton's third law (the action-reaction principle) implies the application of equal and opposite forces by the shaft to the wheel at c. In addition, the unknown friction force $-f\hat{\mathbf{u}}_t$ is applied to the wheel by the floor, with sufficient magnitude to enforce the no-slip constraint. Thus the resultant force on the wheel is

$$\mathbf{F} = P\hat{\mathbf{u}}_t - f\hat{\mathbf{u}}_t + F_r\hat{\mathbf{u}}_r \tag{b}$$

and the resultant moment about c is

$$\mathbf{M}^c = (\tau - PR)\hat{\mathbf{u}}_n - rf\hat{\mathbf{u}}_r + M_t\hat{\mathbf{u}}_t \tag{c}$$

In combination with the kinematical quantities

$$\dot{\mathbf{R}}_c = R\Omega\hat{\mathbf{u}}_t = -r\dot{\theta}\hat{\mathbf{u}}_t \tag{d}$$

and

$$\boldsymbol{\omega} = \Omega\hat{\mathbf{u}}_n + \dot{\theta}\hat{\mathbf{u}}_r = -\frac{r\dot{\theta}}{R}\hat{\mathbf{u}}_n + \dot{\theta}\hat{\mathbf{u}}_r \tag{e}$$

these expressions combine in Eq. (8·99) to produce

$$\begin{aligned} Q_\alpha &= -(P\hat{\mathbf{u}}_t - f\hat{\mathbf{u}}_t + F_r\hat{\mathbf{u}}_r) \cdot r\hat{\mathbf{u}}_t + [(\tau - PR)\hat{\mathbf{u}}_n - rf\hat{\mathbf{u}}_r + M_t\hat{\mathbf{u}}_t] \\ &\qquad \cdot \left(-\frac{r}{R}\hat{\mathbf{u}}_n + \hat{\mathbf{u}}_r\right) \\ &= -Pr + fr - \frac{\tau r}{R} + Pr - rf = -\frac{\tau r}{R} \end{aligned} \tag{f}$$

Substituting Eq. (f) into Eq. (a) yields

$$-\frac{\tau r}{R} = 2\mathfrak{M}r^2\ddot{\theta}\left(1 + \frac{r^2}{4R^2}\right)$$

or

$$-\tau = 2\mathfrak{M}Rr\ddot{\theta}\left(1 + \frac{r^2}{4R^2}\right) \tag{g}$$

The cancellation of the unknown constraint forces and constraint torques (as represented by the scalars P, f, F_r, and M_t) from Q is noteworthy; this is the special feature of Lagrange's equations that makes this approach to dynamic analysis particularly useful. With experience, we can identify the nonworking constraint forces in advance and simply ignore them in calculating generalized forces.

8·6 PERSPECTIVE

This chapter contains a substantial amount of fundamental material on the dynamics of rigid bodies, and it cannot all be assimilated rapidly. The most important equations, however, are $\mathbf{F} = \mathfrak{M}\mathbf{A}^{ci}$ [Eqs. (8·6)] and (for restricted reference points) $\mathbf{M} = \dot{\mathbf{H}}$ [Eqs. (8·15) and (8·16)]; all else in this chapter is either derived from these equations or concerned with the representation of the angular momentum **H**. In the latter category are Eqs. (8·22) to (8·24), which explore the role of the moments and products of inertia in the formation of **H**. The properties of the inertia matrix itself are important; in particular the reference-point transfer theorem [Eq. (8·28)] and the vector-basis transformation [Eqs. (8·35)] deserve special attention, as does the concept of principal axes as presented in §8·4·7 and §8·4·8.

Among the consequences of $\mathbf{M} = \dot{\mathbf{H}}$, the most significant are Euler's equations [Eqs. (8·57)], and its more general matrix counterpart [Eq. (8·70)]. The equations of rotation about a fixed axis [Eqs. (8·74)] also have a special importance. Lagrange's equations [Eq. (8·97)] are particularly useful in application to rigid bodies (or systems of rigid bodies) that are constrained in their motions.

PROBLEMS

8·1 A cart of mass M contains a particle of mass m that rotates with constant angular speed ω on the end of a massless rod of length L (see Fig. P8·1). The cart rolls on smooth, straight, horizontal tracks and is attached to a vertical wall by means of a linear spring with spring constant k. The spring is unstrained when the distance x from the wall to the mass center of the cart equals x_0.

a. Find the acceleration of the mass center of this *system* (cart plus particle) as a function of x and time.

b. Find the equation of motion of the system (a differential equation containing x, its derivatives, and time).

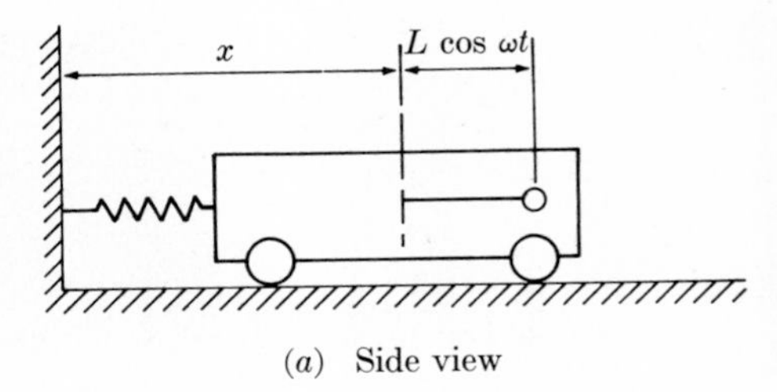

(a) Side view

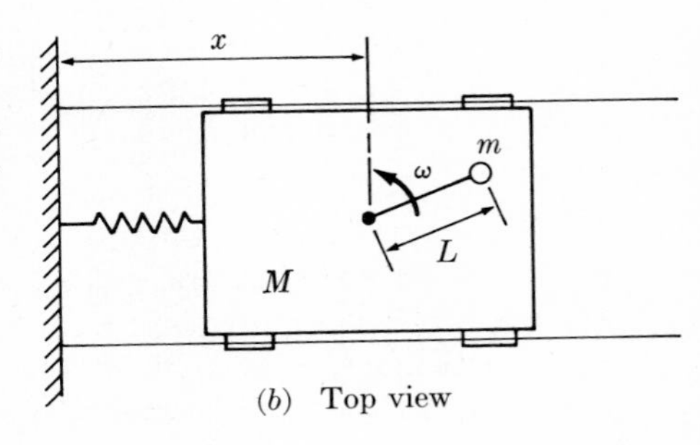

(b) Top view

FIGURE P8·1

8·2 One end of a uniform rod of mass m_1 is attached at point q to a hinge with cartesian coordinates x_1 and x_2 in the plane established by $\hat{\mathbf{u}}_1$ and $\hat{\mathbf{u}}_2$ (see Fig. P8·2). The angle θ is given as Ωt for constant Ω. Point q is fixed also on a block of mass m_2 that can move only vertically with respect to a cart of mass m_3, with the translation of the block resisted by a linear spring with spring constant k. The cart of mass m_3 is free to move in direction $\hat{\mathbf{u}}_1$ on frictionless rollers. Find scalar second-order differential equations in x_1 and x_2.

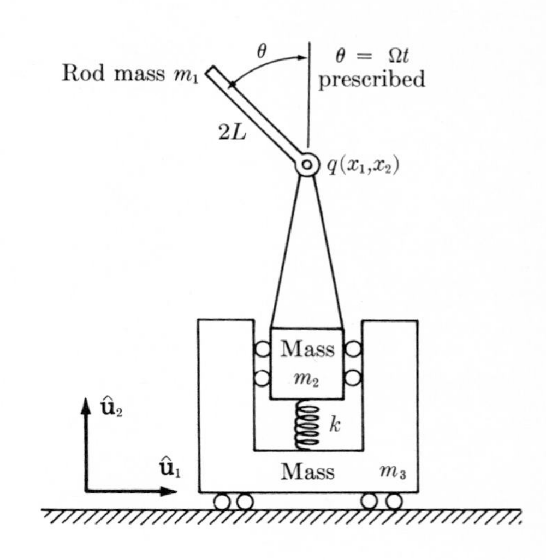

FIGURE P8·2

8·3 The dumbbell shown in Fig. P8·3 consists of two small rigid spheres each of mass m connected by a massless rigid rod of length $2L$. The mass center c moves in the plane of inertially fixed unit vectors $\hat{\mathbf{i}}_1$ and $\hat{\mathbf{i}}_2$, with location relative to the inertially fixed point o given by $\mathbf{R}_c = x_1\hat{\mathbf{i}}_1 + x_2\hat{\mathbf{i}}_2$, and angular velocity given by $\dot{\varphi}\hat{\mathbf{i}}_3$ (planar rotation). In terms of any of the symbols x_1, x_2, $\dot{x}_1$, $\dot{x}_2$, φ, $\dot{\varphi}$, L, m, $\hat{\mathbf{i}}_1$, $\hat{\mathbf{i}}_2$, and $\hat{\mathbf{i}}_3$, express the following:

a. $\mathbf{H}^o$, the angular momentum for point o

b. $\mathbf{h}^o$, the moment of momentum for point o

c. $\mathbf{H}^c$, the angular momentum for point c

d. $\mathbf{h}^c$, the moment of momentum for point c

e. $\mathbf{H}^q$, the angular momentum for point q

f. $\mathbf{h}^q$, the moment of momentum for point q

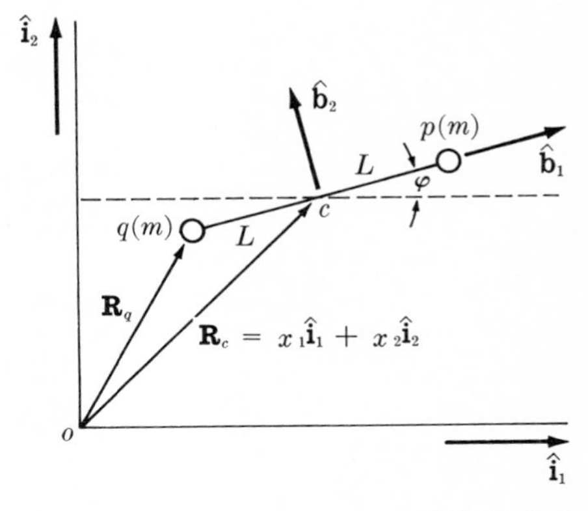

FIGURE P8·3

8·4 In making a soft landing on the moon, the three-legged *Surveyor* vehicle (of mass $\mathfrak{M}$) shown in Fig. P8·4 must be expected to land on one footpad (point p), and point p then either (1) sticks to the lunar surface, (2) slides along it, or (3) bounces off. Write a complete set of vector equations for each of these three cases, making an explicit choice of reference point that minimizes the labors of calculation. Record only those vector equations needed to analyze the motion.

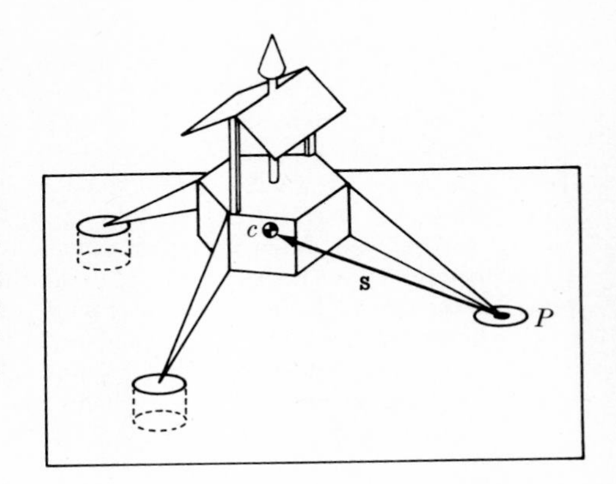

FIGURE P8·4

8·5 A rigid uniform rod of mass m and length L is supported by one end on a frictionless hinge at point q, and q oscillates vertically as prescribed by $y = a \sin \omega t$ (see Fig. P8·5). A uniform gravity force acts in direction $\hat{\mathbf{u}}_v$. Find the scalar second-order differential equation in the angle θ defining the rotation of the rod from the vertical.

a. Use $\mathbf{H}^c$, the angular momentum for mass center c
b. Use $\mathbf{H}^q$, the angular momentum for support point q

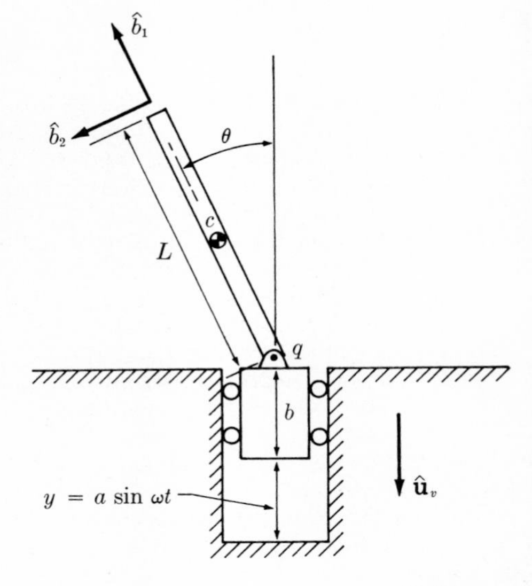

FIGURE P8·5

8·6 Repeat Prob. 8·5, this time requiring instead that point q move in a circle of constant radius a in a manner prescribed by the given (but unspecified) function $\varphi(t)$, as in Fig. P8·6.

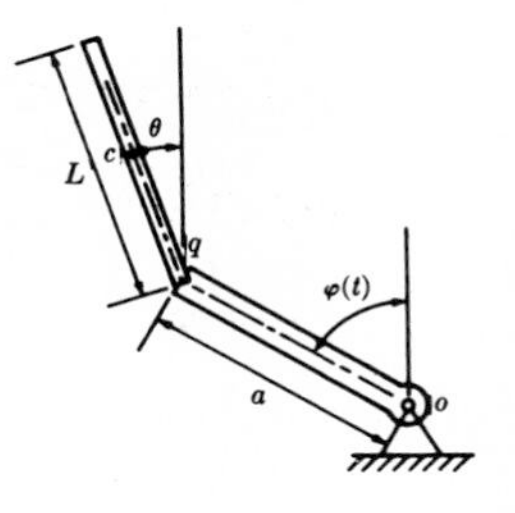

FIGURE P8·6

8·7 Solve Prob. 6·34, modified by the assignment to the teeter-totter board of a uniform mass density of μ and cross-sectional area a, so that the total mass of the board is $2\mu ab$. "Treat the board as a line mass."

8·8 A thin, rigid rod r of mass $\mathfrak{M}$ and length L has one end welded to the center o of a circular disk d that rotates in inertial space with constant angular speed Ω (see Fig. P8·8). Let $\hat{\mathbf{d}}_1$, $\hat{\mathbf{d}}_2$, and $\hat{\mathbf{d}}_3$ be unit vectors fixed in the disk such that $\boldsymbol{\omega}^{di} = \Omega\hat{\mathbf{d}}_2$ and the axis of r lies in the $\hat{\mathbf{d}}_1\hat{\mathbf{d}}_2$ plane and makes a 45-degree angle with $\hat{\mathbf{d}}_1$ and $\hat{\mathbf{d}}_2$. Let $\hat{\mathbf{r}}_1$, $\hat{\mathbf{r}}_2$, and $\hat{\mathbf{r}}_3$ be unit vectors fixed in r such that $\hat{\mathbf{r}}_1$ parallels the rod axis and $\hat{\mathbf{r}}_3 \equiv \hat{\mathbf{d}}_3$.

a. Find the rod's inertia matrix for point o in vector basis $\hat{\mathbf{r}}_1$, $\hat{\mathbf{r}}_2$, and $\hat{\mathbf{r}}_3$. (Assume $I_{11} = 0$.)
b. Find the rod's inertia matrix for point o in vector basis $\hat{\mathbf{d}}_1$, $\hat{\mathbf{d}}_2$, and $\hat{\mathbf{d}}_3$.
c. Find $\mathbf{H}^o$, the angular momentum of the rod relative to point o in vector basis $\hat{\mathbf{r}}_1$, $\hat{\mathbf{r}}_2$, and $\hat{\mathbf{r}}_3$.
d. Find $\mathbf{H}^o$ in vector basis $\hat{\mathbf{d}}_1$, $\hat{\mathbf{d}}_2$, and $\hat{\mathbf{d}}_3$.

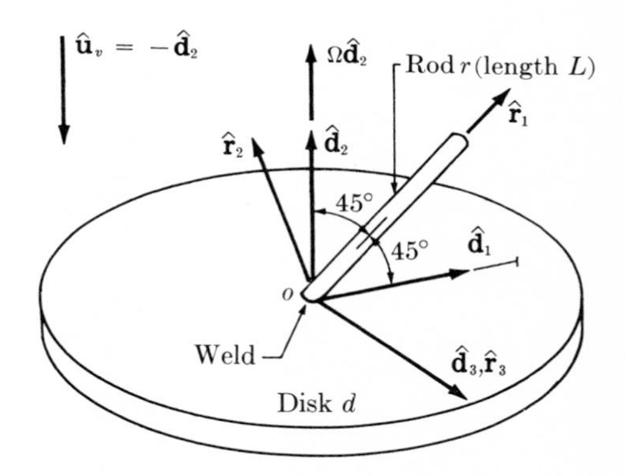

FIGURE P8·8

8·9 A thin rigid rectangular plate of mass $\mathfrak{M}$ and side dimensions a and b rotates about a vertical axis in a gravity field with angular velocity $\boldsymbol{\omega} = \omega\hat{\mathbf{e}}_v$ as shown in Fig. P8·9; the axis of rotation is in the $\hat{\mathbf{n}}_2\hat{\mathbf{n}}_3$ plane, where $\hat{\mathbf{n}}_1$, $\hat{\mathbf{n}}_2$, and $\hat{\mathbf{n}}_3$ are fixed in the plate with $\hat{\mathbf{n}}_3$ normal to the plate and making an angle α with the vertical (the plate is *not* horizontal).

a. Find $\mathbf{H}^c$, the angular momentum relative to the mass center c of the plate, in vector basis $\hat{\mathbf{n}}_1$, $\hat{\mathbf{n}}_2$, and $\hat{\mathbf{n}}_3$.
b. Find $\mathbf{H}^o$, the angular momentum relative to point o at the intersection of the plate and the vertical axis, in vector basis $\hat{\mathbf{n}}_1$, $\hat{\mathbf{n}}_2$, and $\hat{\mathbf{n}}_3$.

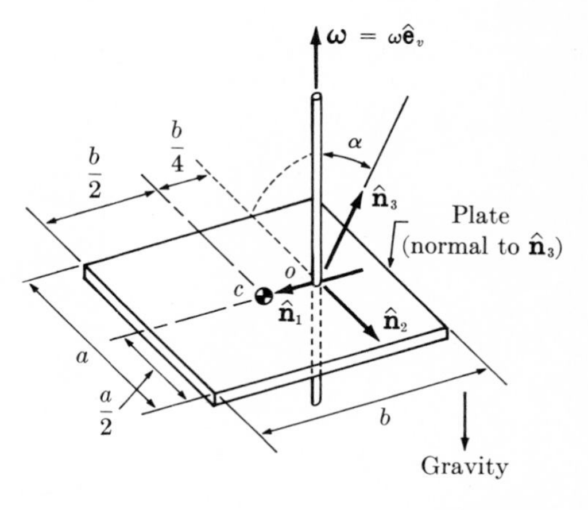

FIGURE P8·9

8·10 Show that if the dyadic $\mathbb{D}$ is symmetric (such that $\hat{\mathbf{u}}_1 \cdot \mathbb{D} \cdot \hat{\mathbf{u}}_2 = \hat{\mathbf{u}}_2 \cdot \mathbb{D} \cdot \hat{\mathbf{u}}_1$) then $\mathbf{V} \cdot \mathbb{D} = \mathbb{D} \cdot \mathbf{V}$ for any vector $\mathbf{V}$. Is it also true that $\mathbf{V} \times \mathbb{D} = -\mathbb{D} \times \mathbf{V}$?

8·11 Given the system of three orthogonally disposed massless rigid rods with particles of differing masses on their ends as shown in Fig. P8·11, with dimensions $d = 10$ in., $a = 5$ in., $m = 1$ lb sec²/in., find the following:

a. The vector $\mathbf{R}$ from the central point o to the mass center c

b. The inertia matrix of the system for point o and vector basis $\hat{\mathbf{b}}_1$, $\hat{\mathbf{b}}_2$, and $\hat{\mathbf{b}}_3$

c. The inertia matrix of the system for point c and vector basis $\hat{\mathbf{b}}_1$, $\hat{\mathbf{b}}_2$, and $\hat{\mathbf{b}}_3$

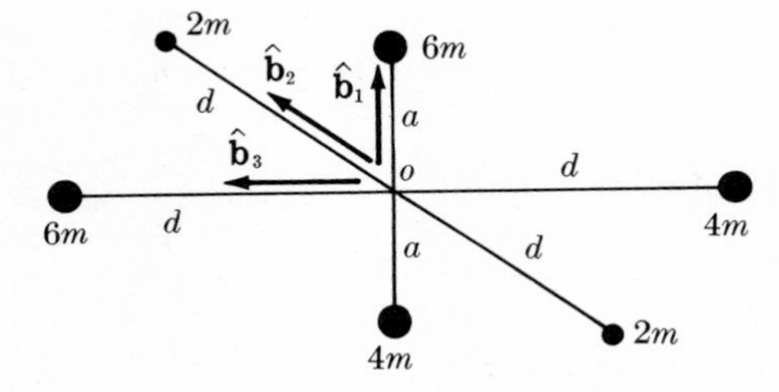

FIGURE P8·11

8·12 Eight small spheres are interconnected as shown in Fig. P8·12 by a lightweight rigid structure to form a cube, each side of which has length $2a$. The symbols $m_1, \ldots, m_8$ denote the masses of the spheres. Assume the connecting structure to be massless and the spheres to be particles, with the numerical values following: $a = 0.5$ ft; $m_1 = m_2 = 1$ lb sec²/ft; $m_3 = m_4 = 2m_1$; $m_5 = m_6 = 3m_1$; and $m_7 = m_8 = 4m_1$.

a. Find the system mass center in terms of a vector from mass point m_1 to c in vector basis $\hat{\mathbf{u}}_1$, $\hat{\mathbf{u}}_2$, and $\hat{\mathbf{u}}_3$.

b. Find the inertia matrix for axes parallel to $\hat{\mathbf{u}}_1$, $\hat{\mathbf{u}}_2$, and $\hat{\mathbf{u}}_3$ through the *geometrical* center c^* of the cube.

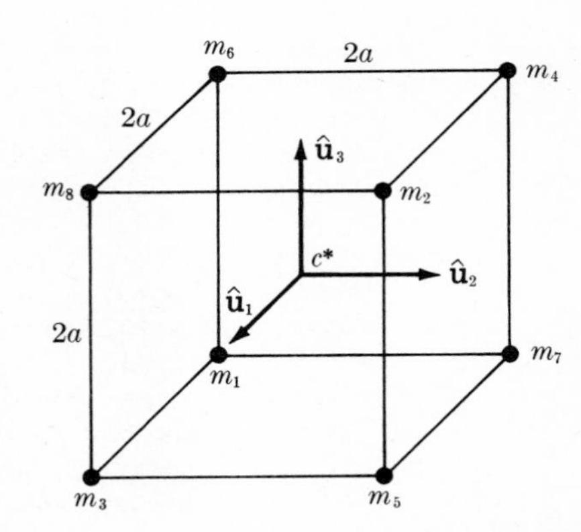

FIGURE P8·12

8·13 By means of the integral definitions in Eqs. (8·22), find the elements of the inertia matrix of the right circular cone of mass density μ in Fig. P8·13 for reference point o and vector basis $\hat{\mathbf{c}}_1$, $\hat{\mathbf{c}}_2$, and $\hat{\mathbf{c}}_3$.

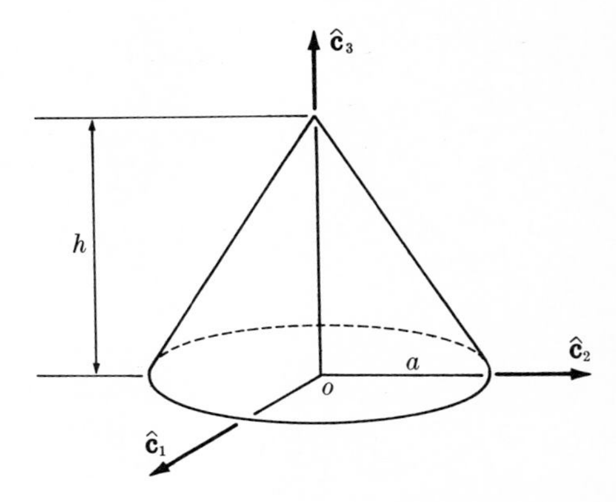

FIGURE P8·13

8·14 Given the homogeneous block of mass $\mathfrak{M}$ and side dimensions a, a, and b shown in Fig. P8·14, use Appendix C and the necessary transfer theorems to find the following:

a. The inertia matrix for mass center c and vector basis $\hat{\mathbf{u}}_1$, $\hat{\mathbf{u}}_2$, and $\hat{\mathbf{u}}_3$

b. The inertia matrix for point q and vector basis $\hat{\mathbf{u}}_1$, $\hat{\mathbf{u}}_2$, and $\hat{\mathbf{u}}_3$

c. The moment of inertia about a line passing through points c and s, where s is at the midpoint of one edge of the block, as shown

d. The moment of inertia about a line passing through the corners q and p (and hence through c)

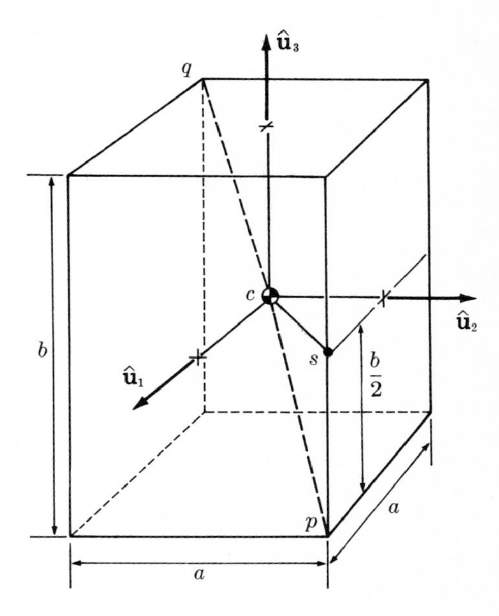

FIGURE P8·14

8·15 Assume that the thin rod of length L and mass m shown in Fig. P8·15*a* is idealized as a *line mass* (a distribution of mass along a *line*, which has no thickness).

a. Find its inertia matrix for midpoint c and vector basis $\hat{\mathbf{e}}_1$, $\hat{\mathbf{e}}_2$, and $\hat{\mathbf{e}}_3$.
b. Find its inertia matrix for point c and vector basis $\hat{\mathbf{u}}_1$, $\hat{\mathbf{u}}_2$, and $\hat{\mathbf{u}}_3$ shown.
c. Use the results of parts (a) and (b) to find the inertia matrix of the welded assembly of thin rods shown in Fig. P8·15*b*, for point o and vector basis $\hat{\mathbf{u}}_1$, $\hat{\mathbf{u}}_2$, and $\hat{\mathbf{u}}_3$.
d. Find the principal axes and principal axis moments of inertia for point o.

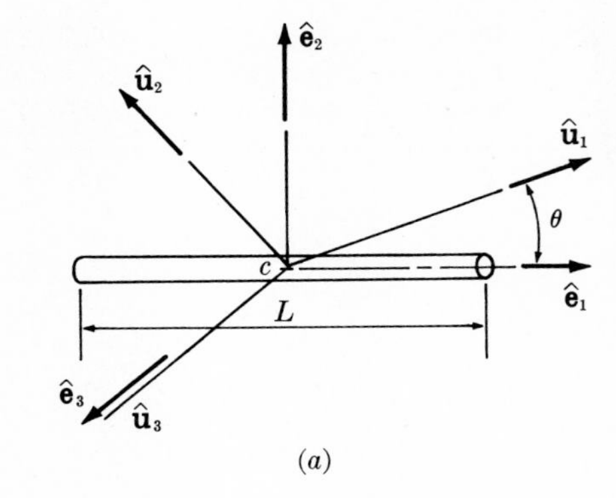

(*a*)

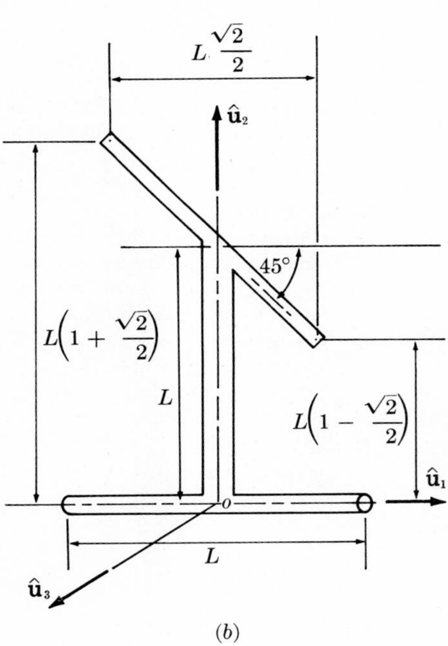

(*b*)

FIGURE P8·15

8·16 For the system shown in Fig. P8·11, using the solution of Prob. 8·11, find:

a. One principal axis for point c
b. An algebraic equation, expressed in determinant form, which *could be* solved for the principal axis moments of inertia for point c.

8·17 For the system shown in Fig. P8·12, using the results of Prob. 8·12, obtain explicitly in simplest form the algebraic equation that *could be* solved for the principal axis moments of inertia for c^*.

8·18 Find the principal axes and the principal axis moments of inertia of the tetrahedron in Fig. P8·18 for point o, using symmetries or eigenvector solutions.

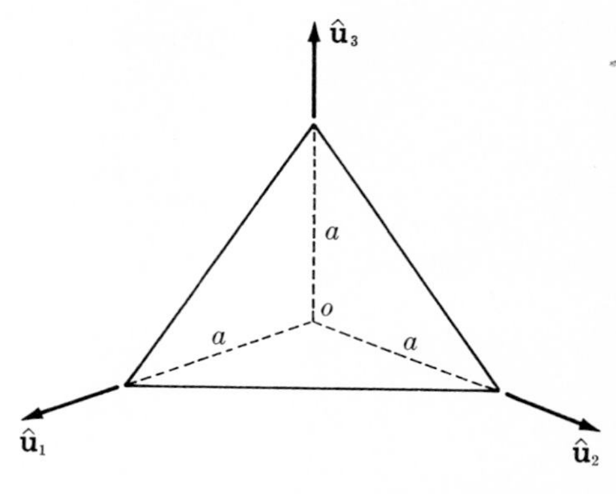

FIGURE P8·18

8·19 The vehicle shown in Fig. P8·19 has principal axes for point c defined by $\hat{\mathbf{e}}_1$, $\hat{\mathbf{e}}_2$, and $\hat{\mathbf{e}}_3$, with corresponding moments of inertia $I_{11} = 500$ ft lb sec², $I_{22} = 400$ ft lb sec², and $I_{33} = 200$ ft lb sec².

In flight, the right wing is sheared off as shown in the figure. The inertia properties *of this wing* for point c and vector basis $\hat{\mathbf{e}}_1$, $\hat{\mathbf{e}}_2$, and $\hat{\mathbf{e}}_3$ are as follows:

$$\begin{bmatrix} I_{11} & I_{12} & I_{13} \\ I_{21} & I_{22} & I_{23} \\ I_{31} & I_{32} & I_{33} \end{bmatrix} = \begin{bmatrix} 50 & 0 & 0 \\ 0 & 40 & -10 \\ 0 & -10 & 20 \end{bmatrix} \text{ ft lb sec}^2$$

a. Write in determinantal form an algebraic equation for the principal axis moments of inertia for point c of the *one-winged vehicle.*

b. Indicate *one* of the principal axes, and the corresponding principal axis moment of inertia for c.

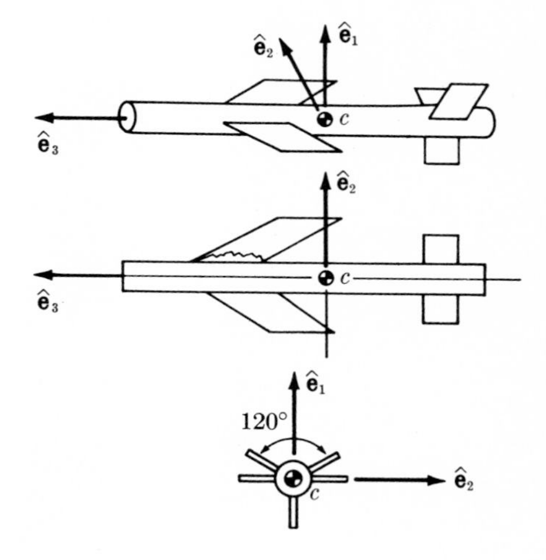

FIGURE P8·19

8·20 Find the smallest possible moment of inertia (about any point) of a body for which the inertia matrix for the mass center and some given vector basis is

$$[I] = \begin{bmatrix} 5 & 0 & 0 \\ 0 & 2 & \sqrt{3} \\ 0 & \sqrt{3} & 4 \end{bmatrix} \text{ ft lb sec}^2$$

8·21 A rigid cylinder of mass $\mathfrak{M}$ and radius R can roll without slipping on a horizontal plane. The center of the cylinder is attached to a vertical wall by means of a linear spring with spring constant k, and x is the extension of the spring (see Fig. P8·21).

a. Find the scalar equation of motion in x.

b. Find the period of oscillation.

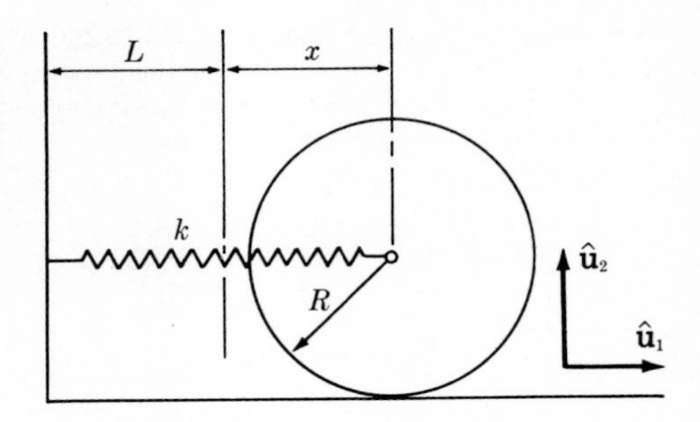

FIGURE P8·21

8·22 Two identical rods of mass m and length l are welded together at a 90° angle as shown in Fig. P8·22. The free end of one rod is then attached to the apex of a cone by means of a frictionless hinge that permits relative rotation only about a horizontal axis. The free end of the other rod rests on the surface of the cone, also assumed frictionless. Next, the entire assembly is made to rotate about a vertical axis with constant angular speed σ.

a. Find the inertia matrix of the two rods for the basis $\hat{\mathbf{b}}_1$, $\hat{\mathbf{b}}_2$, and $\hat{\mathbf{b}}_3$ and the point q as shown. ($\hat{\mathbf{b}}_3$ is perpendicular to $\hat{\mathbf{b}}_1$ and $\hat{\mathbf{b}}_2$ in a right-handed sense.)

b. Determine the directions of the principal axes of the two rods for point q, that is, find three unit vectors $\hat{\mathbf{e}}_1$, $\hat{\mathbf{e}}_2$, and $\hat{\mathbf{e}}_3$ directed along principal axes.

c. Find the angular momentum of the two rods for point q.

d. Find the normal force exerted by the conical surface on the rod at point A.

e. Find the force exerted by the hinge on the rod at point q.

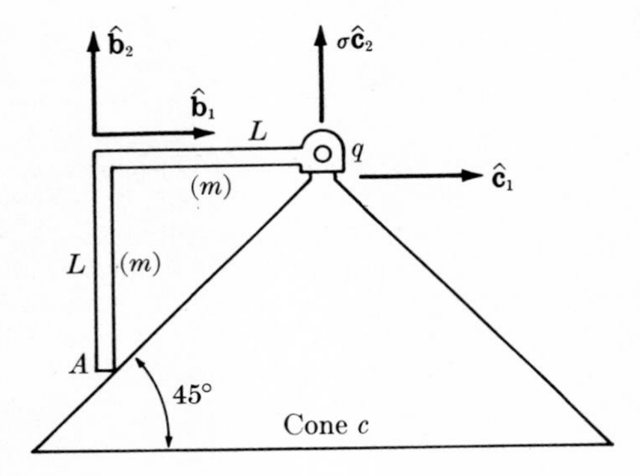

FIGURE P8·22

8·23 Use the results of the example in §8·4·9 to write out Euler's equations for the system shown in Fig. P8·23 to consist of a uniform rigid cylinder of mass $\mathfrak{M}$ supported on a frictionless ball-and-socket joint at point q and subjected to forces applied by jets at point p directed along $\hat{\mathbf{u}}_1$ and $-\hat{\mathbf{u}}_2$, and having magnitude or thrust F. Substitute $\mathfrak{M} = 0.5$ lb sec²/ft; $a = \sqrt{2}$ ft; $h = (a/2)\sqrt{3}$; and $F = \sqrt{3/2}$ lb.

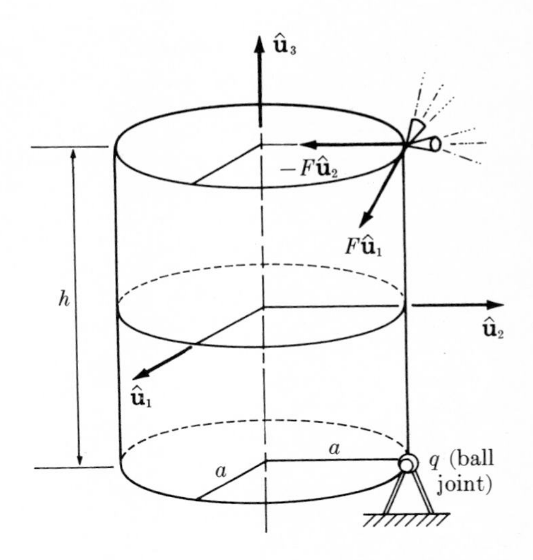

FIGURE P8·23

8·24 A homogenous rectangular parallelepiped is shown in Fig. P8·24 to be rotating on frictionless bearings at constant angular speed Ω about an axis passing through the mass center c and perpendicular to the edges of the body which have dimension b. Find the forces that must be applied at the support bearings A and B (equidistant from c).

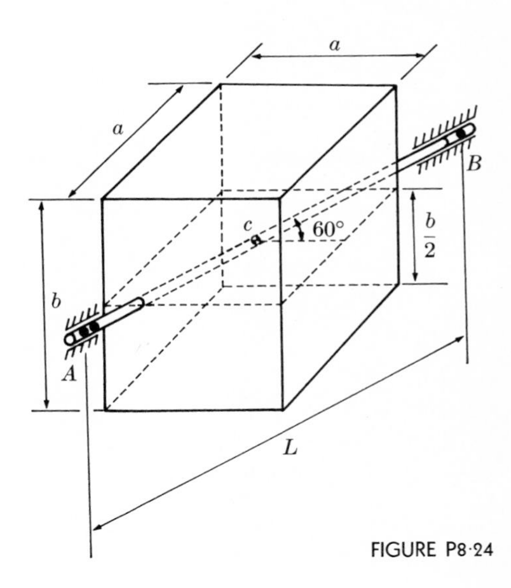

FIGURE P8·24

8·25 A "loop-o-plane" such as you might find among the fun rides in a carnival is shown in Fig. P8·25. Present concern is with the loads on the car bearings shown at A_1, B_1, A_2, and B_2. Orthogonal unit vectors $\hat{\mathbf{a}}_1$, $\hat{\mathbf{a}}_2$, and $\hat{\mathbf{a}}_3$ are fixed relative to the earth, assumed an inertial frame, with $\hat{\mathbf{a}}_3$ vertical. Orthogonal unit vectors $\hat{\mathbf{b}}_1$, $\hat{\mathbf{b}}_2$, and $\hat{\mathbf{b}}_3$ are fixed in the main frame of the machine, with $\hat{\mathbf{b}}_1 \equiv \hat{\mathbf{a}}_1$ and $\hat{\mathbf{b}}_3$ along the long member between the cars. This frame rotates relative to the earth at constant rate Ω. Orthogonal unit vectors $\hat{\mathbf{c}}_1{}^i$, $\hat{\mathbf{c}}_2{}^i$, and $\hat{\mathbf{c}}_3{}^i$ (for $i = 1, 2$) are fixed in car i, with $\hat{\mathbf{c}}_2{}^i \equiv \hat{\mathbf{b}}_2$ and $\hat{\mathbf{c}}_3{}^i$ rotated from $\hat{\mathbf{b}}_3$ an angle φ_i (see figure).

Let m^i be the mass of car i. Let O_i be the geometrical center of car i, lying on the point of intersection of line A_iB_i and line O_1O_2. Denote the elements of the inertia matrix of car i with respect to O_i as $I_{\alpha\beta}{}^i$, where $\alpha, \beta = 1, 2, 3$ correspond to $\hat{\mathbf{c}}_1{}^i$, $\hat{\mathbf{c}}_2{}^i$, and $\hat{\mathbf{c}}_3{}^i$. Let R be the distance from O to O_i, and d be the distance from O_i to A_i or B_i.

a. Car 1 is empty. It is perfectly balanced statically and dynamically (the mass center is at O_i and axes through O_i parallel to $\hat{\mathbf{c}}_1{}^i$, $\hat{\mathbf{c}}_2{}^i$, and $\hat{\mathbf{c}}_3{}^i$ are principal axes of inertia). The car is symmetric, so $I_{33}{}^i = I_{11}{}^i$. Find the forces applied to car 1 at points A_1 and B_1 at its bearings.

b. Car 2 is occupied, but it is otherwise identical to car 1. With the occupants, the car is unbalanced statically and dynamically. Discuss the procedure required to evaluate bearing forces at A_2 and B_2. *Do not* attempt the analysis, but indicate how you might instruct someone else to do so.

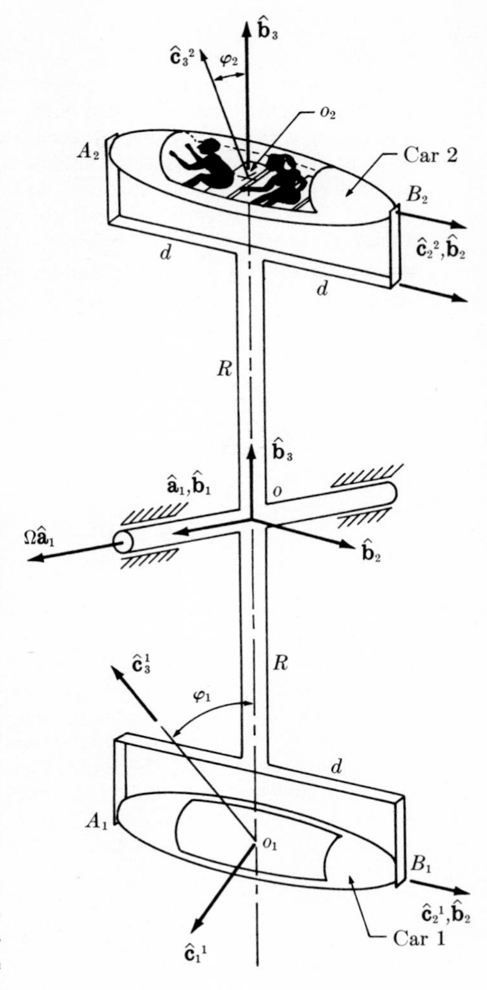

FIGURE P8·25

8·26 For the bar shown in Fig. P8·8 use the results of Prob. 8·8 to find the single force applied at o and the corresponding torque that together are equivalent to the system of forces applied to the bar at point o; the answer is a measure of the load on the welded joint. (Note that gravity forces are also applied to the bar.)

8·27 For the plate shown in Fig. P8·9, use the results of Prob. 8·9 to find the single force applied at o and the corresponding torque that are together equivalent to the system of forces applied to the plate at o; the answer is a measure of the load on the welded joint. (Note that gravity forces are also applied to the plate.)

8·28 Determine the kinetic energy of
a. The bar in Probs. 8·8 and 8·26
b. The plate in Probs. 8·9 and 8·27

8·29 Cubical blocks of ice of side dimension a are fed from a conveyor belt onto a frictionless helical chute (radius R, height h, and constant slope 45°) as in Fig. P8·29, and they accelerate under uniform gravitational attraction until they exit at the bottom.
a. If when the blocks leave the conveyor belt at the top, they are translating horizontally with speed v_0, what is their speed of translation and rotation on leaving the chute?
b. Two cubical blocks freeze together, and go down the chute as a single block. Does this change the answer to part (a)? If so, how?

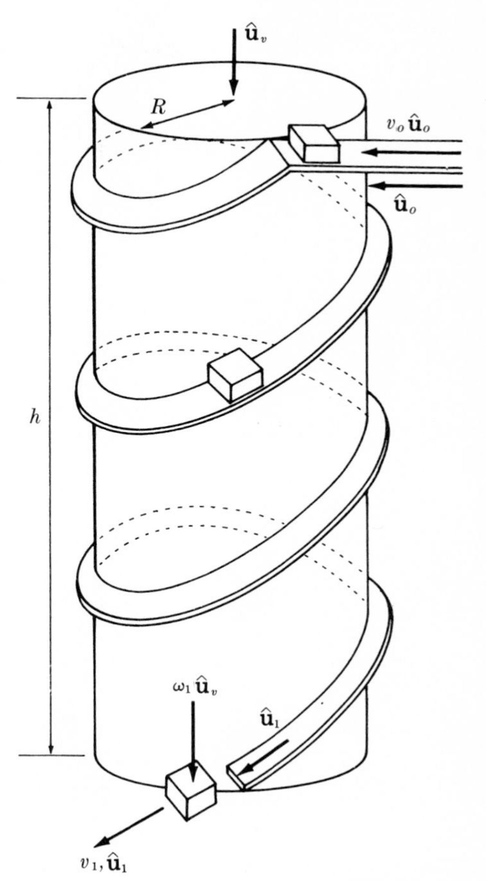

FIGURE P8·29

8·30 A thin, uniform circular disk rolls without slip on a straight horizontal track, maintaining its center c directly above the track, as in Fig. P8·30. Express the constraint equations relating the cartesian coordinates x_1, x_2, and x_3 of c and the angle θ between a vertical through c and a radial line through c and a point q on the wheel's perimeter. (Assume $0 < \theta < \infty$, to allow multiple revolutions.) Are these constraints holonomic or nonholonomic?

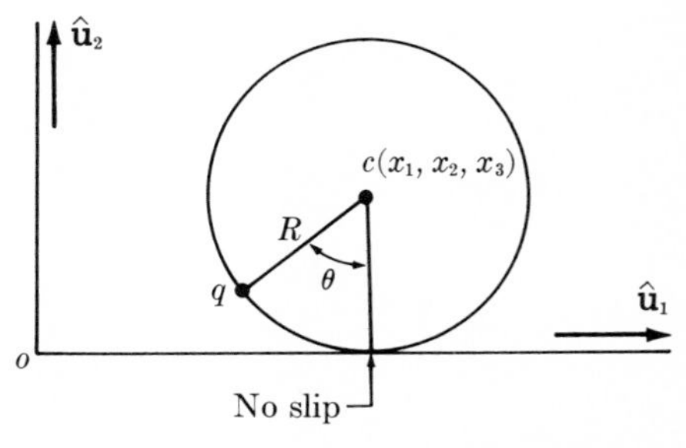

FIGURE P8·30

8·31 A uniform homogeneous disk of mass $\mathfrak{M}$ and radius R is suspended from a massless elastic wire that stretches sufficiently to allow part of the weight of the disk to rest on the rough, concave, circular track below (see Fig. P8·31). Find the natural frequency of oscillation of this modified pendulum,

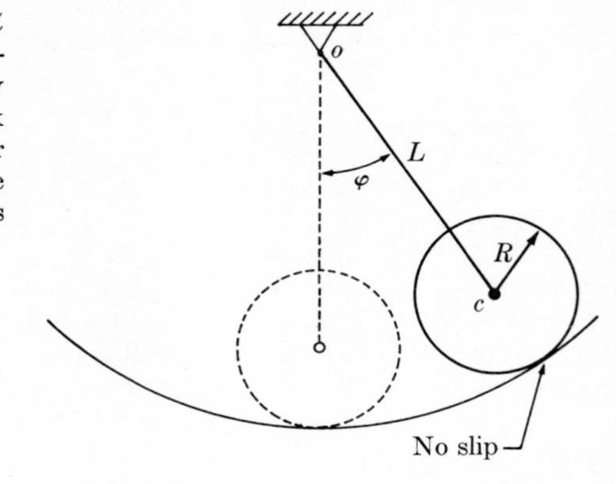

FIGURE P8·31

8·32 The mechanism shown in Fig. P8·32 consists of two uniform thin rigid bars B_1 and B_2 of length a and mass m, connected with frictionless hinges so that the plane in which B_2 rotates relative to B_1 is perpendicular to the plane in which B_1 rotates relative to the stationary base B. Thus the two *axles* shown in the sketch are orthogonal. This assemblage is in a uniform gravity field.

a. Apply Lagrange's equations in appropriate form to obtain nonlinear equations of motion in terms of the generalized coordinates q_1 and q_2. (In Fig. P8·32, q_1 is the angle from the vertical to the centerline of B_1, and q_2 is the angle from the centerline of B_1 to that of B_2.
b. Linearize the equations of motion about a rest state in which $q_1 \equiv q_2 \equiv 0$ (bottom rest position, hanging vertically).

Note 1: Define $p^2 = g/a$, and express results in terms of p when possible.
Note 2: The moment of inertia of a uniform thin rigid bar of mass m and length a is $ma^2/12$ for the mass center and $ma^2/3$ for an end point. (See Appendix C.)

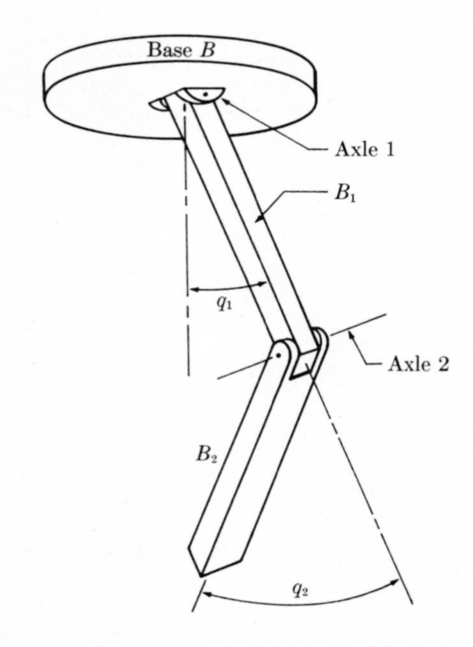

FIGURE P8·32

8·33 Angular coordinates q_1 and q_2 define the position and orientation of a block in its planar motion, as shown in Fig. P8·33. The distance a is constant. The block is subjected to a resultant force

$$\mathbf{F} = C_1(1 + \sin^2 q_2)\mathbf{V} + a\dot{q}_1{}^2\hat{\mathbf{e}}_r$$

where C_1 = constant
$\mathbf{V}$ = inertial velocity of mass center c of the block
$\hat{\mathbf{e}}_n$ = unit vector normal to the paper (fixed inertially), with $\hat{\mathbf{e}}_r$ (radial) and $\hat{\mathbf{e}}_t$ (tangential) completing the orthogonal set

The block is also subjected to a resultant moment about the mass center

$$\mathbf{M} = C_2\boldsymbol{\omega}$$

Find the generalized forces Q_1 and Q_2.

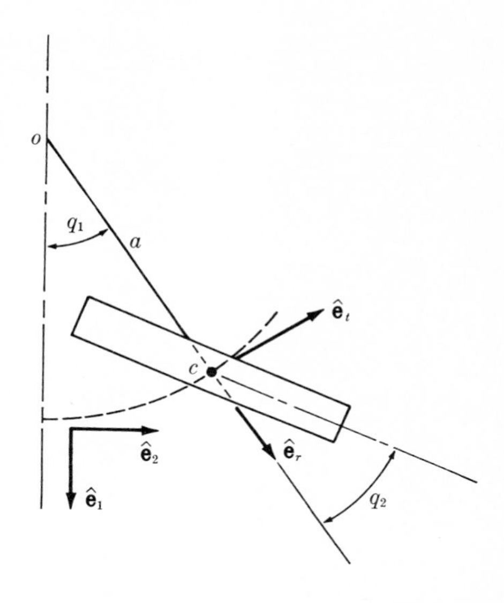

FIGURE P8·33

8·34 A cart is accelerated toward a wall by winding a massless wire (one end of which is attached to the wall) onto a spool on the cart (see Fig. P8·34). The torque applied by the motor turning the spool is $\tau(t)$, a given function of time. The scalar that locates the cart on the straight frictionless track is x, and θ is

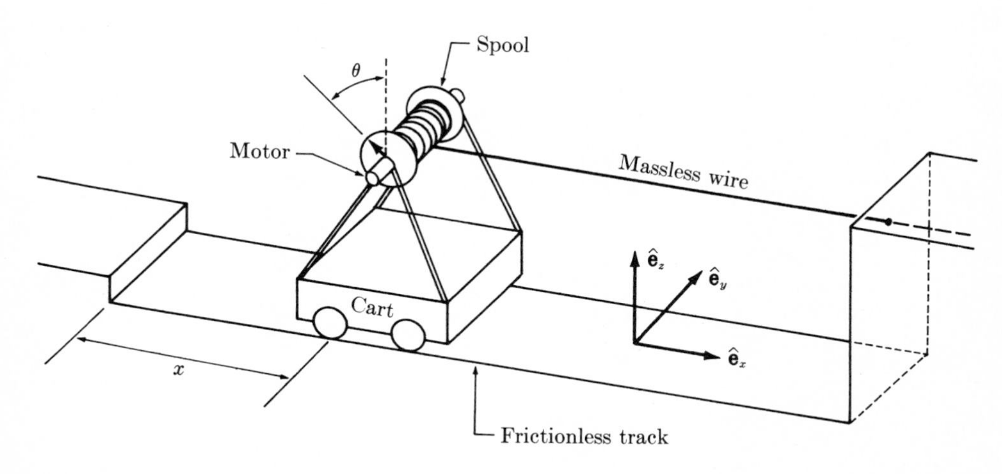

FIGURE P8·34

an angle between the vertical and an arrow painted on the rim of the spool. The radius of the spool is a, I is the moment of inertia of the spool about its axis, and M is the mass of the cart, including the spool. The cart wheels have negligible inertia. Make use of an appropriate form of Lagrange's equations to obtain complete differential equations of motion of this system.

8·35 Derive the equations of rotational motion of a torque-free axisymmetric rigid body, using Lagrange's equations, and choosing as generalized coordinates the angles φ, θ, ψ defined by Fig. 9·3 (see Chap. 9).

CHAPTER NINE RIGID-BODY DYNAMICS: SOLUTIONS

9·1 MOTION OF A FREE RIGID BODY

9·1·1 Angular velocity of an axisymmetric body

The mass center c of a rigid body completely free of external forces (or subject to forces with a zero resultant) must have no acceleration in inertial space, since $\mathbf{F} = \mathfrak{M}\mathbf{A}^{ci}$ [by Eqs. (8·6)]. Since a reference frame that maintains a uniform translational velocity (with no rotation) relative to an inertial reference frame is also an inertial reference frame [see Eq. (3·150)], we can say that *the mass center c of a rigid body under zero resultant force remains fixed in inertial space.* All that remains for such a body is the determination of its rotational motion, which can be established by solving some set of scalar differential equations obtained from the vector equation $\dot{\mathbf{H}}^c = \mathbf{M}^c$ [see Eq. (8·16)]. Euler's equations [Eqs. (8·57)] provide a particularly convenient form, appearing in general as follows:

$$M_1 = I_1\dot{\omega}_1 - \omega_2\omega_3(I_2 - I_3) \tag{9·1a}$$
$$M_2 = I_2\dot{\omega}_2 - \omega_3\omega_1(I_3 - I_1) \tag{9·1b}$$
$$M_3 = I_3\dot{\omega}_3 - \omega_1\omega_2(I_1 - I_2) \tag{9·1c}$$

Recall that the indices 1, 2, and 3 identify unit vectors $\hat{\mathbf{e}}_1$, $\hat{\mathbf{e}}_2$, and $\hat{\mathbf{e}}_3$, which are body-fixed principal axes of inertia for the mass center c or for some point fixed both in the body and in inertial space. For the problem at hand, point c is the required reference point.

If a body has an axis of inertial symmetry through the mass center c (so that the inertia ellipsoid for c as introduced in §8·4·10 has an axis of geometrical symmetry), then that axis is also a principal axis, and the moments of inertia for the other two principal axes are identical. We shall identify $\hat{\mathbf{e}}_3$ with the axis of symmetry, and in Euler's equations we will replace I_3 by I_s (s standing for symmetry axis) and replace both I_2 and I_1 by I_t (t denoting any axis transverse to the symmetry axis). Equations (9·1) then become

$$M_1 = I_t\dot{\omega}_1 + \omega_2\omega_3(I_s - I_t) \tag{9·2a}$$
$$M_2 = I_t\dot{\omega}_2 - \omega_1\omega_3(I_s - I_t) \tag{9·2b}$$
$$M_3 = I_s\dot{\omega}_3 \tag{9·2c}$$

In the special case of a free rigid body, there are no external forces, and hence no external moments, so that $M_1 = M_2 = M_3 = 0$ and Eqs. (9·2) become

$$I_t\dot{\omega}_1 + \omega_2\omega_3(I_s - I_t) = 0 \tag{9·3a}$$
$$I_t\dot{\omega}_2 - \omega_1\omega_3(I_s - I_t) = 0 \tag{9·3b}$$
$$I_s\dot{\omega}_3 = 0 \tag{9·3c}$$

Equation (9·3c) requires that ω_3 maintain a constant value, which we will call Ω. Equations (9·3a and b) can then be written as

$$\dot{\omega}_1 + \lambda\omega_2 = 0 \tag{9·4a}$$
$$\dot{\omega}_2 - \lambda\omega_1 = 0 \tag{9·4b}$$

where $\lambda \triangleq \Omega(I_s - I_t)/I_t$. Differentiating Eq. (9·4a) and substituting (9·4b) yields

$$\ddot{\omega}_1 + \lambda^2\omega_1 = 0 \tag{9·5a}$$

Reversing the process produces

$$\ddot{\omega}_2 + \lambda^2\omega_2 = 0 \tag{9·5b}$$

Equations (9·5) provide very simple examples of differential equations treated in Chap. 5 [see Eq. (5·29)]; these equations have solutions of the form

$$\omega_1 = C_1 \sin \lambda t + C_2 \cos \lambda t \tag{9·6a}$$
$$\omega_2 = C_3 \sin \lambda t + C_4 \cos \lambda t \tag{9·6b}$$

where $C_1, \ldots, C_4$ are to be determined from initial conditions. Here a note of caution is in order. The coupled first-order differential equations (9·4) are the original equations of motion, and their solutions can involve only the *two* initial conditions $\omega_1(0) \triangleq \omega_{10}$ and $\omega_2(0) \triangleq \omega_{20}$. We differentiated Eqs. (9·4) to obtain the more familiar-looking uncoupled second-order differential equations (9·5), but we must not imagine that this manipulation permits us to add new initial-condition information for $\dot{\omega}_1(0)$ and $\dot{\omega}_2(0)$ in order to solve for the four constants $C_1, \ldots, C_4$. Instead we must note from Eqs. (9·4*a*) and (9·6) that these constants are related by $\lambda C_1 \cos \lambda t - \lambda C_2 \sin \lambda t + \lambda(C_3 \sin \lambda t + C_4 \cos \lambda t) = 0$, so that $C_4 = -C_1$ and $C_3 = C_2$. The required solutions now become

$$\omega_1 = C_1 \sin \lambda t + C_2 \cos \lambda t \tag{9·7a}$$
$$\omega_2 = C_2 \sin \lambda t - C_1 \cos \lambda t \tag{9·7b}$$

and the two initial conditions provide

$$\omega_{10} = C_2 \qquad \text{and} \qquad \omega_{20} = -C_1$$

Thus the original equations of motion for a free axisymmetric rigid body [Eqs. (9·3)] have the solutions

$$\omega_1 = -\omega_{20} \sin \lambda t + \omega_{10} \cos \lambda t \tag{9·8a}$$
$$\omega_2 = \omega_{10} \sin \lambda t + \omega_{20} \cos \lambda t \tag{9·8b}$$
$$\omega_3 = \Omega \tag{9·8c}$$

with

$$\lambda \triangleq \frac{\Omega(I_s - I_t)}{I_t} \tag{9·8d}$$

Note that

$$\begin{aligned} \omega_1^2 + \omega_2^2 &= (\omega_{10}^2 + \omega_{20}^2)(\sin^2 \lambda t + \cos^2 \lambda t) \\ &= (\omega_{10}^2 + \omega_{20}^2) \triangleq \omega_0^2 \qquad \text{constant} \end{aligned} \tag{9·9}$$

Hence the *transverse angular-velocity vector* $\omega_1\hat{\mathbf{e}}_1 + \omega_2\hat{\mathbf{e}}_2$ has the constant length ω_0, as established by the initial values of ω_1 and ω_2. Since every axis through c and normal to $\hat{\mathbf{e}}_3$ is a principal axis of inertia, we can without loss of generality assign the vectors $\hat{\mathbf{e}}_1$ and $\hat{\mathbf{e}}_2$ such that $\omega_{20} = 0$ (so $\omega_{10} = \omega_0$); this simplifies the appearance of the solutions in Eqs. (9·8) to

$$\omega_1 = \omega_0 \cos \lambda t \tag{9·10a}$$
$$\omega_2 = \omega_0 \sin \lambda t \tag{9·10b}$$
$$\omega_3 = \Omega \tag{9·10c}$$

With the solutions in this form, it becomes clear that the angular velocity is a vector

$$\boldsymbol{\omega} = \Omega \hat{\mathbf{e}}_3 + \omega_0 \hat{\boldsymbol{\epsilon}} \tag{9·11}$$

where $\hat{\boldsymbol{\epsilon}}$ is a unit vector that rotates relative to the body in the plane of $\hat{\mathbf{e}}_1$ and $\hat{\mathbf{e}}_2$ (normal to the symmetry axis) at the constant angular speed λ. As revealed by Eq. (9·8*d*), λ may be positive or negative, depending on the relative values of I_t and I_s; if $I_s > I_t$ (as for an oblate ellipsoid), then $\hat{\boldsymbol{\epsilon}}$ rotates positively in the body, advancing from coincidence with $\hat{\mathbf{e}}_1$ to coincidence with $\hat{\mathbf{e}}_2$ and around; while if $I_t > I_s$ (as for a prolate ellipsoid), then $\hat{\boldsymbol{\epsilon}}$ rotates negatively, coinciding with $\hat{\mathbf{e}}_1$ and then with $-\hat{\mathbf{e}}_2$, and so on. The geometrical interpretation of these motions is further developed below.

9·1·2 Geometry of motion of an axisymmetric body

Figure 9·1 portrays a typical axisymmetric body (a rectangular parallelepiped with side dimensions a, a, and b), and shows the inertial angular-velocity vector $\boldsymbol{\omega}$ of the body. According to Eq. (9·11),

$$\boldsymbol{\omega} = \Omega \hat{\mathbf{e}}_3 + \omega_0 \hat{\boldsymbol{\epsilon}} \tag{9·12}$$

and the unit vector $\hat{\boldsymbol{\epsilon}}$ rotates steadily around the symmetry axis positively or negatively, depending on the body shape. Here if $b > a$ then (from Appendix C), $I_t > I_s$ and the sense of the rotation is negative. As suggested by Fig. 9·1, the angular-velocity vector $\boldsymbol{\omega}$ then traces out a conical locus fixed in the body.

From the analysis preceding we have full knowledge of the motion of $\boldsymbol{\omega}$ relative to the body, but usually we are more interested in the motion of the body in inertial space. We can fully understand the motion if we observe that the angular-momentum vector $\mathbf{H}$ is [from Eqs. (8·53) and (9·10)] known to be

$$\begin{aligned} \mathbf{H} &= I_t\omega_1\hat{\mathbf{e}}_1 + I_t\omega_2\hat{\mathbf{e}}_2 + I_s\omega_3\hat{\mathbf{e}}_3 \\ &= I_t\omega_0(\hat{\mathbf{e}}_1 \cos \lambda t + \hat{\mathbf{e}}_2 \sin \lambda t) + I_s\Omega\hat{\mathbf{e}}_3 \\ &= I_t\omega_0\hat{\boldsymbol{\epsilon}} + I_s\Omega\hat{\mathbf{e}}_3 \end{aligned} \tag{9·13}$$

When Eq. (9·12) for $\boldsymbol{\omega}$ is compared to Eq. (9·13) for $\mathbf{H}$, we see that these two vectors and the symmetry axis of the body all lie in the same plane (the plane of $\hat{\mathbf{e}}_3$ and $\hat{\boldsymbol{\epsilon}}$ in Fig. 9·1). Moreover, the angular momentum $\mathbf{H}$ is a vector fixed in inertial space, since the moment $\mathbf{M} = 0$ and $\mathbf{M} = \dot{\mathbf{H}}$.

The angle θ between $\hat{\mathbf{e}}_3$ and $\mathbf{H}$ is a constant, as given by

$$\tan \theta = \frac{I_t\omega_0}{I_s\Omega} \tag{9·14}$$

while the angle β between $\hat{\mathbf{e}}_3$ and $\boldsymbol{\omega}$ is a constant given by

$$\tan \beta = \frac{\omega_0}{\Omega} \tag{9·15}$$

Comparing these expressions indicates that $\theta > \beta$ for $I_t > I_s$ (as in Fig. 9·1), while $\beta > \theta$ for $I_s > I_t$ (as for an oblate ellipsoid). In either case, the vectors

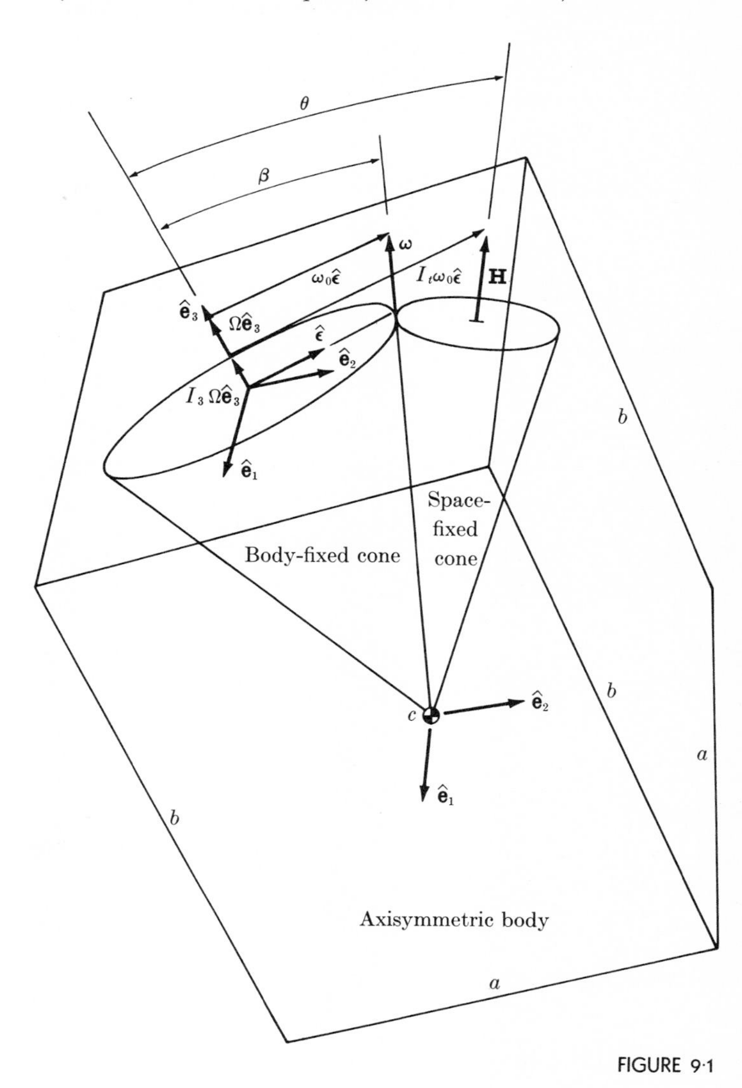

FIGURE 9·1

$\mathbf{H}$, $\boldsymbol{\omega}$, and $\hat{\mathbf{e}}_3$ maintain a fixed relationship to one another; and since only $\mathbf{H}$ is fixed in inertial space, the vectors $\boldsymbol{\omega}$ and $\hat{\mathbf{e}}_3$ must rotate around $\mathbf{H}$. When $\mathbf{H}$ and $\boldsymbol{\omega}$ are drawn through the inertially stationary mass center c (as in Fig. 9·1), then $\boldsymbol{\omega}$ traces out an inertially fixed conical locus with $\mathbf{H}$ directed along the central axis of the cone. Because the inertial velocity of a point of the body located from c by the position vector $\boldsymbol{\rho}$ is given by $\boldsymbol{\omega} \times \boldsymbol{\rho}$, any point of the body on the line through c and parallel to $\boldsymbol{\omega}$ has no inertial velocity. This line, which is called the *instantaneous axis of rotation*, is the line of contact between the space-fixed cone and the body-fixed cone in Fig. 9·1. Since points of the body on this line are instantaneously at rest, while the body rotates about this axis, *the motion of a free axisymmetric rigid body is fully described by the statement that the body-fixed cone rolls without slip on the space-fixed cone.* Figure 9·2*a* and *b* portrays the rolling cones for the two possibilities noted previously, namely, $I_t > I_s$ (as in Fig. 9·2*a*) and $I_s > I_t$ (as in Fig. 9·2*b*).

9·1·3 Euler angle solutions for an axisymmetric body

Although the free-body motions of an axisymmetric rigid body are fully determined in the preceding two articles, it is enlightening to consider the solutions in terms of the Euler angles φ, θ, and ψ first introduced in §3·2·3, and first illustrated in Fig. 3·9. This figure is repeated here as Fig. 9·3, but with the reference base vectors changed to the inertially fixed unit vectors $\hat{\mathbf{i}}_1$, $\hat{\mathbf{i}}_2$, and $\hat{\mathbf{i}}_3$; the body-fixed base vectors changed to the principal axis unit vectors $\hat{\mathbf{e}}_1$, $\hat{\mathbf{e}}_2$, and $\hat{\mathbf{e}}_3$; and the intermediate frame vectors denoted by $\hat{\mathbf{j}}_1$, $\hat{\mathbf{j}}_2$, and $\hat{\mathbf{j}}_3$ and $\hat{\mathbf{k}}_1$, $\hat{\mathbf{k}}_2$, and $\hat{\mathbf{k}}_3$.

The chain rule for angular-velocity construction now produces

$$\begin{aligned}\boldsymbol{\omega} &= \boldsymbol{\omega}^{ei} = \boldsymbol{\omega}^{ek} + \boldsymbol{\omega}^{kj} + \boldsymbol{\omega}^{ji} \\ &= \dot{\psi}\hat{\mathbf{k}}_3 + \dot{\theta}\hat{\mathbf{k}}_1 + \dot{\varphi}\hat{\mathbf{j}}_3 \\ &= \dot{\theta}\hat{\mathbf{k}}_1 + \dot{\varphi}\sin\theta\hat{\mathbf{k}}_2 + (\dot{\psi} + \dot{\varphi}\cos\theta)\hat{\mathbf{k}}_3 \end{aligned} \tag{9·16}$$

If now we make the special choice of the arbitrary inertially fixed unit vector $\hat{\mathbf{i}}_3$ so as to align it with the angular momentum $\mathbf{H}$, then we see that the Euler angle θ is again the angle between $\hat{\mathbf{e}}_3$ and $\mathbf{H}$ (as in the two preceding articles) and again $\dot{\theta} = 0$ and $\tan\theta$ is given by Eq. (9·14). Comparison of $\boldsymbol{\omega}$ as given in Eqs. (9·12) and (9·16) now reveals that

$$\dot{\varphi}\sin\theta\,\hat{\mathbf{k}}_2 + (\dot{\psi} + \dot{\varphi}\cos\theta)\hat{\mathbf{k}}_3 = \omega_0\hat{\boldsymbol{\epsilon}} + \Omega\hat{\mathbf{e}}_3$$

Since $\hat{\mathbf{e}}_3 \equiv \hat{\mathbf{k}}_3$ (see Fig. 9·3), we have

$$\Omega = \dot{\psi} + \dot{\varphi}\cos\theta \tag{9·17}$$

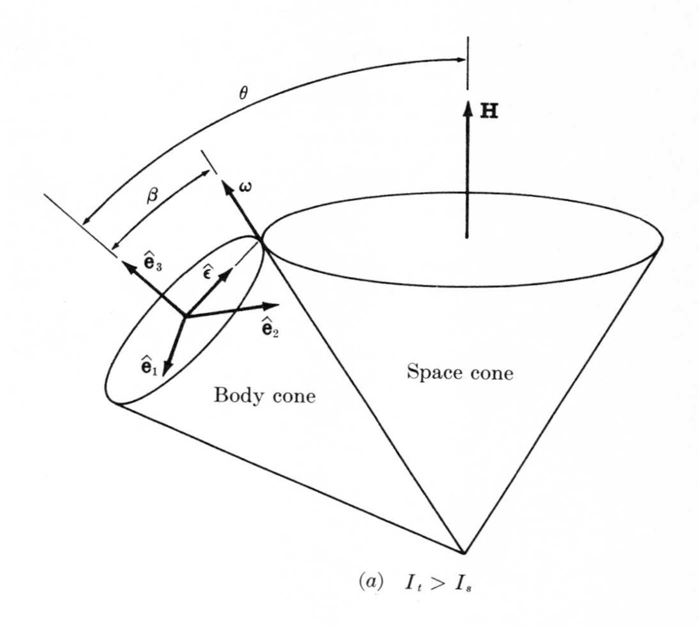

(a) $I_t > I_s$

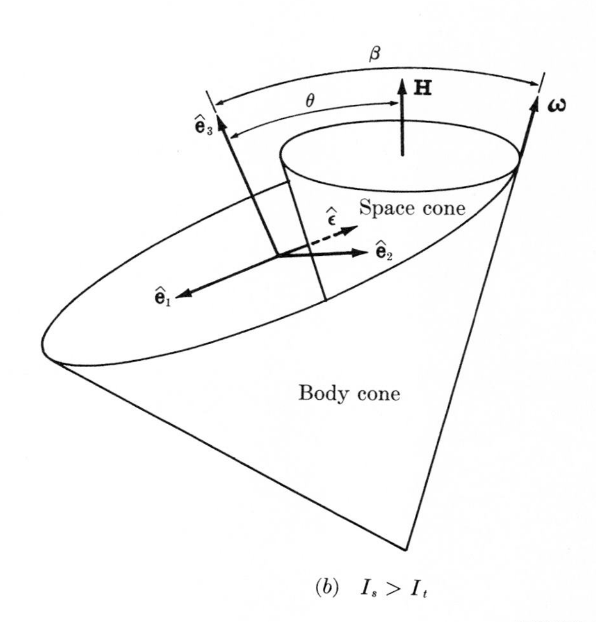

(b) $I_s > I_t$

FIGURE 9·2

and, since $\omega_0 > 0$ and $\dot{\varphi} \sin \theta > 0$, $\hat{\boldsymbol{\epsilon}} \equiv \hat{\mathbf{k}}_2$ and

$$\omega_0 = \dot{\varphi} \sin \theta \tag{9·18}$$

This means that both $\dot{\varphi}$ and $\dot{\psi}$ are constants, and Eq. (9·14) provides

$$\tan \theta = \frac{I_t \dot{\varphi} \sin \theta}{I_s(\dot{\psi} + \dot{\varphi} \cos \theta)} = \frac{I_t \dot{\varphi} \sin \theta}{I_s \Omega}$$

or, if $\sin \theta \neq 0$, then

$$I_s(\dot{\psi} + \dot{\varphi} \cos \theta) = I_t \dot{\varphi} \cos \theta \qquad \text{or} \qquad I_s \Omega = I_t \dot{\varphi} \cos \theta$$

and

$$\dot{\psi} = \left(\frac{I_t}{I_s} - 1\right) \dot{\varphi} \cos \theta = \left(1 - \frac{I_s}{I_t}\right) \Omega \tag{9·19}$$

Evidently $\dot{\psi}\dot{\varphi} > 0$ for $I_t > I_s$, and $\dot{\psi}\dot{\varphi} < 0$ for $I_t < I_s$. Equation (9·18) requires $\dot{\varphi} > 0$ since ω_0 and θ are greater than zero by definition. This means that $\dot{\psi}$ may be positive or negative.

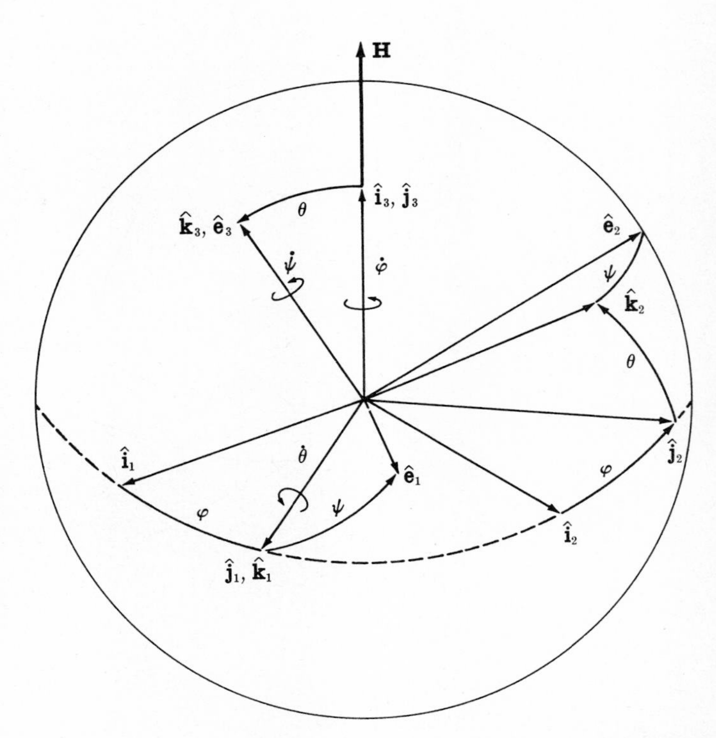

FIGURE 9·3

The utility of the Euler angles stems in part from the simplicity of their solutions for simple cases such as the force-free axisymmetric rigid body considered here. In this case the solutions are summarized by

$$\dot{\psi} = \text{constant} = \left(1 - \frac{I_s}{I_t}\right)\Omega \tag{9·20a}$$

$$\dot{\varphi} = \text{constant} = \frac{I_s\Omega}{I_t \cos\theta} \tag{9·20b}$$

$$\cos\theta = \frac{I_s\dot{\psi}}{(I_t - I_s)\dot{\varphi}} \tag{9·20c}$$

The coning motion established by $\dot{\varphi}$ is classically called *precession* or *free precession,* and the angle θ is classically called the *nutation angle.* In modern engineering practice in gyrodynamics and satellite dynamics the coning motion due to $\dot{\varphi}$ is unfortunately often called *nutation,* so confusion of language has become rampant, and in this book the terms precession and nutation are generally avoided.

9·1·4 Geometry of motion of an arbitrary body

When a force-free rigid body has three unequal principal axis moments of inertia, Euler's equations [Eqs. (9·1)] become

$$I_1\dot{\omega}_1 - \omega_2\omega_3(I_2 - I_3) = 0 \tag{9·21a}$$
$$I_2\dot{\omega}_2 - \omega_3\omega_1(I_3 - I_1) = 0 \tag{9·21b}$$
$$I_3\dot{\omega}_3 - \omega_1\omega_2(I_1 - I_2) = 0 \tag{9·21c}$$

and the closed-form analytical solutions that were so simple in the axisymmetric case [see Eqs. (9·3)] become rather difficult. Integration of Eqs. (9·21) can be accomplished, but the results emerge in terms of a rather unfamiliar class of functions called *elliptic functions* [in contrast to the familiar *trigonometric function* solutions found in Eqs. (9·8)]. More enlightening at this stage than the derivation of the elliptic function solutions is the development of a geometrical interpretation of the motion, as first proposed by Louis Poinsot in 1834.

Poinsot's procedure rests on the observation that two first integrals can be obtained for Eqs. (9·21). By multiplying Eqs. (9·21*a*, *b*, and *c*) respectively by ω_1, ω_2, and ω_3 and then adding, we find

$$I_1\omega_1\dot{\omega}_1 + I_2\omega_2\dot{\omega}_2 + I_3\omega_3\dot{\omega}_3 = 0$$

This equation may be rewritten in the form

$$\frac{d}{dt}\left(\tfrac{1}{2}I_1\omega_1^2 + \tfrac{1}{2}I_2\omega_2^2 + \tfrac{1}{2}I_3\omega_3^2\right) = 0$$

which may be integrated to obtain the energy integral

$$\tfrac{1}{2}I_1\omega_1^2 + \tfrac{1}{2}I_2\omega_2^2 + \tfrac{1}{2}I_3\omega_3^2 = T \tag{9·22}$$

where T is a constant we recognize as the kinetic energy of rotation [see Eq. (8·87)].

Equation (9·22) may be written in the form

$$\frac{\omega_1^2}{2T/I_1} + \frac{\omega_2^2}{2T/I_2} + \frac{\omega_3^2}{2T/I_3} = 1 \tag{9·23}$$

which can be interpreted geometrically as an ellipsoid, called the *energy ellipsoid*, described in a body-fixed cartesian coordinate system with its origin at the mass center. Comparison of Eqs. (9·22) and (8·49) reveals that the energy ellipsoid of a given body has the same shape as its inertia ellipsoid, although a different size (unless $2T = 1$). Since ω_1, ω_2, and ω_3 are the cartesian coordinates of a point on the surface of the energy ellipsoid, the distance from the origin to a given point on the surface is the magnitude of $\boldsymbol{\omega}$ for rotation about the corresponding body-fixed line with the prescribed energy T.

The second first integral of Euler's equations for a free body can be found by multiplying Eqs. (9·21*a*, *b*, and *c*) respectively by $I_1\omega_1$, $I_2\omega_2$, and $I_3\omega_3$, and then adding to obtain

$$I_1^2\omega_1\dot{\omega}_1 + I_2^2\omega_2\dot{\omega}_2 + I_3^2\omega_3\dot{\omega}_3 = 0 \tag{9·24}$$

Integration and cancellation of a factor ½ yields

$$I_1^2\omega_1^2 + I_2^2\omega_2^2 + I_3^2\omega_3^2 = H^2 \tag{9·25}$$

where H^2 is a constant that can be recognized as the square of the angular momentum $\mathbf{H}$ of the body. From Eq. (8·53*c*), $\mathbf{H}$ is given by

$$\mathbf{H} = I_1\omega_1\hat{\mathbf{e}}_1 + I_2\omega_2\hat{\mathbf{e}}_2 + I_3\omega_3\hat{\mathbf{e}}_3 \tag{9·26}$$

Equation (9·25), like (9·22), can be interpreted geometrically as an ellipsoid; the new one is called the *momentum ellipsoid*, and it also is fixed relative to the body.

In order to determine the path of the angular-velocity vector in the body (for comparison with the conical path described as in Fig. 9·1 for the axisymmetric body), we can superimpose the energy ellipsoid and the momentum ellipsoid as in Fig. 9·4, and observe that the tip of the angular-velocity vector emanating from the mass center c of the body must remain simultaneously on the energy ellipsoid *and* the momentum ellipsoid. For a given body (with specified I_1, I_2, and I_3), the specification of initial conditions establishes particular values for $2T$ and H^2, and thereby determines the sizes of the two ellipsoids in Fig. 9·4. The intersections of these ellipsoids (called *polhodes*) represent the only points common to both surfaces, so that the tip of the vector $\boldsymbol{\omega}$ must move along the particular polhode established by the initial conditions. Figure 9·4 shows a pair of polhodes associated with a given set of initial conditions, while Fig. 9·5 shows a whole family of polhodes as they appear for a body with $I_3 = 3I_2/2 = 2I_1$. Since the energy ellipsoid and the inertia ellipsoid (see §8·4·10) are geometrically similar (differing only in size), Fig. 9·5 may be interpreted as portraying the polhodes on either of these two ellipsoids. The indicated "nodding" oscillation between upper and lower bounds on θ is classically called *nutation,* while the coning motion is, as previously noted, classically called *free precession.*

If the body has two identical principal axis inertias, so that all of the noted ellipsoids are axisymmetric, then the polhodes are all circles, and we can

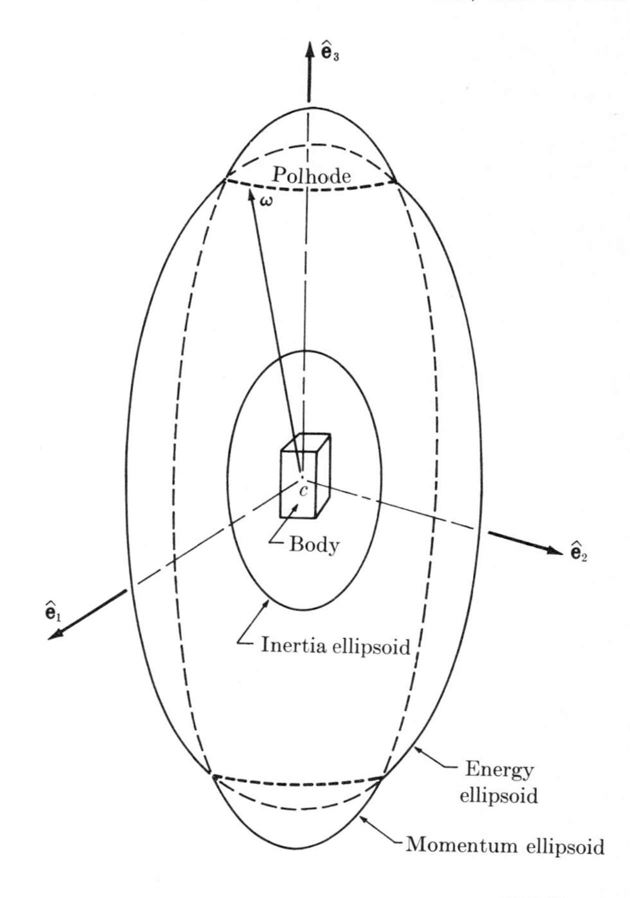

FIGURE 9·4

see again (as in §9·1·2) that the angular-velocity vector traces out a conical locus in the body.

As noted in the discussion of axisymmetric bodies, we are usually more concerned with the rotations of the body in inertial space than we are with the path of **ω** in the body. Understanding of the geometry of motion in the general case requires the observation that the gradient of T in the space of ω_1, ω_2, and ω_3 is

$$\begin{aligned}\nabla T &= \frac{\partial T}{\partial \omega_1}\hat{\mathbf{e}}_1 + \frac{\partial T}{\partial \omega_2}\hat{\mathbf{e}}_2 + \frac{\partial T}{\partial \omega_3}\hat{\mathbf{e}}_3 \\ &= I_1\omega_1\hat{\mathbf{e}}_1 + I_2\omega_2\hat{\mathbf{e}}_2 + I_3\omega_3\hat{\mathbf{e}}_3 = \mathbf{H} \qquad (9{\cdot}27)\end{aligned}$$

from Eqs. (9·22) and (9·26). Geometrically, this means that the normal to the energy ellipsoid at its intersection with the **ω** vector from its origin at c is always parallel to the angular-momentum vector **H**. In the absence of an external moment about c, **H** is fixed in inertial space. If we introduce an inertially fixed plane (called the *invariable plane*) normal to **H**, then we can say that the body moves in such a way that the normal to the body-fixed energy ellipsoid at its intersection with **ω** is always normal to the invariable plane. Moreover, as shown in Fig. 9·6,

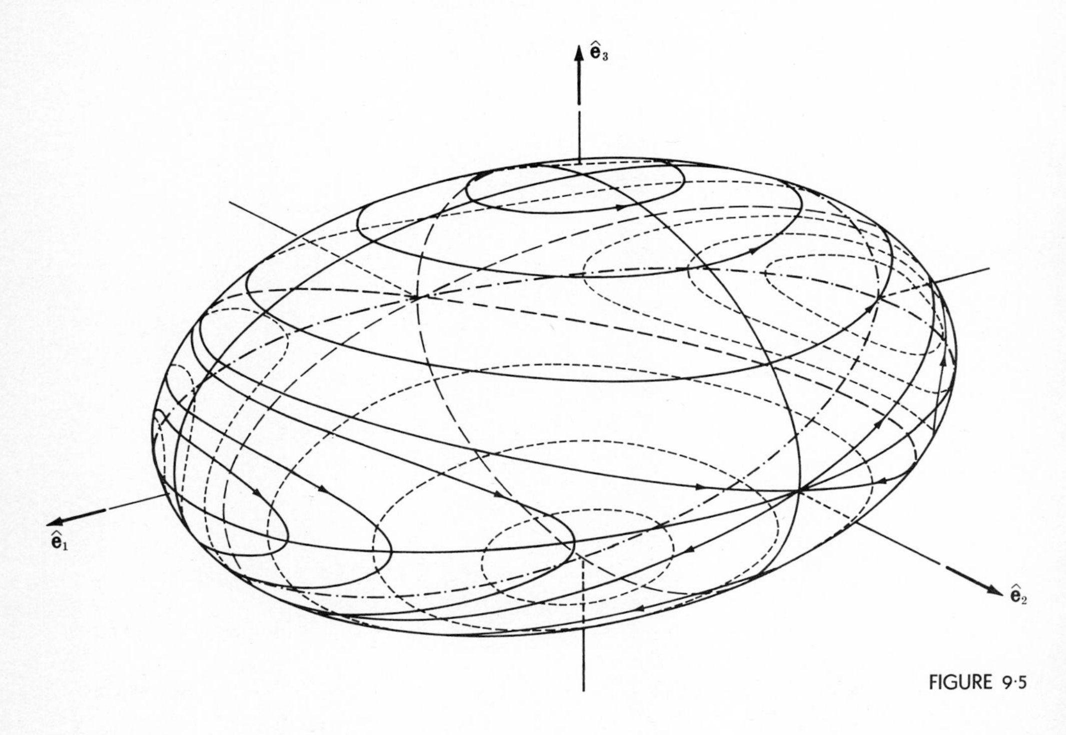

FIGURE 9·5

the distance between c and the invariable plane is given by the constant

$$\boldsymbol{\omega} \cdot \frac{\mathbf{H}}{H} = \frac{2T}{H} \tag{9·28}$$

In the absence of external force, the mass center c is inertially stationary, so that the axis through c and parallel to $\boldsymbol{\omega}$ is the instantaneous axis of rotation, which is at least momentarily at rest both in the body and in inertial space.

Thus we can conclude that *the body-fixed energy ellipsoid moves in inertial space in such a way that it rolls without slip on the invariable plane while its centroid c remains inertially fixed at a distance $2T/H$ above the plane* (see Fig. 9·6).

Whereas the trace of the angular velocity vector $\boldsymbol{\omega}$ on the ellipsoid is called the *polhode*, its trace on the invariable plane is called the *herpolhode*. The polhode is always a closed path on the ellipsoid, while the herpolhode has a periodic geometry that remains concave inward but does not close on itself unless the body is inertially axisymmetric, in which case the geometry of Fig. 9·2 tells the story.

9·1·5 Stability of rotation

The word *stable* has many aspects, and it deserves more mathematical precision in its definition than space permits here. We have already used the word in §6·4·2

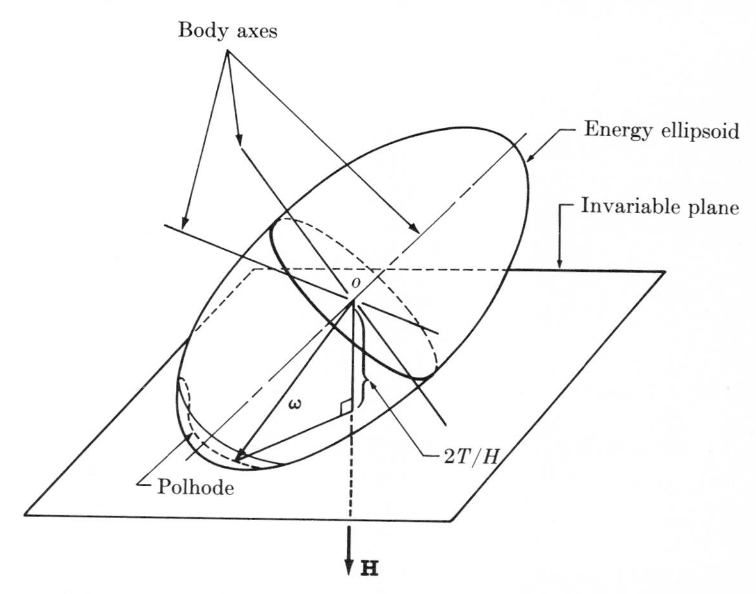

FIGURE 9·6

to describe a state of static equilibrium (rest in inertial space) whenever a choice of initial conditions for generalized coordinates and generalized velocities sufficiently close to the noted equilibrium results in a motion such that these quantities remain for all time arbitrarily close to the values for equilibrium. Generalizing this usage, we will now refer to the rotation of a force-free rigid body about a principal axis as *stable* if any choice of initial conditions sufficiently close to those associated with the nominal spinning motion results in solutions of Euler's equations [Eqs. (9·21)] such that ω_1, ω_2, and ω_3 remain arbitrarily close to their nominal values; a rotation that is not stable is called *unstable*. [The stability concept can be generalized further to apply to a solution of any set of ordinary first-order differential equations, as typified by Eqs. (9·21).]

The stability of rotation of a free rigid body about a principal axis depends on which of the three axes the body is spinning about, as can be established by inspection of Fig. 9·5. For example, if the nominal motion is spin about the principal axis of least inertia, so that nominally $\boldsymbol{\omega}$ is aligned with $\hat{\mathbf{e}}_1$ in Fig. 9·5, then slightly different initial conditions will place the tip of $\boldsymbol{\omega}$ on a neighboring polhode that encircles the $\hat{\mathbf{e}}_1$ axis. Since $\boldsymbol{\omega}$ never departs very far from the nominal location, this motion is *stable*.

On the other hand, if the nominal motion consists of spin about the principal axis of intermediate moment of inertia, so that nominally $\boldsymbol{\omega}$ is aligned with $\hat{\mathbf{e}}_2$ in Fig. 9·5, then shifting to a neighboring polhode causes in most cases a very large eventual change in $\boldsymbol{\omega}$. If the initial conditions place the tip of the $\boldsymbol{\omega}$ vector slightly below the $\hat{\mathbf{e}}_2$ axis in the figure, then the $\boldsymbol{\omega}$ vector eventually encircles the $\hat{\mathbf{e}}_3$ axis; other initial conditions cause $\boldsymbol{\omega}$ to go all the way around the $\hat{\mathbf{e}}_1$ axis. Thus we conclude that *rigid-body spin about a principal axis of intermediate moment of inertia is unstable.*

Finally, Fig. 9·5 reveals that *rigid-body spin about a principal axis of maximum inertia is a stable motion.*

A word of caution is necessary for the proper application of the results of this article. On page 1 we discussed the problems encountered with *Explorer I*, the first artificial earth satellite launched by the United States. This satellite (see Fig. I·1) was an elongated, inertially axisymmetric body designed to spin about its symmetry axis, which was the principal axis of least moment of inertia. According to the results preceding, a rigid body free of external moment should spin stably about such an axis, yet *Explorer I* was observed to be unstable; by the time the satellite had completed its first orbit, the initial spinning motion was transformed into a coning motion (as in Fig. 9·2*a*) with half-angle θ in the neighborhood of 60 degrees!

As noted in the Introduction, the explanation of the behavior of *Explorer I* lay in the flexibility of portions of the vehicle and the resulting energy dissipation. In fact neither *Explorer I* nor any other physical body is truly rigid, and the rigid-body analysis preceding is never *strictly* applicable. Some energy is always dissipated internally, and the kinetic energy T in Eq. (9·22) is not constant, as promised by rigid-body analysis. The utility of the rigid-body idealization lies in its value as a starting point for the dynamic analysis of real physical systems that are *quasi-rigid*, or nearly rigid. In a qualitative sense, we can interpret the loss of kinetic energy due to internal energy dissipation in a quasi-rigid body as requiring a gradual shift of the angular-velocity vector on the inertia ellipsoid of Fig. 9·5 from polhode to polhode, moving continually to a polhode with lower kinetic energy. This process is depicted in Fig. 9·7. Since for a body with a given angular momentum **H**, the state of minimum kinetic energy corresponds to rotation about the principal axis of maximum inertia (and since **H** cannot change in the absence of external moment), it may be expected that the body will continue to dissipate energy internally until it approaches a state of rotation about its maximum inertia axis. *This heuristic argument suggests that a force-free quasi-rigid physical body would rotate stably only about its principal axis of maximum inertia.* Since the surprise performance of *Explorer I* in 1958, quasi-rigid spin-stabilized spacecraft have been designed to spin about the axis through the mass center having the largest moment of inertia. (This restriction is, how-

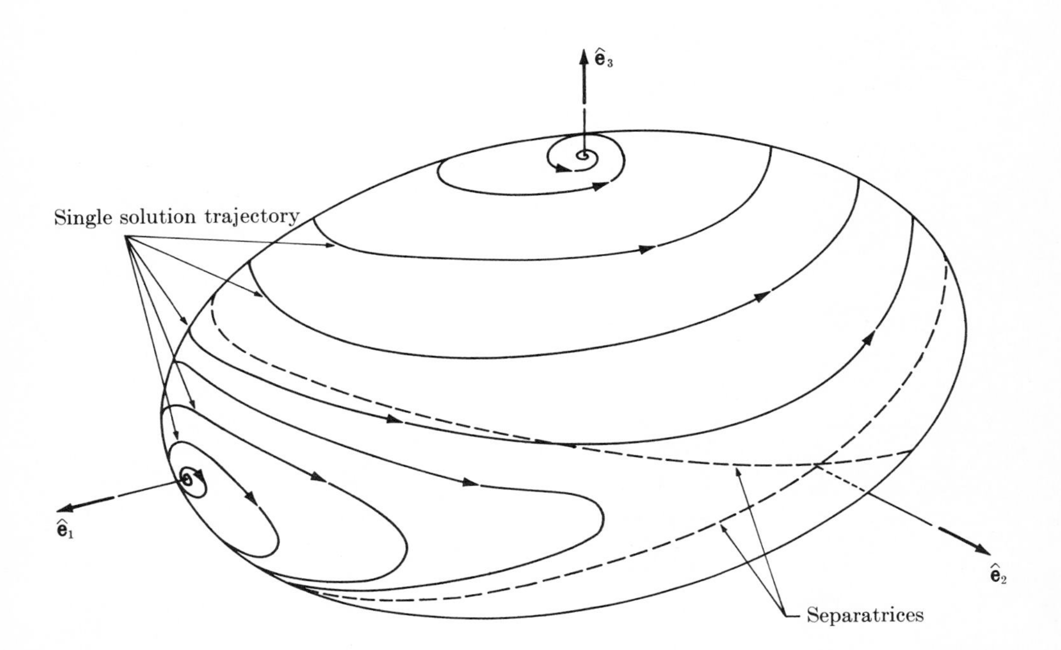

FIGURE 9·7

ever, not always necessary when the spacecraft contains rotating parts, and it is not always sufficient when the body is very flexible.)

9·2 MOTION OF A RIGID BODY WITH ONE POINT FIXED

9·2·1 The general case

Euler's equations, as derived from $\mathbf{M} = \dot{\mathbf{H}}$ in §8·5·1 and repeated in §9·1·1 as Eqs. (9·1), apply to a rigid body with a reference point for $\mathbf{M}$ and $\mathbf{H}$ fixed in the body and either coincident with the mass center or fixed also in inertial space. The indices 1, 2, and 3 in Eqs. (9·1) correspond to body-fixed unit vectors parallel to the principal axes of inertia for the reference point. These equations therefore apply directly to the problem of the rigid body with one point fixed in inertial space.

Of course, the solution of Euler's equations provides only $\omega_1(t)$, $\omega_2(t)$, and $\omega_3(t)$, and thus describes the behavior of $\boldsymbol{\omega} = \omega_1\hat{\mathbf{e}}_1 + \omega_2\hat{\mathbf{e}}_2 + \omega_3\hat{\mathbf{e}}_3$ relative to the body. If we care also about the motion of the body in inertial space, we must introduce some set of attitude variables, such as the Euler angles used in §9·1·3 (and first introduced in §3·2·3). If we elect this option, we must for a complete dynamical description augment the dynamical equations (9·1) by kinematical equations relating ω_1, ω_2, and ω_3 to φ, θ, and ψ and their time derivatives. From Eq. (9·16) and Fig. 9·3, we have

$$\begin{aligned} \omega &= \omega_1\hat{\mathbf{e}}_1 + \omega_2\hat{\mathbf{e}}_2 + \omega_3\hat{\mathbf{e}}_3 \\ &= \dot{\theta}\hat{\mathbf{k}}_1 + \dot{\varphi}\sin\theta\,\hat{\mathbf{k}}_2 + (\dot{\psi} + \dot{\varphi}\cos\theta)\hat{\mathbf{k}}_3 \\ &= \dot{\theta}(\hat{\mathbf{e}}_1\cos\psi - \hat{\mathbf{e}}_2\sin\psi) + \dot{\varphi}\sin\theta(\hat{\mathbf{e}}_2\cos\psi + \hat{\mathbf{e}}_1\sin\psi) \\ &\qquad + (\dot{\psi} + \dot{\varphi}\cos\theta)\hat{\mathbf{e}}_3 \end{aligned} \tag{9·29}$$

so that the required scalar kinematical equations are

$$\omega_1 = \dot{\theta}\cos\psi + \dot{\varphi}\sin\theta\sin\psi \tag{9·30a}$$
$$\omega_2 = -\dot{\theta}\sin\psi + \dot{\varphi}\sin\theta\cos\psi \tag{9·30b}$$
$$\omega_3 = \dot{\psi} + \dot{\varphi}\cos\theta \tag{9·30c}$$

A complete determination of the motion of a rigid body with one point fixed and subject to some specified moment $\mathbf{M} = M_1\hat{\mathbf{e}}_1 + M_2\hat{\mathbf{e}}_2 + M_3\hat{\mathbf{e}}_3$ would require the simultaneous solution of the six first-order nonlinear differential equations recorded here as Eqs. (9·1 *a*, *b*, *c*) and (9·30 *a*, *b*, *c*). Unfortunately, we cannot hope to integrate these equations literally (nonnumerically) even for the simplest nonzero values of $\mathbf{M}$. Even when $\mathbf{M} \equiv 0$ (as in Sec. 9·1), literal integration can be accomplished only with the introduction of elliptic functions and elliptic integrals. Dynamic analysis of a rigid body with one point fixed requires numerical integration with a digital computer, except in a few special cases such as that described in §9·2·2.

9·2·2 The spinning axisymmetric body under the influence of gravity

We shall specialize the general case in §9·2·1 in two ways: (1) The body has an axis of inertial symmetry through the support point o, so that for two principal axes through o, the moments of inertia are the same (say $I_1 = I_2 \triangleq I_t$), and (2) the external moment is due to uniform gravitational attraction, so that the mass center is also the center of gravity and, if $\mathbf{R}_c$ locates the mass center from o, the moment about o is

$$\mathbf{M} = -\mathbf{R}_c \times \mathfrak{M}g\hat{\mathbf{i}}_3 \tag{9·31}$$

where $\mathfrak{M}$ = body mass
g = gravitational constant
$-\hat{\mathbf{i}}_3$ = inertially fixed unit vector in the direction of gravitational attraction

Now we let $\hat{\mathbf{e}}_3$ parallel the axis of inertial symmetry through o, and introduce the Euler angles as shown for the idealized spinning top in Fig. 9·8 (which should be compared to Fig. 9·3). We can see that if a is the distance from o to the mass center then, from Eq. (9·31),

$$\mathbf{M} = -a\hat{\mathbf{e}}_3 \times \mathfrak{M}g\hat{\mathbf{i}}_3 = \mathfrak{M}ga \sin\theta\,\hat{\mathbf{k}}_1 \tag{9·32}$$

If we insist on using Euler's equations as in §9·2·1, or the special case of Euler's equations applicable to axisymmetric bodies [see Eqs. (9·2)], then we

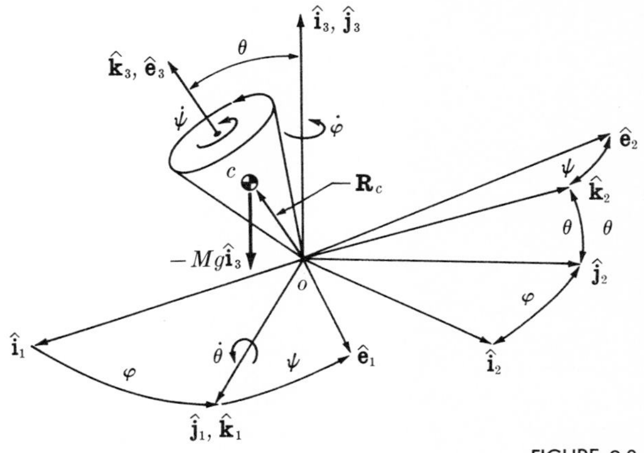

FIGURE 9·8

will be obliged to work with $\mathbf{M}$ as written in terms of body-fixed unit vectors $\hat{\mathbf{e}}_1$, $\hat{\mathbf{e}}_2$, and $\hat{\mathbf{e}}_3$, that is [noting Eq. (9·32)], with

$$\begin{aligned}\mathbf{M} &= \mathfrak{M}ga \sin\theta(\hat{\mathbf{e}}_1 \cos\psi - \hat{\mathbf{e}}_2 \sin\psi) \\ &= M_1\hat{\mathbf{e}}_1 + M_2\hat{\mathbf{e}}_2 + M_3\hat{\mathbf{e}}_3\end{aligned} \tag{9·33}$$

Moreover, we will require $\boldsymbol{\omega}$ as written in the same body-fixed vector basis, as in Eq. (9·29). If we substitute M_1, M_2, and M_3 from Eq. (9·33) and ω_1, ω_2, and ω_3 from Eqs. (9·30) into Euler's equations for the axisymmetric case [Eqs. (9·2)], we find

$$\begin{aligned}\mathfrak{M}ga \sin\theta\cos\psi = I_t(&\ddot{\theta}\cos\psi - \dot{\theta}\dot{\psi}\sin\psi \\ &+ \ddot{\varphi}\sin\theta\sin\psi + \dot{\varphi}\dot{\theta}\cos\theta\sin\psi + \dot{\varphi}\dot{\psi}\sin\theta\cos\psi) \\ &+ (I_s - I_t)(-\dot{\theta}\sin\psi + \dot{\varphi}\sin\theta\cos\psi)(\dot{\psi} + \dot{\varphi}\cos\theta)\end{aligned} \tag{9·34a}$$

$$\begin{aligned}-\mathfrak{M}ga \sin\theta\sin\psi = I_t(&-\ddot{\theta}\sin\psi - \dot{\theta}\dot{\psi}\cos\psi \\ &+ \ddot{\varphi}\sin\theta\cos\psi + \dot{\varphi}\dot{\theta}\cos\theta\cos\psi - \dot{\varphi}\dot{\psi}\sin\theta\sin\psi) \\ &- (I_s - I_t)(\dot{\theta}\cos\psi + \dot{\varphi}\sin\theta\sin\psi)(\dot{\psi} + \dot{\varphi}\cos\theta)\end{aligned} \tag{9·34b}$$

and

$$0 = I_s(\ddot{\psi} + \ddot{\varphi}\cos\theta - \dot{\varphi}\dot{\theta}\sin\theta) \tag{9·34c}$$

We could stare at these equations for a very long time without finding a way to solve them. Perhaps if we give the question some thought (rather than just plunging into Euler's equations), we can formulate the equations of motion more simply.

The special feature of this problem is the fact that $\mathbf{M}$ is directed along the unit vector $\hat{\mathbf{k}}_1$ [see Eq. (9·32) and Fig. 9·8], and because of symmetry $\hat{\mathbf{k}}_1$ always parallels a principal axis through o (even though $\hat{\mathbf{k}}_1$ is not fixed in the body). This suggests that it might be simpler to bypass Euler's scalar equations for the more fundamental vector form $\mathbf{M} = \dot{\mathbf{H}}$, and then write this vector equation in the vector basis $\hat{\mathbf{k}}_1$, $\hat{\mathbf{k}}_2$, and $\hat{\mathbf{k}}_3$.

To proceed with this plan, we can write $\mathbf{H}$ in vector-dyadic form as in §8·3·3, and shift the differentiation of $\mathbf{H}$ from inertial space to the reference frame k established by $\hat{\mathbf{k}}_1$, $\hat{\mathbf{k}}_2$, and $\hat{\mathbf{k}}_3$. The result is

$$\begin{aligned}\mathbf{M} &= \frac{^i d}{dt}\mathbf{H} = \frac{^k d}{dt}\mathbf{H} + \boldsymbol{\omega}^k \times \mathbf{H} \\ &= \frac{^k d}{dt}(\mathbb{I}\cdot\boldsymbol{\omega}) + \boldsymbol{\omega}^k \times \mathbb{I}\cdot\boldsymbol{\omega}\end{aligned} \tag{9·35}$$

where $\boldsymbol{\omega}^k$ is the inertial angular velocity of frame k, which from Fig. 9·8 is given by

$$\boldsymbol{\omega}^k = \dot{\varphi}\hat{\mathbf{i}}_3 + \dot{\theta}\hat{\mathbf{k}}_1 = \dot{\theta}\hat{\mathbf{k}}_1 + \dot{\varphi}\sin\theta\,\hat{\mathbf{k}}_2 + \dot{\varphi}\cos\theta\,\hat{\mathbf{k}}_3 \tag{9·36}$$

Since the axis of inertial symmetry of the body is fixed in frame k, the derivative of the inertia dyadic $\mathbb{I}$ in k is zero. Moreover, since

$$\boldsymbol{\omega} = \boldsymbol{\omega}^k + \dot{\psi}\hat{\mathbf{k}}_3 \tag{9·37}$$

Eq. (9·35) becomes

$$\begin{aligned}\mathbf{M} &= \mathbb{I} \cdot \frac{{}^k d}{dt} (\boldsymbol{\omega}^k + \dot{\psi}\hat{\mathbf{k}}_3) + \boldsymbol{\omega}^k \times \mathbb{I} \cdot (\boldsymbol{\omega}^k + \dot{\psi}\hat{\mathbf{k}}_3) \\ &= \mathbb{I} \cdot [\ddot{\theta}\hat{\mathbf{k}}_1 + (\ddot{\varphi} \sin \theta + \dot{\varphi}\dot{\theta} \cos \theta)\hat{\mathbf{k}}_2 + (\ddot{\varphi} \cos \theta - \dot{\varphi}\dot{\theta} \sin \theta)\hat{\mathbf{k}}_3 + \ddot{\psi}\hat{\mathbf{k}}_3] \\ &\quad + (\dot{\theta}\hat{\mathbf{k}}_1 + \dot{\varphi} \sin \theta \, \hat{\mathbf{k}}_2 + \dot{\varphi} \cos \theta \, \hat{\mathbf{k}}_3) \times \mathbb{I} \\ &\qquad \cdot [\dot{\theta}\hat{\mathbf{k}}_1 + \dot{\varphi} \sin \theta \, \hat{\mathbf{k}}_2 + (\dot{\varphi} \cos \theta + \dot{\psi})\hat{\mathbf{k}}_3]\end{aligned} \tag{9·38}$$

Substituting $\mathbf{M}$ from Eq. (9·32) and

$$\mathbb{I} = I_t\hat{\mathbf{k}}_1\hat{\mathbf{k}}_1 + I_t\hat{\mathbf{k}}_2\hat{\mathbf{k}}_2 + I_s\hat{\mathbf{k}}_3\hat{\mathbf{k}}_3 \tag{9·39}$$

into Eq. (9·38) and multiplying produces

$$\mathfrak{M}ga \sin \theta = I_t\ddot{\theta} + (I_s - I_t)\dot{\varphi}^2 \sin \theta \cos \theta + I_s\dot{\psi}\dot{\varphi} \sin \theta \tag{9·40a}$$
$$0 = I_t(\ddot{\varphi} \sin \theta + \dot{\varphi}\dot{\theta} \cos \theta) - (I_s - I_t)\dot{\varphi}\dot{\theta} \cos \theta - I_s\dot{\theta}\dot{\psi} \tag{9·40b}$$
$$0 = I_s(\ddot{\varphi} \cos \theta - \dot{\varphi}\dot{\theta} \sin \theta + \ddot{\psi}) \tag{9·40c}$$

These equations are much simpler in appearance than Eqs. (9·34), which come directly from Euler's equations, but they must be equivalent. If we compare them carefully, we can see that we could have obtained Eq. (9·40a) by multiplying Eq. (9·34a) by $\cos \psi$ and multiplying Eq. (9·34b) by $-\sin \psi$ and adding, and with similar manipulation of Eqs. (9·34) we could have found Eq. (9·40b); Eqs. (9·34c) and (9·40c) are identical. It is generally better strategy, however, to seek the simpler equations in the course of the dynamic analysis, rather than by purely algebraic manipulations of more complicated equations.

Even Eqs. (9·40) are rather discouraging, and we might explore the possibility of simplifying them further by starting all over again with the derivation of the equations of motion, employing this time Lagrange's equations (see §8·5·9).

The only forces that do work in this problem are gravity forces, the conservative nature of which permits the use of Lagrange's equations in the form [see Eq. (6·46)]

$$\frac{d}{dt}\left(\frac{\partial L}{\partial \dot{q}_\alpha}\right) - \frac{\partial L}{\partial q_\alpha} = 0 \qquad \alpha = 1, 2, 3 \tag{9·41}$$

where the lagrangian L is the kinetic energy T minus the potential energy V. In this problem we will assign the labels $q_1 = \varphi$, $q_2 = \theta$, and $q_3 = \psi$ to the generalized coordinates, and write the potential energy as

$$V = \mathfrak{M}ga \cos \theta \tag{9·42}$$

choosing our datum such that $V = 0$ when the body center of gravity is at the level of the support point o.

The kinetic energy is available from Eqs. (8·89), (9·29), and (9·39) as

$$\begin{aligned} T &= \tfrac{1}{2}[\dot\theta\hat{\mathbf{k}}_1 + \dot\varphi \sin\theta\,\hat{\mathbf{k}}_2 + (\dot\psi + \dot\varphi\cos\theta)\hat{\mathbf{k}}_3] \cdot (I_t\hat{\mathbf{k}}_1\hat{\mathbf{k}}_1 + I_t\hat{\mathbf{k}}_2\hat{\mathbf{k}}_2 + I_s\hat{\mathbf{k}}_3\hat{\mathbf{k}}_3) \\ &\qquad \cdot [\dot\theta\hat{\mathbf{k}}_1 + \dot\varphi \sin\theta\,\hat{\mathbf{k}}_2 + (\dot\psi + \dot\varphi\cos\theta)\hat{\mathbf{k}}_3] \\ &= \tfrac{1}{2}[I_t(\dot\theta^2 + \dot\varphi^2\sin^2\theta) + I_s(\dot\psi + \dot\varphi\cos\theta)^2] \end{aligned} \tag{9·43}$$

so that

$$L = \tfrac{1}{2}I_t(\dot\theta^2 + \dot\varphi^2\sin^2\theta) + \tfrac{1}{2}I_s(\dot\psi + \dot\varphi\cos\theta)^2 - \mathfrak{M}ga\cos\theta \tag{9·44}$$

The remarkable thing[1] about L is its lack of dependence on φ or ψ. This means that Eqs. (9·41) become simply

$$\frac{d}{dt}\left(\frac{\partial L}{\partial\dot\varphi}\right) = 0 \tag{9·45a}$$

$$\frac{d}{dt}\left(\frac{\partial L}{\partial\dot\theta}\right) - \frac{\partial L}{\partial\theta} = 0 \tag{9·45b}$$

$$\frac{d}{dt}\left(\frac{\partial L}{\partial\dot\psi}\right) = 0 \tag{9·45c}$$

The first and third of these equations can obviously be integrated, to obtain

$$\frac{\partial L}{\partial\dot\varphi} = I_t\dot\varphi\sin^2\theta + I_s(\dot\psi + \dot\varphi\cos\theta)\cos\theta = p_{10} \qquad \text{constant} \tag{9·46a}$$

and

$$\frac{\partial L}{\partial\dot\psi} = I_s(\dot\psi + \dot\varphi\cos\theta) = p_{30} \qquad \text{constant} \tag{9·46b}$$

while Eq. (9·45b) expands to

$$I_t\ddot\theta - I_t\dot\varphi^2\sin\theta\cos\theta - I_s(\dot\psi + \dot\varphi\cos\theta)(-\dot\varphi\sin\theta) - \mathfrak{M}ga\sin\theta = 0$$

or

$$\mathfrak{M}ga\sin\theta = I_t\ddot\theta + (I_s - I_t)\dot\varphi^2\sin\theta\cos\theta + I_s\dot\psi\dot\varphi\sin\theta \tag{9·46c}$$

As we attempt to compare Eqs. (9·46) and (9·40), which must be equivalent, we can see immediately that Eq. (9·46c) is identical to (9·40a), and (9·46b) is equivalent to (9·40c); but only with difficulty can we discern that by multiplying Eq. (9·40b) by $\sin\theta$ and adding the result to the product of Eq. (9·40c) with $\cos\theta$ we can get an equation that can be integrated to obtain Eq. (9·46a).

It seems quite clear that we have greatly simplified our problem by attacking it with Lagrange's equations. Now that we have found the simplest

[1] A generalized coordinate that does not appear in the lagrangian is called a *cyclic* or *ignorable coordinate*.

form for the differential equations, we can try to see what they tell us about the motion of the body in Fig. 9·8.

Each of the three approaches pursued here has led us quite easily to the conclusion that the component of $\boldsymbol{\omega}$ along the symmetry axis is a constant, which we will call Ω, so that

$$\dot{\psi} + \dot{\varphi} \cos \theta = \Omega \qquad \text{constant} \tag{9·47a}$$

and

$$I_s \Omega = p_{30} \tag{9·47b}$$

[see Eq. (9·46b) or (9·40c) or (9·34c)]. If we substitute this expression into Eq. (9·46a), we have

$$I_t \dot{\varphi} \sin^2 \theta + I_s \Omega \cos \theta = p_{10}$$

or

$$\dot{\varphi} = \frac{p_{10} - I_s \Omega \cos \theta}{I_t \sin^2 \theta} = \frac{p_{10} - p_{30} \cos \theta}{I_t \sin^2 \theta} \tag{9·48}$$

Equations (9·47) and (9·48) combine to provide

$$\dot{\psi} = \Omega - \frac{(p_{10} - I_s \Omega \cos \theta) \cos \theta}{I_t \sin^2 \theta}$$

or

$$\dot{\psi} = \frac{\Omega(I_t \sin^2 \theta + I_s \cos^2 \theta) - p_{10} \cos \theta}{I_t \sin^2 \theta} \tag{9·49}$$

Thus we have both $\dot{\varphi}$ and $\dot{\psi}$ written explicitly as functions of the angle θ. To obtain information concerning the time behavior of θ, we must obtain another integral of the equations of motion. We could proceed in a purely mathematical fashion, substituting Eqs. (9·48) and (9·49) into (9·46c) to obtain a second-order differential equation involving $\ddot{\theta}$ and θ only, but the result would not be easily integrated. It is better to seek another integral through our understanding of the dynamics of the problem.

As far back as §4·3·4, we noted that if all forces applied to a particle are conservative, then the total mechanical energy $T + V$ is constant [see Eq. (4·67a)]. Extension to the rigid body in the present problem implies that Eqs. (9·43) and (9·42) give us

$$\tfrac{1}{2} I_t (\dot{\theta}^2 + \dot{\varphi}^2 \sin^2 \theta) + \tfrac{1}{2} I_s (\dot{\psi} + \dot{\varphi} \cos \theta)^2 + \mathfrak{M} g a \cos \theta = T_0 + V_0$$

where T_0 and V_0 are the initial values of T and V. Substituting from Eqs. (9·47) and (9·48) produces

$$\tfrac{1}{2} I_t \dot{\theta}^2 + \tfrac{1}{2} I_t \left(\frac{p_{10} - I_s \Omega \cos \theta}{I_t \sin^2 \theta} \right)^2 \sin^2 \theta + \tfrac{1}{2} I_s \Omega^2 + \mathfrak{M} g a \cos \theta = T_0 + V_0$$

or

$$\dot{\theta}^2 = \frac{2(T_0 + V_0)}{I_t} - \frac{I_s}{I_t}\Omega^2 - \left(\frac{p_{10} - I_s\Omega\cos\theta}{I_t\sin\theta}\right)^2 - \frac{2\mathfrak{M}ga\cos\theta}{I_t} \tag{9·50}$$

Now we have in Eqs. (9·48) to (9·50) expressions for $\dot{\varphi}$, $\dot{\psi}$, and $\dot{\theta}$, all of which depend only on θ and the constants p_{10}, T_0, Ω, and V_0 determined by initial conditions. The literal solution of these equations hinges on the integration of Eq. (9·50), which can be accomplished only indirectly and then only by introducing elliptic integrals and elliptic functions. Since our primary interest here is in concepts and methodology rather than explicit formulas for the motion of a top, we will leave this task for a more advanced text, and content ourselves with a qualitative analysis of the motion.

Equations (9·46) are all satisfied by θ identically zero, indicating that the idealized top can remain upright for any spin rate. (We cannot confirm this conclusion by experiment except for relatively high spin rates, since for slow spin the upright motion can be shown[1] to be unstable.)

We can gain some understanding of the motion of the idealized top for $\theta \neq 0$ by establishing those conditions for which $\dot{\theta} = 0$. If the top spins too slowly (for example, if $p_{10} = p_{30} = 0$), there may be no circumstances under which $\dot{\theta} = 0$ when $\theta \neq 0$, but Eq. (9·50) tells us that this is not the case for all initial conditions. If we assume that $\sin\theta \neq 0$, then we can multiply Eq. (9·50) by $\sin^2\theta = 1 - \cos^2\theta$, and in terms of the new symbols

$$u \triangleq \cos\theta \qquad \text{and} \qquad \dot{u} = -\dot{\theta}\sin\theta \tag{9·51}$$

we can rewrite Eq. (9·50) in the more convenient form

$$f(u) = \dot{u}^2 \tag{9·52a}$$

where

$$f(u) \triangleq \{I_t[2(T_0 + V_0) - I_tI_s\Omega^2](1 - u^2) - (p_{10} - I_s\Omega u)^2 - 2I_t\mathfrak{M}gau(1 - u^2)\}/I_t^2 \tag{9·52b}$$

We can plot $f(u)$ versus u, as in Fig. 9·9, to determine those circumstances under which $f(u) = 0$, so that $\dot{u} = \dot{\theta} = 0$. Equation (9·52$b$) indicates that

$$\lim_{u\to\infty} f(u) = \infty$$
$$\lim_{u\to-\infty} f(u) = -\infty$$

and

$$I_t^2 f(1) = -(p_{10} - I_s\Omega)^2 \leqq 0$$
$$I_t^2 f(-1) = -(p_{10} + I_s\Omega)^2 \leqq 0$$

[1] See Leonard Meirovitch, "Methods of Analytical Dynamics," chap. 4, pp. 149–157, McGraw-Hill Book Company, New York, 1970.

In a real physical situation, we can be sure that $\dot{u}^2 > 0$, since $\dot{u} = -\dot{\theta} \sin \theta$ is a real variable. This means that in the physically realizable range of u between -1 and $+1$, we must somewhere have $f(u) > 0$.

With these constraints, the plot of $f(u)$ versus u must appear as in either Fig. 9·9a or b. In the former case, $\dot{\theta} = 0$ when θ reaches either of two distinct values given by

$$\theta_1 = \cos^{-1} u_1 \qquad \text{and} \qquad \theta_2 = \cos^{-1} u_2 \tag{9·53}$$

so that θ oscillates between these two values. In the latter case, θ maintains a constant value θ^* given by

$$\theta^* = \cos^{-1} u_1 = \cos^{-1} u_2 = \cos^{-1} u^* \tag{9·54}$$

as the top spins at the rate $\dot{\psi}^*$ given by Eq. (9·49) with θ replaced by θ^*, and the symmetry axis traces out a conical locus in inertial space at the rate $\dot{\varphi}^*$

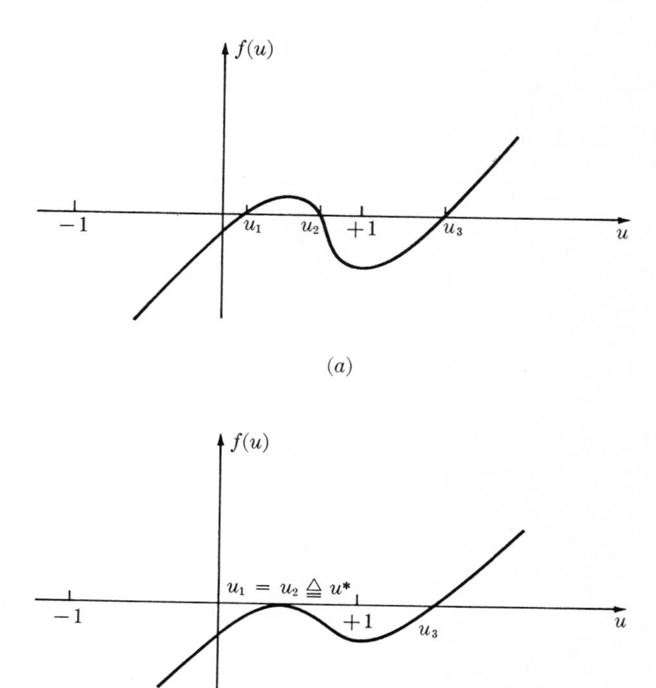

FIGURE 9·9

given by Eq. (9·48) with θ^* for θ. The solution $f(u_3) = 0$ has no physical significance since $u_3 > 1$ and $|\cos \theta| \leqq 1$.

We can obtain new insight into possible values for $\dot{\varphi}^*$ by observing that when θ remains at θ^*, then $\ddot{\theta} = 0$, so that Eq. (9·46c) becomes

$$-I_t\dot{\varphi}^{*2} \sin \theta^* \cos \theta^* + I_s(\dot{\psi}^* + \dot{\varphi}^* \cos \theta^*)(\dot{\varphi}^* \sin \theta^*) - \mathfrak{M}ga \sin \theta^* = 0$$

Substituting for $I_s(\dot{\psi}^* + \dot{\varphi}^* \cos \theta^*)$ from Eq. (9·46b) and dividing by $-\sin \theta^*$ produces the quadratic equation in $\dot{\varphi}^*$,

$$I_t \cos \theta^* \dot{\varphi}^{*2} - p_{30}\dot{\varphi}^* + \mathfrak{M}ga = 0$$

with solutions

$$\dot{\varphi}^* = \frac{p_{30} \pm \sqrt{p_{30}{}^2 - 4I_t \cos \theta^* \mathfrak{M}ga}}{2I_t \cos \theta^*}$$

or

$$\dot{\varphi}^* = \frac{p_{30}}{2I_t \cos \theta^*}\left(1 \pm \sqrt{1 - \frac{4I_t\mathfrak{M}ga \cos \theta^*}{p_{30}{}^2}}\right)$$

that is,

$$\dot{\varphi}^* = \frac{I_s\Omega}{2I_t \cos \theta^*}\left(1 \pm \sqrt{1 - \frac{4I_t\mathfrak{M}ga \cos \theta^*}{I_s{}^2\Omega^2}}\right) \tag{9·55}$$

As long as

$$p_{30}{}^2 > 4I_t\mathfrak{M}ga \cos \theta^* \tag{9·56}$$

there will be two solutions of Eq. (9·56), indicating that for the same constant value $\theta = \theta^*$ given by Eq. (9·54), there will be two different coning rates possible for the symmetry axis, as established by two different sets of initial conditions.

We have pursued the very special case of top motion for which θ has the constant value θ^*. As we have seen, more generally we can expect θ to oscillate between the upper and lower limits θ_1 and θ_2 given by Eq. (9·53). This nodding motion is classically called nutation (as in the moment-free case in Sec. 9·1), and the coning motion due to $\dot{\varphi}$ is called *forced precession.* Although solutions for θ as explicit functions of time are not available, more detailed analysis than is appropriate here can be marshalled to establish the character of the path of the symmetry axis intersection with a unit sphere. This path can be shown[1] to appear as shown in one of the three sketches portrayed in Fig. 9·10, with the choice de-

[1] See D. T. Greenwood, "Principles of Dynamics," Prentice-Hall, Inc., Englewood Cliffs, N.J., 1965; Leonard Meirovitch, "Methods of Analytical Dynamics," McGraw-Hill Book Company, New York, 1970.

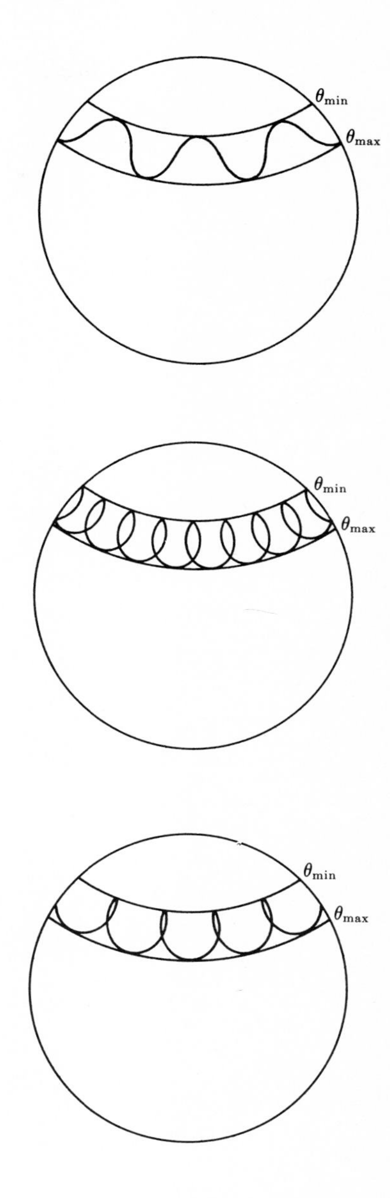

FIGURE 9·10

pending on initial conditions. Thus it is possible with sufficient diligence to obtain a good understanding of the motion of an *idealized* top.

An *actual* top differs in its behavior from the predictions of the preceding analysis in several important respects, due largely to the fact that its support never consists of a single contact point (but involves a contact *area*) and the support is never truly frictionless. As in Sec. 9·1, the idealized analysis is most useful as a starting point in developing an understanding of real physical systems.

9·3 RESPONSE TO MOMENT IMPULSES

9·3·1 Motion of a rigid body after a moment impulse

In §8·2·6 we examined the response of a rigid body to an angular impulse or moment impulse. We noted in Eq. (8·17) that for reference point c (or an inertially fixed reference point), we have

$$\int_{t_1}^{t_2} \mathbf{M}\, dt = \mathbf{H}(t_2) - \mathbf{H}(t_1) \tag{9·57}$$

where the integration is an inertial reference frame integral. Thus if we know $\mathbf{M}$ as a function of time in inertial space, we can evaluate the integral and determine the change in angular momentum in the interval $t_2 - t_1$. In general, we must recognize that in this time interval the body moves, and we might not be able to determine the motion without reverting to numerical integration of equations of motion. But if we assume an *ideal impulsive moment,* so that the interval $t_2 - t_1$ is considered to approach the limit 0 while the integral retains its value, then we can use Eq. (9·57) to calculate an instantaneous change in $\mathbf{H}$, and then use the free-rotation solutions of Sec. 9·1 to determine the ensuing motion. If the reference point for $\mathbf{H}$ is fixed in the body, then from $\mathbf{H} = \mathbb{I} \cdot \boldsymbol{\omega}$ [as in Eq. (8·25*b*)] we can determine $\boldsymbol{\omega}(t_2)$ from $\mathbf{H}(t_2)$, and thereby establish the "initial conditions" for the free-body motions beginning at time t_2. An example of this procedure is given in §8·2·6, although the results of Sec. 9·1 are required for a complete analysis.

9·3·2 Arbitrary moments as impulse trains

It is sometimes helpful for conceptual understanding to consider an applied moment $\mathbf{M}(t)$ to be made up of a sequence of ideal impulsive moments called an *impulse train.* Figure 9·11 illustrates the kind of approximation that might be imagined for a typical scalar component M_α of a moment

$$\mathbf{M} = M_1\hat{\mathbf{i}}_1 + M_2\hat{\mathbf{i}}_2 + M_3\hat{\mathbf{i}}_3$$

where $\hat{\mathbf{i}}_1$, $\hat{\mathbf{i}}_2$, and $\hat{\mathbf{i}}_3$ are inertially fixed unit vectors. After representing the function

$M_\alpha(t)$ by a stair-step pattern of rectangular pulses, we can replace each rectangle in the M_α,t space by an ideal moment impulse at its center in time, with impulse magnitude equal to the area of the rectangle. Then the method of §9·3·1 can be used to evaluate the change in angular momentum $\Delta\mathbf{H}$ due to each ideal impulse, and some attempt can be made to visualize the time variation in $\mathbf{H}$ due to the succession of moment impulses. This procedure is not intended as a computational method (although it bears some resemblance to the superposition integral discussed in §5·1·15); the objective here is simply an aid to visualization.

In the same vein, it is sometimes helpful to imagine that the angular-velocity vector $\boldsymbol{\omega}$ and the angular-momentum vector $\mathbf{H}$ remain aligned, although we have abundant evidence in Sec. 9·2 that this is not always the case. This alignment occurs whenever $\boldsymbol{\omega}$ parallels a principal axis of the body for the reference point, and when this is approximately true (as for a rapidly spinning body), then a general appreciation of the actual motion can often be gained by means of this approximation.

For example, we might improve our understanding of the steady coning or forced precessional motion of the spinning top in Sec. 9·2 by considering Fig. 9·12, in which the moment of the gravity force over a time interval Δt is replaced by the moment impulse $\mathfrak{M}ga \sin\theta\, \Delta t\, \hat{\mathbf{i}}_2$, which causes an increment

$$\Delta\mathbf{H} = \mathfrak{M}ga \sin\theta\, \Delta t\, \hat{\mathbf{i}}_2$$

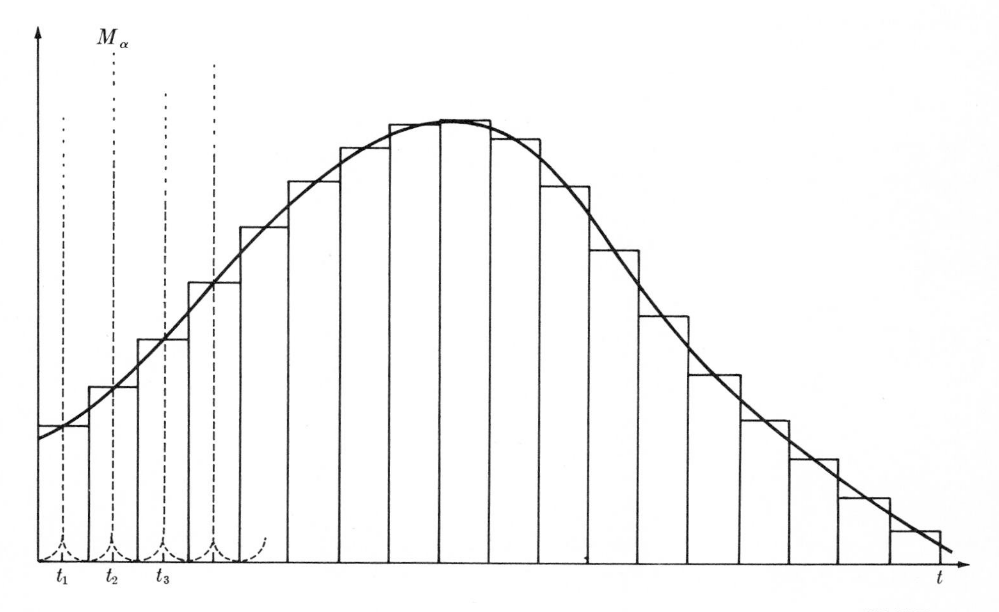

FIGURE 9·11

to be added to the angular momentum $\mathbf{H}_1 \triangleq \mathbf{H}(t_1)$ to obtain the new angular momentum $\mathbf{H}_2 \triangleq \mathbf{H}(t_1 + \Delta t)$. If we assume that the symmetry axis of the top remains aligned with $\mathbf{H}$, then we can in this rough fashion "explain" the steady coning motion discovered in Sec. 9·2. (Without such consideration, we might imagine that the force of gravity would make the center of gravity of the top fall continuously to an increasingly lower elevation, causing an unlimited increase in the angle θ.)

This impulse-train line of argument leads to a general "rule of thumb"

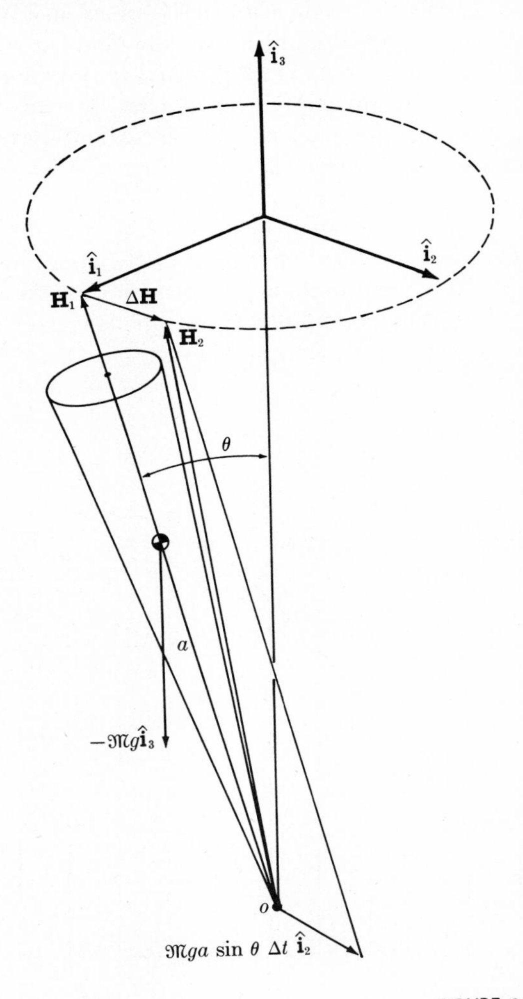

FIGURE 9·12

that is often helpful in understanding the dynamics of a rapidly spinning rigid body with $\boldsymbol{\omega}$ nominally aligned with a body axis called the *spin axis: The spin axis tends to move toward alignment with the external moment vector.*

Conversely, if a spinning body is *constrained* to acquire a new rotation about an axis parallel to some unit vector $\hat{\mathbf{u}}_r$ and normal to the original spin axis defined by $\hat{\mathbf{u}}_s$, then the constraints must apply not only a moment in the direction of the new axis of rotation $\hat{\mathbf{u}}_r$ but also a moment of direction established by $\hat{\mathbf{u}}_r \times \hat{\mathbf{u}}_s$. This moment must develop in order to resist the tendency of the spin axis $\hat{\mathbf{u}}_s$ to align itself with the rotation axis $\hat{\mathbf{u}}_r$.

For example, during retraction of the aircraft landing gear with spinning wheels shown in Fig. 9·13, there will develop a constraint torque $\mathbf{T}$ applied to the strut by the hinge mechanism at A, with the direction of $\mathbf{T}$ established by $\hat{\mathbf{u}}_r \times \hat{\mathbf{u}}_s$, where $\omega_s\hat{\mathbf{u}}_s$ is the wheel spin and $\omega_r\hat{\mathbf{u}}_r$ is the landing-gear rotation during retraction.

9·4 PERSPECTIVE

The most significant aspect of this chapter is its very limited scope. We have been completely successful in obtaining explicit solutions for three-dimensional motions of a rigid body *only* for the case of the free, inertially axisymmetric rigid body. We had some success with the general free rigid body and the inertially axisymmetric rigid body with one point fixed, and in more advanced texts these problems are carried closer to complete solution than we followed them; but centuries of study have revealed only a small catalog of rigid-body dynamics problems for which exact solutions in three-dimensional space can be obtained in closed form. Moreover, our limited investigation has revealed that those few problems for which solutions are completely known may not accurately repre-

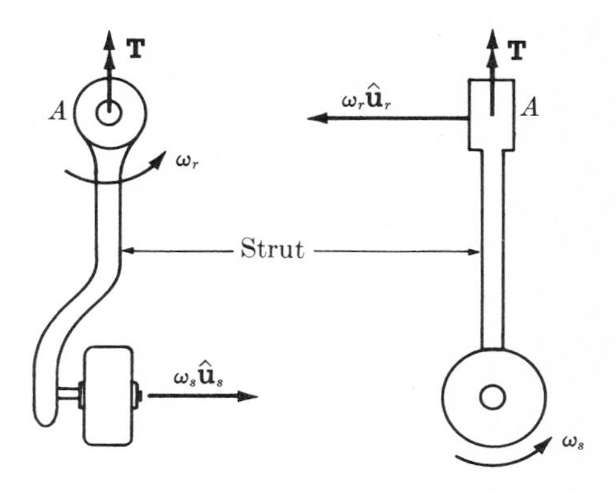

FIGURE 9·13

sent the behavior of physical objects. The lesson of *Explorer I* (see §9·1·5) should not be taken lightly, and the contrast between the solutions in §9·2·2 and the behavior of a child's top should be noted carefully. In both cases there are significant nonconservative forces of energy dissipation present in the physical system but omitted from the mathematical model.

Until the advent of the digital computer, rigid-body dynamics appeared to be essentially closed to further investigation; the possibilities for fruitful investigation appeared to be almost exhausted. Now that the computer has made numerical integration of equations of motion a routine operation, the emphasis in rigid-body dynamics has shifted from the difficult task of *solving* equations of motion of highly idealized problems to the quite different challenge of efficiently formulating equations of motion of realistically complex idealizations of physically significant problems. This is why so much attention has been given in this chapter to the formulation of equations of motion even for the relatively simple problems that permit some measure of closed-form solution. As you review this chapter, you should note carefully the search for a simpler set of equations that we pursued in §9·2·2. In this special case we found Lagrange's equations to be most revealing, while in the general case in §9·2·1 it is probably better, for digital computation, to remain with Euler's first-order equations. In another problem the advantages of these alternative procedures might be reversed, and another approach might be better than either of these. Only with an awareness of the variety of analytical tools at his disposal can the dynamicist develop real competence in selecting the best strategy for formulating the equations of motion of a given dynamical system.

PROBLEMS

9·1 An inertially axisymmetric flashlight is supported at its mass center c in a frictionless gimbal assembly that permits no moment to be applied to the flashlight (see Fig. P9·1). The moment of inertia of the satellite about its symmetry axis is $1/(1+\sqrt{2})$ times the moment of inertia for c about a transverse axis. The flashlight should illuminate on the ground a circular path whose radius R is the same as the height of c above the ground. Find a combination of orientation and angular velocity that, if selected for the initial state of the flashlight, will result automatically in the desired scanning motion.

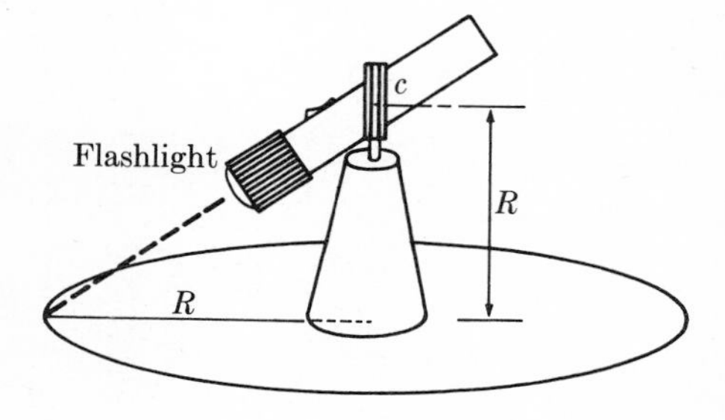

FIGURE P9·1

9·2 After burnout of the final stage of a launch vehicle for a spinning satellite, the axisymmetric combination of expended rocket and attached satellite (see Fig P9·2) is spinning and coning (freely precessing. In terms of the 3-1-3 Euler angles φ, θ, and ψ, the angle θ between the system angular momentum **H** and the symmetry axis $\hat{\mathbf{e}}_3$ is constant, and the rates $\dot{\varphi}$ and $\dot{\psi}$ are constants such that $\dot{\varphi}\dot{\psi} > 0$.

An ideal payload separation takes place; a small impulsive force between the mass centers is applied during separation, with no moments applied about the mass centers. Now the inertially symmetric, oblate satellite spins and cones with constant angle θ between satellite angular momentum and symmetry axis, and the corresponding angular rates $\dot{\varphi}$ and $\dot{\psi}$ for the 3-1-3 Euler angles of this system are constant. But $\dot{\varphi}\dot{\psi} < 0$.

Is there a discontinuity in the angular motion of the satellite indicated by this reversal of sign of $\dot{\varphi}\dot{\psi}$? Explain your answer very carefully, perhaps in terms of sketches of the "rolling cones."

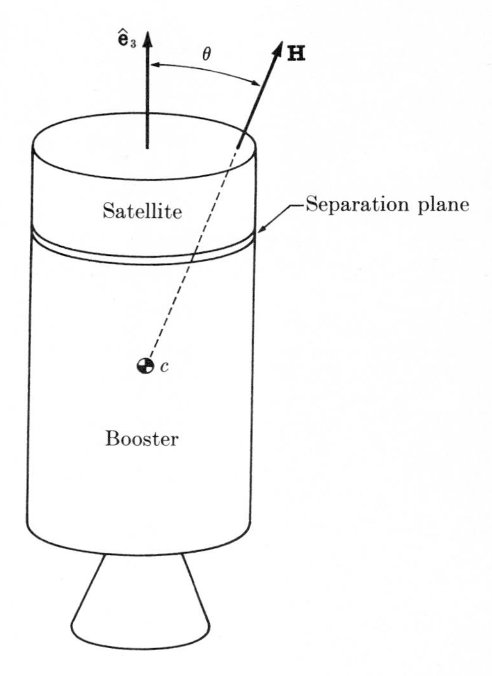

FIGURE P9·2

9·3 A multiple-payload booster carries three satellites, to be spin-stabilized and injected into different orbits. The final stage carries the three payloads as shown in Fig. P9·3; satellites II and III are toroidal, and all parts of the system have a common axis of symmetry (the 3 axis). This assembly is spun-up before the satellites are separated. The engine has a restart capability, so that each satellite can be separated mechanically at a time when the engine is off and all external torques can be neglected. Assume perfect alignment and separation mechanism performance; on separation, the payload and stage will develop a small relative velocity along the axis of symmetry (the 3 axis).

Consider first only the separation between the last satellite (II) and the propulsion stage (I), and assume that their

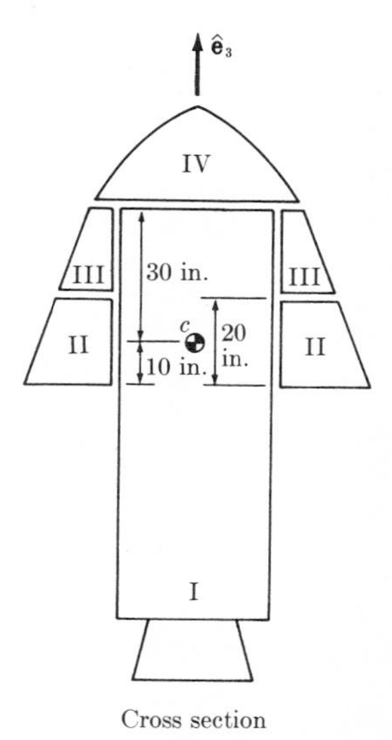

Cross section

FIGURE P9·3

mass centers are *coincident* before separation, and located 10 in. above the bottom of II. Let the height of II be $h_{II} =$ 20 in., and the distance from c to the forward bulkhead of body I be $h_I = 30$ in. Let the moments of inertia of I and II be given by $I_3{}^I = 2000$ ft lb sec^2; $I_1{}^I = 77{,}600$ ft lb sec^2; $I_3{}^{II} = 3000$ ft lb sec^2; and $I_1{}^{II} = 2400$ ft lb sec^2.

If the vehicle just prior to payload-II separation is spinning with $\omega_{30} = 100$ rpm and freely precessing on a cone with half-angle $\theta_0 = 1$ degree, and if the relative velocity of separation is assumed to be arbitrarily small (critical case), how much radial clearance is required between I and II to assure free separation? How would the analysis of separation dynamics of III be qualitatively different from and more involved than that for II?

9·4 For a body with principal axis moments of inertia related by $I_3 = 3I_2/2 = 2I_1$ (as in Fig. 9·5), find the relative size of the three principal dimensions of:

a. The energy ellipsoid
b. The momentum ellipsoid
c. The inertia ellipsoid

9·5 A satellite (see Fig. P9·5*a*) is given a rapid spin about the $\hat{e}_3$ axis while still attached to the final-stage booster. It is then separated from the booster, and for a brief time it spins freely about the principal axis of least inertia. Then the booms shown in Fig. P9·5*b* are telescopically deployed, so that on full deployment the $\hat{e}_3$ axis becomes the principal axis of maximum inertia.

a. For preliminary analysis, adopt the following assumptions:
   i All elements of the system are rigid and nondissipative.
   ii Deployment is perfectly symmetric, so four identical booms deploy identically.
   iii When deployment is initiated, the vehicle is spinning and coning slightly.

On the basis of these assumptions, would you recommend a slow or rapid deployment rate, or does it make any difference?

b. Abandon assumption (i) above, treating the vehicle as quasi-rigid but subject to internal energy dissipation, and answer the two questions again.

c. Be realistic, and recognize the flexibility and limited strength of the booms. Does this influence your recommendation regarding rate of deployment? What will excite boom bending?

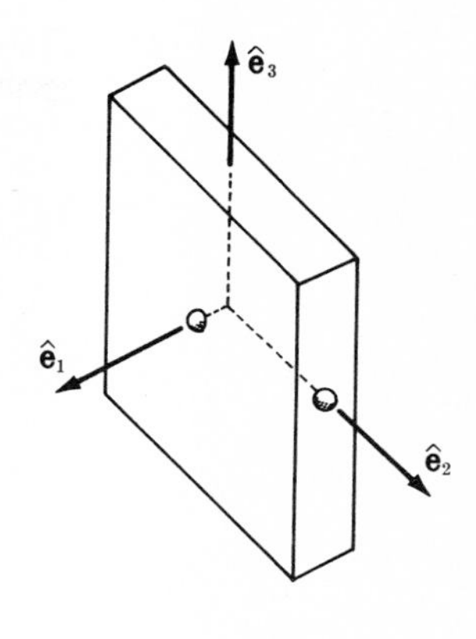

(a) $I_1 > I_2 > I_3$

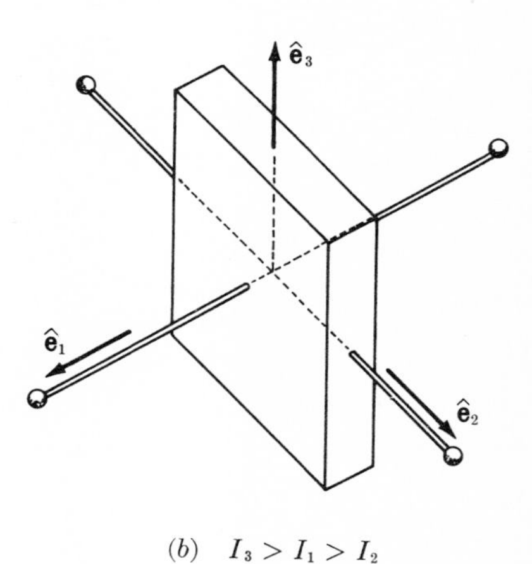

(b) $I_3 > I_1 > I_2$

FIGURE P9·5

9·6 Show that for a rigid body with prescribed angular momentum **H** the minimum kinetic energy state corresponds to rotation about the mass-center principal axis having the greatest moment of inertia.

9·7 An idealization of a dual-spin satellite (see Fig. P9·7) treats the cylindrical portion of the vehicle as rigid and symmetric, and the antenna structure as rigid. The antenna rotates in inertial space once per orbit, so the antenna points toward earth, while the rotor spins rapidly relative to inertial space. Assume that these components can perform free relative rotation about a common axis principal for each. Assume further that the vehicle is free of external torque, and that the equations of mass-center translation are assumed to be uncoupled from rotation.

If, in formulating the attitude equations, we select three Euler angles φ, θ, and ψ to establish the orientation of the antenna in inertial space and in addition an angle γ to fix the rotation of the rotor relative to the antenna, which of these coordinates can be called "cyclic" or "ignorable"? Does any other system of coordinates have more cyclic variables?

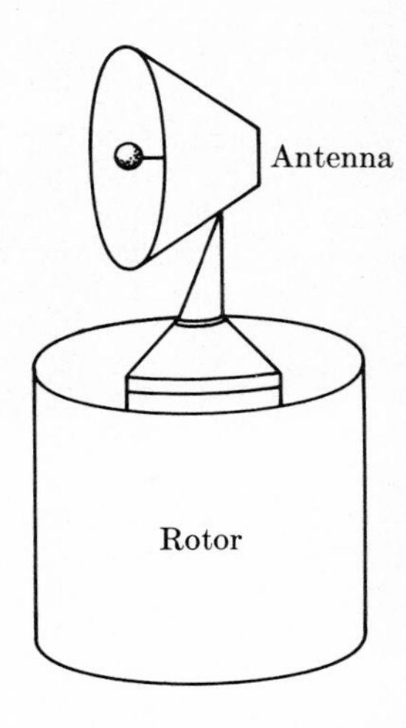

FIGURE P9·7

9·8
a. Derive Eqs. (9·40) from Eqs. (9·34) by algebraic manipulation.
b. Derive Eqs. (9·46) from Eqs. (9·40) by algebraic manipulation and integration.

9·9 Devise and perform some qualitative experiments with a top, and comment on the conceptual differences between the behavior of a top and the predictions of §9·2·2.

9·10 A freely spinning and coning axisymmetric body is subjected to a constant-magnitude torque vector parallel to the symmetry axis, by means of perfectly aligned "spin-up" jets.

a. Determine the time behavior of the angle θ between the angular-momentum vector $\mathbf{H}$ and the body axis of symmetry.
b. Is $\mathbf{H}$ of inertially fixed direction?

Note: The differential equations to be solved for (a) are relatively difficult.

equations tell the whole story, and for a nonrigid continuum these equations as applied to the total system only begin to describe the motion.

In our introduction to the concept of the rigid continuum in §7·2·1, we applied Newton's second law to the differential element of mass as portrayed in Fig. 7·4. In that context each differential element was itself a rigid body, and its relationship to other elements of the total rigid body was fully prescribed by the system geometry. The equations of motion of the differential elements served only as differential quantities to be integrated in order to obtain equations of motion of the total body.

When the continuum is nonrigid, the differential element itself becomes important since its deformations and displacements become unknowns to be determined by the equations of motion. Because these kinematical unknowns are distributed throughout the continuum, rather than identified with discrete objects such as particles and rigid bodies, the equations of motion involve unknown

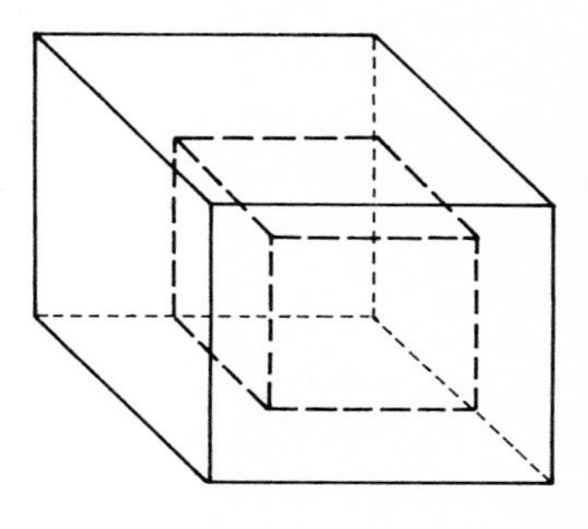

(*a*) Dilatation

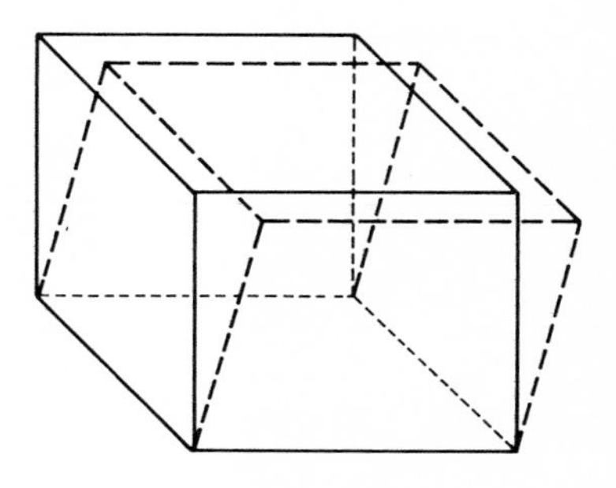

(*b*) Distortion

FIGURE A·1

functions rather than a finite number of unknown *coordinates*. The equations of continuum mechanics are therefore *partial differential equations* rather than the *ordinary differential equations* which describe the motions of particles and rigid bodies.

Continua are classified according to their resistance to deformation or change of configuration, and different classes of continua are often analyzed quite differently, treated in different books, and studied in different courses.

If a differential element deforms in such a way as to change volume without changing shape, as in Fig. A·1*a*, that deformation is called *dilatation*, and if it changes shape without changing volume as in Fig. A·1*b*, then the term *distortion* is applied. If a material is idealized as a continuum that is not subject to dilatation but offers no resistance to distortion, the model is called an *incompressible inviscid fluid*. If the material resists distortion under dynamic loads but offers no resistance under static loads, then a *viscous fluid* model is adopted. If the idealized material sustains both limited dilatations and limited distortions under finite load, but returns on removal of the load to its original state, we call it an *elastic solid*. The term *viscoelastic solid* is applied if the idealized material dissipates energy (transforming mechanical energy into thermal energy) for all time-varying deformations. This list of continuum models is far from complete, but it should serve to explain why continuum mechanics has traditionally been studied as a collection of independent disciplines, often divided into the two areas of *fluid mechanics* and *solid mechanics*.

When a material is idealized as a solid but the distribution of matter is restricted to assemblages of deformable bodies having certain geometries permitting special assumptions and approximations (for example, the geometries of *beams*, *plates*, and *shells*), the study of solid mechanics adopts a more utilitarian aspect, and may bear the label *structural mechanics*.

These are the subjects that lie ahead. While the path is not an easy one, it can be rewarding intellectually and of great practical value as well.

APPENDIX B
MATHEMATICAL FORMALISMS IN MECHANICS: SCALARS, VECTORS, DYADICS, TENSORS, AND MATRICES

Mathematics may be defined as the subject in which we never know what we are talking about, nor whether what we are saying is true.

BERTRAND RUSSELL

Among the many things that we can talk about in the formal language of mathematics are the various quantities that comprise the subject of dynamics, quantities such as force, time, linear and angular velocity and acceleration, mass, and the inertia properties of a rigid body.

In the formulation and solution of problems in dynamics, several different mathematical formalisms are useful. In different treatments of dynamics we encounter *scalars, complex numbers, vectors, quaternions, tensors, dyadics, and matrices.* Although it is well to remember that the mathematical structure of the derivation of equations of motion is secondary for the dynamicist to the end of formulating and solving physically meaningful problems, it is still true that each of these concepts should be sufficiently familiar to permit its use in those applications for which it is the most natural or most efficient analytical tool.

The primary mathematical disciplines used in dynamics are algebra and differential and integral calculus, with trigonometry and geometry (once dominant) playing roles that vary with the analyst. The principal divergence in formalism today is in the dimension of the mathematical entity that is operated on by the rules of algebra and calculus.

The first algebra to which we are exposed is the algebra of scalars—each independent entity is a single scalar quantity. We next consider complex numbers, and construct a double algebra based on these dual-quantity mathematical entities. Sometimes complex numbers are given geometrical interpretation in the "complex plane."

The application of this double algebra to the geometry of a plane suggested to William Rowan Hamilton and others in the 1800s the idea of a triple algebra that could have a similar application to the geometry of three dimensions. Hamilton found he could add and subtract triplets but could not correctly multiply them. In 1843 he observed that geometrical operations in three-dimensional space require not three numbers but four, and he devised the *quaternion* as the quadruple-term entity appropriate for operations in three-dimensional space. A quaternion may be represented as

$$q = a + \mathbf{i}b + \mathbf{j}c + \mathbf{k}d$$

where a, b, c, and d are scalars and $\mathbf{i}$, $\mathbf{j}$, and $\mathbf{k}$ are so-called unit vectors that differ from the modern (gibbsian) unit vectors used in this text in that they have magnitude $\sqrt{-1}$ (so $\mathbf{i}^2 = \mathbf{j}^2 = \mathbf{k}^2 = -1$), and in the definition of vector products so that $\mathbf{ij} = \mathbf{k}$, etc. (permuting).

Hamilton viewed his quaternions as the crowning achievement of his career, and fully expected their adoption as the fundamental analytical tools of physics. They were employed by Maxwell in his famous *Treatise on Electricity and Magnetism* in 1873, but in the 1880s both J. Willard Gibbs and Oliver Heaviside concluded that they could make more efficient use of a modified version of only the triple-term *vector* portion of the quaternion. They each independently devised three-element systems, and in the process abandoned the complex numbers involved in quaternions. The vector analysis we use today is essentially in the form presented by Gibbs.

Among the reasons offered by Gibbs in defense of the triple algebra of vectors was the facility with which it could be extended to higher dimensions, whereas quaternions were confined in application to three-dimensional problems. Although Gibbs was initially ridiculed for his introduction of the fourth dimension into a discussion of physics, the advent of relativity theory made analysis in four dimensions a necessity, and the gibbsian vector was generalized by H. Minkowski and others to meet the needs of modern physics.

The bitter 20-yr controversy that raged over the relative merits of vectors and quaternions was settled only by the gradual acceptance of the former for applications by working physicists and engineers.

The correspondence between geometry and multidimensional algebras may facilitate visualization of algebraic operations, and it may also permit algebraic generalization of concepts originally defined as geometrical constructs. Gibbsian vectors provide an example. We may define (gibbsian) vectors wholly in terms of three-dimensional euclidean geometry: Vectors are directed line segments that add commutatively (so that $\mathbf{u} + \mathbf{v} = \mathbf{v} + \mathbf{u}$) and associatively [so that $\mathbf{u} + (\mathbf{v} + \mathbf{w}) = (\mathbf{u} + \mathbf{v}) + \mathbf{w}$] according to the parallelogram rule (see Fig. B·1). Vectors combine as scalar (dot) products and vector (cross) products defined, respectively, by the following operational rules expressed for two arbitrary vectors $\mathbf{u}$ and $\mathbf{v}$†

$$\mathbf{u} \cdot \mathbf{v} = |\mathbf{u}|\,|\mathbf{v}| \cos \langle \mathbf{u}, \mathbf{v} \rangle \tag{B·1}$$

and

$$\mathbf{u} \times \mathbf{v} = |\mathbf{u}|\,|\mathbf{v}| \sin \langle \mathbf{u}, \mathbf{v} \rangle \hat{\mathbf{e}}_n \tag{B·2}$$

where the unit vector $\hat{\mathbf{e}}_n$ is normal to both $\mathbf{u}$ and $\mathbf{v}$ according to the "right-hand rule"‡; $\langle \mathbf{u}, \mathbf{v} \rangle$ is the angle between $\mathbf{u}$ and $\mathbf{v}$; and the absolute value of a vector is its length. Vector dot multiplication is a commutative operation, so that

$$\mathbf{u} \cdot \mathbf{v} = \mathbf{v} \cdot \mathbf{u}$$

but cross multiplication is anticommutative, so that

$$\mathbf{u} \times \mathbf{v} = -\mathbf{v} \times \mathbf{u}$$

Gibbsian vectors will be discussed here in some detail before the alternative of an *algebraic definition* of vector is introduced.

The rules of vector addition permit the description of any vector in terms of a sum of vectors directed along any three noncoplanar lines; if these lines are

† Gibbsian vectors are printed in boldface; unit vectors (vectors of unit length) are designated by a caret (^).

‡ According to the right-hand rule, the direction of the vector product $\mathbf{u} \times \mathbf{v}$ is the direction in which the thumb of the right hand points when the extended fingers of that hand rotate from alignment with $\mathbf{u}$ to alignment with $\mathbf{v}$, with the fingers rotating toward the palm.

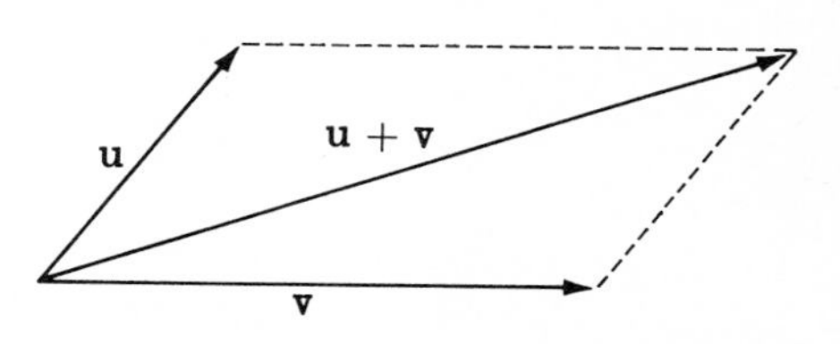

FIGURE B·1

established by three unit vectors $\hat{\mathbf{a}}$, $\hat{\mathbf{b}}$, and $\hat{\mathbf{c}}$, we can write any vector $\mathbf{v}$ in the form

$$\mathbf{v} = v_a\hat{\mathbf{a}} + v_b\hat{\mathbf{b}} + v_c\hat{\mathbf{c}}$$

where v_a, v_b, and v_c are suitable scalars.

It is often convenient to work with *dextral orthonormal triads*, which are sets of three mutually perpendicular vectors of length unity labeled according to the dextral or "right-hand" convention. In terms of the dextral orthonormal triad $\hat{\mathbf{e}}_1$, $\hat{\mathbf{e}}_2$, and $\hat{\mathbf{e}}_3$, any vector $\mathbf{v}$ may be written

$$\mathbf{v} = v_1\hat{\mathbf{e}}_1 + v_2\hat{\mathbf{e}}_2 + v_3\hat{\mathbf{e}}_3 = \sum_{\alpha=1}^{3} v_\alpha\hat{\mathbf{e}}_\alpha \tag{B·3}$$

The vector $\mathbf{v}$ is often written in the abbreviated notation $v_\alpha\hat{\mathbf{e}}_\alpha$, where the summation symbol Σ is omitted, being implied by the repetition of the index α, which is understood to range from 1 through 3. This notation employing the *summation convention* is avoided in this text, although it is very convenient for more advanced analysis. Several expressions in this Appendix are written in terms of the summation convention after more explicit representation, simply to illustrate the procedure. The set of unit vectors $\hat{\mathbf{e}}_1$, $\hat{\mathbf{e}}_2$, and $\hat{\mathbf{e}}_3$ comprise the *vector basis* of $\mathbf{v}$ as written in Eq. (B·3), and the scalars v_1, v_2, and v_3 in that equation are called the *scalar components* of $\mathbf{v}$ for vector basis $\hat{\mathbf{e}}_1$, $\hat{\mathbf{e}}_2$, and $\hat{\mathbf{e}}_3$.

Vector multiplication operations are often facilitated by their expression in terms of the *Kronecker delta* $\delta_{\alpha\beta}$ and the *epsilon symbol* $\epsilon_{\alpha\beta\gamma}$, as defined below:

$$\delta_{\alpha\beta} = \begin{cases} 1 & \alpha = \beta \\ 0 & \alpha \neq \beta \end{cases} \tag{B·4a}$$

$$\epsilon_{\alpha\beta\gamma} = \begin{cases} 1 & \alpha\beta\gamma = \text{cyclic permutation of } 1, 2, 3 \\ 0 & \text{any two indices are the same} \\ -1 & \alpha\beta\gamma = \text{cyclic permutation of } 1, 3, 2 \end{cases} \tag{B·4b}$$

In terms of these symbols, the scalar product of two orthonormal vectors $\hat{\mathbf{e}}_\alpha$ and $\hat{\mathbf{e}}_\beta$ is

$$\hat{\mathbf{e}}_\alpha \cdot \hat{\mathbf{e}}_\beta = \delta_{\alpha\beta} \tag{B·5}$$

as may be confirmed by inspection of the definition of Eq. (B·1). The vector product is available from the definition of Eq. (B·2) as

$$\hat{\mathbf{e}}_\alpha \times \hat{\mathbf{e}}_\beta = \sum_{\gamma=1}^{3} \epsilon_{\alpha\beta\gamma}\hat{\mathbf{e}}_\gamma \triangleq \epsilon_{\alpha\beta\gamma}\hat{\mathbf{e}}_\gamma \tag{B·6}$$

The scalar product of any two vectors **u** and **v** may now be written in the form

$$\mathbf{u}\cdot\mathbf{v} = \sum_{\alpha=1}^{3} u_\alpha \hat{\mathbf{e}}_\alpha \cdot \sum_{\beta=1}^{3} v_\beta \hat{\mathbf{e}}_\beta = \sum_{\alpha,\beta=1}^{3} u_\alpha v_\beta \delta_{\alpha\beta} = \sum_{\alpha=1}^{3} u_\alpha v_\alpha \triangleq u_\alpha v_\alpha \tag{B·7}$$

and their vector product may be expressed as

$$\mathbf{u}\times\mathbf{v} = \sum_{\alpha=1}^{3} u_\alpha \hat{\mathbf{e}}_\alpha \times \sum_{\beta=1}^{3} v_\beta \hat{\mathbf{e}}_\beta = \sum_{\alpha,\beta=1}^{3} u_\alpha v_\beta \epsilon_{\alpha\beta\gamma} \hat{\mathbf{e}}_\gamma \triangleq u_\alpha v_\beta \epsilon_{\alpha\beta\gamma} \hat{\mathbf{e}}_\gamma \tag{B·8a}$$

Equations (B·8) and (B·9) are analytically more convenient than the definitions in Eqs. (B·1) and (B·2), but they are fully equivalent. Yet another way to write Eq. (B·8*a*) is by means of the rules for expanding determinants, as follows:

$$\mathbf{u}\times\mathbf{v} = \begin{vmatrix} \hat{\mathbf{e}}_1 & \hat{\mathbf{e}}_2 & \hat{\mathbf{e}}_3 \\ u_1 & u_2 & u_3 \\ v_1 & v_2 & v_3 \end{vmatrix} = \hat{\mathbf{e}}_1(u_2v_3 - u_3v_2) - \hat{\mathbf{e}}_2(u_1v_3 - u_3v_1) + \hat{\mathbf{e}}_3(u_1v_2 - u_2v_1) \tag{B·8b}$$

Vector multiplication is sometimes facilitated by the following identities

$$\mathbf{u}\times\mathbf{v}\cdot\mathbf{w} = \mathbf{u}\cdot\mathbf{v}\times\mathbf{w} \tag{B·9a}$$

$$\mathbf{u}\times(\mathbf{v}\times\mathbf{w}) = \mathbf{u}\cdot\mathbf{w}\mathbf{v} - \mathbf{u}\cdot\mathbf{v}\mathbf{w} \tag{B·9b}$$

As any vector within the present definition may be written in terms of the vector basis $\hat{\mathbf{e}}_1$, $\hat{\mathbf{e}}_2$, and $\hat{\mathbf{e}}_3$, if we introduce a new vector basis $\hat{\mathbf{e}}_1'$, $\hat{\mathbf{e}}_2'$, and $\hat{\mathbf{e}}_3'$, we can always write

$$\hat{\mathbf{e}}_\alpha' = \sum_{\beta=1}^{3} c_{\alpha\beta}\hat{\mathbf{e}}_\beta \triangleq c_{\alpha\beta}\hat{\mathbf{e}}_\beta \qquad \alpha = 1, 2, 3 \tag{B·10}$$

By forming the scalar product

$$\hat{\mathbf{e}}_\alpha' \cdot \hat{\mathbf{e}}_\gamma = \sum_{\beta=1}^{3} c_{\alpha\beta}\hat{\mathbf{e}}_\beta \cdot \hat{\mathbf{e}}_\gamma = \sum_{\beta=1}^{3} c_{\alpha\beta}\delta_{\beta\gamma} = c_{\alpha\gamma} \tag{B·11}$$

and comparing the result with the geometrically oriented definition of Eq. (B·1), we recognize that $c_{\alpha\gamma}$ is the cosine of the angle between $\hat{\mathbf{e}}_\alpha'$ and $\hat{\mathbf{e}}_\gamma$, that is, the *direction cosine.*

In the study of kinematics (Chap. 3) we are concerned primarily with base vector changes that can be represented physically as rigid rotations, so that the set $\hat{\mathbf{e}}_1'$, $\hat{\mathbf{e}}_2'$, and $\hat{\mathbf{e}}_3'$ can be obtained by rotating the set $\hat{\mathbf{e}}_1$, $\hat{\mathbf{e}}_2$, and $\hat{\mathbf{e}}_3$ without disturbing the dextral orthonormal character of the set. Transformations in this class are known as *proper rotations*, and are characterized by certain relationships among the direction cosines relating $\hat{\mathbf{e}}_1'$, $\hat{\mathbf{e}}_2'$, $\hat{\mathbf{e}}_3'$ and $\hat{\mathbf{e}}_1$, $\hat{\mathbf{e}}_2$, $\hat{\mathbf{e}}_3$. Specifically, due to the preservation of orthonormality, we have

$$\delta_{\alpha\gamma} = \hat{\mathbf{e}}_\alpha' \cdot \hat{\mathbf{e}}_\gamma' = \sum_{\beta=1}^{3} c_{\alpha\beta}\hat{\mathbf{e}}_\beta \cdot \sum_{\delta=1}^{3} c_{\gamma\delta}\hat{\mathbf{e}}_\delta = \sum_{\beta,\delta=1}^{3} c_{\alpha\beta}c_{\gamma\delta}\delta_{\beta\delta}$$

or

$$\delta_{\alpha\gamma} = \sum_{\beta=1}^{3} c_{\alpha\beta}c_{\gamma\beta} \triangleq c_{\alpha\beta}c_{\alpha\beta} \tag{B·12}$$

If we choose to restate Eq. (B·12) in terms of the direction cosine matrix $[C] = [c_{\alpha\beta}]$ and the identity matrix $[U] = [\delta_{\alpha\gamma}]$, we have

$$[C][C]^T = [U] \tag{B·13}$$

where the superscript T indicates the matrix transpose, obtained by interchanging its rows and columns. By definition of the inverse $[C]^{-1}$ of the matrix $[C]$, we have

$$[C][C]^{-1} = [U] = [C]^{-1}[C] \tag{B·14}$$

By premultiplying Eq. (B·13) by $[C]^{-1}$, and then postmultiplying by $[C]$, we obtain

$$[C]^T[C] = [U] \tag{B·15}$$

With vector $\mathbf{v}$ represented in the vector basis $\hat{\mathbf{e}}_1$, $\hat{\mathbf{e}}_2$, and $\hat{\mathbf{e}}_3$, and $\hat{\mathbf{e}}_1$, $\hat{\mathbf{e}}_2$, and $\hat{\mathbf{e}}_3$ related by Eq. (B·10), we have

$$\mathbf{v} = \sum_{\gamma=1}^{3} v_\gamma\hat{\mathbf{e}}_\gamma = \sum_{\alpha=1}^{3} v_\alpha'\hat{\mathbf{e}}_\alpha'$$

The scalar product of $\mathbf{v}$ with $\hat{\mathbf{e}}_\rho$ is therefore

$$\sum_{\gamma=1}^{3} v_\gamma\hat{\mathbf{e}}_\gamma \cdot \hat{\mathbf{e}}_\rho = \sum_{\alpha=1}^{3} v_\alpha'\hat{\mathbf{e}}_\alpha' \cdot \hat{\mathbf{e}}_\rho$$

or

$$\sum_{\gamma=1}^{3} v_\gamma\delta_{\gamma\rho} = \sum_{\alpha,\beta=1}^{3} v_\alpha' c_{\alpha\beta}\hat{\mathbf{e}}_\beta \cdot \hat{\mathbf{e}}_\rho = \sum_{\alpha,\beta=1}^{3} v_\alpha' c_{\alpha\beta}\delta_{\beta\rho}$$

so that

$$v_\rho = \sum_{\alpha=1}^{3} v_\alpha' c_{\alpha\rho} \triangleq v_\alpha' c_{\alpha\rho} \qquad \rho = 1, 2, 3 \tag{B·16}$$

To obtain the primed components explicitly in terms of the unprimed components, multiply each of the three equations implied by Eq. (B·16) by $c_{\beta\rho}$ and sum these three equations. The result, with Eq. (B·12), becomes

$$\sum_{\rho=1}^{3} c_{\beta\rho}v_{\rho} = \sum_{\alpha,\rho=1}^{3} v'_{\alpha}c_{\alpha\rho}c_{\beta\rho} = \sum_{\alpha=1}^{3} v'_{\alpha}\delta_{\alpha\beta} = v'_{\beta}$$

The "free" indices β and the "dummy" indices ρ may, of course, be changed at will to any other letters; if we change β to α and ρ to β, we have for comparison with Eq. (B·10)

$$v'_{\alpha} = \sum_{\beta=1}^{3} c_{\alpha\beta}v_{\beta} \triangleq c_{\alpha\beta}v_{\beta} \qquad \alpha = 1, 2, 3 \tag{B·17}$$

Similarly, a vector equation inverse to Eq. (B·10) may be obtained for comparison with Eq. (B·16); multiply each of the three equations implied by Eq. (B·10) by $c_{\alpha\rho}$ and sum, to provide

$$\sum_{\alpha=1}^{3} c_{\alpha\rho}\hat{\mathbf{e}}'_{\alpha} = \sum_{\alpha,\beta=1}^{3} c_{\alpha\rho}c_{\alpha\beta}\hat{\mathbf{e}}_{\beta} = \sum_{\beta=1}^{3} \delta_{\rho\beta}\hat{\mathbf{e}}_{\beta} = \hat{\mathbf{e}}_{\rho} \qquad \rho = 1, 2, 3 \tag{B·18}$$

Note the direct correspondence between vector-basis relationships and the corresponding relationships among scalar components of the arbitrary vector $\mathbf{v}$.

Although we have begun here with Gibbs' basically geometrical concept of a vector, we now see that, in terms of a specified vector basis, all characteristics of a vector are known if we specify its three scalar components. Thus we could *begin* with a *definition* of a vector as an ordered set of three scalars (a triple), with prescribed rules for operating with scalars and other triples. Having taken this *algebraic* posture, we have no strong predisposition toward restricting the dimension to three, and so we can generalize (although we sacrifice the "vector product" or "cross product" in the process). The word *vector* is usually applied to the generalized n-dimensional concept in books devoted to linear algebra, but in this mechanics book the word vector always means a (three-dimensional) *gibbsian vector*.

Having established the feasibility of replacing the geometrical definition of gibbsian vectors by an algebraic definition in terms of the scalar components v_{α} of a vector $\mathbf{v} = \sum_{\alpha=1}^{3} v_{\alpha}\hat{\mathbf{e}}_{\alpha}$, and having developed the vector-transformation relationship [Eq. (B·17)] corresponding to a linear orthogonal transformation of a vector basis, we are prepared to generalize from these results to establish definitions for *cartesian tensors* of arbitrary order.

Tensors, like gibbsian vectors, were originally conceived in a geometrical context, but may alternatively be defined algebraically. We define a first-order cartesian tensor as a three-element entity whose scalar elements transform under a linear orthogonal transformation by

$$t'_\alpha = \sum_{\beta=1}^{3} c_{\alpha\beta} t_\beta \triangleq c_{\alpha\beta} t_\beta \tag{B·19}$$

If the nine scalars represented by the two-index symbol $t_{\alpha\beta}$ transform under a linear orthogonal transformation according to

$$t'_{\alpha\beta} = \sum_{\delta,\gamma=1}^{3} c_{\alpha\gamma} c_{\beta\delta} t_{\gamma\delta} \triangleq c_{\alpha\gamma} c_{\beta\delta} t_{\gamma\delta} \tag{B·20}$$

then $t_{\alpha\beta}$ are called the components of a *second-order cartesian tensor*. Similarly, three-index symbols $t_{\alpha\beta\gamma}$ are components of a *third-order cartesian tensor* if they transform according to

$$t'_{\alpha\beta\gamma} = \sum_{\delta,\epsilon,\theta=1}^{3} c_{\alpha\delta} c_{\beta\epsilon} c_{\gamma\theta} t_{\delta\epsilon\theta} \triangleq c_{\alpha\delta} c_{\beta\epsilon} c_{\gamma\theta} t_{\delta\epsilon\theta} \tag{B·21}$$

etc., for arbitrary order. Noncartesian tensors expressed in terms of a nonorthogonal vector basis are of little use in classical dynamics, and are ignored here. Even cartesian tensors are bypassed in this book, in favor of equivalent matrix representations.

Thus a vector is a *first-order tensor*. The Kronecker delta and the set of direction cosines of a linear orthogonal transformation are examples of *second-order tensors;* the *inertia tensor* and the *angular-velocity tensor* are additional examples of second-order tensors, although in this book we work instead with the inertia matrix and the angular-velocity matrix. The epsilon symbols transform as third-order tensors *only* under proper rotations; they do not in general qualify as tensors. Since a scalar is not associated with a vector basis, it may be said to be invariant to a basis transformation, and we can then recognize a *scalar* to be a *tensor of zero order*.

It may be emphasized that the tensor as a mathematical entity is comprised of an ordered collection of scalars such as $t_{\alpha\beta}$ or v_α. The value adopted by the scalars depends on the vector basis chosen for the description, but the tensor itself is independent of a vector basis or "coordinate system." We are usually concerned in physics with the *components* of a tensor. These number one for a tensor of zero order, three for a first-order tensor, nine for a second-order tensor, twenty-seven for a third-order tensor, etc. In this sense tensor analysis offers some further examples of multiple-algebra systems (although not all tensors meet the restrictions of the formal definition of a linear algebra).

It may be emphasized that for every explicit description of a tensor by its components there is an underlying vector basis or coordinate system. Since so much work in dynamics involves simultaneous consideration of several different vector bases or corresponding frames of reference, it is very convenient to utilize a notation for a tensor that carries the vector basis along explicitly. This possibility is afforded by the *dyadic* representation of a second-order tensor (or, in general, the polyadic representation of a tensor). Thus the *inertia tensor*, whose elements in a particular vector basis include the scalar moments and products of inertia, may be written in dyadic form as

$$\mathbb{I} = \sum_{\alpha,\beta=1}^{3} I_{\alpha\beta}\mathbf{u}_\alpha\mathbf{u}_\beta \triangleq I_{\alpha\beta}\mathbf{u}_\alpha\mathbf{u}_\beta \tag{B·22}$$

where again in the final representation repeated Greek indices indicate summation from 1 to 3 and I_{11}, I_{22}, I_{33}, $I_{12} = I_{21}$, etc., are (scalar) moments and (negative) products of inertia. The dyadic is not only a convenient representation of a tensor, it may also be defined without recourse to tensor analysis and used in dynamics as a vector operation of great utility. In his basic treatise on vectors, Gibbs defined a *dyad* as an operator to be represented by the juxtaposition of two vectors (for example, **ab**) with no intervening dot or cross, and a *dyadic* as a sum of dyads. Thus the *inertia dyadic* is written as previously

$$\begin{aligned}\mathbb{I} &= I_{11}\hat{\mathbf{u}}_1\hat{\mathbf{u}}_1 + I_{12}\hat{\mathbf{u}}_1\hat{\mathbf{u}}_2 + I_{13}\hat{\mathbf{u}}_1\hat{\mathbf{u}}_3 + I_{21}\hat{\mathbf{u}}_2\hat{\mathbf{u}}_1 + \cdots \\ &= \sum_{\alpha,\beta=1}^{3} I_{\alpha\beta}\hat{\mathbf{u}}_\alpha\hat{\mathbf{u}}_\beta \triangleq I_{\alpha\beta}\hat{\mathbf{u}}_\alpha\hat{\mathbf{u}}_\beta\end{aligned} \tag{B·23}$$

Dyads and dyadics operate on vectors in the most straightforward way; when a dyadic is (pre- or post-) dot- or cross-multiplied by a vector, we simply perform the familiar vector multiplication with those vectors immediately separated by the dot or cross. An expression for the angular momentum **H**, for example, is shown in §8·3·3 to be given in terms of the inertia dyadic $\mathbb{I}$ and the angular-velocity vector $\boldsymbol{\omega}$ by

$$\begin{aligned}\mathbf{H} = H_\alpha\hat{\mathbf{e}}_\alpha &= \mathbb{I}\cdot\boldsymbol{\omega} = \sum_{\alpha,\beta=1}^{3} I_{\alpha\beta}\hat{\mathbf{u}}_\alpha\hat{\mathbf{u}}_\beta \cdot \sum_{\gamma=1}^{3}\omega_\gamma\hat{\mathbf{u}}_\gamma \\ &= \sum_{\alpha,\beta,\gamma=1}^{3} I_{\alpha\beta}\omega_\gamma\hat{\mathbf{u}}_\alpha(\hat{\mathbf{u}}_\beta\cdot\hat{\mathbf{u}}_\gamma) = \sum_{\alpha,\beta=1}^{3} I_{\alpha\beta}\omega_\beta\hat{\mathbf{u}}_\alpha \triangleq I_{\alpha\beta}\omega_\beta\hat{\mathbf{u}}_\alpha\end{aligned} \tag{B·24}$$

or

$$\begin{aligned}\mathbf{H} &= H_1\hat{\mathbf{u}}_1 + H_2\hat{\mathbf{u}}_2 + H_3\hat{\mathbf{u}}_3 \\ &= I_{11}\omega_1\hat{\mathbf{u}}_1 + I_{12}\omega_2\hat{\mathbf{u}}_1 + I_{13}\omega_3\hat{\mathbf{u}}_1 + I_{21}\omega_1\hat{\mathbf{u}}_2 + I_{22}\omega_2\hat{\mathbf{u}}_2 + I_{23}\omega_3\hat{\mathbf{u}}_2 \\ &\qquad + I_{31}\omega_1\hat{\mathbf{u}}_3 + I_{32}\omega_2\hat{\mathbf{u}}_3 + I_{33}\omega_3\hat{\mathbf{u}}_3\end{aligned} \tag{B·25}$$

The final mathematical formalism to be considered here is the matrix formulation, which is the most comprehensive approach available. Since all of the preceding multiple-algebra systems have involved entities composed of ordered collections or arrays of scalars, it may be expected that with proper operational rules these multiple-element systems may be cast in matrix form. Rules for matrix operation are reviewed briefly below, with a notational convention whereby A_{ij} is the scalar in the ith row, jth column of a matrix represented by $[A]$ or $[A_{ij}]$.

Equality:

$$[A_{ij}] = [B_{ij}] \qquad \text{if } A_{ij} = B_{ij} \tag{B·26}$$

Addition:

$$[A_{ij}] + [B_{ij}] = [C_{ij}] \qquad \text{where } C_{ij} = A_{ij} + B_{ij} \tag{B·27}$$

Multiplication:

$$[A_{ij}][B_{jk}] = \sum_{j=1}^{m} A_{ij}B_{jk} \tag{B·28}$$

Matrix transposition:

$$[A_{ij}]^T = [A_{ji}] \tag{B·29}$$

interchanging rows and columns.
Matrix inversion: $[A_{ij}]^{-1}$ is the *inverse* of $[A_{ij}]$ if

$$[A_{ij}]^{-1}[A_{ij}] = [A_{ij}][A_{ij}]^{-1} = [\delta_{ij}] \triangleq [U] \tag{B·30}$$

where $[U]$ is the unit matrix.

$$[[A][B]]^T = [B]^T[A]^T \tag{B·31}$$
$$[[A][B]]^{-1} = [B]^{-1}[A]^{-1} \tag{B·32}$$

Inversion formula:

$$[A_{ij}]^{-1} = \frac{[(-1)^{i+j}M_{ij}]^T}{|[A_{ij}]|} \tag{B·33}$$

where M_{ij} is the minor corresponding to A_{ij}, that is, the determinant of the matrix $[A]$ with the ith row and the jth column deleted.

When the product $[A_{ij}][A_{ij}]^T = [U]$, so that $[A_{ij}]^T = [A_{ij}]^{-1}$, the matrix is called *orthonormal* (some texts use *orthogonal* here) and is subject to the following interpretation: If each column (or row) is treated as the vector to which it corresponds, then these vectors are orthonormal; for 3×3 matrices these vectors give 1 when dot-multiplied by themselves and 0 when dot-multiplied by one another.

Beyond the mathematical elegance of a matrix formulation of the equations of dynamics (which is insufficient reason for the use of matrices by engineers) is the facility with which matrix differential equations are programmed for digital-computer integration. Once the matrix representation becomes sufficiently familiar to seem comfortable to the analyst, the convenience of notation in description of complex systems provides further motivation for its use.

It should again be emphasized that multiple-algebra systems are not necessary in dynamic analysis. No concept beyond the scalar is necessary, and in the final resolution of a problem in dynamics the scalar interpretation is often most useful. Multiple-algebra systems are used solely for convenience and efficiency. It is important to recognize, however, that the *feasibility* of solving a given problem may depend crucially on the efficiency of formulation of the problem. Thus it behooves the modern engineer to develop the full range of mathematical approaches in the study of dynamics.

APPENDIX C
PROPERTIES OF HOMOGENEOUS BODIES

Notation: μ is mass per unit volume of a three-dimensional body
$\bar{\mu}$ is mass per unit area for a lamina, or two-dimensional body
$\bar{\bar{\mu}}$ is mass per unit length for a line mass, or one-dimensional body

Description	*Mass* $\mathfrak{M}$	$\mathbf{R}_c = R_1\hat{\mathbf{u}}_1 + R_2\hat{\mathbf{u}}_2 + R_3\hat{\mathbf{u}}_3$ (vector from o to c)	I^c_{ij} $(i,j = 1,2,3)$	I^o_{ij} $(i,j = 1,2,3)$
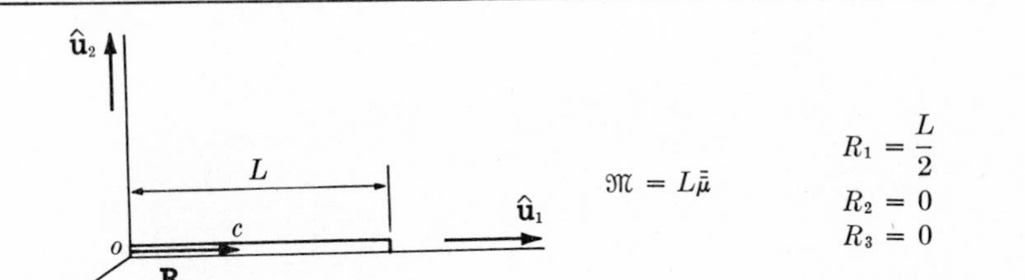 Straight thin rod (line mass)	$\mathfrak{M} = L\bar{\bar{\mu}}$	$R_1 = \dfrac{L}{2}$ $R_2 = 0$ $R_3 = 0$	$I^c_{11} = 0$ $I^c_{22} = I^c_{33} = \dfrac{\mathfrak{M}L^2}{12}$ $I^c_{12} = I^c_{23} = I^c_{31} = 0$	$I^o_{11} = 0$ $I^o_{22} = I^o_{33} = \dfrac{\mathfrak{M}L^2}{3}$ $I^o_{12} = I^o_{23} = I^o_{31} = 0$
Circular thin hoop (line mass)	$\mathfrak{M} = 2\pi a\bar{\bar{\mu}}$	$R_1 = a$ $R_2 = a$ $R_3 = 0$	$I^c_{11} = I^c_{22} = \dfrac{\mathfrak{M}a^2}{2}$ $I^c_{33} = \mathfrak{M}a^2$ $I^c_{12} = I^c_{23} = I^c_{31} = 0$	$I^o_{11} = I^o_{22} = \frac{3}{2}\mathfrak{M}a^2$ $I^o_{3} = 3\mathfrak{M}a^2$ $I^o_{23} = I^o_{31} = 0$ $I^o_{12} = -\mathfrak{M}a^2$

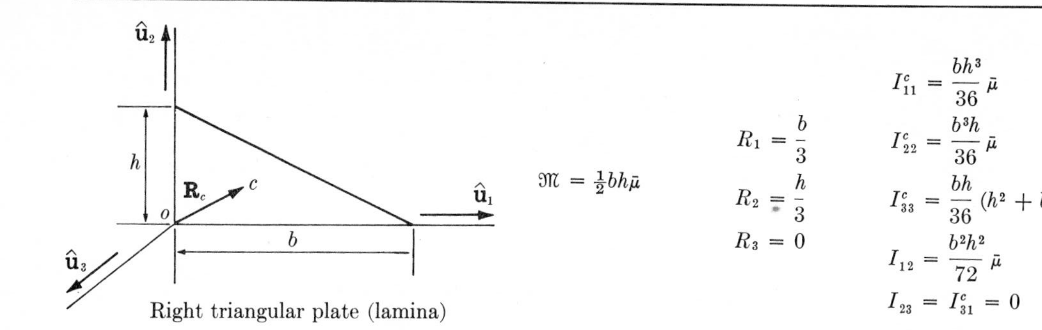

Right triangular plate (lamina)

$$\mathfrak{M} = \tfrac{1}{2}bh\bar{\mu}$$

$$R_1 = \frac{b}{3} \qquad R_2 = \frac{h}{3} \qquad R_3 = 0$$

$$I_{11}^c = \frac{bh^3}{36}\bar{\mu} \qquad I_{11}^o = \frac{bh^3}{12}\bar{\mu}$$

$$I_{22}^c = \frac{b^3h}{36}\bar{\mu} \qquad I_{22}^o = \frac{b^3h}{12}\bar{\mu}$$

$$I_{33}^c = \frac{bh}{36}(h^2 + b^2)\bar{\mu} \qquad I_{33}^o = \frac{bh}{12}(h^2 + b^2)\bar{\mu}$$

$$I_{12} = \frac{b^2h^2}{72}\bar{\mu} \qquad I_{12}^o = -\frac{b^2h^2}{24}\bar{\mu}$$

$$I_{23} = I_{31}^c = 0 \qquad I_{23}^o = I_{31}^o = 0$$

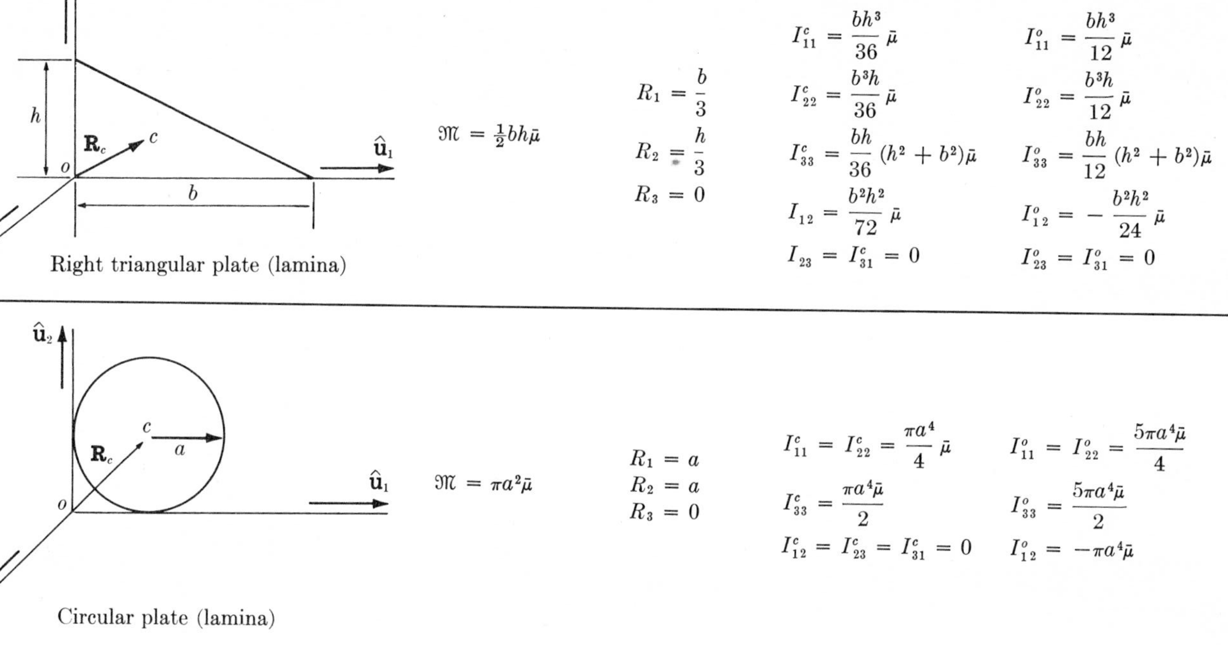

Circular plate (lamina)

$$\mathfrak{M} = \pi a^2\bar{\mu}$$

$$R_1 = a \qquad R_2 = a \qquad R_3 = 0$$

$$I_{11}^c = I_{22}^c = \frac{\pi a^4}{4}\bar{\mu} \qquad I_{11}^o = I_{22}^o = \frac{5\pi a^4\bar{\mu}}{4}$$

$$I_{33}^c = \frac{\pi a^4\bar{\mu}}{2} \qquad I_{33}^o = \frac{5\pi a^4\bar{\mu}}{2}$$

$$I_{12}^c = I_{23}^c = I_{31}^c = 0 \qquad I_{12}^o = -\pi a^4\bar{\mu}$$

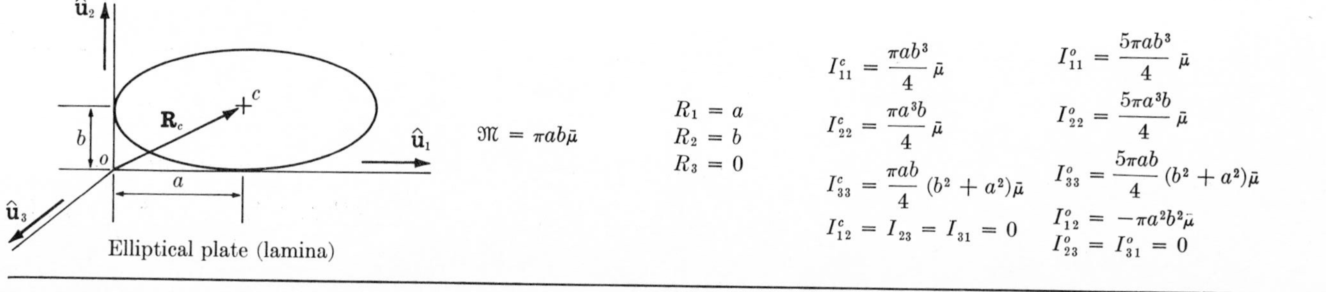

Elliptical plate (lamina)

$$\mathfrak{M} = \pi ab\bar{\mu}$$

$$R_1 = a \qquad R_2 = b \qquad R_3 = 0$$

$$I_{11}^c = \frac{\pi ab^3}{4}\bar{\mu} \qquad I_{11}^o = \frac{5\pi ab^3}{4}\bar{\mu}$$

$$I_{22}^c = \frac{\pi a^3b}{4}\bar{\mu} \qquad I_{22}^o = \frac{5\pi a^3b}{4}\bar{\mu}$$

$$I_{33}^c = \frac{\pi ab}{4}(b^2 + a^2)\bar{\mu} \qquad I_{33}^o = \frac{5\pi ab}{4}(b^2 + a^2)\bar{\mu}$$

$$I_{12}^c = I_{23} = I_{31} = 0 \qquad I_{12}^o = -\pi a^2b^2\bar{\mu}$$

$$I_{23}^o = I_{31}^o = 0$$

Description	Mass $\mathfrak{M}$	$\mathbf{R}_c = R_1\hat{\mathbf{u}}_1 + R_2\hat{\mathbf{u}}_2 + R_3\hat{\mathbf{u}}_3$ (vector from o to c)	I^c_{ij} $(i,j = 1,2,3)$	I^o_{ij} $(i,j = 1,2,3)$
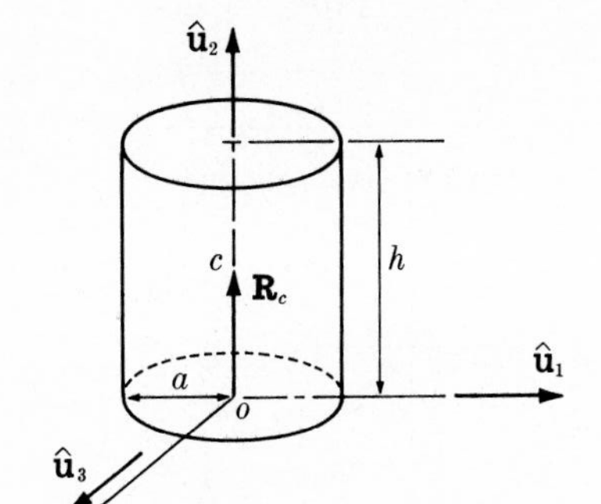 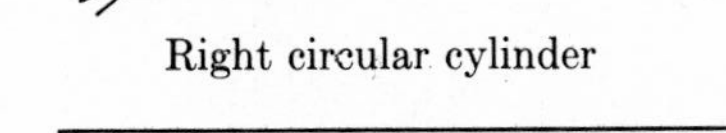 Right circular cylinder	$\mathfrak{M} = \pi a^2 h\mu$	$R_1 = 0$ $R_2 = \frac{h}{2}$ $R_3 = 0$	$I^c_{11} = I^c_{33} = \frac{\mathfrak{M}}{12}(3a^2 + h^2)$ $I^c_{22} = \frac{\mathfrak{M}a^2}{2}$ $I^c_{12} = I^c_{23} = I^c_{31} = 0$	$I^o_{11} = I^o_{33} = \frac{\mathfrak{M}}{12}(3a^2 + 4h^2)$ $I^o_{22} = \frac{\mathfrak{M}a^2}{2}$ $I^o_{12} = I^o_{23} = I^o_{31} = 0$
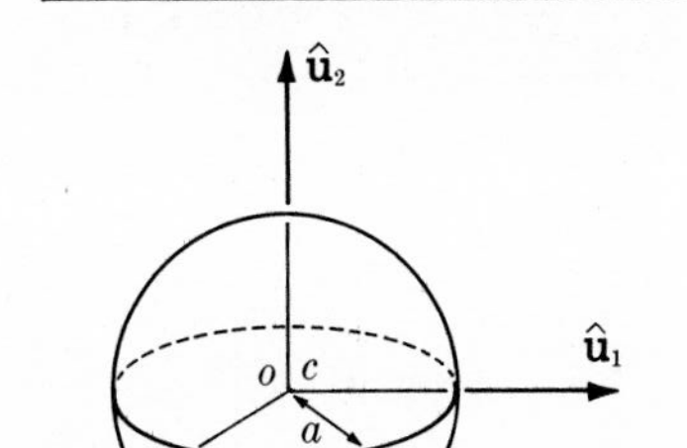 Sphere	$\mathfrak{M} = \frac{4\pi a^3\mu}{3}$	$R_1 = 0$ $R_2 = 0$ $R_3 = 0$	$I^c_{11} = I^c_{22} = I^c_{33} = \frac{2\mathfrak{M}a^2}{5}$ $I^c_{12} = I^c_{23} = I^c_{31} = 0$	$I^o_{ij} = I^c_{ij}$ $(i, j = 1, 2, 3)$

Body	Mass	Center of mass	Moments of inertia about center of mass	Moments of inertia about point o
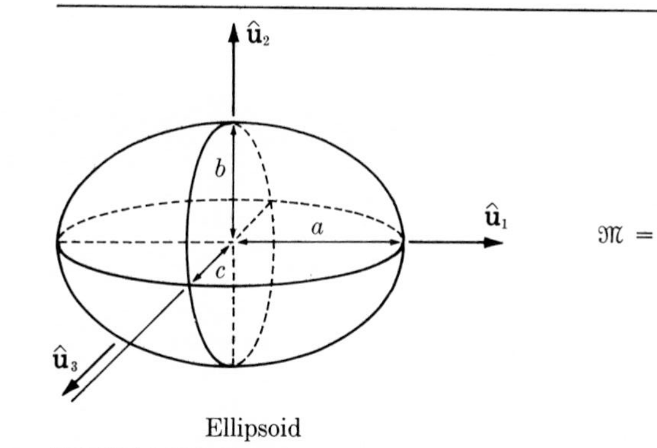 Ellipsoid	$\mathfrak{M} = \dfrac{4\pi abc\mu}{3}$	$R_1 = 0$ $R_2 = 0$ $R_3 = 0$	$I^c_{11} = \dfrac{\mathfrak{M}}{5}(b^2 + c^2)$ $I^c_{22} = \dfrac{\mathfrak{M}}{5}(a^2 + c^2)$ $I^c_{33} = \dfrac{\mathfrak{M}}{5}(a^2 + b^2)$ $I^c_{12} = I^c_{23} = I^c_{31} = 0$	$I^o_{ij} = I^c_{ij}$ $(i, j = 1, 2, 3)$
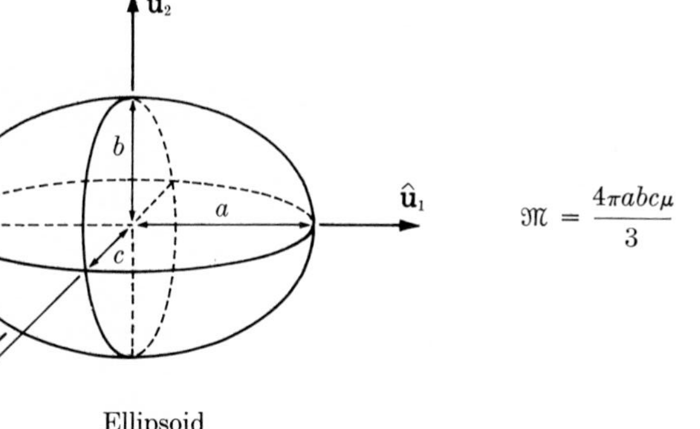 Rectangular parallelepiped	$\mathfrak{M} = abc\mu$	$R_1 = \dfrac{a}{2}$ $R_2 = \dfrac{b}{2}$ $R_3 = \dfrac{c}{2}$	$I^c_{11} = \dfrac{\mathfrak{M}(b^2 + c^2)}{12}$ $I^c_{22} = \dfrac{\mathfrak{M}(c^2 + a^2)}{12}$ $I^c_{33} = \dfrac{\mathfrak{M}(a^2 + b^2)}{12}$ $I^c_{12} = I^c_{23} = I'_{31} = 0$	$I^o_{11} = \dfrac{\mathfrak{M}(b^2 + c^2)}{3}$ $I^o_{22} = \dfrac{\mathfrak{M}(c^2 + a^2)}{3}$ $I^o_{33} = \dfrac{\mathfrak{M}(a^2 + b^2)}{3}$ $I^o_{12} = -\dfrac{\mathfrak{M}ab}{4}$ $I^o_{23} = -\dfrac{\mathfrak{M}bc}{4}$ $I^o_{31} = -\dfrac{\mathfrak{M}ac}{4}$

APPENDIX C (*Continued*)

Description	*Mass* $\mathfrak{M}$	$\mathbf{R}_c = R_1\hat{\mathbf{u}}_1 + R_2\hat{\mathbf{u}}_2 + R_3\hat{\mathbf{u}}_3$ (vector from o to c)	$I^c_{ij}\ (i,j = 1,2,3)$	$I^o_{ij}(i,j = 1,2,3)$
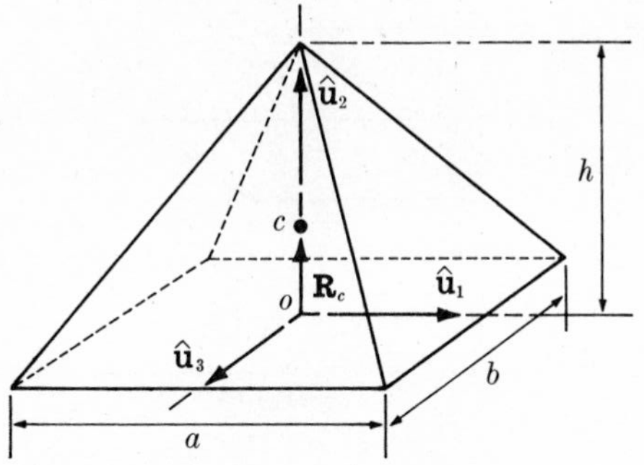 Right rectangular pyramid	$\mathfrak{M} = \frac{abh\mu}{3}$	$R_1 = 0$ $R_2 = \frac{h}{4}$ $R_3 = 0$	$I^c_{11} = \frac{\mathfrak{M}}{80}(4b^2 + 3h^2)$ $I^c_{22} = \frac{\mathfrak{M}}{20}(a^2 + b^2)$ $I^c_{33} = \frac{\mathfrak{M}}{80}(4a^2 + 3h^2)$ $I^c_{12} = I^c_{23} = I^c_{31} = 0$	$I^o_{11} = \frac{\mathfrak{M}}{20}(b^2 + 2h^2)$ $I^o_{22} = \frac{\mathfrak{M}}{20}(a^2 + b^2)$ $I^o_{33} = \frac{\mathfrak{M}}{20}(a^2 + 2h^2)$ $I^o_{12} = I^o_{23} = I^o_{31} = 0$
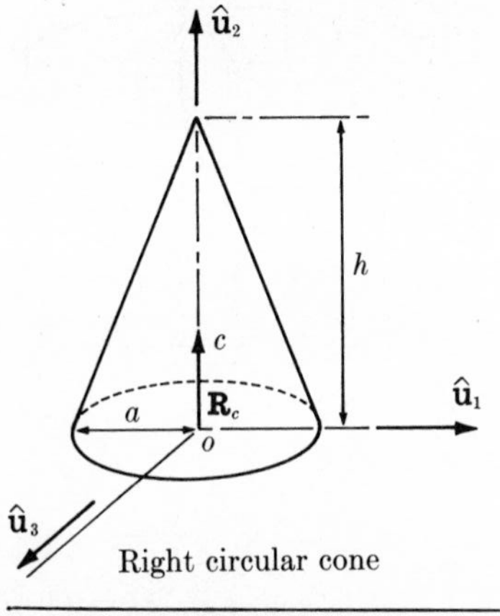 Right circular cone	$\mathfrak{M} = \frac{\pi a^2 h\mu}{3}$	$R_1 = 0$ $R_2 = \frac{h}{4}$ $R_3 = 0$	$I^c_{11} = I^c_{33} = \frac{3\mathfrak{M}}{80}(4a^2 + h^2)$ $I^c_{22} = \frac{3\mathfrak{M}a^2}{10}$ $I^c_{12} = I^c_{23} = I^c_{31} = 0$	$I^o_{11} = I^o_{33} = \frac{\mathfrak{M}}{20}(3a^2 + 2h^2)$ $I^o_{22} = \frac{3\mathfrak{M}a^2}{10}$ $I^o_{12} = I^o_{23} = I^o_{31} = 0$

LIST OF FIGURES

INDEX

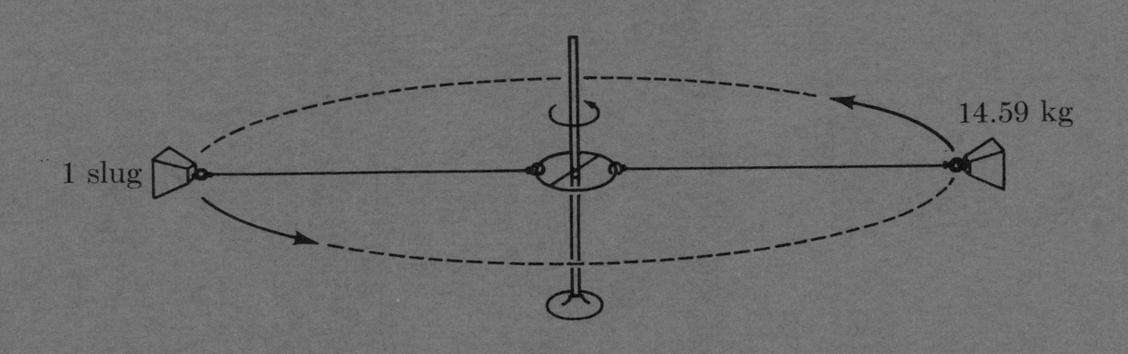

TABLE OF CONVERSIONS

Unit (symbol)		*Factor*		*SI Unit (symbol)*
foot (ft)	×	0.3048 m/ft	≈	meter (m)
inch (in.)	×	0.0254 m/in.	≈	meter (m)
statute mile (mi)	×	1609 m/mi	≈	meter (m)
pound-force (lb$_f$ or lb)	×	4.448 N/lb	≈	newton (N)
slug (slug)	×	14.59 kg/slugs	≈	kilogram (kg)
second (sec)	×	1.0	=	second (s)